Dieter Wöhrle, Anatolii D. Pomogailo

Metal Complexes and Metals in Macromolecules

Related Titles from WILEY-VCH and John Wiley & Sons

Dieter Wöhrle, Anatolii D. Pomogailo

Metal Complexes and Metals in Macromolecules

Synthesis, Structure and Properties

With Contributions from Y. Amao, M. Kaneko,
E. A. Karakhanov, Y. S. Kang, A. L. Maximov, H. Nishide,
T. Ohsaka, I. Okura, R. C. Raj and J. Won

WILEY-VCH

WILEY-VCH GmbH & Co. KGaA

Prof. Dr. Dieter Wöhrle
Institute of Organic and Macromolecular Chemistry
University of Bremen
P.O. Box 330 440
28334 Bremen
Germany

Prof. Dr. Anatolii D.Pomogailo
Institute of Problems of Chemical Physics
Russian Academy of Sciences
Chernogolovka
Moscow Region 142432
Russia

Library of Congress Card No.: applied for

A catalogue record for this book is available from the British Library.

Bibliographic information published by Die Deutsche Bibliothek
Die Deutsche Bibliothek lists this publication in the Deutsche Nationalbibliografie; detailed bibliographic data is available in the Internet at http://dnb.ddb.de

Printed in the Federal Republic of Germany.
Printed on acid-free paper.

Printing: betz-druck gmbH, Darmstadt
Bookbinding: Litges & Dopf Buchbinderei GmbH, Heppenheim

ISBN 3-527-30499-1

Dedicated to G. Manecke in memoriam,
V. Kabanov (Russia), E. Tsuchida (Japan).
To founders of the science on macromolecular metal complexes.

Preface

This book fills a gap in our knowledge of an intensively developing field of science: the chemistry and physics of macromolecular metal complexes (MMCs). This branch of science has appeared at the intersection of macromolecular, organometallic, physical and catalytic chemistry, and physics. On the one hand, it is difficult to generalize numerous data on MMCs because of the need to coordinate the efforts of specialists whose interests are primarily in specific fields of chemistry. Researchers in organic and high molecular branches of chemistry deal with syntheses of metal complexes and metals in macromolecules, whereas specialists in physical chemistry work on problems of structure, the nature of bonds and the catalytic properties of MMCs. On the other hand, as the contents of this book show, the field of MMCs possesses all the features typical of an independent branch of chemistry that operates with specific objects and its own construction principles and methodology. In general, the stage of accumulation of experimental data is almost complete. However, at present many ideas are being elaborated and discussed; some of them (the development of the theory of bonds in macromolecular complexes, the stabilization of thermodynamically unstable structures by polymers, etc.) have still not been formulated. The present state in this area is characterized by efforts to understand the experimental materials and to find fundamental *structure–properties* correlations. Progress in the area of MMCs is a result of intensive work in the fields of catalysis, photochemistry, living chemistry and the science of materials. These investigations have already made a valuable contribution to the solution of general problems in theoretical chemistry. It should be pointed out that the combination of metal complexes and metals with natural macromolecular proteins is the fundamental prerequisite for activity and selectivity in several life functions.

Chapters 1 and 2 of Part A PREFACE introduce into definitions, classifications, history, properties and biological systems of macromolecular metal complexes. Then part B SYNTHESIS AND STRUCTURES contain at first in chapter 3 kinetics and thermodynamics of formation of these complexes. The following chapters 4 till 8 describe in detail the various synthetic routes for the preparation of macromolecular metal complexes. Part C with chapters 9 till 14 is devoted to PROPERTIES. The most important ones are binding of small molecules, physical and optical sensors, catalysis, photocatalysis and electron/photon induced processes. In chapter 15 few closing remarks are made.

We believe that this book will be useful both for researchers who have experience in the synthesis and application of MMCs and for young scientists who wish to deal with this interesting area of science. Experimental procedures and methods for studying the main MMCs are given, and the book contains several references to the most important and fundamental work in this field.

Authors from many countries have worked amicably and productively to create this book. We are pleased and thankful that well-known experts contributed very engaged and in time with chapters of extreme high level. Undoubtedly, Professor D. Wöhrle was the driving force of the author collective; he initiated the publication of the book and expressed the main concepts of its construction.

We wish to thank all the contributors for their creative work on this book. We are also grateful to Rita Fofana and Dr. Gulzhian Dzhardimalieva for their invaluable assistance.

D. Wöhrle (Bremen)

A. Pomogailo (Chernogolovka)

March 2003

Contents

List of Contributors

Amao, Yukata
Department of Applied Chemistry
Oita University
Dannoharu 700, Oita 870-1192, Japan
E-Mail: amao@cc.oita-u.ac.jp

Kaneko, Masao
Faculty of Science
Ibaraki University
Mito, 310-8512, Japan
E-Mail: mkaneko@mx.ibaraki.ac.jp

Kang, Yong Soo
Center for Facilitated Transport Membranes
Korea Institute of Science and Technology
Seoul, Korea
E-Mail: yskang@kist.re.kr

Karakhanov, Edward A.
Department of Chemistry
Moscow State University
Moscow, 119992, Russia
E-Mail: kar@petrol.chem.msu.ru

Maximov, Anton L.
Department of Chemistry
Moscow State University
Moscow, 119992, Russia
E-Mail: max@petrol.chem.msu.ru

Nishide, Hiroyuki
Department of Applied Chemistry
Waseda University
Tokyo 169-8555, Japan
E-Mail: nishide@waseda.jp

Ohsaka, Takeo
Department of Electronic Chemistry
Interdisciplinary Graduate School of Science and Engineering
Tokyo Institute of Technology
4259 Nagatsuta, Midori-ku
Yokohama 226-8502, Japan
E-Mail: ohsaka@echem.titech.ac.jp

Okura, Ichiro
Department of Bioengineering
Tokyo Institute of Technology
4259 Nagatsuta, Midori-ku,
Yokohama 226-8501, Japan
E-Mail: iokura@bio.titech.ac.jp

Pomogailo, Anatolii D.
Institute of Problems of Chemical Physics
Russian Academy of Sciences
Chernogolovka,
Moscow Region, 142432, Russia
E-Mail: adpomog@icp.ac.ru

Raj, Retna C.
Department of Chemistry
Indian Institute of Technology
Kharagpur 721 302, West Bengal, India
E-Mail: crraj@chem.iitkgp.ernet.in

Wöhrle, Dieter
Institute of Organic and Macromolecular Chemistry
University of Bremen
P.O. Box 330 440, 28334 Bremen, Germany
E-Mail: woehrle@chemie.uni-bremen.de

Won, Jongok
Department of Applied Chemistry
Sejong University
98 Kunja, Kwangjin
Seoul, Korea 143-747
E-Mail: jwon@sejong.ac.kr

A OVERVIEW AND BIOLOGICAL SYSTEMS

1 Definitions, Classifications, History, Properties

Dieter Wöhrle and Anatolii Pomogailo

This book provides an overview of possible combinations of metal complexes and metals with organic and inorganic macromolecules (often also named macromolecular metal complexes — MMC [1]). This book covers the formation, synthesis, structure and properties of these exciting and relatively new materials. Metal-containing macromolecules are a fascinating field of science. It is readily understandable that materials with unusual properties are obtained by having a metal complex or metal as part of a macromolecule. Nature shows us the functions of such materials extremely well by the selectivity and activity of, for example, hemoglobin, photosynthesis and metalloenzymes.

After some definitions in Section 1.1, the classifications of metal complexes and metals in macromolecules in four types of combination are given in Section 1.2. A summary of metal complexes in nature and the history of artificial ones is presented in Section 1.3. Then Section 1.4 introduces important properties of these materials. A short review of other inorganic macromolecules – without metal – not dealt with in detail in this book is presented in Section 1.5.

The structure and function of some natural macromolecules containing metal complexes are described in more detail in Chapter 2. The following main Chapters, 3 through 8, include various examples of the formation, synthesis and structures of metal-containing macromolecules. Then Chapters 9 through 14 review some important properties of such materials. Several chapters also contain an experimental part illustrated with selected preparations and investigations of some properties.

It should be pointed out that metal ions, metal complexes and metal clusters are vital for the functioning of living organisms. It is very important to study the structures and working of these natural systems as a first step towards the design of artificial metal-containing macromolecules for various applications. The field of biomimetics is engaged in simulating natural processes [2–4].

1.1 Definitions

Because of their extreme length, organic polymers (macromolecules) give rise to entangled materials with unique properties. They are part of our daily life, and sometimes it is rem arked, based on their broad applications, that we are living now in the so-called "plastic age". Today, macromolecular chemistry is a subject of world-wide importance in education and research at universities, high schools and research institutes, and an extremely important economic factor in different areas of industry. Scientifically and practically, macromolecular chemistry overlaps the disciplines of physics, engineering science and materials science. Besides carbon, the main elements in chain- or network-forming organic macromolecules are oxygen, nitrogen and sulfur. In addition, silicon- and phosphorus-containing macromolecules have become important. Altogether these are only six elements. Organic macromolecules are generally limited in some way, in most cases by their relatively low thermal stability and relatively weak ability to take part in such important processes as binding interactions, activation of small molecules, electron-transfer processes, therapeutic effects, etc. Therefore the development of new materials incorporating others besides these six elements is of fundamental importance.

One of the most important trends in the future process of macromolecular chemistry will be the wide-scale employment of elements of the Periodic Table for the synthesis of macromolecules. Among the around 109 elements of the Periodic Table only the six noble gases and eleven elements of the groups IA and VIIA (including H), which are univalent and can only play the role of terminating groups, are incapable of producing stable macromolecules in solution (although for some elements, especially those of high atomic number, formation of macromolecules is not known). But some of them can contribute to weak chain- or network-forming interactions in the solid state. The remaining around 80 elements are principally capable of functioning as chain- or network-forming elements. The expected and practically unlimited structural possibilities offered by this approach provide an enormous potential to improve greatly the capacity of structural macromolecular materials. Products with new static, dynamic, thermal, electronic, electrical, photoelectrical properties, etc. can be developed. Also there appear to be highly interdisciplinary possibilities for both fundamental studies and applications.

Because many combinations of elements are theoretically possible, a simple subdivision of **main chain and network forming elements** is necessary (Fig. 1-1) [5]:

- **Organic polymers or macromolecules** contain only carbon in the main chain or network. Examples are polyolefines, polystyrene (linear cross-linked), polyacetylenes, etc.
- **Inorganic polymers or macromolecules** contain no carbon but nonmetals, semimetals and/or metals in the main chain or network. Examples are polysiloxanes, polysilanes, polyphosphacenes and several polymers with metal–heteroatom chains (organometallic polymers).
- **Semiorganic polymers or macromolecules** contain at least one other

element besides carbon. Examples in this case are polyesters, polyamides, polyoxyphenylenes, polyaniline, polypyrrol, polythiophenes, polycarbo-siloxanes, polycarbosilanes and several organometallic polymers.

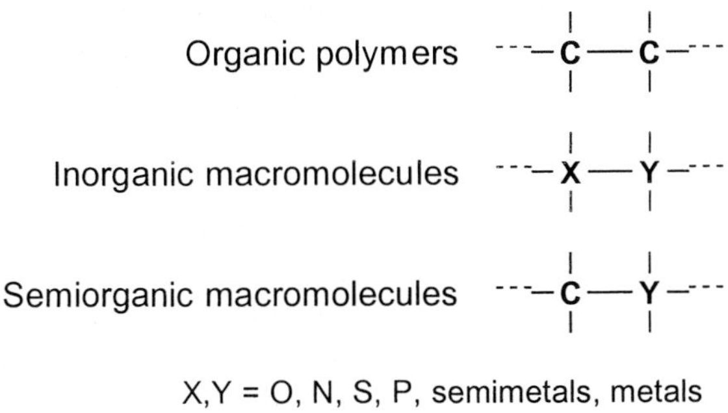

Organic polymers
$$--\!-\overset{\displaystyle |}{\underset{\displaystyle |}{C}}\!-\!\overset{\displaystyle |}{\underset{\displaystyle |}{C}}\!-\!--$$

Inorganic macromolecules
$$--\!-\overset{\displaystyle |}{\underset{\displaystyle |}{X}}\!-\!\overset{\displaystyle |}{\underset{\displaystyle |}{Y}}\!-\!--$$

Semiorganic macromolecules
$$--\!-\overset{\displaystyle |}{\underset{\displaystyle |}{C}}\!-\!\overset{\displaystyle |}{\underset{\displaystyle |}{Y}}\!-\!--$$

X,Y = O, N, S, P, semimetals, metals

Figure 1-1. Main chain macromolecules.

A distinction can also be made between **homochain** and **heterochain macromolecules**. The formation of macromolecules with a homochain structure – one element in the chain – is fairly common with elements such as C, B, Si, Ge, Sn, P, As, Sb, S, Sc and Te. However, a very considerable number of elements can participate in the formation of heterochain macromolecules, and such compounds are more numerous than homochain macromolecules. The ability of an element to be part of a covalently bonded chain depends on the energy of the bonds and the electronegativity difference between the individual elements. Table 1-1 lists examples of dissociation (bond) energies of some main group elements forming homochains or heterochains. Among homochain polymers carbon is prominent, the carbon–carbon bond being much more stable than metal–metal bonds, which are mostly weak (and often sensitive to water and/or air). In heterochain polymers the bond energies generally exceed those of homochain polymers, thus being attractive for applications (Table 1–1). This explains why heterochain macromolecules such as polysiloxanes, glasses, ceramics, etc. are often thermally stable materials whereas homochain compounds such as polysilanes are more easily broken by either thermal or light treatment. Owing to the different electronegativities of elements in heterochain macromolecules, the bonds are polar, resulting in reduced chemical stability. The bonds become more ionic or coordinative in character, and often it is difficult to decide if the interactions between elements are more covalent, ionic or coordinative.

Another large group of polymers or macromolecules is given by modifications of elements in the **side chain of a macromolecule** or simply as substituents of a backbone (Fig. 1-2):

- **Organic side chains** contain only carbon.
- **Inorganic side chains** contain only nonmetals, semimetals or metals.
- **Semiorganic side chains** contain at least one other element besides carbon.

Table 1-1. Dissociation energies D_0 between some elements.

Bond	D_0 (kJ mol^{-1})	Bond	D_0 (kJ mol^{-1})
C–C	346	C–O	358
Si–Si	222	C–N	304
Ge–Ge	188	C–Si	318
Sn–Sn	147	C–Ge	213
Na–Na	72	C–Sn	226
Mg–Mg	129	C–Pb	130
B–B	293	Si–O	452
As–As	146	B–O	536
Sc–Sc	109	B–N	446
La–La	241	As–O	301
Fe–Fe	156	As–C	228
Cu–Cu	190		

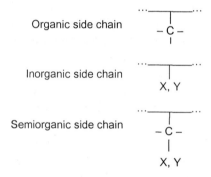

Organic side chain

Inorganic side chain

Semiorganic side chain

X, Y= O, N, S, P, semimetals, metals   **Figure 1-2.** Side chain macromolecules.

In addition, semimetals or metals can be included either as complexes or salts, or an atom/cluster may be physically incorporated in macromolecules (Fig. 1-3). This incorporation gives rise to composite materials with new properties.

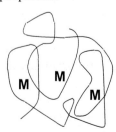

Figure 1-3. Physical incorporation of metals or semi-metals in macromolecules.

This book deals with organic, inorganic or semiorganic macromolecules containing at least one metal either in the main chain, in the side chain or physically incorporated. Macromolecules with semimetals (polysiloxanes, polysilanes, polycarbosilanes, polygermylenes, etc.) are not treated. The same holds for classical inorganic polymers such as polymeric sulfur, selenium and tellurium; polyphosphacenes, borate, silica, silica glasses and titanium dioxide; boron, aluminum, titanium, zirconium, tin, silicon and cerium phosphates (see [6–9], or corresponding keywords in Ullman's *Encyclopedia of Chemical Technology* and *Encyclopedia of Science and Technology*). A short review of some of these inorganic macromolecules is given in Section 1.5.

1.2 Classification of Metal Complexes and Metals in Macromolecules

As indicated in Section 1.1, various combinations of metal complexes or metals with organic, inorganic and semiorganic macromolecules exist. Therefore a strict and clear classification is necessary.

Detailed rules for the **nomenclature of macromolecules** are described on the websites of the IUPAC: www.iupac.org/divisions – then select "IV Macromolecular". For macromolecules with inorganic elements including metals in the backbone, rules are given under "Nomenclature for Regular Single-Strand and Quasi-Single-Strand Inorganic and Coordination Polymers (1984)". The rules are based on the identification, orientation and naming of the constitutional repeating unit. Some examples are given. Because the nomenclature often results in complicated long names it will be not used in this textbook.

1.2.1 Classification by Kind of Metal Complex/Metal Binding

Metal complexes or metals can be part of a macromolecular chain/network as follows: binding at a macromolecule; part of a macromolecule via the ligand; part of a macromolecule via the metal; physically incorporated into a macromolecule. This classification, first given in 1996 [1], is used throughout this book because from the numerous possibilities a metal complex or metal is easily classified by the kind of interaction it has with a macromolecule.

Type I: A metal ion, metal complex or metal is bound to a chain of a linear or cross-linked organic or inorganic macromolecule via a covalent (at the metal), a coordinative (at the metal), a complex (at the ligand of a complex), an ionic or a π-bond (so-called "Macromolecular Metal Complexes", Fig. 1-4). Additional possibilities exist for different kinds of binding at the surface of a carrier or the end group of a macromolecule. Examples are described in Chapters 4 and 5.

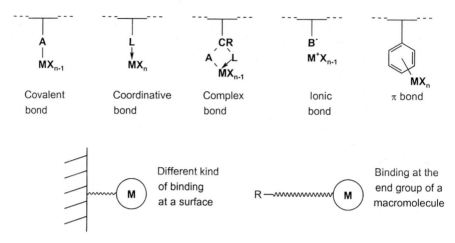

Figure 1-4. Type I: Metals bound at macromolecules.

Type II: The ligand of a metal complex is part of a macromolecular chain or network (so-called "Ligand Macromolecular Complexes", Fig. 1-5). Besides the direct synthesis of macromolecular metal complexes from low molecular precursors, a macromolecular ligand can be prepared first which is then metallated in a second step. Examples of this type are given in Chapter 6.

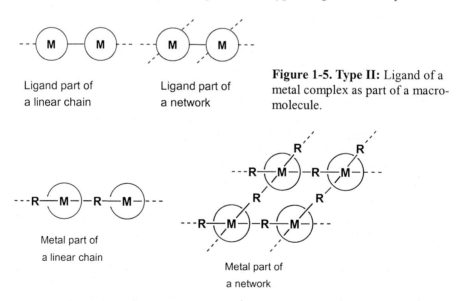

Figure 1-5. Type II: Ligand of a metal complex as part of a macro-molecule.

Figure 1-6. Type III: Metal as part of a macromolecule.

Type III: Several examples of compounds exist in which the metal of a metal complex or another metal derivative is directly part of a macromolecular chain or network (so-called "Metal Macromolecular Complexes", Fig. 1-6). In type III the macromolecular chain can consist only of metal atoms. But in most cases the metal is connected with another element such C, N, O, S via a covalent, a coordinative, an ionic or a π–bond. Chapter 7 contains examples of type III compounds.

Type IV: The physical incorporation of metal (and also semiconductor) clusters or metal complexes in macromolecules has become an important field (so-called "Macromolecule Incorporated Metal Complexes and Metals", Fig. 1-7). By stabilization of metal clusters in a macromolecular environment new composite materials have been synthesized. Chapter 8 concentrates on some aspects of metal clusters in macromolecules. This chapter also describes the monomolecular or aggregated distribution of metal complexes in macromolecules.

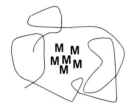 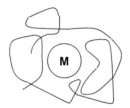

Physical incorporation
of metal nano-particles

Physical incorporation
of metal complexes

Figure 1-7. Type IV: Physical incorporation of a metal or metal complex.

Further examples of type I, II, III and IV metal complexes and metals in macromolecules are given in Chapters 9 through 14.

Chapter 3 introduces the complicated thermodynamics and kinetics of formation of macromolecular metal complexes. Formation is in general thermo-dynamically favored by a negative free energy. An additional contribution to the free energy in comparison to the known "chelate effect" for the formation or stabilization of a low molecular weight metal complex/metal is given by the so-called "polymer chelate effect": change of the free energy due to the addition of a metal derivative to a macromolecular ligand for polymer binding, the reaction of a metal with a bi/higher-functional precursor for the formation of a macromolecular chain/network or the stabilizing macromolecular environment for a metal complex/metal. Compared to low molecular weight metal com-plexes, the situation for macromolecular analogues is more complicated due to the local, molecular and supramolecular organization of different kinds of macromolecules. This influences kinetic parameters, formation constants, cooperative effects, reorganization of structured arrangements and other transformations as will be shown in Chapter 3. For all these aspects it is important to consider the formation and stereochemistry of low molecular weight coordination compounds, the chemical and physical behavior of macromolecules and aspects of physical and theoretical chemistry.

1.2.2   Classification by Connectivities

In 1978 a classification by connectivities was introduced for inorganic polymers [6]. This is defined by the number of atoms attached to a defined atom which is part of a polymer chain or side chain. Some simple examples are mentioned (for more details see [10]).

A connectivity of 1 is given for side chain metal-containing polymers as **1** in which the metal complex is connected via one bond to the main chain [11]. Additional examples for other connectivities are (Fig. 1-8): connectivity 2 in metal(yne)s **2** [12]; connectivity 3 in arsenic(III) sulfide **3**; connectivity 4 in a polymeric methyl rhenium oxide of the formula $\{H_{0.5}[(CH_3)_{0.92}ReO_3]\}_\infty$ **4** [13] or polymeric phthalocyanines **5** [14]; connectivity 8 in lanthanide complexes of bis(tetradendate) Schiff base bridging ligands **6** [15].

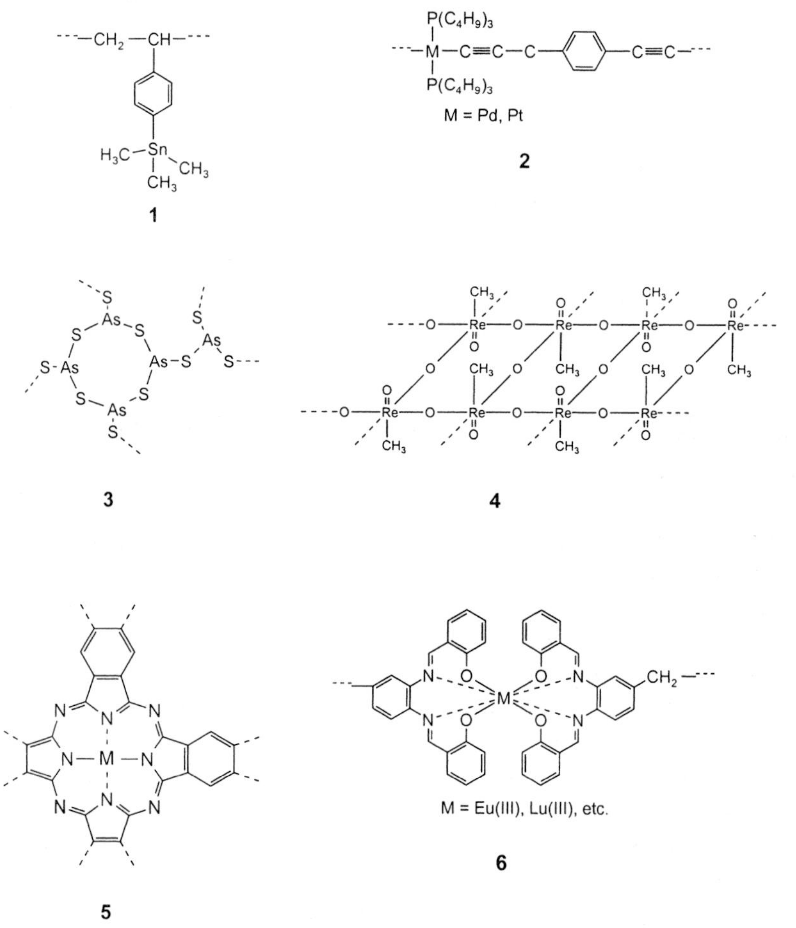

Figure 1-8. Classification by connectivities.

1.2.3   Classification by Dimensionality

This classification is described in [10,16]. A linear chain as shown for **2** in Fig. 1-8 is classified as a one-dimensional (1-D) polymer. An example of a two-dimensional (2-D) polymer is given by **5** in Fig. 1-8. A simple example of a three-dimensional (3-D) macromolecule is quartz (SiO_2). Reaction of 2,4,6-tri(4-pyridyl)-1,3,5-triazine **7** with $HgClO_4$ yields the 3-D structure **8** (Fig. 1-9) [17].

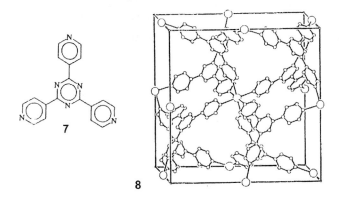

Figure 1-9. Unit cell of a 3-D coordination polymer **8** from **7** and $HgClO_4$ (C and N atoms as small circles and Hg as large circles)

1.3   Metals in Nature and the History of Artificial Metal Complexes and Metals in Macromolecules

The variety of metals used by biological systems is very large, ranging from the alkali metals to the transition metals (for more details and references see Chapter 2). They play an essential role in living systems, in both growth and metabolism. *Some metal ions are necessary in gram quantities, others are trace elements* and essential beneficial nutrients at low levels but metabolic poisons at high levels, and some metal ions are called *detrimental metal ions* because they are toxic and impair the regular course of life functions in all concentrations. In living organisms metal ions can coordinate to a variety of biopolymers in ionic, coordinative and also covalent bonds:

- In proteins at the (C=O) or (N–H) bonds and especially at N, O, S donor atoms of side chains.
- In nucleic acids at basic N atoms or at phosphate groups.
- In carbohydrates and lipids at (C–O) and (P–O) groups.

- In solid bones, teeth and kidney stones.

Metal ion interactions also play an important role in the binding of nucleic acids with proteins, influencing denaturation stability and protein synthesis and thus keeping the inherited information.

Metal complexes are involved in various processes essential for life:

- Hemoglobin, myoglobin → gas transport.
- The mechanism of photosynthesis → energy conversion.
- Metalloenzymes → catalysis.
- Cofactors → electronic interactions.

Also metal or metal compound clusters fulfill necessary functions as catalysts or for electron transfer in living systems. Two examples are:

- Mn clusters for the four electron oxidation of water to oxygen in the mechanism of photosynthesis.
- Electron-rich Fe–S clusters for electron transport in the respiratory chain.

Metals also play an important role in non-living nature. Humic acids originating from dead plant materials are heteropolycondensates with a molecular weight between 30,000 and 100,000 $g\,mol^{-1}$ and consist of a polycyclic cross-linked backbone with bound polysaccharides, proteins and phenols [18]. Metal ions of, for example, iron, copper, zinc or manganese are bound at carboxylic acid and carbonyl groups. The complexation of humic acids with metal ions is involved in many biological, biochemical and geochemical processes. For example, the interaction of humic acids with metal compounds is a necessary part of soil formation. A "deposit" of bioelements is formed which regulates the supply for plants, depending on environmental conditions, as well as the processes of migration and delivery of biogenetic materials to biological systems. Moreover, the leaching of metal ions from humic acids plays an important role in ore formation, including uranium- and gold-containing ores and bauxites. Partly dissolved in natural water, humic acids can bind toxic metal ions. Altogether, the complexation of humic acids is important for the soil, for biological systems and for solving ecological problems.

The interaction of metal ions with natural inorganic sorbents influences the redistribution of some elements in geological deposits and soils. Sorption of metal ions, especially iron, copper, manganese and gold by microorganisms has been widely studied [19]. For example, the interaction of microorganisms with gold leads to bacterial enrichment and aggregation of gold particles on a geological time scale [20]. Microbiological geotechnology uses the extraction of metal ions from ores, rocks and solutions by microorganisms [21,22].

Humans used some metals in macromolecules in ancient times with no knowledge of the kinds of interaction between these compounds. From the historical point of view it is worth mentioning that the oldest synthetic inorganic polymer (in fact, the oldest of all synthetic polymers) was a simple alkali silicate glass used during the Baldarian period in Egypt (around 12,000 B.C.) as a glaze in ornamental amulets and beads [23]). Through the centuries subsequent improvements in glass manufacture reached a stage of perfection during the time of Tutankhamen (1300 B.C.) [6]. Around 3000 years later the next important discovery of inorganic polymers occurred when sodium polyphosphates in crystalline and amorphous forms were prepared [24]. Their potential importance, academically and industrially, can be seen today by considering the well-known polysiloxanes. Other ancient processes using metal ions are the tanning of hides (consisting of proteins) to get leather and the mordanting of (protein) wool fibers for fixing dyes (see keywords in *Encyclopedia of Science and Technology* and Ullmann's *Encyclopedia of Technical Chemistry*). It is worthwhile mentioning that the production of metals played a decisive role in human history.

Systematic studies of metal complexes and metals in macromolecules – often called macromolecular metal complexes [1] – have a short history of about sixty years.

During World War II, the United States Air Force established a research program to develop polymers which were stable up to 600 °C [25]. Undoubtedly, this promoted studies in the field of metals in polymers. The first publication of the results obtained in terms of the program appeared in the late Fifties and early Sixties [26,27]. About this time investigations of chelate-forming resins and their metal complexes were started [28] in connection with aspects of the nuclear-power industry and hydrometallurgy, and also including analytical aspects and the separation of metals, especially rare elements and isotopes. It should be mentioned that the initiatives for intensive studies of metals in polymers were always determined by the demands of some branches of industry. Thus, in the early Seventies an increasing number of studies were concerned with the catalytic properties of polymer-bound metal complexes with the aim of heterogenizing homogeneous catalysts for various chemical reactions [29,30]. Most likely, the first attempt to use metal complexes bonded with the polymer ligand as catalyst (the product of the interaction of K_2PdCl_4 with an ion-exchange resin) was in 1969 (US patent 3668271). As early as 1975 British Petroleum and Mobil Oil Corporation had received about 2000 patents on immobilized catalysts for hydrolysis, carbonylation, isomerization, polymerization and other reactions [31].

At the end of the Seventies and beginning of the Eighties the results of the first intensive studies of the binding of small molecules and the conversion of solar energy using polymer metal complexes were published [32,33]. Also around this time reviews and the first books summarized the developing interest in the field of metal complexes and metals in macromolecules [32–37].

Today several summarizing reports exist in this field covering such themes as macromolecular metal complexes, organometallic polymers, metal-containing monomers, metal-containing polymeric materials, polymer-immobilized metal complexes, macromolecular complexes and macromolecular metal chelates in different ways [1,10,11,14,16,29,38–67]. In addition, International Conferences on Macromolecular Metal Complexes (MMC) initiated in 1985 by E. Tsuchida (Waseda University, Tokyo) have been held regularly every second year since 1985 under the sponsorship of the IUPAC [68–75]. Since the Seventies/Eighties the field has increasingly developed. Now in several countries scientific groups are active to develop this field further in many directions for the future. A great number of macromolecules involving practically all metals of the Periodic Table of elements have been synthesized, the kinetic and thermodynamic characteristics of complex formation have been obtained and several aspects of the structural or supramolecular organization of macromolecular metal complexes have been elucidated. But it is important to notice that we are only partially able to mimic natural systems based on specific structural organizations. Today, as well as their thermal stability and catalytic activity, macromolecules containing metals are of increasing interest as conductors, photoconductors, sensors, magnetic materials and medical materials, as are their interactions with small molecules and their role in electron/photoelectron induced processes, (see Section 1.4 and in more detail Chapters 9–14). The most recent development is the formation and stabilization of metal nanoparticles/clusters (size 1–100 nm) in macromolecules (Chapter 8). This aspect is extensively developed now and has led to new composite materials with several interesting properties.

Thus, the field of metals in macromolecules covers all aspects of chemistry, especially polymer chemistry, analytical chemistry, coordination chemistry, theoretical chemistry and bioinorganic/organic chemistry, and overlaps with biology, physics, medicine and engineering science.

1.4   Properties and Potential Applications of Metal Complexes and Metals in Macromolecules

The combination of a specific metal with a definite macromolecule (organic or inorganic, either linear or cross-linked) gives rise to both active and selective properties, with high performance and stability. Therefore the first step must be the planning and carrying out of a successful synthesis. The necessary second step is a detailed analysis by different instrumental analytical techniques. The complex molecular arrangements on different levels must consider:

- primary structure → composition of the metal-containing macromolecule,
- secondary structure → steric orientation of a metal-containing macro-molecular unit,
- tertiary structure → orientation of a whole metal-containing macromolecule,

- quarternary structure → interaction of different metal-containing macro-molecules.

Nature shows how complicated it is to construct metal-containing macromolecules that are active but also selective with respect to a specific property. Natural systems do not need to exhibit a high stability towards storage and heat because the active materials are readily replaced. On the other hand, artificial systems must be more stable over time and towards heat. Therefore extreme demands are placed upon artificial metal-containing macromolecules. It is important to point out that the fundamental behaviors of low molecular weight metal complexes will be shown also in macromolecular analogues. But these behaviors, and thus the properties, are strongly influenced by the kind of macromolecular environment or the kind of incorporation into a macromolecule.

Type I describes the binding of a metal compound at a macromolecule (Sections 1.1.1, 4, 5). In linear organic polymers the secondary binding forces of a bound metal to other parts of the chain are weak (coordination bonds, hydrogen bonds, charge-transfer interactions, hydrophobic interactions). But because they are multiple, these binding forces cooperatively play an important role. These dynamic and multiple weak secondary forces are often affected by dynamic conformational changes of the macromolecular molecular environment. Electronic interactions and transfer processes to guest molecules can occur which can lead also to molecular conversions (Fig. 1-10). Binding of metal compounds at rigid cross-linked organic polymers (e.g. cross-linked polystyrene), at the surface of organic polymers (e.g. polyethylene) or at the external and, in the case of porous materials, at the internal surface of inorganic macromolecules (e.g. silica, zeolites) results in more "free" metal compounds, not firmly bound to the matrix by strong interactions. Thus, the properties of such compounds are more comparable with those of analogous low molecular weight metal complexes. Characteristic properties and uses of type I metal-containing macromolecules, which may or may not involve photoexcitation interactions with each other or guest molecules, include:

- Binding of small molecules (gas transport, gas separation, sensor).
- Catalysis, photocatalysis.
- Photoinduced electron or energy transport.
- Electrocatalysis, photoelectrocatalysis.
- Biosensors and biochips.
- Smart materials.
- Polymer electrolytes.
- Ion-exchange resins.
- Polymer coatings.

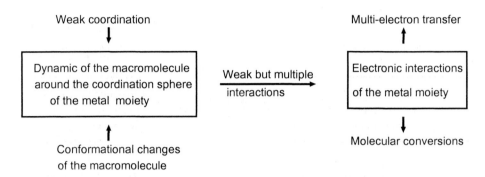

Figure 1-10. Dynamic and fixed situations in type I metal-containing macromolecules.

An important use of polymer ligands is the extraction of trace amounts of metals, e.g. gold, uranium, from seawater. For example, an industrial extraction of uranium from seawater has been developed using polymers with amidoxime groups [76]. Also the role of macromolecular metal complexes in Green chemistry is very significant, for example their efficient use in the neutralization of effluents from chemical companies.

Type I macromolecules have become more and more important as drugs in medicine:

- They can supply iron for iron deficiency anemia [77].
- They can remove toxic metal ions (Pb^{2+}, Cd^{2+}, Cu^{2+}) [78].
- Mn^{2+}, Ce^{4+} radioactive isotopes as well as molybdate ions bound to polyoxiranes can be used for treatment of pulmonary disease [79].
- Products of the reaction of K_2PtCl_4 with polymers are used similarly as the antineoplastic drug *cis*-dichlorodiamminoplatinum, for example for treatment of leukemia [80].

In **type II and III** metal-containing macromolecules the ligand or the metal is part of a cross-linked macromolecule. The solubility of such often rigid macromolecules is enhanced by introducing bulky substituents, which also improve the processibility. The alignment of complex moieties along the macromolecule matrix or within the macromolecule results in a material with integrated electronic processes (Fig. 1-11). Charge interaction/transfer, induced by an electrical field or by photoexcitation, yields materials with the following uses:

- Conductors, photoconductors.
- Photovoltaic cells, light emitting diodes.
- Electrochemical, photoelectrochemical cells.

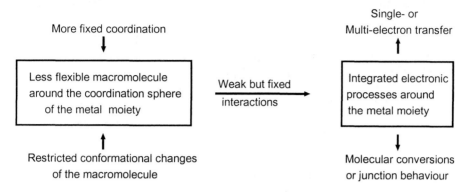

Figure 1-11. Electronic interactions in type II or III metal-containing macromolecules.

In **type II** macromolecules the metal complex moieties can also be connected to each other in a chain via covalent single bonds, which results in a more flexible backbone. In principle, this can give rise to the properties mentioned for type I macromolecules.

Type IV metal-containing macromolecules involve metal nanoparticles or metal complexes physically incorporated in macromolecules. The properties are general influenced greatly by the kind of macromolecular environment:

- Metallo–polymer nanohybrids.
- Polymer–inorganic composites.
- Intercalation metallopolymers.
- Magnetic nanocomposites.
- Bioinorganic nanocomposites.
- Metallized plastics.

Metal complexes can be incorporated in a macromolecule distributed either monomolecularly or in an aggregated state. These materials are characteristically used in thin film devices as:

- Conductors, photoconductors.
- Photoinduced electron or energy transporters.
- Electrocatalysts, photoelectrocatalysts.

1.5   Examples of Inorganic Polymers

Inorganic polymers or macromolecules which contain no carbon in the chain or network are not treated in detail in this book. They are only mentioned in the following chapter when a metal is bound at the chain, the network or on the surface of an inorganic material.

Classical "inorganic macromolecules" such as silica, silica glasses, borate, inorganic molecular sieves, polyphosphates, polymeric sulfur, etc. are treated for example in [6–10]. A short overview of Si, P, N and S macromolecules is given now because such macromolecules are intermediate between organic polymers and metal-containing macromolecules (in addition to the cited references see also keywords in Ullmann's *Encyclopedia of Industrial Chemistry* and *Encyclopedia of Chemical Technology*).

Silicones (polysiloxanes) contain $(-Si(R_x)-O-)_n$ structural elements [8,81–83]: **9a** in linear silicones, **9b** in cross-linked silicones and **9c** as possible end groups in silicones (Fig. 1-12). R is commonly $-CH_3$. **9d** is the moiety in silica and sol–gel based silica materials. The common synthesis, e.g. of a linear silicone, is the acid-catalysed controlled hydrolysis of a dichlorodialkylsilane to cyclosiloxane trimers and/or tetramers (Eq. 1-1). Then a ring-opening polymerization either thermally or using an anionic or cationic initiator leads to the linear **9a**. Silicones exhibit the advantage of low glass transition temperature (T_g for **9a** at $-123\,°C$), high oxidative stability ($\sim300\,°C$), high volume resistivity and good dielectric strength/loss. Silicone elastomers with excellent mechanical properties are obtained by different cross-linking reactions. The properties are modified by preparing block copolymers or graft copolymers of linear silicones with organic polymers.

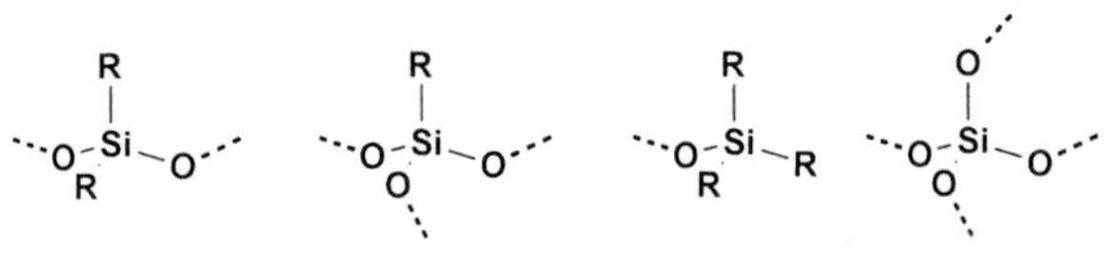

Figure 1-12. Structural units of silicones 9 (R = −alkyl, −aryl, −H).

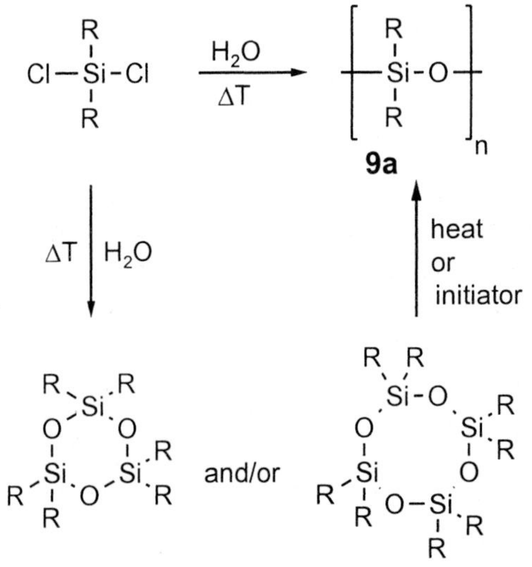

$(1-1)$

Polycarbosilanes contain carbon and silicon in the chain and commonly consist of $(-Si(R_x)-CH_2-)_n$ (R = $-H$, $-CH_3$) [9,10,82,84,85]. These polymers are prepared by thermal rearrangement of poly(dimethylsilane) e.g. to the idealized structure **10** (Eq. 1-2), coupling of CH_2X_2 or as shown for **11** of dilithium acetylides with R_2SiX_2 (Eq. 1-3) and ring-opening polymerization of 1,3–disilacyclobutane with H_2PtCl_6 as catalyst to **12** (Eq. 1-4). **10** has a melting temperature of ~25 °C and a glass temperature of about −135 °C. Polycarbosilanes are used as preceramic polymers for silicon carbide (SiC) fibers.

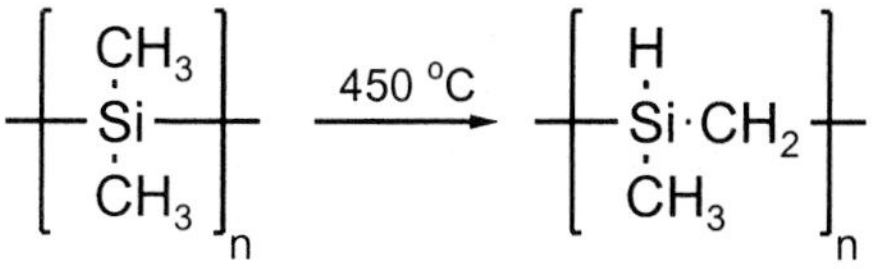

$$
\left[\begin{array}{c} CH_3 \\ | \\ Si \\ | \\ CH_3 \end{array} \right]_n \quad \xrightarrow{450\ °C} \quad \left[\begin{array}{c} H \\ | \\ Si \cdot CH_2 \\ | \\ CH_3 \end{array} \right]_n
$$

$$(1\text{-}2)$$

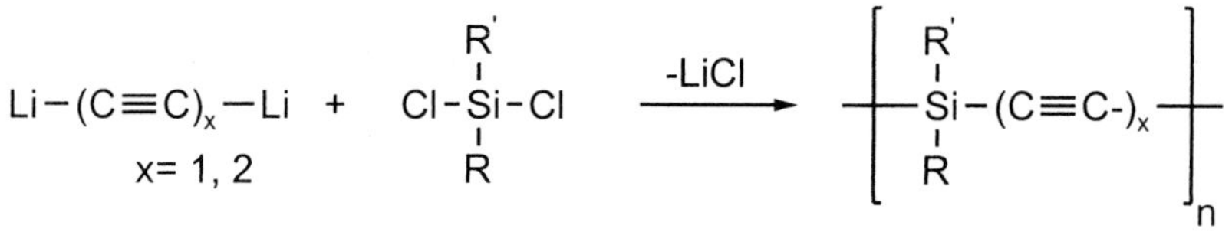

10

$$
Li-(C{\equiv}C)_x-Li \; + \; \begin{array}{c} R' \\ | \\ Cl-Si-Cl \\ | \\ R \end{array} \quad \xrightarrow{-LiCl} \quad \left[\begin{array}{c} R' \\ | \\ Si-(C{\equiv}C\text{-})_x \\ | \\ R \end{array} \right]_n
$$
$$x = 1, 2$$

$$(1\text{-}3)$$

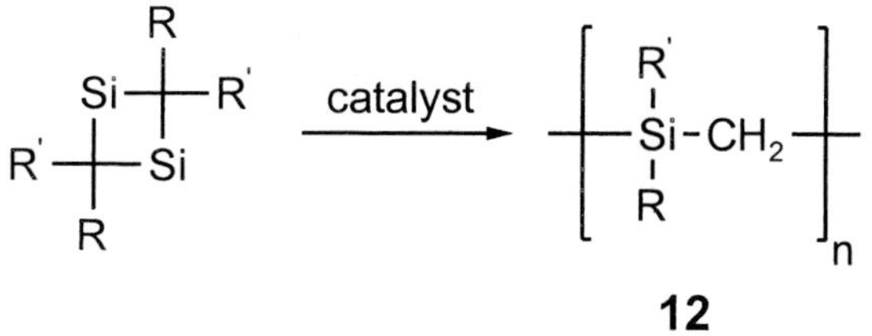

11

$$
\begin{array}{c} R \\ Si-\!\!\!\!\!\!+\!\!\!-R' \\ R'-\!\!\!+\!\!\!-Si \\ R \end{array} \quad \xrightarrow{catalyst} \quad \left[\begin{array}{c} R' \\ | \\ Si-CH_2 \\ | \\ R \end{array} \right]_n
$$

$$(1\text{-}4)$$

12

A Si–Si bond with the structural element $(-SiR_2-)_n$ (R = alkyl, aryl, silyl) is realized in polysilanes (polysilylenes) **13** [8–10,81,86,87]. The main route of synthesis uses the dehalogenation of diorganodichlorosilanes with finely dispersed sodium (Eq. 1-5). The metal-catalyzed dehydrogenation of diorganosilanes is an alternative route for the synthesis of **13**. Polysilynes $(RSi)_n$ are prepared by reduction of alkyl- or aryltrichlorosilanes. Depending on the substituents R in polysilanes **13** the glass transition temperatures are between −75 °C and 120 °C. Because these polymers are photochemically degraded by UV light, their most promising technical application are as positive photoresists in microlithography.

The structural element of polysilazanes **14** consists of $(-SiR_2-NR-)_n$. They are prepared by the reaction of halogenosilanes with ammonia or amines (Eq. 1-6) [9,10,84]. High molecular weight polysilazanes are obtained by anionic ring-opening polymerization of cyclosilazanes. These materials are interesting as preceramic polymers for silicon nitride (Si_3N_4).

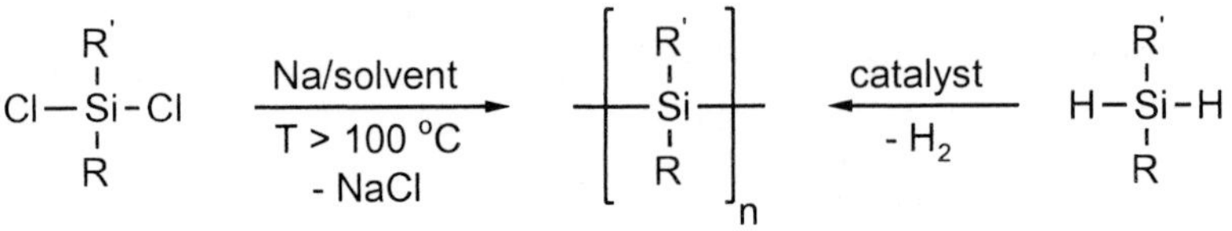

$$
\begin{array}{c}
R' \\
| \\
Cl-Si-Cl \\
| \\
R
\end{array}
\quad
\xrightarrow[\substack{T > 100\,°C \\ -\,NaCl}]{Na/solvent}
\quad
\left[
\begin{array}{c}
R' \\
| \\
Si \\
| \\
R
\end{array}
\right]_n
\quad
\xleftarrow[-\,H_2]{catalyst}
\quad
\begin{array}{c}
R' \\
| \\
H-Si-H \\
| \\
R
\end{array}
\qquad (1\text{-}5)
$$

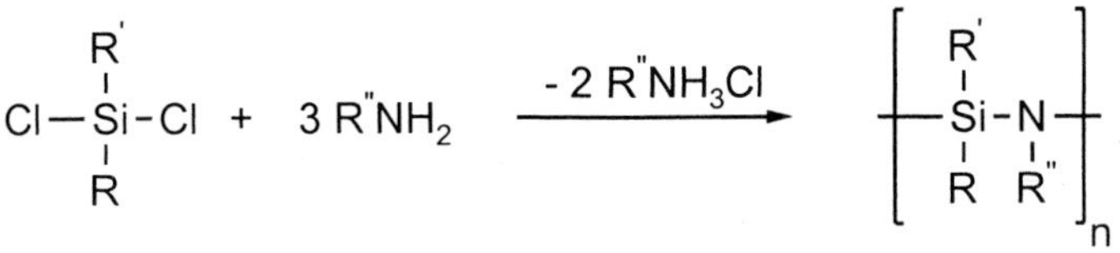

13

$$
\begin{array}{c}
R' \\
| \\
Cl-Si-Cl \\
| \\
R
\end{array}
+\ 3\,R''NH_2
\quad
\xrightarrow{-\,2\,R''NH_3Cl}
\quad
\left[
\begin{array}{c}
R' \\
| \\
Si-N \\
| \quad | \\
R \quad R''
\end{array}
\right]_n
\qquad (1\text{-}6)
$$

14

The backbone of polyphosphazenes **17** contains ($-PR_2=N-$). (R = $-OR'$, $-NR''_2$) structural units [8–10,88]. The first step in the most generally used method of synthesis is the ring-opening polymerization of hexachloro-cyclotriphosphazene **15** (obtained from PCl_5 and NH_4Cl) to poly(di-chlorophosphazene) **16** at raised temperature or in the presence of $AlCl_3$ as catalyst. Then reaction with different nucleophiles leads to **17a–c**. Alkyl or aryl polyphosphazenes ($-PR_2=N-$) (R = alkyl, aryl) are prepared by thermal condensation or in the presence of a catalyst of N-silylphosphoranimines, $Me_3SiN=PR_2X$ (X = $-OCH_2CF_3$, $-OC_6H_5$). A wide-range of properties (e.g. crystallinity vs. amorphous behaviour and hydrophilicity vs. hydrophobicity) can be introduced into the polyphosphazenes through variation of the substituents. Glass transition temperatures vary between around $-100\,°C$ and around $+100\,°C$. All of this leads to a wide variety of practical uses.

Some other inorganic polymers may be mentioned [8,10]: polyoxothiazenes **18** (prepared by thermal condensation of sulfon imidates, $RO(O)S(R)=NH$); highly conducting polythiazyl (poly(sulfur nitride)) **19** (prepared from S_2Cl_2 and NH_3 via a cyclic alternating nitride tetramer $(SN)_4$); polycarboranes and polycarbosilanes (e.g. ($-CB_{10}H_{10}C-Si(CH_3)_2-O-)_n$) containing a m-carborane moiety.

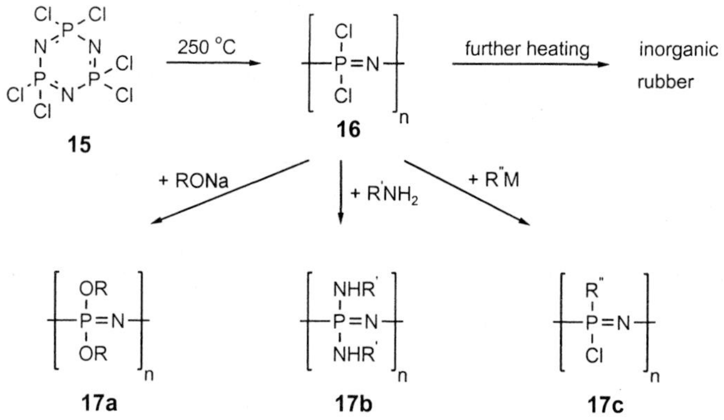

$$(1-7)$$

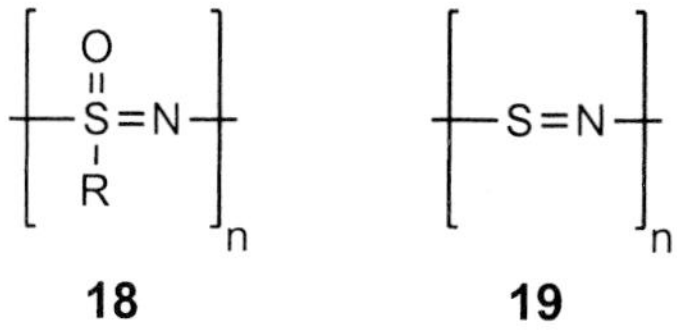

1.6 References

1. F. Ciardelli, E. Tsuchida, D. Wöhrle, *Macromolecule–Metal Complexes*, Springer, Berlin, 1996.
2. S. Mann, J. Webb, R. Williams, *Biomineralization: Chemical and Biochemical Perspectives*, VCH Publishers, Weinheim, 1989.
3. S. Mann (Ed.), *Biomimetic Materials Chemistry*, VCH Publishers, Weinheim, 1996.
4. M. Sarikaya, I.A. Aksay, *Biomimetics. Design and Processing of Materials,* AIP Press, New York, 1995.
5. B.P. Block, P.M. Thomas, K.M. Donovan, *J. Chem. Soc.* **1969**, *9*, 242.
6. N.H. Ray, *Inorganic Polymers*, Academic Press, New York, 1978.
7. F.G.A. Stone, W.A.G. Graham (Eds.), *Inorganic Polymers*, Academic Press, New York, 1962.
8. U. Schubert, N. Hüsing, *Synthesis of Inorganic Materials*, Wiley-VCH, Weinheim, 2000.
9. J.E. Mark, H.R. Allcock, R. West, *Inorganic Polymers*, Prentice Hall, Englewood Cliffs, 1992.
10. R.D. Archer, *Inorganic and Organometallic Polymers*, Wiley-VCH, Weinheim, 2001.
11. A.D. Pomogailo, V.S. Savost'yanov, *Synthesis and Polymerization of Metal Containing Monomers*, CRC Press, Boca Raton, 1994.
12. N. Hagihara, K. Sonogashira, S. Takahashi, *Adv. Polym. Sci.* **1981**, *41*, 149.
13. M.R. Mattner, W.A. Hermann, R. Berger, C. Gerber, J.K. Gimzewski, *Adv. Mater.*

1996, *8*, 654.

14. D. Wöhrle, *Macromol. Rapid. Commun.* **2001**, *22*, 68.
15. H. Chen, R.D. Archer, *Macromolecules* **1995**, *28*, 1609; ibid. **1996**, *29*, 1957.
16. J.E. Sheats, C.E. Carraher, C.U. Pittman, M. Zeldin, B. Currell (Eds.), *Inorganic and Metal-Containing Polymeric Materials*, Plenum Press, New York, 1990.
17. S.R. Batten, B.F. Hoskins, R. Robson, *Angew. Chem.* **1995**, *107*, 884.
18. F. Flaig, in: *Soil Components*, H. Gieseking (Ed.), Springer, New York, 1975, vol. 1, p. 1–121.
19. R.B. Frankel, R.P. Blakemore, *Iron Biominerals*, Plenum Press, New York, 1991.
20. S.A. Marakushev, *Geomicrobiology and Biochemistry of Gold*, Nauka, Moscow, 1991.
21. G.A. Ozin, *Acc. Chem. Res.*, **1977**, *30*, 17.
22. A.S. Monin, A.P. Lisitsyn, *Biogeochemistry of Ocean*, Nauka, Moscow, 1983.
23. A.K. Dey, *J. Indian Chem. Soc.* **1986**, *63*, 357.
24. T. Graham, *Philos. Trans.* **1833**, *123*, 253.
25. J.C. Bailar, in: *Organometallic Polymers*, C.E. Carraher, J.E. Sheats, C.U. Pittmann (Eds.), Academic Press, New York, 1978.
26. C.S. Marvel, N. Tarkoy, *J. Am. Chem. Soc.* **1957**, *79*, 6000; ibid. **1958**, *80*, 832.
27. H.A. Goodwin, J.C. Bailar, *J. Am. Chem. Soc.* **1961**, *83*, 2467.
28. J.R. Millar, *Chem. and Ind.* **1957**, *606*.
29. F.R. Hartley, *Supported Metal Complexes. A New Generation of Catalysts,* Reidel Pub. Co., Dordrecht, 1985.
30. A.D. Pomogailo, *Catalysis by Polymer-Immobilized Metal Complex*, Gordon and Breach Publ., Amsterdam, 1998.
31. A.L. Robinson, *Science* **1976**, *194*, 1261.
32. E. Tsuchida, H. Nishide, *Adv. Polym. Sci.* **1977**, *24*, 1.
33. M. Kaneko, A. Yamada, *Adv. Polym. Sci.* **1984**, *55*, 1.
34. D. Wöhrle, *Adv. Polym. Sci.* **1983**, *50*, 45.
35. C.E. Carraher, J.E. Sheats (Eds.), *Organometallic Polymers*, Academic Press, New York, 1978.
36. C.E. Carraher, J.E. Sheats, C.U. Pittmann (Eds.), *Advances in Organometallic in Inorganic Polymer Science*, Marcel Dekker, New York, 1982.
37. J.E. Sheats, C.E. Carraher, C.U. Pittmann (Eds.), *Metal-Containing Polymeric Systems*, Plenum Press, New York, 1985.
38. E. Tsuchida (Ed.), *Macromolecular Complexes, Dynamic Interactions and Electronic Processes*, VCH Publishers, New York 1991.
39. A.D. Pomogailo, I.E. Uflyand, *Macromolecular Metal Chelates*, Nauka, Moscow, 1991.
40. C.U. Pittmann, C.E. Carraher, M. Zeldin, J.E. Sheats, B.M. Culbertson, *Metal Containing Polymeric Materials*, Plenum Press, New York, 1996.
41. D. Wöhrle, A.D. Pomogailo, Metal-Containing Macromolecules, in: *Advanced Functional Molecules and Polymers*, Vol. 1, H.S. Nalwa (Ed.), Gordon & Breach Science Pub., Amsterdam, 2001.
42. D. Wöhrle, G. Schnurpfeil, Porphyrins and Phthalocyanines, in: *The Porphyrin Handbook*, Vol. 17, K.M. Kadish, K.M. Smith, R. Guilard (Eds.), Academic Press, San Diego, 2002.

43. Scientific Journal: *J. Inorg. Organomet. Polym,.* Plenum Press, New York and London.
44. J. Reedijk (Ed.), *Bioinorganic Catalysis*, Marcel Dekker, New York, 1993.
45. A.D. Pomogailo, *Polymeric Immobilized Metallocomplex Catalysts*, Nauka, Moscow, 1988.
46. Y.I. Yermakov, B.N. Kuznetsov, V.A. Zakharor, *Catalysis by Supported Complexes*, Elsevier, Amsterdam, 1981.
47. Y.I. Yermakov, V. Likholobov, *Homogeneous and Heterogeneous Catalysis*, VNU Science Press, Utrecht, 1987.
48. E.A. Bekturov, S.E. Kudaibergenov, *Catalysis by Polymers*, Hüthig & Wepf–Verlag, Heidelberg, 1996.
49. P.A. Jacobs, N.I. Jaeger, L. Kubelkova, B. Wichterlova (Eds.), *Zeolite Chemistry and Catalysis, Studies in Surface Science and Catalysis*, Vol. 69, Elsevier Science Pub., Amsterdam, 1991.
50. H. Karge, J. Weitkamp (Eds.), *Zeolites as Catalysts, Sorbents and Detergent Builders, Studies in Surface Science and Catalysis*, Vol. 46, Elsevier Science Pub., Amsterdam, 1989.
51. V. Ramamurthy (Ed.), *Photochemistry in Organized and Constrained Media,* VCH Publishers, New York, 1991.
52. V.V. Korshak, N.M. Kozyreva, *Russ. Chem. Rev.* **1983**, *54*, 1091.
53. M. Hanack, M. Lang, *Adv. Mater.* **1994**, *6*, 819; ibid, *Chemtracts – Org. Chem. Adv.* **1995**, *8*, 131.
54. G.A. Ozin, C. Gil, *Chem. Rev.* **1989**, *89*, 1749.
55. D.C. Sherrington, *Pure Appl. Chem.* **1988**, *60*, 401.
56. D. Wöhrle, Polymers with Metals in the Backbone, in: *Handbook of Polymer Synthesis*, Vol. B, H. Kricheldorf (Ed.), Marcel Dekker, New York, 1992.
57. D. Wöhrle, Phthalocyanines in Polymer Phases, in: *Phthalocyanines, Properties and Applications*, Vol. 1, C.C. Leznoff, A.B.P. Lever (Eds.), VCH Publishers, New York, 1989.
58. A.D. Pomogailo, *Russ. Chem. Rev.* **1992**, *61*, 133.
59. A.D. Pomogailo, I.E. Uflyand, *Adv. Polym. Sci.* **1990**, *97*, 61.
60. M. Rehahn, *Acta Polym.* **1998**, *49*, 201.
61. M. Kaneko, D. Wöhrle, *Adv. Polym. Sci.* **1988**, *84*, 141.
62. M. Hanack, S. Deger, A. Lange, *Coord. Chem. Rev.* **1988**, *83*, 115.
63. M. Biswas, A. Mukherjee, *Adv. Polym. Sci.* **1994**, *115*, 89.
64. I. Manners, *Angew. Chem.* **1996**, *168*, 1713.
65. L. Oriol, J.L. Serrano, *Adv. Mater.* **1995**, *7*, 348.
66. S.R. Batten, R. Robson, *Angew. Chem.* **1998**, *110*, 1558.
67. A.D. Pomogailo, Polymer Immobilized Nanoscale and Cluster Particles, *Russ. Chem. Rev.* **1997**, *66*, 679.
68. MMC II, *J. Macromol. Sci.–Chem.* **1988**, *A25*, Vol. 10 and 11.
69. MMC III, *J. Macromol. Sci.–Chem.* **1990**, *A26*, Vol. 2 and 3; **1990**; *A27*, Vol. 9–11.
70. MMC IV, *Macromol. Chem., Macromol. Symp.* **1992**, *59*.
71. MMC V, *Macromol. Chem., Macromol. Symp.* **1994**, *80*.
72. MMC VI, *Macromol. Chem., Macromol. Symp.* **1996**, *105*.

73. MMC VII, *Macromol. Chem., Macromol. Symp.* **1998**, *131*.
74. MMC VIII, *Macromol. Chem., Macromol. Symp.* **2000**, *156*.
75. MMC IX, *Macromol. Chem., Macromol. Symp.* **2002**, *in press*.
76. F. Vernon, T. Shah, *React. Polym.*, **1983**, *1*, 301.
77. F.T. Chuk, M. Koleva, S. Mikhailova, *Farmacy*, **1988**, *38*, 16.
78. A. Malovikova, R. Kogan, *Collect. Czech. Chem. Commun.* **1983**, *48*, 3154.
79. D.M. McPhillips, T.A. Armer, D.R. Owen, *J. Biomed. Mater. Res.*, **1983**, *17*, 993.
80. H.R. Allcock, R.W. Allen, J.P. O'Brien, *J. Amer. Chem. Soc.* **1977**, *99*, 3984.
81. S. Patei, Z. Rappaport (Eds.), *The Chemistry of Organic Silicon Compounds*, Wiley, New York, 1989.
82. Z. Rappaport, Y. Apeloig (Eds.), *The Chemistry of Organic Silicon Compounds*, Wiley, New York, 1998.
83. H.R. Kricheldorf (Ed.), *Silicon in Polymer Synthesis*, Springer, Heidelberg, 1996.
84. M. Birot, J.–P. Pillot, J. Dunogues, *Chem. Rev.* **1995**, *95*, 1443.
85. R.M. Laine, F. Babonneau, *Chem. Mater.* **1993**, *5*, 260.
86. K. Matyjaszewski, D. Greszta, J.S. Hrkach, H.K. Kim, *Macromolecules* **1995**, *28*, 59.
87. R.D. Miller, *Angew. Chem. Adv. Mater.* **1989**, *101*, 1773.
88. C.W. Allen, *Coord. Chem. Rev.* **1994**, *130*, 137.

2 Macromolecular Metal Complexes in Biological Systems

Dieter Wöhrle and Masao Kaneko

The functions of biological systems are based on five main components:
- polymers such as proteins, carbohydrates, nucleic acids
- metal ions/complexes/clusters (see below)
- water
- gases such as carbon dioxide, oxygen and, to some extent, nitrogen
- photons (in photosynthesis).

Super- or supramolecular arrangements of *protein polymers* and the metal part are essential for biologically important reactions such as gas transport, catalysis and photocatalysis. An understanding of the reactivity of biological reaction centers is of fundamental importance for the construction and optimization of artificial systems. In modern textbooks of biochemistry a lot of information is available on the construction and function of biological systems. This chapter first outlines a general view of elements in life, and then gives examples of the interaction and function of metals in biological materials. It is very important to realize that activity and selectivity only arise from the combination of a specific metal in a specific ligand surrounding with a specific natural macromolecule. References [1–7] contain selected examples of metals in biological matter.

It should be noted that the interdisciplinary field of bioinorganic chemistry draws on the strength of both inorganic chemistry and the biological sciences. Investigations go beyond the influence of metals ions, metal cluster and other inorganic compounds on the function of biological matter. Other subjects such as biomineralization and biomimetic materials chemistry are part of this field. Biological macromolecular metal complexes can be treated as part of bioinorganic chemistry.

2.1 Elements Essential for Life

Life is believed to have originated spontaneously some $4 \cdot 10^9$ years ago from a mixture of 15–20 elements in the sea. Around 2 to $1 \cdot 10^9$ years ago an important change occurred when the atmosphere, which had been a reducing one, changed to an oxidizing one because light could be used to reduce CO_2 and at the same time to liberate O_2. The following development of a "fuel cell" from O_2 + reduced carbon allowed life to escape from a static (plant-like) existence to a mobile animal-like scavenger condition.

A look at the Periodic Table shows that almost every group is represented in living matter (Fig. 2-1).The variety of metals used by biological systems is very large, ranging from the alkali metals to the transition metals. They play an essential role in living systems, both in growth and metabolism.

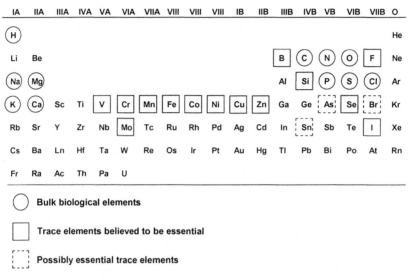

Figure 2-1. Elements essential for life (after Williams in [5]).

Table 2-1 gives a survey of metals essential for life and summarizes the amounts of metals in the human body. The question is, what makes the function of these around twenty-five elements so valuable in making life. Some metals are necessary in gram quantities. Other trace elements such as Mn, Mo, Co, V, W, Ni and Cr are essential beneficial nutrients at low levels but metabolic poisons at high levels. Some metal ions such as Pb, Cd are called "detrimental metal ions" because they are toxic and impair the regular course of life functions at all concentrations.

Elements essential for life are now classified in four groups:

H, C, N, O, P, S: The elements H, C, N and O make the best polymers such as proteins, with high kinetic stability. The elements P in phosphate esters or anhydrides and S in thiols or thioethers increase the reactivity of small molecules and polymers. Clusters of S with Fe or Cu are important electron carrier systems.

Table 2-1. Important metals in biology.

Metal	Concentration in human in mg	Human daily required amount in mg	General role in biology
Ca	10^6	800	Bones, teeth, muscle activity
K	$1.5 \cdot 10^5$		Nerve axon potential, osmotic cell pressure
Na	$1 \cdot 10^5$		Nerve axon potential
Mg	$3 \cdot 10^4$	500	Photosynthesis, nucleic acid
Fe	$4.5 \cdot 10^3$	15	Enzymes, respiratory
Zn	$2 \cdot 10^3$	12	Enzymes, nucleic acid
Cu	100	3	Enzymes, O_2-transport
Mn	20	3	Photosynthesis, enzyme activity
Mo	5	0.2	Redox enzymes
Co	1.5	0.3	Vitamin B12
Cr	1	0.06	Glucose tolerance factor
V	<0.1	<0.1	Role in bromooxidase in algae
Ni	<0.1	0.4	Role in hydrogenase in bacteria

Cl, Na, K: The anion Cl^- and the metal ions Na^+, K^+ do nothing chemically in living systems but they are important in fundamental life processes. Na^+ and K^+ control the osmotic pressure and therefore the integrity and shape of living cells. In addition, Na^+ and K^+ neutralize the negative charges of intracellular anions and serve as charge carriers in neurophysiological processes. The most important biochemical devices are the energy-consuming pumping and passive transport across biomembranes, both catalyzed by integral membrane proteins. Na^+-, K^+-ATPase is involved in the enzyme activity for the osmotic balance of cells as well as in the generation of nerve axon potentials, using Na^+, K^+ and Mg^{2+} for its activity. The action of this enzyme involves the hydrolysis of one ATP with the export of three Na^+ and the import of two K^+ through the cell membrane. Therefore living cells accumulate K^+ relative to their exterior medium and maintain in membranes at low K^+ concentration. Toxic effects for these functions result mainly from Tl^+, Ba^{2+} and VO_3^-. In addition, transition metal ions such as Co^{2+}, Ni^{2+}, Cu^{2+}, Zn^{2+} and CrO_4^{2+} interfere with Na^+-, K^+-ATPase.

Mg, Ca: Mg^{2+} is necessary for nearly every reaction in the production and consumption of biochemical energy which means it serves as a cofactor in the biochemistry of phosphates. For these reactions, Mg^{2+} is bound to oxygen atoms in phosphates, and bridges these to the side groups of enzymes. Mg^{2+} functions therefore as a weak Lewis acid catalyst in phosphate regeneration. Biological functions of Mg^{2+} are disrupted by Li^+, Be^{2+}, Al^{3+}, Mn^{2+} and other transition metal ions. Another important location of Mg^{2+} is in the center of chlorophyll. This porphyrin derivative is essential in the harvesting of light and as a photosensitizer in the photosynthesis mechanism. Ca^{2+} is active as a signal and regulator of nearly every major biological process in multicellular organisms. The processes include control of intermediary metabolism, cell proliferation and differentiation, motion and membrane transport. The concentration of Ca^{2+} is controlled by the balance between Ca^{2+} pumping out of the cytoplasm by Ca^{2+}-ATPase and Ca^{2+} influx through controlled channels. In response to extra-cellular signals, Ca^{2+} in the cytoplasm can activate enzymes by binding first to special proteins and then to specific regulator sites. Binding of Ca^{2+} occurs mainly at oxygen atoms, with the highest affinity for uncharged oxygen as in alcohols or carbonyl compounds. In contrast to Mg^{2+}, Ca^{2+} can form complexes with up to eight ligands with irregular geometry: the ionic radius is relatively large (~ 0.1 nm); ($Mg^{2+} \sim 0.067$ nm), the character of the ion is very hard and ligand exchange rate constants are very high ($\sim 10^8$ to 10^9 s^{-1}). Typical toxic metal ions which interfere with the metabolism of Ca^{2+} are Pb^{2+}, Cd^{2+} and Hg^{2+}.

So far we have briefly discussed eleven elements: H, C, N, O, P, S, Cl, and the metal ions Na, K, Mg and Ca. These elements are the basis of life.

Zn, Fe, Cu, Mn, Mo, Co, (V), Ni, Si, Se, (B): Life also requires many of these ten other elements. Some living systems also require elements such as Sr, Ba, W and I, so now almost every group of the Periodic Table is represented. Metal ions are active in redox chemistry as catalysts and photocatalysts for the use of raw materials such as H_2O, CO_2, N_2 and in some cases also CH_4. For example, no N_2 is fixed without Mo, and very little C/H/O redox chemistry proceeds without Fe. Some examples are listed below. Moreover, as well as redox activity, stability constants and kinetics transition metal ions modify structural and dynamic properties in living systems. Zn^{2+} is necessary for all forms of life and is active in around 200 enzymes, a dozen regulatory proteins and some peptide toxins. Zn^{2+} is an ideal Lewis acid catalyst due to its small ionic radius (~ 0.074 nm) and high charge density. This metal ion binds to N-, S- and O-containing functional groups, forming tetrahedral, pyramidal and octahedral coordination with four or six ligands. The broad variation in biochemical function of Zn^{2+} also correlates with different ligand exchange rates in proteins: the half-life for ligand exchange varies between one hour in all S-coordination (e.g. metallothioneine) and a few days with N- and O-coordination. In many hydrolases (e.g. protein catalyzing enzymes) and dehydrogenases (e.g. alcohol dehydrogenase) Zn^{2+} is responsible for the folding of various proteins. In transcription factors Zn^{2+} folds the protein structure by coordinating four amino acid side chains with cystein sulfur and histidine nitrogen atoms. Toxic metal ions interfering with Zn^{2+} functions are mainly Co^{2+}, Cd^{2+} and Pb^{2+}.

Fe^{2+}/Fe^{3+} and Cu^{2+} are involved in biochemical catalysis especially in electron transfer under physiological conditions. FeS-proteins are active in the redox reactions of mitochondria and chloroplasts. Fe-porphyrins as cytochromes carry electrons in the respiratory chain of mitochondria. They are in the cytoplasm as catalytically active groups in monooxygenases to detoxify hydrocarbons by hydroxylation. Fe^{2+} is the reversible O_2 carrier in hemoglobin and myoglobin. Special Fe^{3+}-transporting proteins (transferrin) are used for transmembrane transport of iron. Little is known about the interference of toxic metal ions with iron. Some heavy metal ions do not compete with the cytochrome iron but inactivate the enzyme protein component. Pb^{2+} can interfere with the insertion of Fe^{2+} into protoporphyrin. Cu replaces Fe as the active part for electron transfer in some enzymes and in photosystem II. Because free Cu^{2+} is highly toxic, it is transported in the blood stream to a protein (ceruloplasmin). Excess of Cu^{2+} in tissues is reduced to Cu^+, bound to metallothionein and then no longer bioavailable.

It is beyond the scope of this review to mention in detail the functions of other transition metals or their ions. Detailed information is available in [1–7].

We need to know how the nonmetallic elements C, H, N, O, P and S combine with the metal ions. It is possible to divide the metal elements roughly into three groups (Table 2–2) [5]. The cations Na and K combine only with water, the cations Mg, Ca (S, Ba) with water and oxygen anions, while transition metal ions Mn, Fe, Co, Ni, Cu, Zn and Mo combine with N and S donors. Exceptions such as Mg in chlorophyll do exist. Na, K, Mg, Ca and Mn preferentially form ionic bonds, and the transition metal ions form bonds intermediate between ionic and covalent bonds. Toxic metals such as Hg, Cd, Pb, which impair life functions at all concentrations, form covalent bonds to N and RS^- ligand groups. Also the so-called HSAB theory [8] can be used to discriminate between the preferences of certain metals for certain ligands (Reedijk in [9]).

Table 2-2. The character of the major combinations between metal ions and ligands in living matter (after Williams in [5]).

Element	Combination
Na^+, K^+, Cl^-	H_2O
Mg^{2+}, Ca^{2+}	Combined O in C–O, P–O compounds
	H_2O
(Sr) (Ba)	H_2O, SO_4^{2-}
Mn^{2+}/Mn^{3+}	N, O ligands
Fe^{2+}/Fe^{3+}	RS^-, S^{2-}, N, O ligands
Co^{2+}/Co^{3+}	N (vitamin B_{12})
Ni^{2+}/Ni^{3+}	N, RS^-
Cu^+/Cu^{2+}	N, RS^-
Zn^{2+}	N, RS^-, combined O
Mo^{6+}/Mo^{5+}	RS^-, S^{2-}, O^{2-}

We also need to know how metals are transported into cells and stored. Iron has been investigated most intensively. Mammals bind and transport iron by the serum protein transferrin and store it in ferritin. One protein can bind around 4500 Fe^{3+} ions. Copper is taken up by the serum protein ceruloplasmin. Also albumin can bind and transport metal ions. The cystein-rich protein metallothioneine is formed in cells if toxic Cd^{2+} and PS^{2+} are present. This aprotein protects cells against the toxic effects of such metal ions.

2.2   Some Functions of Metals

It is necessary now to classify the metals with regard to their function in biological systems: metals as cofactors of proteins, metalloenzymes, communicative functions of metals, interaction of metal ions with polynucleotides, biometal-organic chemistry (e.g. metals in medicine).

2.2.1   Metals as Cofactors of Proteins, Metalloproteins

- Oxygen binding and transport
  Metalloproteins are involved in respiration. In these proteins a sensitive equilibrium exists for binding of O_2 to a metal center without irreversible oxidation of the metal. The three systems are hemoglobin/myoglobin, hemerythrin and hemocyanin. In the hemoglobin/myoglobin system, oxygen binds reversibly at an iron porphyrin complex which is embedded in the hydrophobic pocket of a protein. In hemerythrin, a protein of some marine invertebrates, the oxygen molecule is coordinated at an Fe_2-unit. The hemocyanin of molluscs and arthropods contain the oxygen molecule between two copper ions.

- Energy transfer followed by electron transfer
  In the photosynthesis of green plants, photosystems I and II (PS I, PS II) contain chlorophyll a, a Mg(II)-porphyrin, as an antenna system for light absorption and energy transfer to the reaction centers of PS I and PS II. PS II consists of a dimeric chlorophyll a as reaction center, pheophytin a, a metal-free chlorophyll a as electron transfer system to PS I and – on the other side – a water-oxidizing Mn cluster. The electron connection between PS II and PS I is carried out by a cytb_6/f complex (heme complexes and an FeS protein). The reaction center of PS I is also a dimeric chlorophyll (perhaps together with other chlorophylls), and chlorophyll a_o and several FeS proteins for electron transfer.

- Electron transfer
  Electron transfer centers exist in cytochromes and FeS clusters. The various cytochromes are structurally related to the Fe porphyrins in hemo-

globin/myoglobin. Iron–sulfur clusters of the general composition Fe_nS_m contain 4, 3, 2 and 1 iron atoms and carry charges between -4 and -1.

- Structural functions of metal ions
  Several proteins regulating the expression of mammalian genes contain Zn^{2+} ions. In these proteins the presence of the metal ions has structural consequences [2,5,10]. The regulatory structure of the genes consists of a sequence coding for a protein, a proximal upstream promoter sequence, which binds the general transcription factors. These are proteins and combine with the RNA polymerase, binding short DNA sequences. The transcription factors share common structural motifs. One of the most frequent is zinc fingers. Zn^{2+} ions are present in most DNA and RNA polymerases.

2.2.2   Metalloenzymes

Metalloenzymes are metalloproteins with catalytic functions, converting a substrate chemically.

- Hydrolytic enzymes (hydrolases)
  These proteins catalyze the addition or elimination of water to or from a substrate molecule. The carboanhydrases catalyze the hydration of CO_2, peptidases and esterases the hydrolysis of carboxylic acid compounds, and phosphatases the cleavage of phosphoric esters. The active centers of hydrolytically active enzymes often contain Zn^{2+} ions. Other metal ions in some of these enzymes are Mn^{2+}, Ni^{2+}, Ca^{2+} and Mg^{2+}.

- Redox enzymes
  Several metalloenzymes catalyze the oxidation and reduction of substrates. Often these in general two-electron redox processes are combined with a reaction of the substrate with another molecule. Examples of the addition of oxygen are: oxidation of hydrocarbons to alcohols by cytochrome P450 containing an Fe porphyrin (oxygen from an O_2 molecule), *ortho*-hydroxylation of phenolic substrates by tyrosinase containing a dimeric Cu complex (oxygen from an O_2 molecule); oxidation of sulfite to sulfate by sulfite-oxidase containing an Mo atom (oxygen from a water molecule). Other metalloenzymes such as ribonucleotide-reductase remove an oxygen atom from a substrate. Probably a dimeric Fe center is responsible for this reaction. The enzyme nitrate-reductase contains a Mo unit and is responsible for reduction of NO_3^- to NO_2^- as a key intermediate for the assimilation of nitrate by plants. Dehydrogenases remove two electrons and two protons. The alcohol-dehydrogenase in the liver converts acetaldehyde to ethanol and contains Zn^{2+} ions in an active center. Other hydrogenases are complex systems containing FeS clusters or often nickel.

- Redox reactions with several electron pairs

  These redox reactions involving more than two electrons in one reaction are of central importance in living organisms.

  One reaction is the reversible conversion of two molecules water to one molecule oxygen with exchange of four electrons (!): $O_2 + 4H^+ + 4e^- \leftrightarrows 2H_2O$. Cytochrome c oxidase, consisting of a complex enzyme with two copper and two Fe–heme centers, catalyzes the reduction of O_2 to H_2O which is finally important for the phosphorylation of ADP to ATP. The reverse reaction, the oxidation of H_2O to O_2, has a central role in the oxygen-evolving complex containing four Mn of photosystem II in photosynthesis.

  The enzyme dinitrogenase converts N_2 from air to NH_3 with eight electrons (!): $N_2 + 8H^+ + 8e^- \rightarrow 2NH_3 + H_2$. This enzyme contains one protein with a cluster of Mo, Fe and S atoms as the active center and a second protein with a [4Fe–4S]-cluster.

  The nitrate-reductase contains an Fe–porphyrin complex and catalyzes the reduction of NO_2^- to NH_3 with six electrons (!): $NO_2^- + 8H^+ + 6e^- \rightarrow NH_4^+ + 2H_2O$. A related enzyme is active in the conversion of sulfite to hydrogen sulfide with six electrons (!): $SO_3^{2-} + 8H^+ + 6e^- \rightarrow H_2S + 3H_2O$.

- Rearrangements

  In several biological reactions the oxidation state is not changed but rearrangements occur. Vitamin B_{12}, an alkyl–cobalt(III) complex of a substituted corrin, is a cofactor of enzymes catalyzing 1,2–carbon rearrangements. Another important reaction is the conversion of citrate to isocitrate in the citric acid cycle catalyzed by the enzyme acotinase containing a FeS-cluster.

2.2.3   Interference of Detrimental Metal Ions with Normal Cell Metabolism

Metal ions can be metabolic poisons. Some metal ions are essential at low levels but can be toxic at more elevated doses. Also, the chemical forms of the metal ions are important. For example, Cr^{3+} is necessary for mammals in micrograms per day, but Cr(VI), which is anionic, causes cancer. Cu^{2+} is highly toxic for mammals as free ion but when chelated by natural carriers this ion is essential. The intracellular distribution is also crucial. Fe^{2+}/Fe^{3+} taken up by humans in milligrams per day reaches millimolar concentrations in the cytoplasm of blood and liver cells. But micromolar concentrations of free Fe^{2+}/Fe^{3+} in the cell nucleus can catalyze the formation of reactive oxygen species which damage genetic material.

Detrimental metals are toxic at all concentrations because they impair the regular course of life functions. Detrimental effects of metals in biological systems occur primarily through metal–protein interactions. Toxic metals combined to enzymes, membrane receptor proteins, nuclear regulatory proteins and structural proteins cause significant alterations in their normal functions (Squibb, Fowler in [5]). These effects come about by chemical and structural changes caused by the binding of metals to critical ligands in the protein molecules. Examples are Pb^{2+} and Cd^{2+} which bind to sulfhydryl groups of proteins and enzymes. The most sensitive biochemical pathway for Pb^{2+} is the biosynthesis of heme where this ion inhibits the dimerization of 5-aminolevulinic acid (ALA) to a pyrrole derivative called porphobilinogen or PBG. This reaction is catalyzed by the zinc enzyme porphobilinogen synthase (also called 5-aminolevulinic dehydratase). The enzyme is an octameric protein containing eight zinc ions. Four proteins contain Zn^{2+} bound to tyrosine, histidine and cysteine (Zn_A sites). In each of the other proteins Zn^{2+} is coordinated by cystein (Zn_B sites) (Fig. 2–2). The active sites are Zn_A-ones but they only work in cooperation with the Zn_B sites. Pb^{2+} will substitute the Zn ion in the sulfur-rich Zn_B site with the result that PBG synthesis is inhibited. This example clearly shows the interplay between different centers and the sensitivity towards interference.

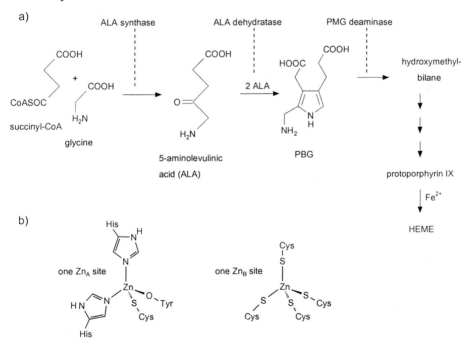

Figure 2-2. a: Synthesis of PBG as precursor of porphyrins by the enzyme ALA-synthase and -dehydratase. b: Structure of ALA-dehydratase bearing Zn_A and Zn_B sites.

2.2.4   Communicative Functions of Metals in Biology

Metal ions function as magnetic compasses, as initiators of specific cell functions and for the regulation of gene expression. Magnetotactic bacteria use magnetite (Fe_3O_4) as an internal compass to orientate in relation to the terrestrial magnetic field. Alkali and alkaline earth metals such as Na^+, K^+ and Ca^{2+} can actuate cell reactions, e.g., the regulation of intracellular functions by the Ca containing protein calmoduline. Gene expression is regulated by zinc finger proteins (as mentioned above).

2.2.5   Interactions of Nucleic Acids with Metal Ions

Nucleic acids are *polyanions*, and the compact structure is stabilized by the interaction of negatively charged phosphate groups with cations such as Na^+, K^+, Ca^{2+} and cationic polyamines. The stabilization of DNA and RNA with Na^+ and Mg^{2+} is often not specific, although in some cases a specific interaction has been found. K^+ stabilizes the DNA telomere structure at the end of a double helix of chromosomes by interaction with eight guanine bases as shown in Fig. 2–3. The relative stability depends on the cellular concentration of Na^+ and K^+.

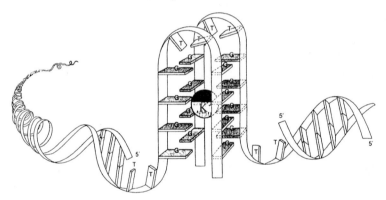

Figure 2-3. Postulated structure of a guanine tetrad stabilized by K^+.

2.2.6   Biometal-Organic Chemistry

The interaction of artificial metal ions/complexes with peptides/proteins [11], nucleic acids/DNA [12,13], enzymes [14], steroids [15] and carbohydrates [16] forms a bridge between natural and artificial macromolecular metal complexes. Biometal-organic chemistry concentrates on such complexes [17]. The reason for the increasing interest in this field lies in medical applications of metal complexes (cancer, photodynamic therapy of cancer, immunoassays, fluorescence markers, enantioselective catalysis, template orientated synthesis of peptides, etc.). Figure 2-4 presents an overview of metals in medicine [18]. Some examples are given below.

Stable metal complexes can be employed as markers for biochemical and biological systems in immunoassays, in radiographic and electron microscopic investigations of active centers and as radio pharmaceuticals. A stable covalent linkage is essential. One simple possibility is the functionalization of peptides and proteins by acylation of e.g. lysine side chains using succinimyl esters [11]. Modification of this reactive unit with transition metal complexes such as cyclopentadienyl complexes, sandwich complexes or alkinyl clusters leads to activated carboxylic acid derivatives, which can be isolated and reacted with the free amino group of lysine units in peptides and proteins. Fourier-transform-infrared spectroscopy (FT–IR) at 1900–2100 cm^{-1} allows the detection of the bound carbonyl complexes down to a detection limit in the picomole region. A carbonyl-metallo-immunoassay (CMIA) has the advantage that no radioactive compounds are necessary and by using different metal-organic markers several immuno assays can be carried out simultaneously. Other possibilities are reviewed in [11].

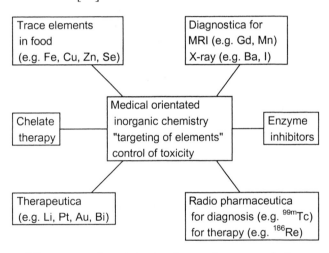

Figure 2-4. Overview of the applications of metals in medicine.

The use of cytotoxic drugs is one of the major approaches to the chemotherapy of cancer. Most cytotoxic anticancer drugs are only anti-proliferative which means that the process of cell division is interrupted. *cis*-Diamminedichloroplatinum(II) (*cis*-Pt(Cl_2)(NH_3)$_2$, nicknamed *cisplatin*) is used today routinely against testicular and ovarian cancer [2,5]. If we are to develop new more selective and active anticancer drugs based on platinum, we need to understand the interaction of the active model compound cisplatin with DNA. Structural data have shown that the binding of cisplatin to DNA occurs preferentially at the N7 position of adjacent guanines [13,19]. This binding leads to local denaturation of DNA, inhibits the replication process and kills the tumor cells. Because cisplatin possesses two reactive Cl-groups, intrastrand and interstrand cross-linking can occur. Figure 2-5 shows the intramolecular interaction of an oligonucleotide as model compound at the N7 positions of the guanines. Both guanines are orientated anti with respect to the sugar ring and are perpendicular to each other because of the square planar geometry around the platinum center.

The interactions of several ruthenium complexes with proteins, cytochromes and nucleic acids have been investigated [20]. The intention is to use these Ru-complexes as luminescence sensors (e.g. optical O_2 sensor), to trigger electron transfer and photoinduced electron transfer in proteins and DNA. For example, electrogenerated chemoluminescence (ECL) of Ru(phen)$_3^{2+}$ (phen: 1,10-phenanthroline) can be used to detect the presence of double-stranded DNA (for details see [20], p. 642). Ru(phen)$_3^{2+}$ binds strongly to double-stranded DNA and minimal binding is observed in the presence of single-stranded DNA. If a given single-stranded DNA sequence is immobilized on an electrode, treatment with a suitable target DNA may generate double-stranded DNA which allows the binding of the Ru-complex and the detection of ECL by electrode reactions.

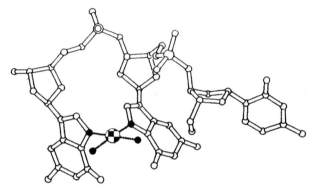

Figure 2-5. Solid-state structure of an oligonucleotide containing two guanines after reaction with cisplatin.

Synthetic metal complexes can preferentially bind to and so be used to recognize nucleic acid structures. After binding, nucleic acid photocleavage reactions have served as an important assay to examine nucleic acid recognition which is dominated by steric and symmetry factors (Sitlani, Barton in [5]). Shape selection, where the three-dimensional structure of the metal complexes is matched in shape and symmetry to a nucleic acid site, serves as a primary recognition element and distinguishes the global features of nucleic acid tertiary and local structures associated with a single pair of bases. Enantiomeric isomers of metal (phenanthroline) complexes bind to DNA with different affinities. An example of the binding of a Ru-complex is shown in Fig. 2-6.

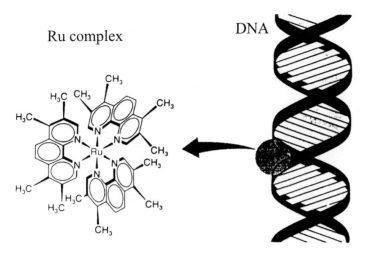

Figure 2-6. Binding of a chiral Ru-complex at a DNA.

The following sections show the structure and functions of some metalloproteins and metalloenzymes in more detail in order to explain the interaction of the metal centers with the macromolecular protein environments.

2.3   Heme Proteins

Heme (Fe-protoporphyrin IX, **1**), whether modified or not, is the active part in several proteins. The architecture of the hemoproteins is essential for the reactivity of heme in the macromolecular protein environment, as exemplified by three examples in Fig. 2-7:

- Heme in hemoglobin/myoglobin relatively deep inside a hydrophobic pocket. At Fe(II) one axial position is more strongly and the other axial position more weakly coordinated by a Lewis base of an amino acid residue. The weak axially coordinated side is free for reversible binding of O_2 (see Section 2.3.1).
- Heme in cytochrome c with covalent binding of the former vinyl groups at the protein as shown in **2**. Coordination of two different Lewis bases of amino acid residues to Fe(III) deeply inside a pocket allows electron transfer but not coordination to other species (see Section 2.3.3).
- Heme in cytochrome P-450 with the Fe(III) porphyrin in a hydrophobic environment and coordinated by a Lewis base but near the surface of the protein. Therefore binding of e.g. oxygen can occur at the free axial position of the Fe(III) and a substrate peripheral (see Section 2.3.2).

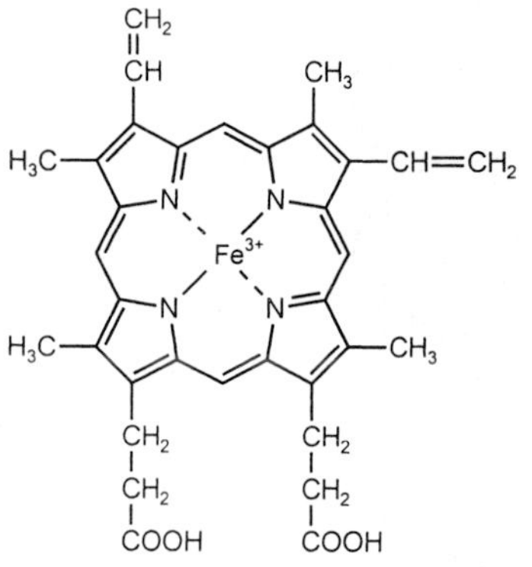

1

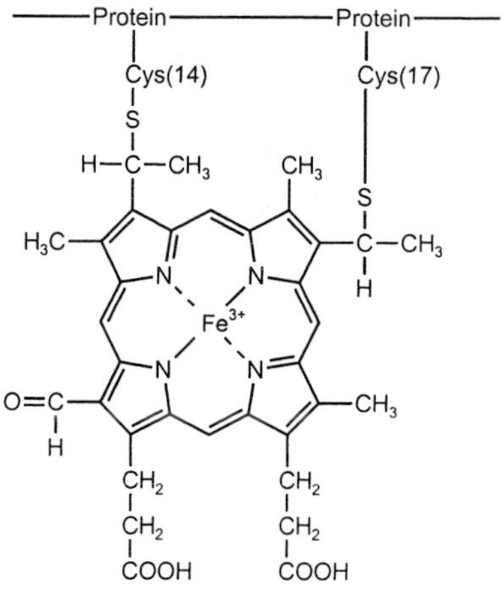

2

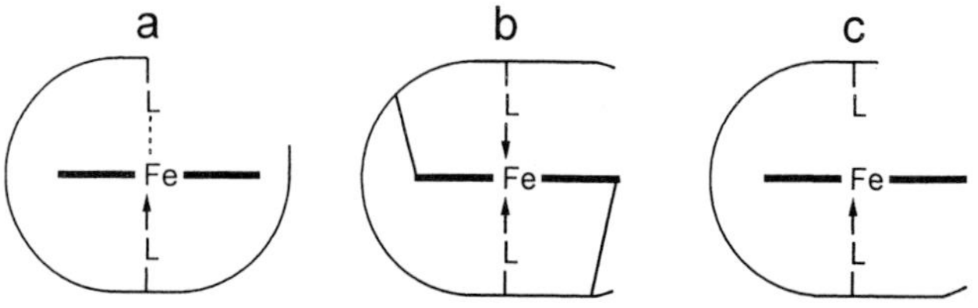

Figure 2-7. Scheme of location of heme in (a) myoglobin/hemoglobin, (b) cytochrome *c*, (c) cytochrome P-450.

2.3.1   Oxygen Transport Hemes

The oxygen carrier systems in vertebrates are the proteins hemoglobin and myoglobin: hemoglobin in the erythrocytes is responsible for O_2-transport in blood (also CO_2 and proton transport) and myoglobin for storage and transport in muscle. A human being has between 700 and 900 g hemoglobin in their blood. Because hemoglobin contains 0.35% Fe, 3 g or 70% of a human's Fe is concentrated in this protein. Each hemoglobin (molecular weight 64,500 Da) is approximately four times the size of a myoglobin unit with four heme **1** units as active centers for O_2 transport. Details of the function of hemes in proteins are available in modern textbooks of biochemistry and it is useful to read the excellent textbook *The Colours of Life* by L.R. Milgrom [21]. Some examples are discussed now.

Heme **1** is embedded in the protein matrix globin. The Fe(II) interacts coordinatively with two imidazole moieties of histidines of globin (one imidazole above and close to the Fe(II) and one below but further away), so that it can reversibly bind oxygen (A in Fig. 2-8) either for transport around the body (hemoglobin) or for storage in the muscle tissues (myoglobin). Surprisingly, the Fe(II) in these globins does not irreversibly change its oxidation state to Fe(III) upon oxygen binding (D via C and B in Fig. 2-8). The porphyrin ligand and the macromolecular protein environment slightly reduce the oxidizability of Fe(II) to Fe(III): E° for the Fe(III)/Fe(II) couple is +0.77 V vs. NHE with free Fe^{2+} and +0.82 V vs. NHE with Fe(II) in hemoglobin. Also complexing with water at Fe(II) is hindered in the more hydrophobic pocket.

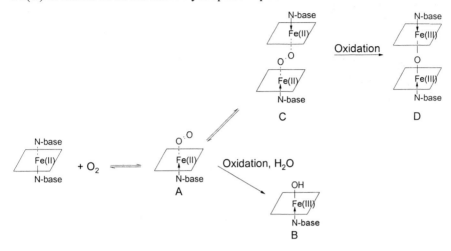

Figure 2-8. Possible reactions of heme with O_2.

A cartoon in Fig. 2-9 taken from [21] shows schematically steps in the development of myoglobin and then hemoglobin. The free heme (step c in Fig. 2-9), without the globin environment, will bind oxygen irreversibly with formation of Fe(III)-OH inside the ligand (B in Fig. 2-8). Myoglobin (step d in Fig. 2-9), with a single protein chain of 152 amino acids and the heme embedded within the hydrophobic heart of the polymer globin (Fig. 2-10), allows reversible O_2-binding.

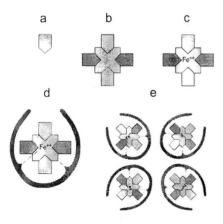

Figure 2-9. A cartoon (taken from [21]) shows the steps in the construction of a macromolecular metal complex. (a) pyrrole, (b) porphyrin, (c) heme, (d) myoglobin, (e) hemoglobin.

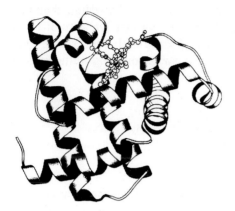

Figure 2-10. Structure of myoglobin [2].

The dependence of the saturation with O_2 on the O_2 pressure in myoglobin is described by a rectangular hyperbola. In hemoglobin (step e in Fig. 2-9), with two α-globin chains (each 141 amino acids) and two β-globin chains (each 146 amino acids), the hemes show a sigmoid relationship between their affinity for the gas and the O_2 pressure. Figure 2-11 shows that the oxygen affinity of hemoglobin increases as oxygen is taken up. In the muscles tissues hemoglobin can transfer O_2 to myoglobin. In hemoglobin four subunits cooperate with each other during O_2 uptake (a so-called cooperative interaction).

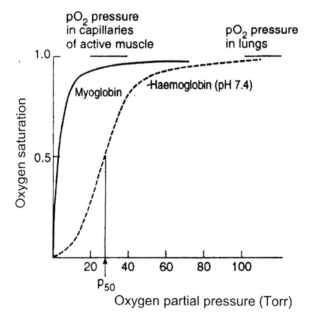

Figure 2-11. Oxygen binding curves for hemoglobin and myoglobin.

In myoglobin and hemoglobin 70–80% of the amino acids as monomer units in the globin form helices stabilized by hydrogen bonding, with 3.6 steps required for a complete turn. Oxygen-free deoxyhemoglobin contains all Fe(II) cations in a paramagnetic high-spin configuration with four unpaired electrons. The Fe(II) cations are too big to fit into the hole of the porphyrin and sit 0.05–0.1 nm above the ligand planes directly linked to the proximal imidazolyl ring of the proximal histidines (Fig. 2-12). Oxygen binding converts the Fe(II) cations into a diamagnetic low spin d^6 configuration which, because of a 13% reduction in volume, fits snugly into the hole of the porphyrin. This small motion triggers a series of structural changes. These have a large effect on all of the globins (Fig. 2-13), which can be seen as molecular amplifiers.

In each of the four globin chains of hemoglobin the penultimate amino acid is tyrosine. In the so-called "spring loaded" hemoglobin the phenolic side chain of tyrosine fits into a cleft between two helices, the F and H helices. When Fe(II) moves by oxygenation into the core of the porphyrin, it also pulls down a proximal histidine. This motion moves the F helix a little closer to the H-helix, and the phenolic side chain is squeezed out of the two small tyrosine-containing pockets. This motion results in the breaking and movement of hydrogen bonds holding the chains together, initially arginines in α-chains and then histidine in β-chains, which means a stepwise oxygenation. This reversible hydrogen bond breaking on oxygenation and reformation on deoxygenation leads to the release and uptake of protons in different parts of the globin chains.

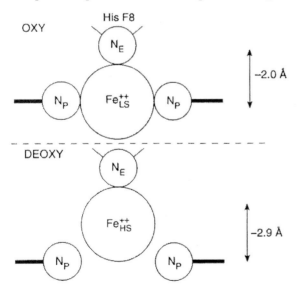

Figure 2-12. Positions of Fe(II) in heme of deoxyhemoglobin and oxyhemoglobin (N atom of a proximal histidine also shown).

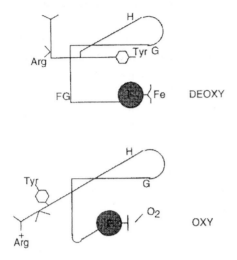

Figure 2-13. Schematic view of the conformational changes of the globin chain around heme during oxygenation and deoxygenation.

Small changes in the macromolecular protein chain show how sensitive the function of the heme is. Changes in the amino acid sequence of the globin around the heme site caused by genetic disorders can result in the irreversible oxidation of Fe(II) of heme to Fe(III) (B in Fig. 2-8) now called methemoglobin. In one of several cases the histidines whose imidazole side chains coordinate to Fe(II) are replaced by tyrosines which have a phenol side chain. Now the oxygen of the phenolic group coordinates to the Fe(II). This results in a weaker ligand field around the Fe(II) so that the low-spin configuration is less accessible. The coordinated oxygen now oxidizes Fe(II) to Fe(III) which is less amenable to reduction by enzyme-based reductive mechanisms. Also water now finds its way into the pocket, and this leads to the formation of Fe(III)-OH. In this methemoglobinaemia only one or two of the globin chains of hemoglobin are affected. Therefore a relatively normal life can be led. The presence of methemoglobin leads to skin pigmentation (cyanosis). A sickle-cell gene found mainly in people of African origin results in an important abnormal variant of hemoglobin, which is called hemoglobin S. The red blood cells have a tendency to collapse into a sickle shape when the oxygen concentration in the blood is low and crystallize out in the intracellular fluid. On one hand the deoxyhemoglobin S is less able to pick up oxygen than the normal deoxyhemoglobin and on the other hand it causes membrane damages. The reason for the crystallization of hemoglobin S and low oxygen concentration is a variation of one or more of the amino acids in the β-globin chains. One possibility is the replacement of the hydrophobic glutamic acid residue at a specific position in the β-chains by a hydrophobic valine residue. An important result is that deoxyhemoglobin S is less soluble in the intracellular fluid and its molecules can fit together more easily. These two examples show how sensitive the behavior of a macromolecular metal complex is to minor changes of the composition of the macromolecular chain.

We have shown that the reversible oxygen uptake of myoglobin and hemoglobin is very sensitive to the kind and sequence of α-amino acids in the macromolecular protein chain. Now it is interesting to speculate if it is possible to substitute the heme in, for example, myoglobin by other porphyrin-type compounds without loss of reversible oxygen function. This topic forms a bridge between fully natural and fully artificial/synthetic systems.

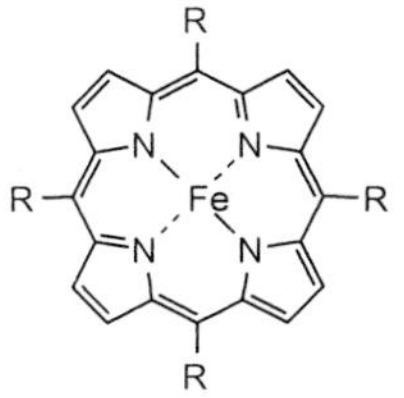

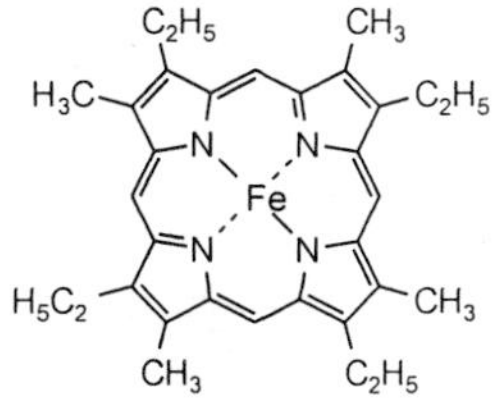

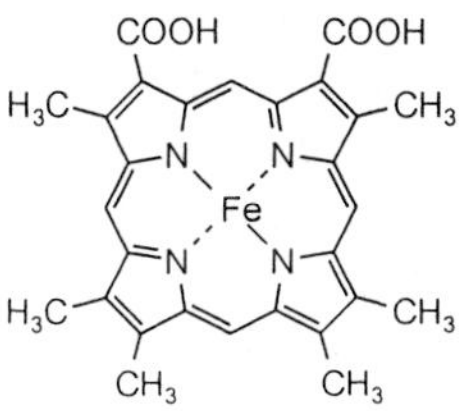

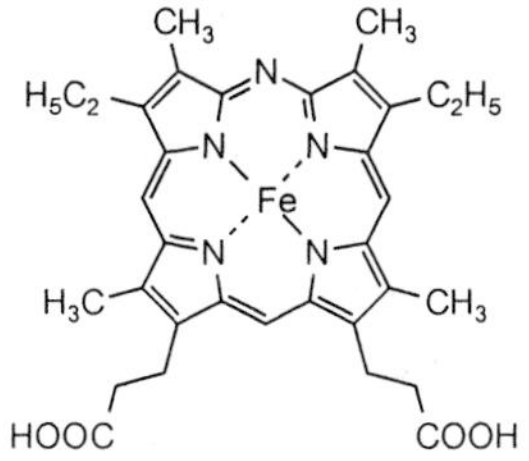

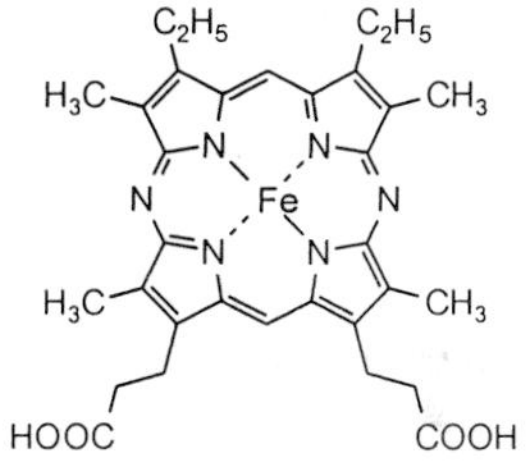

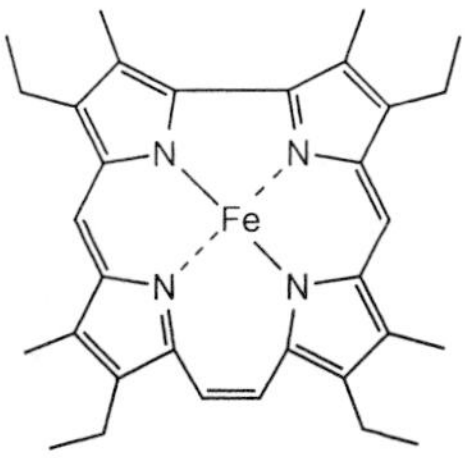

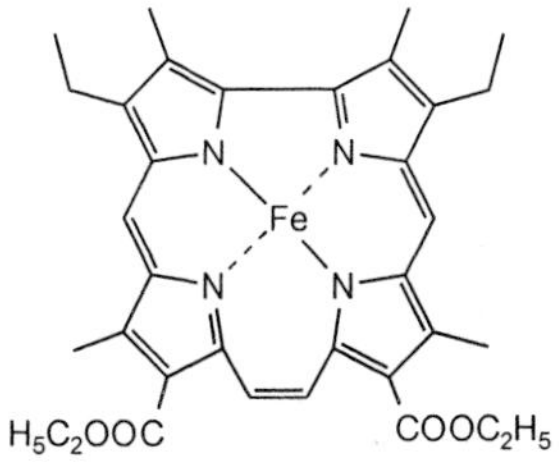

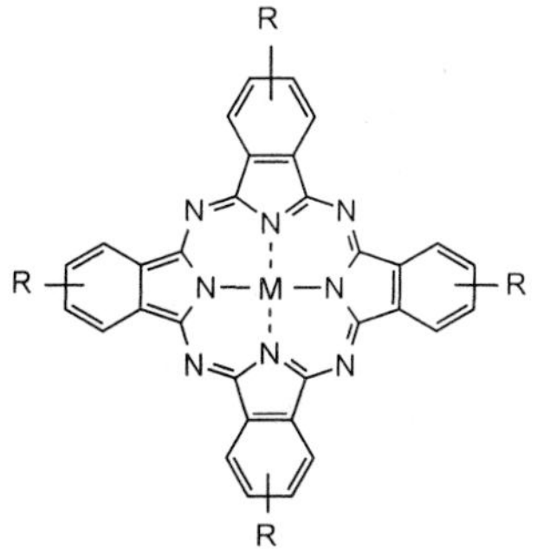

In order to study the influence of a synthetic porphyrin on the properties of a reconstituted "myoglobin" the following parameters are interesting: the size of the porphyrin-type ligand and its electron density; the kind of substituents (size, hydrophilic or hydrophobic). The following porphyrin-type compounds have been used to reconstitute myoglobin: Fe(II)-porphyrin **3** (R = –H) [22], Fe(III)-5,10,15,20-tetraalkylporphyrins **3** (R = –CH$_3$, –C$_2$H$_5$, –C$_3$H$_7$) [23], Fe(II)-octamethylporphyrin **4** (R = –CH$_3$) [24], Fe(III)-octaethylporphyrin **4** (R = – CH$_2$CH$_3$) [25], Fe(II)-etioporphyrin **5** [24], Fe(II)-6,7-dicarboxyporphyrin **6** [26], monoazahemin **7** [27], diazaheme **8** [27], Fe(II)-etiocorrphycene **9** [28], Fe(II)-corrphycene **10** [29,30], and phthalocyanines such as Fe(II)-2,9,16,23-tetrasulfophthalocyanine **11** (R = –SO$_3$H) [31–33]. Also the reconstitution of a hemoglobin has been described [36].

The introduction of porphyrin-type compounds was generally carried out as follows (Fig. 2-14). Heme-free apomyoglobin was first prepared from myoglobin by the acid/methylethylketone method [34,35]. Then in the case of **6** the apoprotein was mixed with a 1.2-fold molar excess of a porphyrin dissolved in 50 mM aqueous NaOH [26]. In the case of water insoluble **4**, **5** the porphyrins were dissolved in DMSO and mixed with an apomyoglobin solution (excess of porphyrin 1.2–1.5) [24]. Purification was achieved by dialysis or ion exchange chromatography. The incorporation of the synthetic porphyrin-type compounds into apomyoglobin was followed by spectrophotometry. A 1:1 complex formation was confirmed. Incorporation into the heme cavity to form the original Fe–N (His-F8) bond was demonstrated by NMR spectroscopy. Therefore "myoglobin" was reconstituted. X-ray crystallographic structure analysis also confirmed the results [22]. IR and EPR were used as other analytical methods. The binding of small molecules was investigated. For example, the Fe(II)-porphyrin **8** in the globin exhibits strong binding for CO, pyridine, imidazole, cyanide and azide, and reacts moderately with NH$_3$ [27]. In particular, the interaction with oxygen is interesting. For the green-colored complex of globin and the phthalocyanine derivative **11** neither O$_2$ nor CO-binding was found because the distal histidine blocks the coordination site, and the bulky substituted phthalocyanine significantly deforms the compact heme pocket [32,33]. Another reason is that the imidazole affinity for Fe(II) increases rapidly with increasing numbers of meso-nitrogens, thus leading to binding of proximal and distal imidazoles at the Fe(II) of the phthalocyanine. The porphyrin derivative **4** (R = –CH$_2$CH$_3$) binds O$_2$ similarly to native myoglobin [25]. The electron withdrawing carboxylate groups in **6** lowers the O$_2$ affinity by 3-fold compared to native myoglobin [26]. Surprisingly, the O$_2$ affinity of the monoazaheme **7** in globin is 50 times larger than that of the native one [27]. Also the diazaheme **8** is capable of strong oxygen binding [27]. Fe(II)-corrphycene derivatives such as **9** show reversible O$_2$ and CO binding [28]. But the equilibrium constants are one or two orders of magnitude lower than those for native myoglobin. It should be pointed out that reversible oxygen binding can be achieved with totally artificial porphyrin derivatives [37]. Different blood substitutes are described in detail later in this book.

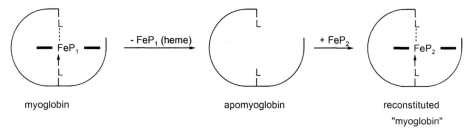

myoglobin   apomyoglobin   reconstituted
                                                             "myoglobin"

Figure 2-14. Schematic representation of the conversion of myoglobin into porphyrin-free apomyoglobin and its reconstitution after introduction of a new porphyrin FeP_2.

2.3.2 Oxygen Transfer Hemes

Various heme-containing proteins catalyze reactions in which oxygen is transferred to a substrate and inserted into (C–H) or (C–C) bonds. Examples are monooxygenases using only one atom of oxygen such as cytochrome P-450 or dioxygenases involving dioxygen such as peroxidases and catalases. Cytochrome P-450 in the liver catalyzes various reactions in which dioxygen is activated by two electrons and two protons with the insertion of one oxygen into a substrate and formation of water as shown in Eq. 2-1. P-450-dependent monoxygenases are of cconsiderable importance in the biosynthesis and catabolism of endogeneous compounds such as fatty acids, steroids, prostaglandins and leukotrienes. They are also involved in the oxidative metabolism of exogenous compounds such as drugs and other environmental products.

$$RH + O_2 + 2e^- + 2H^+ \rightarrow ROH + H_2O \tag{2-1}$$

The X-ray structure of cytochrome P-450 from *Pseudomonas putida* grown on camphor has been determined [5]. Heme is attached to the protein through a cystein residue (Cys357, Fig. 2-15). The heme pocket where dioxygen and substrate molecules are bound is entirely composed of the lipophilic residues of the amino acids Leu, Val and Phe. The hydrophobicity of the active site stabilizes the binding of a nonpolar hydrocarbon substrate to the protein as well as the binding of O_2 to iron. The active site is only accessible to the substrate and O_2 by transient broadening of the protein channels (for the mechanism of oxidation see modern textbooks of biochemistry).

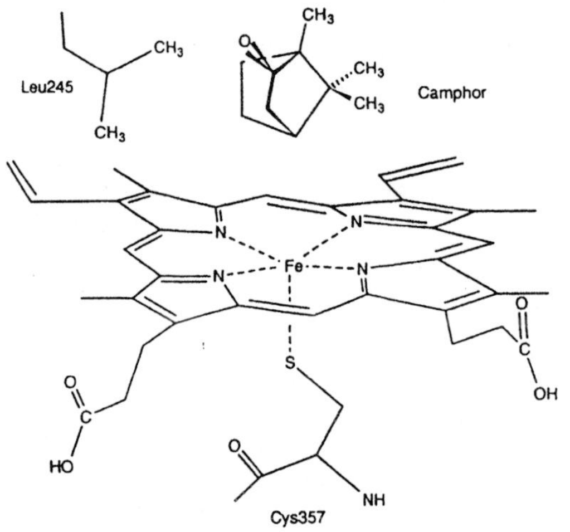

Figure 2-15. View to the active center of cytochrome P-450 with camphor bound.

The reactions with peroxides or hydrogen peroxides are shown in Eqs. 2-2 to 2-5. In the active centers the heme coordinates with histidine for cytochrome *c* peroxidase, with cysteine for chloroperoxidase and with tyrosine for catalase.

Cytochrome *c* peroxidase: $ROOH + 2H^+ + 2e^- \rightarrow ROH + H_2O$ (2-2)

Horse radish peroxidase: $Ar–H_2 + H_2O_2 \rightarrow Ar + 2H_2O$ (2-3)

Chloroperoxidase: $Ar–H + X^- + H^+ + H_2O_2 \rightarrow Ar – X + 2H_2O$ (2-4)

Catalase: $2H_2O_2 \rightarrow 2H_2O + O_2$ (2-5)

The structure of beef liver catalase is shown in Fig. 2-16 [2]. The heme site is accessible by a channel 3 nm long. The specificity is largely determined by the ability of a substrate to form H-bonding interactions to some amino acids [2]. Unlike peroxidases, the fifth ligand of catalase is tyrosine. The phenyl ring is tilted 42° inwards to the heme plane with an iron–heme distance of 0.22 nm. This suggest that the phenolic side chain is deprotonated and has a localized charge. The reaction cycle with Fe(III) and Fe(IV) is shown in Eq. 2-6.

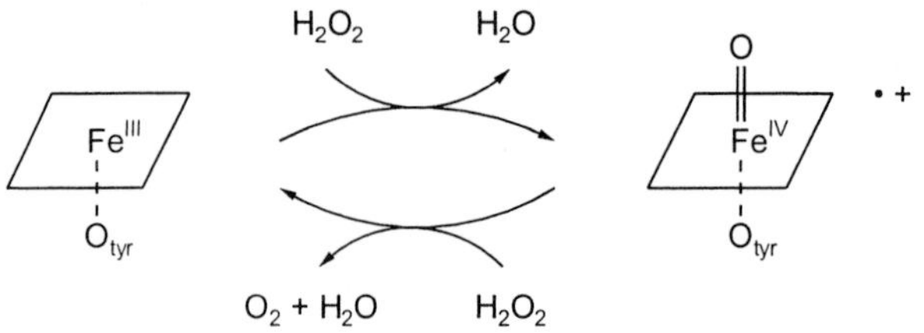

(2-6)

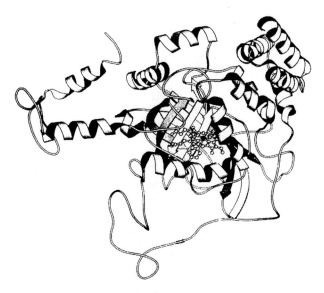

Figure 2-16. The structure of beef liver catalase with heme as the active center.

Enzymes can be used as natural catalysts for the transformation of artificially made organic compounds [38]. Either microorganisms or isolated enzymes can be employed. The mild, environmentally acceptable conditions and the high selectivities achieved by using natural enzymes are advantageous. The sensitivity of enzymes and the fact that several reactions will not work in water, where enzymes display their highest catalytic activity, are disadvantageous. Examples are given for dioxygenases which normally contain bound iron either in a heme complex or in a related environment. Typical dioxygenase reactions, during which two oxygens are transferred onto a substrate, are shown in Eqs. 2-7 to 2-9. In all cases, a highly reactive and unstable peroxospecies is formed, e.g. a hydro- or *endo*-peroxide. In general, these latter intermediates are not isolated but further reduced to yield the more stable corresponding hydroxyl derivatives.

- Alkenes may be transformed into an allyl hydroperoxide which, upon reduction (e.g. by sodium borohydride), yields an allylic alcohol (Eq. 2-7). In living systems, the formation of lipid peroxides is believed to be involved in some serious diseases and malfunctions including arteriosclerosis and cancer.

- Alternatively, an *endo*-peroxide may be formed, whose reduction leads to a diol (Eq. 2-8). The latter reaction resembles the cycloadditon of singlet-oxygen onto an unsaturated system and occurs in the biosynthesis of prosta-glandins and leukotrienes. Both intermediate hydro- or *endo*-peroxide species are highly reactive and may be subject to further transformations such as (enzymatic or non-enzymatic) reduction or rearrangement.

• In prokaryotic cells such as bacteria the initial step of the metabolism of aromatic compounds is the cycloaddition of oxygen catalyzed by a dioxygenase. In living microbial cells, the resulting *endo*-peroxide (dioxetane) is then enzymatically reduced to yield synthetically useful *cis*-glycols (Eq. 2-9).

$$\text{Sub-H} + \text{O}_2 \xrightarrow{\text{dioxygenase}} \underset{\text{hydroperoxide}}{\text{Sub-O-O-H}} \xrightarrow[\text{e.g. NaBH}_4]{\text{reduction}} \text{Sub-OH} \qquad (2\text{-}7)$$

$$\text{Sub} + \text{O}_2 \xrightarrow{\text{dioxygenase}} \underset{\text{endo-peroxide}}{\text{Sub}\overset{O}{\underset{O}{|}}} \xrightarrow[\text{e.g. NaBH}_4]{\text{reduction}} \text{Sub}\overset{OH}{\underset{OH}{<}} \qquad (2\text{-}8)$$

$$\text{(benzene)} + \text{O}_2 \xrightarrow{\text{dioxygenase}} \text{dioxetane} \xrightarrow[\text{[H}_2]]{\text{reduction}} \text{cis-glycol} \qquad (2\text{-}9)$$

2.3.3 Electron Transfer Hemes

In the electron transfer cytochromes the iron of heme is bound in a hexacoordinate low-spin state, with two protein ligands (typically histidine and/or methionine) above and below the heme plane. They serve as electron carrier proteins in mitochondria and endoplasmatic organelles as well as in bacterial redox chains. At least three classes of cytochromes, *a*, *b* and *c*, are known. They can alternate between an oxidized Fe(III) low-spin state with a single unpaired electron and a reduced Fe(II) low-spin form with no unpaired electrons. Since iron remains low spin, electron transfer is greatly facilitated. The best characterized family are the *c* cytochromes.

Cytochrome *c* (see **2** and Fig. 2-7) in the mitochondria is part of the chain of electron carrier proteins that ultimately produce ATP from ADP (oxidative phosphorylation). In principle, the heme in cytochrome *c* is the same as that in hemoglobin. But in detail, the vinyl groups, after conversion to thioethers, are covalently linked to cysteine amino residues of the protein chain [2,5]. The fifth coordination site of iron is occupied again by the imidazole N-atom of a histidine, but the sixth position is now coordinated to the S-atom of a methionine (Figs. 2-7, 2-17). The redox potential of low-spin Fe(III)/Fe(II) with E°= +0.25 V vs. NHE is drastically altered compared to heme. Now this heme iron is much more able to act in the electron-transporting chain by redox cycling. The mechanisms of electron transfer are available in modern textbooks of biochemistry. It is interesting to note that the macromolecular protein chain in cytochrome *c* contains a variable number of amino acids: 103–104 in some fishes and earthbound vertebrates, up to 112 in some green plants. Because the differences between simple yeasts, plants, insects, higher mammals and even humans are not huge, cytochrome *c* from one organism will also work in another.

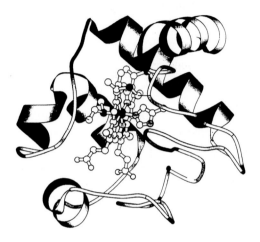

Figure 2-17. Structure of cytochrome *c* of tunny fish with a modified heme as active center (see also **2** and Fig. 2-7).

2.4   Non-Heme Proteins

Iron-containing clusters function as electron carriers, catalytic or storage sites, or as structural determinants. Two distinct types of iron-containing clusters are mentioned: iron–sulfur clusters composed of iron and inorganic sulfide and coordinated in most cases to cystein ligands; diiron–oxo clusters having oxygen-containing bridging ligands and an additional ligand sphere composed of N and/or O from proteins. Mononuclear iron stabilized by ligand groups from amino acids is also known. Some examples are given below.

2.4.1   Oxygen Transport Non-Heme Proteins

The non-heme counterpart of hemoglobin/myoglobin is hemerythrin, a reversible O_2 carrier found in marine invertebrates. This has one diiron site per subunit, coordinated by several histidines of the protein chain (Fig. 2-18) [2,5]. The X-ray structure shows a triply bridged high-spin diiron(II) center (Fe–Fe distance 0.332 nm). As well as two carboxylato bridges (aspartic acid, glutamic acid) there is also a bridging hydroxyl group. One of the Fe(II) ions is five-coordinated with two histidine ligands, whereas the other is six-coordinate. The desoxy-form binds O_2 reversibly by a two-electron oxidative addition process, generating a peroxodiferric moiety, present in a hydroperoxide form and hydrogen bound to a bridging oxo atom.

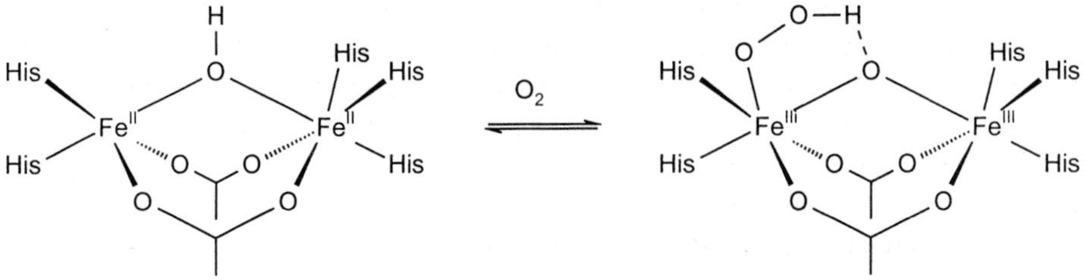

Figure 2-18. The active center of hemerythrin.

2.4.2   Oxygen Transfer Non-Hemes

A heterogeneous collection of enzymes and proteins contain iron or require it for their activity in the non-heme form. Some of them can be classified as mono- and dinuclear Fe proteins, and further characterized as tightly bound Fe(III) centers and loosely bound mononuclear Fe(II) centers: oxidases, dioxygenases, monoxygenase, dismutases, hydratases. One example is the mononuclear Fe(II) superoxide dismutase of *Escherichia coli* catalyzing the reaction shown in Eq. 2-10. The catalytic ferric ion is coordinated by four protein ligands [5]. The Fe coordination geometry forms an approximate trigonal bipyramid with a bound water completing the trigonal iron coordination in the equatorial plan (Fig. 2-19).

$$2O_2^- + 2H^+ \rightarrow H_2O_2 + O_2 \tag{2-10}$$

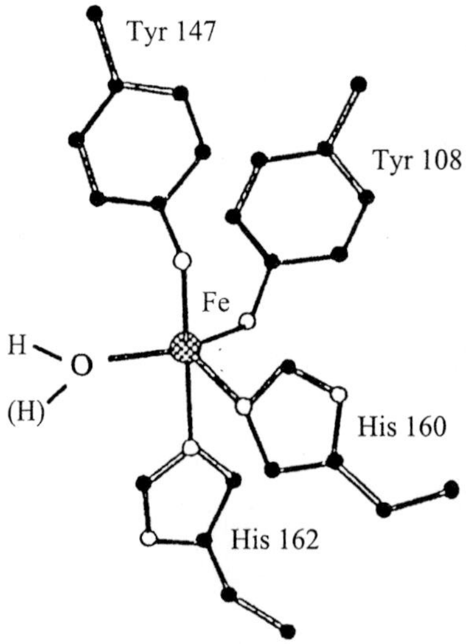

Figure 2-19. The active center of Fe superoxide dismutase (Fe–SOD).

The non-heme enzyme methane monooxygenase (MMO) from methanotropic bacteria catalyzes the hydroxylation of methane to methanol. Methane is most difficult to hydroxylate and cytochrome P-450 cannot perform this reaction. MMO consists of three components. Component A is a dimer with subunits of dinuclear iron with monooxygenase activity. Components B and C are electron donor and transfer sites. Like cytochrome P-450, a high valent iron–oxo complex is proposed for component A in MMO. This species abstracts a H atom from CH_4 to generate a $CH_3\cdot$ radical.

2.4.3   Electron Transfer Non-Hemes

A variety of iron–sulfur clusters responsible for electron transfer are found in proteins where the ligands to the metal of the cluster are provided by the amino acid residues of a polymer protein. The structure of some Fe_nS_m units and their bioinorganic redox states are shown in Fig. 2-20 [2,5].

A distorted tetrahedral environment of sulfur ligands was found for rubredoxin, which contains [1Fe–OS] and is an intermediate electron carrier in the ω-hydroxylation of fatty acids in *Pseudomonas oleovorans*. Fe cycles between the Fe(II) and Fe(III) states with E°= + 20 mV vs. NHE in the protein environment. (E°= + 770 mV vs. NHE for free Fe^{3+}/Fe^{2+} in water at pH 7.5.)

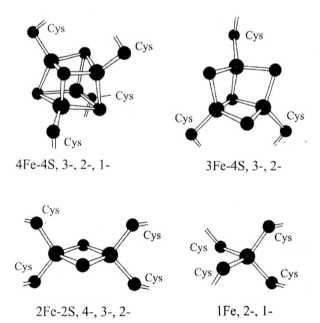

4Fe-4S, 3-, 2-, 1-   3Fe-4S, 3-, 2-

2Fe-2S, 4-, 3-, 2-   1Fe, 2-, 1-

Figure 2-20. Structure of some iron–sulfur clusters, and their whole cluster charge.

The [2Fe–2S]-ferredoxins serve as electron carriers in many systems such as the chloroplasts of higher plants. The iron atoms are linked by two sulfide bridges which provide an efficient pathway for superexchange (antiferro-magnetic coupling). Each Fe(III) in the cluster $[2Fe-2S]^{2+}$ is d^5 high-spin ($S = 5/2$), but the whole cluster has a ground state with $S = 0$. Upon reduction one iron site becomes high-spin d^6 ($S = 2$) and the whole system yields a ground state with $S = 1/2$. The plant-type ferredoxin of cyanobacteria has $E° = -400$ mV vs. NHE, whereas $E°$ for the ferredoxins of mitochondrial complexes vary from ~ 0 to 300 mV vs. NHE.

Clusters with a [4Fe–4S] core are found in a large variety of proteins their main function being intra- and interprotein electron transport. Mostly the clusters have tetracysteinate coordination. Changes in the protein environment allow the redox potential to vary from -650 to $+450$ mV vs. NHE. A $[4Fe-4S]^{2+}$ cluster contains formally 2 Fe^{2+} and 2 Fe^{3+} and 4 S^{2-} atoms with a spin $S = 0$. A one-electron reduction yields the $[4Fe-4S]^{1+}$ state with a spin $S = 1/2$ whereas a one-electron oxidation yields the $[4Fe-4S]^{3+}$ state with a spin also of $S = 1/2$.

The enzyme acotinase transforms citrate to isocitrate in the Krebs' cycle, catalyzing successive reactions of dehydration and rehydration. Detailed X-ray analysis shows that acotinase contains iron–sulfur clusters with the composition [3Fe–4S]. In the presence of Fe^{2+} and a reducing agent this cluster converts into the cluster [4F–4S]. In acotinase a solvent-derived hydroxyl moiety acts as a fourth ligand to provide distorted tetrahedral coordination in the [3Fe–4S] cluster. Site-differentiated iron–sulfur clusters may also be involved in the regulation of gene expression in iron metabolism.

Iron–sulfur clusters are also found with other cofactors [2,5]. For example, the active site of sulfite reductase, which catalyzes the reduction of sulfite to sulfide, is a heme group and a [4Fe–4S] cluster. They are bridged by a cysteine sulfur atom. The iron–sulfur cluster plays the role of electron transport. Sulfite probably binds to the sixth position of the heme iron atom in the first step of reduction. Fe-containing nitrogenases, reducing N_2 to NH_3, are a complex of several proteins, the main parts being dinitrogenase–reductase, Fe-containing protein and dinitrogenase with Fe/Mo. For the reduction of N_2 to NH_3 eight electrons are needed: two electrons for the formation of H_2 and six electrons for the formation of NH_3. The Fe/Mo-protein stores the electrons till reduction by transport to the Fe-protein occurs. Details for this interesting process with eight electrons are not known. Mg-ATP is involved as an energy-rich component (Eq. 2-11).

$$N_2 + 8H^+ + 16 \text{ Mg-ATP} + 8e^- \rightarrow 2NH_3 + H_2 + 16 \text{ Mg-ADP} + 16 \text{ P} \qquad (2\text{-}11)$$

2.5 Copper Proteins

One of the important roles of metalloproteins is electron transport between functional molecules in biological systems [39]. Copper proteins are involved in electron transfer, redox reactions and the transport and activation of dioxygen. They are classified into Types I, II and III, and their properties are as follows:

Type I: One copper is involved in one unit. The copper has a strong absorption around 600 nm and small hyperfine coupling constants in ESR. It is called Blue copper protein.

Type II: One copper is involved in one unit. The copper has an absorption similar to a conventional copper complex and hyperfine coupling constants in ESR.

Type III: Two or three coppers are involved in one unit. Two copper atoms show antiferromagnetic interaction, and have a strong absorption around 300 nm.

2.5.1 Type I Copper Proteins

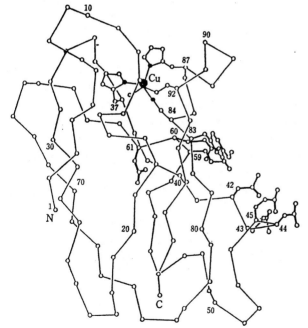

Figure 2-21. Structure of poplar plastocyanine. O represents the α-carbons of amino acid residues.

Plastocyanine, azurine, and pseudoazurine are of this type. It shows a very strong absorption around 600 nm (molar extinction coefficient $\varepsilon \sim 3000$–5000 M^{-1} cm^{-1}) and a large redox potential of about 0.2–0.8 V vs. NHE. In the ESR spectrum the hyperfine coupling constant is very small ($A_{//} = 0.006$ in comparison to that of a conventional square planar copper complex, 0.015–0.02) showing a distorted coordinated structure close to a tetrahedral structure. Figure 2-21 shows the structure of a poplar plastocyanine [40]. In this complex, the coordinating amino acid residues are His-84(S), Met-92(S), His-37(N) and His-87(N); they form a distorted CuN_2S_2 structure. This distorted structure stabilizes Cu(I) making the redox potential larger than that of any other copper complex. It is important that the three-dimensional structure of the protein makes such a distorted structure possible, which is not easily modelled by a synthetic low molecular weight metal complex.

2.5.2   Type II Copper Proteins

Superoxide dismutase (SOD) of Cu, Zn-type and galactose oxidase are of this type. The SOD catalyzes disproportionation of O_2^- to O_2 and H_2O_2 by successive reactions such as Eqs. 2-12 and 2-13 where M is a metal ion.

$$O_2^- + M^{n+} \rightarrow O_2 + M^{(n-1)+} \tag{2-12}$$

$$O_2^- + M^{(n-1)+} + 2H^+ \rightarrow H_2O_2 + M^{n+} \tag{2-13}$$

Its physiological role is to protect the organism against biohazardous action of O_2^-. SOD involving Cu and Zn (Cu, Zn-type) forms a dimeric structure and four His residues coordinate to Cu forming a square planar structure distorted towards a tetrahedral one (Fig. 2-22) [41]. The Zn^{2+} is located at 0.67 nm distance from the Cu, and three His and one Asp are coordinated to the Zn. The Zn plays an important role in maintaining the structure of the SOD.

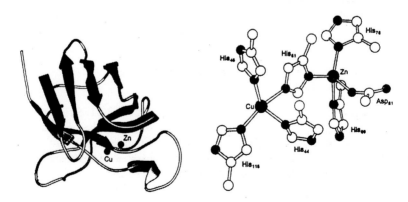

Figure 2-22. Structure of Cu, Zn superoxide dismutase and its catalytically active center.

In the galactose oxidase of *Dactylium dendroides* Cu^{2+} is coordinated by two His and one Tyr in an equatorial plane and by one Tyr at the apical position [42].

All these specific coordination structures are also caused by the three-dimensional structure of the protein.

2.5.3   Type III Copper Proteins

Hemocyanine and tyrosinase are of this type. Hemocyanine is the oxygen carrier in molluscs and arthropods. Two Cu ions are involved in a subunit and bind one O_2. A deoxy structure of a lobster hemocyanine from *Panulitrus interruptus* has been reported [43] for which both the Cu are monovalent. Three His coordinate to each Cu, and two Cu and four His are located in the same plane (Fig. 2-23). The two copper ions are not equivalent and no bridging ligand exist between them. Tyrosinase oxidizes phenols to orthoquinones. O_2 is coordinated between two Cu ions as a μ-peroxo structure.

Using a number of dicopper(II) complexes of dinucleating ligands it has been shown that reversible O_2 binding is also possible with artificial complexes at low temperature. Copper–dioxygen complexes $[Cu(TPA)_2(O_2)]$ (TPA = tris(2-pyridylmethyl)amine) contains two copper bridged by two oxygen in a μ-peroxo group [44]. Another copper–dioxygen complex $[HB(3,5-R_2pz)_3Cu(O_2)]$ (HB = 3,5-*iso*-propylpyrazoyl, R = *iso*-propyl) contains O_2 in an η^2,η^2-peroxo structure as in hemocyanine [45].

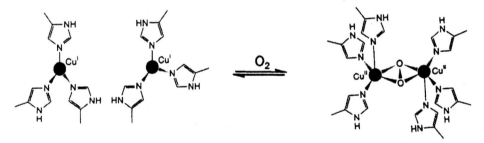

Figure 2-23. Structure of the active center of hemocyanine and oxy-hemocyanine containing O_2 as η^2,η^2-peroxide.

2.5.4   Mixed Type

This type involves a mixture of types I, II and III coppers. Ascorbic acid oxidase, laccase and celluloplasmin are of this type. These enzymes oxidize substrates with O_2, forming two H_2O molecules from one O_2.

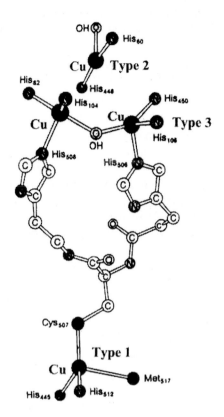

Figure 2-24. Structure of the active center of ascorbic acid oxidase with four Cu centers (N atoms of His, S atoms of Cys and Met).

Ascorbic acid oxidase oxidizes ascorbic acid to dehydroascorbic acid. In a pumpkin, two His, Cys and Met coordinate to type I copper, two His and OH^- (or H_2O) to type II copper, and three His to each two type III copper (Fig. 2-24) [46]. Type II and Type III coppers form a triangular cluster, which presumably plays an important role in the four electron reduction of O_2. Ascorbic acid oxidase and laccase have similar structures composed of three domains, but celluloplasmin is composed of six domains.

Nitrous acid reductase is a trimer [47]. The type I copper at the N-terminal end accepts an electron from pseudoazurin and transports it to the type II copper. This type II copper is coordinated by two His, another His belonging to another monomer unit and water, and reduces NO_2^- to NO.

2.6   Manganese Proteins

Manganese proteins are important in the photosynthesis of green plants where a Mn-cluster (in a stable state most probably a mixture of 3^+ and 4^+ Mn ions) complex oxidizes water to O_2 and provides electrons to the whole system. Other Mn proteins are pyruvic acid carboxylase, arginase, phosphatase, catalase and superoxide dismutase (SOD).

2.6.1   Oxygen Evolving Center (OEC) of Photosynthesis

In photosynthesis electrons are abstracted from water by a four electron process (Eq. 2-14) catalyzed by a Mn cluster, and the electrons are provided to the whole photosynthetic system.

$$2\,H_2O \rightarrow 4e^- + 4H^+ + O_2 \tag{2-14}$$

The electrons are transferred from the OEC to the photosystem II reaction center where they are excited to higher energy by solar visible light, and then acquire energy again at photosystem I. The high energy electrons are stored in photosynthetic products by reducing carbon dioxide; the main product is carbohydrate, which provides almost all the energy resources of living things as well as fossil fuels. This reaction is therefore of primary importance for biological activities.

The exact structure of the Mn cluster is not yet certain. At least four Mn ions are involved in the OEC to realize four electron oxidation of two water molecules wherein a histidine residue is inferred to work also as an oxidation site. A Mn cluster model is proposed as shown in Fig. 2-25.

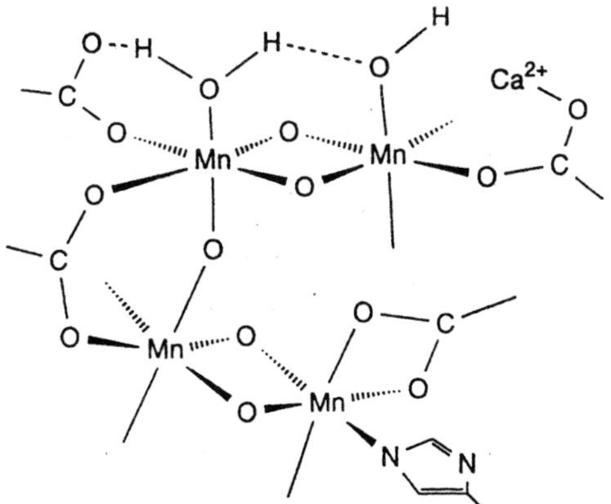

Figure 2-25. A model of the OEC Mn cluster.

The crystal structure of the photosystem II of cyanobacteria (*Synechococcus elongates*) has been presented at a resolution of 0.38 nm and the location of four Mn ions has been proposed [48], but the coordination structure of the protein is not elucidated yet. Many Mn cluster model complexes have been investigated [49,50], but most of them do not show catalytic activity for water oxidation [51].

2.6.2   Mn-Superoxide Dismutase (Mn-SOD)

The amino acid sequence of Mn-SOD is similar to that of Fe-SOD (see Section 2.4.2), and their three-dimensional structures also resemble each other. Mn is coordinated by His 28, His 83, Asp 166 and His 170, for which His 28 is the axial ligand and the other residues are planar ones forming a trigonal bipyramidal structure with a probable fifth aquo ligand. Although the ligands of Mn-SOD and Fe-SOD are the same and the protein structures closely resemble each other, substitution of one metal ion by another brings an end to enzymatic activity, showing that the specificity of the metal ion for the activity is very high. Such a large difference may be caused by a very slight difference in the coordination environment. It is inferred that a difference of amino acid residues located at the active center affects the state of the fifth ligand, resulting in a difference of the electronic state of the metal center.

2.7   Magnesium Proteins – Chlorophyll

Magnesium porphyrin complexes called chlorophyll (Chl, **12**) play important roles in photosynthesis in green plants and some bacteria. In green plants the chloroplasts of the thylakoid membrane contain macromolecular anisotropically oriented protein complexes with photosystems II and I (Fig. 2-26).

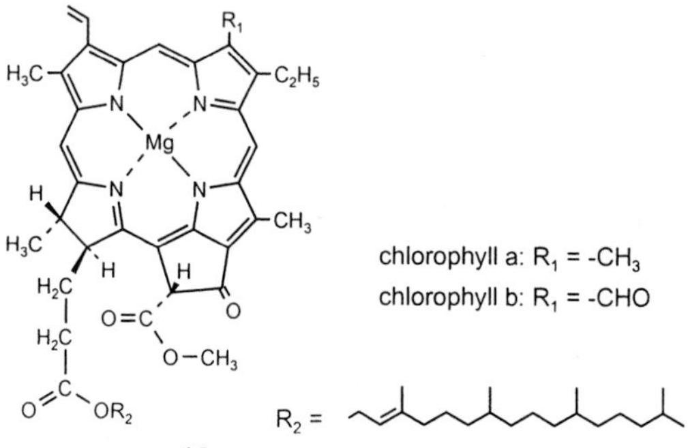

chlorophyll a: R_1 = -CH$_3$
chlorophyll b: R_1 = -CHO

12

    In each photosystem two kinds of chlorophyll exist, i.e. antenna (light-harvesting, LH) and reaction center (RC) chlorophylls. The LH Chl exists in excess with respect to the RC Chl, and harvests solar energy by excitation energy migration between the LH Chls. This excitation energy reaches the RC Chl and brings about charge separation (see also Chapter 13). A complete analysis of the structure of photosystems I and II of green plants involving Chls has not yet been achieved, but the structures of the LH and RC Chls with proteins for photosynthetic bacteria have been published [51].
    In the light reaction of photosynthesis solar light is converted into chemical

energy. Water, NADP$^+$, ADP and phosphate are converted into the energy-rich products O$_2$, NADPH and ATP (NADP$^+$/NADPH: oxidized and reduced form of nicotinamide adenine dinucleotide phosphate; ATP/ADP: adenosine tri- or diphosphate) (Fig. 2-26, Eq. 2-15 and Section 13.1). In the dark reaction, reduction of CO$_2$ by NADPH leads to carbohydrates (Calvin cycle).

$$2H_2O + 2NADP^+ + 3ADP^{3-} + 3HPO_4^{2-} + H^+$$
$$\xrightarrow{\;h\nu\;} O_2 + 2NADPH + 3ATP^{4-} + 3H_2O$$

(2-15)

Figure 2-27 shows the structures of the RC complex of the bacterium *Rhodopseudomonas* (Rps.) *viridis* (a) and the extracted functional groups (prosthetic groups) (b). L and M subunits located in the center surround the photosynthetic dyes, and cytochrome and H subunits are located above and below these, respectively. In the RC bacterial Chl (BC) forms a dimer (special pair), which is sandwiched by monomeric BCs, and two bacteriopheophytins (BP) exist in contact with the BC under which iron and two kinds of quinone compounds (menaquinone (MQ) and ubiquinone (UQ)) are located. These RC cofactors bound to the L and M subunits are arranged in two arms, and they show a high degree of two-fold symmetry with the symmetry axis perpendicular to the membrane plane. Only one branch is active in the electron transfer and the final electron acceptor is UQ. After accepting photon energy from LC, the excited special dimer reduces BP forming P$^+$BP$^-$, and the electron is subsequently transferred to the UQ via MQ [52].

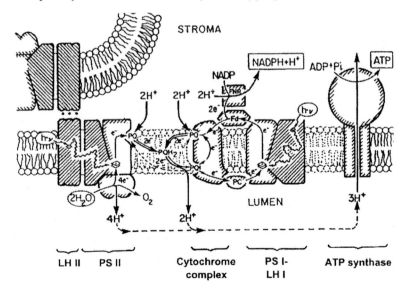

Figure 2-26. Functional organization of protein complexes in the thylakoid membrane for photosynthesis.

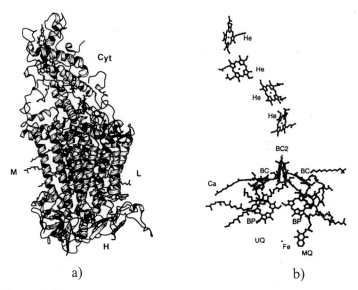

Figure 2-27. Structure of the reaction center complex of *Rps.viridis*. a) Whole structure. b) Prosthetic groups extracted from the whole structure (He = heme iron complex).

In this electron transfer system nearly 100% selective unidirectional electron transport is possible, with only a small fraction of back electron transfer. The proteins surrounding the prosthetic groups (functional molecules) play important roles in establishing such efficient electron flow by maintaining the location of the groups most suited for electron hopping.

Photosynthetic bacteria contains two types of LH Chl, that is, LH1 and LH2. A schematic representation of these structures is shown in Fig. 2-28, where LH Chl molecules are arranged to form a wheel-like structure.

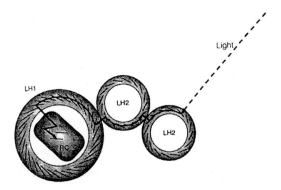

Figure 2-28. Schematic representation of light-harvesting chlorophylls.

The photon energy harvested by the antenna complexes is transferred to the RC with almost 100% efficiency by energy migration between the Chl molecules (singlet–singlet energy transfer after the Förster mechanism). The structure of the LH1 of *Rhodospirillum rubrum* associated with RC is shown in Fig. 2-29 [53]. The LH Chls are so beautifully arranged like a wheel by the protein molecules that efficient energy migration between the Chl molecules is possible. Inside the LH1, an RC complex exists which contains a special pair of Chls.

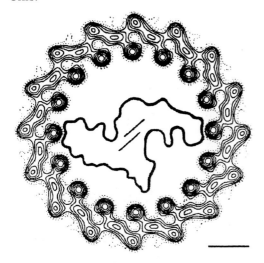

Figure 2-29. The image extracted from a projection map of two-dimensional crystals from LH1 of *Rhodospirillum rubrum* with RC complex inside[39].

2.8   Zinc Proteins

Zinc is the second most abundant metal *in vivo* next to iron (see Table 2-1). Most of the zinc exists as zinc proteins, and they function as (1) carboxy peptidase, (2) carbonic anhydrase, (3) alkaliphosphatase, (4) alcohol dehydrogenase, and (5) DNA, RNA polymerase with divalent Zn. In this section only the carboxy peptidase is described.

There have been strong arguments over the mechanism of peptide hydrolysis by carboxypeptidase. One possible mechanism is the anhydride mechanism [54] in which the oxygen of a carbonyl group is coordinated to a Zn ion, activation of an amide takes place, nucleophilic attack of $-COO^-$ of the glucose residue of the protein occurs to form an acid anhydride, and then the anhydride reacts with water by being activated by the Zn to result in hydrolysis (Fig. 2-30).

Cooperative action of the functional groups of the enzyme is important for this enzymatic catalysis as for many other enzymatic reactions.

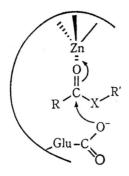

Figure 2-30. The catalysis mechanism of carboxypeptidase (anhydride mechanism, X = O, NH).

2.9   References

1. H. Sigel (Ed.), *Metal Ions in Biological Systems*, Marcel Dekker, New York, 1999, over 37 Volumes.
2. J.S. Lippard, M.J. Berg, *Principles of Bioinorganic Chemistry*, University Science Books, Mill Vallay Co., USA, 1994.
3. J. Reedijk (Ed.), *Bioinorganic Catalysis*, Marcel Dekker, New York, 1993.
4. R.J.P. Williams and J.R.R. Frausto da Silva, *The Natural Selection of the Chemical Elements*, Oxford University Press, Oxford, 1995.
5. G. Berthon (Ed.), *Handbook of Metal–Ligand, Interactions in Biological Fluids*, Marcel Dekker, New York, 1995.
6. Deutsche Forschungsgemeinschaft, *Bioinorganic Chemistry, Transition Metals in Biology and their Coordination Chemistry*, Wiley-VCH-Verlag, Weinheim, 1997.
7. *Metal-Based Drugs, An International Journal*, Freund Publishing House, Tel Aviv, Israel, Vol. 5, 1998.
8. R.G. Pearson, *J. Am. Chem. Soc.* **1963**, *85*, 3533; ibid, *Science* **1966**, *151*, 172.
9. F. Ciardelli, E. Tsuchida, D. Wöhrle (Eds.), *Macromolecule–Metal Complexes*, Springer-Verlag, Berlin, 1996.
10. D. Beyersmann, Regulation of Mammalian Gene Expression, in: *New Approaches to Drug Development*, P. Jolles (Ed.), Birkhäuser Verlag, Switzerland, 2000.
11. K. Severin, R. Bergs and W. Beck, *Angew. Chem.* **1998**, *110*, 1722.
12. H. Chen, S. Ogo and R.H. Fish, *J. Am. Chem. Soc.* **1996**, *118*, 4993.
13. M.J. Bloemink and J. Reedijk, in: *Organic Photochemistry and Biology*, CRC Press, Boca Raton, FL, 1995.
14. A.D. Ryabov, *Angew. Chem.* **1991**, *103*, 945.
15. G. Jaouen, A. Vessieres and I.S. Butler, *Acc. Chem. Res.* **1993**, *26*, 361.
16. S. Krawielitzki and W. Beck, *Chem. Ber.* **1997**, *130*, 1659.
17. A. Sigel and H. Sigel (Eds.), *Metal Ions in Biological Systems*, Marcel Dekker, New York, 1996.
18. Z. Guo, P.J. Sadler, *Angew. Chem.* **1999**, *111*, 1610.
19. J. Reedijk, *Chem. Commun.* **1996**, 801.
20. Several articles in *J. Chem. Educ.* **1997**, *74*, 633–651.
21. L.R. Milgrom, *The Colours of Life*, Oxford University Press, Oxford, 1997.
22. S. Neya, N. Funasaki, T. Sato, N. Igarashi, N. Tanaka, *J. Biol. Chem.* **1993**, *268*,

8935.

23. S. Neya, N. Funasaki, *J. Biol. Chem.* **1987**, *262*, 6725.

24. Y. Mie, K. Sonoda, M. Kishita, E. Krestyn, S. Neya, N. Funasaki, I. Tanagushi, *Electrochim. Acta* **2000**, *45*, 2903.

25. S. Neya, N. Funasaki, K. Imai, *J. Biol. Chem.* **1988**, *263*, 8810.

26. S. Neya, N. Funasaki, N. Igarashi, A. Ikezaki, T. Sato, K. Imai, N. Tanaka, *Biochem.* **1998**, *37*, 5487.

27. S. Neya, T. Kaku, N. Funasaki, Y. Shiro, T. Iizuka, K. Imai, J. Hori, *J. Biol. Chem.* **1995**, *270*, 13118; ibid. *J. Biochem.* **1997**, *121*, 654.

28. S. Neya, M. Nakamura, K. Imai, N. Funasaki, *Chem. Pharma. Bull.* **2001**, *49*, 345.

29. S. Neya, N. Funasaki, H. Hori, K. Imai, S. Nagatomo, T Iwase, T. Yonetani, *Chem. Lett.* **1999**, 989.

30. S. Neya, M. Tsubaki, H. Hori, T. Yonetani, N. Funasaki, *Inorg. Chem.* **2001**, *40*, 1220.

31. J. Przywarska-Boniecka, L. Trynda, E. Antonini, *Eur. J. Biochem.* **1975**, *52*, 567.

32. D.V. Styness, S. Liu, H. Marcus, *Inorg. Chem.* **1985**, *24*, 4335.

33. L. Tyagi, *Inorg. Chim. Acta* **1988**, *151*, 29.

34. F.W.J. Teale, *Biochem. Biophys. Acta* **1988**, *35*, 543.

35. T. Asakura, *Methods Enzymol.* **1978**, *52*, 447.

36. S. Neya, N. Funasaki, *Biochem.* **1986**, *25*, 1221.

37. E. Tsuchida (Ed.), *Artificial Red Cells*, John Wiley & Sons, Chichester, 1995.

38. K. Faber, *Biotransformation in Organic Chemistry*, Springer-Verlag, Berlin, 2000.

39. C. Nicolini (Ed.), *Biophysics of Electron Transfer and Molecular Bioelectronics*, Plenum Press, New York (1998).

40. P.M. Coleman, H.C. Freeman, J.M. Guss, M. Murata, V.A. Norris, J.A.M. Ramshaw, M.P. Venkatappa, *Nature* **1978**, *272*, 319.

41. E.D. Getzoff, J.A. Tainer, P.K. Weiner, P.A. Kollman, J.S. Richardson, D.C. Richardson, *Nature* **1983**, *306*, 287.

42. N. Ito, S.E. Phillips, C. Stevens, Z.B. Oquel, M.J. McPherson, J.N. Kenn, K.D.S. Yadov, P.F. Knowles, *Nature* **1991**, *350*, 87.

43. A.Volbeda, W.G.J. Hol, *J. Mol. Biol.* **1989**, *209*, 249.

44. K.D. Karlin, Z. Tyeklar, A.D. Zuberbuehler, in: *Bioinorganic Catalysis*, J. Reedijk (Ed.), Marcel Dekker, New York, 1993, p.261.

45. N. Kitajima, K. Fujisawa, A. Nakamura, *J. Am.Chem. Soc.* **1992**, *114*, 1277.

46. A. Messerschmidt, A. Rossi, R. Ladenstein, R. Huber, M. Bolognesi, G. Gatti, A. Marchesini, R. Petruzzeli, A. Finazzi-Agro, *J. Mol. Biol.* **1989**, *206*, 513.

47. J.W. Godden, S. Turley, D. Teller, E.T. Adman, M.Y. Liu, W.J. Payne, J. LeGall, *Science* **1991**, *253*, 438.

48. A. Zouni, H.-T. Witt, J. Kern, P. Fromme, N. Krauss, W. Saenger, P. Orth, *Nature* **2001**, *409*, 739.

49. W. Ruettinger, G.C. Dismukes, *Chem. Rev.* **1997**, *97*,1.

50. M. Yagi, M. Kaneko, *Chem. Rev.* **2001**, 101, 21.

51. J. Deisenhofer, O. Epp, M. Miki, R. Huber, H. Michel, *Nature* **1985**, *318*, 618.

52. J. Deisenhofer, H. Michel, *EMBO J.* **1989**, *8*, 2149.

53. S. Karrasch, P.A. Bullough, R. Ghosh, *EMBO J.* **1995**, *14*, 631.

54. J. Suh, T.H. Park, B.K. Hwang, *J. Am. Chem. Soc.* **1992**, *114*, 5141.

B SYNTHESIS AND STRUCTURES

3 Kinetics and Thermodynamics of Formation of Macromolecular Metal Complexes and Their Structural Organization

Anatolii Pomogailo

Chapter 3 concentrates on different aspects of the formation of macromolecular metal complexes (MMCs) by interaction of a macromolecular ligand with a metal compound MX_n. Such metal-containing macromolecules are classified in Section 1.2.1 as **type I**. Chapter 4 concentrates on polymerizations of metal-containing monomers. Examples of the binding of MX_n at macromolecular ligands are given in more detail in Chapter 5. The kinetics and thermodynamics of formation of macromolecular metal complexes of type II, type III and type IV are not known in detail. A few aspects are included in Chapters 6, 7 and 8.

The quantitative data (reaction constants, functions of formation, thermodynamic parameters, etc.) of the interactions of metal compounds (MX_n) with macroligands are important for the evaluation of the complexation ability of polymeric ligands. Well-directed attempts to synthesize MMCs should be governed on the one hand by the necessity to obtain products with homogeneous structures, and on the other hand by their practical use as metallopolymer composite materials and metal complex catalysts. Moreover, the study of complexation reactions in systems containing a macroligand is of further interest. First, for coordination chemistry, it is necessary to clarify the influence of the macromolecular chain on the character of the coordination centers formed, their structure and the nature of the bonds. From the point of view of the chemistry of high molecular weight compounds it is interesting to study the peculiarities of the reactivity of macromolecules. So far, kinetic and thermodynamic parameters have only been estimated using data on the average composition of the MMC formed. In many respects, homogeneous complexation (sometimes called ideal complexation) resembles well-known polymer-analogous reactions proceeding in dilute solutions of macromolecules [1–4]. This is the simplest model of the interaction of metal complexes with macroligands. However, the ideal regime is rarely realized, so more complicated models of complexation based on statistical physics are currently being developed, in which the mutual influence of coordination centers or cooperative effects are considered. The methods of quantitative analysis of such systems are similar to those for describing the complexation reactions of low molecular weight compounds, taking into account the specific character of polymer ligands. Two extreme possibilities should be considered: complexation in dilute polymer solutions, i.e. in solutions in which association of chains can be neglected (homogeneous complexation); and interactions with cross-linked or insoluble polymers (heterogeneous complexation). We shall briefly discuss these variants as well as some intermediate cases.

3.1   Complexation in Dilute Polymer Solutions

Mathematical models of MX_n complexation with polymers are based on the law of equilibrium and take into consideration the macromolecular nature of the ligand.

The reactions of polymers in dilute solutions have some peculiarities caused by their chain structure. The most essential ones are the following [5]:

- the effect of "neighbor" groups
- the effect of the "chain"
- the change in chain conformation in the course of the reaction
- the cooperative effect.

As a rule, any description of chemical reactions involving the participation of polymers is based on the principle that all the sites of the polymer chain are of equal reactivity. This principle is applied only to complexation in so-called θ-solvents. However, it is usually difficult to select a θ-solvent for both the starting polymer and the macrocomplex formed.

The effect of "neighbor" groups consists of the change in the free energy of formation depending upon whether a neighbor group of the reaction site has already reacted or not, in other words, the constant of equilibrium depends on the location of the reacted group. For a *macromolecular chain, MX_n,* system in the simplest case there are six possible values for the free energy of formation of the central unit of a triad (three values for a reacted group (•) and three for an unreacted group (o)) (G_i is the free energy of formation for the central unit of the triad) (Fig. 3-1).

o—o—o   (G1)   o—•—o   (G4)

•—o—o   (G2)   •—•—o   (G5)

•—o—•   (G3)   •—•—•   (G6)

Figure 3-1. Schematic representation of six sets of free energies G in a triad for a macromolecular chain – MX_n system.

Usually the simplest models – the effect of one or two reacted groups – are analyzed. Nevertheless, even in such a simplified variant calculation of the kinetic parameters is very complex because it is necessary to take into account a whole series of distributions:

- of the lengths of reacted blocks (concentrations of i-length units m_i),
- of the lengths of unreacted blocks (concentrations of i-length units n_i),
- of the structure at the given composition,
- of the composition at constant molecular weights of macroligands,
- of the compositions at different molecular weights of polymers.

Distributions of the lengths of reacted and unreacted blocks are estimated by finding the minimum of the free energy of the system or by solving the equation for the rates of the forward and reverse reactions of formation of blocks of different length [6]. Let us consider an example in Eq. 3-1 for the calculation of minimum distributions of the free energy of formation in a general case.

$$G = G_3 n_1 + G_4 m_1 + \sum_{i=2}^{i=\infty} [2G_2 + (i-2)G_1]_{n_1} + \sum_{i=2}^{i=\infty} [2G_5 + (i-2)G_6]_{m_1} - RT \ln \frac{(\sum_{i=1}^{\infty} n_i)!}{\prod_{i=1}^{\infty} (n_i!) \prod_{i=1}^{\infty} (m_i!)} \tag{3-1}$$

The last term of the equation involves the additional entropy caused by different structures of the chain at the same composition. Minimizing the equation is carried out for chains of infinite length taking material balance equations into consideration. In general, if there are q metal ions on the polymer chain containing m active centers, it is necessary to consider the combinatorial analysis of complexes formed having different structures, i.e. the number of combinations (t) of q molecules on m centers of the macroligand $t = C^q{}_m$.

The "neighbor" effect in complexation (mostly an inhibiting influence due to interactions between adjacent metal centers as well the statistical character of complexation) is revealed in a change in the local rigidity, which is increased at a high degree of conversion (α, the ratio of the actual number of metal ions bonded with one chain and the maximum possible number) as well as in the change in the valency of metal ions. According to the Landau model [7] for a flexible thread, local rigidity is the resistance to bending of a chain segment. At complexation, the free energy changes are caused by both the chemical reaction and the change in the linear sizes of the chain (so-called work of stretching A_s). The sizes are determined by the locations of flexible and rigid segments of the chain. The reaction sites will consequently have different equilibrium constants due to different A_s values. Therefore, the probability of reaction in this zone will be higher at less stretching of the chain. This leads to a new type of distribution – distribution by "site" – which influences other types of distribution including that by length of reacted and unreacted sequences. It is important that an increase in the rigidity of chain leads to the longer chains reacting.

The "chain" effect in complexation consists of a change in macromolecular shape without any change in the contour length and is caused by a high local concentration of ligand groups in the solution. Therefore, the stability constant is determined not by the mean concentrations of reacting components but by their local concentrations in coils (acting as microreactors). A metal ion in a zone of high concentration of functional groups forms coordination-saturated complexes at once, and the remaining free groups (according to the different distributions mentioned above) may participate in the formation of a small number of coordination-unsaturated structures. The observed permanency of composition of complexes at different ratios of reacting components is a feature of the reactions of macroligands. For macromolecules, the formation of coordination-saturated complexes (even in small quantities) is thermo-dynamically more favorable than that of unsaturated complexes because the first addition of a metal ion into a polymer chain is accompanied not only by a direct chemical interaction but also by the loss of most of the entropy, because of the change in local rigidity of the chain at the point of the MX_n attachment, and all following stages of addition proceed with minimal changes in the entropy, as they are intramolecular cyclizations (Fig. 3-2). This leads to an increase in the reactivity of unreacted functional groups (for example, the oxygen atoms of poly(ethylene glycol) (PEG)).

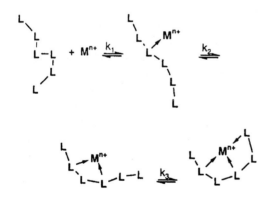

Figure 3-2. Schematic representation of the interaction of a metal ion with a polymer chain containing the ligand L.

In other words, coordination produces bending of the polymer chain and its conformation becomes more favorable for continuing the reaction. In this case the stability constant may be represented as the product of the corresponding stages, where $\sigma = K_i/K_{i-1}$ is the parameter of the cooperativity which indicates that the value of the stability constant of each subsequent addition is greater than that of previous one (Eq. 3-2).

$$K = \prod_{i=N}^{i=1} \int K_i = \sigma K_i \qquad (3\text{-}2)$$

All subsequent constants of formation should be the same except the constant of the first stage and $\sigma = K_2/K_1 = 10^{-4} \div 10^{-8}$. Therefore, the interaction with one chain proceeds until all potential reaction sites have been used completely. In this case, both the complexes with maximum amounts of bonded functional groups (the maximum coordination number) and uncoordinated macroligands exist simultaneously in the system. In other words, in such systems a widely extended biological principle "all or nothing" is realized. This is supported by numerous experimental facts. Thus, using a labeled atom technique with tritium, an interaction of PE-grafted-poly(acrylic acid) with nickel acetate was found to be an equilibrium and reversible reactions as shown in Eqs. 3-3 and 3-4 take place.

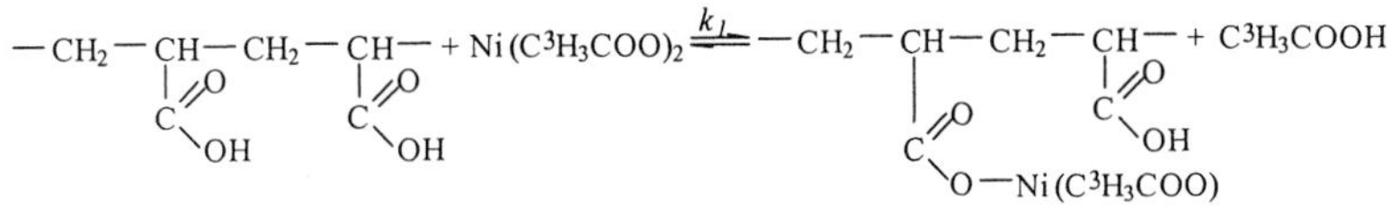

$$ -CH_2-CH-CH_2-CH- + Ni(C^3H_3COO)_2 \xrightleftharpoons{k_1} -CH_2-CH-CH_2-CH- + C^3H_3COOH \tag{3-3} $$

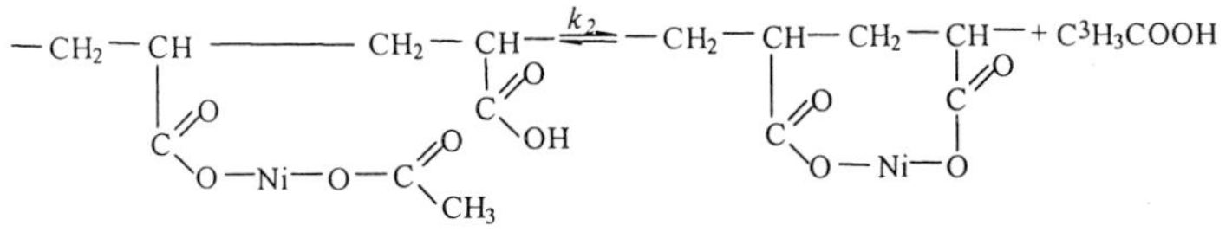

$$ -CH_2-CH \underline{\hspace{2cm}} CH_2-CH- \xrightleftharpoons{k_2} -CH_2-CH-CH_2-CH- + C^3H_3COOH \tag{3-4} $$

The first stage of the reaction is the formation of a monosubstituted species C_1, then at the second stage a disubstituted species C_2 arises with equilibrium constants K_1 and K_2, respectively. Apparently, the second reaction should not occur if two adjacent units have already reacted. Therefore, it is necessary to take into account the probability of the presence of reacted neighboring groups. The probability is equal to $p = [COOH]/[COOH]_0$ (the equilibrium concentrations of unreacted groups and initial concentrations of functional groups on the chain, respectively) if no units react, and the probability of a reversible process is $1 - p$. Two variants are possible: no both neighboring units or only one of them are reacted. A system of equations which describes this equilibrium can be presented as follows:

- $K_1[Ni(CH_3COO)_2][COOH] = C_1[CH_3COOH]$,
- $2K_2C_1p^2 + K_2C_1p(1-p) = C_2[CH_3COOH]$,
- $[Ni^{2+}]_0 = [Ni^{2+}] + C_1 + C_2$,
- $[COOH]_0 = [COOH] + C_1 + 2C_2$,
- $[CH_3COOH] = C_1 + 2C_2$.

The validity of the proposed scheme was confirmed by a comparison of the experimental and calculated dependences of the amount of bonded Ni^{2+}-ions on the initial ratio $[Ni^{2+}]/[COOH]_0$ (Fig. 3-3) [8]. Using the temperature dependence of the ratio, $\Delta H = 7300$ J mol^{-1}, $K_1 = 1.49 \cdot 10^{-2}$, $K_2 = 1.22 \cdot 10^{-2}$ mol g^{-1} at 333 K were found. The equilibrium of the reaction is shifted to the formation of the species C_2 because the entropy of the system is increased, the value ΔS is equal to 104 ± 20 J mol^{-1} deg^{-1}, which corresponds to the formation of a new molecule of CH_3COOH in each part of the reaction. Depending on the reaction conditions, the content of the species C_1 varies from 0.2 to 8.2 mol %. For comparison note that complexation of Cu^{2+} ions with this macroligand proceeds by the same mechanism with $K_1 = 16 \cdot 10^{-2}$ and $K_2 = 2.54 \cdot 10^{-3}$ mol g^{-1} (at 333 K), and the amount of the C_1 form is more than that for the Ni^{2+}-system and reaches 16 mol % [9].

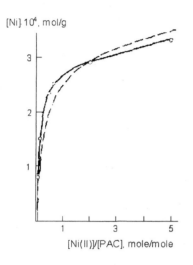

Figure 3-3. Experimental (solid line) and calculated (dotted line) dependences of bound Ni^{2+} ions on the initial ratio $[Ni^{2+}]/[COOH]$.

Finally, for many polymers at high degrees of complexation a change in the chain conformation is observed because the effect of the intramolecular interactions of valence non-bound atoms becomes important. Here different conformations differ in flexibility and a drastic increase in the equilibrium constant is observed for a small change in the degree of conversion. Configurational (a change of the spatial location of the adjacent unit, "neighbor effect") and conformational (the rearrangement of a macro-molecule during the reaction with MX_n) effects take place. In addition, conformational transformations may occur at each act of MX_n binding as well as at certain degrees of conversion involving different numbers of monomer units. Metal ions usually interact by forming two or more bonds with the nearest active centers forming five- and six-membered metal rings. To form larger rings requires more energy as entropy losses rise [5]. If the neighbor of the reacted group is a conformationally inactive group (the "neighbor effect" extends right and left of the units entering the coordination sphere of the transition metal), no reaction occurs with such a group. For example, Co^{2+} ethylene diamine complexes have rather a large volume (about 10 Å^3). Therefore in their coordination, even at closest packing, many units of poly(4-vinylpyridine)

(PVPy) are uncoordinated for purely steric reasons. The calculated maximum degree of conversion for PVPy amounts to 0.63, which is close to the experimental value [10]. It should be noted that the change in rigidity of a chain, which depends on the number of bonded functional groups, influences the MMC stability: most stable complexes are formed at low degrees of conversion (η), only a few at high degrees. Owing to the high mobility of the polymer ligand, its conformation changes at the point of binding according to the electron configuration of the transition metal. In addition, both transitions of the helix-coil typical of charged chains and changes in the preferred conformation of a statistical coil are possible. The presence of hydrogen bonds and steric effects hinder bending and rotation, i.e. conformational optimization. This may explain why many functional groups do not participate in complex formation.

It should also be mentioned that complex formation often results in the stabilization of atypical polymer conformations and tautomeric forms. A classical example is the molecule of poly(ethylene glycol) (PEG), having a helix structure with alternating *trans* and *gauche* C–O, O–C and C–C bonds, respectively (conformation T_2G, helix 7_2) [11]. If it is treated with $HgCl_2$, complexes of regular composition $HgCl_2(OCH_2CH_2)_4$ with an elementary unit composed of four molecules of $HgCl_2$ and 16 units of (OCH_2CH_2) [12] is obtained. Three of each four G-forms are transformed to T-, T-, $\overline{G}$-forms (G and $\overline{G}$ mean dextro- and levorotatory *gauche* forms) and the resulting chain carrying bound complexes is in $T_5GT_5\overline{G}$ conformation with an increased proportion of turned-off configurations (*gauche*), each elementary unit of PEG changing from monoclinic to orthorhombic. These transformations cause a change in the linear orientation of the $HgCl_2$ molecule (the angle ClHgCl is equal to 176°). The more noteworthy change occurs in the repeating period of the macromolecule. It is composed of four monomer units instead of seven for the uncomplexed polymer, though each unit of the complex is more elongated. Structural studies [13] show that the MMC properties in PEG–alkali metal ion complexes are better described in terms of the double-helix model.

Metal ions may be bonded to one or more polymer chains forming intramolecular or intermolecular complexes, respectively. Intermolecular interactions between metal complexes and macroligands rise significantly when going from dilute polymer solutions to concentrated ones or a matrix. During intramolecular complex formation the first addition may be considered as a second-order reaction (the first one on each component), and all the following intramolecular interactions are first-order reactions. The overall interaction of MX_n with one chain is a second-order reaction. Because the rates of complex formation are high, it is difficult to isolate the successive stages of the process experimentally. Only in rare cases has it been possible to resolve this problem. The graduated stability constant of the *i*-complex (the actual or characteristic K_i)

was found to increase with conversion in contrast to low molecular weight analogues (Table 3-1).

Table 3-1. Stability constants and thermodynamic data for the M^{2+}–macroligand complexes.

M^{2+}-Macroligand	B_2	K_2	$-\Delta G$, kJ·mol^{-1}	ΔH, kJ·mol^{-1}	ΔS, J mol·grad^{-1}	References
Cu-PMA	$2.5 \cdot 10^{-4}$	$4.0 \cdot 10^{9}$	55.2	21.3	254.6	[25]
Zn-PMA	$6.3 \cdot 10^{-6}$	$1.0 \cdot 10^{8}$	46.0	14.6	187.7	[25]
Cd-PMA	$4.0 \cdot 10^{-5}$	$6.3 \cdot 10^{8}$	50.6	10.9	204.4	[25]
Ni-PMA	$2.5 \cdot 10^{-6}$	$4.0 \cdot 10^{7}$	43.5	4.6	161.3	[25]
Co-PMA	$2.0 \cdot 10^{-6}$	$3.2 \cdot 10^{7}$	43.0	4.6	158.4	[25]
Mn-PMA			2.1			[25]
Ca-PMA			2.1			[25]
Mg-PMA	$6.3 \cdot 10^{-7}$	$1.0 \cdot 10^{7}$	40.1	0.6	135.8	[25]
Cu-PMAAc (syndiotactic)		$4 \cdot 10^{9}$	55.2	21.3	255	[27]
Cu-PMAAc (isotactic)		$1.2 \cdot 10^{10}$	58.1	15.9	246.8	[27]
Mg-PMAAc (syndiotactic)		$1 \cdot 10^{7}$	40.1	0.6	135.8	[27]
Mg-PMMAAc (isotactic)		$0.4 \cdot 10^{7}$	38.0	3.3	137.9	[27]
Eu(III)-PAA		$7.9 \cdot 10^{10}$				[10]
Co-PVAl		$1.1 \cdot 10^{11}$	66.9			[28]
Ni-PVAl		$6.6 \cdot 10^{11}$	66.9			[28]
Zn-PVAl		$3.7 \cdot 10^{13}$	80.4			[28]
Cu-PVAl		$8.5 \cdot 10^{15}$	94.1			[28]
Ag-PVIA	10^{-2}	10^{8}				[29]
Zn^a-PVIA		10^{4}				[4]
Cu^a-PVIA	10^{-7}	10^{11}–10^{15}				[4]
Cu–PVPy		$5 \cdot 10^{9}$				[4]

Note: PAC, poly(acrylic acid); PVAl, poly(vinyl alcohol); PMAAc, poly(methacrylic acid); PVIA, poly(N–vinylimidazol); PVPy, poly(4–vinylpyridine). ^{a} coordination number (CN) = 4.

This is caused by the cooperative effect arising from the change in the macromolecular form during complex formation. The cooperative interactions in the system macroligand–MX_n in solution also produce a change in the charge of the chain.

Formal thermodynamic approaches can be elaborated to describe complex formation using different complexation schemes (see, for instance [14]), where c_M, c_L, c_{ML} are the molar concentrations of free metal ions, and the formal species free and bound sites, s is the maximum number of metal ions bound with one macromolecule, c_j is the molar concentration of the real species M_jP (j = 1, ...s), and θ is the coverage or fraction of occupied sites in a macromolecule (Eq. 3-5).

$$K_c = \frac{c_{ML}}{c_M c_L} = \frac{1}{c_M} \frac{\sum_{j=1}^{s} j c_j}{\sum_{j=0}^{s} (s-j) c_j} = \frac{1}{c_M} \frac{\theta}{(1-\theta)} \tag{3-5}$$

Activity coefficients for bound and free sites are used to describe the complex formation. Independently of the complexation model, the concentration equilibrium relationship for free and occupied sites tends to K when the concentration of MX_n tends to vanish (ligand excess), and the total equilibrium constant of complexation is the product of all k_i. The distribution function of the equilibrium constants ($\rho(k)$) is embedded in the model of a continuous distribution of the constants and represents that proportion of the functional groups that corresponds to an individual value of K (Eq. 3-6).

$$\int_0^{\alpha} \rho(k) dk = 1 \tag{3-6}$$

The physical meaning of the function $\rho(k)$ is the density of probability of the stability constant k and is connected with experimental data, i.e. the change in the concentration of metal ions in solution and unreacted functional groups, or the concentration of surface sites occupied by metal ions $f([M])$. At $\rho(k) \geq 0$ the equilibrium constant can be presented as an average equilibrium function depending on the molar concentration, c_j, of the real species M_jP (j = 1, ...s) and on the fraction of occupied sites in a macromolecule, θ.

The total constant K is obtained as the limit of k_i for $c_M \to 0$, taking into account the intrinsic constant of complexation for each site of the chain (k) and the probability distribution of the microscopic stability constants $\rho(k)$ (Eq. 3-7).

$$K = \lim_{c_M \to 0} \frac{\int_0^{\infty} \rho(k) \dfrac{k}{1+kc_M} dk}{1 - \int_0^{\infty} \rho(k) \dfrac{k}{1+kc_M} dk} = \int_0^{\infty} \rho(k) k \, dk \equiv \langle k \rangle \tag{3-7}$$

The concentration of surface sites occupied by metal ions can be obtained using the expression of Eq. 3-8.

$$f([M]) = \int \frac{\rho(k)K[M]dk}{(1+K[M])} \qquad (3-8)$$

This expression is used to characterize the complexation properties of macroligands such as humic acids, proteins, etc. (see for instance [15–17]). There are many formalized models of complexation [18–20] in which expressions for the constant of formation considering the different forms of complexes are suggested. However, they are inconvenient for practical use and require a multiplex calculation procedure.

The interaction of metal ions with macroligands is a very fast process with rates (about 10^8 mol L^{-1} s^{-1}) comparable to those of low molecular weight compounds. Usually the reaction proceeds over several seconds [21]. It should be mentioned that the kinetic analysis of these reactions has been developed for a long time (see [1–4]).

The data on constants of complexation (k_k) and dissociation (k_d) are rather limited. Thus, complexation of Ni(II)-ions with neutral and protonated poly(4-vinylpyridine) (PVPy) in water–methanol solution was studied using the stopped stream method [22]. The values of the reaction rate constants of formation of both the polymer complex and its monomer analogue are close (4500 and 3500 mol L $^{-1}$ s^{-1}, respectively) but the constant of the reverse reaction rate is greater for the polymer complex than for the monomer one (6.6 and 2.3 s^{-1}, respectively). This is also observed for Cu^{2+} and PAC complexes, for which k_k is equal to $\approx 10^9$ mol L $^{-1}$ s^{-1}. Complexes with monomer analogues are characterized by similar parameters. Interesting kinetic peculiarities were observed for the complexation of Co^{2+} ions with a copolymer of styrene and 4-vinylpyridine, 4-VPy, (the content of latter monomers is 2 to 95 mol %) using the stopped stream method along with spectrophotometry [23]. The dependence of k_k on the content of 4-VPy monomers is described by a bell-shaped curve with a maximum at 50 mol % of 4-VPy, and for all polymer complexes the constant of formation is higher than that for the complex with 4-VPy. Both this and the character of the curve are caused by lower values of k_d. One of the reasons for such a dependence may be also a combination of two contrary effects – a negative, inhibitory "neighbor" effect and a positive "chain" effect.

Polymers containing many active complexing groups are polydentate ligands, and this determines their characteristics in complexation. Binding of metal ions to several functional groups leads to cross-linking of the macromolecular coil with a decrease in the viscosity of the solution, and, finally, insoluble complexes form at high concentrations of MX_n.

Two different principal approaches are used for theoretical considerations of specific features of the equilibrium binding of macromolecules with metal ions in solutions [23]. According to one of them, the central species is a metal ion with which a few functional groups of the chain are successively or simultaneously reacted (Eq. 3-9). Simultaneous reactions are assumed to be independent of each other.

$$M + nL \rightleftarrows ML_n \tag{3-9}$$

The validity of this model is limited because it fails to take into account the changes in macromolecule conformation during complexation as well as the inaccessibility of parts of the functional groups involved in complexation.

In another model the central species is a macromolecule as matrix with a certain number of binding sites. In this case the formation of complexes in solution can be described by sets of equilibria [24], where P is the coil of macromolecule and N is the maximum number of metal ions capable of binding with one coil (Eq. 3-10).

$$PM_{n+1} + M \xrightleftharpoons{k_n} PM_n \, (n=1,2,\dots N) \tag{3-10}$$

The equilibrium constant can be written as $K_j = C_{M_{j-1}} / C_{M_{j-1}P} . C_M$.

Although such a model satisfactorily describes complexation with proteins or DNA, its use for flexible chain polymers is restricted because there are no definite "binding sites" – one metal ion coordinates with a few arbitrarily located functional groups.

In a soluble macromolecular complex, there may be 10–20 units of the chain for every metal ion, i.e. the majority of the polymer units do not participate in complexation, apparently for steric reasons as well as the above-mentioned polymer effects. With a great excess of metal ions the macrocomplex is precipitated accompanied by a small additional binding of the metal. In these cases intermolecular cross-linking of polymer coils by metal ions probably occurs (see below).

Different methods are used to study complexation in solution. Potentiometric titrations are often applied to determine the equilibrium in complexes of polymer acids and bases. When a polymer reacts with a metal ion, the pH of the solution is substantially decreased (by 1 or 2 units) either because of proton elimination or because protons cannot add to the complexed unit. Thus, the interaction of polymer acids with metal ions proceeds according to Eqs. 3-11, 3-12 and 3-13 [25].

$$2(RCOO^-, H^+) + M \rightarrow (R{-}COO)_2M + 2H^+ \tag{3-11}$$

$$2A^- + M^{2+} \xrightleftharpoons{K} MA_2 \tag{3-12}$$

$$2AH + M^{2+} \xrightarrow{B_2} MA_2 + 2H^+ \tag{3-13}$$

K and B_2 are expressed by the relationship $B_2 = K \cdot k_a^2$, where k_a is the observed dissociation constant of the polymer acid (for instance, for syndiotactic poly(methacrylic acid) $k_a = 2.5 \cdot 10^7$). The values of K for MA_2 complexes are known (Table 3-2).

Table 3-2. The stepwise constants of formation of macrocomplexes and their low molecular weight analogues [4].

Macroligand	M	$\lg K_1$	$\lg K_2$	$\lg K_3$	$\lg K_4$	$\lg \overline{K}$
PAC	Sm(III)	5.7	5.4	-	-	11.1
	Eu(III)	5.7	5.2	-	-	10.9
	Pr(III)	5.6	5.2	-	-	10.8
	Cu(II)	4.8	4.2	-	-	9.0
	Ni(II)	3.9	3.4	-	-	7.3
	Co(II)	3.7	3.1	-	-	6.8
PE-gr-PAC	Sm(III)	6.2	5.9	-	-	12.1
	Tb(III)	3.86	-	-	-	-
	Cu(II)	5.9	5.2			11.1
Propionic acid	Sm(III)	3.9	3.2	2.9	-	10.0
	Eu(III)	3.8	3.2	2.9	-	9.9
	Pr(III)	3.7	3.1	2.9	-	9.7
PVPy (DP = 108)	Cu(II)	3.4	2.9	-	-	6.3
	Cu(II)	1.1	1.6	2.7	4.4	9.7
	Cu(II)	2.5	1.9	1.3	0.8	6.5

Note: PAC, poly(acrylic acid); PE–gr–PAC, polyethylene grafted poly(acrylic acid); PVPy, poly(4–vinylpyridine); DP, degree of polymerization.

The titration curve of a polymer in the presence of a metal salt is lower than that for a free polymer. There is a well-developed technique which takes into account this difference of pH and allows us to calculate the average number of polymer units bound with metal ions (the modified Bjerrum method). Although in recent years more modern computing techniques have been suggested [26], the major limitation of this method is the assumption that non-bound polymer units in free macromolecules and metal complexes are of equal acidity. But the conformation of the macromolecule changes during complexation, in particular it becomes more compact, and this should influence the acid–base properties of the unbound groups. Therefore the assumption that the properties of ligand units are independent of the degree of complexation restricts the applicability of this method. Nevertheless, potentiometric titration is still widely used in comparative analysis.

MX_n binding is appreciable influenced by hydrophobic, hydrophilic, and electrostatic interactions between reagents. The formation of charged chains during M^{n+} binding results in a straightening of the macromolecule. Moreover, it may contain positively or negatively charged groups which either facilitate or hinder its complexation with the metal ion. Therefore electrolyte solutions are specially added to the reaction medium to neutralize the electrostatic repulsions. The formation constants and thermodynamic parameters of formation of divalent metal complexes with some polymers (poly(methacrylic acid) (PMMA), poly(-N-vinylimidazole) (PVIA)) are listed in Table 3-2. Some kinetic parameters of complexation are: PVIA–Ag (coordination number (CN) for Ag is 2) $K_2 = 10^8 \, \mathrm{mol}^{-2} \, \mathrm{L}^{-2}$, $B_2 = \sim 10^2$; PVIA–Zn (CN = 4), $K_4 = 10^3 \, \mathrm{mol}^{-4} \, \mathrm{L}^{-4}$; PVIA–Cu (CN = 4) $K_4 = 10^{11}\text{–}10^{15} \, \mathrm{mol}^{-4} \, \mathrm{L}^{-4}$ depending on the ionic force and polymer concentration.

For the following reaction (Eq. 3-14) the equilibrium constant $B_4 = 10^{-7}$, whereas for the monomer analogue its value is 10^{-16}.

$$4Cu^{2+} + 4 \, ImH^+ \rightleftarrows (CuIm^{2+})_4 + 4H^+ \tag{3-14}$$

More accurate data can be obtained by direct measurements of the equilibrium concentration of the free ion-complexing agent in solution. Ion-selective electrodes, equilibrium dialysis when the solution of a complex is in equilibrium with water or base electrolyte through a membrane permeable only to low molecular weight ions and other methods are used for this purpose [30,31]. If the equilibrium constants of free metal ions are known, we can estimate the kinetic parameters of complexation [32]. The difficulty of such an approach is that the equilibrium concentrations of free metal ions in the case of highly stable complexes are very small. Therefore even small errors in their determination will change the values of the stability constant and other parameters. Occasionally sedimentation methods are used to study the polymer–metal system. The phenomenon of disproportionation was discovered using this method [24]: that is, the coexistence in solution of coils of the macromolecules containing both low numbers and the maximum number of metal ions. Electron spin resonance is widely used to study low molecular weight coordination compounds [33] because of its high sensitivity to the electronic state of the metal ion, the number and nature of ligands and the geometry and topography of metal complexes, in other words, to its high information density. This method is also applicable to determine the fine structure of an MMC.

3.2 Complexation with Insoluble Polymers (Heterogeneous Complexation)

Kinetic analysis of reactions in which insoluble polymers take part requires us to consider difficulties arising from both diffusion (transfer of functional groups at the glass point of the system) and topology (in practice a complete lack of translational diffusion of the functional groups grafted into the polymer). Because there is a diffusion process, the surface functional groups react first, and the penetration of MX_n into the network is limited (although sometimes it can be large, see below). Kinetic and thermodynamic analysis of the processes must also consider the polymer effects described above. Distribution of the types of coordination species is also a specific problem in complexation but has no physical meaning in the study of adsorption. This also applies to conformational transformations and transfer or accumulation processes in MMC systems.

Binding isotherms described by the modified Langmuir equation are often used to provide an estimate of interactions in such systems and to calculate the formation constants. The reason for observed deviations of the experimental dependences from the theoretical curves is a departure from the major postulates of the Langmuir equation, namely: one active site of the surface and one molecule of the low molecular weight compound should participate in the reaction, and all unreacted sites have the same reactivity. Obviously, this postulate is violated in complexation. At low conversion of functional groups, concentrations of unreacted and inactive groups (from unreacted blocks and containing a smaller number of functional groups than required for coordination with MX_n) can be neglected. Then the modified equation of the localized adsorption isotherm for evaluation of formation constants is as shown in Eq. 3-15 [34], where f_{max} is the constant characterizing the limit binding (limit adsorption) of metal by functional groups of the polymer, (i.e. $[M]_b/[L]_o$).

$$[M]/[M]_b = 1/K + (1/f_{max}[M]) \tag{3-15}$$

This approach is comparatively widely used, for example to evaluate the formation constants of Cu(II) and Ni(II) complexes with polyethylene-grafted-poly(acrylic acid) (PE-gr-PAA) (Table 3-2), Mn(II)-poly(acrylic acid), for studying sorption of low molecular weight compounds by polymers and complexation with modified silica, etc. [35–37]. It was found that one third of the carboxylic groups failed to react with copper ions in the PE-gr-PAA-Cu(II)-system ($K = 300$ L mol^{-1}, $f_{max} = 0.35$). For Cu(II) complexes with polystyrene-grafted-poly(vinylpyridine) of different types the values of K and f_{max} vary from ~2500 to 9100 L mol^{-1} and ~0.15 to 0.37, respectively [38].

This equation is mathematically like the one suggested by Langmuir for describing physical adsorption on an energy heterogeneous surface. The term "energy heterogeneity" means that the binding of metal ions by different surface sites is characterized by different adsorption heats. One can distinguish biographical (pre-history) heterogeneity of stoichiometrically identical sites and evolutionary heterogeneity – the change in stability constants during surface filling by metal. These calculations are based only on the degree of surface filling by metals and concentration equilibrium constants [18]. The prehistory heterogeneity of a polymer surface can be caused by differences in the topochemistry of bonded metal complexes (see below), the unit variability of the polymer chains, unequal solvation of functional groups and complexes, the distribution of pore sizes, etc. Energy heterogeneity means the difference between the standard chemical potentials of the reagents bound to the surface and having the same stoichiometric composition.

The principal differences between descriptions of equilibrium complexation both in solution and on a surface also involve the number of independent variables, which unambiguously determine the state of equilibrium. If a knowledge of the total concentrations of ligands and metals and the equilibrium pH value (altogether five independent variables, see above) is sufficient to estimate all the equilibria in a solution and the case of complexation on the surface [39], such an equilibrium is influenced also by a variable surface potential ψ_s. The variable involving this factor is the degree of neutralization (α). The dependence of binding degree (η) on reaction conditions indicates that such systems are non-equilibrium, so the constants obtained with equations for equilibrium systems have a formal character. It is difficult to analyze and compare the values of K at least from conventions of their dimensionalities (for example, the formal constant of formation of binding of four ligand groups of a polymer in solution has units of $mol^{-4} L^{-4}$). The approaches are more widely applied to calculate the constants using the assumption that all functional groups are bound to one molecule of MX_n and are considered as a one n-dentate ligand.

The composition and stability of macrocomplexes are also affected by other factors, especially the porosity of polymers. As a rule, transition from a nonporous (gel) to a porous structure is accompanied by a decrease in the average constant of formation of complexes determined by a greater density of packing of chains and lower mobility of the functional groups. This leads to appreciable losses of free energy for complexation. The change in local rigidity arising from conversion also has an influence on the stability of complexes. The most stable complexes (in the case of cross-linked polymers) are formed at small values of the ratio MX_n:$[L]_o$ and the less stable ones at high values. Thus, at high degrees of conversion, complexes of the composition 1:2 and even 1:1 were found and, for example, [M]:[L] = 1:3, for Eu(III) with poly(n-benzoylacetylstyrene) [40]. As a matter of fact, "neighbor" effects in cross-linked polymers should be more pronounced than in dilute solutions. But the rates of reaction in such systems are substantially lower than those in solution owing to diffusion restrictions. In such systems formation of external sphere complexes and their transformation to internal ones with change of the position of the ligands occur. Therefore, due to lower concentrations of MX_n in the network, the average composition of complexes differs from those obtained in solution. In view of this, carrying out reactions on the surfaces of polymer ligands (for instance, functionalization of a polymer surface by means of grafted polymerizations) results in an increase in both the reaction rate of complex formation and the constant of formation.

Equation 3-16 has been successfully used to estimate the value of the binding constant (K^s) in a heterogeneous system macroligand, MX_n [39], where $\tilde{n}$ is the average number of ligands bound with M^{n+}, i.e. the mean denticity of binding; c^s_M and c^s_{RH} is the total concentrations of metal and ligand groups on the polymer surface (mol g^{-1}); K_{RH} is the characteristic dissociation constant of ligand groups RH, V is the volume of the solution phase, g is the weight of polymer.

$$K^s = \frac{\tilde{n}c^s_M [H^+]^n (c^s_{RH}\alpha + \dfrac{[H^+]V}{g} - \tilde{n}c^s_M)^{n-1}}{(K_{RH})^n [M^n](c^s_{RH}(1-\alpha) - \dfrac{[H^+]V}{g}^n)} \tag{3-16}$$

The mean denticity of binding is expressed by Eq. 3-17, where c^i_M is the concentration of particles of corresponding composition on the surface of the polymer in mol g^{-1}.

$$\tilde{n} = \frac{\displaystyle\sum_{i=1}^{m} i[c^i_M]}{c^s_M} \tag{3-17}$$

The correctness of this model has been confirmed in many cases. Thus, lg K^s is 3.4 to 2.5 for the binding of Cu(II) with carboxymethylcellulose depending on the salt concentration (0.1 ÷ 1.0 mol L^{-1} NaCl) and the mean denticity of binding is 1.8 to 2.0. Such a model is also in agreement with the sorption of molybdenum(VI) at carboxymethylcellulose and phosphatecellulose.

In classic coordination chemistry the parameter n is known as the formation function, showing the number of functional groups directly bound with metal ions, whereas in the coordination chemistry of macromolecular complexes this parameter corresponds to the average number of functional groups which are excluded from the total number of ligands for binding of one metal ion. Moreover, for complexation of MX_n with cross-linked polymers the total concentration of macroligands is constant and does not depend on the dilution of the suspension in a solvent.

3.3   The Structural Organization of MMCs

Going from a dilute solution to a polymer matrix, intermolecular interactions increase both between segments of the polymer chains (where they are stronger than for low molecular weight analogues) and in complex formation. For macromolecular complexes the interactions seem to be stronger in comparison with linear polymers due to the decrease of chain flexibility, as well as to interactions with other chains. The parameters of intermolecular interactions in MMCs are influenced by the reactivity of functional groups and their distance from the backbone, which determine the stability of the complex and the magnitude of intermolecular interactions between chains under otherwise equal conditions. They are also affected by the presence and size of other non-complexing groups. Because of these factors, MMCs often form more perfect structures than the initial macroligands. The valence of metals also has a significant influence on the stability of complexes. Thus, the stability of poly(keto–enol) chelates formed on previously oxidized films of poly(vinyl alcohol) increases as follows: $M^+ > M^{2+} > M^{3+}$ [40].

An increase in the degree of cross-linking results in a decrease in the stability of macrocomplexes because, on one hand, the stress of a twisted chain occurs during coordination with the second and subsequent coordination sites and, on the other hand, the character of the intermolecular interactions alters. Moreover, at a high degree of cross-linking of the macroligands the structure of the MMC may change: for instance, stable square-planar complexes of Cu(II) become distorted tetrahedral. Other factors, in particular the temperature, may also affect intermolecular processes and result in changes in the structure and morphology of the polymer.

The molecular weight characteristics of the starting polymer play an important role in intermolecular interactions. For "weak" intermolecular interactions between macromolecules, association of chains occurs at a certain chain length. In the case of polymers with vigorous intermolecular interactions changes in both the morphology of the macromolecules and their cross-linking are observed. For example, poly(vinyl alcohol) (PVAl) with a degree of polymerization over 260 forms stable intramolecular complexes with Cu(II) via coordination of four hydroxyl groups of one molecule, whereas polymers with a degree of polymerization of less than 160 yield intermolecular complexes by cross-linking PVAl chains with Cu(II) ions [4].

Changes in permolecular structure during complexation become more abrupt in the case of cross-linked polymers. Thus, reorganization of a copolymer of styrene with divinylbenzene caused by its alkylation with 5-chloromethyl-8-hydroquinoline in nitrobenzene in the presence of AlCl$_3$ proceeds on the basis of covalent fixation of the complex into the polymer [41]. The formation of stable complexes is accompanied by micronet rearrangements and curling of the polymer. As a rule, such processes take place in highly elastic polymers (in a swollen state) at high concentrations of the metal ion. During the interaction of rigid polymer chains with MX$_n$, the interchain cross-links change their permolecular organization considerably, and separate structural elements (microfibrils, fibrils, etc.), and cross-linked chains form the network.

The structure of the reacting units of the polymer changes according to the steric demands of the metal-complexing agent. In all cases the ability of chain molecules to curl depends on the rigidity of the chain and the nature of the solvent. Using these factors one may reduce or, on the contrary, reinforce the curling or stretching of the chain. As the structure of MMCs is determined by the conditions of their synthesis, thermodynamically non-equilibrium products are apparently formed (a distribution of different complexes along the chain is observed) and we should make allowance for a relaxation time and a rearrangement of the complexes into the most thermodynamically stable form. However, in many cases no such rearrangement is observed, and the MMC has an inhomogeneous composition. Since the complexation reaction is very rapid, there is no time to reach equilibrium and a product of uncertain composition precipitates. Rearrangement of an MMC is also hindered by the rigidity of the macromolecular skeleton, whereas in the case of linear polymers in solution, as a rule, homogeneous coordination-saturated MMCs are formed.

As mentioned above, metal complexes cause an increase in the chain rigidity which becomes apparent in relaxation transitions of the chains and their segments. Evidently, the α-relaxation associated with the micro-Brownian motion of long segments of the chains near the glass point is more subject to such influences. For low degrees of MX_n binding the overall restriction in mobility of the chains is slight but with a subsequent incorporation of metal ions a great number of cross-links between the adjusted chains is formed, which leads to an increase in the glass point (T_g). Let us consider the following example [42]. In the complexes of copolymers of ethylene and acrylic acid (9.3%) with Ni(II) or Mn(II) acetate, α-relaxation does not vary significantly if up to 14% of the acrylic acid units are neutralized. With higher numbers of reacted units T_g rises as a consequence of the toughening of individual segments, and decreases the mobility of an entire chain. In the region of β-relaxation the glass point is almost independent of the metal content. There are many similar examples.

3.4   Main Approaches to Describing the Thermodynamics of MMC Formation

The above-mentioned factors essentially influence the thermodynamic parameters of MMC formation. One of the "neigbor" effects apparent in metal–polymer systems is that thermodynamic parameters do not depend on the length of the chain (i.e. the molecular weight of the macroligand). Using the temperature dependence of the constant of formation K, the changes in the free energy (ΔG) and entropy (ΔS) for complexation are described by the well known Eq. 3-18.

$$\Delta G = RT\ln K = \Delta H - T\Delta S \tag{3-18}$$

It is useful to examine three levels of the spatial organization of an MMC [43]:

- The local level, which reflects the chemical structure of a single chelate unit in the chain molecule (the nature of the metal-complexing agent and functional groups, the ring size and structure, etc.).
- The molecular level, which is determined by the chemical structure of the polymer chain (chain length, the elemental composition of repeat units, the shape and conformation of the chain, etc.).
- The supermolecular level, which reflects the nature of the intermolecular interaction of macromolecules and their degree of mutual order.

Taking these three levels of the spatial organization of the MMC into account, the free energy of formation of the MMC may be represented (on the assumption that its components are additive [44]) by Eq. 3-19, where ΔG_1, ΔG_2 and ΔG_3 are the free energy changes for the local, molecular and supermolecular levels, respectively.

$$\Delta G = \Delta G_1 + \Delta G_2 + \Delta G_3 \tag{3-19}$$

In some cases such an analysis becomes simpler. For instance, in dilute solutions the association of the chains (the supermolecular level, $\Delta G_3 \rightarrow 0$) may be neglected (Eq. 3-20).

$$\Delta G = \Delta G_1 + \Delta G_2 \tag{3-20}$$

Under conditions of infinitely long chains ($l \rightarrow \infty$) and low degrees of conversion ($\eta \rightarrow 0$) the complexation reaction may not exert a significant influence on the form and conformation of the macromolecule, so that $\Delta G_2 \rightarrow 0$ (Eq. 3-21).

$$\Delta G = \Delta G_1 \tag{3-21}$$

Therefore, the macromolecule behaves similarly to its low molecular weight analogue and the familiar Flory principle then operates. When the components of the model reaction are chosen correctly, the reactivity of binding centers is independent of whether they belong to the polymer chain or to the low molecular weight analogue. This makes it possible to exclude from consideration the molecular and supermolecular levels and to analyze only the local units. The entropy and enthalpy components of ΔG for MMCs are determined mainly by the same contributions as for low molecular weight ligands. Numerous experimental data confirm these conclusions. For example, the thermodynamic functions of the formation of Cu(II) complexes with poly(amidoamines) and their low molecular weight analogues are similar ($\Delta G \approx 51$–57, $-\Delta H = 30$ kJ mol^{-1}, $\Delta S \approx 72$–90 J mol^{-1} K^{-1}) [45]. This indicates a similar spatial pattern of coordination sites in these metal complexes.

Formally the MMC formation can be divided into several hypothetical elementary stages and the entropy change can be expressed by Eq. 3-22, where ΔS^0_1 is the experimental entropy of the reaction in solution, ΔS^0_4 is the calculated theoretical entropy of the reaction in the gas phase, ΔS_{solv} is the entropy of solvation of the metal ion (ΔS^0_2), of the ligand (ΔS^0_3), and of the complex formed (ΔS^0_5), which is determined as the difference between the entropies in solution and in the gas phase.

$$\Delta S^0_1 = \Delta S_{solv} + \Delta S^0_4 = \Delta S^0_2 + \Delta S^0_3 + \Delta S^0_5 \qquad (3\text{-}22)$$

ΔS^0_4 represents the sum of the entropies of transfer (ΔS_t), internal rotation ($\Delta S_{int.rot}$), symmetry (ΔS_{sym}), isomerization (ΔS_{isom}), vibration (ΔS_{vib}), and rotation (ΔS_{rot}) [46]. ΔS_t, $\Delta S_{int.rot}$, ΔS_{sym}, and ΔS_{isom} are positive contributions to the entropy component whereas ΔS_{vib}, ΔS_{rot}, and ΔS_{solv} are negative contributions.

For a high degree of conversion and/or small length of polymer chains it is necessary to take into account the molecular level of the spatial organization of the MMC. A change of chain conformation involving different numbers of monomer units is most frequently found after reaching a certain degree of conversion. It is likely that a helix-coil conformation transition, characteristic of charged chains, will occur as well as the formation of preferred conformations by statistical coil changes. The chain conformation is determined by many factors (for example, the nature of the solvent). Thus, in the interaction of poly(2-methyl-N-vinylimidazole) with the CoCl$_3$(ethylenediamine)$_2$ system in a H$_2$O–C$_2$H$_5$OH mixture, an increase in the ethanol content from 0 mass % to 10 mass % entails a decrease in thermodynamic parameters: from 46.5 to 43.6 kJ mol^{-1} in ΔG, from 97.2 to 41.5 kJ mol^{-1} in ΔH, and from 164.3 to 47.3 J K^{-1} mol^{-1} in ΔS [10].

On passing from a dilute to a concentrated solution and further to a massive specimen, there is a significant increase in the extent of the intermolecular interactions of both the initial polymers and the metal complexes based on them, which makes it is necessary to take into account the third (supermolecular) level of spatial organization of the MMC. Studies have shown that the reactions of metal ions with concentrated solutions of macroligands, cross-linked or graft polymers are distinguished primarily by greater free energy changes in the process. In concentrated solutions the polymeric nature of the ligands considered should be more evident than in dilute solutions, and the characteristics of the complexation process with polymer character become more pronounced. Therefore, considering the kinetic and thermodynamic features of these reactions on the supermolecular level, we have to take into account all the principal effects of the polymer chain. Macromolecular ligands of the graft type (a readily swelling functional coating on an insoluble substrate) occupy an intermediate position between soluble and cross-linked polymers. The interaction of metal ions with such macroligands simulates to a large extent complex formation in solutions. As a rule, the free energy changes are similar in both cases. Intramolecular rather than intermolecular cyclizations are characteristic for grafted macroligands, as they are for soluble polymers [47].

At high concentrations of macroligand solutions there is an appreciable increase in the probability of the formation of the intermolecular (and not the intramolecular) MMC [48]. By varying the degree of conversion, it is possible to observe the reversible transition between intramolecular and intermolecular complexes. This is shown in Fig. 3-4 for the formation of complexes between Fe^{3+} and poly(hydroxamic acids) [49]. At high polymer concentrations the hydroxamic groups from different polymers are closer to one another than the functional groups in the same chain. This increases the probability of formation of intermolecular bonds (path 1 in Fig. 3-4), so that cross-linking becomes a significant feature of the process. The cross-linking of the polymer chain is accompanied by an increase in the viscosity of solution. As a result of further addition of iron ions, the cross-linked complexes of composition [L]:[M] = 3:1 are converted into 1:1 complexes (path 2) and, by virtue of the rupture of the cross-links, each iron is bound to only one hydroxamic group (path 4 also leads to the same products). At low concentrations of the macroligand, where the neighboring hydroxamic groups in the same chain are closer to each other than the groups from different polymer chains, the iron complex is formed intramolecularly (path 3) which diminishes the limiting size of the disordered polymer coil and hence diminishes the viscosity. With increase in polymer concentrations the change in viscosity becomes more appreciable, since the density of cross-links increases. It is noteworthy that the molecular mass has a greater influence on the viscosity than the content of the metal centers. The cross-linked polymers are "kinetic" products, which after a time are converted into thermodynamically more stable intramolecular complexes (path 5). The transition between the intramolecular and intermolecular complexes is reversible.

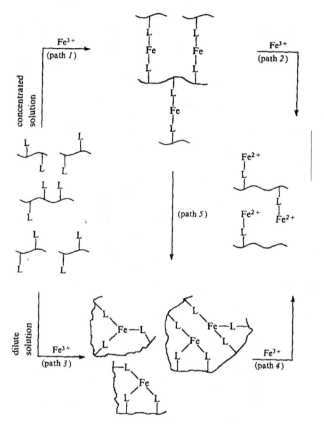

Figure 3-4. Possible interactions between Fe^{3+} and poly(hydroxamic acid).

Since the chemical bond is the same in both cases, the enthalpy of the transition should be zero and hence the transformation is determined by entropy factors. The conversion of intermolecular complexes into intramolecular ones is a first-order reaction, the rate constants for which are $5.21 \cdot 10^{-6}$ s^{-1} at 303 K and $49.7 \cdot 10^{-6}$ s^{-1} at 333 K. The activation energy and the entropy are 63.3 kJ mol^{-1} and 4.2 J mol^{-1} K^{-1} respectively.

The following two parameters are most important for the formation reactions of intermolecular metal complexes: the critical polymer concentration c^0 and the temperature T_t of the liquid $\Leftrightarrow$ gel transition (Eq. 3-23).

$$T_t = \frac{\Delta H}{(\Delta S - R \ln[(c - c^0)/c^0])} \tag{3-23}$$

The quantity c^0 represents the concentration above which the intermolecular interaction becomes significant (Fig. 3-5). The transition temperature is determined mainly by the thermodynamic characteristics associated with the formation of metal–polymer bonds and is obtained from the classical theory of gel formation. The results of the calculation [49] indicate that the hydroxamic acid groups of the polymer mentioned above react in a "disordered" fashion and form infinite networks only when the probability of the intermolecular binding of the metal exceeds 50 %.

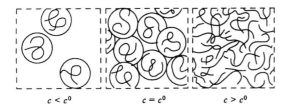

$c < c^0$ $c = c^0$ $c > c^0$

Figure 3-5. Schematic representation of intermolecular interactions at different concentrations of a polymer.

Analysis of the three levels of structural organization of an MMC shows that the characteristics of their formation differ significantly from those for low molecular weight analogues. For the latter, the second and third levels are not as a rule considered. On the most thoroughly investigated local level, important information about the mechanism of the formation reaction can be obtained precisely by comparing the thermodynamic parameters.

Examination of the second level leads to the conclusion that, for calculation of the thermodynamic parameters, account must be taken of the principal polymer effects, as for the kinetics of chemical reactions with participation of macromolecular reagents. Only this approach permits quantitative estimation of the formation processes of macromolecular complexes. In ideal cases, it is possible to estimate the contribution of each effect (or at least to discover which is the dominant one) for the thermodynamics of the process, which helps in understanding the characteristic features of the formation and structure of the MMC.

The third, supermolecular, level has been least investigated. This is associated, on one hand, with the dependence of its parameters on the degree of conversion in complexation and, on the other hand, with the possibility of the occurrence of so called "stationary states" characteristic of long-chain molecules. The properties of the MMC obtained differ appreciably from the equilibrium ones. Unfortunately, not only the properties of the system in a specified "stationary state" but also the conditions of the appearance of such a state still remain uninvestigated.

In general, the observed equilibrium constant can be represented to a first approximation as the product of the equilibrium constants for the different levels. Unfortunately, the number of kinetic studies on MMCs is still insufficient for more specific generalizations.

3.5   The Problem of the Topochemistry of MMCs

For many purposes we need to know not only the kinetic and thermodynamic parameters of MMC formation and the structure of the metal centers, but their positioning relative to each other (topography or topochemistry, if physicochemical interactions occur between the centers). Moreover, a knowledge of topochemistry of the MMC permits us to understand their fine structural organization in greater detail, and deliberately to change the properties of such systems by creating favorable conditions for the realization of one or another spatial formation. In particular, a spin labeling technique is applied to this end. Using this method [50] it has been shown that the functional groups of the grafted layer of PE-grafted-poly(allylamine), PE-grafted-poly(allyl alcohol), and PE-grafted-poly(acrylic acid) and other macromolecular ligands exhibit a high reactivity with respect to MX_n. The external groups are the first to react. Then, like "chromatography", MX_n penetrates into deeper layers (the thickness of which amounts to 100–300 Å). The scheme of $TiCl_4$ binding followed by spin labeling of the complex formed can be represented by Eq. 3-24 [51].

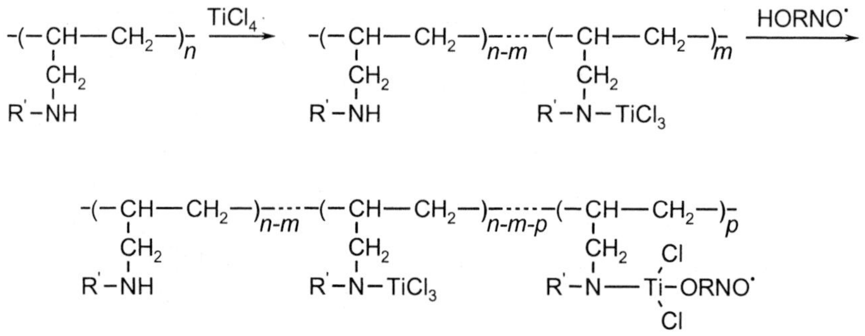

$$-(-\overset{\underset{\displaystyle CH_2}{\mid}}{\underset{\displaystyle R'-NH}{}}{CH}-CH_2-)_{\overline{n}} \xrightarrow{TiCl_4} -(-\overset{\underset{\displaystyle CH_2}{\mid}}{CH}-CH_2-)_{\overline{n-m}}\cdots-(-\overset{\underset{\displaystyle CH_2}{\mid}}{CH}-CH_2-)_{\overline{m}} \xrightarrow{HORNO^\bullet}$$

$$(3\text{-}24)$$

In the range of concentration of the grafted fragments $(1.3–5.7)\cdot10^{-1}$ mol g^{-1} the content of bonded Ti(IV) is proportional to the number of grafted functional groups. The average ratio of [–N=]/[Ti] is equal to 3 in the same concentration range. The shape of the ESR spectra of spin-labeled Ti-containing polymers depends on the [RNO$^\bullet$]/[Ti] ratio, and at values of 0.04–0.2 the spectrum is singlet. This is the result of a spin–spin exchange between –N–O–groups caused by their close proximity ($r \leq 12$ Å) on the polymer. The analysis shows that such MMCs include two types of Ti(IV) complex: some complexes relatively homogeneously located on the front of the grafted layer (isolated Ti(IV) ions) and "aggregations" in the external layer close together with little penetration into the layer.

More detailed information may be obtained using paramagnetic ions as labels bonded with polymers. A characteristic example is the binding of VCl$_4$ with PVPy [52]. At low concentrations of VCl$_4$ (up to $0.4 \cdot 10^{-4}$ mol g^{-1}) a well-defined eight-component ESR spectrum characteristic of isolated V(IV) complexes is observed, but at higher concentrations a wide signal appears and grows with the same g-factor, representing the spectra of the same complexes widened by the dipole–dipole interaction. Therefore, at high concentrations the bound part of VCl$_4$ should be considered as a cooperative system of cluster formations containing a relatively large number of the particles. Estimates show [53,54] that the upper limit of the local concentrations of isolated complexes C_{isol} (assuming that the distribution of isolated complexes of V^{4+} is chaotic and homogeneous in the polymer) is equal to $2 \cdot 10^{-4}$ mol cm^{-3}. The same estimate of the local concentration of the vanadium ions in cluster formations results in the value $C_{clus} \approx (2–6) \cdot 10^{-3}$ mol cm^{-3}.

These values permit us to estimate the mean distances between both isolated complexes and particles which form a cluster: $r_{isol} \geq 22$ Å, $r_{clus} \approx 6.8–9$ Å. As the energy of dipole–dipole interactions at a distance of $\sim 7–10$ Å is much less than kT, we can suppose that for cluster formation it is not the dipole–dipole interactions themselves that are important but a favorable spatial location of V(IV) ions. Cluster formation seems to occur around a VCl$_4$ molecule which joins two polymer chains. This increases the probability of coordination of other VCl$_4$ molecules nearby. For vanadium ions in a coordination environment of six such a hypothetical model can be represented as shown in Fig. 3-6.

Interesting results were obtained in a study of the topochemistry of Cu(II) complexes on polyethylene-grafted-poly(acrylic acid) [55]. A superposition of two different states of copper was observed. Even at concentrations of bound copper ions ~ 0.1 mmol g^{-1} more than half are localized in exchange-bonded cluster aggregations, and at concentrations > 0.15 mmol g^{-1} virtually all copper is bonded into aggregations. The local concentrations of Cu(II) in these formations amount to $(6.0–40) \cdot 10^{20}$ cm^{-3} (Table 3-3). One may picture the following pattern of distribution of copper complexes in the grafted layer of the macroligand. There are isolated complexes with a mean distance between them $\bar{r} \geq 15$ Å (type I complexes), but the major part of the copper is localized in cluster-type aggregations with $\bar{r} \leq 7$ Å, in which copper complexes are apparently not bound chemically (type II complexes).

Finally, a significant part of the copper is bound into clusters with strong exchange interactions (exchange-bonded clusters, type III) determined by chemical bonding of copper complexes with each other. The content of the latter is appreciable increased with an increase in concentration of bound Cu(II) complexes: the accumulation of complexation sites along a macromolecule leads to the formation of polynucleic complexes. The macromolecule segments seem to be large enough and their motion slow enough that they are unable to average the local environment of the complexes. Inhomogeneous distribution of copper on the polymer is also influenced by presorption treatment. A rapid treatment of carboxyl-containing fibers results in the formation of regions with non-equilibrium high C_{loc} value ($r_{agr} \leq 9.1$ Å) even at low concentrations of copper whereas for a slow introduction of copper ions the formation of Cu^{2+} aggregations is observed, with a higher content of bound copper [56].

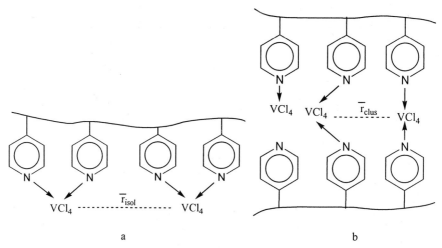

a   b

Figure 3-6. Coordination model of VCl_4 at poly(4-vinylpyridine).

Recently [57], a computer has been used for the mutual subtraction of the experimental spectra associated with different copper concentrations, and three types of individual signals have been identified in the poly(methacrylic–Cu(II)) system (Table 3.3 and Fig. 3-7). The ESR spectrum of complex I ([Cu]/[COOH] = 1:60) is a weak asymmetric signal characteristic of mononuclear tetragonal Cu(II) compounds with the structures A and B shown in Fig. 3.8.

The intensity of the signals are in agreement with the total content of incorporated copper which increases up to [Cu]/[COOH] = 1:20. A wide symmetrical signal corresponding to the complex II appears and increases. ESR spectra of this type are characteristic of magneto-concentrated polynucleic aggregations (clusters) of Cu(II) ions in which exchange and dipole–dipole interactions occur.

Table 3-3. The ESR data for isolated ions and dipole-bound aggregations in the polyethylene-grafted-poly(acrylic acid)–Cu(II) system [9,55].

Total content of Cu, %	Content, %		Density of Cu in clusters 10^{20} cm^{-3}	Total content of Cu, %	Content, %		Density of Cu in clusters 10^{20} cm^{-3}
	isolated ions	clusters			Isolated ions	clusters	
0.2	100	0	-	1.51	1	99	19.4
0.29	100	0	-	2.48	2.5	97.	31.4
0.55	16	84	6.0	3.18	2.0	5	40.6
0.94	4	96	11.7	18.4[a]	0	98	17.2
1.37	2	98	17.4			100	

[a] For Cu(II) complex with homopolymeric poly(acrylic acid).

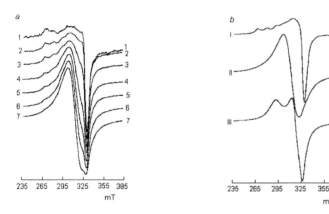

Figure 3-7. a: The ESR spectra of the PMMA-Cu(II) systems at initial ratios [Cu^{2+}]/[COOH] equal to 1:60 (1), 1:40 (2), 1:20 (3), 1:10 (4), 1:8 (5), 1:6 (6), 1:4(7). b: The individual ESR signals I, II, III.

They appear with an increase in binding of the copper complexes and only after the formation of mononuclear tetragonal complexes I. The steady form of symmetrical signal II over a wide range of Cu(II) ion content in the polymer (1:20 < [Cu]/[COOH] < 1:4) indicates the permanence of a high local concentration of copper ions in clusters. This indicates that the embedding of the first copper ions is a limited stage of the formation of each cluster, giving rise to the conditions for cluster growth. Embedding of copper ions and the formation of carboxylates should result in an additional stabilization of polymer coils and, as mentioned above, in the cross-linking of macromolecules. The ESR signal appearing at [Cu]/]COOH] > 1:4 is usually referred to physically adsorbed Cu(II) complexes III, although such a conclusion is ambiguous, in our opinion. Interestingly, the destruction of polynuclear clusters and the formation of dimer complexes in the reaction of poly(methacrylic acid) with NaOH have already occurred at the degree of neutralization $\alpha = 0.1$.

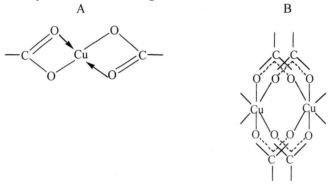

Figure 3-8. Structures of the poly(methacrylic acid–Cu) system.

Thus, the synthesis of MMCs is a dynamic process, and their structural organization depends on many factors. Knowledge of these factors enables us, within limits, to control both the composition of coordination units and the structural organization of macrocomplexes in general, and to design MMCs with specific properties. Of course, such a problem becomes immeasurably more complicated when two or more different metals are participating in the complexation. Such investigations are only now being developed.

3.6   References

1. M. Kaneko, E. Tsuchida, *J. Polymer Sci., D, Macromol. Rev.* **1981**, *16*, 397.
2. S.L. Davydova, N.A. Plate, *Coord. Chem. Rev.* **1975**, *16*, 195.
3. N.A. Plate, A.D. Litmanovich, O.V. Noah, *Macromolecular Reactions,* John Wiley & Sons, Chichester, 1995.
4. A.D. Pomogailo, *Polymer-Immobilized Metal Complexes Catalyst,* Nauka, Moscow, 1988 (and the references cited therein).
5. E.F. Vainstein, Izv. Vuzov. Khimiya i Khim. Tekhnol. **1993,** 36, 35, 41.
6. E.F. Vainstein, G.E. Zaikov, in: *Polymer Yearbook,* R.A. Pethric (Ed.), Harwood Academic Publishers, London, 1993, Vol.10, p.231.
7. L.D. Landau, E.M. Lifschitz. *Statistical Physics*, Nauka, Moscow, 1966.
8. N.M. Bravaya, A.D. Pomogailo, E.F. Vainstein, *Kinetika i Kataliz* **1984**, *25*, 1140.
9. A.D. Pomogailo, N.D. Golubeva, *Kinetika i Kataliz.* **1985**, *26*, 947.
10. E. Tsuchida, *Macromolecular Complexes. Dynamic Interactions and Electronic Processes*, VCH Publishers, New York, 1991.
11. H. Todokoro. Y.Chatani, T. T. Yoshihara, *Macromol. Chem.* **1964,** *73*, 109.
12. R. Iwamoto, Y. Saito, H. Ishikawa, H. Iadokoro, *J. Polym. Sci. A-2* **1968**, *6*, 1509.

13. J.M. Parker, P.V. Wright, C.C. Lee, *Polymer*, **1981**, *22*, 1305.
14. J.L. Garces, F. Mas, J. Puy, J.Galceran, J.Salvador, *J. Chem. Soc., Faraday Trans.* **1998**, *94*, 2783.
15. B. Leunberger, P.W. Schindler, *Anal. Chem.* **1986**, *58*, 1471.
16. M. Fukushima, S. Tanaka, K. Hasebe, *Anal. Chim. Acta* **1995**, *302*, 365.
17. J.C.M. De Wit, W.H. Van Riemsdijk, L.K. Koopal, *Environ. Sci. and Technol.* **1993**, *27*, 2015.
18. Yu. V. Kholin, *The quantitative physical-chemical analysis of complexation in solution and on the surface of chemical modified silica: substantial models, mathematical methods and their applications*, Kharkov, Pholio, 2000, 288 p (in Russian).
19. A.M. Evseev, L.S. Nikolaeva, *Zh. Phys. Khimii* **1993**, *67*, 508.
20. L.S. Nikolaeva, Yu. F. Kir'yanov, *Zh. Phys. Khimii* **1997**, *71*, 746.
21. C. Fenn-Barraba, A. Pohlmeier, W. Knoche, *Colloid Polym. Sci.* **1998**, *276*, 627.
22. T. Okubo, A. Enokida, *J. Chem. Soc., Faraday Trans. I* **1983**, *79*, 1639.
23. A.D. Pomogailo, D. Wöhrle, in: *Macromolecule-Metal Complexes,* F .Ciardelli, E. Tsuchida, D. Wöhrle (Eds), Springer, Berlin, Heidelberg, 1996, p.11.
24. A.S. Polynskii, V.S. Pshezhetskii, V.A. Kabanov, *Vysokomol. Soedin.* **1985**, *A27*, 2295.
25. M. Morcellet, *Polym. Bull.* **1984**, *12*, 127.
26. J.C. Benegas, R.F.M. Cleven, M.A.G. van den Hoop, *Anal. Chim. Acta* **1998**, *369*, 109.
27. M. Morcelletett, *J. Polymer Sci., Polym. Lett. Ed.* **1985**, *23*, 99.
28. T. Suzuki, H. Shirai, F.Hojo, *Polym. J.* **1983**, *15*, 409.
29. B.C. Cornilsen, K. Nakamoto, *J. Inorg. Nucl. Chem.* **1974**, *36*, 2467.
30. K. Yamaoka, T. Nasujima, *Bull. Chem. Soc. Japan* **1979**, *52*, 1819.
31. T. Miyajima, K. Yoshida, Y. Kanegae, *React. Polym.* **1991**, *15*, 55; *The Science of the Total Environment*, **1992**, *117–118*, 129.
32. M.T. Vaconcelos, C.A. Gomes, *Eur. Polym. J.* **1997**, *33*, 631.
33. Yu. V. Rakitin, G.M. Larin, V.V. Minin, *Interpretation of the ESR Spectra of Coordination Compounds,* Nauka, Moscow, 1993.
34. L. Wang, J. Smid, *J. Am. Chem. Soc.* **1977**, *99*, 5637.
35. D.Wöhrle, A. Pomogailo, in: *Advanced Functional Molecules and Polymers,* H.S. Nalwa (Ed.). Gordon & Breach Science Publishers, New York, 2001, Vol.1, p.87.
36. A.D. Pomogailo, *Catalysis by Polymer-Immobilized Metal Complexes*, Gordon & Breach Science Publishers, Amsterdam, 1998.
37. K. Hikichi, T. Hiraoki, N. Ogata, *Polym. J.* **1984**, *16*, 437.
38. H. Nishide, N. Shimidzu, E. Tsuchida, *J. Appl. Polym. Sci.* **1982**, *27*, 4161.
39. A.P.Filippov, *Theoretical and Experimental Chemistry* (*Transl. Teoreticheskaya i Eksperimental'naya Khimiya*), **1986**, *21*, 693.
40. A.Z. El-Sonbati, A.A. El-Binderi, M.A. Diab, *Polymer* **1994**, *35*, 647.
41. A. Warshawsky, R. Kalir, *J. Appl. Polym. Sci.* **1979**, *24*, 1125.
42. S. Yano, Y. Yamashita, M. Matsushita, *Colloid Polym. Sci.* **1981**, *259*, 514.
43. A.D. Pomogailo, I.E. Uflyand, E.F. Vainshtein, *Russ. Chem. Rev.* **1995**, *64*, 857 (Engl. Transl.).
44. I.E. Uflyand, E.F. Vainshtein, A.D. Pomogailo, *Zh. Obshch. Khim.* **1991**, *61*, 1790.

45. R. Barbucci, M.J.M. Campbell, M. Casolaro, M.Nocentini, G. Reginato, P. Ferruti, *J. Chem. Soc., Dalton Trans.* **1986**, 2325.

46. C.-S. Chung, *Inorg. Chem.* **1979,** *18*, 1321; ibid, *J. Chem. Educ.* **1984**, *61*, 1062.

47. I.E. Uflyand, A.D. Pomogailo, *J. Inorg. Organomet. Polym.* **1992**, *2*, 373.

48. L.V. Miroshnik, A.V. Aleksandrov, V.N. Tolmachev, *Ukrain. Khim. Zh.* **1987**, *53*, 1148; *Izv. Vuzov. Khimiya i Khim. Tekhnol.* **1987**, *30*, 81.

49. J.W. Rosthauser, A. Winston, *Macromolecules* **1981**, *14*, 538.

50. N.M. Bravaya, A.D. Pomogailo, F.S. D'yachkovskii, *Vysokomolek. Soedin. A.* **1979**, *21*, 1781.

51. N.M. Bravaya, A.D. Pomogailo, *J. Inorg. Organomet. Polym.* **2000**, *10*, 1.

52. A.D. Pomogailo, A.T. Nikitaev, F.S. D'yachkovskii, *Kinetika i Kataliz.* **1984**, *25*, 166.

53. A.D. Pomogailo, *J. Inorg. Organomet. Polym.*, in press.

54. A.D. Pomogailo, *Macromol. Chem. Symp.* **2002**, *186*, 15.

55. A.D. Pomogailo, N.D. Golubeva, *J. Inorg. Organomet. Polym.* **2001**, *11*, 67.

56. A.I. Kokorin, M.P. Zhilenko, A.P. Rudakov, *Kh. Phis. Khim.* **1995**, *69*, 1796.

57. V.V. Annenkov, Thesis, Irkutsk, University, 2001.

4 Polymerization of Metal-Containing Monomers (MCMs) as a Method for Incorporating Metals in Macromolecules

Anatolii Pomogailo

There are three main approaches to the binding metal ions or metal complexes into polymers. The first is by a physical interaction. A relatively developed surface of the polymer is sufficient to bring this about (occasionally specific porosity and mechanical properties of the polymer are necessary). Physical binding may be carried out by different methods: impregnation, precipitation and coprecipitation, sorption, spraying, sublimation and incorporation into a polymer matrix, microcapsulation, dispersion, absorption of appropriate metal compounds followed by decomposition, etc. For instance, mixtures of polymers with highly dispersed metal particles, sometimes of colloidal dimensions (metallofilled polymers), belong to such systems. Partial formation of a chemical bond between metal and polymer may also be observed in such processes. However, in most cases the metal component reacts with the polymer by physical forces whose specific characteristics are usually little known (see Chapter 8).

The chemical method of synthesis of macromolecular metal complexes (MMCs) is more widespread. It is based on the chemical binding of a metal compound MX_n with a polymer. In this case the polymer should have corresponding reactive functional groups. The type of bond formed (donor–acceptor, covalent, ionic, etc.) is determined by the nature of the reacting components. Such a bond, as a rule, possesses a high stability (see Chapter 5).

The construction principles of MMC formation are almost the same in all cases: polymer carrier → its functionalization → the chemical reaction of binding → the separation of chemically unbound reagents. Some variants are known when fewer stages are used and the formation of an MMC is carried out using a shortened scheme. For example, a chemist may first attach a binding ligand to the polymer and then bind it to a metal complex or the first stage may be the binding of a metal complex to a ligand and then the attachment of the complex to a polymer, or combined synthesis of the metal complex with the required ligand and its binding to a polymer.

Finally the third method – polymerization transformations of metal-containing monomers – is to some extent like the second, but instead of polymer the appropriate monomer is used.

Metallomonomers are metal complexes in a ligand environment which includes

- at least one group with a multiple bond capable of polymerization,
- a ring capable of rupture,
- an exosphere group capable of entering into condensation reactions with other compounds or metallomonomers

The last type of metallomonomer will not be considered in this chapter because it is the basis of the subject of the coordination type of metallopolymers (see Chapter 6).

Thus, polymerization conversions of metal-containing monomers is a direct method for the synthesis of metallopolymers. Some aspects of the topics we cover are analyzed in the first monographs [1–4] and reviews [5–7].

4.1 MCM Classification

From our point of view [8], the most effective classification is based on the type of bond between the metal and the organic part of the monomer molecule. On this principle, MCMs can be classed into the following major types: monomers with a covalent, ionic, donor–acceptor, chelate and π-bound metal (Fig. 4-1).

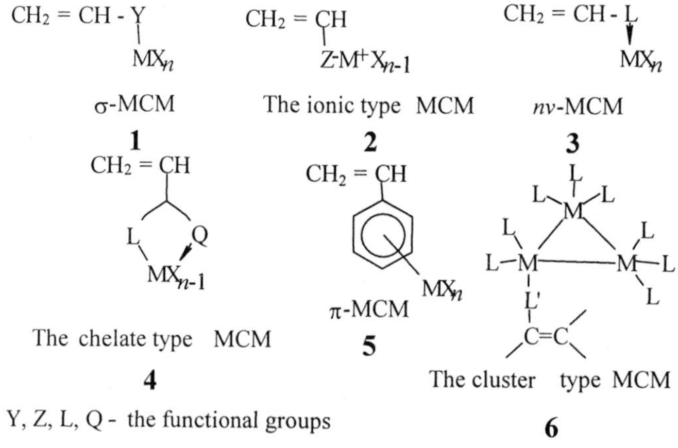

Figure 4-1. Classification of MCMs.

The groups capable of polymerization may be of different types, namely, vinyl, allyl, styryl, diene, allenic, acetylenic, acrylic, alkoxide, etc. Also, different rings bonded with metal such as epoxy, ethyleneimine, and others may be used. It goes without saying that multiple bond can be present not only as double bonds but also as triple, allene, diene bonds as well as their combinations. The correlation between the electronic structure and the properties of MCMs as well as their ability to polymerize is one of the most fundamental problems in the chemistry of MCMs. In the MCM of the donor–acceptor type (*nv*), functional groups, including heteroatoms with an unshared pair of electrons, take the part of n-donors, while unsaturated hydrocarbons act as π-donors. The former present unshared electron pairs for the formation of coordination bonds, and the latter give the π-electronic system. Such MCMs are more often formed in the case of transition metal compounds. As a rule, π-MCMs are characteristic of the transition metals of groups VIA, VIIA and VIII of the Periodic Table of the elements. MCMs of the ionic type are more characteristic of nontransition metals, MCMs that are true organometallic compounds (with a metal–carbon bond) occupy only a small place in MCM chemistry.

4.2 Overview of the Different Methods of Synthesis of MCMs

All the typical procedures of preparative inorganic and organometallic chemistry are used for MCM synthesis although some methods are elaborated especially for MCMs. Both transition and nontransition metals are used. The chemistry of metals such as Sn, Zn, Ba, Mg, etc. has been most widely developed. Moreover, the methods of synthesis of metallomonomers have been successfully applied to the corresponding derivatives of metalloids (especially for silicon and boron). At present over 75 elements of the Periodic Table are involved in the synthesis of high molecular weight compounds of different types. Although over 5000 new organometallic polymers are reported every year, on the whole the fraction of polymers obtained by polymerization and co-polymerization of MCMs is very small. This situation is caused, first, by the need to synthesize MCMs of new types, and also because the scientific interests of researchers specializing in the related fields of organometallic synthesis, coordination chemistry and polymerization processes frequently do not overlap. Here we consider the main tendencies and perspectives of developments in this field of science.

4.2.1 Synthesis of σ-MCMs of the Organometallic Type

The nature and properties of a metal–carbon bond are analyzed in the literature at some length (see, for example, a review of the calculation of the chemical bond in transition metals by *ab initio* methods [9]). Here we only note that in the compounds MRX_{n-1} (where M is transition metal, R is a hydrocarbon radical, X is a halide, alkoxide, cyclopentadienyl, etc.) the properties of a M–C σ-bond are rather contradictory. On the one hand, the average energy of dissociation of the bond D(M–C) is comparable with D(M–Cl) even for the least-stable alkyl derivatives of transition metals; for instance, in the case of $(CH)_3TiCl_3$ they are equal to 264 and 343 kJ mol^{-1}, respectively. On the other hand, the M–C bond is immeasurably more reactive than the M–Cl bond. This is determined by many factors [10]. The basic technique for the preparation of σ-type organometallic monomers consists of vinylation of metal halides by magnesium (or less commonly by Li or Na) organic compounds (Normant's reagents or their derivatives, which, in turn, are obtained by reaction of vinyl chloride or vinyl bromide with magnesium in THF to **7**) (Eq. 4-1).

$$CH_2=CHMgBr + MCl_2 \rightarrow (CH_2=CH)_2M + MgClBr \qquad (4\text{-}1)$$

7

Arylvinyl derivatives of transition and nontransition metals, especially, α- and β-styryl ones, are of considerable interest for polymerizations. As a rule, they are more stable than vinyl derivatives. To enhance their stability additional stabilizing ligands of electron donor type (Py, Dipy, PPh$_3$, etc) are also used (examples are trans-dichlorobis(triphenylphosphine)nickel(II), Pd(II), Pt(II)).

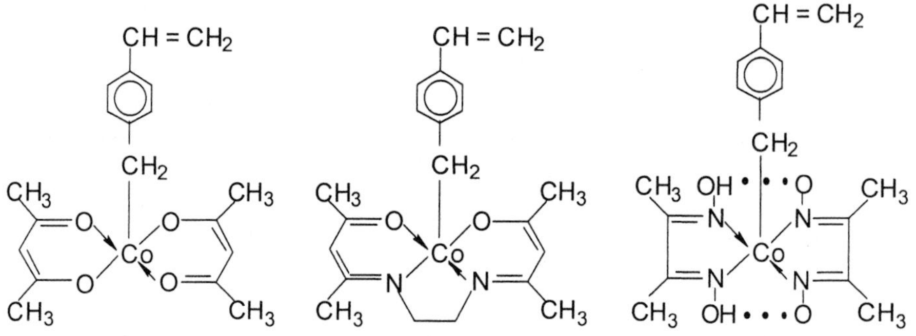

8   9   10

Mono- and polynuclear complexes of these metals, stable enough for polymerization, have been prepared from the corresponding metal halides and BuMgC≡CH by the same method [14].

Unsaturated organocobalt compounds (**8–10**), for instance, are stable only if they contain bulk chelate ligands of dimethylglyoxime type, etc. [11–13].

One convenient method for the preparation of alkene (**11–14**) and alkyne (**15–16**) derivatives of transition metals (Ni(II), Pd(II), Pt(II)) is the interaction of the corresponding metal halides with Grignard reagents [15–17] through a series of simple consecutive reactions(Eqs. 4-2, 4-3). The MCMs obtained are stable enough to be characterized in detail (Table 4-1).

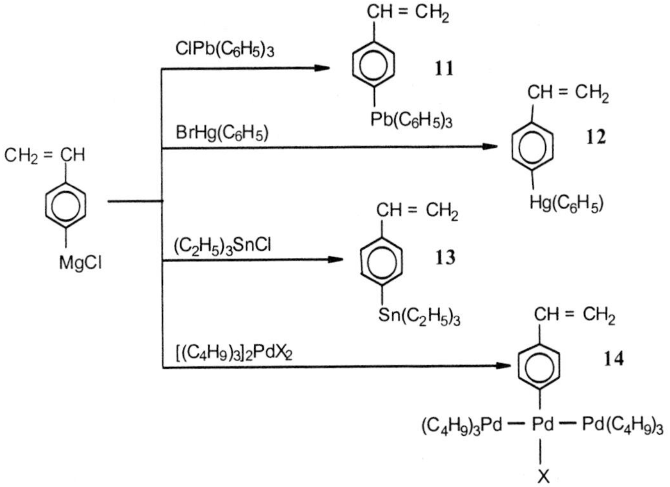

$$(4\text{-}2)$$

Table 4-1. Characteristics of metal–polyine monomers.

MCM	$\nu \equiv CH$	$\nu C \equiv C$	λ_{max} (nm)	lg ε	δ (NMR $^{31}P)^a$ (ppm)
trans- Pt(PBu$_3$)$_2$(C≡C–C≡CH)$_2$	3307	2147	318	4,39	−4,7
*cis-*Pt(PBu$_3$)$_2$(C≡C–C≡CH)$_2$	3315	2153	301	4,22	+3,7
*trans-*Pt(PBu$_3$)$_2$(C≡C–⟨⟩– 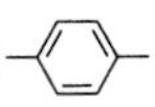 –C≡CH)$_2$	4295 3280	2145 2098	338	4,73	−3,3
*cis-*Pt(PBu$_3$)$_2$(C≡C–⟨⟩– 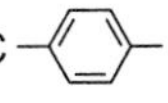 –C≡CH)$_2$	3268	2120	308	4,75	+3,3
ClPd(PBu$_3$)$_2$C≡C–C≡ ≡C–Pd(PBu$_3$)$_2$Cl	–	2130 2000 2280	299	3,35	−9,83
PhC≡C–Pd(PBu$_3$)$_2$Cl	–	1978 2240	314	1,55	−10,74

^{a}With respect to H$_3$PO$_4$ (85%).

$$(PBu_3)_2NiBr_2 + BrMgC \equiv CH \longrightarrow trans\text{-} (PBu_3)_2Ni(C \equiv CH)_2 \quad \textbf{15}$$

$$\downarrow HC \equiv C - C \equiv CH$$

$$trans\text{-} (PBu_3)_2Ni(C = C - C = CH)_2 \quad \textbf{16}$$

$$(4\text{-}3)$$

Data on σ-bonded diene compounds are practically absent. However, we note reports [18] concerning the synthesis of 1,4-digermabutadiene $H_2Ge=HC-CH=GeH_2$ (**17**) and the peculiarities of enhanced conjugation in its molecule [19, 20]. However no data on its polymerization conversions are known.

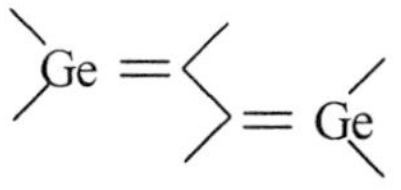

17

4.2.2　σ-Type Organoelemental Monomers

σ-Type MCMs with a strong covalent M–O bond (or occasionally with M–N or M–S bonds) received some attention in polymerization practice. Exchange reactions of alkoxyderivatives of corresponding metals with alcohols, alcoholates, glycols are comparatively useful methods for their preparation. Unsaturated organotitanium compounds – ethers of orthotitanium acid (**18**) – have received much attention as potential monomers for a long time [21].

$$(RO)_nMX_{n-m} + M'OR' \rightarrow (RO)_nM(OR')_{n-m}$$

M = Sn(IV), Ti(IV), V(IV), V(V), Zr(IV)
M′ = H, Li, Na
R′ = the residue of the unsaturated alcohol
X = OR, halogen

18

Alkoxytitanium compounds [22] and the esters of orthovanadium acid [23] have been prepared by the same method and characterized (R = $CH_2CH_2OCOC(CH_3)=CH_2$, $CH_2C\equiv CH$, $CH_2CH=CH_2$, $C(CH_3)_2C\equiv C-CH=CH_2$) (see Experiment 4-2, Section 4.6).

The following approaches are often used:

- direct interaction of alcohols with metal halides
- interaction of alcohols (or sodium alcoholate) with metal halides or alkoxyhalides,
- substitution of alkoxy groups by unsaturated ones (interesterification).

The reactions proceed with a high yield. In the case of halide derivatives evolved HCl is bound by Py and/or NH_3. The transesterification methods are applicable to those cases when the substitution is conducted using an alcohol with a boiling point that is higher than that of the alcohol to be substituted. By this method mixed titanates have also been prepared (Fig. 4-2). The released alcohol can be removed from the reaction mixture either directly during the synthesis or as an azeotropic solution with solvent (most often with benzene).

Such potential MCMs occur widely. For instance we can note titanium enolates as $(\eta^5\text{-}C_5H_5)_2Ti(OCH=CH_2)_2$ **(19)** $(\eta^5\text{-}C_5H_5)_2(CH_3)TiOC(CH_3)=CH_2$ **(20)** [24] as well as $(\eta^5\text{-}C_5H_5)_2Zr(OC(CH_3)_2CH_2CH_2CH=CH_2)^+$ **(21)** as model intermediates in metallocene-based olefin polymerization catalysis [25], etc. A series of zirconium aryloxide complexes containing pendent vinyl groups have also been described. For instance [26], the interaction of zirconium alkoxide with two equivalents of eugenol gives rise to $\{Zr(O\text{-}iPr)_2[(OC_6H_3(2\text{-}OMe)(4\text{-}CH_2CH=CH_2)](\mu\text{-}O\text{-}i\text{-}Pr)\}_2$ **(22)**. A similar method has been applied to the synthesis of $(C_5H_5)_2U[OC(CH_3)_2CH_2CH=CH_2]_2$ **(23)** [27].

One of the interesting specific methods of MCM synthesis is the interaction of metal alkoxy derivatives with enolizing aldehydes and ketones [28]. It has been useful in producing tetrastyryloxy-, dihexadienyloxy-, and divinyloxy-diisopropoxy titanates. The scheme of the reaction can be represented as shown in Eq. 4-4.

$$Ti(OR)_4 + R_2''CHCR'O \Leftrightarrow Ti(OR)_{4\text{-}n}(OCR'=CR'')_n$$

24   (4-4)

(R′ and R″ = H, alkyl, aryl)

It is important that the transesterification method is convenient for producing MCMs and that their polymers contain optically active groups (Fig. 4-2) [29].

To judge from the first publications, the following development of the field of synthesis and polymerization of optically active MCMs is expected. Thus, the optically active σ-type MCM $Ti(OBu)_2[OCH_2CH_2OCOC(CH_3)=CH_2]-[OC^*H(CH_3)CH_2CH_3]$ **(31)** has been synthesized from alkoxyderivatives of Ti(IV), using the monomethylacrylic ester of ethyleneglycol residue as the polymerizable group and secondary optically active butanol, 1-menthol, ephedrine, cinchonidin as chiral substituents. The same method was previously used to make *nv*-type MCMs – complexes of $PdCl_2$ with R-,S-(4-vinylphenyl)ethylamine [30] **(28)**. Polymers and copolymers based on such MCMs are chiral catalysts for different reactions; in particular, they reveal a double asymmetric induction effect during hydrogenation of optical isomers of ethylamides of acetaminocinnamic acid. The range of such investigations will undoubtedly be extended.

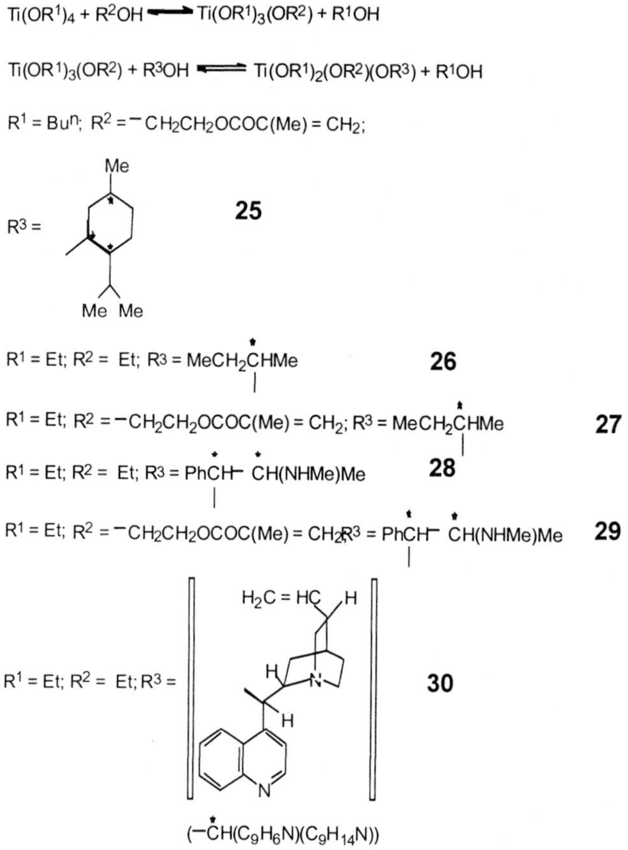

$Ti(OR^1)_4 + R^2OH \rightleftarrows Ti(OR^1)_3(OR^2) + R^1OH$

$Ti(OR^1)_3(OR^2) + R^3OH \rightleftarrows Ti(OR^1)_2(OR^2)(OR^3) + R^1OH$

$R^1 = Bu^n;\; R^2 = {}^-CH_2CH_2OCOC(Me) = CH_2;$

$R^3 =$

25

$R^1 = Et;\; R^2 = Et;\; R^3 = MeCH_2\overset{*}{C}HMe$ **26**

$R^1 = Et;\; R^2 = {}^-CH_2CH_2OCOC(Me) = CH_2;\; R^3 = MeCH_2\overset{*}{C}HMe$ **27**

$R^1 = Et;\; R^2 = Et;\; R^3 = Ph\overset{*}{C}H{-}\overset{*}{C}H(NHMe)Me$ **28**

$R^1 = Et;\; R^2 = {}^-CH_2CH_2OCOC(Me) = CH_2;\; R^3 = Ph\overset{*}{C}H{-}\overset{*}{C}H(NHMe)Me$ **29**

$R^1 = Et;\; R^2 = Et;\; R^3 =$

$H_2C = HC$

$(-\overset{*}{C}H(C_9H_6N)(C_9H_{14}N))$

30

Figure 4-2.
Scheme for the synthesis of optically active monomers.

Similar approaches are used in the production of organometallic epoxides. Recently, much attention has been paid to tin epoxides (**32**) obtained by the reaction of allylglycide ester with Et$_3$SnH [3].

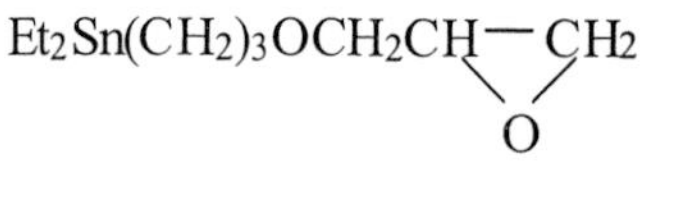

$Et_2Sn(CH_2)_3OCH_2CH{-}CH_2$ **32**

O

As will be shown in Chapter 12, metallopolymer catalysts based on zirconocene are used in the catalysis of polymerization processes. In this context, the σ-type of zirconocene monomers [31], namely, alkoxy derivatives and carboxylates are of great interest.

Covalent type MCMs with a metal–nitrogen bond (for instance, the products of the interaction of metal halides and alkoxides with allyl- and diallylamine, alkaline metal amides, etc.), a metal–sulfur bond and others have no wide uses in polymerization practice.

4.2.3   Ionic-type MCMs

The salts of unsaturated mono- and dicarboxylic acids have received the most attention among the MCMs of ionic type. The ionic bond in a pure state is only present in salts of alkaline and alkaline earth metals. In other compounds, especially in the case of transition metal compounds, it is complicated by an admixture of covalent bonding. The general method for their synthesis, for examples **33**, comes by the interaction of salts, (hydr)oxides and (hydro)carbonates of the metals or their mixtures as well as alkyl(aryl)-derivatives with unsaturated mono- and dicarboxylic acids or their anhydrides (Eq. (4–5) (see Experiment 4-1, Section 4.6).

$$CH_2 = CH - COOH \xrightarrow[M(CO_3)_{n/2} \cdot xM(OH)_n \cdot yH_2O]{M(OH)_n} M(OCOCH = CH_2)_n \qquad (4\text{-}5)$$

33

In such MCMs the carboxylic group can take the part of mono-, bi-, tri- and even tetradentate ligand during coordination with metal ions to form **34–40**.

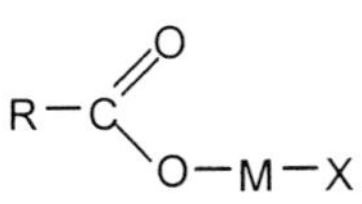

$$R-C \underset{O-M-X}{\overset{O}{<}} \qquad R-C \underset{O-M-O}{\overset{O}{<}} \underset{}{\overset{O}{>}}C-R \qquad \left(R-C \underset{O^-}{\overset{O}{<}} \right)_3 M$$

34 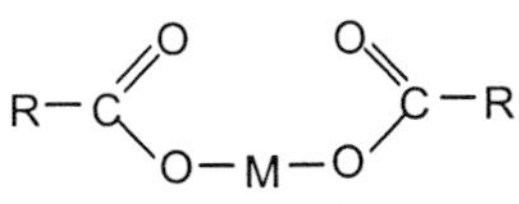   **35** 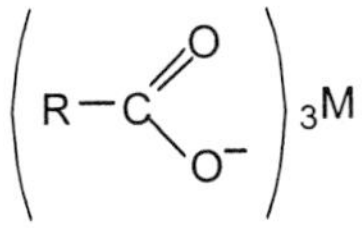   **36**

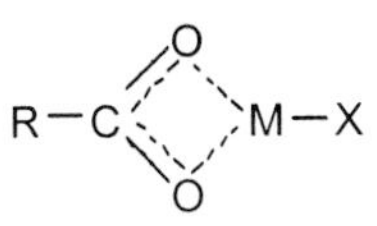

$$R-C \underset{O}{\overset{O}{<}} M-X \qquad R-C \underset{O}{\overset{O}{<}} M \underset{O}{\overset{O}{>}}C-R \qquad \left(R-C \underset{O}{\overset{O}{<}} \right)_3 M$$

37 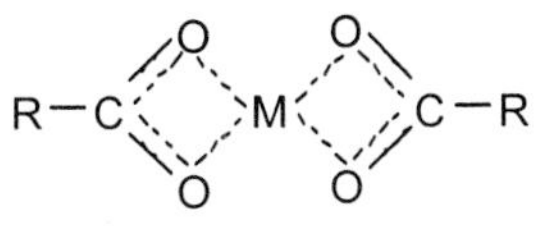   **38** 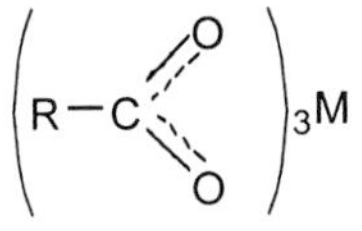   **39**

$$R-C \underset{O \ \ \ S}{\overset{O \ \ \ S \ \ \ O}{<}} M \underset{O}{\overset{O}{>}}C-R$$

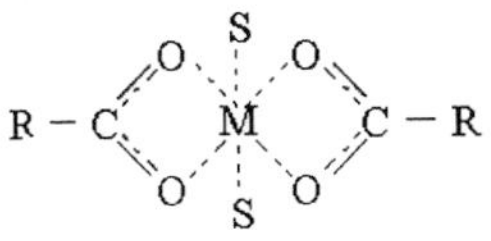

40

S - solvent

Besides acrylic (Ac) and methacrylic acids, maleic, fumaric, itaconic, ethyleneglycolmethacrylatephthalic and acetylenedicarboxylic acids are used. Different complications, which are important for polymerization processes, are often observed during these syntheses. Among them should be noted the possibility of the formation of mono- and dihydrates (which are sometimes crystalline), the formation of "acid" salts with additional coordination of acid molecules, the occurrence of phase transitions, connected with a change of crystalline structure (for example in K and Rb acrylates), processes of aggregation, conjugation of multiple bonds and carboxylate groups, the formation of salts with metals of different valence (for example, $Fe(Ac)_2$ and $Fe(Ac)_3$, etc.).

Similar methods are applicable to the synthesis of ionic-type MCMs with metallocene derivatives of Ti, Zr, and V (**41–46**). The most typical reactions of this type are given in Fig. 4-3 [4].

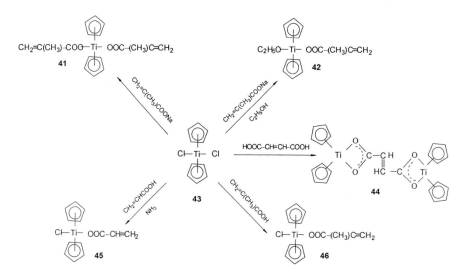

Figure 4-3. Synthesis of metallocene carboxylates.

Data on salts of unsaturated aromatic acids are rather limited. One example is uranyl vinylbenzoate $UO_2(VBA)_2$ [32].

4.2.4   *nv*-Type MCMs

The formation of *nv*-type MCMs proceeds through filling of the low-energy vacant d- (or f-) orbitals of transition metals IVA-VII or VIII groups (*v*-acid) by an unshared pair of electrons from the heteroatoms of amines, esters, phosphines etc. (*n*-bases) (Fig. 4-1, **3**). Thus, the different vinylpyridines (VPy) – 2-VPy, 2-Me-5-VPy, 4-VPy bipyridyls (in particular, 4-vinyl-4'-methyl-2,2'-bipyridydine) – possess pronounced basic properties owing to the presence of an unshared pair of electrons. The complexes formed quite often have the composition $MX_n(VPy)_2$ (M = Co, Ni, Zn, Ti, V, Pd, Cu, etc.) (see Experiment 4-3, Section 4.6). MCMs of the same composition are formed from vinylimidazoles, vinyl derivatives of tri- and tetraazols. This class of coordination-type monomers is most interesting and fairly well studied. This can be attributed to the interesting properties of their metallopolymers. It has been noted [33] that coordination of MX_n with vinylimidazoles (VIA) occurs via the N(3) – "pyridine" – nitrogen atom and is accompanied by a decrease of conjugation between the vinyl and imidazole groups. In the complexes of vinylpyrazole and its derivatives the N(2) atom participates in such a process. This refers also to vinyl-1,2,4-tetrazole (**47–51**) (for examples see Fig. 4-4).

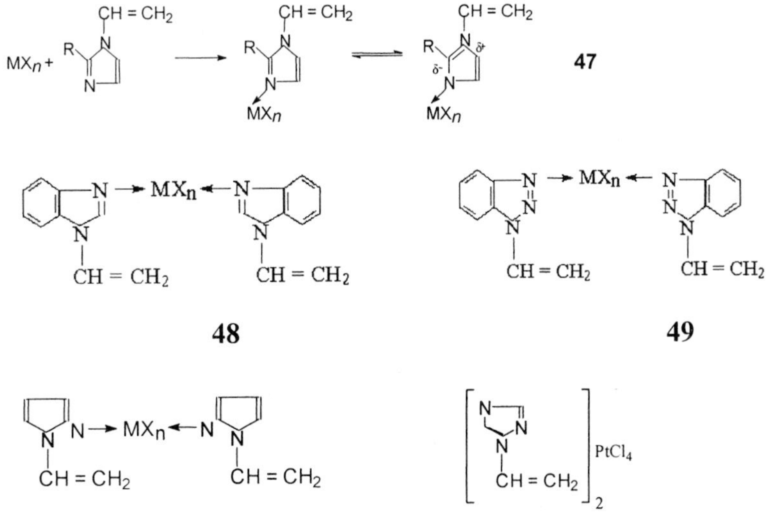

Figure 4-4. Metal-containing monomers formed from polyazoles.

In the series aliphatic amine-nitrogen to heterocycle to nitrile the carbon atom hybridization changes from sp³ to sp (**52**) with the result that the negative charge gets concentrated at this atom and the nitrogen basicity decreases as shown in Eq. 4-6 [34].

(4-6)

Unsaturated nitriles, in principle, have two centers of coordination: an unshared pair of electrons on the nitrogen atoms and the π-electrons of double and triple bonds, sometimes, both π- and n-electrons take part in coordination with MX_n simultaneously. There are representatives of all types of the complexes, but for the synthesis of metallopolymers the *nv*-type MCM complexes are of greatest interest. For instance, acrylonitrile and methacrylonitrile are widely used ligands for the synthesis of this type of MCM, as is acrylamide (AAm). The existence of the double bond in AAm markedly affects the electronic state of the amide group. Assuming the resonance of the C=C and C=O groups with the participation of NH$_2$, three different resonance structures I–III are shown in Eq. 4-7) [35].

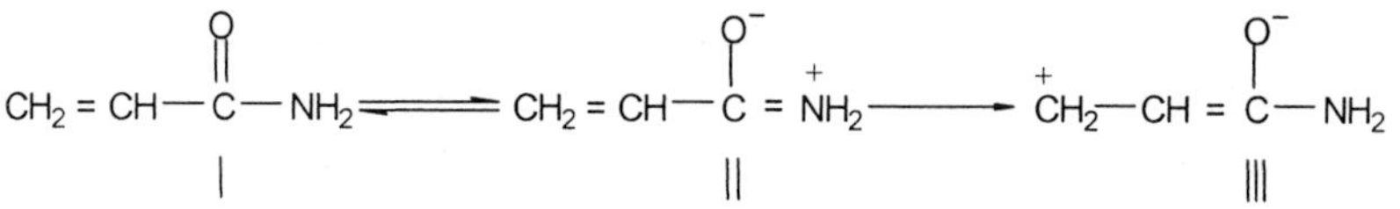

$$CH_2 = CH-\overset{\displaystyle O}{\overset{\|}{C}}-NH_2 \rightleftharpoons CH_2 = CH-\overset{\displaystyle O^-}{\overset{|}{C}} = \overset{+}{N}H_2 \longrightarrow \overset{+}{C}H_2-CH = \overset{\displaystyle O^-}{\overset{|}{C}}-NH_2 \quad (4\text{-}7)$$

$$\qquad\qquad\qquad \text{I} \qquad\qquad\qquad\qquad \text{II} \qquad\qquad\qquad\qquad \text{III}$$

If complexing is via the oxygen atom, structure II is preferred, whereas structures I and III are expected to exist in the case of N-coordination. X-ray structure studies of the [Co(AAM)$_4$(H$_2$O)$_2$](NO$_3$)$_2$ (**53**) [36] suggest that AAm is coordinated via the oxygen atom. The MCM structure is composed of octahedral [Co(AAm)$_4$(H$_2$O)$_2$]$^{2+}$ cations and NO$_3^-$ anions united by through a three-dimensional system of hydrogen bonds (see Experiment 4-4, Section 4.6).

Methacrylate (MMA) and vinylphosphines, which have received widespread attention for these purposes should also be noted. Note here the Co(II), Ni(II), Pd(II), and Pt(II) complexes which are prepared as shown in Eq. 4-8 [37].

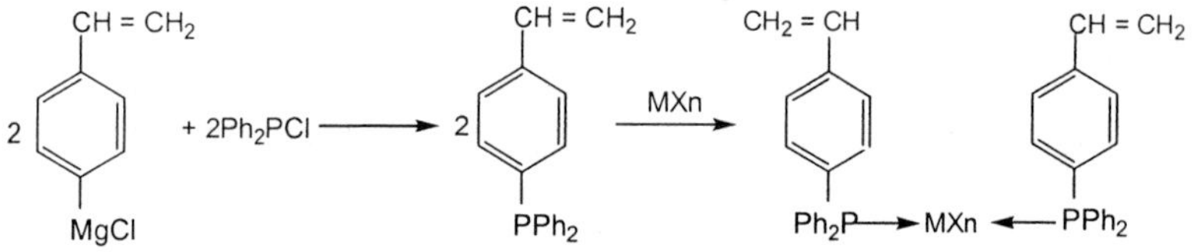

$$\text{(4-8)}$$

$$M = Co(II), Ni(II), Pd(II), Pt(II)$$

54

55

The coordination chemistry of saturated analogues of these ligands is well known.

Cyclic metalloimines, although they are outside the limits of the topic we are considering, are readily polymerized with opening of the ring [38]. The rings of ethylimine and its N-substituted derivatives are often used for this purpose. Complexation significantly alters the electronic structure of small heterocycles. Their structure can be represented as shown in **56**.

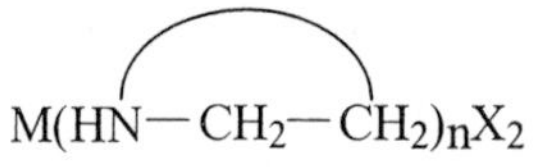

M(HN—CH$_2$—CH$_2$)$_n$X$_2$

56

(M=Zn(II), Co(II), Ni(II), Pd(II), X=-Cl, -Br)

4.2.5   π-Type Metal-Containing Monomers

In π-type MCMs, metal bonding is accomplished through a system of π-electrons, which are equidistant from the metal, so the metal's valence state is unusual. There is no doubt that this is responsible for the unusual properties of such monomers in comparison with the monomers of other types. π-Complexes are only formed by transition metals. π-type MCMs can be divided into vinylcyclopentadienyl-, vinylarene-, (meth)acrylate substituted-, divinyl- and other types. Two basic methods of synthesis of such MCMs have received the greatest attention. In the first method the cyclopentadienyl (or other aromatic) ring is completed by a vinyl group (or another group capable of polymerization). The first representative of this class of MCM – vinylferrocene (**57**) was prepared by this method about 50 years ago [39] (see Experiment 4-5, Section 4.6), then vinyl cymantrene (η^5-(vinylcyclopentadienyl)tricarbonyl-manganese), vinylcynichrodene (η^5-(vinylcyclopentadienyl)dicarbonylnitrosyl-chromium) (**58**), vinylrutheniumcene, etc. were synthesized. The second method, which is more useful for the preparation of vinylarene monomers, consists of the introduction of a metal grouping into a template capable of polymerization. For example, (η^6-styrene)tricarbonylchromium (**59**) (Fig. 4-1, **5**) is prepared by the interaction of triaminetricarbonylchromium with styrene (see Experiment 4-6, Section 4.6). Variations of these methods can lead to the preparation of substituted methacrylate, for example, ferrocenylmethyl-methacrylate) (**60**), and divinyl compounds, such as 1,1′-divinylruthenocene (**61**) [40] and vinyldicyclopentadienyltitaniumchloride (**62**).

4.2.6   Polymerizable Chelate-Type MCMs

This type of MCM is essential for the combination of the classes of MCMs considered above (Table 1), mostly of σ- and *nv*-types and more rarely of ionic and π-types [41]. Their distinctive feature is that chelate nodes and multiple bonds capable of polymerization are located on the same ligand molecule. Sometimes a chelate grouping is the only ligand environment of the metal, especially in the case of MCMs in pure organometallic compounds (Section 4.2.1). Examples of these MCMs are **63–65**.

Let us briefly consider MCMs with chelate nodes of the traditional type. As we know, the chelate effect consists of the formation of more stable complexes than those with non-chelate ligands. This is a result of saving of the Gibbs' free energy (ΔG) due to the addition of a bidentate (ΔG_n) ligand to MX_n instead of binding two monodentate (ΔG_m) ones to the same reacting groups (Eq. 4-9).

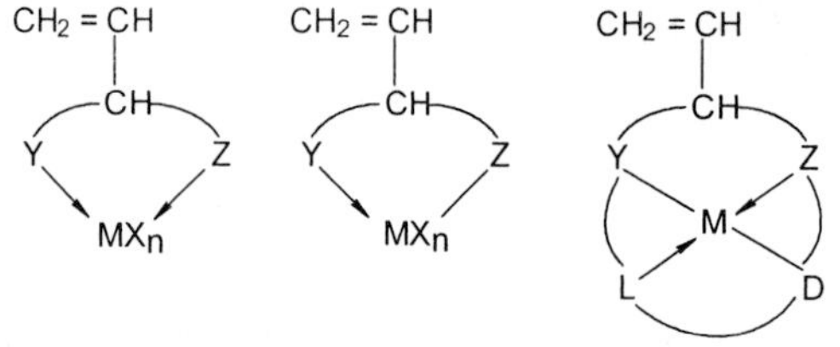

$$\Delta G_{chel} = \Delta G_n - \Delta G_m \tag{4-9}$$

The change of the enthropy (ΔS) makes the maximum contribution to the chelate effect. The enthalpy component of the free energy (ΔH) is mainly determined by the nature of the chelate unit and the metal. The well-known Chugaev's "ring rule" applies to MCM chelates, i.e. six- and five-membered rings are the most stable. In fact, MCMs with all types of chelate nodes are known. These monomers are largely represented by dioxo-, diaza- and dithiochelates as well as MCM chelates with combined asymmetric nodes – oxoaza-, phosphoazachelates, etc. A bidentate (rarely monodentate) binding of metals is observed in metal chelate monomers, whereas a tetradentate mode is characteristic of MCMs formed from porphyrins and phthalocyanines.

While the types of MCMs described earlier have already received comparatively wide popularity in polymerization practice, polymers based on metal-containing monomers of the chelate type have only been prepared more recently. The methods of "assembly" of such MCMs, i.e. the simultaneous formation of the ligand and the corresponding complex, have been substantially developed. The synthesis of MCMs from p-aminostyrene, 2-formylpyrrole and Cu(II) or Co(III) salts is an example of such a method. The last approach is especially characteristic of the preparation of MCMs with macrocyclic chelate nodes, in particular, from porphyrins, phthalocyanines and other macrocycles with exocyclic multiple bonds. It is worth noting that traditional methods of chelation are used for preparing MCMs when scientists want to ensure strong multicenter fixation of metals into monomer molecules, and, thus, into (co)polymers.

Vinyl ethers of ethanolamine **66** may coordinate with metal ions via both the ester oxygen atom and the amine group to form two five-membered rings.

The interaction of Co(OCOCH$_3$)$_3$ with a Schiff base – the product of condensation of 4-(p-vinylbenzyloxy)salicylaldehyde – leads to the formation of a classical chelate ring (**70**).

Let us consider some typical examples **66–69** [42].

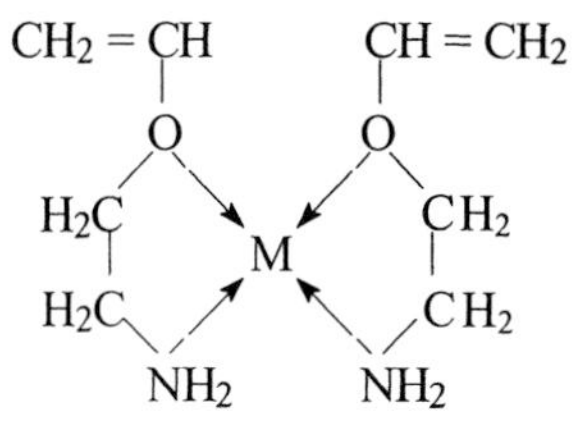

CH$_2$ = CH   CH = CH$_2$

H$_2$C   CH$_2$

M

H$_2$C   CH$_2$

NH$_2$ NH$_2$

66

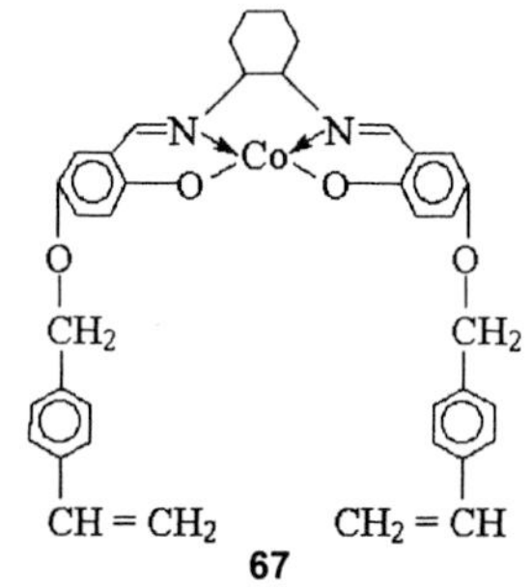

=N   Co   N=

O   O

CH$_2$ CH$_2$

CH = CH$_2$ CH$_2$ = CH

67

CH$_2$=CH   CH$_2$=CH   CH$_2$=CH

CH$_2$

N

HC   M

N

]$_n$

N

HC   M

N

]$_n$

C—O   O=C

CH   Eu   CH

C=O   O—C

]$_2$

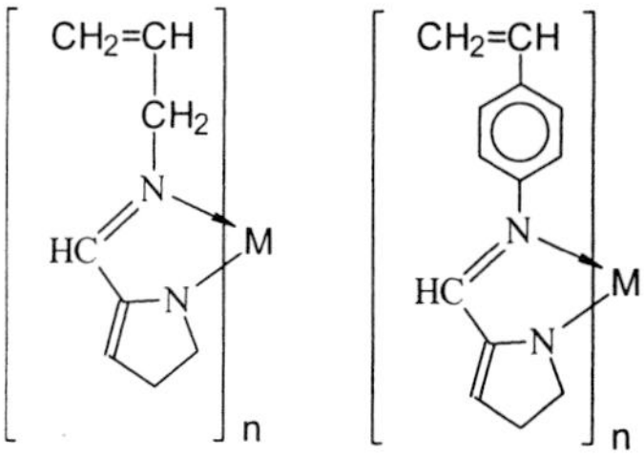

68 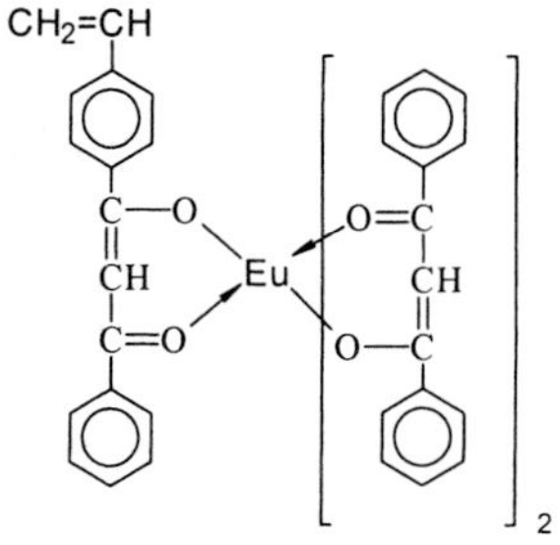   **69**

Using the method of chemical "assembly" from allylamine or *p*-amine-styrene and 2-formylpyrrole in the presence of Co^{2+} or Cu^{2+} we may obtain the corresponding Schiff bases – MCMs (**71,72**) (see Experiment 4-8, Section 4.6).

MCMs formed from (*p*-vinylphenyl)-3-phenyl-1,3-propandione (**73**) have been prepared from EuCl$_3$ (in the presence of Py) by means of the keto–enol tautomerism of 1,3-propandiones.

Chelate complexes with methacroylacetone are synthesized through the interaction of transition metal hydroxides with methacroylacetone [43,44].

4.2.7   MCMs with Macrocyclic Chelate Nodes

These are the same chelate-type MCMs, but with tetradentate coordination (Table 4-1). Moreover, MCMs with a small strained structure (vinylimine, epoxide) have already been considered above. By considering MCMs with macrocycles in a separate section, we wish to focus attention on monomers of this type because of their widespread use in polymerization practice, especially, porphyrin and phthalocyanine derivatives. As we know, such macrocycles are π-conjugated planar tetradentate ligands and capable of forming rather stable chelates with almost all metals. A basic method of MCM synthesis is based on the incorporation of metal ions into a "window" of macrocycles completed by an exocyclic multiple bond [45–48]. Examples are **74–76**.

Metalloporphyrin complexes of Co(II), Fe(III), Fe(II), Mg(II), Zn(II), Cu(II), Ni(II) , etc. having two or more vinyl groups have also been synthesized [49,50]. Typical representatives are **77–79** below.

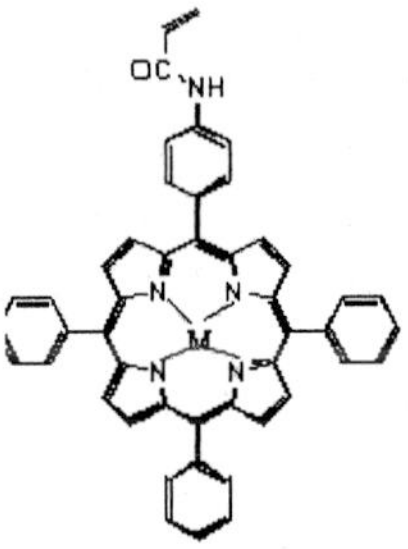

74 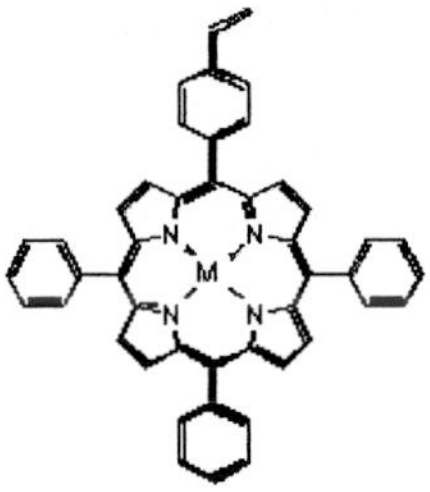   **75** 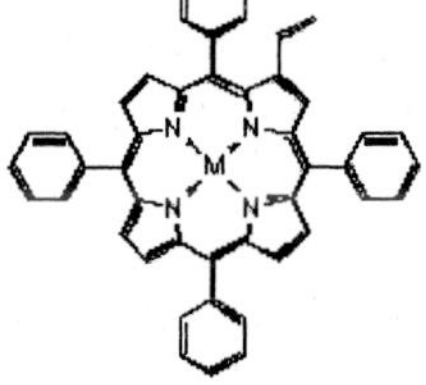   **76**

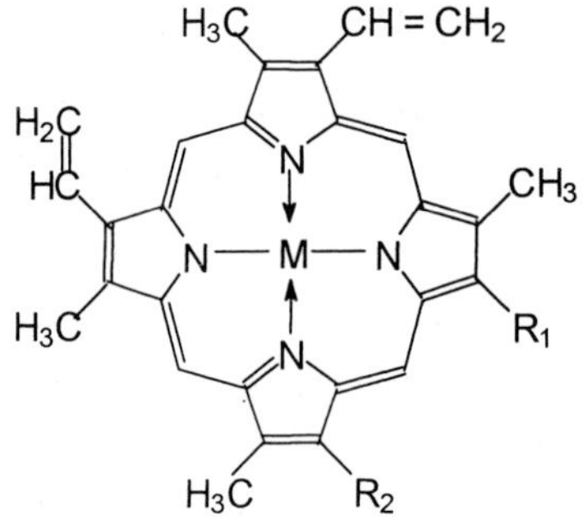

77

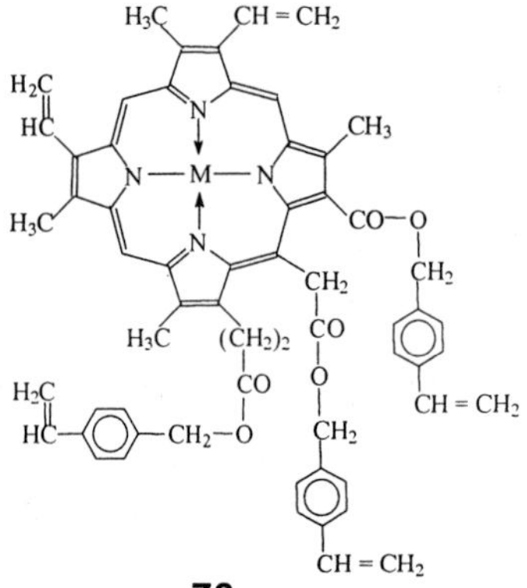

78

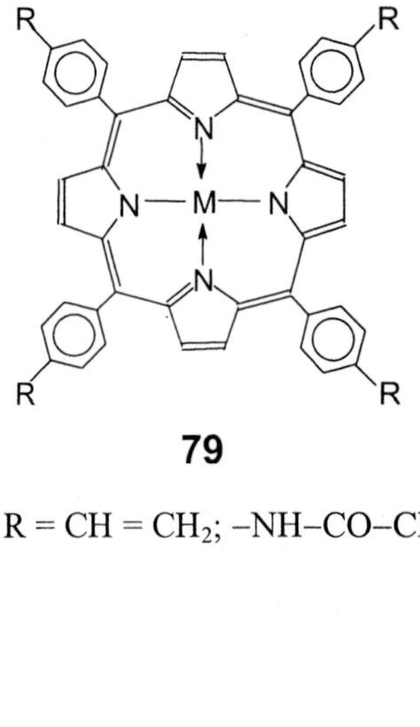

79

R = CH = CH$_2$; –NH–CO–CH = CH$_2$

Metalloporphyrin monomers are largely synthesized through interaction of reactive peripheral groups of transition metal porphyrinates (e.g. chlorophyll, chlorine, Mn(II) tetra-*n*-aminoporphyrinate, etc.) with monomers (e.g. *n*-chlormethylstyrene or the chloranhydride of acrylic acid). Carboxyl and amine groups represent the most widely used peripheral species of this kind. A direct "assembly" method for the synthesis of such monomers at the stage of formation of macrocycles (as is typical for conventional metall-ophthalocyanines) would be of interest. However, no such experimental details have been published.

Among another macrocyclic MCMs we can note the complexes of Zn(OTf)$_2$ and Hg(OTf)$_2$ with *N*-(4-vinylbenzyl)-1,4,7-triazacyclononane (**80**), which are used as templates [51, 52].

Crown ethers with exocyclic multiple bonds and a cavity filled by metal ions would be a potential monomer of this type. The cavity can be considered as a polydentate chelate node. The nature of the cation and the size of athe macrocyclic cavity play an important role (the "key–lock" principle) [53]. There are reports of the production of such monomers from alkali metals and 4'-vinylmonobenzo-18-crown-6, (**81**), 4'-vinyldibenzo-18-crown-6 (**82**), and 4'-vinylbenzo-15-crown-5 (**83**).

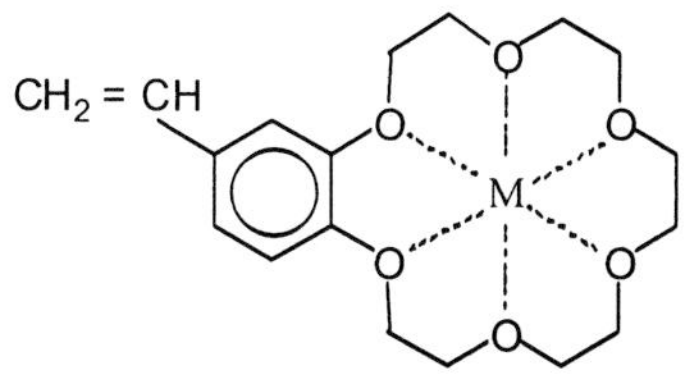

81

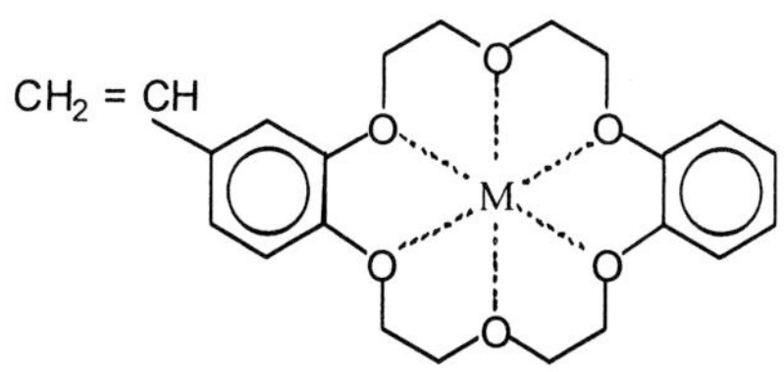

82

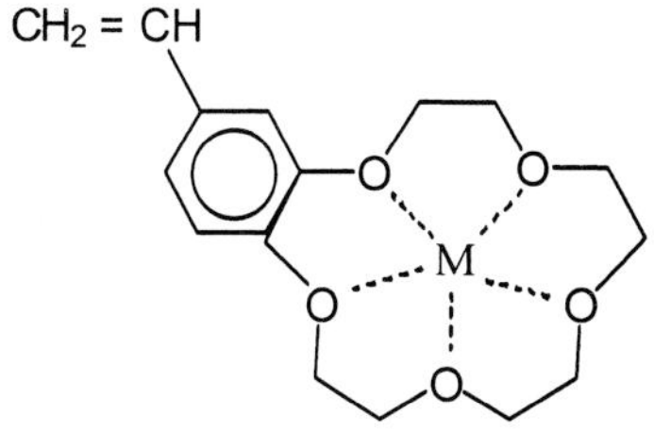

83

4.2.8   Polynuclear and Cluster MCMs

So far we have considered mononuclear MCMs. However, polynuclear MCMs, i.e. aggregates with assemblies of interacted metal ions bound in definite proportions and structural ratios, are of substantial interest. The earliest reports related to the trivalent metals V(III), $V_3O(OCOCH=CH_2)_6 \cdot 3CH_2=CHCOOH$ (**84**), and Fe(III), Cr(III) **85** acrylates. For instance, for the antiferromagnetic polynuclear complex **84** it has been established that the structure has no direct metal–metal bond [54].

It is probable that the cyclopentadienyl derivatives of Ti(IV), $[(C_5H_5)_2Ti(\mu$-$OCOC\equiv COCO)]_2$ (**86**) and $[(C_5H_5)_2Ti(\mu$-$OCOC\equiv COCO)]_5 \cdot 5L$ (**87**) (where L is a solvent molecule, dioxane or methylene chloride), obtained in a two-phase system – water/organic solvent) are related to this class of MCM [55]. The compound $Zr_4[OCOC(CH_3)=CH_2]_{10}O_2(OR)_2 \cdot n$ H_2O (n = 2, 4) (**88**) has been obtained by boiling zirconium alkoxides with acrylic acid [56].

Heterometallic structures of this type and structures including the ions of one metal in different valence states (for example, Fe(III) and Fe(II)) can also be synthesized. Not only carboxylate groups but also other polymerizable ligands may possibly be involved, and the polynuclearity of such formations can, in principle, be more prominent.

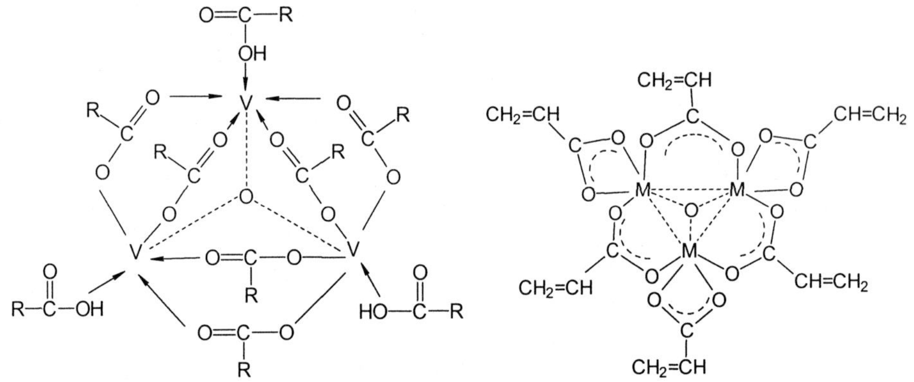

84   **85**

$M = Fe(III), Cr(III); R = CH=CH_2$

Wider possibilities are also available with cluster-type MCMs. They are molecular compounds that include a framework of metal atoms, which are separated by short distances (not more than 3.5 Å) permitting direct metal–metal interaction. Such a framework is enclosed within a set of polymerizable ligand groups. The first report in this regard relates to MCMs based on $Co_2(CO)_8$ or $Fe_2(CO)_9$ and the methyl ester of p-vinyldithiobenzoic acid as well as N-cyclohexyl-4-vinylthiobenzamide [57]. MCMs based on a carbonyl cluster of osmium **89**, ruthenium [58] and rhodium [59] with 4-VPy **90** and allyldiphenylphosphine **91** are stable compounds (see Experiments 4-9 and 4-10, Section 4.6) [59,60].

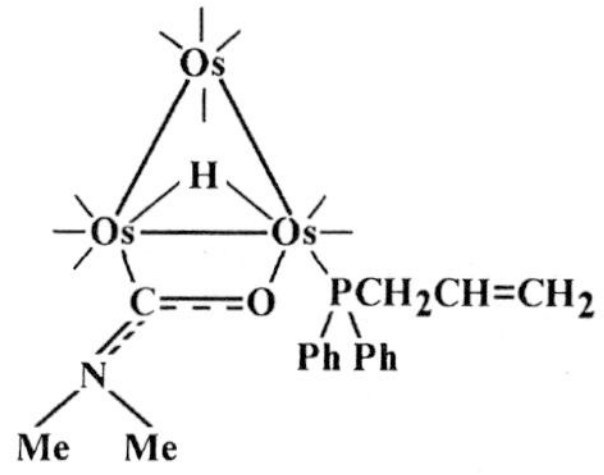

89 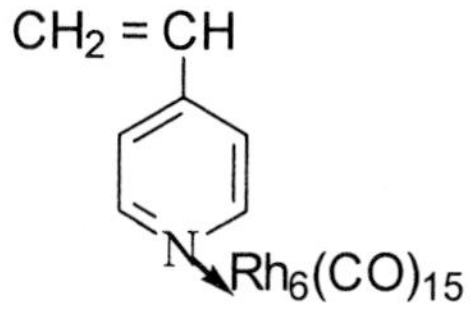   **90** 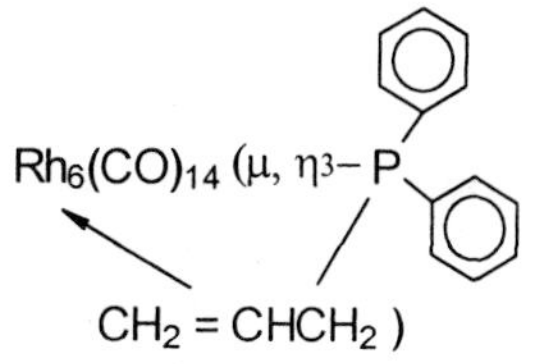   **91**

At the same time, independent developments in different fields of chemistry (as mentioned above, MCMs are of interest for specialists in organometallic, coordination and physical chemistry) mean that specific features of the chemistry of high molecular weight compounds in such systems are not considered. Most importantly for polymerization, transformation properties, e.g. the state of multiple bonds in MCMs, the effect of metal ions on them and how they can be activated, remain largely insufficiently studied. Thus, it is important to obtain data on the composition and structure of MCMs in solution, the kinetics and thermodynamics of formation of intermediate complexes (especially in copolymerized systems), and a knowledge of which have metal–ligand and which have multiple bonds in the inner coordination sphere, etc.

As can be seen from the previous analysis, the scale of scientific research into methods of MCM synthesis is broad enough. At present methods have been developed for the preparation of MCMs based on practically all metals including lanthanides and actinides, as well as of all possible valence states.

We imagine that in the near future therte will be substantial progress in the synthesis of pure organometallic MCMs. Probably, new types of polymerizable groups as well as new stabilizing ligands (most likely of the chelate type) will be developed.

Thus, the synthesis of metal-containing monomers is a well-developed field of the science which has appeared as a branch of inorganic, coordination and organometallic chemistry. The chemist can practically obtain all types of MCM. Characteristic representatives are given in Fig. 4-5.

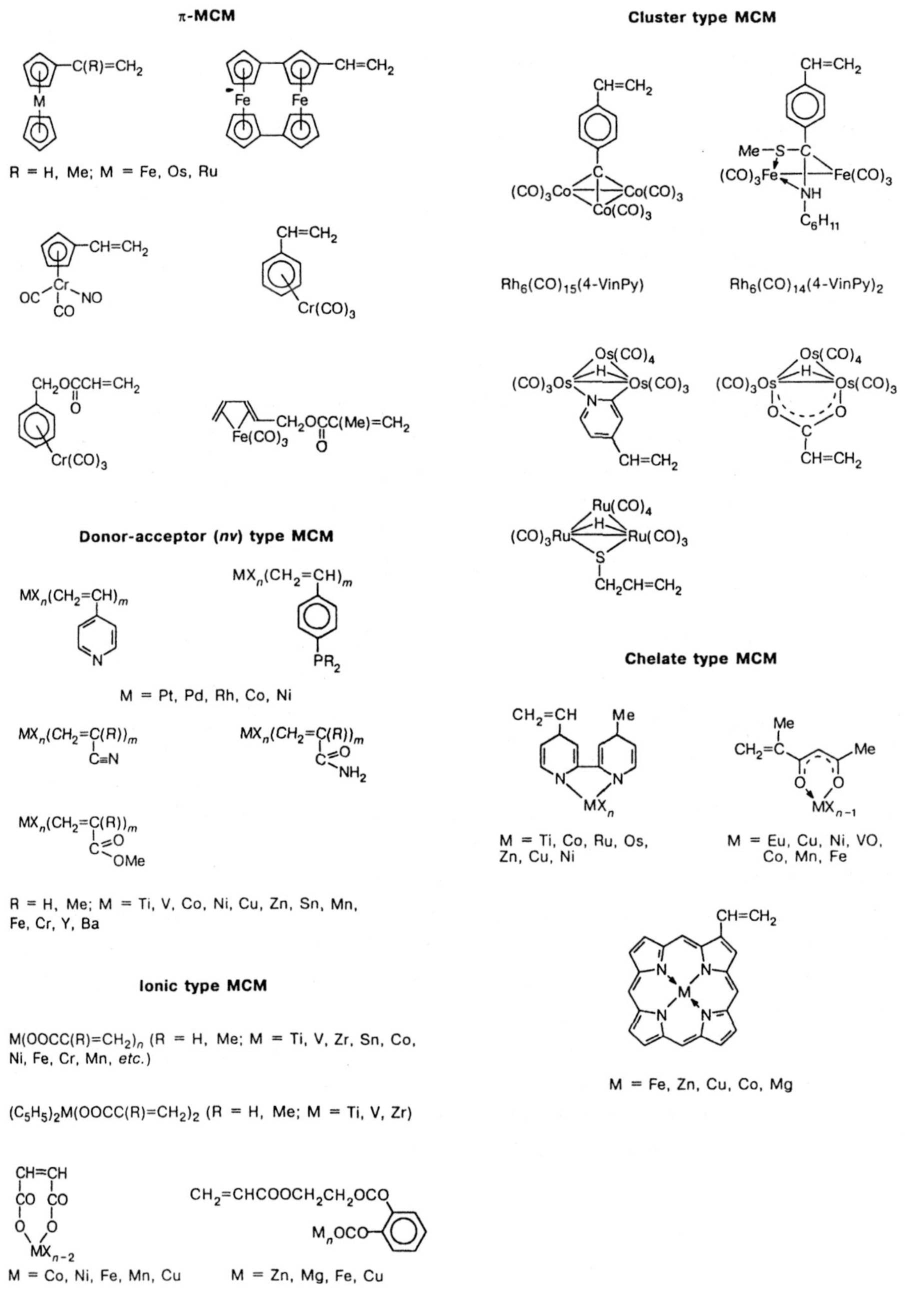

Figure 4-5. Representatives of the main types of MCM.

4.3   Homopolymerization of Metal-Containing Monomers

Although homopolymerization of MCMs is usually carried out according to traditional schemes, the basic stages of the polymerization process are characterized by their own features. Firstly, this is related to the initiation and chain termination reactions.

The reactivities of MCMs vary extremely widely, i.e. from monomers that will polymerize as soon as they are made to monomers incapable of polymerization even under very extreme conditions.

Radical polymerization of MCMs has only been studied to a limited extent. One of the difficulties of this method is the selection of an appropriate solvent. The polymer formed is often precipitated even at a small degree of conversion. For many reasons, quantitative data on the polymerization of such monomers are extremely scarce. Only a few papers consider the determination of effective rates and of kinetic orders, monomer concentrations and activation energies of polymerization. The reasons are the many complications and side-reactions that accompany MCM polymerizations.

A metal-containing group as a specific substituent affects all the elementary stages (especially during copolymerization, see below).

4.3.1   The Specific Character of the Homopolymerization of True Organometallic Monomers

We should note first that the literature contains no systematic data on the homopolymerization of the MCMs under discussion. Among the scant data, we note the homopolymerization of *p*- and *m*-vinylbenzylmagnesium chlorides that results in the formation of a Grignard polymer reagent [61]. At 0 °C in THF these monomers undergo spontaneous polymerization with a low rate. The rate of the process shows a drastic increase as the temperature is raised to 60 °C, but in this case an insoluble polymer is formed. The reactivity of the *m*-isomer is higher than that of the *p*-isomer. The polymer obtained in hexamethyl phosphoramide (HMPA) as solvent is very branched, which is related to the enhanced activity of the residual benzylmagnesium chloride groups (Eq. 4-10). This activity is sufficient for MCM polymerization (Table 4-2).

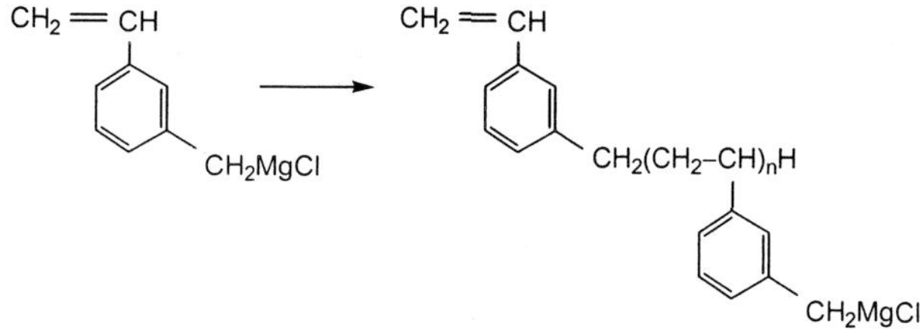

$$(4\text{-}10)$$

Table 4-2. Homopolymerization of vinylbenzylmagnesium chlorides.

MCM[a]	Solvent	T, °C	Time (d)	Yield,%	$\overline{M}_n$	$\overline{M}_w/\overline{M}_n$
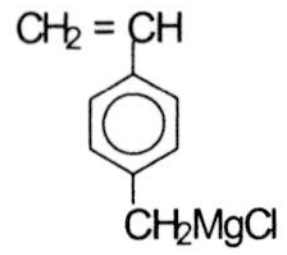 CH₂ = CH ⬡ CH₂MgCl	(C₂H₅)₂O	0	7	0	–	–
	THF	0	3	2.5	7800	3.3
	THF	60	1	11	–	
	HMPA/ toluene	0	3	78	4900	–3.7
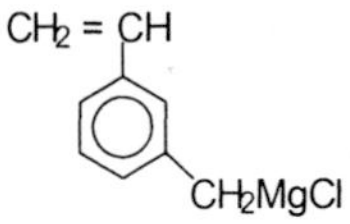 CH₂ = CH ⬡ CH₂MgCl	THF	0	3	11	3300	1.8
	THF	60	1	33[b]	–	–
	Toluene	0	3	0	–	–

[a] [MCM]= 1 mol L⁻¹; [b] Polymer insoluble in most organic solvents.

We mention here an interesting peculiarity of one more organometallic monomer. The polymerization of 1-(trimethyltin)alkylmethacrylate [62] is distinguished by chain transfer to an organotin monomer and polymer due to the presence of a labile H atom in the α-position relative to the Sn(CH₃)₃ group. Such a reaction is accompanied by a decrease in the polymerization rate, since the new radical is incapable of participating in the initiation process. Note that the presence of a metal atom in the monomer molecule is necessary for the degenerate chain transfer, because such a process is not characteristic of "free metal" analogues. The chain-transfer process goes more readily to the polymer than to the MCM, since the growing radical abstracts the hydrogen atoms in a five-membered transition state (Eq. 4-11).

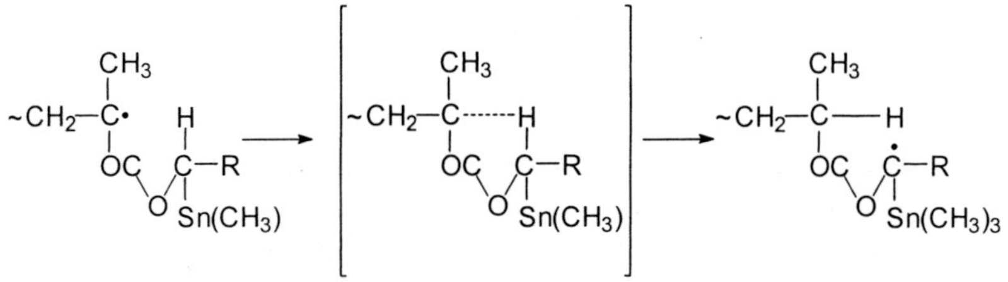

$$(4-11)$$

Such an intramolecular process is facilitated as compared with an intermolecular hydrogen abstraction because the chain is transferred to the MCM. Cases of the mutual influences of initiating (R˙) or growing radicals and the metal of monomers are frequent in the systems we are considering.

Finally, we should mention the homopolymerization of organometallic styrene Pd(II) and Pt(II) derivatives (Eq. 4-12).

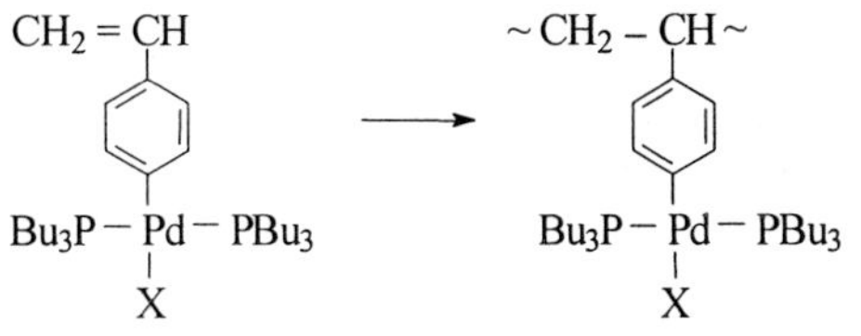

$$CH_2 = CH \qquad \sim CH_2 - CH \sim$$

$$Bu_3P - Pd - PBu_3 \qquad Bu_3P - Pd - PBu_3 \qquad\qquad (4\text{-}12)$$
$$\quad\; X \qquad\qquad\qquad\quad\; X$$

14

Table 4-3. Homopolymerization of MCMs – organometallic compounds of Pd(II) and Pt(II).

Amount of MCM, g	AIBN, g	Temperature, °C	Time, h	Yield,%	$\overline{M}_n$	$\overline{P}_n$
		$CH_2{=}CHC_6H_4Pd(P(Bu_3)_2Cl$				
1.36	0.033	60	4	15.2	3610	5.5
0.413	0.016	60	3	8.2	4110	6.3
1.295	0.032	50	48	53.2	**	**
0.669	0.016	100	24	25.0	7550	11.6
1.25	0.032	55	54	57.5	8210	12.6
2.598	0.133	55	72	41.0	8450	13.0
2.120	0.197	55	96	61.0	8050	12.4
1.945	0.295	55	24	28.0	10900	16.8
2.577	0.534	55	72	65.0	10000	15.4
3.399	0.170	55	48	68.9	10900	16.8
3.264	0.164	55	72	72.0	9350	14.3
		$CH_2 = CHC_6H_4Pd(PBu_3)_2Br$				
2.710	0.132	55	50	58.0	6460	9.3
0.334	0.032	55	72	23.0	7500	10.7

Pd(II)-containing monomers (X = Cl or Br) and also a Pt(II)-containing monomer are readily polymerizable in the presence of azobisisobutyronitrile (AIBN). At the same time, Pd(II)-containing monomers (X = CN and C_6H_5) decompose at polymerization temperatures of 55 and 60 °C; however, they are capable of effective polymerization at 0 to 15 °C when the process is initiated with BBu_3O_2. The molecular mass of the resulting polymer is not high ($P_n = 5$–17) (Table 4-3).

Attempts to conduct a cationic polymerization of these MCMs (BF_3 as the catalyst) have not been successful, apparently because of the interaction of the palladium MCM with BF_3. Moreover, the polymerization of organometallic monomers can be complicated by a whole series of side-reactions.

Thus, for MCMs that are pure organometallic compounds of transition metals, the most important problem is how to avoid the elimination of a metal hydride during polymerization (Eq. 4-13) [64]. Such transformations are not experimentally observed for MCMs with polymerizable groups of other types (styrene, methacrylate, etc.).

$$
CH_2=CH \xrightarrow{\text{initiation}} \sim CH_2\text{-}CH \sim \longrightarrow \sim CH=CH \sim + MX_{n\text{-}1}H \tag{4-13}
$$
$$
\underset{MX_{n\text{-}1}}{|} \qquad\qquad \underset{MX_{n\text{-}1}}{|}
$$

Nevertheless, reactions of MCMs with the initiator (especially peroxide types) during polymerization are very widespread. We shall analyze some cases for varied types of MCMs. There are many ways of avoiding side-reactions, or reducing them to a minimum during polymerization of MCMs.

4.3.2   Homopolymerization of Organoelemental Monomers

Radical polymerization of MCMs involves the same elementary steps as with ordinary vinyl-type monomers, and the stationary rate (w_p) of the reaction is described by the classical kinetic equation of radical polymerization, the order with respect to the initiator being 0.5 and with respect to the MCM close to 1.0.

The choice of initiators for polymerizations is an important problem. MCMs can affect the initiator dissociation and initiation rate, and sometimes a chemical interaction of the MCM with an initiator occurs. Thus, MCMs based on Ti(IV) effectively interact with peroxide initiators, with the formation of titanium acrylates. In this connection the polymerization of $(BuO)_3TiOCOC(CH_3)=CH_2$ does not compete with the acylation reaction, while in the presence of azobisisobutyronitrile (AIBN) it can be avoided completely. Such examples are numerous, as are variants in which the MCM serves as a "co-initiator", increasing or decreasing the polymerization rate.

A very curious phenomenon was discovered during radical polymerization of $(BuO)_3TiOC(CH_3)_2C\equiv C-CH=CH_2$ under the action of AIBN. The total equation representing the polymerization rate is presented in Eq. 4-14 [65].

$$
W_n = k[M]^{0.5}[I]^{0.5} \tag{4-14}
$$

The polymerization rate (W_n) is proportional to the square root of the concentration not only of the initiator (I), but also of the monomer (M). Such a relationship is rarely observed in radical polymerization and is specific for this MCM in particular, due to the interaction of primary radicals with Ti(IV). The kinetic scheme is described by equation Eq. 4-15.

$$I \xrightarrow{k_i} 2R^\bullet$$

$$R^\bullet + M \xrightarrow{k_c} R^\bullet M$$

$$R^\bullet + M \xrightarrow{k_p^1} RM^\bullet$$

$$RM^\bullet + M \xrightarrow{k_p} RM_n^\bullet \qquad\qquad (4\text{-}15)$$

$$RM^\bullet + RM^\bullet \xrightarrow{k_t} P$$

$$RM^\bullet + R^\bullet \xrightarrow{k_t'} P$$

$$R^\bullet + R^\bullet \xrightarrow{k_t''} P$$

In Eq. 4-15 k_c is the constant of the complex formation reaction of the monomer with the radical, k_p^1 and k_p are the rate constants of reaction of chain growth with participation of radicals coordinated and uncoordinated with Ti(IV), k_t, k_t', k_t'' are the rate constants of the corresponding chain termination reactions and P represents reaction products. The chain termination proceeds according to a bimolecular mechanism, and the radicals coordinated with monomer (R M) play a substantial role. The concurrent reactions of initiation and formation of coordinated radicals at polymerization of Ti(IV) monomers can occur as shown in Eq. 4-16.

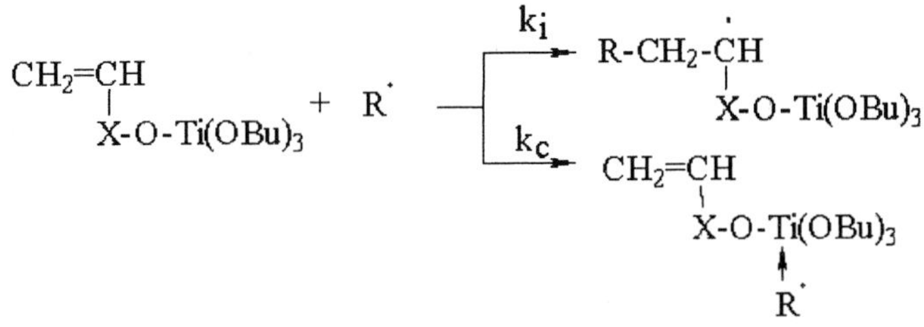

$$
\begin{array}{c}
CH_2{=}CH \\
\;\;\;\;| \\
X\text{-}O\text{-}Ti(OBu)_3
\end{array}
+ \; R^\bullet
\quad
\begin{cases}
\xrightarrow{k_i} & \begin{array}{c} R\text{-}CH_2\text{-}\overset{\bullet}{C}H \\ | \\ X\text{-}O\text{-}Ti(OBu)_3 \end{array} \\[2em]
\xrightarrow{k_c} & \begin{array}{c} CH_2{=}CH \\ | \\ X\text{-}O\text{-}Ti(OBu)_3 \\ \uparrow \\ R^\bullet \end{array}
\end{cases}
\qquad (4\text{-}16)
$$

Thus, Ti(IV)-coordinated radicals take part both in chain growth and in chain termination. In other words, the metal ion is an original trap for radicals. The same process can have an intramolecular nature, when it proceeds in growing chains and leads to their termination (by a type of degenerate chain transfer, Eq. 4-17).

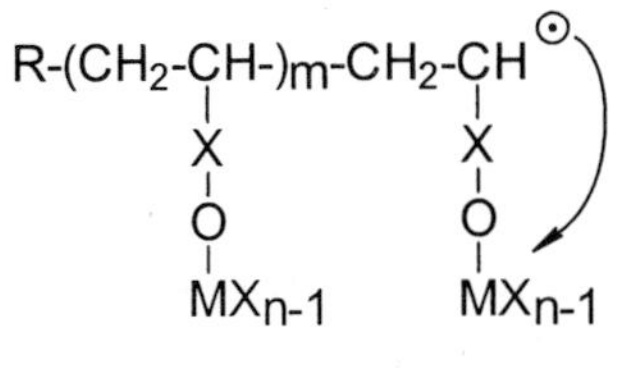

$$
\left.
\begin{array}{c}
R\text{-}(CH_2\text{-}CH\text{-})_m\text{-}CH_2\text{-}CH^{\ominus} \\
\;\;\;\;\;\;| \;\;\;\;\;\;\;\;\;\;\;\;\;\;\;\;\; | \\
\;\;\;\;\;\;X \;\;\;\;\;\;\;\;\;\;\;\;\;\;\;\;\; X \\
\;\;\;\;\;\;| \;\;\;\;\;\;\;\;\;\;\;\;\;\;\;\;\; | \\
\;\;\;\;\;\;O \;\;\;\;\;\;\;\;\;\;\;\;\;\;\;\;\; O \\
\;\;\;\;\;\;| \;\;\;\;\;\;\;\;\;\;\;\;\;\;\;\;\; | \\
\;\;\;MX_{n\text{-}1} \;\;\;\;\;\;\;\; MX_{n\text{-}1}
\end{array}
\right)
\qquad (4\text{-}17)
$$

The nature of both the polymerized system and the metal ion is important for the way in which such processes proceed.

4.3.3   Radical Polymerization of Salts of Unsaturated Carboxylic Acids

The polymerization of this class of monomers has been fairly well studied. Many studies have been carried to quantitative completion. Polymerizations of alkali and alkaline earth metal salts based on unsaturated acids occur, as a rule, in aqueous or aquo-organic environments. Under these conditions the MCMs are virtually dissociated [66]. One of the first qualitative studies of polymerization of Mg(II), Sr(II), Ba(II) and Ca(II) acrylates was carried out in 1955 [67]. The influence of different factors on the rate of polymerization of these monomers has been studied (Fig. 4-6).

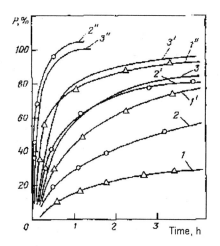 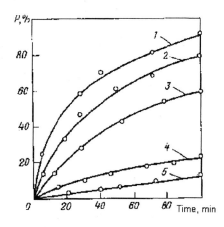

Figure 4-6. Kinetic curves of homopolymerization of nontransition metal acrylates: Ca(II) (1), Sr(II) (2), Mg(II) (3); [AIBN] = 0.25 (mol.-%), [MCM] (mol L^{-1}): 0.2 (1, 2, 3), 0.5 (1′, 2′, 3′), and 0.8 (1″, 2″, 3″).

Figure 4-7. Kinetic curves of homopolymerization yield for AA(*1*); and metal acrylates: Co(II) (*2*), Ni(II) (*3*), Fe(II) (*4*), Cu(II) (*5*). [MCM] = 0.9 mol L^{-1}. [AIBN] = 2.5·10^{-2} mol L^{-1}, ethanol, 78 °C.

The rate of polymerization increases in this series of metals: Mg(II) < Sr(II) < Ba(II) < Ca(II). The nature of the cation is likely to have a significant effect on the kinetics of the polymerization of salts of unsaturated acids in ionizing environments [68–70]. These differences are attributed to a different charge density at the macroradical anion, which influences the rate of interaction in the "propagating macroradical–monomeric anion" system. In comparable conditions the rate of radical polymerization of transition metal acrylates is lower than that of acrylic acid (AA) and decreases in the series (Fig. 4-7) [71]: AA > Co(II) > Ni(II) > Fe(III) > Cu(II) (see Experiment 4-1, Section 4.6). The resulting metallopolymers are insoluble in any organic solvent, which indicates the formation of cross-linked structures. However, they dissolve in a methanol:HCl mixture to give polyacids.

It should be taken into account that the salts of monocarboxylic acids and alkaline earth or transition metals comprise divinyl-type species with unconjugated double bonds. Such MCMs can be homopolymerized in two ways: with the opening of one or both multiple bonds and with the appearance of either linear or cross-linked polymers (Eq. 4-18).

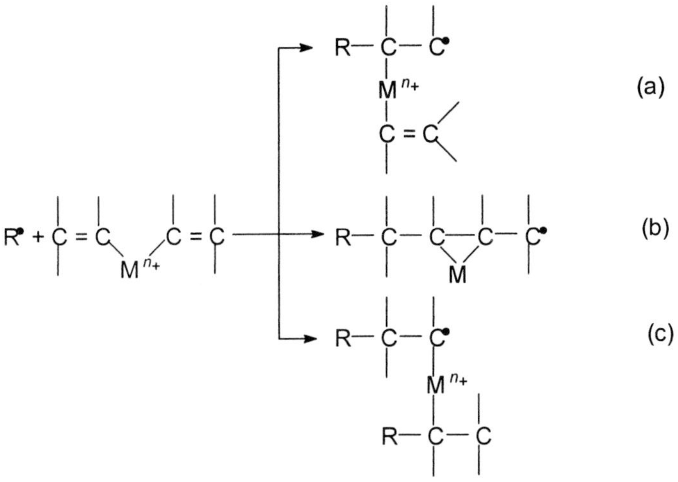

$$ (4\text{-}18) $$

The insolubility of the corresponding polymers may suggest a significant role of reaction (c) leading to the presence of structures cross-linked through a metal ion. The mechanism of cyclic unit formation seems to be close to the polymerization of unconjugated divinyl monomers (diene-type, methacrylic anhydride, etc.). It arises from the alternation of the inter- and intramolecular chain propagation reactions (Eq. 4-19).

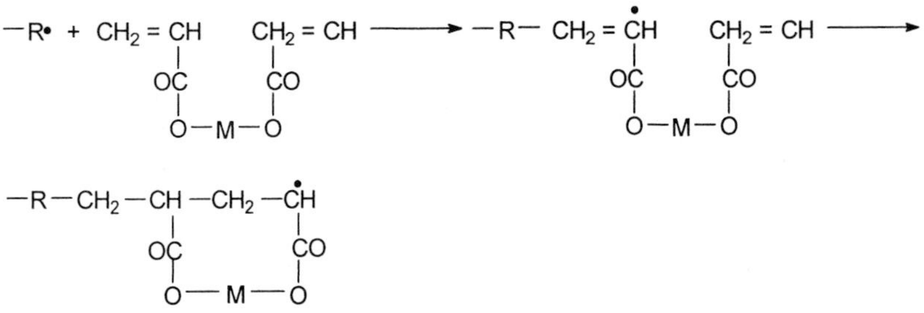

$$ (4\text{-}19) $$

Metallocene mono- and dimethacrylates are fairly polymerizable (80 °C, DMFA, 0.3–0.5% benzoyl peroxide as initiator) to give the polymers **93** and **94** [72].

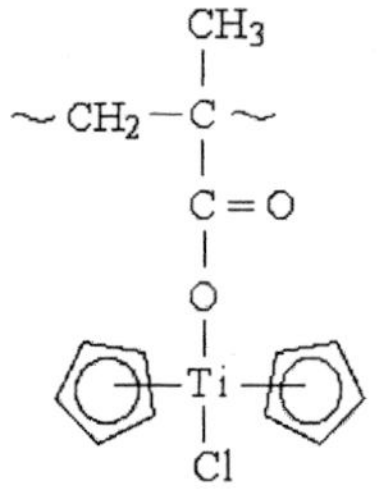

$$
\begin{array}{c}
CH_3 \\
| \\
\sim CH_2-C \sim \\
| \\
C=O \\
| \\
O \\
| \\
\text{(Cp)}-Ti-\text{(Cp)} \\
| \\
O \\
| \\
C=O \\
| \\
\sim CH_2-C \sim \\
| \\
CH_3
\end{array}
$$

$$
\begin{array}{c}
CH_3 \\
| \\
\sim CH_2-C \sim \\
| \\
C=O \\
| \\
O \\
| \\
\text{(Cp)}-Ti-\text{(Cp)} \\
| \\
Cl
\end{array}
$$

93 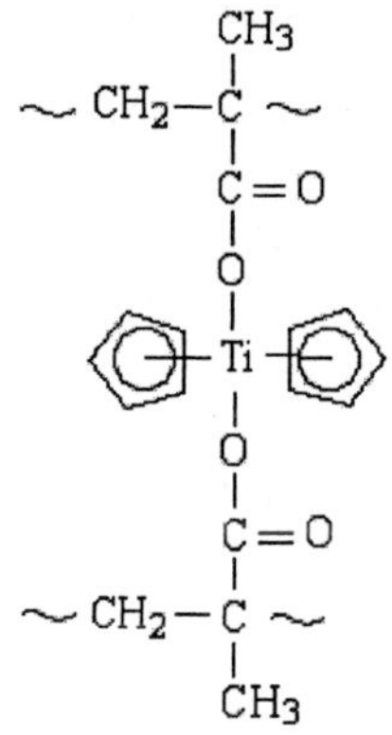   94

The correlation between the polymerization rate of metal diacrylates and the metal's electronegativity has been determined experimentally. Electron density on the double bond and the polymerization rate decrease as the metal electronegativity increases. The limited polymerization of Cu(II) and Fe(III) acrylates is a consequence of intermolecular termination of the chain (Eq. 4-20).

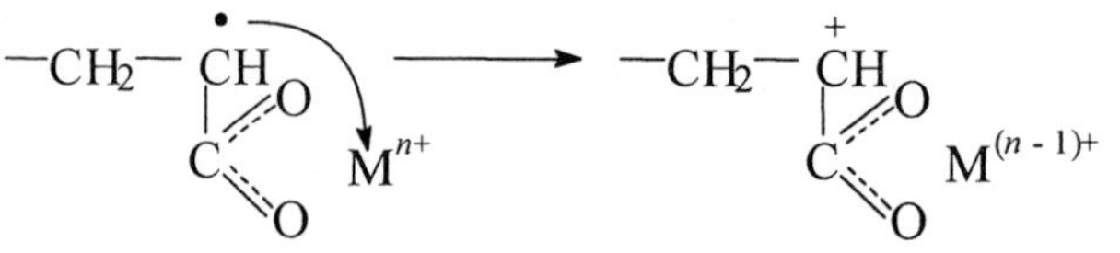

$$
\begin{array}{ccc}
-CH_2-\overset{\bullet}{CH} & & -CH_2-\overset{+}{CH} \\
\underset{C}{\big|}\diagdown O & \longrightarrow & \underset{C}{\big|}\diagdown O \\
\quad \diagdown O \ M^{n+} & & \quad \diagdown O \ M^{(n-1)+}
\end{array}
\qquad (4\text{-}20)
$$

This process is favored by the fairly high standard electrode potentials of the copper and iron ions: $E^0(Cu^{2+}/Cu^+) = 0.15$ and $E^0(Fe^{3+}/Fe^{2+}) = 0.77$ V vs. normal hydrogen electrode.

Thus, the polymerization of transition metal acrylates can be accompanied by redox reactions that involve metal ions and cause termination of the propagating chain.

We note here new tendencies in polymerizations of this type of monomer:

- low-temperature radical MCM polymerization,
- polymerization in the solid phase,
- thermal polymerization,
- solid-phase ultraviolet- and radiation-induced polymerizations,
- solid-phase polymerization under pressure,
- solid-phase polymerization under high-pressure/shear strain or mechano-chemical initiation conditions.

These methods and others have been described in detail in monographs [4,114] and are therefore not considered here.

4.3.4   Homopolymerization of *nv*-Type MCMs

The first studies that illustrated the effect of complexing additives on the radical polymerization of conventional monomers marked the starting point for investigations of the polymerization behavior of *nv*-type MCMs. Thus, complexation of MMA with MX_n was a basis for establishing the fruitful concept of complex-radical polymerization. It is based on the assumption that a transition state is formed in the presence of MX_n in which both the propagating radical and the monomer molecules are coordinated at the same metal atom (Eq. 4-21):

$$(4\text{-}21)$$

The formation of such a state involves special steric conditions which favor monomer addition to the radical, with the result that the polymerization process takes place in the coordination sphere of one metal atom. Along with the steric factors, a large role is played by an energetic contribution. Owing to a significant delocalization of the lone electron in the intermediate complex, E_a decreases. All this leads to an increasing k_p. Indeed, the complexation of vinylpyridines with $ZnCl_2$, for example, leads to a noticeable process acceleration (by 1.5 to 5 times) as compared with the case when an uncoordinated ligand is used [79]. The other types of initiation were ineffective in this case: anionic initiation because of a substantial contribution by side-reactions with MCM, and cationic initiation ($SnCl_4$, THF, -78 °C) because of high positive values of the e parameter of such MCMs. A steric double bond accessibility and the nature of the complexing metal and ligands are important factors for the activation of the exocyclic C=C bond. This also applies to the polymerization of vinylimidazole complexes. It is probable that because of a strong dynamic conjugation of the vinyl group and heterocycle, coordination with MX_n leads to a higher polarity of the double bond and a greater parameter e [76]. The complexing gives rise to larger $(k_p/k_t)^{1/2}$ ratios (0.116 and 0.348 $mol^{1/2}$ $s^{1/2}$ for vinylimidazole and the complex, respectively), which is likely due to a smaller k_t value as a result of the lower molecular mobility of the growing complex-bonded polymer chains. Complexes of the vinyl derivatives of different types of azoles, namely, imidazoles, triazoles, tetrazoles, containing 2, 3 and 4 nitrogen atoms in the ring, are capable of polymerization. Studies of the effect of the ring nature and MX_n as well as the polymerization conditions will have to be undertaken.

This also applies to the polymerization of MX_n with unsaturated nitriles, although homopolymerization of acrylonitrile complexes with Zn(II) halides (in

1:1 and 2:1 ratios) has been comparatively well studied [77]. The rate of polymerization of methacrylonitrile proceeds 300 times slower than that of a complex-bonded one. The resulting products have a high molecular mass (degree of polymerization is 1600 to 14,000). However, in polar solvents (THF-type) partial dissociation of the complexes is frequently observed.

Addition of complexing agents to the systems with allyl monomers affects at least two factors: first, strengthening the C–H bond in the monomer (with a decrease in the monomer degradative chain transfer rate); and second, enhancing the probability of attaching the resulting complex-bonded allyl radical to the double monomer bond (i.e. to the conversion of the degradative chain transfer into an effective one) (Eq. 4-22):

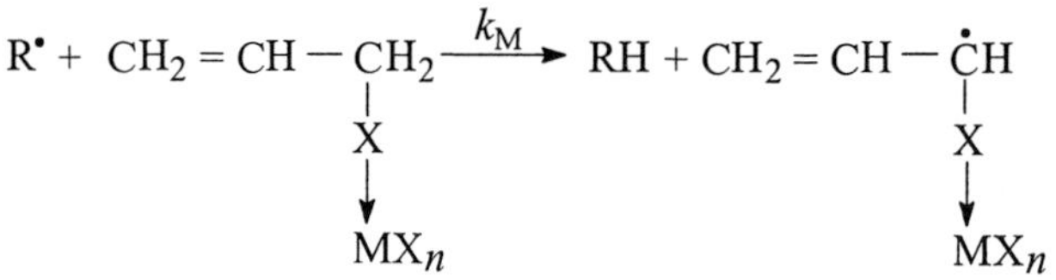

$$R^{\bullet} + CH_2 = CH - CH_2 \xrightarrow{k_M} RH + CH_2 = CH - \overset{\bullet}{C}H \qquad (4\text{-}22)$$

with X and MX_n substituents below the CH_2 and $\overset{\bullet}{C}H$ groups.

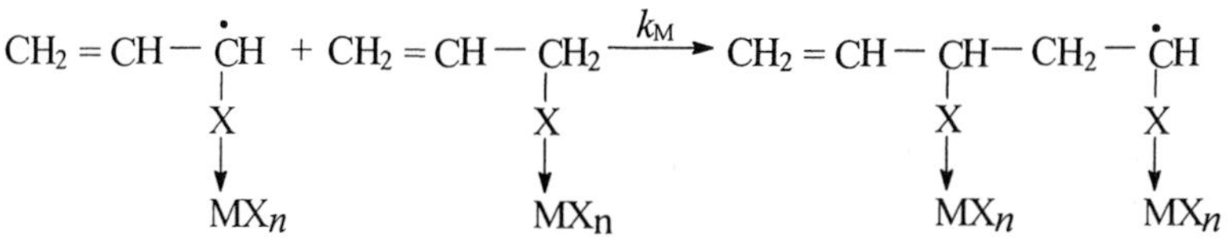

$$CH_2 = CH - \overset{\bullet}{C}H + CH_2 = CH - CH_2 \xrightarrow{k_M} CH_2 = CH - CH - CH_2 - \overset{\bullet}{C}H \qquad (4\text{-}23)$$

with X and MX_n substituents.

Note the most interesting examples. In particular, as $ZnCl_2$ is added, we can observe an accelerated radiation polymerization of allyl acetate and allylamine by 6 and 30 times, respectively. When MX_n complexes include conventional monomers, the polymerizations are also typical in the case of $ZnCl_2$. The same is observed for $CoCl_2$, $NiCl_2$, and $PdCl_2$ complexes with ethylene and triethylene imines and their derivatives (ML_nCl_2 with $n = 1$ to 4). Owing to the formation of a coordinate bond, these complexes are distinguished by a rearrangement of the electronic density at the atoms and bonds of the heterocycle, with the result that the ring stress increases appreciable (e.g. for ethylene imine by 42 kJ mol^{-1}) [78]).

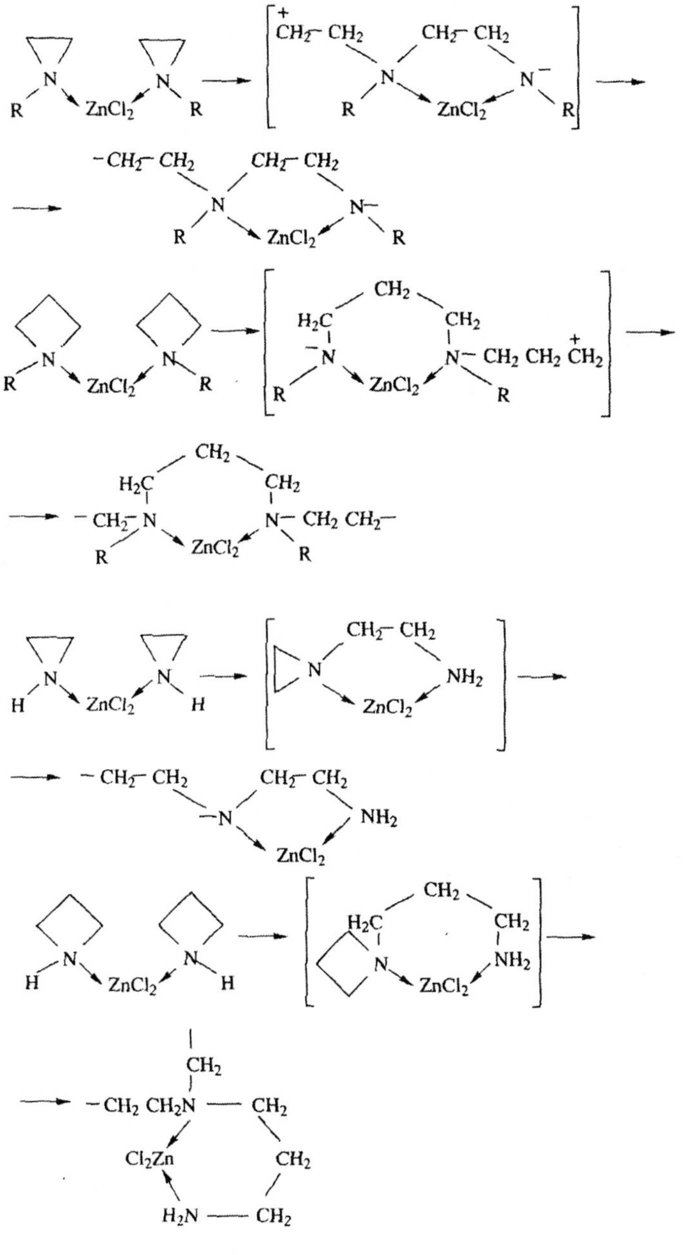

(4-24)

An enhanced heterocycle stress in the coordination sphere leads, in turn, to a sharp drop in the endocyclic bond strength. This causes heterocyclic ring opening to give bipolar ions with the coordination bonds remaining unchanged.

Polymerizations are carried out in block or aprotic solvents at 70 to 130 °C. Polymerization of complex-bonded heterocycles can be described as shown in Eq. 4-24.

Thus, complex-bonded monomerscan be polymerized using a variety of experimental approaches, both conventional (polymerization in solution, in bulk) and comparatively specific:

* in the solid phase,
* in electrochemical processes,
* at devitrification of an irradiated matrix,
* at a thermal front wave, etc.

4.3.5   Frontal Polymerization of Acrylamide Complexes

This method has become popular in recent years and is therefore discussed in detail. The process involves the conversion of a monomer into a polymer in a localized reaction zone that propagates in a layer mode of the bulk medium. Self-heating of the polymerizing system in a condensed phase under certain conditions can lead to a change of the bulk polymerization into an anisotropic self-propagating frontal mode. Such a process can be considered as an analogue of self-propagating high-temperature synthesis [80]. The heat evolved during polymerization of one monomer layer is consumed in the activation of an other layer. The stationary propagation of the thermal wave proceeds in a narrow temperature range close to the adiabatic heating of the reaction medium. Frontal polymerization is realized under a specific combination of thermo-physical characteristics of the polymerization medium; i.e. high exothermicity of the reaction and low coefficient of monomer–polymer heat conductivity. Up to now only one example of polymerization of MCMs in front of a thermal wave is known. The ability of acrylamide (AAm) complexes of transition metal nitrates such as those of Mn(II), Co(II), Ni(II) and Zn(II) in the composition $M(AAm)_4(NO_3)_2(H_2O)_2$ to polymerize in frontal conditions is their unique property [81–83]. It should be noted that frontal polymerization of these MCMs proceeds under the mildest possible conditions that are known for such processes: namely, at atmospheric pressure and thermal initiation in the absence of chemical initiators or activators. The reaction is accompanied by melting of the monomer with a corresponding change of its color from pink to dark-violet for Co(II), light-green to green for Ni(II), white to light-pink for Mn(II) and white to light-yellow for Zn(II). Studies of the effect produced by a complexing metal on the reaction behavior indicate that the rate decreases in the following series of metals: Co(II) > Ni(II) > Mn(II) > Zn(II) (see Experiment 4-4, Section 4.6).

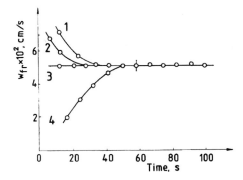

Figure 4-8. Front propagation velocity in the polymerization of Co(NO₃)₂·(AAm)₄·2H₂O versus 'starting' temperature and process duration, d = 4 mm, ρ = 1.02 g cm⁻³, T_{st} (K): (1) 573, (2) 523, (3) 453, (4) 393.

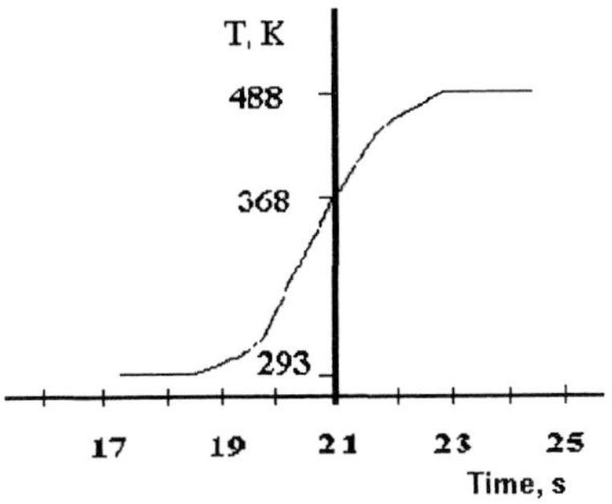

Figure 4-9. Temperature profile of front propagation in the polymerization of Co(NO₃)₂·(AAm)₄·2H₂O (T_{st} = 413 K, d = 1.2 cm, ρ = 1.38 g cm⁻³).

The kinetic curves have two zones corresponding to the non-steady-state and steady-state modes of front propagation. Depending on the starting temperature, T_{st}, the w_{fr} values in the pre-stationary state conditions may be either higher or lower than the stationary rate of front propagation (Fig. 4-8). The stationary rates of the process for the monomers studied apparently indicate an insignificant role for free convection in the polymerization medium. The initiation of the reaction was optimal at 413–493 K. Below this temperature range no polymerization wave appeared. Moreover, at higher temperatures a front of intense oxidation, with the formation of metal oxides or carbides, occurred. It was found that changing the temperature of initiation does not significantly affect the rate of frontal propagation or the maximum frontal temperature (Fig. 4-9, Table 4-4).

Table 4-4. The dependence of the reaction velocity (w) versus T_{max} of frontal polymerization of Co(NO₃)₂·4AAm·H₂O on initiation temperature (d = 1.2 cm).

T_{st}, K	$w \cdot 10^2$, cm s⁻¹	T_{max}, K
413	4.8	488
433	4.8	483
353	4.1	483
473	4.2	483

The frontal temperatures for Ni- and Cd-AAm complexes are 488 and 483 K, respectively, while the rates of propagation are different: 1.4·10⁻² and 7.0·10⁻² cm s⁻¹. It is also clear that the nature of the metal atom affects the reactivity of double bonds in the AAm ligand. For example, for the Fe(III)–

AAm complex the oxidation front occurs at 453 K with $w = 2.4 \cdot 10^{-2}$ cm s^{-1} and $T_{max} = 597$ K. By comparison, all experimental attempts to polymerize the Cu(II)–AAm complex in a frontal mode were unsuccessful. It appears that, under these conditions, there are preferred structural considerations for frontal polymerization; i.e. a significant distance between double bonds in the spatial lattice prevents self-sustaining frontal propagation. A low polymerization activity of the monomer may also be caused by redox processes owing to a fairly high standard electrode potential for the copper ions. The mechanism for the initiation of frontal polymerization is probably quite complicated. For example, the formation of NO_2 is possible as a result of the hydrolysis of nitrate complexes during hydration. It is more likely, however, that a process involving elimination of water molecules and formation of NO_2 occurs in a synchronized manner (Eq. 4-25, a and b); e.g. NO_2 occupies a vacant coordination site in the Co(II) coordination sphere as one H_2O molecule is removed. This is followed by changes in the coordination sphere of the six-coordinated metal, which reduces the energy consumption that is needed for the rupture of the double bond of an acrylamide group (Eq. 4-25, c). The chain growth process proceeds via a preliminary π-coordination of the double bond of the monomer with the Co atom (Eq. 4-25, c and d). This also leads to a six-coordination state for the metal. The sequence of successive transformations that involve neighboring structurally organized molecules of the metallo-monomer results in the formation of a metallopolymer chain (Eq. 4-25, e). The details of such a concerted mechanism and the state of water in the metallopolymer that is formed require further study.

There are several modes for the frontal polymerization of metal-containing monomers. One such mode is polymerization in a high-temperature burning regime. This method is followed by thermolysis of the products to obtain metal-containing composites that include nano-sized materials. The kinetic peculiarities of high-temperature pyrolysis of the Co(II) acrylamide complex have been studied [84]. The rate of the process is approximately satisfied by a first-order equation of autocatalysis (Eq. 4-26), where $k = 4.2 \cdot 10^7 \exp[-24,000/(RT)]$ (s^{-1}), $\xi_0 \approx 1.9 \cdot 10^{-2}$.

Study of the material's topography by electron microscopy demonstrated that this product is a single-phase material consisting of shapeless glassy particles with a uniform electron density. The magnetic properties of the resulting products appear only in the case of thermolysis at $T_{therm} > 680$ K. In other words, at relatively low temperatures of thermolysis, very small particles are formed which are stabilized by the polymer matrix. By contrast, at higher temperatures the particles become larger and the resulting domains exhibit magnetic properties.

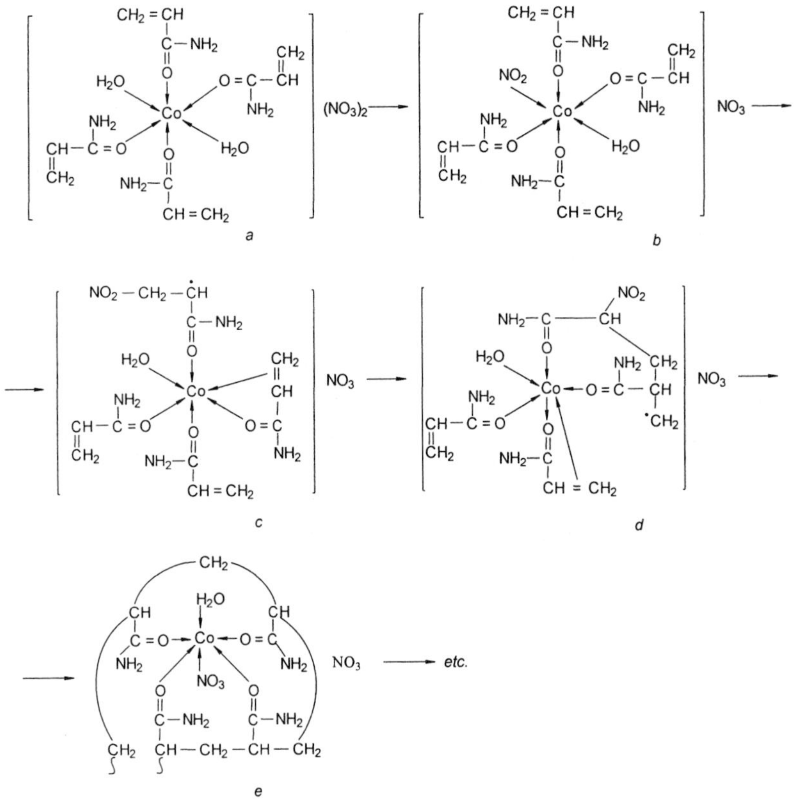

$$ (4\text{-}25) $$

$$ W = d\eta/dt = k(1 - \eta)(\eta + \xi_0) \qquad (4\text{-}26) $$

4.3.6   Homopolymerization of π-Type MCMs

Much progress has been achieved in the study of homopolymerization of π-type MCMs, and a number of such monomers are known. The first well-known representative is vinylferrocene, polymerizable under appropriate conditions by any mechanism including the Ziegler–Natta type. The choice of initiation method is determined, in the first place, by the type and structure of the π-MCM and then by the required properties of the polymer formed [4,8 and references therein]. Vinyl arene derivatives of Cr and Mo, for example, are decomposed by acids, so cationic initiation is not possible in those cases. Vinyl cyclopentadienyl MCMs, namely vinyl ferrocene, vinylcymantrene, vinyl-cynichrodene, etc., and MCMs with $Fe(CO)_3$-groups are incapable of anionic polymerization. Similarly, radical polymerization of η^5-(vinylcyclopentadienyl)-dicarbonylcobalt proved to be inefficient. This is because stable radicals form when the initiating radical becomes attached to the vinyl group of the MCM

(Eq. 4-27).

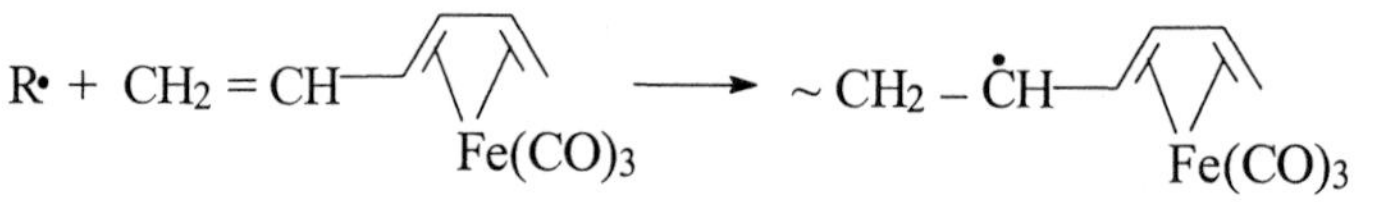

$$R\bullet + CH_2 = CH\!\!-\!\!\underset{Fe(CO)_3}{\wedge} \longrightarrow \ \sim CH_2 - \overset{\bullet}{C}H\!\!-\!\!\underset{Fe(CO)_3}{\wedge}$$

(4-27)

Other reasons are also possible. For example, X-ray analysis of styryltricarbonylchromium shows that coordination occurs between the vinyl group and a benzene ring from a neighboring molecule of MCM [85]. Such a location of the double bond and the formation of dimeric structures can prevent the polymerization. The susceptibility of π-MCMs to a particular type of initiation may be changed by introducing some substitutents into the vinyl group or bridging groups between the cyclopentadienyl (or arene) groups and multiple bonds. Examples are known of the polymerization of such monomers in the presence of Ziegler–Natta catalysts, apparently by the anionic-coordination mechanism. Soluble polyconjugated polymers ($M = 300{,}000$) have been obtained by a ring-opening metathesis polymerization of ferrocenophane (1,1′-(1-*tert*-butyl)-1,3-butadienylene)ferrocene by using of a tungsten-based metathesis catalyst (Eq. 4-28) [86] (see Experiments 4-5 and 4-6, Section 4.6 for polymerization of π-type MCMs).

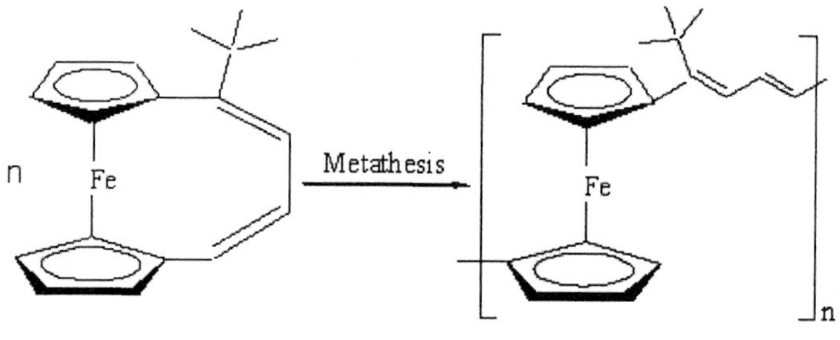

(4-28)

95   96

The increase in the variety of possible initiating systems is probably determined by the higher stability of the metal bond of π-monomers in comparison with σ- or *nv*-bonded MCMs. However, in spite of abundant experimental data on the homopolymerization of π-MCMs there is no data concerning the rate constants of elementary stages, in particular, k_p. The ratios of $k_p/k_t^{1/2}$ have only been estimated for radical initiation. Deviations from the kinetic scheme of polymerization are often observed, for example, when an MCM molecule participates in initiation, ($W = k[M]^{3/2}[I]^{1/2}$) or if there are chain termination peculiarities ($W = k[M]^{1.12}[I]^{1.11}$) such as the intramolecular transfer of an electron from a Fe atom to the terminal radical. The latter mechanism is particularly characteristic of the polymerization of vinylferrocene.

Intramolecular chain termination occurs as shown in Eq. 4-29.

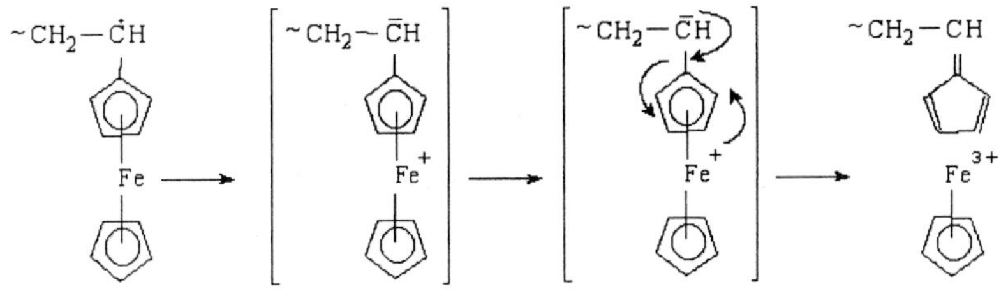

$$(4\text{-}29)$$

The ferrocene cation which is formed by the transfer of an electron from the iron atom undergoes a regrouping to give a high-spin $3d^5$ Fe(III) complex. This process is facilitated by the conjugated system of ferrocene. Thus, we can note some analogy to the mechanism of chain termination in the polymerization of MCMs of the π-, σ- and ionic types. In addition, higher values are observed for the constant of chain transfer to the monomer ($k_m/k_p = 8 \cdot 10^{-3}$ at 60 °C, whereas for styrene $k_m/k_p = 6 \cdot 10^{-5}$), and chain transfer to the polymer plays a greater role than with other monomers (Eq. 4-30):

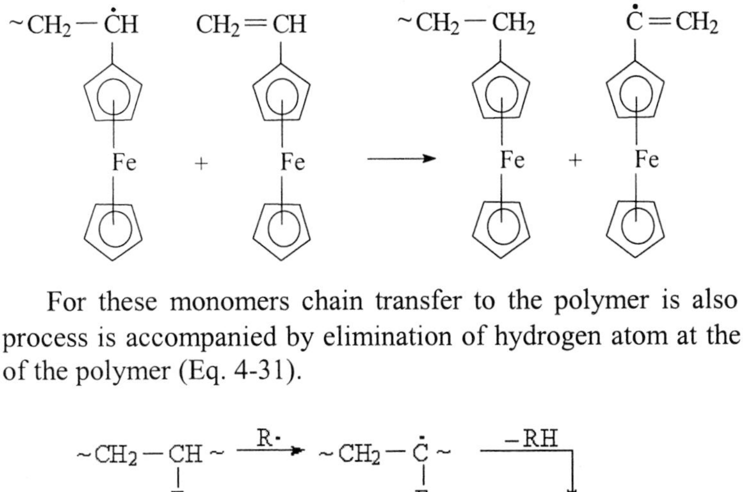

$$(4\text{-}30)$$

For these monomers chain transfer to the polymer is also observed. This process is accompanied by elimination of hydrogen atom at the α-carbon atom of the polymer (Eq. 4-31).

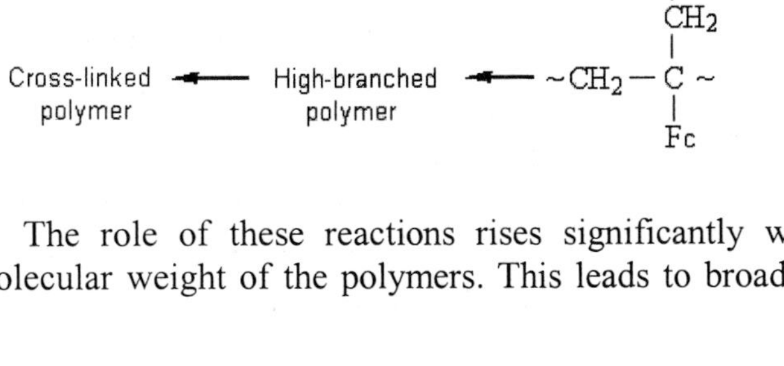

$$(4\text{-}31)$$

The role of these reactions rises significantly with an increase of the molecular weight of the polymers. This leads to broadening of the molecular-weight distribution of the polymers, and in some cases bi- and trimodal distributions can be observed. It should be noted that, as a rule, investigations of MCM polymerization are limited to polymerization in solutions. Polymerization in bulk is rarely studied and data on their solid-phase polymerization is scarce. Among π-MCM polymerization conversions the majority of studies deal with Fe-containing monomers while monomers of other metals are less studied. However, note that it has been possible to polymerize styryltricarbonyl-chromium and its analogues (n-methylstyryl- and α-methylstyryltricarbonyl-chromium) in the presence of a complexing agent [87]. Carrying out the process in a polar solvent such as ethylacetate, which is capable of forming a complex with an arenetricarbonylchromium, or introducing complexing additives (such as triisobutylboron) gives the polymers in yields as high as 60% and molecular weights of several thousand.

4.3.7   Polymerization of Chelate-Type MCMs

Information concerning the polymerization conversions of metallochelate monomers has appeared in the last few years. Radical polymerization has been applied to the production of metal poly(p-acryloxyphenyltriphenylporphyrinates [4 and references therein]. Also, Cu(II), Ni(II), Cd (II) and UO$_2$(II) have been polymerized with 5-vinylsalicylidene aniline by boiling in dimethylolpropionic acid in the presence of AIBN [88]. In the case of the nitrate or acetate of UO$_2^{2+}$ the polymer **97** were obtained:

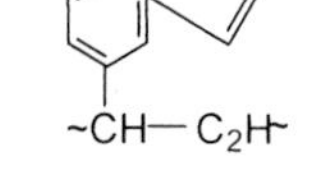

97

More detailed consideration has been give to the electrochemical reductive polymerization of the perchlorate and mixed-ligand hexafluorophosphate chelate monomers of Ru(II), Os(II) and Fe(II) that contain chelating fragments of the Dipy, Phen and 2,2′,2″-tripyridyl types and also a monofunctional monomer (namely, 4-VPy and trans-4-stilbazol) [89]. The polymerization was performed by repeated cyclic changes in the electrode potential around a value corresponding to ligand reduction. As this takes place, the electrode surface gets covered with a uniform polymeric film. A similar process occurs when polymerizing monomers in which the vinyl group is bonded to the chelating fragment, e.g. Zn(II), Ru(II) and Ir(III) complexes with 4-vinyl-4'-methyl-2,2'-dipyridyl [90].

Careful design of the radical polymerization conditions of monovinyl-porphyrins led to homopolymers with high molecular weight. Homopolymers of the following porphyrin derivatives have been prepared and metallation could be carried out afterwards without problems: 5-(4-acryloxyphenyl)-, 5-(4-methacryloxyphenyl)-, 5-(4-vinylphenyl)-, 5-(4-acrylamidophenyl)- and 5-(4-methacrylamidophenyl)-10,15,20-triphenylporphyrin, etc. [91–96]. Examples are shown in the structures **74–79**. Also homopolymers containing paramagnetic metal ions (Cu(II), Ag(II), VO(IV), Co(II)) have been prepared, and their magnetic properties investigated by ESR and magnetic susceptibility measurements [97,98]. It was found, for example, that Ag-porphyrin moieties in the polymer chain interact with oxygen at low temperature to form reversible ESR-active species [99].

Liquid-crystalline phthalocyanines have been prepared as asymmetric monomers. They contain one methoxy, six dodecyloxy and one acryloyloxy or methacryloyloxydodecyloxy groups [100]. The last-mentioned compounds have been polymerized under mild conditions. Highly crystalline monolayers have been prepared by the Langmuir–Blodgett technique [101]. Phthalocyanines bearing eight or two terminate olefin groups have been polymerized in dichloromethane by olefin metathesis using a ruthenium initiator [102]. The cross-linked insoluble polymers prepared from **98** do not contain a regular ordered structure. In contrast, the linear polymers prepared from **99** show an ordered rodlike nanostructure.

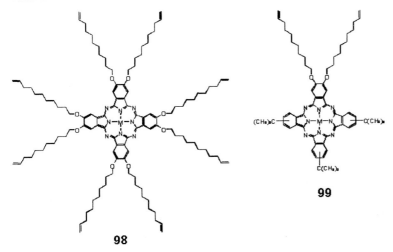

98

99

The homopolymerization of porphyrins and phthalocyanines substituted by four methacryloyloxy or 2,4-hexadienoyloxy groups, such as the zinc complexes of 2,9,16,23-tetrakis(4-methacryloyloxy)- and 2,9,16,23-tetrakis(4-(11-methacryloyloxyundecylox)-phenoxy)phthalocyanines (**100**, **101**) has been studied (see Experiment 4-7, Section 4.6) [103].

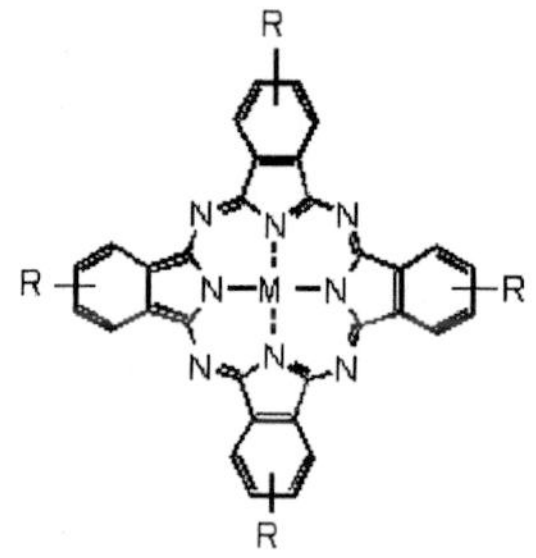

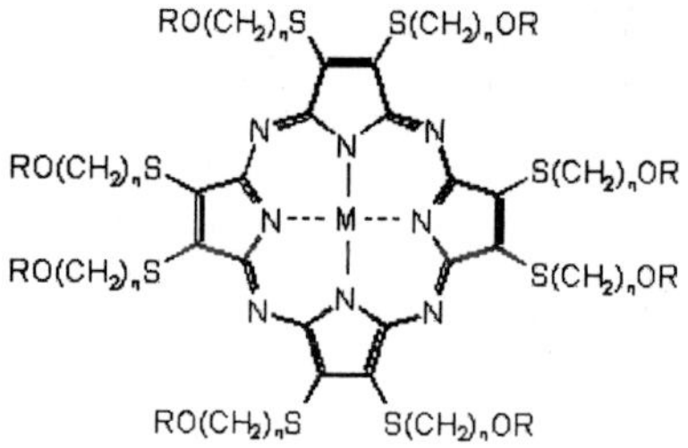

100 R = ─O─⟨⟩─O-C-C=CH$_2$
              CH$_3$

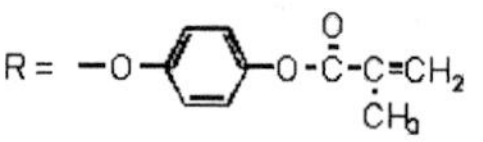

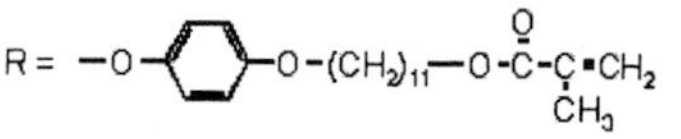

101 R = ─O─⟨⟩─O-(CH$_2$)$_{11}$─O-C-C=CH$_2$
              CH$_3$

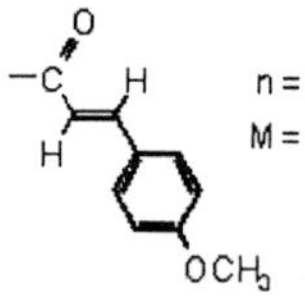

102 R = ─C   H   n = 3,6,8,11
                 H   M = 2H, Mg, Zn, Cu, Ni
            OCH$_3$

Photopolymerization of drop-coated films on KBr plates has also been investigated. 2,3,7,8,12,13,17,18-Octakis[(4-methoxycinnamoyl)oxyalkylthio]-tetraazaporphyrins (**102**) with different methylene chain lengths were reacted through [2+2]-photocycloaddition by UV irradiation in solution and as spin-coated films in the solid state [104–106]. The spin-coated films are optically transparent, do not scatter light and are not birefringent, suggesting that they are amorphous or microcrystalline.

Surprisingly, the electrooxidation of metal complexes of protoporphyrin-IX dimethyl ester, possibly via the vinyl groups, leads to the deposition of electroactive porphyrin films on the electrode surface [107–109]. The electrochemical polymerization of pyrrole, thiophene and amine metal-substituted complexes is described in more detail in Section 6.3.

4.4 MCM Copolymerization

Compared to homopolymerization the copolymerization of MCMs has been investigated in more detail, as it is often used to modify the properties of traditional polymers. Equally, joint polymerization gives additional opportunities to study the statistical processes and factors that influence the reactivity of the multiple bonds of monomers, as well as to expose latent effects intrinsic to MCMs.

The canonical equation for copolymer composition is shown in Eq. 4-32, where $r_2 = k_{22}/k_{21}$ is the copolymerization constant characterizing the relative activity of the MCM towards "own" or "strange" radicals, m_2 and M_2 are its content in the polymer and monomer mixture, $r_1 = k_{11}/k_{12}$, m_1 and m_2 are those related to a "free metal" comonomer.

$$\frac{[m_1]}{[m_2]} = \frac{[M_1]}{[M_2]} \frac{r_1[M_1] + [M_2]}{[M_1] + r_2[M_2]} \tag{4-32}$$

As we know, the ability to copolymerize is enhanced by an increase in the difference of resonance stabilization between the adding monomer and the forming radical (Eq. 4-33).

$$\sim CH_2 - \overset{\bullet}{C}H + CH_2 = CH \longrightarrow \sim CH_2 - CH - CH_2 - \overset{\bullet}{C}H \tag{4-33}$$
$$\quad\quad\quad | \quad\quad\quad\quad | \quad\quad\quad\quad\quad\quad\quad | \quad\quad\quad\quad | $$
$$\quad\quad\quad X \quad\quad\quad\quad Y \quad\quad\quad\quad\quad\quad\quad X \quad\quad\quad Y$$

In the $Q - e$ scheme, Q characterizes the resonance stabilization of a monomer during copolymerization and e is the factor reflecting the degree of the polar effect of substituents at a multiple bond. These parameters are associated with the constants of relative activity of monomers by the known expressions (Eq. 4-34).

$$r_1 = (Q_1/Q_2)\exp[-e_1(e_1 - e_2)]; \; r_2 = (Q_2/Q_1)\exp[-e_2(e_2 - e_1)] \tag{4-34}$$

As MCMs have two reactive groups taking part in polymerization the "double" mole concentrations of the MCM should be considered. Then the copolymerization equation will be as shown in Eq. 4-35.

$$\frac{[m_1]}{[m_2]} = \frac{M_1}{2[M_2]} \left(\frac{r_1[M_1] + 2[M_2]}{[M_1] + 2r_2[M_2]} \right) \tag{4-35}$$

The copolymerization constants are estimated by the Mayo–Lewis method using the different means of linearization and optimization (most often the Fineman–Ross and Kelen–Tudos methods). We shall now consider the principal problems in the field of copolymerization of MCMs under the same conditions as for their homopolymerization.

4.4.1 Organometallic Monomers in Copolymerization Reactions

Among organometallic MCMs σ-bonded derivatives of *trans*-$Pd(PBu_3)_2(C_6H_4CH=CH_2)X$ are of interest [63]. Experimental data on the copolymerization of the MCM on the base of Pd^{2+} (X = Cl) with styrene are given in Table 4-5.

Table 4-5. Copolymerization of *trans*-Pd(PBu$_3$)$_2$(C$_6$H$_4$CH=CH$_2$)Cl with styrene.[a]

M$_1$ (mmole)	M$_2$ (mmole)	AIBN (mmole)	Yield of copolymer (%)	m$_2$ of copolymer (mol-%)
0.8	3.62	0.2	7.2	71.54
1.4	3.1	0.22	6.5	52.25
1.6	2.51	0.17	21.0	46.60
2.2	2.17	0.15	6.4	36.57
2.87	1.61	0.22	9.7	27.22
6.7	1.13	0.22	11.2	21.61
7.7	1.45	0.25	6.5	11.11

[a]The conditions of the reaction: 55 °C, 3 h, benzene, 1 mL (in two last experiments 2 mL).

The values of the copolymerization constants (r_1= 1,49 ±0,01, r_2 =0,45 ±0,002) indicate that the MCM is a less-active monomer in copolymerization than styrene. Note the large negative *e* value of −1.4. For common vinyl monomers polymerizing by a radical mechanism *e* is equal to −1 to + 1.5. Since this parameter characterizes a polarity measure of the substituent at the double bond the correlation of *e* and its Hammet's constant (σ_x) [110] for the monomers CH$_2$=CHX is observed (Fig. 4-10).

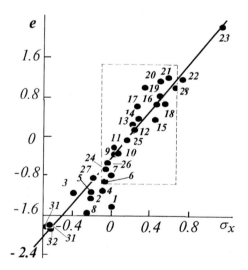

Figure 4-10. Correlation of the polarity parameter (*e*) of the X substituent in CH$_2$=CHX and its Hammet's constant (σ_x), where X equal 1 - S-CH$_3$; 2 - p-C$_6$H$_4$N(CH$_3$)$_2$; 3 - -(CH$_3$)$_2$; 4 - p-C$_6$H$_4$0CH$_3$; 5 - CH$_3$, C$_6$H$_5$; 6 - p-C$_6$H$_4$CH$_3$; 7 - C$_6$H$_5$; 8 - OC$_2$H$_5$; 9 - p-C$_6$H$_4$Cl; 10 - p-C$_6$H$_4$I; 11 - p-C$_6$H$_4$Br; 12 - Cl; 13 - p-C$_6$H$_4$CN; 14 - C$_6$H$_4$NO$_2$; 15 - CH$_3$, COOCH$_3$; 16 - Cl, Cl; 17 - CH$_2$Cl; 18 - COOCH$_3$; 19 - CO(CH$_3$), 20 - CH$_3$, CN; 21 - CN, 22 - SO(CH$_3$); 23 - CN, CN; 24 - Co(OCOCH=CH$_2$)OCO; 25 - Cu(OCOCH=CH$_2$)OCO; 26 - (C$_4$H$_9$)$_3$SnOCO; 27 - (C$_4$H$_9$)$_3$SnOCO, CH$_3$; 28 - (CH$_3$)$_3$Sn; 29 - C$_5$H$_3$N(CH$_3$)-ZnCl$_2$; 30 - (η^5-C$_5$H$_4$)Fe(η^5-C$_5$H$_5$); 31 - (η^5-C$_5$H$_4$)Cr(CO)$_2$NO; 32 - (η^5-C$_5$H$_4$)Mn(CO)$_3$.

From the data of this correlation analysis, we may suppose that the σ-bonded $Pd(PBu_3)_2$ group possesses strong electron-donor properties. The value of the parameter e close to $n\text{-}N(CH_3)_2$ or $n\text{-}OCH_3$ substituents of the aromatic derivatives, i.e. the electron density in the MCM, is heavily shifted to the C=C bond. On the other hand, the higher the parameter e the more reactive the monomer in anionic polymerization (in cationic polymerization the reverse is true). However, attempts at cationic copolymerization of these monomers (etherate BF_3 catalyst) were not successful, as mentioned above, because of the interaction of the catalyst with the Pd of the MCM. Vinyl and allyl Sn(IV) and Pb(IV) derivatives incapable of homopolymerization under ordinary conditions also produce an inhibiting effect on the radical polymerization of other monomers such as styrene, methyl methacrylate and vinyl acetate [4]. This can be attributed to the fact that the free radicals appearing in the system are trapped and bonded to the metal. As this takes place, the inhibiting effect increases with the number of unsaturated groups in the MCM molecule, the allyl compounds being more efficient inhibitors than the vinyl ones. In terms of the inhibiting action, unsaturated Sn(IV) compounds are arranged as follows: $Sn(CH_2CH=CH_2)_4 > Sn(CH=CH_2)_4 > (CH_3)_3Sn(CH_2CH=CH_2) > (C_6H_5)_3Sn(CH_2CH=CH_2) > (C_6H_5)_3Sn(CH=CH_2)$. Copolymerization of such MCMs is distinguished by close to zero r_2 values (Table 4-6).

The process occurs in low yields, and the final products normally contain a relatively small amount of MCM. Styrene Sn(IV) and Pb(IV) derivatives are much more active in these processes. As shown in Table 4-6, the activity of the Pb(IV)-containing monomer is close to that of the styrene- but lower than that of the Sn(IV)-based monomer. In terms of activity in copolymerization with styrene, the organotin monomers decrease as follows: $(C_6H_5)_3SnC_6H_4CH=CH_2 > (C_6H_5)_3SnC_{12}H_4CH=CH_2 > (C_6H_{11})_3SnC_6H_4CH=CH_2$. It is evident that an increase in the size of a substituent and a weaker double bond conjugation lead to a lower activity of these MCMs. Some interesting features are typical of the copolymerization of another organotin monomer, 1-(trimethyl)methyl methacrylate, with MMA [62].

The process is distinguished by the fact that the composition of the copolymer is close to that of the parent monomeric mixture ($r_1 = r_2 = 1.0$). In other words, in this case one can observe an "azeotropic" copolymerization. Thus, the substitution of an organometallic group by a hydrogen atom does not affect the double bond reactivity, which is probably due to the presence of a three-atom bridge.

Table 4-6. Relative reactivity constants and parameters Q-e of copolymerizable systems based on Sn(IV)- and Pb(IV)-containing covalent-type MCM [4].

MCM (M_2)	Comonomer Parameter	Styrene	MMA
$(CH_3)_3Sn(CH=CH_2)$	r_1	44.8	25.1
	r_2	0.001	0.03
	Q_2	0.005	0.036
	e_2	0.96	0.93
$(C_4H_9)_3Sn(CH=CH_2)$	r_1	16.0	27.9
	r_2	0.005	0.03
	Q_2	0.017	0.03
	e_2	0.82	0.82
$(CH_3)_3SnCH_2OCOC(CH_3)=CH_2$	r_1	-	1.0
	r_2	-	1.0
 $(C_6H_5)_3Sn-\langle\bigcirc\rangle-CH=CH_2$	r_1	0.826	-
	r_2	2.86	-
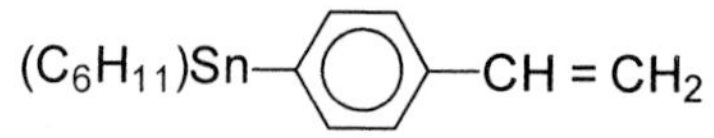 $(C_6H_{11})Sn-\langle\bigcirc\rangle-CH=CH_2$	r_1	0.96	-
	r_2	1.6	-
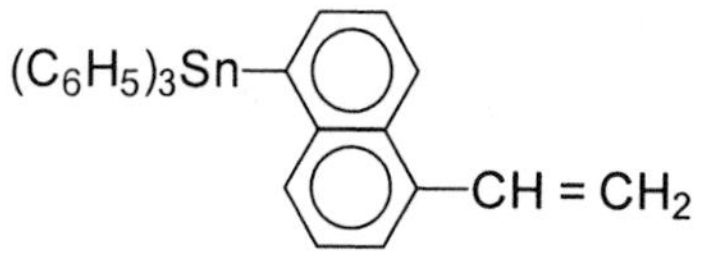 $(C_6H_5)_3Sn-\langle\text{naphthyl}\rangle-CH=CH_2$	r_1	0.897	-
	r_2	4.7	-
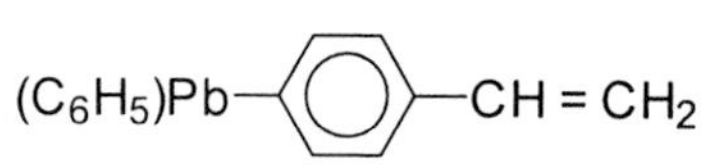 $(C_6H_5)Pb-\langle\bigcirc\rangle-CH=CH_2$	r_1	0.978	-
	r_2	1.22	

4.4.2   Copolymerization of Ionic-Type MCMs

Copolymerization of ionic-type MCMs, alkali and alkaline earth metal salts, with various comonomers has been studied quite well. A detailed review of the basic studies has been given in a monograph [4] and they will not be analyzed in this book. Studies of the mechanism of copolymerization of such types of MCM were carried out in the case of organotin and -lead monomers. The widest recognition among these MCMs used as comonomers has been gained by trimethyl-, tributyl-, and triphenyltin methacrylates (Table 4-7) [4 and references therein].

Thus, trimethyltin acrylate readily copolymerizes with styrene [111]. A negative value of e (-0.5 to -0.88) for this MCM is due to the ionic nature of the Sn–O bond and COO$^-$ group specify as compared with COOH or COOR.

An increase in the parameter Q (Q = 0.23 to 0.78) of Sn-organic monomers in comparison with the corresponding alkylmethacrylates testifies to the influence of electron-acceptor $-SnR_3$ groups (possessing a strong positive inductive effect) on the conjugation of multiple bonds and carbonyl groups. Dimethyltin dimethacrylate with styrene affords a cross-linked copolymer. Under identical conditions to tributyltin methacrylate (TBTM) copolymerization, the yield of the products depended on the type of comonomers used and was 90% (MMA), 40% (AN), 30% (methacrylonitrile) and 25% (styrene) [111]. The main features of TBTM copolymerization with alkyl methacrylates reside in the fact that these comonomers are randomly distributed in the chain and the tendency to alternate increases with alkyl chain length in the case of both acrylates and meth-acrylates. The same is true of the copolymerization of other tributyltinalkyl acrylates. The rate of this reaction is higher than that in the case of "metal-free" analogues. Moreover, an increase in the bulk of the side groups and formation of coordinate bonds between the monomer units appears to hinder macroradical interaction and to create favorable conditions for chain propagation.

The contribution of complex formation to MCM copolymerization is more substantial than it is to homopolymerization. The coordinational unsaturation of central metal atoms, for example the pentacoordination state of Sn(IV), plays a definite part. The transfer of an electron from an MCM (electron-donor monomer) to a multiple bond of the comonomer is comparatively easily carried out in the transition state [112] as shown in Eq. 4-36, for example for maleic aldehyde the complex formation constant K = 0.17 ± 0.002 L mol^{-1}.

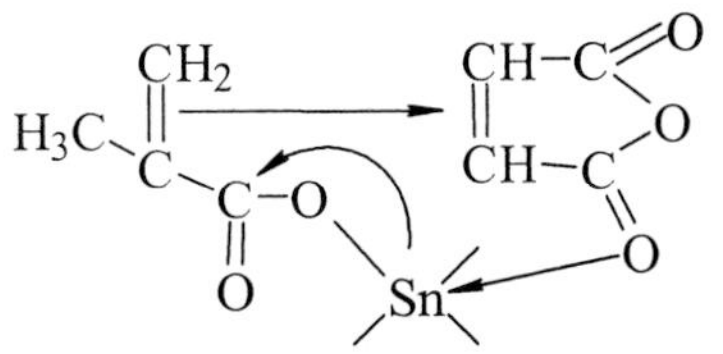

$$(4\text{-}36)$$

The redistribution of electron density on all ligand bonds, including the vinyl group, proceeds during this process. Chain propagation can take place through three mechanisms, namely, consecutive addition of free monomers to macroradicals, addition of complex-bonded monomer pairs and simultaneous participation of the free and complex-bound monomers. As a rule, the reactivity of complex-bound monomers is higher than that of free ones. They are observed to make a major contribution to the reaction with growing radicals, irrespective of the nature of terminal unit.

Table 4-7. Copolymerization parameters of Sn(IV)-containing ionic-type monomers.

MCM (M₂)/Q-e parameters	Parameter	Styrene	MMA	Vinyl-acetate	Methyl-acrylate	BA	BMA	AN	VPr	MMA	Allyl-methacrylate	Glycydyl-methacrylate
$(C_4H_9)_3SnOCOCH=CH_2$ $Q = 0.235$, $e = 0.401$	r_1	0.21	2.2	0.06	0.82	0.70	1.65	1.01	0.610	–	0.20	4.29
$(CH_3)_3SnOCOC(CH_3)=CH_2$ $Q = 0.45$, $e = -0.37$ $Q = 0.31$, $e = -0.64$	r_2	1.91	0.4	2.57	0.03	0.15	0.20	0.24	0.513	–	2.26	0.34
	r_1	1.57	–	–	–	–	–	–	–	0	–	–
$(CH_3)_2Sn[OCOC(CH_3)=CH_2]_2$ $Q = 1.36$, $e = 0.41$	r_2	0.53	–	–	–	–	–	–	–	0.22	–	–
	r_1	0.28	–	–	–	–	–	–	–	–	–	–
$(C_4H_9)_2SnOCOC(CH_3)=CH_2$ $Q = 0.852$, $e = 0.197$ $Q = 0.18$, $e = -0.88$ $Q = 0.78$, $e = -0.38$	r_1	0.83	–	–	–	–	–	–	–	–	–	–
		1.14	1.04	0.017	0.66	0.57	0.68	0.47	0.44	0	1.01	0.79
		0.51	1.0	0.028	–	–	1.65	–	–	–	–	–
		1.10	–	0.008	–	–	–	–	–	–	–	–
	r_2	0.40	0.93	4.408	1.75	0.85	0.65	0.47	3.16	0.05	2.30	0.75
		0.49	0.79	0.76	–	–	0.20	–	–	–	–	–
		0.26	–	0.61	–	–	–	–	–	–	–	–
$(C_6H_5)_3SnOCOC(CH_3)=CH_2$	r_1	0.47	1.09	–	0.63	0.35	0.78	0.16	0.36	–	–	–
	r_2	0.76	1.08	–	2.48	1.17	0.64	0.69	1.22	–	–	–
$(C_4H_9)_3SnOCOC(=CH_2)CH_2$ $COOSn(C_4H_9)_3$	r_1	0.68	1.64	–	0.67	0.96	–	0.43	–	–	–	–
	r_2	0.23	0.22	–	0.39	0.52	–	0.44	–	–	–	–

Table 4-7. Copolymerization parameters of Sn(IV)-containing ionic-type monomers.

MCM (M$_2$)/Q-e parameters	Parameter	Comonomer										
		Sty-rene	MMA	Vinyl-acetate	Methyl-acrylate	BA	BMA	AN	VPr	MMA	Allyl-metha-crylate	Glycyd-ylmetha-crylate
CH$_2$=CHC(O)OC$_6$H$_4$COOSn (C$_4$H$_9$)$_3$ Q = 0.045, e = 1.39	r_1	1.34	1.72	–	1.05	2.11	2.83	–	–	–	–	–
	r_2	0.12	0.16	–	0.10	0.02	0.01	–	–	–	–	–
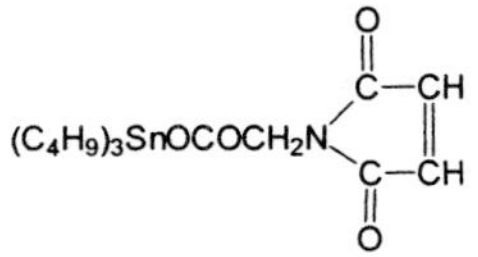 (C$_4$H$_9$)$_3$SnOCOCH$_2$N<ring>	r_1	1.65	–	–	–	–	–	–	–	0.02	–	–
	r_2	0.004	–	–	–	–	–	–	–	0.16	–	–
(C$_4$H$_9$)$_3$SnOCOCH=CHC(O) OCH$_2$CH=CH$_2$ Q = 1.11, e = 1.67	r_1	0.12	–	–	–	–	–	–	–	–	–	–
	r_2	0.018	–	–	–	–	–	–	–	–	–	–
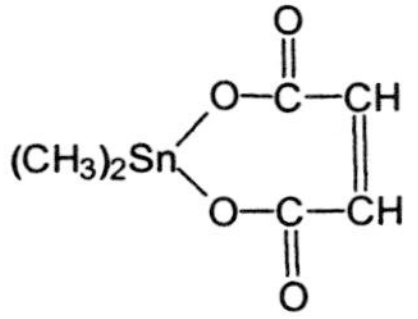 (CH$_3$)$_2$Sn<ring> Q = 0.24, e = 0.08	r_1	–	–	–	2.08	–	–	–	–	–	–	–
	r_2	–	–	–	0.22	–	–	–	–	–	–	–

The rate of the radical copolymerization of styrene or acrylonitrile with transition metal acrylates increases with an increase of MCM content in the monomer mixture with the exception of $Cu(OCOCH=CH_2)_2$, which distinctly inhibits the process even at 2 mol%. The negative values of e (-0.22 ± -0.85) (Table 4-8) point to an electron-donor character of the metallocarboxyl group in the MCM. The drop in this value is correlated with an increase of the metal's electronegativity.

Table 4-8. Copolymerization parameters of styrene and acrylonitrile with transition metal acrylates [113–115].

M_1	Acrylate (M_2)	r_1	r_2	$r_1 \cdot r_2$	$1/r_1$	Q_1	e_2
Styrene	Co(II)	1.74	0.56	0.97	0.58	0.51	−0.64
Styrene	Ni(II)	1.83	0.53	0.97	0.55	0.48	−0.63
Styrene	Cu(II)	5.94	0.12	0.71	0.17	0.11	−0.22
Styrene	Zn(II)	1.10	0.90	0.99	0.91	0.84	−0.70
AN	Co(II)	0.14	0.16	0.022	7.1	0.42	−0.75
AN	Ni(II)	0.10	0.17	0.017	10	0.59	−0.85
AN	Cu(II)	0.21	0.08	0.017	4.8	0.26	−0.82
AN	Zn(II)	0.41	0.24	0.106	2.44	0.24	−0.30

The copolymerization of diacrylates can involve some unsaturated bonds being retained in the product. For such MCMs copolymerization can occur by two different mechanisms. The propagation chain interacts with one of the acrylate residues only, followed by comonomer addition; at a certain stage the process includes the second acrylate residue to give rise to a cross-linked chain (Eq. 4-37).

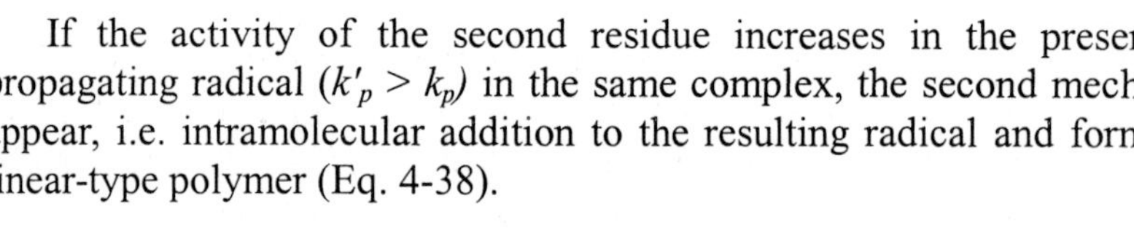

$$- M_1^{\cdot} + C=C \underset{M^{n+}}{\overset{}{\underbrace{}}} C=C \xrightarrow[M_1]{k_p} - M_1 - C - C^{\cdot} \underset{M^{n+}}{\overset{}{\underbrace{}}} C=C \xrightarrow[M_1]{k_p'}$$

$$(4\text{-}37)$$

$$- M_1 - C - C \underset{M^{n+}}{\overset{}{\underbrace{}}} {}^{\cdot}C - C - M_1$$

If the activity of the second residue increases in the presence of the propagating radical ($k'_p > k_p$) in the same complex, the second mechanism will appear, i.e. intramolecular addition to the resulting radical and formation of a linear-type polymer (Eq. 4-38).

$$- M_1 - C - C \cdot \quad \underset{M^{n+}}{\bigcap} \quad C = C \quad \xrightarrow[M_1]{k_p''} \quad - M_1 - C - C \quad \underset{M^{n+}}{\bigcap} C - C \cdot \quad \xrightarrow[M_1]{k_p}$$

(4-38)

$$- M_1 - C - C \quad \underset{M^{n+}}{\bigcap} C - C - M_1^-$$

The contribution of each of these mechanisms seems to depend on a number of factors and can only be assessed after studying the structures of the resulting polymers. Special studies indicate that the products appearing in the initial copolymerization stages are soluble in organic solvents [113]. Consequently, in this case no intermolecular cross-linking due to the vinyl groups takes place, and the majority of these groups interact with the two acrylate groups. However, at higher conversions unsaturation becomes more pronounced. Thus, the number of unreacted double bonds in the copolymers grows in the acrylates of these metals as follows: Zn(II) (4 to 35%) < Co(II) (14 to 39%) < Ni(II) (22 to 49%) (unsaturation is 50% if no more than one acrylate group reacts).

The distribution of MCM units in polymeric chain is an important characteristic of the copolymerization products, and defines many of their properties. In such systems the tendency to alternation is often pronounced. For example, in copolymers of styrene and nickel acrylate 46 mol% of the total number of acrylate units belong to strictly alternate units [116].

4.4.3 Peculiarities of Copolymerization of Donor–Acceptor Type MCMs

Two different possibilities can arise during copolymerization of nv-type MCMs, which stem from the ability of a metal ion to take part in additional coordination (extracoordination) with the comonomer. The first case is typical of co-monomers with properties approximating to those of the MCM-contained ligands. In principle, this case may involve situations favorable for a complex-radical polymerization. The limiting case of such a process is copolymerization of the MX_n complex with an excess ligand. However, in most studies involving this approach, the complexes taking part in the process were not identified.

Although the second case seems to be simpler, there is a certain probability of an additional polarization of the complexed comonomer, with participation of the π-electron cloud and corresponding strengthening of the double bond's electron-acceptor properties. The formation of such triple complexes results in the appearance of alternating copolymers.

The majority of data concerning copolymerization of nv-type MCMs with conventional monomers relate to complexes incorporating a nitrogen-containing ligand. Thus, in the copolymerization of Zn(II) 2-methyl-5-vinylpyridine (MVPy) complexes with styrene [79] we can observe a significant strengthening of the polar factor e as compared with an uncoordinated MVPy (Table 4-9) (see Experiment 4-3, Section 4.6).

Table 4-9. Copolymerization of MVPy complexes (M$_2$) with styrene.[a]

M$_2$	r_1	r_2	Q_2	e_2
MVPy	0.72	1.20	1.03	-0.42
Zn(MVPy)$_2$Cl$_2$	0.93	0.08	0.30	+0.81
Zn(MVPy)$_2$Br$_2$	0.33	0.38	0.97	+0.63
Zn(MVPy)$_2$I$_2$	0.60	0.55	0.70	+0.21

[a]DMFA, 60 °C, [AIBN] = $2.0 \cdot 10^{-3}$ mol L^{-1}, [M$_2$] = 0.98 mol L^{-1}.

This effect, as in the case of homopolymerization, appears to be due to the electron-acceptor nature of M^{n+}, the coordination with which involves an electronic density distribution at the ligand bonds including those of the vinyl group. In this case, a remarkable role is played by the nature of the anion, which decreases the electron affinity of the metal ion in a different way in that the e values are more positive as I$^-$ is replaced by Cl$^-$ (Fig. 4-11).

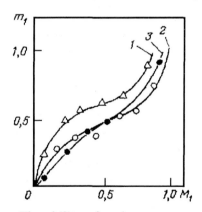

Figure 4-11. Diagram of the composition in the copolymerization of the complex of MVPy with styrene at 60 °C (DMF; [AIBN] = $2.0 \cdot 10^{-3}$ mol L^{-1}; [MCM] = 0.98 mol L^{-1}): 1 – Zn(MVPy)$_2$Cl$_2$; 2 – Zn(MVPy)$_2$Br$_2$; 3 – Zn(MVPy)$_2$I$_2$.

The ability of such monomers as 4-vinylpyridine (VPy), acrylonitrile (AN) and methylmethacrylate (MMA) to coordinate can be used in the anionic-coordination copolymerization with ethylene using a catalytic system of MX$_n$·L –Al(C$_2$H$_5$)$_2$Cl. The complexes of transition metal halides such as V(AN)$_n$Cl$_4$, V(VPy)$_2$Cl$_4$, V(MMA)Cl$_4$, Ti(VPy)$_2$Cl$_4$ (n = 2, 3) are both comonomers and catalytic components [117,118]).

4-Vinylpyridine is copolymerized with the Ru(II) complex of 4-vinyl-4′-methyl-2,2′-bipyridine by a radical mechanism to polymers **103** (Eq. 4-39) [119].

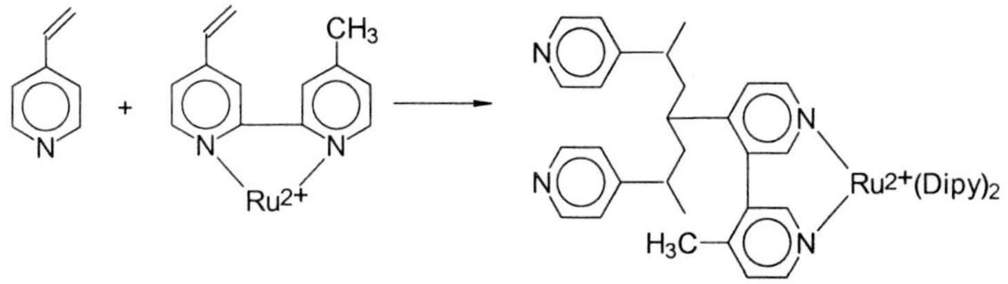

(4-39)

103

Among the copolymerizations of vinylazole complexes, a successful process with $ZnCl_2$–1-vinylbenzimidazole (VBI) or 1-vinylbenztriazole (VBT) adducts (styrene as the comonomer) (Table 4-10) may be mentioned. It is evident that the activity of the VBI and VBT complexes in the interaction with a styrene macroradical ($1/r_1$) is lower than that for an uncoordinated ligand.

Among other vinylazole-based MCMs it would be interesting to investigate $M(VIA)_4(OCOCH_3)_2$, where M = Co(II), Ni(II) or Zn(II). Ternary co-polymerizations of these complexes with N-vinylpyrollidone and divinyl-benzene have been conducted in an aqueous buffer solution under radiation initiation [121].

Table 4-10. Relative reactivity constants of vinylazoles and their $ZnCl_2$ complexes (M_2) in the copolymerization with styrene [120].

M_2	r_1	r_2	$1/r_1$
VBT	2.5	0.20	0.4
$Zn(VBT)_2Cl_2$	4.6	0.27	0.22
VBI	2.82	0.36	0.35
$Zn(VBI)_2Cl_2$	4.0	0.24	0.25

Earlier we pointed out possible changes in MCM composition during homopolymerizations. The products of such a process can be regarded as off-beat MCM copolymers with a smaller number of coordinated ligands and with a free ligand. This can be exemplified by polymerization of methylvinyltetrazole complexes in ethanol (Eq. 4-40) [4].

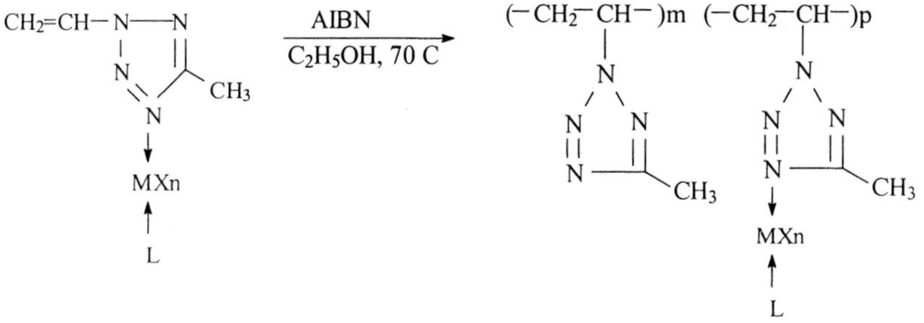

(4-40)

The ability of AAm complexes with Cr(III), Er(III) and other metal nitrates to involve other comonomers, such as acrylic or methacrylic acid and maleide anhydride, in spontaneous copolymerization in aqueous solutions has been reported [122]. It was possible to copolymerize transition metal nitrate AAm complexes with maleic acid under frontal conditions by thermal initiation [83].

Products with enhanced photophysical properties were obtained by copolymerization of a ruthenium complex of 4-vinyl-4′-methyl-2,2′-bipyridine with coumarin-2 (route A) or by a polymer analogous reaction of the appropriate copolymer with Ru(bpy)₃ (route B). Using route A, copolymers containing up to 25% bipyridine monomer units can be produced (Eq. 4-41) [123].

Graft polymerization of covalent-, ionic and donor–acceptor-type MCMs will not be considered in this book.

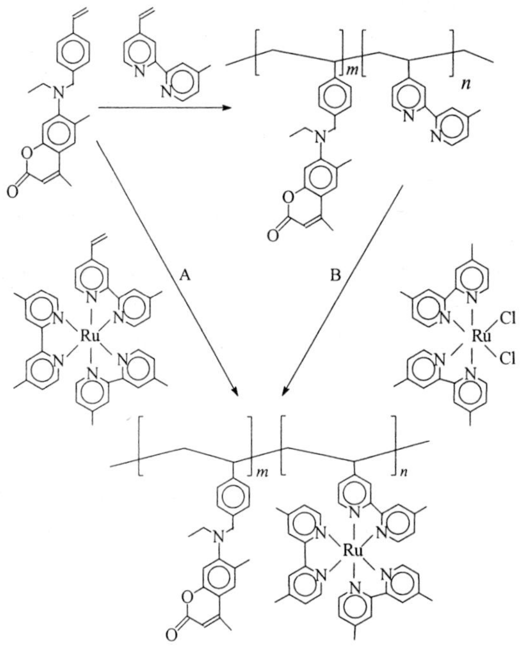

$$(4\text{-}41)$$

104

4.4.4 Copolymerization of π-MCMs

The copolymerization of π-type MCMs has also been studied extensively because a much greater number of monomers can be used for copolymerization than for homopolymerization. There are known cases of MCMs that cannot be polymerized by any mechanism, but which enter into copolymerization according to a specific mechanism. The effect of steric factors due to the presence of metal rings in such MCMs, for example, is also much lower in copolymerization than in homopolymerization. Thus, styrenetricarbonyl-chromium which cannot be homopolymerized has been copolymerized with styrene and MMA (AIBN initiation, $r_2 = 0$). Another non-polymerizable monomer, cinnamoylferrocene, is capable of copolymerization with styrene, acrylonitrile or MMA in bulk or solution [124 and references therein]. It is obvious that copolymerization of such MCMs can be regarded as a method for obtaining the copolymers with statistical (not block) distribution of MCM units. Polymerizations of vinylferrocene (VFe) including homopolymerization have been well studied (Table 4-11) (see Experiments 4-5 and 4-6, Section 4.6).

VFe copolymerized effectively with different monomers including electron-saturated and electron-deficient ones. This monomer has the largest value of the parameter e (−2.1) (comonomer is styrene), i.e. VFe is a fairly strong electron-saturated monomer. From the data of Table 4-11 we may suppose that the Q-e scheme probably does not operate if the electron-deficient monomers take part in copolymerization (i.e. vinylacetate, etc.). This is confirmed by the values of the parameter e_2. When VFe copolymerized with MMA ($e_1 = +2.25$), fumaronitrile ($e_1 = +1.96$) and acrylonitrile ($e_1 = +1.2$) alternating copolymers were obtained. We may conclude that the formation of charge transfer complexes takes place in such systems. Thus, the copolymerization rate dependence on the concentration of VFe in the initial mixture is characterized by a well-defined maximum at 50% content of VFe. Indeed, if the charge transfer complexes enter a growing chain the Q-e scheme should not be applied because of the suggestion that the reactivity of the growing chain depends on the nature of the end units.

Table 4-11. Activity of VFe (M_2) in copolymerization with conventional monomers according to the Q–e scheme.

M_1	e_1	r_1	r_2	e_2
N-vinylcarbazole	-1.40	0.20	0.60	-2.4
n-N,N-Dimethylaminestyrene	-1.37	3.8	0.15	-2.2
Butadiene	-1.05	3.5	0.3	-1.05
	-1.05	3.97	0.14	-1.80
	-1.40	3.97	0.14	-2.1
Vinylpyrrolidone	-0.90	0.33	0.67	-2.1
Styrene	-0.80	2.70	0.08	-2.04
	-0.80	2.63	0.07	-2.1
MMA	+0.40	1.22	0.52	-0.29
	+0.40	1.25	0.56	-0.20
MA	+0.58	0.63	0.82	-0.21
	+0.58	0.61	0.73	-0.32

Table 4-11. Activity of VFe (M_2) in copolymerization with conventional monomers according to the Q–e scheme.

M_1	e_1	r_1	r_2	e_2
Acrylonitrile	+1.20	0.16	0.15	-0.73
	+1.20	0.17	0.11	-0.81
Diethylfumarate	+1.25			
Fumaronitrile	+1.96			
		Alternate copolymer		
MAA	+2.25	0.19	0.02	-0.1

Radical copolymerization of another π-type MCM has also beenstudied in detail. The constants r_1 and r_2, and parameters of the Q–e scheme were estimated. These data are summarized in the monograph [124]. Copolymerization of styrenetricarbonyl chromium (**59**) with MMA ($r_1 = 0.71$, $r_2 = 0$), styrene and vinylacetate has also been studied [125,126].

The availability of a large amount of data on radical copolymerizations of π-type MCMs in contrast to other types allows us to carry out some comparisons of their reactivity [124]. Thus, the values of the polar factor e for different vinylcyclopentadienyl MCMs turned out to be very similar (from −1.97 to −2.1). Therefore, the presence of different metals (from Cr to Ir and W) in such monomers as well as electron-donor carbonyl and nitrosyl groups bound to the metal does not affect the electronic saturation of the vinyl group. Also the similarity of the values of r_1 and r_2 in copolymerizations of many π-type MCMs with styrene should be noted. The high value of Q, from 3.11 to 4.1, is another common property of these MCMs, which indicates a high delocalization of the electrons of cyclopentadienyl radicals. The activities of Fe-containing π-type MCMs (ferrocenylethylmethylmethacrylate, ferrocenylmethacrylate, ferrocenyl-methylmethacrylate) in their copolymerizations with traditional monomers have been compared. The introduction of a methylene group between the Fe-nucleus and the vinyl group increases the process rate, but it is lower compared to MMA or MA. The positive value of e highlights the electron-acceptor character of these MCMs. Many data concerning non-radical copolymerizations of π-type MCMs are available. However, in recent years investigations in this field have not been as intensive as they were in the seventies.

4.4.5   Chelate-Type Monomers in Copolymerization with Conventional Monomers

The behavior of this type of monomer in homo- and copolymerizations as well as the stability of the products formed are the same as those of the π-type MCMs discussed above. As has been mentioned, their homopolymerization proceeds with difficulty; many of them are not polymerizable. Such a situation stimulates investigations into copolymerization of these MCMs. One of the classical cases is the copolymerization of 4-vinylbenzoxy-2-benzoate-di-(2-benzoylbenzoate) europium **105** (or its analogue – Eu(III)-(*p*-vinylphenyl)-3-phenyl-1,3-propanedione) with styrene and MMA [127].

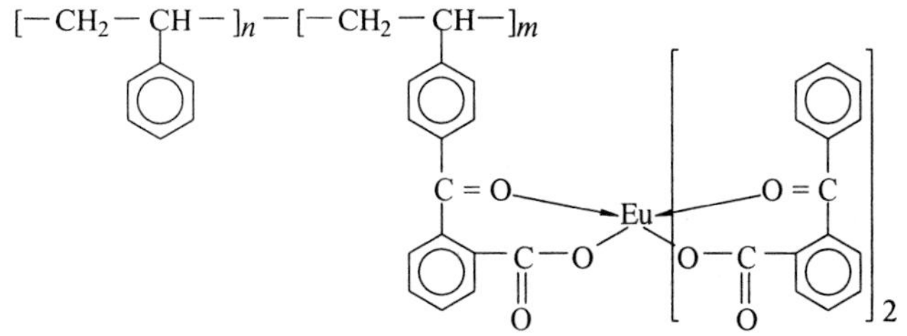

$$[-CH_2-CH-]_n-[-CH_2-CH-]_m$$

105

The products formed contain as much as 4 wt % Eu^{3+} and display a high fluorescence efficiency. The fluorescent intensity is proportional to the content of the metal. Interesting processes occur in the case of the copolymerization of conventional monomers (styrene, acrylonitrile, MMA, acrylic acid) with copper and cobalt chelates containing Schiff bases with vinyl groups in pyrrolyl-methyleneiminostyrene or pyrrolylmethyleneiminopropene (see Experiment 4-8, Section 4.6) [13]. The mole fraction of the monomer M_2 in the resulting copolymer is higher for the MCM than for its "metal-free" analogue (Table 4-12).

In this case, the monomer reactivity increases as it accepts a metal, which is a fairly rare situation in MCM polymerizations. A Co(III) chelate was copolymerized with the participation of no more than one or two multiple bonds of the three available. For a Cu(II) chelate the process typically involves both vinyl groups, but the initiation mechanism for the polymerization of this MCM is more intricate (Eq. 4-42).

Table 4-12. Copolymerization of chelate-type MCMs.

M_1	M_2	r_1	r_2
Acrylonitrile	$CH_2=CH-CH_2N=CH-C_4H_4NH$	2.30	0.40
Acrylonitrile	$(CH_2=CH-CH_2N=CH-C_4H_4N)_3Co$	0.86	8.6
Styrene	$CH_2=CH-C_6H_4N=CH-C_4H_4NH$	0.50	1.8
Styrene	$(CH_2=CH-C_6H_4N=CH-C_4H_4N)_2Cu$	0.45	6.9

The primary radical attacks the vinyl group in the monomer first and then the free electron is transferred from this group to the central metal ion. This process is facilitated by the presence of a conjugated chain (a long coplanar π-electronic system). Chain propagation does not start until after the complete reduction of Cu(II) to Cu(I). A concurrent stage consists of β-elimination of a proton from the vinyl group attacked by the primary radical and its addition to the pyrrolyl nitrogen atom.

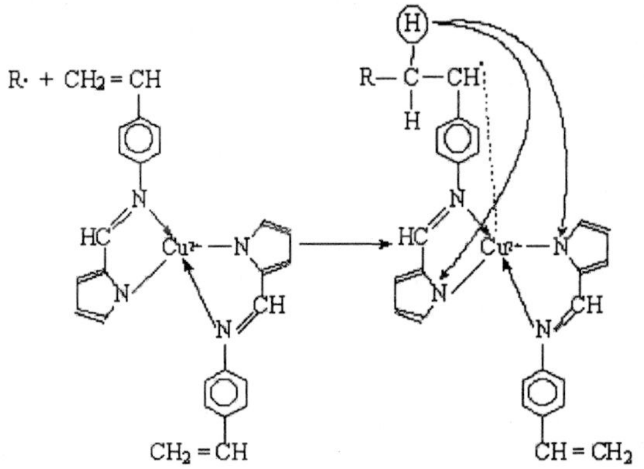

$$\text{(4-42)}$$

Numerous pieces of evidence were obtained for such a mechanism (low molecular weights of the copolymers, study of the interaction of stable radicals with such copper complexes, etc.). A triple radical copolymerization of an MCM based on a Co(II) complex of a Schiff base – the product of condensation of 4-(n-vinylbenzyloxy)salicylaldehyde with 1,2-diaminocyclohexanone – gave a product in which one unit of MCM accounts for 20 units of styrene and 4 units of divinylbenzene [128]. There are reports of quantitative studies in the copolymerization of a Cu(II)-containing chelate monomer based on ethyl-α-(acetoacetoxymethyl)acrylate with styrene and benzyl methacrylate (40 °C, THF) [129]. The resulting copolymers contained 52 and 72 mol% of MCMs (comonomers: styrene and benzyl methacrylate, respectively).

It is evident that macrocyclic MCMs including porphyrin and phthalocyanine derivatives are of great interest. Because attempts at copolymerization of hemin with monomers such as vinylpyrrolidone or vinylimidazole were unsuccessful, the parameter Q should be taken into consideration. Thus, monomers with high Q values (such as styrene, MMA and π-conjugated species) can be effectively copolymerized with hemin (Eq. 4-43).

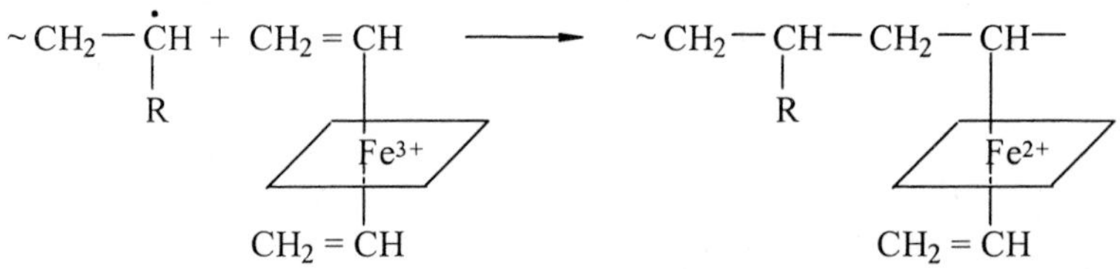

$$\text{(4-43)}$$

At the same time, copolymerization that involves unconjugated monomers with a low Q value (N-vinylpyrrolidone, 1-vinylimidazole) is not effective because primary radicals add to hemin to give stable products.

However, hemin copolymerization with participation of unconjugated monomers can be conducted in the presence of a third conjugated monomer to give, for example, a ternary hemin-IX-vinylimidazole-styrene product.

Radical copolymerization of tetraphenylporphyrin monomers with the vinyl group in a benzene ring or pyrrole ring and their Cu(II), Co(II) and Zn(II) complexes with styrene or methylmethacrylate has been studied [47,48]. Compared to the homopolymerization of styrene, the copolymerization decreases both the overall polymerization rate and the molecular weight of the polymers formed. For example, the rate of chain transfer in the methylmethacrylate Co(II) porphyrin system is 10^7 times greater than the homopolymerization of the methacrylate.

Table 4-13. Radical copolymerization of styrene with Fe(III) protoporphyrin-IX dimethylester (FePP) or hemine (FeH).

Porphyrin type	Porphyrin in monomer mixture (mol %)	Porphyrin in polymer (mol %)	Polymer yield (%)	$\overline{M} \cdot 10^{-4}$	π_{sp}/c $(\cdot 10^2)$ (dL g^{-1})
-	0	0		12.3[a]	
FePP	0.025	0.1		8.3[a]	
FePP	0.05	0.12		5.9[a]	
FePP	0.125	0.31		3.0[a]	
FeH	0.49	0.53	75.6		9.96[b]
FeH	1.95	1.94	60.7		8.66[b]
FeH	3.44	3.31	41.7		7.65[b]

[a] Copolymerization in bulk with AIBN at 60 °C. [b] Copolymerization in pyridine with AIBN at 80 °C.

Chain termination occurs by an intramolecular electron transfer in a redox reaction. It is assumed that a proton of the former vinyl group is eliminated and added to the N-atom of the porphyrin ring. According to ESR spectra, the copper centers in the copolymers containing Cu(II) porphyrins are fairly well separated from each other. But fluorescence spectra show that the quantum yield of fluorescence decreases with increasing molar fraction of porphyrin groups in the copolymers.

Natural protoporphyrins containing vinyl groups are suitable for polymerization. Fe(III) protoporphyrin-IX dimethylester has been copolymerized with styrene or methylmethacrylate in bulk or with acrylamide in methanol using radical initiators [130]. Increasing the porphyrin/styrene ratio decreases the molecular weight, indicating chain transfer. On the other side, the content of covalently incorporated porphyrin increases.

As usual, the content of porphyrins in copolymers were determined by UV-Vis spectroscopy taking the molar extinction coefficient of the non-aggregating monomers as reference.

The content of hemin in its copolymer with styrene generally corresponds with the monomer composition (Table 4-13) [131]. The molecular weight characteristics of the copolymers also depend on the location of double bonds in the ring (benzene or pyrrol) (Table 4-14). But a reduced polymer yield and viscosity again indicate chain termination.

Table 4-14. Molecular weight parameters of copolymers of styrene and their metal complexes.

Sample (content of copolymer (mol %))	$\bar{M}_n$	$\bar{M}_w$	$\bar{M}_w/\bar{M}_n$	Proportion of high – MW fraction (%)
Styrene	145,000	380,000	2.62	–
Styrene – **76** (0.1)	143,000	678,000	4.74	Absent
Styrene – **76** (0.5)	Insoluble			
Styrene – **76** (1.0)	Insoluble			
Styrene – **75**-Cu^{2+} (0.5)	77,000	248,000	3.22	0.6
Styrene – **75**-Cu^{2+} (1.0)	68,000	300,000	4.41	~6
Styrene – **75**-Cu^{2+} (5.0)	60,000	180,000	3.00	~40
Styrene – **75**-Zn^{2+} (0.1)	106,000	290,000	2.73	Absent
Styrene – **75**-Zn^{2+} (0.5)	45,000	120,000	2.66	Absent
Styrene – **75**-Zn^{2+} (1.0)	50,000	120,000	2.40	Absent
Styrene – **76**-Cu^{2+} (0.1)	140,000	390,000	2.78	Absent
Styrene – **76**-Cu^{2+} (0.5)	Insoluble			
Styrene – **76**-Cu^{2+} (1.0)	Insoluble			
Styrene – **76**-Zn^{2+} (0.1)	128,000	470,000	3.67	Absent

In general, it seems that metal-free porphyrin comonomers are more suitable for polymerizations because with metal-containing porphyrins the metal ion in the core can be involved in some chain terminations. Surprisingly, 5,10,15,20-tetra(4-arylamidophenyl)-porphyrin (M = Mn(III)) could be homo- and copolymerized (with methylmethacrylate or divinylbenzene) in DMF to soluble or insoluble polymers [132].

4.4.6   Mutual MCM Copolymerization

The mutual copolymerization of various MCMs allows the production of polymers containing different metals. Thus, radical copolymerization of dicyclopentadienyltitanium dimethacrylate with Ni(II) and Cu(II) acrylates takes place relatively rapidly in both liquid (DMFA, AIBN) and solid (high-pressure initiation in combination with shear strains) phases [133,134]. Although these systems are distinguished by relatively similar reactivity constants, the *e* values do not correspond to those of the conventional monomers in that the metal-containing groups behave as strong electron-donor substituents (e between -1.96 and -2.65). The MCM units are distributed over the chain randomly. Copolymerization parameters are listed in Table 4-15.

The absence of any significant differences between the copolymerization parameters of the processes in the liquid phase and plastic flow conditions may suggest that mechanisms of the two processes are fairly similar.

Table 4-15. Parameters for copolymerization of transition metal acrylates (M_1) with $(C_2H_5)_2Ti(OCOC(CH_3)=CH_2)_2$ [134].

M_1	r_1	r_2	Q_1	Q_2	e_1	e_2
Ni(II) acrylate	0.95	0.56	0.53	0.56	-2.63	-1.9
Cu(II) acrylate	1.09	0.89	0.05	0.56	-2.13	-1.9

The application of high-pressure and shear strains has been used to effect tercopolymerization of Y(III), Ba(II) and Cu(II) acrylates in 1:2:3 molar ratio [135].

There are also reports of mutual copolymerizations of π-type MCMs although they are scarce [124 and references therein]. Thus, copolymerization of vinylcymantrene with VFe gives a product with $\overline{M}_n = 7700$ and $\overline{M}_w = 13{,}800$. The copolymerization parameters are: Q_1 and $Q_2 = 0.436$ and 0.485, $e_1 = -2.6$, $e_2 = -2.1$, respectively. η^6-(Benzylacrylate)tricarbonylchromium was readily copolymerized with ferrocenylethylacrylate to the polymer **106**. The structure of heterometallic copolymers can be shown as follows:

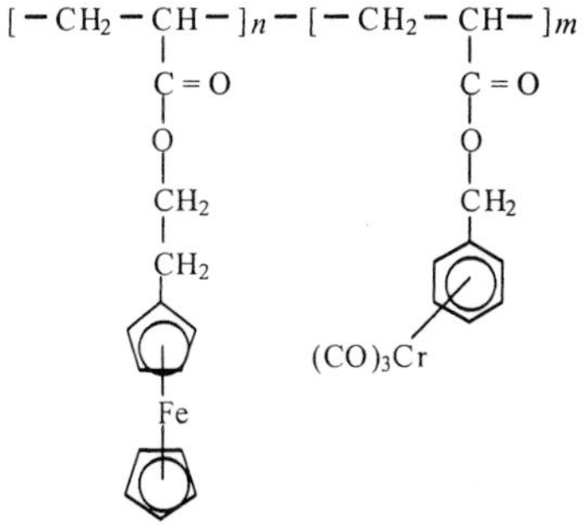

106

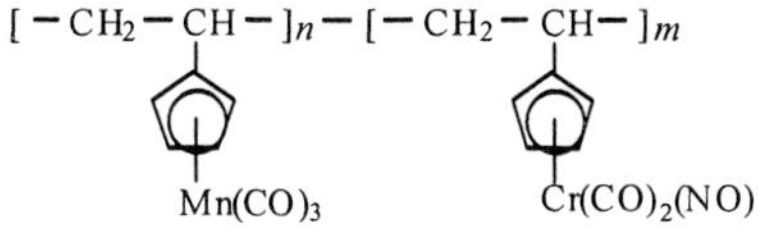

107

The copolymer **107** is formed only at a narrow molar ratio of η^5-vinylcyclopentadienyltricarbonylmanganese and η^5-vinylcyclopentadienyldicarbonylnitrosylchromium comonomers.

4.4.7 Copolymerization of Cluster-Containing Monomers

At present copolymerization of cluster-type monomers is the only way to involve them in polymerizations because all attempts to homopolymerize these monomers have been unsuccessful. Thus, copolymerization of such Co(II)-

containing monomers with styrene or *N*-[*tris*-1,1-(hydroxymethyl)methyl]-acrylamide gave products incorporating as much as 1.5 mol% of clusters [57]. Osmium cluster-containing monomers differing in the nature of the unsaturated ligand (4-vinylpyridine, allyl amine and acrylic acid) $(\mu\text{-H})Os_3(CO)_{10}(\mu\text{-}NC_5H_3CH=CH_2)$ **(89)**, $(\mu\text{-H})Os_3(CO)_{10}(\mu\text{-CONHCH}_2CH=CH_2)$ **(90)**, $(\mu\text{-H})Os_3\text{-}(CO)_{10}(\mu\text{-COOCH}=CH_2)$ **(91)**) have been copolymerized with styrene and acrylonitrile [136]. It was found that each copolymer chain contained on average one cluster unit and one terminal double bond, i.e. cluster monomer addition restricted chain propagation. These observations suggested that cluster monomer addition to the growing polymer chains terminates propagation by fast chain transfer (Fig. 4-12) (for copolymerizations of cluster-containing monomers see Experiments 4-9 and 4-10, Section 4.6).

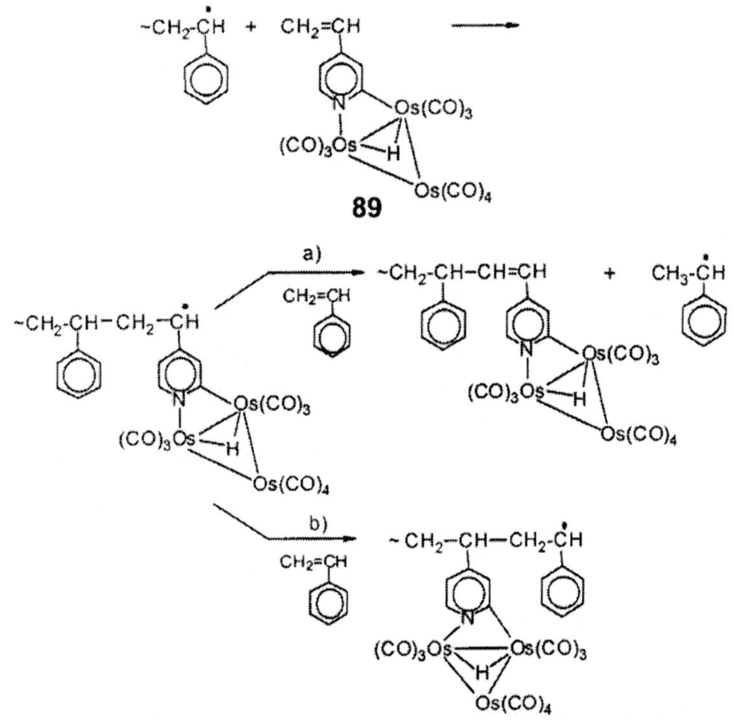

Figure 4-12. Scheme of chain transfer in copolymerizations of cluster-containing monomers.

Styrene is the most probable chain transfer agent (route a). The estimated value of $C_s = K_s/K_p$ – the relative rate constant for chain propagation in the presence of a chain termination agent – is equal to ~0.2, the absolute value of this process K_s is within the range 40–500 at 60 °C. The presence of cluster-containing monomers leads to a decrease of the copolymer molecular weights (Table 4-16).

Table 4-16. Copolymers of styrene (St) with cluster-containing monomers **97–99**. Comparison of their number average molecular weights, $\overline{M}_n$, average number of polymer chains per gram of polymer, q, average number of double bonds (mol g$^{-1}\cdot 10^5$), and cluster units content, CM, (mol g$^{-1}\cdot 10^5$).

Molar ratio styrene:CM	$\overline{M}_n$	[q]	[C=C]	[CM]
Polystyrene[a]	31,000	3.2	2.8	
St:89 0.005[a]	31,000	3.2	2.2	1.2
St:89 0.01[a]	37,000	2.7	3.1	2.8
St:89 0.02[a]	35,000	2.9	6.5	6.0
St:89 0.05[b]	6000	11.6	12.3	19.2
St:90 0.1[b]	5000	19.1	17.6	32.1
St:90 0.005[a]	29,000	3.4	4.2	0.42
St:90 0.01[a]	21,000	4.8	5.0	0.43
St:90 0.005[a]	15,000	6.7	3.2	1.5
St:91 0.01[a]	12,000	8.3	4.0	3.5

[a] Polymerization in bulk, 60 °C, [AIBN] = 0.5 mol %. [b] Polymerization in toluene solution, 60 °C, [AIBN] = 0.5 mol %.

Cluster units cause chain transfer, most probably to styrene molecules. This conclusion is also confirmed by the correspondence between the cluster content and the specific number of double bonds in the copolymers obtained.

It should be pointed out that an alternative way of producing polymeric clusters, i.e. through interactions of clusters such as $Co_2(CO)_8$ with a polymer (a macroligand) involves some difficulties such as chain cross-linking, cluster adsorption, formation of colloidal metal species, etc. In other words, copolymerization of such an MCM is the best way of producing polymeric clusters characterized by high purity and structural homogeneity. This is also confirmed by radical copolymerization of common monomers with Rh_6- [137] and $[Mo_6Cl_8]$- [138] containing monomers.

4.5 The Future Development of Polymerization and Copolymerization of MCMs

Despite intensive investigations in this field, the problem of correlating MCM structures and reactivities is far from being solved. First of all, it relates not only to the influence of the metal bond type in the MCM, but also to its valence state, ligand environment and polynuclearity. It seems that very interesting results would be obtained by studies of polymerization of MCMs based on polyvalent metals such as Mn(VII), Mn(IV) or Cr(VI), Cr(III), Cr(II), as well as during copolymerization of MCMs based on metals in different valence states. This is especially true of the copolymerization of MCMs with each other in order to prepare polymers of different metals (heterometallic copolymers). This aspect is only beginning to be developed.

The copolymerization of polynuclear and cluster-containing monomers is the most promising approach to obtaining a high degree of uniformity and integrity of polynuclear formations in the polymers. This method is expected to be one of the ways of preparing novel structurally organized catalysts for different reactions.

The intensive development of chelate-type MCMs can be taken into account. Fragmentary data (most often without either quantitative data or study of the reaction mechanism) on radical copolymerizations of MCMs – Schiff bases with vinyl groups, macrocyclic complexes (hemins, phthalocyanines etc.) – point us in this direction. Probably, polymerizable crown ethers with exocyclic groups and metal ions in the ring (cryptands, podands etc.) will be synthesized.

In our opinion, detailed investigation of the structural organization of metallopolymers is urgently needed, in particular the stereoregulation of the course of radical polymerizations of MCMs. The orienting influences of coordinate bonds, noted above, and polar effects can exert a substantial influence on the structure of the metallomonomer unit. Indeed, radical polymerization of zinc, barium, magnesium and lead acrylates leads to the formation of polymers with regular structures, characterized by primarily (up to 80 wt.%) syndiotactic addition of monomer units. In addition to the classical factors such as the structure of active centers, the temperature and the kind of solvent, steric hindrances caused by the forming network structure of the metallopolymer affect the stereoregularity of addition of acrylate groups of such bifunctional monomers as $M(OCOCH=CH_2)_2$. This is related to an increase of internal (shrinking) strengthening at high degrees of transformation, which lead to an atactic structure of the chain. The results obtained so far seem to suggest that we should change our understanding of stereoregulation during radical polymerization.

It is expected that investigations on homo- and copolymerization of optically-active MCMs will receive a new impetus. This particularly relates to the influence of the end unit on the stereoconformation of the growing chain. It is reasonable to expect considerable advances in chiral catalysis using such metallopolymers. A metallocenter may activate a substrate and generate an asymmetric induction simultaneously.

Nontraditional methods of polymerization (low temperature radical poly-merization initiated by organometallic compounds, ionic and ion-coordination polymerization, electro- and photochemical initiation, etc.) with nontraditional monomers (which are undoubtedly MCMs) will in future receive much more attention. At present, intensive research into solid-phase polymerization is on-going. The reason is to avoid the problems that appear during MCM polymerization, namely, those connected with the selection of an appropriate solvent and initiator, with precipitation of the polymeric product and with removal of residual reaction-medium components from the product.

4.6   Experimental

At present, several thousand metallopolymers have been described, although the part of those obtained by polymerization and ☐opolymerisation of metal-containing monomers is not great. Nevertheless, the procedures of synthesis of over 200 metallomonomers are given in the monograph [124]. In this book we restrict ourselves to preparative methods of synthesis only of some main representatives of metal-containing monomers and their polymers.

Experiment 4-1:   Polymerization and Copolymerization of Acrylate Cobalt(II) 33 (Section 4.2.3) [139]

Synthesis of 33: A solution of acrylic acid (18 g, 0.25 mol) in 20 mL of ethanol was added dropwise at room temperature to a suspension of $CoCO_3$ (10 g, $8.4 \cdot 10^{-2}$ mol) in 150 mL of ethanol. The reaction mixture was stirred for 3 h at room temperature. After removal of two-thirds of the solvent the product was precipitated with ether. The precipitate was filtered, washed and dried in vacuum. Yield 15.2 g ($6.9 \cdot 10^{-2}$ mol, 84%). $CoC_6H_8O_5$ calcd. C 32.9, H 3.7, Co 26.9; found C 32.6, H 3.7, Co 27.5; IR (KBr): 1640 (C=C), 1065 (=CH–C), 1360, 1560 (COO), 280 cm^{-1} (M–O); λ: 18760 (v_2), 20,500 cm^{-1} (v_3).

Homopolymerization of 33: 1 g of cobalt(II) acrylate, 0.02 g of AIBN, and 15 mL of ethanol were placed and evacuated in a glass tube. The tube was sealed under vacuum. Polymerization was carried out at 70 °C for 3 h. The polymer precipitate was filtered, washed and dried in vacuum. Yield 0.96 g (96%). $CoC_6H_8O_5$ calcd. Co 26.9; found Co 26.8; IR (KBr): 1640 (C=C), 1360, 1560 (COO), 280 cm^{-1} (M–O); λ: 18,760 (v_2), 20,700 cm^{-1} (v_3). The molecular weights were determined by thermomechanical spectroscopy. The following molecular weights were determined: $M_n = 71.9 \cdot 10^3$, $M_w = 301.5 \cdot 10^3$ Da.

Copolymerization of styrene with 33: The copolymerization was carried out in DMF, containing 0.46 mol L^{-1} of each comonomer and $1.22 \cdot 10^{-2}$ mol L^{-1} of AIBN as initiator. Metal acrylate and AIBN were placed in a glass tube and evacuated. The monomer and the solvent were then frozen in the reactor tube and the tube was sealed under vacuum. The polymerization was carried out at 70 °C for 3 h. The precipitated copolymer was filtered, washed and dried in vacuum. The yield was 60%.

Experiment 4-2: Synthesis, Polymerization and Copolymerization of Metal Etherate 18 OV(O-i-C$_3$H$_7$)$_2$(OCH$_2$CH$_2$OC(O)C(CH$_3$)=CH$_2$) (Section 4.2.2) [23,140]

Synthesis of the metallomonomer: Using a dropping funnel, 3.82 g (0.029 mol) of 2-oxymethylmethacrylate in 10 mL benzene was added to a solution of 6.79 g of OV(O-i-C$_3$H$_7$)$_3$ (0.029 mol) in 40 mL freshly distilled benzene in specially designed equipment that allowed us to carry out simultaneously the synthesis, azeotropic solution distillation and addition of solvent in an atmosphere of dry Ar. The duration of the reaction with simultaneous azeotropic distillation of the evolved isopropyl alcohol alcohol was 3 h. At the end the reaction mixture was vacuum distilled. The yellowish product contained 16.3% (calcd. 16.2%) V and 1 mol of double bonds per mol. The density (ρ) was 1.1352 g mL^{-1}, IR: 1005 ($v_{V=O}$), 1140, 1175 (skeletal vibrations of i-Pr group), 1640 ($v_{C=C}$), 1720 cm^{-1} ($v_{C=O}$). ^{1}H NMR (δ, ppm): 1.31–1.33 (i-Pr, Me, J = 6 Hz), 1.84–1.86 (R', Me), 4.21 (R', C–CH$_2$–C), 4.76–4.91 (R', O–CH$_2$–C), 5.12–5.15 (i-Pr, CH, J = 6.0 Hz), 5.27 and 6.20 (R', C=C(H$_A$) (H$_B$) (R is the fragments of 2-oxymethylmethacrylate).

Homopolymerization of 18: The monomer (0.2 mol L^{-1}) azo-bisisobutyronitrile (0.5 to 2.0 wt. %) and benzene (40 mL) were placed in a glass tube and evacuated. The tube was sealed under vacuum. Under the experimental conditions (75 °C, 1% of the initiator, 100 min) the yield of the polymer was 40%. The polymer was precipitated by pentane, and its intrinsic viscosity was equal to 0.16 (in DMSO). The content of V(IV) in the polymer was found to be ~1%.

Copolymerization of 18 with styrene and acrylonitrile: In the range of 40 to 60 mol % of the metallomonomer the copolymer composition does not depend on the monomeric mixture. The copolymers are soluble in benzene; the molecular weights are about 10^4 Da. Under copolymerization conditions (75 °C, benzene, 1% of the initiator) of **18** with acrylonitrile (25 mol %) a light-yellow product containing ~12% vanadium was obtained. The yield was 15%, the product is soluble in DMFA and DMSO, and its intrinsic viscosity was 0.11 (DMSO, 30 °C). IR: 1720 ($v_{C=O}$), 2245 cm^{-1} ($v_{C\equiv N}$); the ratio of the intensities of the absorptions $I(C=O)/I(C\equiv N)$ = 13:1. This method can also be used for the synthesis and polymerization of optically active metallomonomers.

Experiment 4-3: Synthesis, Polymerization and Copolymerization of Vinylpyridine Complexes 3, Co(4-VPy)$_2$Cl$_2$ and Zn(4-VPy)$_2$(SCN)$_2$ (Section 4.2.4) [141]

Synthesis of the monomer 3: Method I: 4-VPy was added to a hot ethanol solution of CoCl$_2$; after cooling the crystal adduct was isolated and dried over NaOH in a desiccator.

Method II: To an aqueous-alcohol or aqueous solution of metal halide the stoichometric amount of the ligand was added. Isopropyl alcohol was used to precipitate the complex because it was very soluble in ethanol.

Method III: A solution of 4-VPy (2.1 g, 0.02 mmol) in 10 mL of ethanol was added to a solution of CoCl$_2$·6H$_2$O (2.4 g, 0.01 mmol) in 5 mL of ethanol. After 1 h the blue crystals were filtered, washed with ether containing a few drops of 4-VPy, and dried in vacuum for 1 h. The yield was 2.6 g (59%). Methods I and II were used for the synthesis of Co(2-VPy)$_2$Cl$_2$, and method II was also used for the synthesis of Zn^{2+}, Ni^{2+} and 4-VPy adducts followed by precipitation with ethanol.

Polymerization of vinylpyridine complexes 3: Homopolymerization of Zn(4-VPy)$_2$(SCN)$_2$ was carried in DMF using AIBN at 70 °C. The rate of reaction was 24.4·10^5 mol L^{-1} s^{-1}. The molecular weight of the polymer was 2.5·10^4 Da.

Experiment 4-4: Synthesis and Frontal (Co)polymerization of Acrylamide Complex 53, Co(AAm)$_4$(NO$_3$)$_2$ (Section 4.2.4) [81, 83]

Synthesis of 53: A five-fold molar excess of acrylamide was mixed with Co(NO$_3$)$_2$·6H$_2$O with the evolution of crystal water. The product obtained was then dried in an desiccator, washed in hot benzene and ether, and dried under vacuum at room temperature. The product was a pink crystal powder. The melting point is 78–79 °C, ρ = 1.53 g mL^{-1}. IR: 3190(ν_sNH), 3290 (ν_{as}NH), 1665 (νCO), 1580, 1590(δNH), 1445 (δ(CH$_2$)), 1280 (νCH), 1385 (ν(NO$_3$)), 354 cm^{-1} (νMO).

Homopolymerization of 53: After preliminary compression at 50 MPa, the powdered acrylamide complex in the form of a cylindrical tablet (d = 1.2 cm, ρ = 1.38 g mL^{-1}) was placed in a glass ampoule. The bottom of the ampoule was then heated in a bath (with an alloy of low-melting metal) at T_{st} = 413 K for 10 s until the sample was observed to undergo a marked color change from pink to dark violet. The frontal propagation rate was determined visually by following the movement of the boundary color change of the melting sample with time. The polymer obtained (M = 61.5·10^3 Da) was insoluble in organic solvents.

Copolymerization of 53: The copolymerization of Co(NO$_3$)$_2$·(AAm)$_4$·2H$_2$O with maleic acid was studied for a wide range of initial compositions (17.0 to

83.0 mol. % of **53**) at an initiation temperature of 433 K. Pressed samples of the monomeric mixture as cylindrical tablets ($d = 1.2$ cm, $\rho = 1.38$ g mL^{-1}) were used for the reaction. The reaction procedure was the same as for the homopolymerization of **53**. The degree of conversion was 40–50%. Unreacted compounds were removed by extraction with methanol in a Soxhlet apparatus, the products obtained were dried under vacuum at room temperature. The copolymers contain 5.73 to 9.72% of Co.

Experiment 4-5:   Synthesis and Polymerization Conversions of Vinylferrocene 57 (Section 4.2.5) [40,124]

Synthesis of the metallomonomer 57: A number of fairly convenient methods for synthesis of this monomer have been developed. We consider only some of them.

Method I, Dehydration of ferrocenylethanol: 9.9 g (0.043 mol) of 1-ferrocenylethanol was mixed with 30 g of activated Al$_2$O$_3$ and placed in a vacuum sublimator supplied by a water-cooled probe. The surface of the mixture was covered by a small quantity of Al$_2$O$_3$. The system was cooled and placed in an oil bath at 155 °C and 270 Pa for 6 h. After cooling and slow admission of air, orange crystals of **57** were isolated from the probe. The yield was 7.83 g (85%).

Method II, The Meerwein–Ponndorf–Verley reaction: A solution of 3 g of acetylferrocene in 300 mL of isopropyl alcohol was passed through a tube of diameter 25 mm and length 700 mm filled with activated Al$_2$O$_3$ for 3 h at a pressure of 900 to 1300 Pa and 200–210 °C. From the condensate 2.3 g of vinylferrocene was isolated and chromatographed via a column with Al$_2$O$_3$, hexane was used as the eluent. After evaporation of the solvent, 1.95 g of the metallomonomer was obtained (70% of the yield).

Method III, The Wittig reaction: Ferrocenylaldehyde was reacted with methyltriphenylphosphorane in boiling ether for 10 h. The precipitate was then isolated, dissolved in ether, with water and dried. After evaporation of the ether the product was purified by chromatography with SiO$_2$ and sublimed at 13 Pa.

Polymerization: Vinylferrocene was polymerized in bulk using AIBN at 80 °C. $M_n = 17,200$. Polymerization at 70 °C leads to formation of the polymer with $M_n = 23,000$, $M_w = 40,000$ Da.

Experiment 4-6:   Synthesis and Copolymerization of Styrenetricarbonylchromium 59 (Section 4.2.5) [142, 143]

Synthesis of 59: 20 g of triaminechromiumtricarbonyl, 10 mL of styrene and 500 mL of dioxane were placed in a three-neck flask. The mixture was heated for 2.5 h under Ar and then cooled in a water bath. The solution was filtered through a glass filter, and the residue was repeatedly washed with dioxane to extract **59**. The solvent was then evaporated at reduced pressure and the precipitate was recrystallized from hexane. The yield was 60% of orange crystals. The same method can be used to obtain π-MCMs from *p*- and *m*-methylstyrenechromiumtricarbonyl.

Homopolymerization of 59: A few years ago this homopolymerization was assumed to be impossible. However, it has since been found that the use of complexing additives (for instance, triisobutylboron) leads to the formation of the homopolymer of **59** with yields of 60% after 6 h (M_w = 18,000 Da, M_w/M_n = 3.3 to 3.8). The same procedure can be used for the polymerization of other arenechromiumtricarbonyls.

Copolymerization of 59: The copolymerization of the monomer (10 mol %) with methylmethacrylate was carried out in ethylacetate at 50 °C in the presence of AIBN (0.5 wt. %). The conversion was 13.1%, M_w = 168,000, M_n = 91,000 Da, M_w/M_n = 1.8. Under the same conditions **59** as well as other arene-chromiumtricarbonyls (*p*- and *m*-methylstyrenes) can be copolymerized with styrene, methylacrylate, butylacrylate and vinylchloride.

Experiment 4-7:   Synthesis, Homopolymerization and Copolymerization of the Tetramethacrylic acid Ester 100 of a Phthalocyanine (Section 4.3.7) [103]

Synthesis of the monomer 2,9,16,23-tetrakis(4-(methacryloyloxy)phenoxy)-phthalocyanine zinc(II) complex: 192 mg (0.15 mmol) of zinc complex of 2,9,16,23-tetrakis(4-hydroxyphenoxy)phthalo-cyanine and 0.7 mL (5 mmol) of triethylamine were dissolved under dry inert gas in 10 mL dry THF. Using a dropping funnel 0.6 mL (6 mmol) of methacryloyl chloride dissolved in 5 mL dry THF was added slowly. The mixture was stirred for 7 days, and the reaction was followed by thin layer chromatography (TLC) (SiF, CHCl₃). The hydrochloride of triethylamine was filtered off, and the solvent was removed under vacuum at 30 °C. The purification of the dried monomer was carried out by column chromatography on SiO_2 in CHCl₃/acetone (95:5). Yield: 113 mg (60%) of a green powder. IR (KBr): 1735 (ν(C=O)), 1237 (ν(Ar–O)), 1185, 1026 cm⁻¹ (ν(Ar–O–CO)). UV/Vis in DMF at λ (ε/L mol⁻¹ cm⁻¹): 674 (150,000), 608, 354 nm. ¹H NMR (DMSO): δ = 2.1 (dd; –CH₃), 6.0 (m; vinyl), 6.4 (m; vinyl), 7.3–7.8 ppm (m; aromatic H). Intensity ratio 12:4:4:28. MS (DCl, negative, NH₃): *m/z* 1282 (M⁻).

Homopolymerization: 0.02 mmol of the phthalocyanine monomer in a bomb vessel was dissolved in 2 mL DMF containing 0.2 mol L⁻¹ AIBN (0.5 mol-% AIBN per vinyl group). The polymerization was carried out as described below for the copolymerization for 12 h at 60 °C. The products were precipitated with methanol. Yield ~35% of a green powder soluble in DMF, THF and CHCl₃.

Copolymerization with styrene: The copolymerizations were carried out in

DMF containing 10% of the mixtures of monomers. AIBN (0.5 mol % corresponding to the whole amount of styrene and porphyrin) was used as the initiator. 0.005 mmol of the metallomonomer and 2 mg AIBN were dissolved in 3 mL DMF in a bomb vessel. Freshly distilled styrene (0.26 g, 2.5 mmol) was added, and the bomb vessel was then degassed using three freeze–thaw cycles at 10^{-3} mba. The sealed vessel was heated at 60 °C for 4 h. Dilution with 50 mL methanol precipitated the polymer, which was purified further by three reprecipitations from DMF/methanol (yield ~10%). The amount of porphyrin moieties in the copolymers was determined by quantitative electronic spectra taking the absorption coefficients of the monomer as standard. The molecular weights were determined by GPC in THF using polystyrene as standard. The employed molar ratio of styrene : the metallomonomer was 500. The polymer contains the comonomers styrene and metallomonomer in a molar ratio of 729 (0.013 mmol per g polymer = 1.6 wt. %). The following molecular weights were determined: M_n = 13,000, M_w = 23,100 Da. Other porphyrin-type monomers and other monomer compositions employed for co-polymerizations are described in [103].

Experiment 4-8:   Synthesis and Polymerization of the Chelate Monomer $(CH_2=CH–CH_2N=CHC_4H_4N)_2Cu$ (Section 4.2.6) [13]

Synthesis of the monomer. Bis[p-pyrrolymethyleneamino)propen-1]Cu(II) was obtained from allylamine (3 mL, 0.05 mol) by adding 25 mL of an aqueous solution containing $CuCl_2$ (1.4 g, 0.01 mol). The reaction mixture was stirred for 15 min at room temperature, an aqueous solution of 2-formylpyrrole was then added and the mixture was stirred for 1 h at 60 °C. The color of the solution changed from green-blue to dark brown. The product was precipitated by addition of 10 mL of 3N solution of NaOH and was recrystallized from methanol. The yield was 87%. Using this ligand (L) and $CoCl_3$ the metallo-monomer CoL_3 can be obtained. Bis[p-(2-pyrrolylmethyleneimino)styryl]Cu(II) was obtained from $Cu(OCOCH_3)_2$ and p-(2-pyrrolylmethyleneimino)styrene. Yield: 89% of dark brown crystals (melting point 201–202 °C)

Polymerization: The chelate monomer was polymerized by a radical iniator (AIBN) in a sealed tube. The reaction mixture was poured into diethyl ether after the polymerization at 70 °C and the precipitate was filtered and dried under vacuum. The product was further purified by Soxhlet extraction with THF to remove unreacted monomer.

Experiment 4-9: Synthesis and Copolymerization of Cluster Monomer 90 Rh$_6$(CO)$_{15}$(4–C$_6$H$_4$NCH=CH$_2$) (Section 4.2.8) [137]

Synthesis of the monomer: 4-Vinylpyridine (1 mmol) was added to a solution of Rh$_6$(CO)$_{16}$ (0.1065 g, 0.1 mmol) in chloroform (150 mL). A solution of (CH$_3$)$_3$NO·2H$_2$O (0.15 mmol) in ethanol (15 mL) was added dropwise to the reaction mixture with vigorous stirring. The reaction mixture was then evaporated to 15 mL under reduced pressure, diluted with heptane (50 mL), and concentrated in vacuo to one-half of its original volume. The concentrate was chromatographed on a silica-gel (Silpearl) column using hexane–ether (1:3) as eluent. The eluate was evaporated in a rotary evaporator, and heptane was added periodically to the pot residue to induce crystallization. The precipitated crystals were filtered, washed with pentane and dried. The yield of the monomer was 0.045 g (42%). Rh, found 53.9% (calcd. 54.07%). IR (KBr): 2104, 2068, 2038, 2010, 1788, 1619 cm^{-1}; ^{1}H NMR (CDCl$_3$): δ = 8.88 (2H-β), 7.39 (2H-α), 6.68 (1H, –CH=CH$_2$, J = 10.74 Hz), 6.10 (1H, –CH=CH$_2$, J = 17.54 Hz), 5.68 ppm (1H, –CH=CH$_2$, J = 10.81 Hz).

A disubstituted derivative, Rh$_6$(CO)$_{14}$(4-VPy)$_2$, as well as the cluster monomer **91**, Rh$_6$(CO)$_{14}$(μ,η^2-P(C$_6$H$_5$)$_2$CH$_2$CH=CH$_2$) were obtained in the same way.

Copolymerization of the cluster monomer **90** *with styrene*: Radical copolymerization of **90** with styrene was carried out in bulk or in toluene in the presence of AIBN (0.1 to 0.6 mol. %). Rh$_6$(CO)$_{15}$(4-VPy) (0.06 g, 0.052 mmol), AIBN (1 wt. %), styrene (1.4 g, 10 mmol), benzene (18 mL) were placed in a glass ampoule. The mixture was degassed by freeze–thaw cycles to remove the oxygen from the reaction mixture completely. The ampoule was then sealed in vacuum and heated at 70 °C for 6 h. The copolymer formed was precipitated with ethanol from benzene solution, and dried. The yield: 0.57 g of the copolymer (3.27% Rh, M_n = 180,000 Da).

Experiment 4-10: Synthesis and Copolymerization of the Cluster Monomer 89 (μ-H)Os$_3$(μ-OCNMe$_2$)(CO)$_9$P(C$_6$H$_5$)$_2$CH$_2$CH=CH$_2$ (Section 4.2.8) [137]

Synthesis of 89: A cluster complex of osmium, (μ-H)Os$_3$(μ-OCN(CH$_3$)$_2$(CO)$_{10}$, was synthesized [145] by the interaction of Os$_3$(CO)$_{12}$ with liquid dimethylamine, which led to the carbonylation of the amine by the cluster followed by its coordination to form the carbamoyl ligand μ-OCN(CH$_3$)$_2$. A methanol solution of (CH$_3$)$_3$NO·2H$_2$O (9 mL, 0.081 mmole) was added dropwise to (μ-H)Os$_3$(μ-OCN(CH$_3$)$_2$(CO)$_{10}$ (0.041 g, 0.44 mmol) in ether (5 mL) for 20 min. The mixture was diluted with hexane (1.5 times) and passed through a short column with silica-gel to remove the excess $(CH_3)_3NO$. The complex was obtained as an orange powder (0.04 g, 0.042 mmol). Allyldiphenylphosphine $P(C_6H_5)_2CH_2CH=CH_2$ (0.05 mL, 0.23 mmol) was added to a solution of the complex $(\mu\text{-}H)Os_3(\mu\text{-}OCNMe_2)(CO)_9NCH_3$ (0.196 g, 0.2 mmol) in methylene chloride (5 mL) and stirred. After ~15 h two fractions of the products, I (the main fraction) and II, R_f = ~0.35 were found. The separation was carried on Silufol plates and the fractions desorbed with CH_2Cl_2. A yellow crystalline compound, $(\mu\text{-}H)(\mu\text{-}OCNMe_2)(CO)_9PPh_2CH_2CH=CH_2$ (0.195 g) was obtained.

Copolymerization of 89 with styrene: The monomer (0.02 g, 0.018 mmol), styrene (0.26 g, 2.5 mmol), and AIBN (0.002 g, 1 wt. %) were placed in an ampoule, repeatedly degassed by freeze–thaw cycles and sealed in vacuum. The copolymerization was carried out for 5 h at 70 °C. The product was precipitated with ethanol. Yield: 0.18% of a yellow copolymer product (1.85% Os, M_n = 65,000 Da). In the IR spectra the absorption bands characteristic for polystyrene and **89** are observed.

4.7 References

1. J.E. Sheats, *J. Macromol. Sci.-Chem.* **1981**, *A15*, 1173.
2. C.E. Carraher, J.E. Sheats, C.U. Pittman Jr. (Eds.), *Metal-containing Polymer System,* Plenum, New York, 1985.
3. V.F. Mishchenko, Z.M. Rzaev, V.A. Zubov, *Biostable Tin-containing polymers,* Khimiya, Moscow, 1995.
4. A.D. Pomogailo, V.S. Savost'yanov, *Synthesis and Polymerization of Metal-Containing Monomers,* CRC Press, Boca Raton, FL, 1994.
5. J.E. Sheats, in: Kirk-Othmer *Encyclopedia of Chemical Technology,* Vol 15, Wiley, New York, 1982, 184.
6. A.D. Pomogailo, V.S. Savost'yanov, *Russ. Chem. Rev.* **1983**, *52*, 973; ibid, **1991**, *60*, 762.
7. A.D. Pomogailo, V.S. Savost'yanov, *J. Macromol.Sci.-Rev.* **1985**, *C25*, 375.
8. A.D. Pomogailo, in: *The Polymeric Materials Encyclopedia,* J.S. Solomone (Ed.), Vol. 6, CRC Press, Boca Raton, FL, 1996, 4123.
9. G. Frenking and U. Pidum, *J. Chem. Soc., Dalton Trans.* **1997**, 1653.
10. M.L.H. Green, *Organometallic Compounds. The Transition Elements*, Methuen & Co, London, 1968.
11. G.N. Schrauzer, R.J. Windgassen, *J. Am. Chem. Soc.* **1967**, *89*, 1999.
12. M. Numberg, N.V.K. Duong, A. Graudemer, *J. Organomet. Chem.* **1970**, *25*, 231.
13. T. Tomono, K. Honda, E. Tsuchida, *J. Polymer Sci.: Polym. Chem. Ed.* **1974**, *12*, 1243.
14. K. Sonogashira, K. Ohga, S. Takagashi, N. Nagihara, *J. Organomet. Chem.* **1980**, *188*, 237.
15. K. Sonogashira, S. Takagashi, N. Nagihara, *Macromolecules* **1977**, *10*, 879.

16. S. Takagashi, E. Murata, K. Sonogashira, N. Nagihara, *J. Polymer Sci.: Polym. Chem. Ed.* **1980**, *18*, 661.

17. S. Takagashi, Y. Onyama, E. Murata, *J. Polymer Sci.: Polym. Chem. Ed.* **1980**, *18*, 349.

18. M.A. Chaubon-Deredempt, J. Eescudie, C. Couret, *J. Organomet. Chem.* **1994**, *467*, 37.

19. C. Jouany, S. Mathieu, M.A. Chaubon-Deredempt,G. Trinquier, *J. Am. Chem. Soc.* **1994**, *116*, 3973.

20. C. Jouany, G. Trinquier, *Organometallics* **1997**, *16*, 3148.

21. R. Feld, P.L. Cowe, *The Organic Chemistry of Titanium*, Laporte Titanium Ltd. Butterworth, London, 1965.

22. G.I. Dzhardimalieva, A.D. Pomogailo, A.N. Shupik, *Bull. Russ. Acad. Sci. USSR, Div. Chem. Sci.* **1985**, *34*, 411.

23. A.D. Pomogailo, N.D. Golubeva, A.N. Kitaigorodskii, *Izv. Akad. Nauk, Ser. Khim.* **1991**, 482.

24. C.P.Gibson, D.S. Berm, *J. Organomet. Chem.* **1991**, *414*, 23.

25. J.-F. Carpentier, Z. Wu, C.W. Lee, S. Strömberg, J.N. Christopher, R. Jordan, *J. Am. Chem. Soc.* **2000**, *122*, 7750.

26. W.J. Evans, M.A. Ansari, J.W. Ziller, *Inorg. Chem.* **1999**, *38*, 1160.

27. A. Berton, M.Porchia, G. Rossetto, P. Zanella, *J. Organometal. Chem.* **1986**, *302*, 351.

28. Pat. USA, 2708205.

29. E.I. Klabunovskii, E.I. Karpeiskaya, G.I. Dzhardimalieva, N.D. Golubeva, A.D. Pomogailo, *Russ. Chem. Bull.* **1999**, *48*, 1717.

30. E.I. Karpeiskaya, L.F. Godunova, E.S. Levitina, M.R. Lyubeznova, E.I. Klabunovskii, E.S. Shapiro, G.N. Baeva, E.D. Lubuzh, L.V. Krentsel', A.D. Litmanovich, N.A. Plate, *Bull. Russ. Acad. Sci., Div. Chem. Sci.* **1992**, *41*, 1858.

31. Z.M. Dzhabieva, G.I. Dzhardimalieva, A.D. Pomogailo, Yu. M. Shulga, N.M. Bravaya, in: *XIIth FECHEM Conference on Organometallic Chemistry. August 31 – September 5, 1997. Book of Abstracts,* Institute of Chemical Process Fundamentals Academy of Sciences of the Czech Republic, Praga, 1997, p. A26.

32. A. Kimaro, L.A. Kelli, G.M. Murray, *Chem. Commun.* **2001**, 1282.

33. G.G. Skvortsova, E.S. Domnina, Yu. N. Ivlev, *Zh. Obshch. Khim.* **1972**, *42*, 168; 596; A.G. Starikov, G.I. Dzhardimalieva, I.E. Uflyand, T. K. Goncharov, A.D. Pomogailo, *Russ. Chem. Bull.* **1993**, *42*, 66.

34. D.C. Ayres, *Carbanions in Synthesis,* Oldbourne Press, London, 1965.

35. M.F. Farona, W.T. Agers, B.G. Ramsey, and J.G. Grasseli, *Inorg. Chem. Acta* **1969**, *3*, 503.

36. V.S. Savostyanov, V.I. Ponomarev, A.D. Pomogailo, B.S. Selenova, I.N. Ivleva, A.G. Starikov, L.O. Atovmyan, *Bull. Acad. Sci. USSR, Div. Chem. Sci.* **1990**, *39*, 674.

37. C.A. Fyfe, H.C. Clark, J.A. Davies, *J. Am. Chem. Soc.* **1983**, *105*, 6577.

38. V.N. Perchenko, G.A. Sytov, L.E. Ledina, *Dokl. AN USSR* **1981**, *261*, 114.

39. C.U. Pittman, Jr., P.L. Grube, O.E. Ayers, *J. Polymer Sci., A-1* **1972**, *10*, 379.

40. M.D. Rausch, A. Siegel, *Organometal. Chem.* **1968**, *11*, 317.

41. A.D. Pomogailo, I.E. Uflyand, *Adv. Polym. Sci.* **1990**, *97*, 61.

42. A.D. Pomogailo, I.E. Uflyand, *Macromolecular Metallochelates,* Khimiya, Moscow, 1991.
43. T.I. Movchan, A.G. Starikov, I.N. Ivleva, I.S. Voloshanovskii, A.D. Pomogailo, *Bull. Russ. Acad. Sci., Div. Chem. Sci.* **1992**, *41*, 545.
44. T.I. Movchan, I.I. Zheltvai, I.S. Voloshanovskii, A.D. Pomogailo, *Izv. Akad. Nauk, Ser. Khim.* **1992**, 2066.
45. Yu.S. Avlasevich, O.G. Kulinkovich, V.N.Knyukshto, K.N. Solov'ev, *J. Appl. Spectroscopy* **1999**, *66*, 597; ibid. **2000**, *67*, 663.
46. Yu.S. Avlasevich, O.G. Kulinkovich, V.N. Knyukshto, A.P. Losev, K.N. Solov'ev, *Polym. Sci., Ser. A* **1997**, *39*, 1155.
47. A.D. Pomogailo, N.M. Bravaya, V.F. Razumov, I.S. Voloshanovskii, N.A. Kitsenko, V.V. Berezovskii, A.I. Kuzaev, A.G. Ivanchenko, *Russ. Chem. Bull.* **1996**, *45*, 2773.
48. A.D. Pomogailo, V.F. Razumov, I.S. Voloshanovskii, *J. Porphyrins Phthalocyanines* **2000**, *4*, 45.
49. J.H. Schutten, P.Piet, A.L. German, *Macromol. Chem.* **1979**, *180*, 2341.
50. H. Kamogawa, J. Polym. Sci.: Polym. Lett. Ed. **1972**, *10*, 711.
51. H. Chen, M.M. Olmstead , R.L. Albright, J. Devenii, R.H. Fish, *Angew. Chem. Int. Ed. Engl.* **1997**, *36*, 642.
52. H.C. Lo, H. Chen, R.H. Fish, *Eur. J. Inorg.Chem.* **2001**, 2217.
53. S. Kopov, T.E. Hogen Esch, J. Smid, *Macromolecules* **1971**, *4*, 359; ibid, **1973**, *6*, 133.
54. Yu. M. Shulga, O.S. Roshchupkina, G.I. Dzhardimalieva, I.V. Chernushevich, A.F. Dodonov, Yu. V. Baldochin, P.Ya. Kolotyrkin, A.S. Rozenberg, A.D. Pomogailo, *Russ. Chem. Bull.* **1993**, *42,* 1661; ibid, **1994,** *93,* 983.
55. T. Güttner, U. Thewalt, *J. Organomet. Chem.* **1988**, *350*, 235.
56. Pat. Germany, 3137840, 1984.
57. J.C. Gressier, G. Levesque, H.Patin, in: *Metal-Containing Polymer Systems*, J.E. Sheats, C.E. Carraher, Jr. (Eds), Plenum Press, New York, 1985, p. 291; J.C. Gressier, G. Levesque, A.Patin, *Polymer Bull.* **1982**, *8*, 55.
58. V.A. Maksakov, V.P. Kirin, S.N. Konchenko, N.M. Bravaya, A.D. Pomogailo, A.V. Virovets, N.V. Podberezskaya, I.G. Barakovskaya, S.V. Tkachev, *Russ.Chem. Bull.* **1993**, *42*, 1236.
59. S.P. Tunik, S.I. Pomogailo, G.I. Dzhardimalieva, A.D. Pomogailo, I.I. Chuev, S.M. Aldoshin, A.B. Nikolskii, *Russ. Chem. Bull.* **1993**, *42*, 937.
60. V.A. Ershova, A.V. Golovin, L.A. Sheludyakova, P.P. Semyannikov, S.I. Pomogailo, A.D. Pomogailo, *Izv. Akad. Nauk, Ser. Khim.* **2000**, 1455.
61. M. Steinbach, C. Wegner, *Makromol. Chem.* **1977**, *178*, 1671.
62. J.W. Labadie, S.A. McDonnald, C.G. Willson, *Am. Chem. Soc., Polym. Prepr.* **1986**, *27*, 216.
63. N. Fujita, K. Sonogashira, *J. Polym. Sci.: Polym. Chem. Ed.* **1974**, *12*, 2845.
64. C.U. Pittman,Jr., *Organometallic Reactions,* Vol. 6, Marcel Decker, NewYork, **1977**, 1.
65. G.I. Dzhardimalieva, A.O. Tonoyan, A.D. Pomogailo, S.P. Davtyan, *Bull. Acad. Sci. USSR, Div. Chem. Sci.* **1987**, *36*, 1612.
66. V.A. Kabanov, D.A. Topchiev, *Polymerization of Ionizing Monomers,* Nauka,

Moscow, 1975.

67. R.P. Hopkins, *Ind. Eng.Chem.* **1965**, *47*, 2258.
68. G.I. Dzhardimalieva, A.D. Pomogailo, Macromol. Symp. **2002**, *186*, 147; *Russian Chem. Rev.* **2003**, *72* (in press).
69. V.F. Kurenkov, A.K. Vagapova, V.A. Myagchenkov, *Eur. Polym. J.* **1982**, *18*, 763.
70. V. F. Kurenkov, O.A. Zaitseva, N.V. Isaeva, *Zh. Prikl. Khim.* **2001**, *74*,1684.
71. G.I. Dzhardimalieva, A.D. Pomogailo, S.P. Davtyan, V.I. Ponomarev, *Bull. Acad. Sci. USSR, Div. Chem. Sci.* **1988**, *37*, 1352.
72. R.Ralea, G. Ungurenasu, I. Maxim, *Rev. Roum. Chim.* **1967**, *12*, 523.
73. C.H. Bamford, A.D. Jenkins, R. Jonston, *Proc.Roy.Soc., A* **1957**, *241*, 364.
74. V.A. Kargin, V.A. Kabanov, V.P. Zubov, *Vysokomolek. Soedin.* **1960**, *2*, 765.
75. M. Imoto, T. Otsu, S. Shimizu, *Makromol. Chem., B* **1963**, *65*, 174.
76. L.I. Skushnikova, E.S. Domnina, G.G. Skvorstova, *Vysokomolek. Soedin. B* **1982**, *24*, 11.
77. T. Maekawa, S. Mah, S. Okamura, *J. Macromol. Sci.-Chem.* **1978**, *12*, 1.
78. V.H. Perchenko, G.A. Sytov, G.L. Kamneva, N.C. Nametkin, *Vysokomolek. Soedin. A* **1983**, *25*, 1723.
79. S. Tazuke, M. Sato, S.Okamura.,*J. Polym. Sci., A-1* **1966**, *4*, 2461.
80. J. Pojman, D. Fortenberry, V. Ilyashenko, *Int. J. Self-Propag. High-Temp. Synth.* **1997**, *6*, 355.
81. V. S. Savostyanov, A. D. Pomogailo, B. S. Selenova, D. A. Kritskaya, A. N. Ponomarev, *Bull. Acad. Sci. USSR, Div. Chem. Sci.* **1990**, *39*, 680; *J. Polym. Sci.: Part A: Polym. Chem.* **1994**, *32*, 1201.
82. V. V. Barelko, A. D. Pomogailo, G. I. Dzhardimalieva, S. I. Evstratova, A. S. Rozenberg, I. E. Uflyand, *Chaos* **1999**, *9*, 342; *Dokl. Akad. Nauk* (Dokl. Chem. Engl. Transl.), **1999**, *365*, 201.
83. G. I. Dzhardimalieva, A. D. Pomogailo, V.A. Volpert, *J. Inorg. Organomet. Polym.* **2003**, *12* (in press).
84. A. S. Rozenberg, A. V. Raevskii, E. I. Aleksandrova, O. I. Kolesova, G. I. Dzhardimalieva, A. D. Pomogailo, *Russ. Chem. Bull., Int. Ed.* **2001**, *50*, 901.
85. G.M. Brown, C.C. Frazier, *Acta Cryst. C* **1989**, *45*, 1158.
86. R.W. Heo, F.B. Somoza, T.R. Lee, *J. Am. Chem. Soc.* **1998**, *120*, 1621.
87. D. F. Grishin, L.L. Semenycheva, I.S. Il'ichev, Yu. A. Kurskii *Vysokomolek. Soed., A* **2000**, *42*, 594; ibid, **2001**, B, *43*, 1416.
88. M.A. Diab., A.Z.El-Sonbati, A.A.Aggour, *J. Therm. Anal.* 1990, **36**, 957.
89. I.M Calvert., R.H Schemhl., B.P. Sullivan, J.S Facci., T.J. Meyer, R.W. Murray, *Inorg. Chem.* 1983, **22**, 2151.
90. P.A.Ennis, J.M.Kelly, C.M.O'Connel, *J. Chem. Soc. Dalton Trans.* **1986**, 2485.
91. M. Kamachi, H. Akimoto, W. Mori, M. Kishita, *Polym. J.* **1984**, *16*, 23.
92. S. Nozakura, M. Kamachi, *Makromol. Chem. Suppl.* **1985**, *12*, 255.
93. M. Kamachi, X.S. Cheng, T. Kida, A. Kajiwara, M. Shibasaka, S. Nagata, *Macromolecules* **1987**, *20*, 2665.
94. H. Aota, Y. Morishima, M. Kamachi, *Photochem. Photobiol.* **1993**, *57*, 989.
95. H. Aota, S. Araki, Y. Morishima, M. Kamachi, *Macromolecules* **1997**, *30*, 4090.
96. Y. Morishima, S. Nomura, T. Ikeda, M. Seki, M. Kamachi, *Macromolecules* **1995**, *28*, 2874.

97. A. Kajiwara, K. Aramata, M. Kamachi, K. Sumi, *Polym. J.* **1994**, *26*, 215.
98. N. Watanabe, X.S. Cheng, A. Harada, M. Kamachi, H. Nakayama, W. Mori, M. Kishita, *Polym. J.* **1989**, *21*, 633.
99. Y. Morishima, M. Tsuji, M. Seki, M. Kamachi, *Macromolecules* **1993**, *26*, 3229.
100. J.F. van der Pol, E. Neeleman, R.J.M. Nolte, J.W. Zwikker, W. Drenth, *Makromol. Chem.* **1989**, *190*, 2727.
101. C.F. van Nostrum, R.J.M. Nolte, M.A.C. Devillers, G.T. Oestergetel, M.N. Teerenstra, A.J. Schouten, *Macromolecules* **1993**, *26*, 3306.
102. M. Kimura, K. Wada, K. Ohta, K. Hanabusa, H. Shirai, N. Kobayashi, *Macromolecules* **2001**, *34*, 4706.
103. H. Eichhorn, M. Sturm, D. Wöhrle, *Makromol. Chem. Phys.* **1995**, *196*, 115.
104. H. Eichhorn, M. Rutloh, D. Wöhrle, J. Stumpe, *J. Chem. Soc., Perkin Trans.* **1996**, *2*, 1801.
105. H. Eichhorn, D.W. Bruce, D. Wöhrle, *Adv. Mater.* **1998**, *10*, 419.
106. H. Eichhorn, J. Porphyrins Phthalocyanines **2000**, 4, 88.
107. V.C. Dall'Orto, C. Danilowicz, S. Sobral, A. LoBalbo, I. Rezzano, *Anal. Chim. Acta* **1996**, *336*, 195.
108. V.C. Dall'Orto, C. Danilowicz, J. Hurst, A. LoBalbo, I. Rezzano, *Electroanalysis* **1998**, *10*, 127.
109. R. Paolesse, C.D. Natale, V.C. Dall'Orto, I. Rezzano, M. Mascini, A. D'Amico, *Thin Solid Films* **1999**, *354*, 245.
110. J.Furukawa, T. Tsuruta, *J. Polym. Sci.* **1959**, *36*, 275.
111. N.E.Ikladious, N.N.Messiha, A.F.Shaaban, *Eur. Polym. J.* **1984**, *20*, 625
112. Z.M. Rzaev, S.G. Medyakova, S.G. Mamedova, *Vysokomol. Soed., B* **1983**, *25*, 111.
113. A.Gronowskii, Z.Wojtczak, *Acta Polym.* **1985**, *36*, 59.
114. Z.Wojtczak, A.Gronowskii, *Makromol. Chem.* **1985**, *186*, 139.
115. T. Czerniawskii, Z.Wojtczak, *Acta Polym.* **1984**, *35*, 443.
116. G.I. Dzhardimalieva, A.D. Pomogailo. In: *Metal-Containing Polymeric Materials* (C.U. Pittman, Jr., C.E. Carraher, Jr., M. Zeldin, J.E. Sheats, B.M. Culbertson, Eds.), Plenum Press, New York, 1996, p. 63.
117. P.E. Matkovskii, I.D. Leonov, G.A. Beikhold, A.D. Pomogailo, Yu.V. Kissin, N.M. Chirkov, *Izv. Akad. Nauk SSSR, Khim.* **1970**, 1311; **1968**, 930.
118. A.D. Pomogailo, E. Baishiganov, F.S. Dy'achkovsky, *Vysokomolek. Soedin. A* **1981**, *23*, 230.
119. P.M. Ennis, J.M. Kelly, C.M. O'Connell, *J.Chem. Soc., Dalton Trans.* **1986**, 2485.
120. E.N. Danilovtseva, A.I. Skushnikova, E.S. Domnina, A.V. Afonin, *Vysokomolek. Soedin., B* **1989**, *31*, 777.
121. M. Kato, H. Nishide, E. Tsuchida, *J. Polym. Sci., Polym. Chem. Ed.* **1981**, *19*, 1803.
122. V.S. Savost'yanov, V.N. Vasilets, O.V. Ermakov, E.A. Sokolov, A.D. Pomogailo, D.A. Kritskaya, A.N. Ponomarev, *Izv. Akad. Nauk, Ser. Chem.* **1992**, 2073.
123. X. Schulze, J. Serin, A. Adronov, M.J. Frechet, *Chem. Commun.* **2001**, 1160.
124. A.D. Pomogailo, V.S. Savostyanov, *Metal-Containing Monomers and Their Polymers*, Khimiya, Moscow, 1988.
125. L.L. Semyonycheva, D.F. Grishin, I.S. Illichev, A.N. Artèmov, *Polymery* **2001**, *46*, 178.

126. L.L. Semyonycheva, A.N. Artemov, I.S. Illichev, E.V. Sazonova, *Khim. i Khim. Tekhnol.* **1999**, *42*. 96.

127. Y. Ueba, K.J. Zhu, E. Banks, Y. Okamoto, *J.Polym. Sci., Polym. Chem. Ed.* **1982**, *20*, 1271.

128. Y.Fujii, K. Kikuchi, K. Matsutani, *Chem. Lett.* **1984**, 1487.

129. T. Sato, N. Morita, I. Kamiya, H. Tanaka, T. Ota, *Makromol. Chem. Rapid Commun.* **1990**, *11*, 553.

130. H. Nishide, K. Shinohara, E. Tsuchida, *J. Polym. Sci., Polym. Chem. Ed.* **1981**, *19*, 1109.

131. E. Kokufuta, H. Watanabe, I. Nakamara, *Polym. Bull.* **1981**, *4*, 603.

132. A.B. Solovieva, E.A. Lukashova, A.V. Vorobiev, S.F. Timashev, *Reactive Polymers* **1991**, *16*, 9.

133. G.I. Dzhardimalieva, V.A. Zhorin, I.N. Ivleva, A.D. Pomogailo, N.S. Enikolopyan, *Dokl. Akad. Nauk SSSR* **1986**, *287*, 654.

134. G.I. Dzhardimalieva, A.D. Pomogailo, *Izv. Akad. Nauk SSSR, Ser. Khim.* **1991**, 352.

135. V.S. Savostyanov, V.A. Zhorin, G.I. Dzhardimalieva, A.D. Pomogailo, V.A. Dubovitskii, V.N. Topnikov, M.K. Makova, A.N. Ponomarev, *Dokl. Akad. Nauk SSSR* **1991**, *318*, 278.

136. N.M. Bravaya, A.D. Pomogailo, V.A. Maksakov, V.P. Kirin, V.P. Grachev, A.I. Kuzaev, *Russ. Chem. Bull.* **1995**, *44*, 1062.

137. S.I. Pomogailo, G.I. Dzhardimalieva, V.A. Ershova, S.M. Aldoshin, A.D. Pomogailo, *Macromol. Symp.* **2002**, *186*, 155.

138. O.A. Adamenko, G.V. Lukova, N.D. Golubeva, V.A. Smirnov, G.N. Boiko, A.D. Pomogailo, I.E. Uflyand, *Dokl. Akad. Nauk* **2001**, *381*, 360.

139. G.I.Dzhardimalieva, A.D.Pomogailo, *Macromol. Symp.* **1998**, *131*, 19.

140. A.D. Pomogailo, N.D. Golubeva, *Russ. Chem. Bull.* **1994**, *43*, 2020.

141. N.H. Agnew, M.E. Brown, *J. Polym. Sci., Polym. Chem. Ed.* **1974**, *12*, 1493.

142. M.D. Rausch, G.A. Moser, E.J. Zaiko, A.L. Lippman, Jr., *Organometal. Chem.* **1970**, *23*, 185.

143. I.S. Il'ichev, *Thesis*. Nizhnii Novgorod State University by N.I. Lobachevskii, Nizhnii Novgorod, 2001.

144. D.F. Grishin, L.L. Semyonycheva, I.S. Il'ichev, *Izv. Akad. Nauk, Ser. Khim.* **2000**, 1319.

145. Y.C.Lin, A. Mayr, C.B. Knobler, H.D. Kaesz, *J. Organomet. Chem.* **1984**, *272*, 207; R.Szostak, C.E. Strouse, H.D. Kaesz, *J. Organomet. Chem.* **1980**, *191*, 243.

5 Binding of Metal Ions and Metal Complexes to Macromolecular Carriers

Dieter Wöhrle

A macromolecule bearing suitable ligand groups or substituents can interact with a metal part (metal ion, metal complex) by covalent, coordinative, ionic or π–complex interactions as schematically shown in Fig. 5-1 [1–5]. A great variety of macromolecules can be employed for the binding of the metal part (Section 1.2.1):

- Linear organic polymers bearing suitable ligand groups or substituents are soluble in organic solvents or water. The interaction with a metal part can be studied using methods employed for analogous low molecular weight metal complexes. The resulting polymeric metal complexes are often insoluble materials.
- Cross-linked insoluble organic copolymers consist of a monovinyl compound bearing the group for interaction with a metal part and a divinyl comonomer for cross-linking (often 1,4-divinylbenzene or 1,2-ethanedioldimethacrylate). With a low degree of cross-linking the resins are capable of swelling and with a high degree of cross-linking macroporous resins have a permanent pore structure. Cross-linked networks can also be obtained by polycondensation reactions, for example, of phenol derivatives with formaldehyde.
- Either different monomers for polymerization or reactive molecules bearing groups for interaction with the metal part can be grafted onto the surface of organic polymers such as polyethylene or poly(tetrafluoroethylene) by γ–irradiation, accelerated electrons, low pressure gas discharge or chemical treatment.
- Inorganic macromolecules/carriers especially different kinds of silica (sol–gel, monosphere, glass surfaces, etc.) can be modified through their surface –OH groups with the effect of giving functionalities for interaction with the metal part.

The advantages are that well-defined macromolecules are employed and the kind of binding or complex formation can be studied in analogous reactions with low molecular weight compounds. Three sections of this chapter will cover the binding of metal ions, metal complexes (or their ligands) and some π–complexes/organometallic compounds. In references [1–3] the literature in this field up to around 1988 is reviewed from the viewpoint of synthesis and the use of these polymeric metal complexes as catalysts. Another important property for technical application is the use of polymeric ligands as metal ion exchangers and for selective metal ion binding. Some examples are given below, as are some other properties. Examples are given here of ways in which the reaction of a metal ion with a polymer system can bring about interactions that do not occur in analogous low molecular weight compounds. Stability constants and the kinetics and thermodynamics of formation of polymeric metal complexes are described in more detail in Chapter 3.

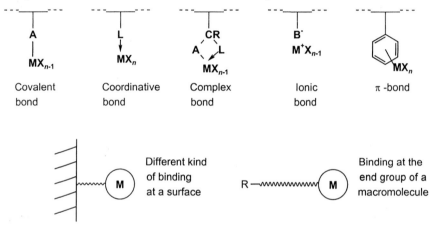

Figure 5-1. Schematic representation of the binding of metal ions, metal complexes or π-complexes at macromolecular carriers.

The interactions of a metal compound MX_n with an organic polymer ligand may occur either through monodentate binding (a) (when MX_n possesses only one coordination vacancy or group for interaction with the polymer ligand) or by either intra-(b) or intermolecular (c) polydentate binding (see Eq. 5-1). In the case of linear or branched organic polymers the macromolecular metal complexes (a) are, as a rule, soluble in organic solvents and their structure can be identified rather easily. The solubility of the bridged macromolecular metal complexes (b) decreases; they are more stable and have a less-defined structure. The complexes (c) with the intermolecular bridge bonds are insoluble and difficulty to characterize.

As an example, it has been shown for hydroxamic acid copolymers that infinite networks are formed when the probability of intermolecular binding of metal ions exceeds 50% (see also Section 3.4) [6].

Complex formation on the surface of inorganic carriers preferentially occurs by the intramolecular types (a) and (b).

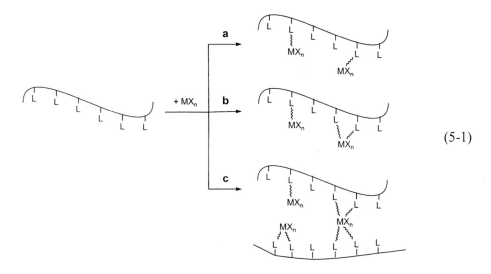

$$(5\text{-}1)$$

The interaction of a polymer ligand with metal ions in aqueous solutions can be explained in more detail. Figure 5-2 shows the dependences of the change of the hydrogel swelling coefficient (K_{sw}) of poly(ethyleneimine) (PEI) and poly(allylamine) (PAA·HCl) hydrochloride hydrogel, and the reduced viscosity (η) of its linear polymer, on the concentration of copper sulfate (C_{CuSO_4}) in aqueous solution (curves 1, 2) and the ratio of polymer functional groups (C_p) to metal ion concentration (curves 3, 4) [7]. The strong compression of hydrogel volume with increasing amount of the metal ion is a characteristic of both investigated systems. Attention must be paid to the influence of the degree of macroligand ionization on the character of the conformational changes of linear segments of the gel. It may be seen that at high pH the swelling coefficient of the PEI gel passes through a maximum in the gel–metal ion systems at a concentration of $CuSO_4$ equal to $8 \cdot 10^{-3}$ mol L^{-1} for Cu^{2+} (molar ratio of Cu^{2+}:PEI = 0.25). The increase in the degree of hydrogel swelling following complexation with metal ions at high pH can be explained in terms of additional charges in the slightly charged gel from bivalent metal ions coordinated with the amino groups of PEI. The latter increases the electrostatic energy of the system resulting in an increased swelling coefficient. For PAA-HCl hydrogel a decrease of the swelling coefficient is observed, caused by intramolecular chelation between metal ions and polyligands. This results in additional cross-linking in the network due to the donor–acceptor bonds and compaction of linear parts of the polymers between covalently cross-linked points. At low pH values the complexation proceeds by a substitution mechanism, in which the hydrogen ions of protonized nitrogen atoms in the gel are replaced by metal ions, thus avoiding the stage at which the polymer chain acquires a charge as was observed at a high pH value (see Fig. 5-2, gel PEI-Cu^{2+}, curves 1 and 2). An appropriate correlation between changes of K_{sw} and the reduced viscosity of the gel and linear polymer is observed (Fig. 5-2, curves 1 and 2).

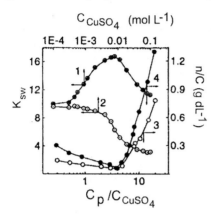

Figure 5-2. Dependences of the swelling degree and reduced viscosity for PEI gels (1, 2), PAA • HCl gel (3) and its linear polymer (4) on the $CuSO_4$ concentration at pH 8.3 (1, 3, 4) and 6.5 (2).

In most cases the structure of the local chelate unit in macromolecular complexes is the same as in the low molecular weight analogues. But the polymer chain may provide a significant influence. For salicylaldime ligands the structure of the complex units are different in low molecular weight and macromolecular ones: planar for the low molecular weight complex and distorted tetrahedral for the macromolecular complex [8]. The influence of ring size on the mechanism of binding of metal ions by polymers can be illustrated in relation to the formation of complexes between $TiCl_4$ and copolymers of styrene and the diallyl esters of dicarboxylic acids [9]. For $n = 1$ or 2, a mixture of complexes with the *cis*-disposition of **1** and *trans*-disposition of **2** of the carbonyl groups is formed. An increase in the size of the separating bridge ($n > 3$) precludes the formation of type 1 complexes (see also Sections 3.3 and 3.5).

Complex formation can also be influenced by the nature of the connecting bridge between the complexing unit and the polymer chain. For example, the transfer of Cu^{2+} from the aqueous to the organic phase (chloroform, toluene) for the formation of a complex with a hydrophobic low molecular weight ligand (compound **3a**) occurs readily. In contrast, complexation by the polymeric analogue **3b** is ineffective. Only the replacement of the short and hydrophobic methylene bridge in compound **3b** by the long hydrophilic ethylenediamine (compound **3c**) or methylamine (compound **3d**) units leads to appreciable hydrophilicity and spatial mobility of the complexing unit. This results in the diffusion of ions in the polymeric medium and allows the ligands bound to the polymer to be more mobile [10]. Steric hindrance of the macromolecular chain often prevents the formation of a multidentate complex. If polystyrene is substituted by bipyridyl groups, the formation of a monodentate complex **4** is observed and not the expected trisbipyridyl complex [11].

3a: R = CH₂, R₁ = CH₃
3b: R = CH₂, R₁ = polystyrene
3c: R = -NHCH₂CH₂NH-, R₁ = polystyrene
3d: R = -CH₂NH-, R₂ = polystyrene

4

Close packing in a polymer chain may lead to uncoordinated ligand groups. Poly(4-vinylpyridine) dissolved in an ethanol/water mixture with Co(II)–acetylacetonate results in a degree of complexation of ~0.7. For steric reasons, the rate of formation of the Co(II) complex with R of partly quarternized poly(4-vinylpyridine), decreases as follows: R = –CH₃ > CH₂–C₆H₅ [12].

Another important point of stereochemical recognition with metal ions called the "template" or "memory" effect is worth mentioning. A template effect is observed during the formation of the complexes of corresponding ions with some copolymers followed by cross-linking of the chains [13–16]. The structure of the macrocomplex formed during interaction of the metal ion with the ligand is strictly determined by their nature. If then the metal ion is removed while the formed stereostructure of the polymer is preserved, the remaining polymer ligand has a "pocket" (template) that will fit the same metal ions which were removed from the polymeric matrix (Eq. 5-2). The selectivity and value of the template effect depend on the spatial organization, on the nature of the complexing ligands and the stabilities of the formed complexes. Examples are complexes of poly(4-vinylpyridine) cross-linked with 1,4-bromobutane or complexes of poly(ethyleneimine) cross-linked with N,N'-methylene-diacrylamide [16]. For the cross-linked poly(ethyleneimine) the distribution coefficients of Cu ions between the non-prearranged polymer and the Cu(II)-prearranged polymer is ~6.25 and for Ni ions using Ni(II)-prearranged polymer it is ~0.9 which shows the different selectivity in metal ion uptake. Catalytic activities for oxidation reactions were investigated in detail.

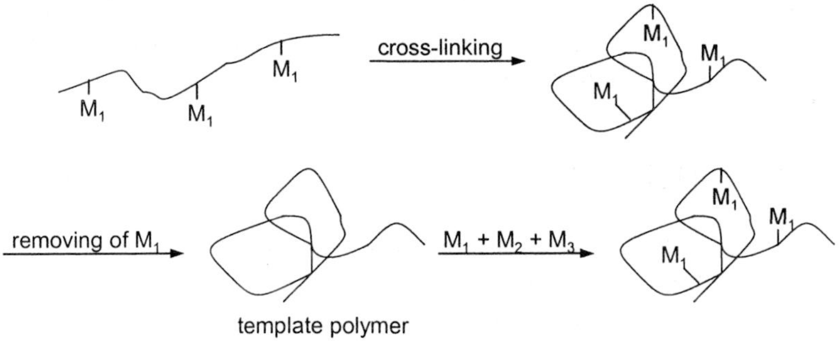

$$(5-2)$$

template polymer

Another possibility for realizing a template effect uses the copolymerization of metal complex vinyl monomers. Copolymerization of Ni(II), Co(II) or Cr(III) complexes of bis[di-(4-vinylphenyl)]dithiophosphinates with styrene and ethyleneglycoldimethacrylate yields cross-linked polymers, which, after removal of the metal ion, exhibit some degree of selectivity towards their "own" metal ion [16,17]. Copolymerization of the Zn(II)-complex of 1,4,7-triazacyclononane with divinylbenzene (molar ratio ~1:3) results in a macroporous copolymer containing sandwich complexes **5** of the Zn(II) complex [18]. After removal of Zn(II) the prearranged copolymer shows a selectivity of Cu(II):Zn(II) up to 157:1. This means that in this case the thermodynamic stability of the new complex formation dominates over the template effect. This template effect of Zn(II) for Cu(II) results also in a high selectivity of Cu(II) against other transition metal ions such as Fe(III). Altogether, prearrangement effects existing are difficulty to predict and further research is necessary.

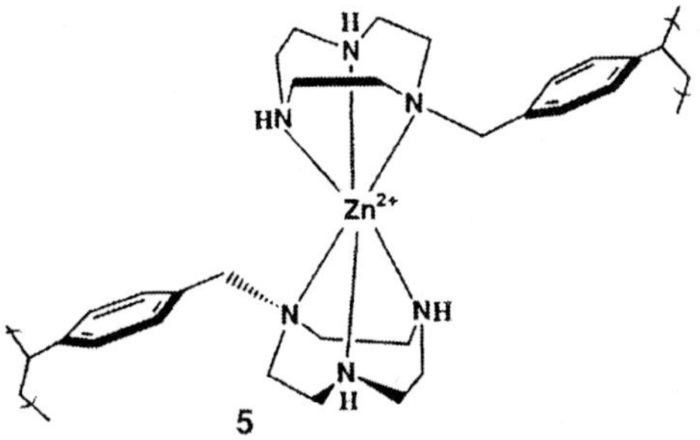

5

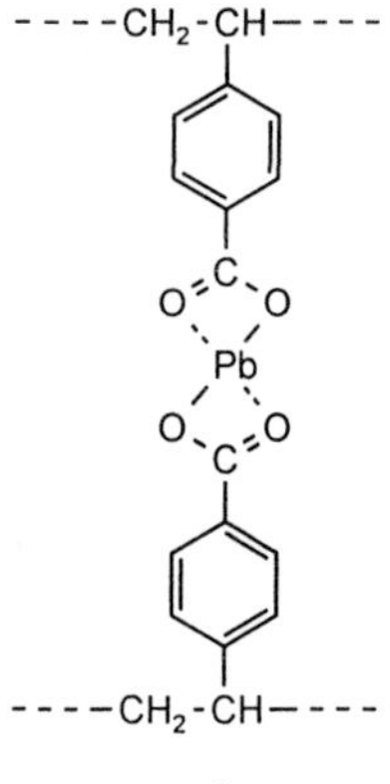

6

Styrene has been copolymerized with lead(II) vinylbenzoate to form a template resin **6** which was then stripped of the Pb ions [19]. The resin had a memory effect and was selective for Pb(II) over Cd(II) and Cu(II) at pH 2.6–2.8. The selectivity was increased by a factor of 1.7 over Cd(II) and 3 over Cu(II) compared to the untemplated resin.

5.1 Binding of Metal Ions

In this case a macromolecule containing ligand groups for metal ion binding is synthesized first. Immobilized ligand groups contain, for example, oxygen, nitrogen, sulfur and phosphorus. Often chloromethylated polystyrenes either linear or with a different degree of cross-linking are employed as starting material to immobilize ligand groups for metal ions. Water-soluble or insoluble cross-linked hydrophilic polymers with complex binding abilities are formed by derivatization of, for example, poly(vinylamine), poly(methacrylic acid), poly(acrylic acid) or poly(N-vinylpyrrolidone). In addition, modification of inorganic macromolecules has become important in recent years. The best-known example is the modification of surface Si–OH-groups with ligand groups. Reversible metal ion binding, especially at organic polymer ligands, has found broad application in ion exchangers. Further, metal ion-loaded carriers are interesting as catalysts, or in the case of alkali metal ion-binding as conducting electrolytes. Several properties of polymeric metal complexes are mentioned in subsequent chapters: binding of small molecules in Chapter 9, physical and optical sensors in Chapter 10, catalytic properties in Chapters 11 and 12, photo-electron- and photo-energy-induced processes in Chapter 13. We shall now discuss the properties of polymer ligands for use as ion exchangers.

The most common applications of ion exchangers are (1) process water deionization or softening; (2) rinse water recovery; (3) recovery of metals from rinse water; (4) final effluent purification to ensure compliance; (5) bath purification [1,20–24]. Another application of increasing interest is obtaining rare metals such as gold or uranium from sea water [1,2,24]. A scheme for pollution prevention in a metal finishing process is illustrated in Fig. 5-3. To achieve zero or near-zero discharge, water must be recycled, metals or metal salts must be recycled in the process or recovered and contaminants must be removed from the process bath. The reversible nature of ion exchange and sorption processes make them particularly suited to these applications. Over 80% of industries use de-ionized or softened water for rinsing. Besides ion exchange, solvent extraction is also used for removal of metal ions from water [25].

Ion-exchange materials often contain charged groups (usually sodium or hydrogen for cation exchangers and chloride or hydroxide for anion exchangers), and three key factors determine the effectiveness of the ion-exchange process: selectivity for the target metal ion, the number of active sites (capacity) and regenerability. It is necessary to concentrate the target metal ion 100–10,000 times in the regeneration effluent. Commercially available organic ion exchangers have a particle size ranging from 0.3 to 1.2 mm and a pore size between 5 nm (micropores) and more than 50 nm (macropores).

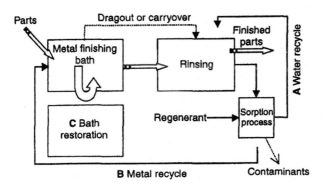

Figure 5-3. Scheme of generic metal finishing process identifying in-process pollution.

Ion-exchange affinity refers to the equilibrium constants for metal ions between the solid resin phase and the mobile liquid phase. A significant factor for the affinity of a resin for a particular ion is pH. Qualitatively, the following criteria determine the affinity of a chelating polymer for a given metal: ionic charge, hydrated ionic radius and ligand bonding with exposed electron pairs on nitrogen, oxygen, phosphorus or sulfur. Ion-exchange selectivity is defined as the ratio of the equilibrium constants between the target ion and a reference ion. Commercially available strong acidic, weak acidic, strong basic and weak basic ion-exchange materials containing sulfonic acid, carboxylic acid, quarternary amine or tertiary amine groups, respectively, are not very selective and exhibit a dry capacity between 1 and 10 eq. kg^{-1} [23]. Selective metal ion binding has been reported for special ligand groups and definite pH [22–24]. From the tremendous body of literature some examples of more recent papers are given below. Several ion exchangers are available from different chemical companies. Commercial names of ion exchangers are, for example, Amberlite®, Amberlist®, Dowex®, Nafion®.

5.1.1   Metal Ion Binding at Polymeric Oxygen Donor Ligands

As just mentioned, strong or weak acidic ion exchangers contain sulfonic acid or carboxylic acid groups in the H$^+$ or alkali metal salt form and consist of 1,4–divinylbenzene cross-linked polystyrene or poly(acrylic acid) as shown in formulae **7a** and **7b**. Another example is the sulfonated phenol–formaldehyde condensation polymer **8**. The preparation of ion-exchange resins and the determination of their capacities are described in Section 5.4, Experiments 5-1 and 5-2.

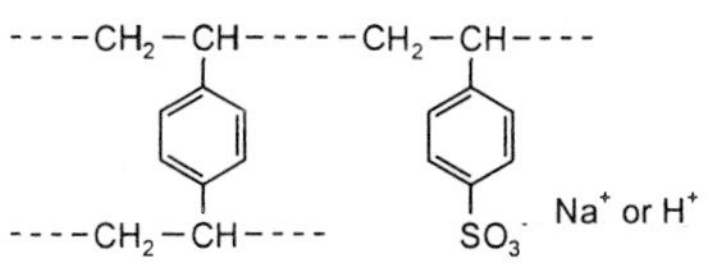

7a 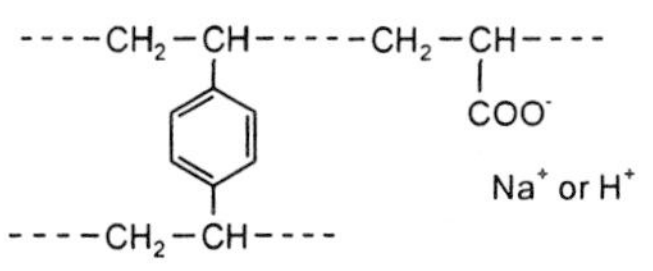   **7b** 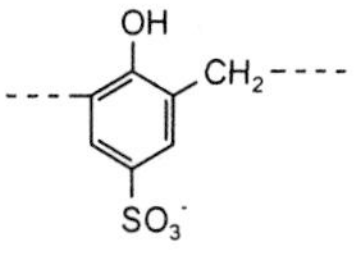   **8**

Completely water-soluble copolymers from 2-arylamido-2-methylpropane sulfonic acid and N-acryloyl-N-methylpiperazine have been synthesized by radical polymerization in yields >90% [26]. Their binding capacities for metal ions were studied at different pH and filtration factors, and the metal ion recovery was studied using ultrafiltration. The metal ion retention increased with increasing pH and content of sulfonic acid moieties in the copolymer. To increase the adhesion to silver, polyamide surfaces were modified by sulfonation reactions [27]. A maximum rate of modification was attained via both the gaseous SO_3 and concentrated H_2SO_4 in the presence of Ag_2SO_4 procedures. Excellent adhesion of Ag^+ was found. Platinum-containing catalysts were obtained by ion exchange of the complex $[(dppb)Pt(\mu-OH)_2]^{2+}$ (dppb = 1,2-bis-diphenylphosphinobutane) with commercially available sulfonated styrene–divinylbenzene copolymers [28]. These polymeric complexes exhibit a good catalytic activity in the Baeyer–Villinger oxidation of methylcyclohexanone with H_2O_2 to methyl-substituted lactones.

Polyester **9a**, which has amino sulfonic acid moieties, was obtained by a liquid/solid biphase polycondensation of terephthalyl chloride and N,N-bis(2-hydroxyethyl)-2-aminoethane sulfonic acid in trimethylphosphate using triethylamine as an acid acceptor [29]. A thin film of blends was prepared by casting a solution containing **9a**, poly(vinyl alcohol) and Ni^{2+} or Cu^{2+} salts. **9a** and the metal salts formed an interpenetrating network as shown in **9b**. The blends exhibit an electrical conductivity up to 10^{-6} S cm^{-1} (for an experiment see Section 5.4, Experiment 5-3).

9a 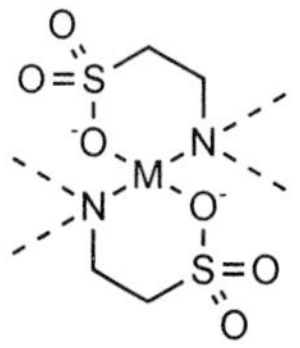   **9b**

Insoluble polychelates separate when an aqueous solution of poly(acrylic acid-co-acrylamide) is added to an aqueous solution of Cu(II), Co(II) and Ni(II) [30]. Tetrahedral arrangements were found for Co(II) and Ni(II). But homogeneous complex solutions can also be obtained as was shown by controlling the composition of an inter-polyelectrolyte of poly(acrylamide-co-acrylic acid) and poly(acrylamide-co-dimethyldiallylammonium chloride) with metal ions [31]. Cross-linked terpolymers prepared by radical polymerization of [2-(methacryloyloxy)ethyl]trimethyl ammonium chloride, acrylic acid and N,N-methylene-bis-acrylamide were tested as adsorbents for various metal ions at different pH values (Table 5-1) by a batch equilibrium procedure (for a preparation see Experiment 5-4, Section 5.4) [32]. The resin shows a high sensitivity for U(VI) with respect to all other metal ions and a recovery of over 65% by H_2SO_4 and Na_2CO_3.

Table 5-1. Sorption in percentage and in parenthesis in meq of metal ions on a terpolymer (see text) at different pH values. Concentration of metal ion: 1.0 g L^{-1}.

Metal ion	pH			
	1	2	3	5
Cu(II)	0.0		0.0	0.0
Cd(II)	3.9 (0.069)		0.0	0.0
Zn(II)	14.0 (0.367)		0.0	0.0
Pb(II)	8.6 (0.083)		2.0	5.2 (0.05)
Hg(II)	0.0	17 (0.169)		
Cr(II)	0.0		19.6 (1.133)	25.0 (1.445)
U(VI)	0.0		65.0 (1.638)	74.0 (1.865)

Copolymerization of N-acryloyldiethyl iminodiacetate with acrylic acid was carried out, and the resulting copolymer **10** complexed over 97% ^{152}Eu(III) from solution at pH 4 [33]. Metal ion adsorption was studied by introducing iminodiacetic acid ($-R-N(CH_2COOH)_2$) in a weak acid type ($-R(COOH)-)_n$ ion-exchange resin [34]. The luminescence properties of lanthanide ions with polycarboxylates were investigated in detail and are interesting for applications as luminescence sensors [35]. When water-soluble polymers having pendent carboxylic acid residues and powdered metal oxides containing leachable Ca(II), Al(III) etc. ions are mixed in the presence of controlled amounts of water, metal cation–carboxylate anion salt-bridges are generated which bring about curing or hardening of the formulation [36]. These so-called glass-ionomers are applied as dental biomaterials. An example is a terpolymer based on acrylic acid, itaconic acid and methacrylglutamic acid **11** hardened with Ca(II) or Al(III).

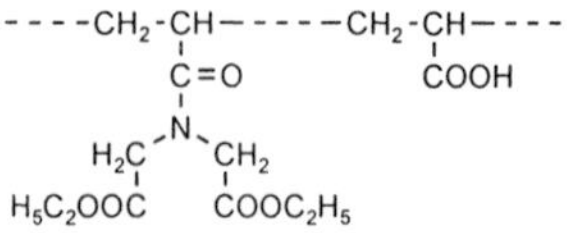

10 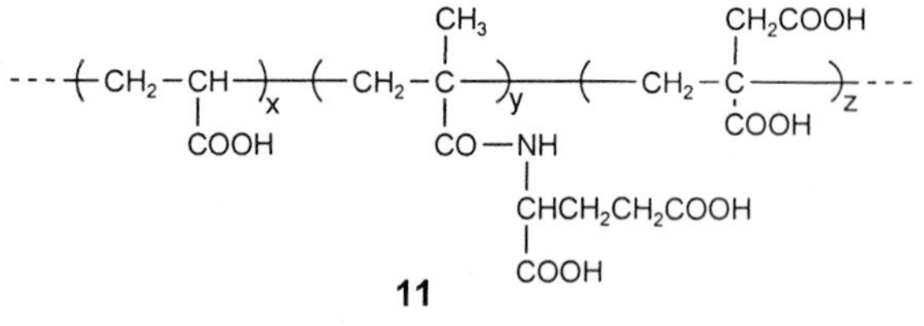   **11**

The interaction of Tl(I), Ca(II) and Tb(III) with copolymers of methacrylic acid and methyl methacrylate containing 9-vinylanthracene as a fluorescence label (formula **12**) has been studied [37]. With Tl(I), quenching of the fluorescence was observed. The cation is held in the polyelectrolyte domain. Stronger binding was found with Ca(II) and Tb(III), and in this case an increase in fluorescence intensity was observed on complexation due to the anthracene group being in a more hydrophobic region as a result of conformational changes in the polymer chain. In all cases, the extent of binding increases as a function of the charge of the polymer with either increasing fraction of carboxylic acids or the pH. A new interesting property was shown with membranes prepared from methyl methacrylate–methacrylic acid copolymer ionically cross-linked with Fe(III) and Co(II) [38]. Permeation and separation studies for a benzene/cyclohexane mixture indicate both permselectivity and permeability for benzene. Differences between the two metal ions are discussed.

12

Some recent papers report on the metal ion binding at homo- and copolymers of maleic acid: a) Insoluble polymer metal complexes have been prepared from Cu(II), Co(II), Ni(II) and Zn(II) with poly(maleic acid) in water. For Ni(II) and Co(II) complexes a distorted octahedral symmetry has been proposed (see Section 5.4, Experiment 5-5) [39]. b) The kinetics and equilibrium of the complexation of Al(III) with a random copolymer of maleic acid and acrylic acid has been studied [40]. In the first step of complexation an outer sphere complex is formed, followed in the second rate-determining step by the formation of an inner sphere complex. c) Insoluble complexes have been obtained from poly(maleic acid–co–olefin) with Cu(II), Co(II), Ni(II) and Zn(II) [41]. For the Zn(II) complexes a distorted octahedral geometry has been proposed.

We give two examples for the introduction of carboxylic acid groups into polymers. A glycine function can be incorporated by transamidation in cross-linked poly(acrylamide)s to obtain **13** (for a preparation see Experiment 5-6, Section 5.4) [42]. The metal uptake varied in the order: Cu(II) > Ni(II) ≈ Co(II) > Fe(II). The polymer **14** obtained via a Friedel–Crafts reaction of polystyrene with phthalic anhydride exhibits a high affinity for Cu(II) compared with other metal ions [43]. Alginic acid and hyaluronic acid are examples of natural polymers containing carboxylic acid groups. The removal of Fe(III) from aqueous solution wasstudied using membrane filtration. At first the water soluble alginic acid was used to bind the metal ion, and this was followed by batch ultrafiltration in a membrane of poly(methacrylic acid) modified with poly(ethylene glycol) [44]. The coordination of Zn(II) at a hyaluronate-modified electrode has also been studied [45].

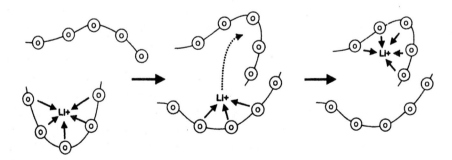

13

Poly(oxyethylene) combinations with various other comonomers [46], are of interest as solid polymer electrolytes after complex formation with Li(I) (complexation with Na(I), K(I), Mg(II), Ba(II), etc. has also been studied) [1,5,46–48]. The synthesis is carried out by direct interaction of the ligand and metal ions in solution or, if cross-linked poly(oxyethylene) is employed, by immersing the polymer ligand into a solution of the metal salt. Poly(oxypropylene), modified polysiloxanes, cross-linked phosphate esters and ethers [46,49,50], and structurally different ligands such as 2,5-dimercapto-1,3,4-thiadiazol-polyaniline [51] have also been used as polymer ligands. The developments in this field are reviewed in [46]. In this review the segmental motion of Li(I) in a poly(oxyethylene) is described as shown in Fig. 5-4.

Figure 5-4. Scheme of the segmental motion of Li(I) in a poly(oxyethylene) matrix. The circles represent the ether oxygen of the polymer.

Crown ethers have been immobilized within polymer supports and used in metal ion separation. Crown ether moieties at cross-linked polystyrene have been prepared by the reaction of cross-linked chloromethylated polystyrene with hydroxyl-substituted crown ethers [52]. The resins **15** formed from the condensation of dibenzo—18-crown-6 and formaldehyde were modified with phosphonate monoethylester and phosphonic acid groups [53]. Resin **15** demonstrates selectivity for K(I) over the other alkali metal ions at pH < 8, and selectivity for Li(I) at pH > 8. A resin **16** prepared from the condensation of a substituted crown ether with formaldehyde is selective for Na(I) over the other alkali metal ions [54].

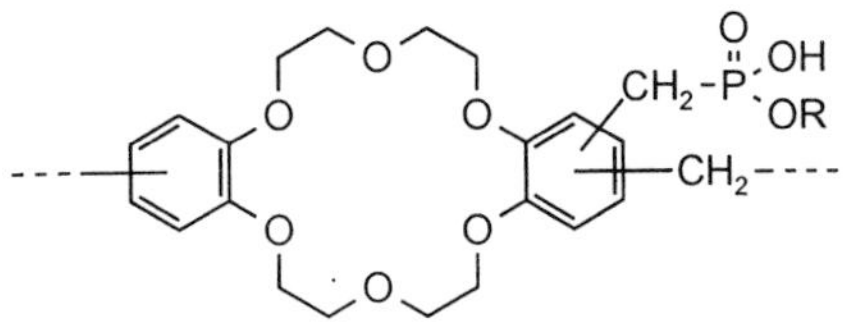

15 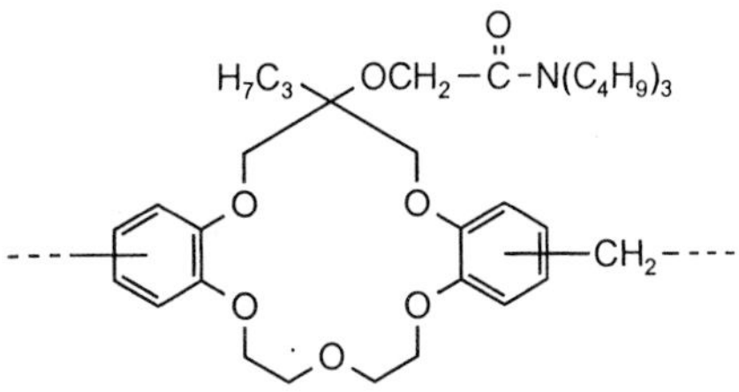   **16**

Polymeric calix[4]crown-4 and calix[4]arenes have been investigated for metal ion binding [55]. For example, the copolymer **17** was prepared by the copolymerization of styrene and the acryloyloxy-substituted calix[4]arene. Complexation studies were carried out using the liquid–liquid extraction procedure. Polymer **17**, like its monomer, exhibits high selectivity towards the toxic Hg(II) compared to alkali, alkaline earth and transition metal ions. Catechol, phenol, resorcinol, hydroxyacetophenone and other substituted aromatic compounds have been employed in polycondensation reactions with formaldehyde [24]. They exhibit different good selectivities: Li(I) against other alkali metal ions [56]; U(VI) against other transition metal ions [57]; Mg(II) against other transition metal ions [58]. An idealized structure of a polymer of the 2,4-dinitrophenylhydrazone derivative of 2-hydroxyacetophenone and 2-hydroxybenzoic acid with formaldehyde is shown in **18** [58]. They have a selectivity order at pH 8 of Cu(II) > Ni(II) > Zn(II) > Mg(II) > Mn(II) > Co(II).

Poly(2-hydroxy-4-acryl or methacryloyloxybenzophenone) was prepared via radical polymerization of the corresponding monomers [59]. They form polychelates in a metal-to-ligand ratio of 1:2 by coordination through the oxygen of the keto group and the oxygen of the phenolic –OH group. Polychelates of Cu(II) are square planar, of Ni(II), Mn(II) and Co(II) are octahedral and Ca(II), Cd(II) and Zn(II) are tetrahedral.

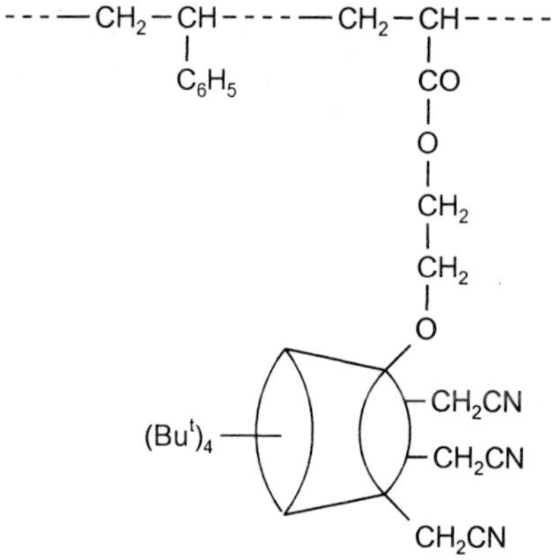

17

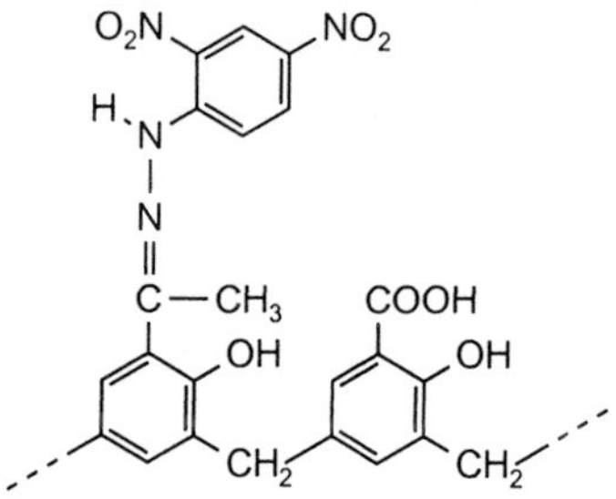

18

5.1.2   Metal Ion Binding at Polymeric Nitrogen Donor Ligands

As mentioned above, most commercially available strong basic or weak basic ion-exchange materials contain quarternary or tertiary amine groups, often bound to cross-linked polystyrene **19a** or poly(methacrylic acid ester) **19b**. These positively charged resins are not suitable for binding of positively charged metal ions, but can be used for the ionic binding of negatively charged metal complexes (see below in this section and Section 5.2.3).

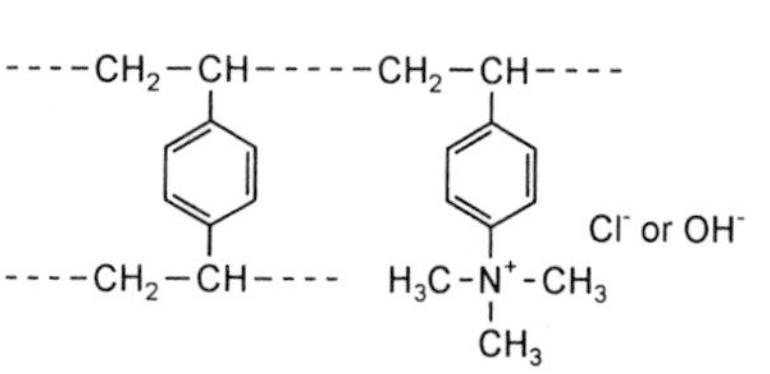

19a

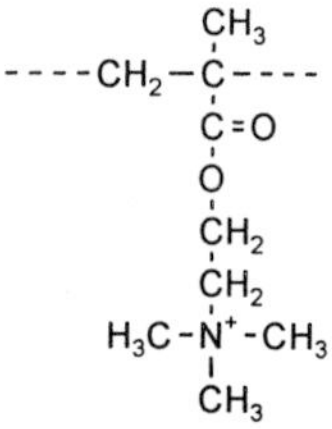

19b

Open-chain and cyclic amines can coordinate with various metal ions. Poly(ethyleneimine) prepared from 2-methyloxazoline by ring-opening polymerization wasinvestigated for Na(I) binding [60]. Various open-chain amines and amides, cyclic amines and amides such as **20** were synthesized starting from cross-linked chloromethylated polystyrene [61]. The modified polymers contained up to 2.7 mmol g^{-1} amine or amide group and were investigated for the reversible binding of Co(II), Ni(II) and Cu(II) (see Experiment 5-7, Section 5.4). Resin **21**, formed from the reaction of poly(ethyleneimine) with poly(vinyl benzaldehyde), is selective for Fe(III) over Cu(II), Ni(II), Co(II), Fe(II), Mn(II) and Zn(II) [62]. Wofaxit Mk 51, a polystyrene-based resin, which contains methylaminoglucitol, is capable of removing stannate, germanate, tellurite, molybdate and tungstate from brine

[63]. Soluble poly(allylamine) metal complexes of Co(II), Cu(II) and Ni(II) have been prepared by mixing the two components in different molar ratios and at different pH values in aqueous solution [64]. All aqueous solutions showed a pseudoplastic behavior.

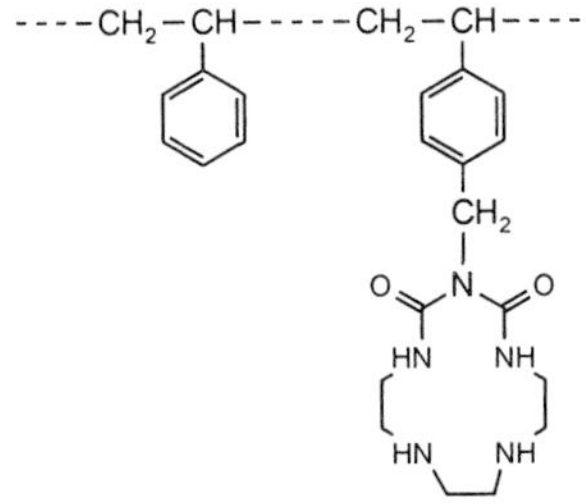

20

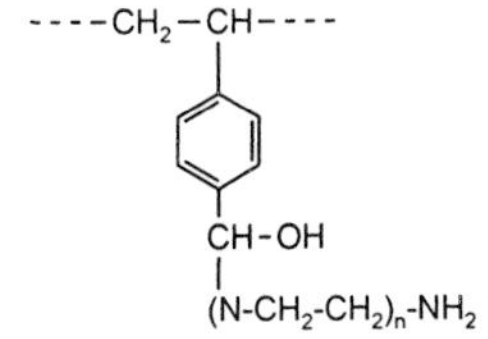

21

Platinum drugs such as *cisplatin* (*cis*–diamminodichloroplatinum(II)) are firmly established as a prominent class of anticancer agents available to clinical oncologists (for reviews see [65,66]). In order to overcome the limitations of the low molecular weight Pt compounds, such as short circulation residence time and powerful toxic side-effects, reversible binding of the active drug species to macromolecular carriers has recently been developed. Linear and water-soluble polymeric platinum complexes have been developed containing ethylene-diamine segments as side chain or main chain components [67–69]. The polymer ligands were prepared first and then reacted in water with specified equivalents of the tetrachloroplatinate(II) anion, followed by purification by dialysis. Two examples of prepared Pt complexes are shown in **22** and **23**. *In vitro* tests against the HeLa human cervic carcinoma cell line found that these two polymeric Pt complexes exhibit the best cytotoxicity [70].

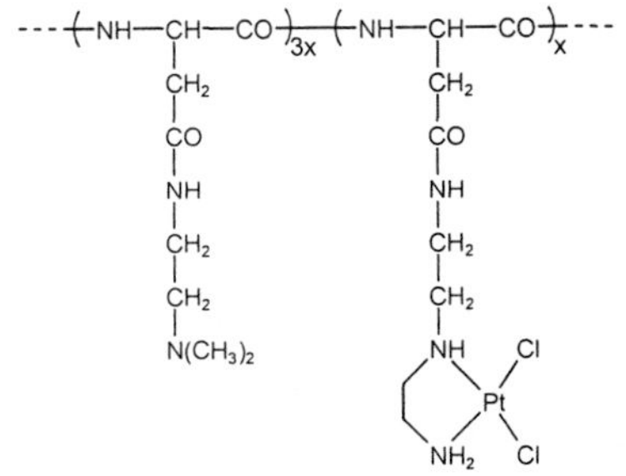

22 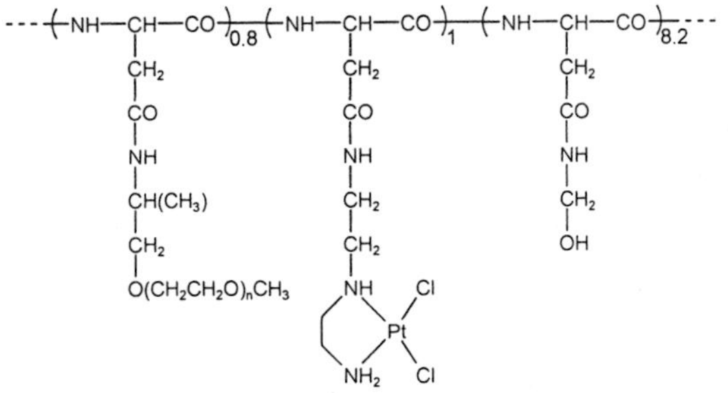   **23**

Linear and cross-linked poly[2-(methacryloyloxy)ethyl]trimethylammonium chloride **19b** have been investigated in aqueous solution for their interaction with $K_3[Fe(CN)_6]$ and $K_4[Fe(CN)_6]$ [71]. In Fig. 5-5 the dependences of the equilibrium swelling degree of different cross-linked hydrogels on the concentration of $K_4[Fe(CN)_6]$ are plotted. At a salt/gel ratio of 0.25 the gels attain a completely collapsed state which corresponds to the connection of one $[Fe(CN)_6]^{4-}$ ion with four monomer units. The positively charged o-phenylenediamine hydrochloride was covalently bound at cross-linked polystyrene [72]. The resin is very selective for $AuCl_4^-$ ions in strongly acidic solutions (Fig. 5-6) and can be regenerated with thiourea solution.

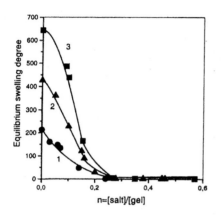

Figure 5-5. Swelling of cross-linked poly(methacryloyloxy) trimethylammonium gel in aqueous solution containing $K_4[Fe(CN)_6]$.

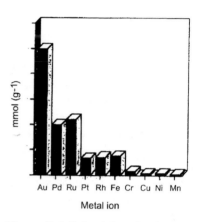

Figure 5-6. Selectivity of polymer bound o-phenylenediamine for $AuCl_4$ ion. 25 mg resin were shaken for 16 h with aqueous acidic solution containing 1 mM $KAuCl_4^-$, K_2PtCl_4, K_2PdCl_4, K_2RuCl_5, $Na_3Rh_2Cl_6$, MCl_n.

Cross-linked polystyrene–divinylbenzene copolymer has been function-alized with an aldehyde group at the benzene ring of the styrene moiety [73]. The aldehyde group was reacted with 1,2-diaminoethane to get a polymer bound Schiff-base ligand $R–N=CH–CH_2–CH_2–NH_2$. Immobilization of Fe(III) to this ligand resulted in catalytically active metal complexes. These supported catalysts are active in the epoxidation of cis-cyclooctene and styrene in the presence of $tert$-butylhydroperoxide. Solutions of undoped polyaniline in 1-methylpyrrolidone have been treated with Cu, Fe and Pd salts [74]. A bathochromic shift of the absorption of polyaniline at $\lambda\sim$ 640 nm is attributable to charge transfer from the benzoid to the chinoid form of the polymer. The complexes **24** are effective in the dehydrogenative oxidation of, for example, cinnamoyl alcohol. A Mo-complex bound at a silica surface **25** wasinvestigated as an effective catalyst for the oxidation of alcohol in the presence of $tert$-butylhydroperoxide [75]. The Si–OH groups were first silicated with octadecyldimethyl[3-(trimethoxysilyl)propyl] ammonium chloride and then reacted with MoO_3 (see Experiment 5-8, Section 5.4). Similar Fe(II) and Cu(II) catalysts for the oxidation of hydrocarbons have been prepared.

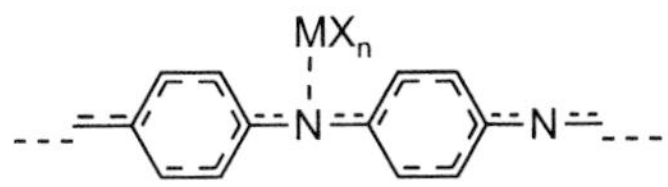

24

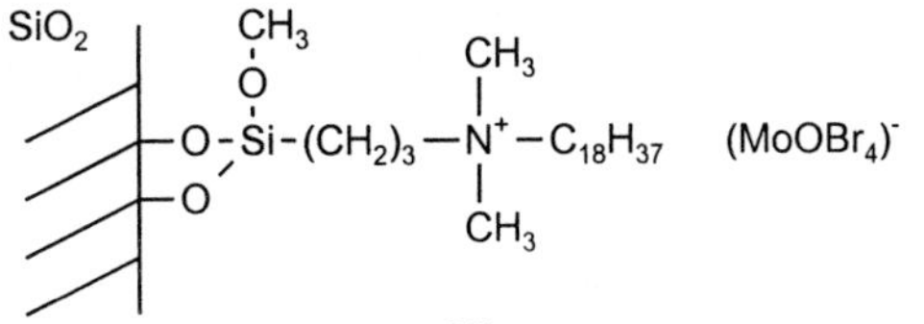

25

The excellent complexing ability of the pyridine group led to several investigations of the coordination of polymers bearing pyridyl or bipyridyl groups with metal ions such as Pt(II), Ru(II), Re(II), Co(II), Cu(II), Cr(III) and others [1–5,76,77]. Polymer metal complex formation of different poly(vinylpyridine)s in solution in hydrogels and at interfaces have been investigated [78]. In aqueous solution, interaction of H_2PtCl_6 with linear or cross-linked poly-(vinylpyridine)s results in reduced viscosities and reduced swelling coefficients, respectively. Complexation leads to molecular bridges and folding of the polymer. Film formation has been observed at the interface of poly(2-vinylpyridine) dissolved in benzene and metal salts dissolved in water. Chloromethylated styrene–divinylbenzene copolymer beads have been chemically modified by reaction with 2-aminopyridine to get the polymer bound 2-aminomethylenepyridine ligand [79]. Complexation with $RuCl_3$ led to coordination of the Ru(III) ion at the nitrogen of the pyridine ring. The materials are effective catalysts in the epoxidation of cyclic olefins and styrene in the presence of alkyl hydroperoxide as oxidant. Different tridendate bis(2-pyridylalkylamines) have been coupled to 3-(glycidyloxypropyl)trimethoxysilane and subsequently grafted onto silica to get **26** [80]. Most polymer ligands selectively absorb Cu(II) from aqueous solutions containing a mixture of different metal ions. Silica has been modified with 2-(phenylazo)pyridine for the synthesis of a macromolecular ligand which is a good ligand for Ru(III) [81]. The macromolecular Ru complex **27** is an excellent catalyst for the epoxidation of *trans*-stilbene.

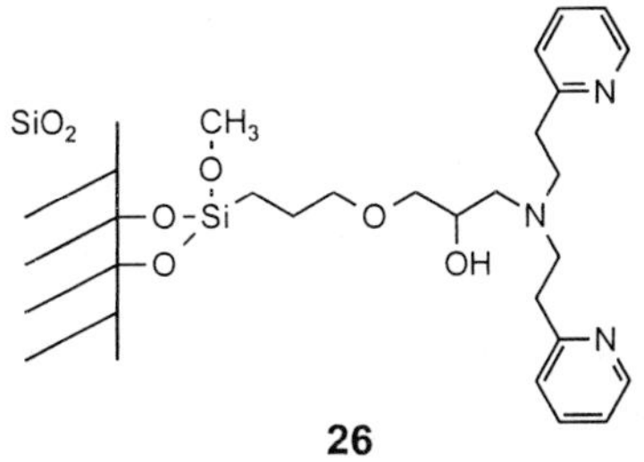

26

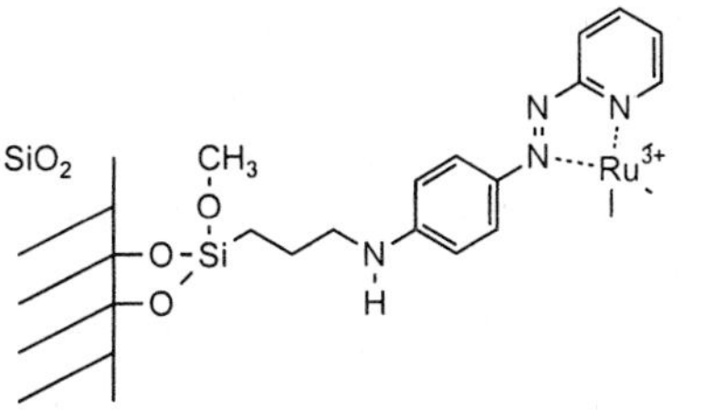

27

The synthesis of poly(dimethyl siloxanes) and polyisoprenes, substituted by *N*-pyrid-2-ylmethyl amide groups, has been described in a series of papers [82].

Two examples are shown in **28** and **29**. The polymeric ligands form 1:1 complexes with $ZnCl_2$ linked to the pyridyl group. Although the polymer ligands are liquid at room temperature, the melting point of the complexes has been observed to be ~140 °C. The polymeric complexes are thermoelastic and exhibit elastic properties.

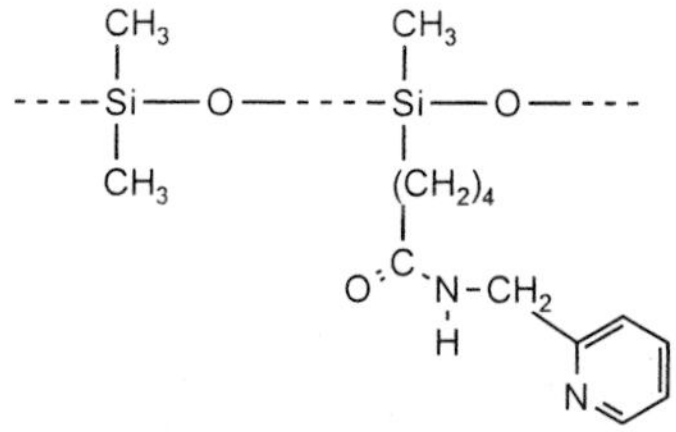

A polymer complex containing $Ru(bpy)_3^{2+}$ pendent groups was obtained by the reaction of a lithium-substituted polystyrene with 2,2'-bipyridyl followed by interaction with *cis*-$Ru(bpy)_2^{2+}$ [83]. Another example is binding of 4,4'-dicarboxy-2,2'-bipyridyl to a copolymer of *p*-aminostyrene followed by reaction with *cis*-$Ru(bpy)_2^{2+}$ to form **30** [84]. Other copolymers with pendent $Ru(bpy)_3^{2+}$ bound via a spacer or containing additional bound 4,4'–bipyridyl have also been prepared. These materials are interesting as sensitizers for visible light energy conversion. A series of polymeric ligands were prepared by radical polymerization of vinylpyridines and 4-methyl-4'-vinyl-2,2'-bipyridine with different dimethacrylate cross-linkers [85]. The polymeric ligands are selective for Cu(II) over Co(II) or Ni(II), for Hg(II) over Cd(II), for Au(III) over Ag(I) and for Pd(II) over Pt(II). Different phenylene–ethynylenethienylene-linked polymers with pendent oligopyridine groups, for example **31**, soluble in organic solvents and highly emissive, have been described [86]. These polymers are interesting as optical sensor materials. For terpyridine as substituent 5% initial emission quenching by transition metal ions such as Ni(II) in a concentration as low as $4 \cdot 10^{-8}$ M was measured. A polymer having fluorescent pyridyl-benzoxazol groups and soluble in organic solvents was prepared for the purpose of detecting of different transition metal ions [87]. Among the metal ions investigated, Fe(III) was shown to be the most effective in fluorescence quenching ability.

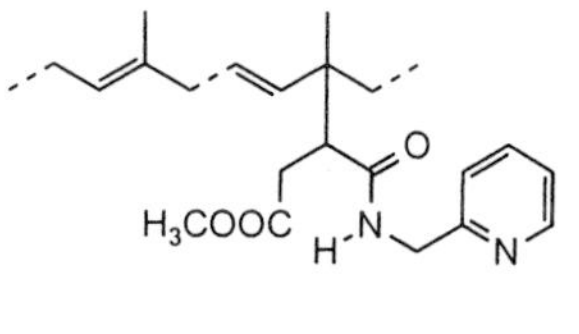

Other polymeric heterocyclic amines such as imidazole, benzimidazole pyrazole, triazole or tetrazole are also excellent ligands for transition metal ions [1,2,23,24]. A detailed study on the complexation of Ag(I) and Cu(II) with poly(*N*-vinylimidazole) gave insights into the intrinsic complexation equilibria [88].

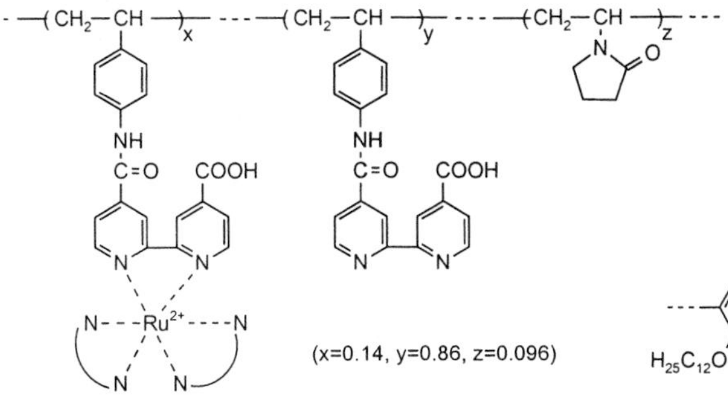

30   (x=0.14, y=0.86, z=0.096)   **31**

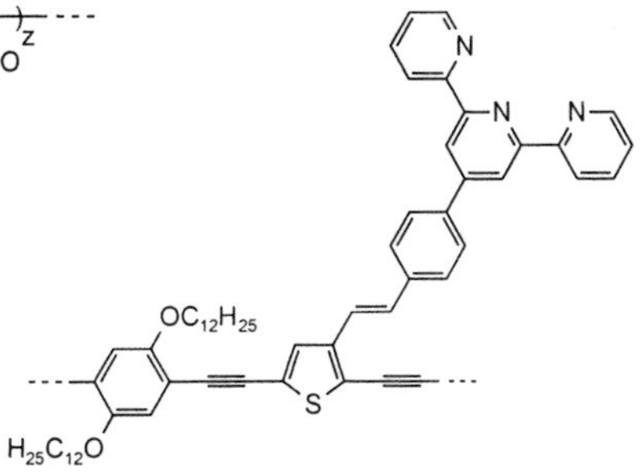

Poly(glycidyl methacrylate) resins were modified with pyrazole, imidazole, and 1,2,4-triazole as shown in **32** and also with bis(benzimidazole)s [89]. All resins selectively remove Cu(II) from a solution also containing Cd(II), Co(II), Ni(II) and Zn(II) at pH >2.5.

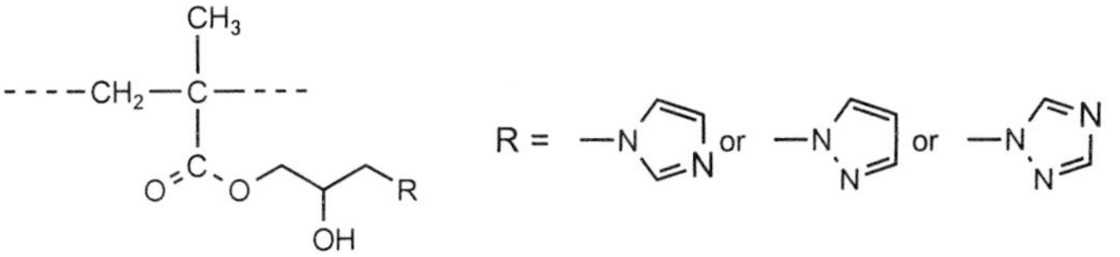

32

5.1.3 Metal Ion Binding at Polymeric Oxygen/Nitrogen Donor Ligands

Most of these polymeric ligands contain a =N bond and an additional C–O or C–N bond, or a C=O bond and an additional C–N bond for metal ion coordination. Remarkable selectivities for a specific metal ion were often reported.

A cross-linked polystyrene with amidoxime groups **33** was prepared from cross-linked polystyrene by cyanoethylation and reaction with hydroxylamine [90]. This polymer shows a good selectivity for the separation of U(VI) from sea water. Amideoxime polymers and their complexes with Cu(II) were also prepared from macroporous acrylonitrile divinylbenzene copolymers by reaction with NH$_2$OH [91]. Polymer fibers prepared from the grafting of acrylonitrile to polypropylene through irradiation were modified with hydroxylamine to give amidoxime ligands [92]. The polymer complexes Cu(II) from solution at pH 6 as well as U(VI) from sea water. Polyacrylonitrile grafted on poly(vinylalcohol) fibers and modified with NH$_2$OH removes Au(III) from solutions containing different transition metal ions [93]. The determination of metal ions in synthetic and environmental samples using chromatographic separation of Cu, Zn, Ni, Co, Fe and Fe ions on poly(methacrylohydroxamic acid) has been described [94]. The polymer ligands were prepared by the reaction of poly(methacryloyl chloride) with several *N*-arylhydroxylamines.

A *N*-isopropylacrylamide-bound hydroxamic acid copolymer **34** was prepared by the reaction of poly(*N*-isopropyl acrylamide)-*co*-(2-acryl-oxysuccinimide) with 6-aminohexanehydroxamic acid [95]. This water-soluble copolymer quantitatively separates from aqueous solution after Fe^{3+} uptake and its concentration is reduced from 15.5 ppm to 116 ppb. Poly(hydroxamic acid) beads prepared from polymethacrylate strongly complex Cu(II), Fe(III) and Pb(II) at pH 5 [96]. In addition, Co(II), Cu(II) and Fe(II) were selectively eluted from the resin at pH 4, and Pb(II) and Ni(II) with 4 M HNO_3.

33

34

A chelating ion-exchange resin 1-hydroxy-2-naphthaldoxime polymer **35** was synthesized by polycondensation from 1-hydroxy-2-naphthaloxime with formaldehyde [97]. A heterogeneous membrane of **35** was obtained by casting a THF solution with poly(vinyl chloride) as binder material. The polymer exhibits a high selectivity for Ni(II) in comparison to various other metal ions, which is expressed as a IUPAC-defined selectivity coefficient. The soluble polymeric ligand **36** was obtained from the corresponding acrylate monomers [98]. Polychelates with Cu(II) and Ni(II) (coordination at the −OH and =N groups in the ratio 1:2 metal to ligand) are insoluble. Polymer Schiff-base anchored Ru(III) complexes synthesized by sequential attachment of ethylenediamine, salicylaldehyde and $RuCl_3$ to cross-linked chloromethylated polystyrene were investigated as catalysts for the oxidation of benzyl alcohol and hydrogenation of styrene [99]. Polymer-bound catalysts were found to be more active than their homogeneous counterparts. The polymer **37** containing cinnamaldehyde anthranilic acid moieties forms spontaneous endothermic and entropically favored metal complexes [100]. The order of stability constants was found to be Th(IV), U(VI), Ce(III) > La(III) > Cu(II) > Ni(II) > Co(II) > Mn(II). A hint of polymeric semicarbazone ligands with chelate-forming properties is given in [101,102]. There are also reports of some other polymeric ligands which have been investigated for their metal ion binding: polyethers with 8-hydroxy-5-quinolinyl end groups [103], polysulfone functionalized by 8-hydroxyquinoline and other groups [104], poly(2-hydroxyethyl)-DL-aspartamide [105] and copolymers of 1-vinyl-2-pyrrolidone and acrylamide derivatives [106].

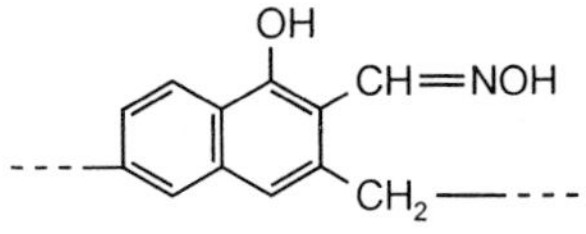

35

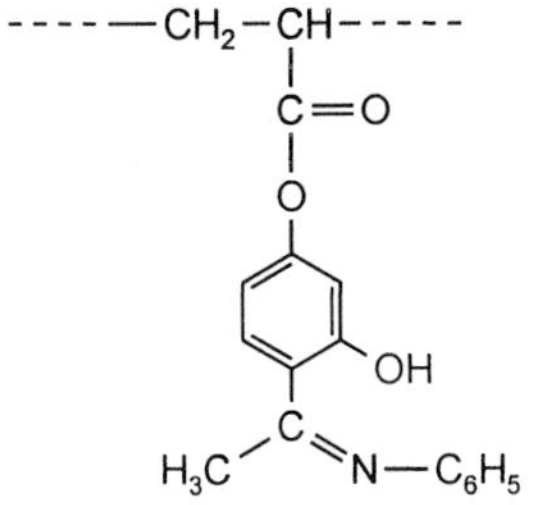

36

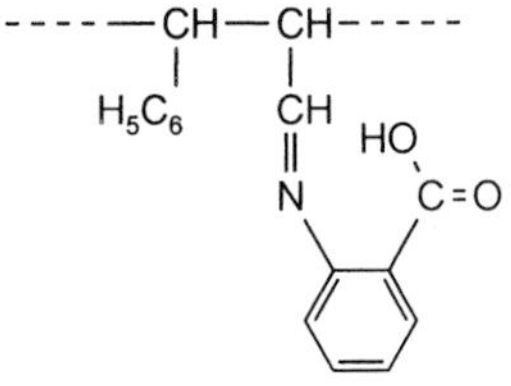

37

Grafting on polymers to obtain functional groups for binding of metal ions has been reported: polyethylene-graft-poly(methylvinylketone)/Schiff base with 2-aminophenol **38** [1,107] or salicylaldehyde hydrazide [108] (for other grafted ligands see [109]).

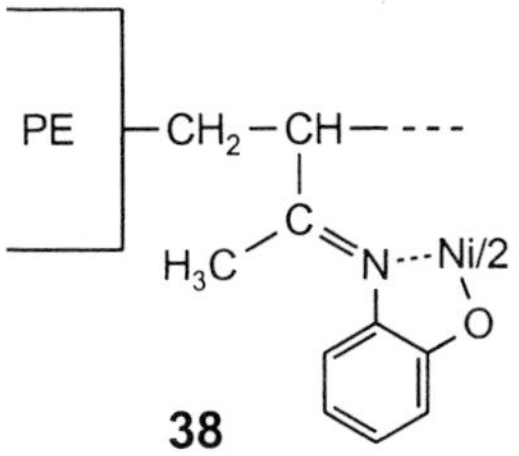

38

5.1.4 Metal Ion Binding at Polymeric Sulfur or Phosphorus Ligands

Thiacrown ether polymers have been prepared by copolymerization of 6-(4'-vinylbenzyloxy)-1,4,8,11-tetrathiacyclotetradecane with styrene and N-vinyl-pyrrolidone [110]. The thiacrown ether copolymers are selective for Ag(I) and Hg(II) over other transition metal ions. Thiacrown polymers **39** were obtained by the copolymerization of the corresponding 4-vinyl-substituted amino-methylthiacrowns with divinylbenzene [111]. **39** ($n = 1$) efficiently removes the toxic Hg(II) ions from acidic aqueous solution. This resin was also shown to be selective for Hg(II) over a variety of other metal ions (Table 5-2). In addition, Hg(II) recovery and polymer regeneration has been demonstrated. Chlorosulfonated styrene–divinylbenzene copolymers were modified to obtain polymer-bound N-sulfonylethylenebis(dithiocarbamate)s **40** with good metal binding abilities [112]. They show very high selectivities for Ag(I) and Hg(II) ions and moderate selectivities for Cu(II), Pb(II) and Cd(II). Phosphine sulfide-type chelating polymers bearing different spacer arms and good selective adsorption of Au(III) and Pd(II) have been reported [113].

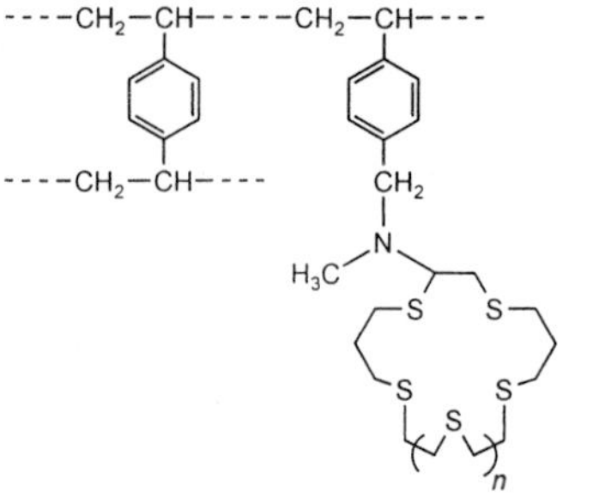

39   (n = 0, 1)

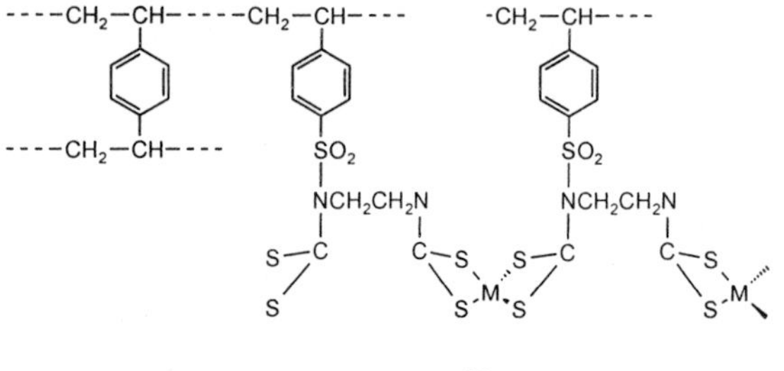

40

Table 5-2. Ion selectivity data for the resin **36** (n= 1).

Metal	pH	M^{n+}:Hg^{2+} molar ratio	Percent extraction[a]
Pb^{2+}	1.40	5	98
Cd^{2+}	1.51	5	97
Fe^{3+}	1.53	10	98
Al^{3+}	1.62	100	95

[a] extraction (%)= 100–([remaining Hg^{2+}]/[starting Hg^{2+}] x 100).

A styrene/divinylbenzene copolymer was phosphorylated and converted to a polymer with phenylphosphinic acid groups **41** [114] (see Experiment 5-9, Section 5.4). Ion-exchange coordination to Cu(II), Cd(II), Ni(II), Zn(II) and Eu(III) was studied. Resins with carboxylic groups in the α- and β-positions display higher metal ion uptake than those with –COOH in the γ-position. Macroporous polymers **42** with a functional group based on triisobutyl-phosphine sulfide have been synthesized and characterized for the selective adsorption of Au(III) and Pd(II) [115].

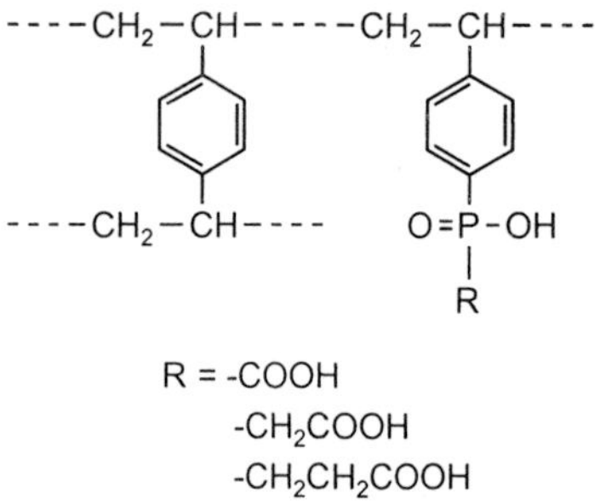

R = -COOH
   -CH₂COOH
   -CH₂CH₂COOH

41

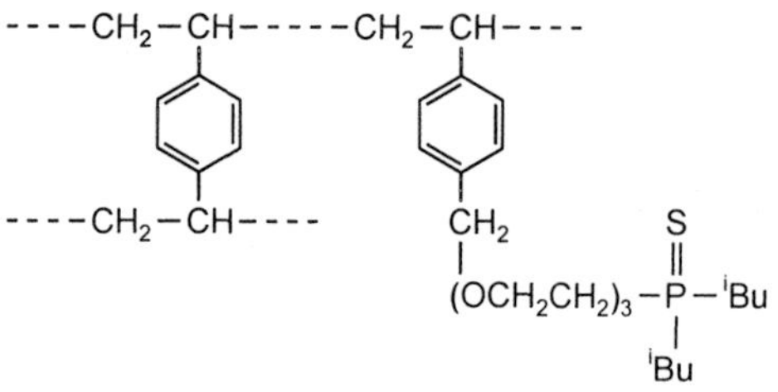

42

5.2   Binding of Metal Complexes

Different kinds of bonds can be formed when a macromolecule reacts with a metal complex: covalent bonds by the reaction of a substituted macromolecule with a substituted metal complex, ionic bonds by interaction of a charged macromolecule with an oppositely charged metal complex and coordinative bonds by interaction of a macromolecule which is substituted by donor atoms with the metal ion of a metal complex. The aim of this work is to obtain an analogy to natural macromolecular metal complexes for binding of small molecules and with catalytic or photocatalytic properties. Also the coating of electrodes to produce electrocatalysts or photoelectrocatalysts is interesting. Very recently macromolecules containing bound bipyridine and terpyridine metal complexes were reviewed: U.S. Schubert, C. Eschbaumer, *Angew. Chem. Int. Ed.* **2002**, *41*, 2892.

5.2.1   Covalent Binding of Metal Complexes

The reaction of cross-linked polystyrene with 5-chloromethyl-2-hydroxybenzaldehyde followed by interaction with the Co(II) chelate of the Schiff-base from 2-hydroxybenzaldehyde with diaminomaleonitrile yields the polymer chelate **43** (content 0.2 mmol g^{-1} chelate centers) [116]. This polymer complex was investigated as a catalyst for the conversion of quadricyclane to norbornadiene. Cross-linked chloromethylated polystyrene was reacted with N_2O_3–Schiff-base ligands. The resulting macroligands were investigated for the binding of Co(II), Mn(II) and Fe(II) (formula **44**) [117]. Also cyclic Schiff-base chelates have been synthesized [118]. Gel-type and macroporous versions of a chiral Mn(III)-salen complex **45** (R = –H) were prepared by the reaction of poly[4-(4-vinylbenzyloxy)salicylaldehyde] [119]. These polymers are very active catalysts in the asymmetric epoxidation of alkenes.

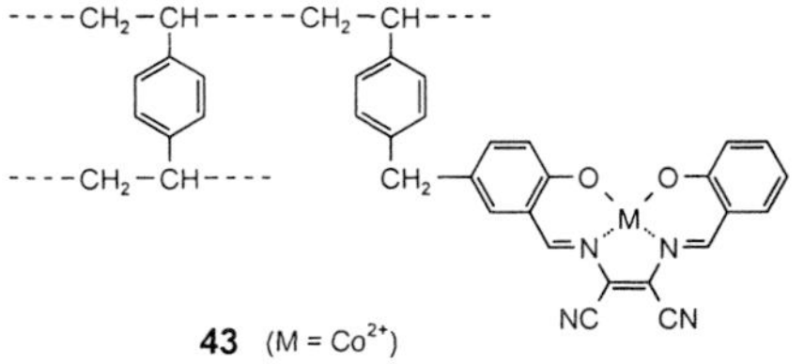

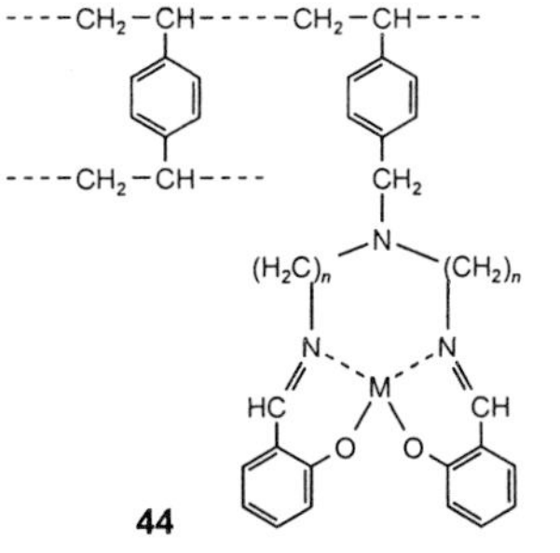

Analogously **45** (R = –OH)has been synthesized [120]. The Mn-catalyst produces the epoxides of 1,2-dihydronaphthalene, styrene and *cis*-β-methyl-styrene with enantiomeric excesses of 46.9 and 79%, respectively. Chiral Mn-salens have also been prepared by polymerization and copolymerization of chiral Mn divinylsalens [121,122]. Finally, a polymeric salen ligand **46** has been reported which was obtained by the reaction of cross-linked chloromethylated polystyrene with the carboxylic acid-substituted Schiff-base ligand [123]. The complexes with Cu(II) and Co(II) have a paramagnetic square-planar structure, and with Ni(II) a diamagnetic square-planar structure; with Fe(III) they are paramagnetic and octahedral, with Zn(II) and Cd(II) they are diamagnetic and tetrahedral and with U(VI) they are diamagnetic and octahedral.

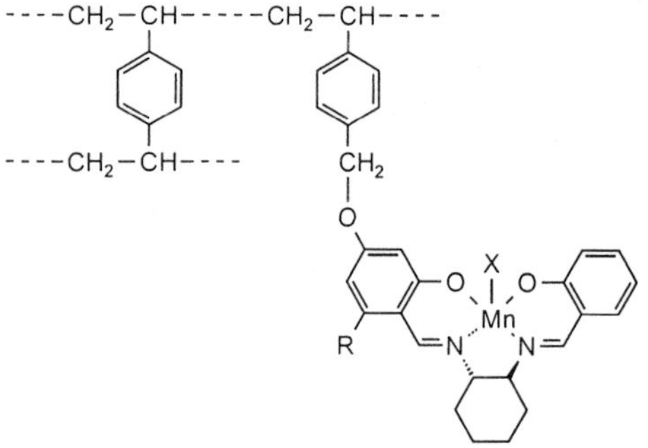

45 (R = -H, -OH) 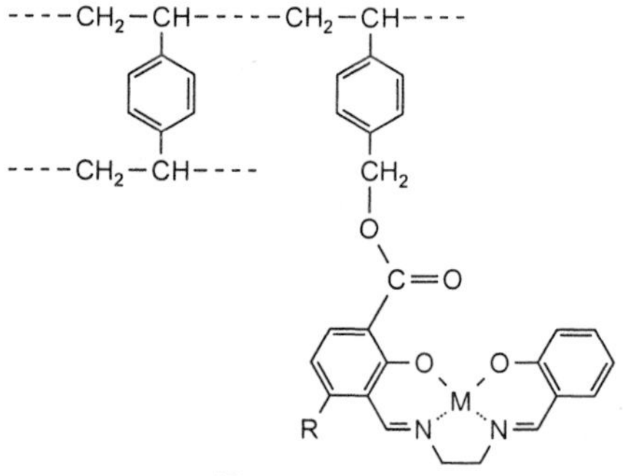   **46**

Metal complexes of various porphyrins **47**, phthalocyanines **48** and naphthalocyanines **49**, with intense absorptions in the visible region of light, have also been covalently bound to macromolecular carriers. Early work in this field was reviewed in 1977 and 1983 [124,125]. These reviews, for example, reported the fixation of metal-free and metal-containing protoporphyrin-IX, chlorophyll, mesoporphyrin-IX and substituted 5,10,15,20-tetraphenyl-porphyrins. The binding of substituted phthalocyanines to substituted polymers has also been reviewed [125,126] and covers mainly the binding of carboxylic or sulfonic acid-substituted phthalocyanines to polystyrene.

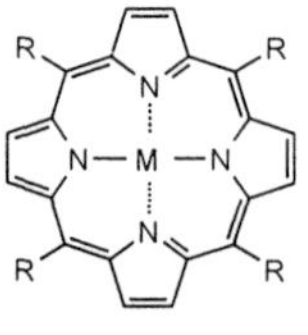

47 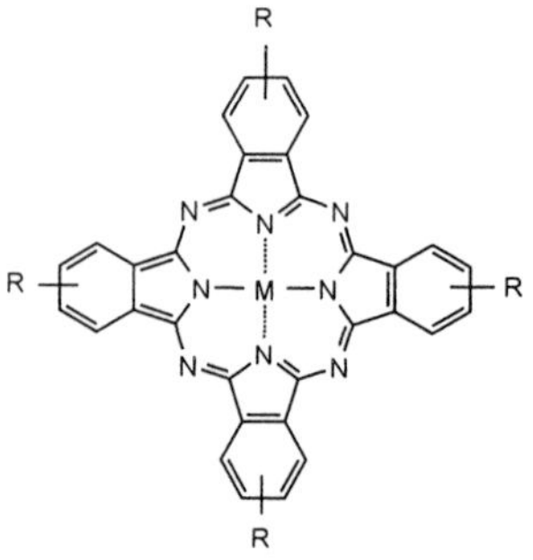   **48** 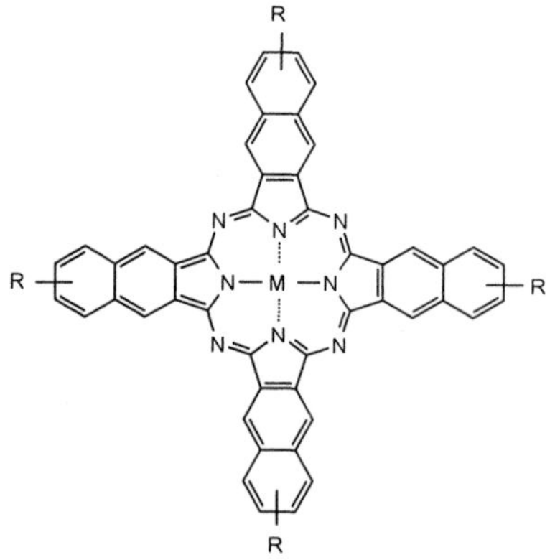   **49**

Recently, water-soluble tetraarylporphyrin-labeled polymers have been synthesized [127–129]. The radical copolymerization of *N*-isopropylacrylamide with 4-bromobutylacrylate was carried out to give the copolymer **50**. Subsequent reaction with 5-(4-pyridyl)-10,15,20-tri(4-methoxyphenyl)porphyrin in THF led to the porphyrin-containing copolymer **51** (Eq. 5-3). The porphyrin content of **51** is around 10^{-5} mol g^{-1}. The reason for synthesizing such copolymers is to use the Zn(II)-porphyrin derivatives as a fluorescence labels. In methanol the quantum yields are consistent with low molecular weight tetraarylporphyrins, but in water the quantum yields decrease by a factor of around 40 and the kinetics are bifunctional. Therefore, in water the macromolecule forms compact globules containing monomolecular unquenched, partially aggregated quenched and absolutely non-fluorescent porphyrins.

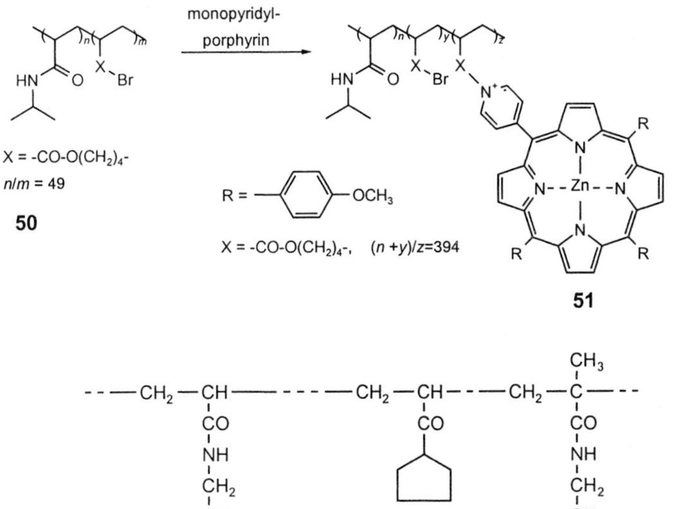

$$(5\text{-}3)$$

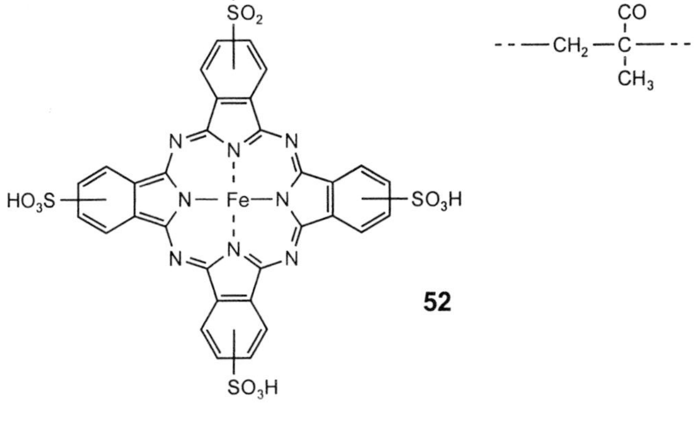

Bis(3,4-dicarboxybenzoyl)phthalocyanines with M = Co(II) or Cu(II) have been covalently bound to linear polystyrene by a Friedel–Crafts reaction [130]. The polymers contained 0.13 mmol g^{-1} (12.4 wt. %) phthalocyanine moieties. The sensitivity of the polymers toward toxic gases were investigated by quartz balance transducers. The sensitivities are $6 \cdot 10^{-7}$ m^3 mL^{-1} for NO$_2$ and $2 \cdot 10^{-8}$ m^3 mL^{-1} for chloroform and perchloroethylene, respectively. **52** (R = SO$_2$Cl, M = Fe) has been covalently bound to the amino groups of a cross-linked, hydrophilic polymer consisting of N-acryloyl-β-alanine-(aminoethyl-ene)-amide, N-acryloylpyrrolidine and N,N-bis(methacryloyl)-1,2-diamino-ethane as comonomers [131]. The polymer obtained, **52**, contains 20 µmol phthalocyanine units per g. The catalytic oxidation of 2,4,6-trichlorophenol or 3,5-di-tert-butylcatechol in the presence of H$_2$O$_2$ or KHSO$_5$ as oxidant has been studied. Around 80% of the trichlorophenol was catalytically oxidized within 2 h at pH 7.

Some years ago a general route for the binding of different porphyrin-type compounds was described [132–135]. The substituted porphyrins **47** (R = –C$_6$H$_4$–NH$_2$), **48** (R = –NH$_2$) and **49** (R = –NH$_2$) contain nucleophilic amino groups of similar reactivities. Therefore, an identical synthetic procedure can be applied to achieve covalent binding to a polymer with reactive sites. Besides the binding of one porphyrin, the addition of different porphyrins to the reaction mixture allows the fixation of two or three porphyrins in one polymer system in a one-step procedure. A method was selected in which a diluted solution of the polymer was added dropwise to a diluted solution to the porphyrins. If the reaction of poly(4-chloromethylstyrene) is carried out in the presence of an excess of triethylamine, the covalent binding of the porphyrin and a quarternization reaction occur simultaneously (see Experiment 5-10, Section 5.4). Positively charged polymers **53** soluble in water were obtained [133]. Figure 5-7 gives an example of a UV-vis spectrum of polymer **53** with combined moieties of all porphyrins **47**, **48** and **49**. It can be seen that by the protection of the polymer domain the porphyrins are monomolecularly distributed. In addition to a porphyrin, viologen as an electron relay has also been covalently bonded to positively charged polystyrene [135]. Also negatively charged and uncharged polymers **54** and **55** containing porphyrin moieties have been prepared [133,135]. The polymers prepared contained porphyrin moieties up to ~10 wt. %.

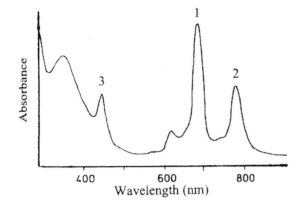

Figure 5-7. UV-vis spectrum of the positively charged **53** in water containing covalently bound moieties of **47**, **48** and **49**.

The properties of the systems as photosensitizers in photo-induced electron-transfer reactions were investigated. A suitable system consists of a photosensitizer in low concentration ($\sim 10^{-6}$ mol L^{-1}), the acceptor methylviologen MV^{2+} (1,1'-di-methyl-4,4'-bipyridinium dichloride, $\sim 10^{-3}$ mol L^{-1}) and the donor 2-mercaptoethanol ($\sim 10^{-1}$ mol L^{-1}) (see Experiment 5-10, Section 5.4). Irradiation with visible light under inert gas (to exclude energy and electron transfer to oxygen) leads to the excited triplet state of the photosensitizer, and oxidative quenching and reduction of the acceptor followed by oxidation of the donor can occur. The influence of the surrounding polymer on the photo-induced electron-transfer processes was studied using the zinc(II) complexes of polymer-bound porphyrins in water [134,135]. Low molecular weight Zn(II)-complexes of positively charged tetramethylpyridiniumporphyrin and negatively charged tetrasulfophenylporphyrin were used as low molecular weight photosensitizers for comparison. Taking the low molecular weight positively charged porphyrin with 1.0 as standard, the photoredox activities of the porphyrins increase as follows: negatively charged polymer **54** 1.4 < positively charged polymer **53** 6.6 < neutral polymer **55** 15.6. The surrounding polymer strongly influences the electron-transfer reactions. The highest reactivity within the uncharged polymer shows good separation of all charged species from the polymer coil. The photoredox reactions were also studied using a polymer-bound MV^{2+} with comparable results [135].

Water-soluble polymers **53–55** containing the covalently bound **47** (R = -C$_6$H$_4$–NH$_2$) with M = Mn(III)OH have been employed to study electron transfer from sodium dithionite as reducing agent [136]. The reduction of a Mn(III) porphyrin can be followed by electronic spectra, using the charge-transfer band of Mn(III) porphyins at ~ 470 nm and the Soret-band of Mn(II) porphyrins at ~ 440 nm. With an excess of dithionite analogous charged low molecular weight Mn(III) porphyrins exhibit kinetically linear pseudo-first-order plots. The curved first-order plots of polymer-bound Mn(III) porphyrins can be separated into two parallel first-order reactions and are due to the influence of the polymer chains. Porphyrins located in the outer sphere of the coil are reduced more easily. A change in conformation of the polymer coil with time makes the inner-sphere-located porphyrins accessible also. As expected, the positive charges of the polymer porphyrin **53** enhance reactions with the anionic dithionite in comparison to the negatively charged polymer porphyrin **54** by a factor of around 50, whereas the uncharged polymer porphyrin **55** exhibits an intermediate reaction rate. Polymer electrolytes are more efficient in the acceleration and retardation of reactions than monomeric electrolytes.

Covalent binding can be also achieved to inorganic carriers. The heterogeneous systems thus obtained are very suitable for investigating catalytic, photocatalytic or optical properties. Because of the stable bond, no leaching out of the immobilized metal complex from the carrier systems occurs. At first the surface of silica is modified by a substituted trialkoxysilane and then binding of a metal complex is carried out. Binding will occur at the external and internal surfaces of the silica.

For modification of the surface different silicas (all from Merck AG), such as macroporous Lichrosorb (surface area ~300 m^2 g^{-1}), macroporous Lichrosphere (surface area ~40 m^2 g^{-1}) were employed [137–139]. In the first step, the silica surfaces were modified to obtain chemically active positions for the attachment of a substituted phthalocyanine. Functionalization was achieved, for example, by reacting the 3-aminopropylsilyl groups of **56** with **48** (R = –COCl) to synthesize **57** (Eq. 5-4) (see Experiment 5-11, Section 5.4). The loadings are between 10^{-3} and 10^{-4} mol g^{-1} with substituted silyl groups and between 10^{-5} and 10^{-6} mol g^{-1} with phthalocyanines. The binding of **48** (R = SO$_2$Cl, M = Fe) at aminopropyl-modified silica has also been described [139]. Also other surfaces such as glass can be modified. A glass surface was first activated with 3-aminopropyltriethoxysilane, and then in a second step a porphyrin was bound covalently [140]. In a second route, substituted porphyrins may be modified at first with triethoxysilyl-groups and then employed in the hydrolysis with tetraethoxysilane to give silica-gel-bound metal complexes [141,142]. The metallated materials obtain with M = Mn(III) and Fe(III) are very active in the epoxidation of cyclooctene.

$$(5\text{–}4)$$

Covalent binding at mesoporous silica molecular sieves of the Si-MCM-41 type with a pore diameter of 3.1 nm, for example, is necessary to overcome the leaching of metal complexes or chromophores. In order to achieve covalent binding only at the interior surface the following method was developed [143]. In a first step the external surface of the mesoporous silica was passivated by reaction with dichlorodiphenylsilane. Then, in a second step, aminopropyl-groups were bound at the inner pore walls (Eq. 5-5). In a third step the 3-aminopropylsilyl-MCM-41 was reacted with the carboxylic acid or sulfonic acid chlorides of different chromophores such as **48** (R = –SO_2Cl, M = Zn) to get the modified mesoporous silica **58**. All steps were confirmed by X-ray diffraction, nitrogen physisorption and FTIR and UV-vis reflectance spectroscopy. Loadings of ~10 μmol g^{-1} were determined.

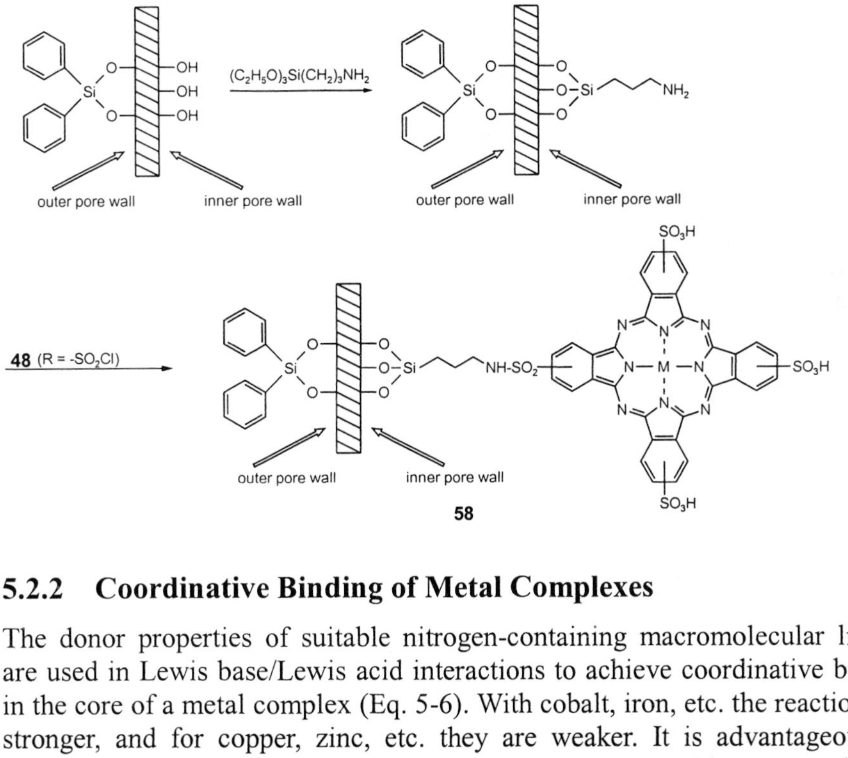

$$(5–5)$$

5.2.2   Coordinative Binding of Metal Complexes

The donor properties of suitable nitrogen-containing macromolecular ligands are used in Lewis base/Lewis acid interactions to achieve coordinative binding in the core of a metal complex (Eq. 5-6). With cobalt, iron, etc. the reactions are stronger, and for copper, zinc, etc. they are weaker. It is advantageous for binding at macromolecules that the formation constants with macromolecular ligands are between 10^1 and 10^3 times higher than with low molecular weight ligands [5,124,144]. Any linear or cross-linked organic polymer or inorganic macromolecule that has been modified with a strong Lewis base group can be used for binding. Coordinative bonding of a donor base to the metal ion leads to the properties being altered as described for natural porphyrins in Section 2.3.

$$(5-6)$$

Because coordinatively bound complexes are quite easy to prepare, numerous papers on this subject have been published. Soluble complexes are prepared by dissolving stoichiometric amounts of polymer and metal complex in an organic solvent or water. To obtain solid materials, the solvent is removed or a film is cast. Cross-linked insoluble polymers are suspended in a solvent and a solution of a metal complex is added. The equilibrium of the coordinatively polymer-bound metal complex with the unbound one in solution is not favorable (Eq. 5-6), so an excess of strong low molecular weight donor base can destroy the polymer coordinative bond. Even the heme in myoglobin can be cleaved to give the apomyoglobin, as described in Section 2.3.1, and another porphyrin derivative subsequently coordinatively bound.

Two older reviews summarize work in this field [124,125]. The following derivatives have been employed as porphyrins: Fe(II,III) protoporphyrin-IX (heme, hemin), Fe(II,III) or Co(II) protoporphyrin-IX-diester, chlorophyllins with different metal ions in the core, Fe(II) tetraphenylporphyrin, Mg(II) or Fe(II,III) octaethylporphyrin, Fe(II,III) tetrakis[o-(alkylamido)phenyl]-porphyrin. Polymers with N-donor groups are based on proteins such as poly(L-lysine), poly(L-histidine), poly(γ-benzyl-L-glutamate) or synthetic polymers such as homopolymers and copolymers with vinylpyridine, N-vinylimidazole or ethyleneimine.

The metal complexes at the polymer ligand have been analyzed by spectroscopic methods (UV-Vis, ESR), determination of the coordination number (formation of pyramidal or octahedral complexes (Eq. 5-6)), and formation constants [124]. Very often the coordination numbers of low molecular weight or polymer ligands are the same. In most cases only one base is coordinated. Using poly(L-lysine) [124] or poly(1-vinylimidazole) [145] with heme two units of these bases are coordinated. The coordination number of Co(II) protoporphyrin investigated by ESR changed from five to six by increasing the amount of poly(4-vinylpyridine-co-styrene) in solution [146]. ESR measurements show that the spin of the Fe(III) porphyrin complexes with poly(1-vinyl-imidazole-co-styrene) depends on the amount of imidazole units in the copolymer [144]. Copolymers with more than 10% imidazole units show low-spin adducts typical for a coordination number of six. With less than 1% vinylimidazole in the copolymer, high-spin indicates five as the coordination number. Using 1-vinylimidazole substituted in position 2, steric hindrance keeps the pyramidal coordination even at a high concentration of the imidazole in the polymer. Five-coordinating complexes are active for O_2 uptake, so with artificial polymer ligands the situation may be similar to the coordinating property of globin. The formation constants of porphyrin complexes with polymer ligands are 10^1 to 10^3 times higher than with low molecular ligands which is explained by a higher concentration of the ligand now fixed in the polymer domain [124,144].

Coordinatively polymer-bound Co salenes and Co or Fe porphyrins have mainly been investigated for reversible binding of oxygen, which is of commercial interest for artificial blood and air separation by membrane processes (for a detailed discussion of these materials see Chapter 9) [5,124,147]. A chiral Mn(III) Schiff-base complex was bound to the insoluble cross-linked terpolymer of styrene, 4-vinylpyridine and divinylbenzene (formula **59**) [148]. The materials have been used successfully as catalysts in the presence of iodosylbenzene as oxidant for the enantioselective epoxidation of styrenes, as can be seen from Table 5-3.

Table 5-3. Enantioselective epoxidation of styrene derivatives with the catalyst **59** in dichloromethane at 4 °C.

Substrate	Conversion (%)	Time (h)	ee product (%)	Configuration
Styrene	56	3.5	22	R
4–chlorostyrene	39	24.0	30	R
4–methylstyrene	48	4.5	20	R
4–nitrostyrene	25	24.0	40	R

Fe(III) tetraphenylporphyrin **47** (R = $-C_6H_5$) and Fe(III) deuteroporphyrin-IX dimethylester have been incorporated by coordinative interaction into films of the hydrogel of the polymer **60**, containing imidazole pendent arms [149,150]. The films were studied on ITO and platinum electrodes by cyclic voltammetry in contact with acetonitrile or an aqueous electrolyte. The Fe(III)/Fe(II) redox couple was clearly seen and the films exhibited Nernstian behavior [149]. Similarities with cytochrome c_3 were pointed out in the redox behavior.

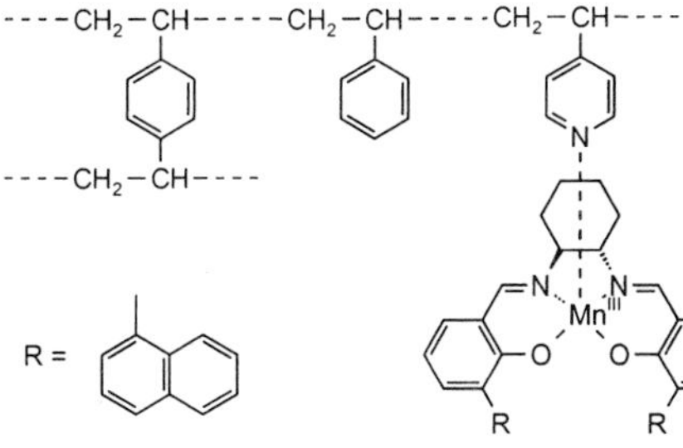

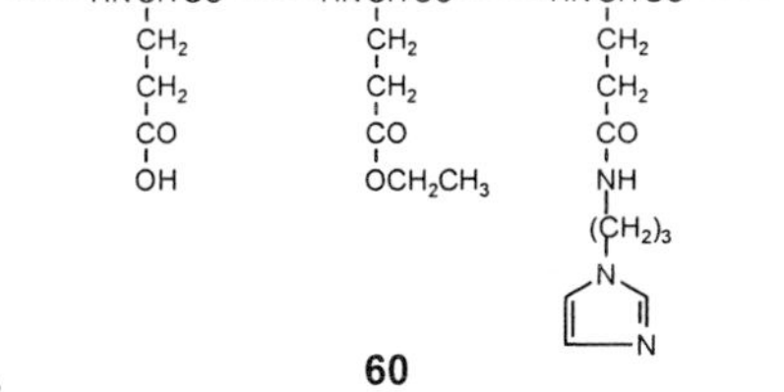

The coordinative binding of cobalt phthalocyanines **48** with R = –COOH or R = –SO$_3$H taking polymer ligands such as poly(ethyleneimine), poly(vinylamine), amino group-modified poly(acrylamide) or modified silica gel was reviewed some years ago [125,126].

Recently, the electrochemical properties of cobalt phthalocyanines included by coordinative binding in membranes of poly(4-vinylpyridine) [151] or poly(4-vinylpyridine-co-styrene) were investigated [152] (see also Chapter 13). The membranes were prepared by dissolving **48** (R = –H, M = Co(II)) in DMF in the presence of poly(4-vinylpyridine). The coordinative interaction of the metal complex to the pyridyl group strongly enhances the solubility of the phthalocyanine in DMF. The film formed on a carrier after casting and evaporation of the solvent is homogeneously blue. The pattern in the UV-vis spectra of the films is comparable to the Co phthalocyanine dissolved in pyridine, thus showing homogeneous monomeric distribution of the metal complex in the polymer. In contrast, the film of the cobalt complex cast from a pyridine solution shows a strong resonance broadening of the long wavelength band, indicating its crystallinity.

For coordinative binding to inorganic carriers the surface has to be modified. In a one-step procedure for the preparation of silica **61** modified on the surface with imidazoyl-groups, different silica materials as mentioned in Section 5.2.1 were treated with a mixture of 3-chloropropyltriethoxysilane and an excess of imidazole in *m*-xylene (Eq. 5-7). Subsequent treatment with different kinds of substituted cobalt phthalocyanines, naphthalocyanines and porphyrins **47, 48** and **49** in DMF led to the modified silica. Examples are shown with **48** (R = –H) and **61** for **62**. The silica contains ~0.8–12 µmol g^{-1} metal complex moieties [153,154]. These materials are very efficient heterogeneous catalysts in the oxidation of thiols R–SH to disulfides R–S–S–R which is an important process for the treatment of petroleum distillates (MEROX process, Eq. 5-8) (see Experiment 5-12, Section 5.4). Whereas the soluble low molecular weight tetrasulfonic acid **48** (R = –SO$_2$H, M = Co) used industrially exhibits a turnover number (TON) of 822 min^{-1}, the heterogeneous catalysts **62** modified with **48** (R = –NO$_2$, M = Co) or **49** (R = –H, M = Co) show a TON of 1981 and 2458 min^{-1}, respectively [153].

$$(5-7)$$

61   62

$$4R\text{–}S^- + O_2 + H_2O \rightarrow 2R\text{–}S\text{–}S\text{–}R + 4OH^- \tag{5-8}$$

Ruthenium phthalocyanine **49** (R = –H, M = Ru(II)) monolayers were obtained by self-assembly on pyridyl-functionalized SiO_2 or Al_2O_3 substrates by coordination between Ru and the pyridino group to obtain **63** [155]. The pyridino-functionalized metal oxides were dipped into a solution containing the more soluble benzonitrile derivative of the phthalocyanine $RuPc(NC\text{–}C_6H_5)_2$, and ligand exchange reactions led to **63**. Strategies have been described to immobilize a second layer of a Ru-phthalocyanine or a Co tetraphenyl-porphyrin.

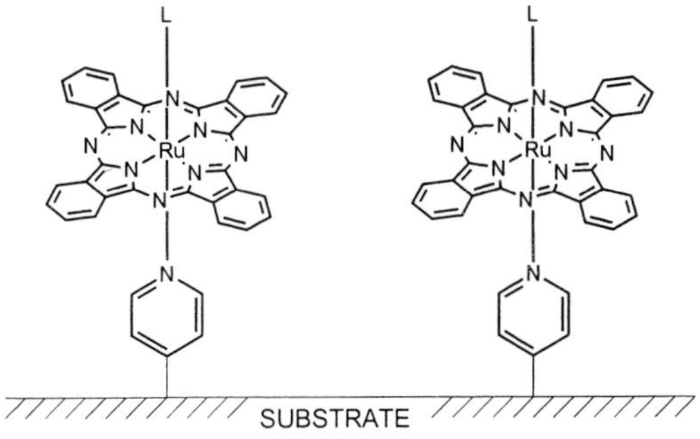

63

5.2.3   Ionic Binding of Metal Complexes

Ionic bonding by electrostatic interactions between charged macromolecular carriers and oppositely charged metal complexes is easily carried out. Either charged linear or cross-linked organic polymers or inorganic carriers bearing charged groups are employed as macromolecular carriers. Then to a solution or a suspension of the macromolecular compound a solution of an oppositely charged porphyrin or phthalocyanine is added with stirring. The complexes obtained may be investigated in solution, as suspensions or after casting a film from solution with evaporation of the solvent. Some examples of these macromolecular complexes are now discussed.

Interaction of the tetraanionic 5,10,15,20-tetrakis(4-sulfonatophenyl)-porphyrin (**47**, R = $-C_6H_4\text{–}SO_3H$) and the tetracationic 5,10,15,20-tetrakis(4-N-methylpyridyl)porphyrin (**47**, R = $-C_5H_4N^+\text{–}CH_3$) with poly(lysine) and poly(glutamic acid) has been studied in detail [156–158]. The tetraanionic **47** (M = 2H, R = $-C_6H_4\text{–}SO_3H$) exists in aqueous solution at pH 5–12 in a monomeric form, and its fluorescence is not pH dependent. In contrast, polymer complexes prepared from a 1 µM solution of this porphyrin and a 100 µM poly(lysine) solution, have a fluorescence intensity vs. pH which shows a sigmoidal profile [156,157]. Also, circular dichroism and fluorescence light scattering indicate an extensive polymer-induced self-aggregation of the porphyrin [156]. In the pH range 3.2–4.8, the cationic 47 (M = Cu(II); R = – $C_5H_4N^+$–CH_3) interacts with poly(L-glutamate), forming a kinetically labile chiral binary complex. The addition of the anionic 47 (M = 2H; R = –C_6H_4– SO_3H) induces significant changes in the absorption, fluorescence and CD spectra, indicating the formation of chiral ternary complexes. Interestingly, these complexes are very stable even at pH values as high as 12. At this pH the poly(glutamate) exists in a random-coil conformation, and the porphyrin assemblies retain their original chirality even when the matrix loses it (for details see [158]).

The tetraanionic 47 (M = Cu, Zn, Ag, Cd; R = –C_6H_4–SO_3H) has been incorporated into different polymers including linear and cross-linked chloromethylated polystyrenes partially quarternized with pyridine (ionic interactions with the metal porphyrin), poly(4-vinylpyridine) (coordinative interactions with the metal porphyrin via the metal) and poly(ethylene glycol) (dipolar interactions with the metal porphyrin) [159,160]. Spectra of the samples were measured in solution or in the solid state. Red-shifts of both Q- and B-bands in the electronic spectra of the porphyrin occurred in the order: poly(ethylene glycol) < poly(4-vinylpyridine) < quarternized poly(styrene). The greatest interaction occurred in ionic bonds between the carrier and the porphyrin [159].

Multilayered conducting polymer heterostructures have been prepared by the self-assembly method as follows [161]. At first a positively charged surface of a quartz plate was immersed in a methanol solution of poly(thiophene acetic acid) 64. Then the substrate was immersed in an equimolar solution of poly(N-methylpyridinium vinylene) 65 and the tetracationic porphyrin 47 (M = 2H; R = –$C_5H_4N^+$–CH_3). By repeating these procedures, multilayered films were obtained on the substrate. These multilayers show remarkable photocurrent enhancement, which can be explained by improved charge separation at the differently charged interfaces. The photogeneration of singlet oxygen has been studied by excitation of porphyrins immobilized in polymeric supports [162]. Compound 47 (M = 2H; R = –C_6H_4–SO_3H) was added to an aqueous solution of the cationic poly(dimethyldiallylammonium chloride), and a film was cast on quartz glass. 5-(4-Amino)-phenyl-10,15,20-triphenylporphyrin was immobilized in a strong acidic film of Nafion. The energy-rich singlet oxygen was generated by excitation of the porphyrins in the presence of oxygen, and measured in the gas phase. The amount of singlet oxygen is one to two orders of magnitude higher compared to porphyrins deposited without polymer on a quartz plate. Positively charged polymers such as ionenes [–$N^+(CH_3)_2$–$(CH_2)_x$– $N^+(CH_3)_2$–$(CH_3)_2$–$(CH_2)_y$–]$_n$ form stoichiometric complexes in their interaction with tetrasulfonated 48 (R = –SO_3^-; M = Co(II)) having the composition N^+/CoPc(SO_3^-)$_4$ of 4:1 [163–166].

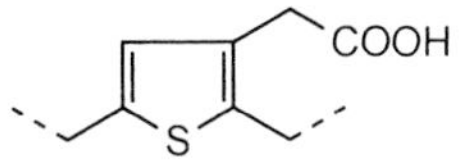

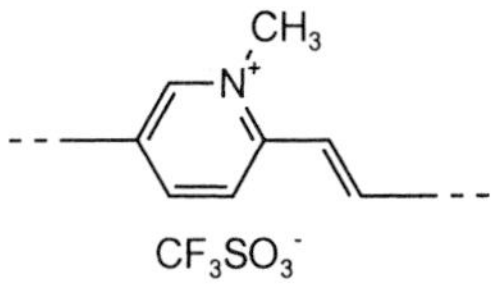

COOH

CH$_3$

CF$_3$SO$_3^-$

64   **65**

Ionic binding at charged cross-linked polymers was easily obtained by shaking an aqueous solution of a positively charged ion exchanger, for example, Amberlite, with the negatively charged **48** (R = –SO$_3^-$; M = Zn(II), Al(III)(OH), Si(IV)(OH)$_2$), resulting in blue-colored polymeric complexes **66** containing a monomeric distribution of the MPc (see Experiment 5-13, Section 5.4) [167,168]. Employing 2.5 g Amberlite and 10 μmol of the anionic phthalocyanine, a quantitative loading of 4 μmol phthalocyanine per g carrier is obtained. Polymers **66** with M = Zn(II), Al(III)OH or Si(IV)(OH)$_2$ are excellent photosensitizers in the photooxidation of sulfide to sulfate, thiols to sulfonic acids and phenol under partial mineralization to carbonate (see Section 13.4). Manganese(III) complexes of tetraanionic and tetracationic porphyrins have been immobilized on counter-charged, surface-modified silica supports [169]. These supported Mn porphyrins are very active catalysts in epoxidation reactions.

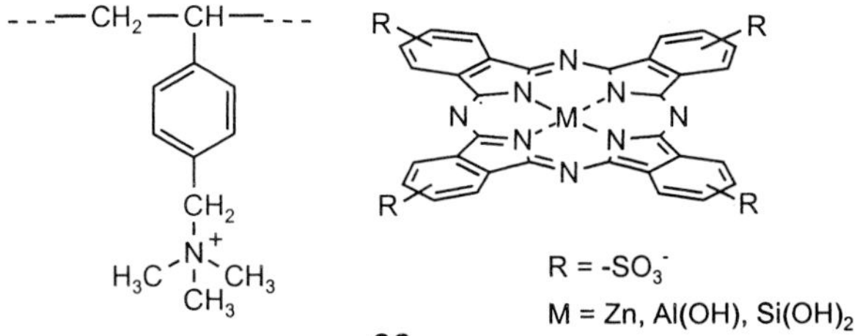

---CH$_2$—CH—---

CH$_2$

H$_3$C–N$^+$–CH$_3$
CH$_3$

R = -SO$_3^-$
M = Zn, Al(OH), Si(OH)$_2$

66

Porphyrins have been incorporated into a positively charged poly-(*N*-alkylpyrrole) film prepared by oxidative electropolymerization of the alkyl-ammonium pyrrole monomer on ITO [170]. Incorporation of the tetraanionic **47** (M = Mn(III), Zn(II), Fe(III); R = –C$_6$H$_4$–COOH or –C$_6$H$_4$–SO$_3$H) was achieved by soaking the polymer deposited on the electrode in an aqueous solution of these porphyrins. The reactivity of the electrodes containing Fe(III) porphyrin against nitric oxide was proven. Metalloporphyrin/polypyrrole films have been prepared by oxidative polymerization of pyrrole in the presence of carboxylated or sulfonated porphyrins [170–172] or phthalocyanines [172–176] as counter-ions in the electrolyte instead of inorganic counter-ions such as perchlorate. These films are interesting for their electrical and chemical sensing properties.

5.3 Examples of Organometallic Compounds and π-Complexes

In this section a few examples are given for other interactions of a metal compound with a macromolecule not previously mentioned.

π-Complexes are formed between aromatic groups or unsaturated parts of a macrochain as π-donors and a transition metal compound. By interaction of the phenyl group of polystyrene with $TiCl_4$ or $VOCl_3$ π-complexes of the composition monomer unit: $MX_n = 1:1$ are formed [177]. The interaction of graphite with metal compounds such as gaseous $TiCl_4$ is well known [178]. Treatment of cross-linked styrene bearing silylnaphthyl groups with Li or Na biphenylide in THF results in highly colored paramagnetic polymers **67** [179]. Supported naphthalide complexes **68** on silica, alumina and titania have also been prepared [179].

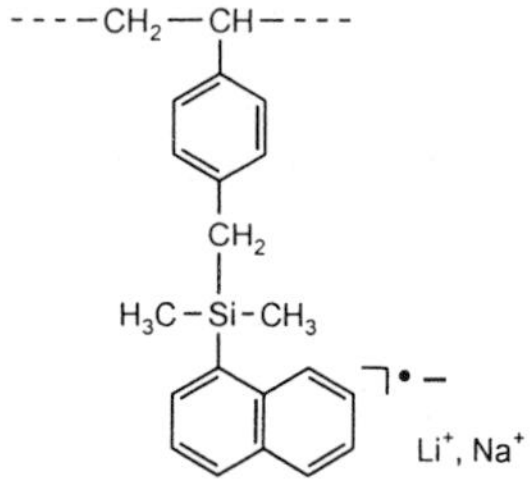

67

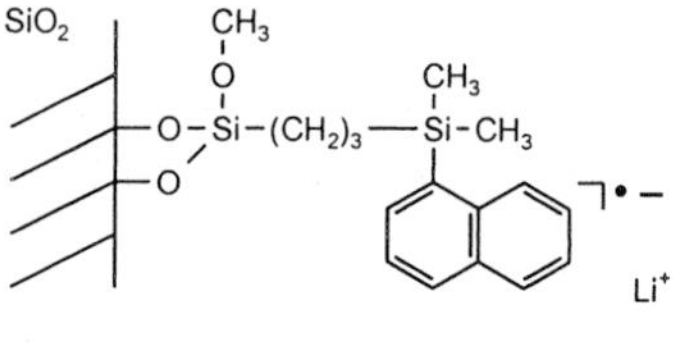

68

Different polystyrene-bound biscyclopentadienyl complexes have been prepared. One example is shown as **69** in Eq. 5-9 [180]. Treatment of soluble copolymers of styrene and methacrylic acid with dichlorotitanocene resulted in soluble titanium-containing copolymers **70** (Eq. 5-10] [181]. Finally polymeric organometallic mesogens **71** with [1,3-(diethynyl)cyclobutadiene]cyclopentadienyl Co(II) moieties were prepared by Heck coupling of a corresponding cyclobutadiene monomer with 2,5-diiodothiophene [182].

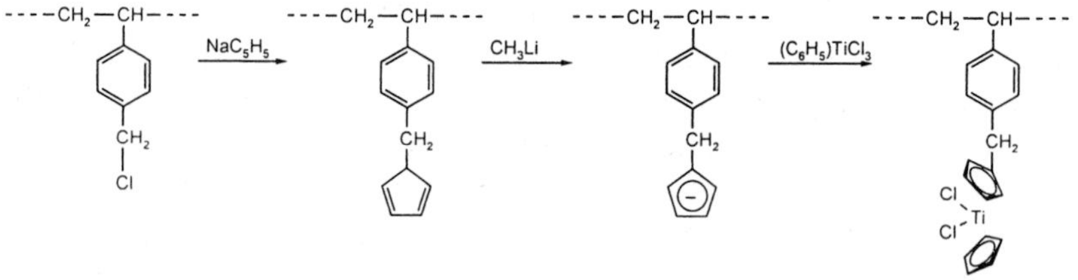

(5-9)

69

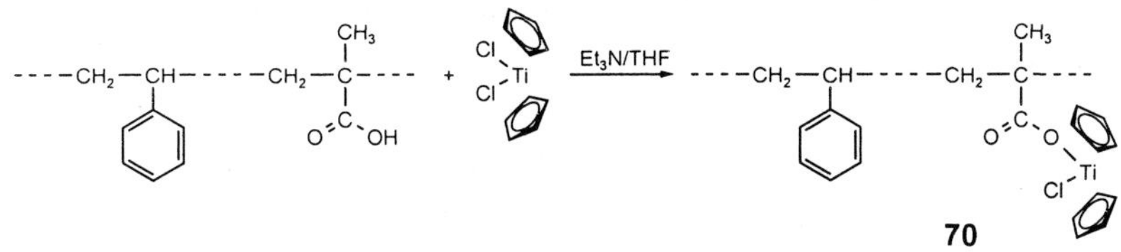

(5-10)

70

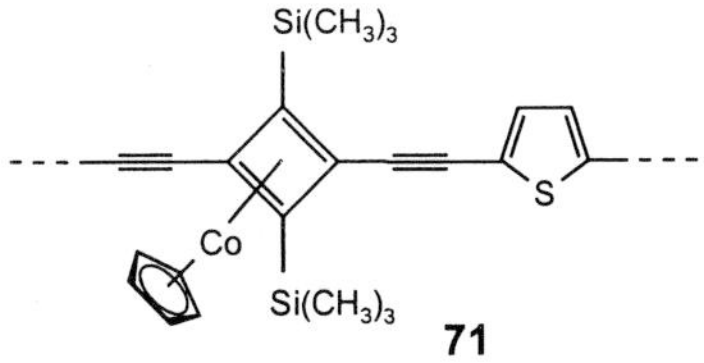

71

5.4 Experimental

Experiment 5-1: Preparation of Cation Exchanger 7a by Sulfonation of Cross-linked Polystyrene (Section 5.1.1)

Reference: D. Braun, H. Cherdron, H. Ritter, *Polymer Synthesis: Theory and Practice; Fundamentals, Methods, Experiments*, Springer-Verlag, Berlin, Heidelberg, 2001.

Radical copolymerization of styrene with 1,4-divinylbenzene in aqueous suspension (cross-linking copolymerization): Monomeric styrene is freed from phenolic inhibitors by shaking twice with 10% sodium hydroxide solution, washing three times with distilled water, drying over calcium chloride or silica gel and distilling into a receiver under reduced pressure of nitrogen (b.p. 82 °C/100 torr, 46 °C/20 torr). It is stored in a refrigerator until required. 1,4-Divinylbenzene is used as received or freed from inhibitors as described for styrene but without distillation.

A three-necked flask, fitted with stirrer (preferably with a revolution counter), thermometer, reflux condenser and nitrogen inlet, is evacuated and filled with nitrogen three times. 250 mg of poly(vinyl alcohol) are placed in the flask and dissolved in 150 mL of deaerated water at 50 °C. A freshly prepared solution of 0.25 g (1.03 mmol) of dibenzoyl peroxide (or twice the amount if the divinylbenzene used contains an inhibitor) in 25 mL (0.22 mmol) of styrene and 2 mL (7 mmol) of 1,4-divinylbenzene are added with constant stirring so as to produce an emulsion of fine droplets of monomer in water. This is heated to 90 °C on a water bath while maintaining a constant rate of stirring and passing a stream of nitrogen through the reaction vessel. After about 1 h (about 5% conversion) the cross-linking becomes noticeable (gelation). Stirring is continued for another 7 h at 90 °C, the reaction mixture then being allowed to cool to room temperature with stirring. The supernatant liquid is decanted from the beads which are washed several times with methanol and finally stirred for another 2 h with 200 mL of methanol. The polymer is filtered off and dried overnight in vacuum at 50 °C. Yield: practically quantitative. The cross-linked copolymer of styrene and 1,4-divinylbenzene so obtained is used for the preparation of ion-exchange resins.

The swellability is determined by placing 1 g of the dried polymer in contact with toluene for three days in a closed 250 mL flask. The swollen beads are collected on a sintered glass filter (porosity G 1), suction is applied for 5 min and the filter immediately weighed. The percentage increase in the weight of the beads can then be calculated. The size of the swollen and unswollen beads can be compared under the microscope.

Sulfonation of the cross-linked polystyrene: Insoluble polystyrene cross-linked with divinylbenzene can easily be converted by sulfonation to a usable ion exchanger. For this purpose a mixture of 0.2 g of silver sulfate and 150 mL of concentrated sulfuric acid are heated to 80–90 °C in a 500 mL three-necked flask fitted with stirrer, reflux condenser and thermometer. 20 g of the polymer of styrene and divinylbenzene are then introduced with stirring; the temperature rises spontaneously to 100–105 °C. The mixture is maintained at 100 °C for 3 h, then cooled to room temperature and allowed to stand for some hours. Next the contents of the flask are poured into a conical flask that contains about 500 mL of 50% sulfuric acid. After cooling, the mixture is diluted with distilled water, and the gold-brown colored beads are filtered off on a sintered glass filter and washed carefully with water.

Determination of the ion-exchange capacity: To determine the ion-exchange capacity, the sulfonated polymer is washed into a glass tube closed at one end with a stopcock (chromatographic column) above which is a plug of glass wool. 100 mL of 2 M NaCl solution are allowed to run through the column, followed by 100 mL of 2 M HCl. Finally, the column is washed with distilled water, the washings being collected in 10 mL portions and titrated with 0.1 M NaOH using phenolphthalein as indicator. When the concentration has fallen below 0.001 M the washing is stopped. The water remaining in the column is allowed to run off and the damp ion exchanger is poured into a beaker. Three samples, each of 2 g, are weighed into weighed 100 mL conical flasks as quickly as possible. One is heated to constant weight in an oven at 110 °C in order to determine the water content of the sample. 50 mL of 0.1 M NaOH are added to each of the other two flasks and vigorously shaken. After 30 min the mixtures are filtered, the resin is washed with a little water, and the filtrate back-titrated with 0.1 M HCl. The ion-exchange capacity is expressed in meq g^{-1} of dry exchanger.

Experiment 5-2: Preparation of Cation Exchanger 8 by Sulfonation of a Phenol-Formaldehyde Condensation Polymer (Section 5.1.1)

Reference: D. Braun, H. Cherdron, H. Ritter, *Polymer Synthesis: Theory and Practice; Fundamentals, Methods, Experiments*, Springer-Verlag, Berlin, Heidelberg, 2001.

Acid-catalysed phenol-formaldehyde condensation: Because of the formation of gaseous formaldehyde all reactions must be carried out under a closed hood. 65 g (0.69 mol) of phenol, 46 g of a 37% aqueous formaldehyde solution (0.57 mol), 7.5 mL of water and 1 g of oxalic acid dihydrate are placed in a 250 mL three-necked flask, fitted with stirrer and reflux condenser, and heated under reflux on an oil bath for 1.5 h. 150 mL of water are then added, stirred briefly, and allowed to cool, whereon the condensation polymer settles out. The aqueous layer is separated and the residual water distilled off at 50–100 torr while the temperature is slowly raised to 150 °C. This temperature is maintained (at most for 1 h) until test samples solidify on cooling. The resin is poured out while still warm and solidifies to a colourless, brittle mass, soluble in alcohol. The Novolak obtained is used to fabricate a molding by mixing with sawdust and hexamethylenetetramine as follows: 25 g of finely ground Novolak, 25 g of dry sawdust, 3.5 g of hexamethylenetetramine, 1 g of magnesium oxide (to trap residual acid from the condensation) and 0.5 g of calcium stearate (as lubricant) are thoroughly mixed (best in a ball mill or analytical mill) and then heated in a mold at 140 bar for 5 min at 160 °C. The resulting molding is infusible and insoluble.

Sulfonation of the phenol-formaldehyde condensation polymer: 40 g of the uncross-linked phenol-formaldehyde condensation polymer are gradually warmed to 140 °C in 120 g of 95% sulfuric acid in a 250 mL round-bottomed flask fitted with a reflux condenser. As soon as the resin has completely dissolved, the solution is cooled to room temperature, poured into an iron dish and 30 mL of 37% aqueous formaldehyde solution are stirred in with a spatula as quickly as possible. The metal dish is then placed in an oil bath at 110 °C and the resin allowed to harden for 2 h. The cooled product is washed with water, broken up with a hammer, and ground down to pieces of 1–3 mm size in a mortar. The particles are then washed with water until the washings are clear.

Determination of the ion-exchange capacity: The exchange capacity is determined as described above for sulfonated cross-linked polystyrene.

Experiment 5-3: Polyester 9a with Sulfonic Acid Groups (Section 5.1.1) [29]

Synthesis of the polyester 9a: 2-[Bis(2-hydroxyethyl)-amino]-ethane sulfonic acid (0.02 mol), triethylamine (0.04 mol) and 0.15 g tetra-*n*-butylphosphonium bromide were added to 60 mL dimethylacetamide. The mixture was heated at 70 °C for 20 min until the sulfonic acid was dissolved. Then at 20 °C 0.02 mol terephthaloyl chloride in 20 mL dimethylacetamide was added dropwise at 20 °C over 10 min. The resulting slurry was poured into ethanol to precipitate the polyester 9a (yield 80%).

Preparation of films from the polyester and their metal complexes: Films of 9a with poly(vinylalcohol) (PVA, M_w ~100,000 Da) were prepared by dissolving 9a in an aqueous 10% PVA solution and casting the solution onto a glass plate. Metal complex films of the polyester were prepared by adding 0.09 g of KOH and 0.18 g of Ni(CH₃COO)₂ or 0.09 g of CoCl₂ to the aqueous solution of 0.5 g of 9a and PVA and casting the solution.

Preparation of metal complexes of 9a: 9a (1.5 g) was added to the solution containing 0.27 g of KOH and 0.54 g of Ni(CH₃COO)₂ or 0.27 g of CoCl₂ in 20 mL of methanol. The reaction was continued for 1 h, and the reaction mixture was poured into 100 mL of diethyl ether to precipitate the metal complex 9b (M = Co, Ni).

Experiment 5-4: Cross-linked Terpolymer from [2-(Methacryloyl-oxy)-ethyl]trimethyl Ammonium Chloride, Acrylic Acid and *N,N*-Methylene-bis-acrylamide (Section 5.1.1) [32]

Synthesis of terpolymer: The cross-linked terpolymer was prepared by radical polymerization using a mixture of [2-(methacryloyloxy)ethyl]trimethylammonium chloride (0.05 mol, 8.72 g), acrylic acid (0.05 mol, 3.60 g), *N,N'*-methylene-bis acrylamide (6 mol %, 462.5 mg) and ammonium peroxydisulfate (0.5 mol %, 57.05 mg) at 80 °C for 2 h.

Metal-uptake experiments: The metal-uptake experiments were performed using standard metal salts (Cu(NO₃)₂, CdCl₂, HgCl₂, Zn(NO₃)₂, Pb(NO₃)₂, Cr(NO₃)₃ and UO₂(CH₃COO)₂) 1 g L⁻¹ at pH range 1–5 depending on the metal ion. All experiments were performed in flasks mounted on a shaker at 25 °C. The capacities for Cu(II), Cd(II), Zn(II), Hg(II), Pb(II), Cr(III) and U(VI) under non-competitive and competitive conditions were determined as a function of the pH. Batches of 0.1 g resin were used, together with a mixture of 10 mL of metal solution. After a shaking time of 1 h, the samples were filtered, washed with water, and dried in vacuum at 50 °C. The results after metal analysis are shown in Table 5-1.

Experiment 5-5: Metal Complexes of Poly(maleic acid) (Section 5.1.1) [39]

Synthesis of metal complexes of poly(maleic acid): Poly(maleic acid) (from Polysciences, M ~1000 Da) was dissolved in distilled water and an aqueous solution of a metal halide was added with stirring until no further precipitate was formed. The colored complexes were isolated at pH 6 (for the Ni, Co and Zn chelates) or pH 8 (for the Cu chelate). After elemental analysis the composition of the polymer chelates are: poly(maleic acid):meta l= Cu(II) 1:1.5; Ni(II) 1:1.0; Co(II) 1:1.0; Zn(II) 1:1.0.

Experiment 5-6: Polymer 13 by Incorporation of Glycine Functions in Cross-Linked Poly(acrylamide) (Section 5.1.1) [42]

Synthesis of poly(acrylamide-co-N,N'-methylene-bis-acrylamide): For the preparation of 2 mol % cross-linked poly(acrylamide) 14 g acrylamide, 0.6 g *N,N'*-methylene-bis-acrylamide and 100 mg potassium persulfate were dissolved in 100 mL water and heated at 80 °C for 30 min.The isolated polymer was washed with water, ethanol, methanol and dried at 70 °C.

Synthesis of polymer 13: 10 g of cross-linked poly(acrylamide) was treated with the sodium salt of glycine and refluxed at 90–100 °C for 20 h. The unreacted glycine and NaOH were removed by washing with distilled water, ethanol, methanol, and the residue was treated with acetone in a Soxhlet apparatus and dried at 50 °C.

Determination of carboxyl capacity: For the determination of carboxyl content 100 mg of the resin was equilibrated with HCl (0.2 mol L^{-1}, 10 mL) for 24 h with magnetic stirring. The resin samples were filtered, washed with distilled water to remove unreacted HCl and the filtrate was titrated with NaOH (0.2 mol L^{-1}) to a phenolphthalein end point. The polymer contains ~3.3 mmol g^{-1} carboxylic acid groups.

Complexation of metal ions with the carboxyl resin: Complexation of the carboxyl resin was carried out towards Fe(II), Fe(III), Co(II), Ni(II) and Cu(II) ions by a batch equilibration technique. A quantity (100 mg) of the resin was stirred in an aqueous metal salt solution (0.05 mol L^{-1}, 50 mL) for 24 h. The complexed resins were collected by filtration and washed with distilled water to remove uncomplexed metal ions. The concentrations of the metal salt solutions were estimated by UV spectrophotometric methods.

Experiment 5-7: Polymer-Bound Amide 20 (Section 5.1.2) [61]

At first a macroporous poly(styrene-*co*-divinylbenzene) is chloromethylated.

Then this cross-linked chloromethylated polystyrene is converted with the sodium salt of malonic diester to an ester-modified polymer. This polymer-bound malonic diester is reacted with a diamine to yield polmer **20**.

Chloromethylation of the cross-linked polystyrene: This synthesis must be carried out with rubber gloves under a hood, preferably in a glove box. 20 g of a macroporous cross-linked poly(styrene–co–divinylbenzene) (Aldrich) and a solution of chloromethyl methyl ether (alkylating agent, very poisonous!) in 40 mL of tetrachloroethylene are placed in a 250 mL three-necked flask fitted with a stirrer and reflux condenser (with drying tube attached), the third neck being closed with a ground-glass stopper. This mixture is stirred for 30 min at room temperature, causing the beads to swell somewhat. 10 g of anhydrous zinc chloride are added at 40–50 °C over a period of 60 min with continuous stirring; stirring is continued at this temperature for another 2 h. The unconverted chloromethyl methyl ether is now destroyed by careful addition of water; the beads are washed several times with water and dried in vacuum at 50 °C. The chlorine content of the beads should be around 15 wt. % (4.1 mmol g^{-1}).

Poly[styrene-co-divinylbenzene-co-chloromethylstyrene-co-4-[2,2-bis-(ethoxycarbonyl)ethyl]styrene]: 2.3 g (100 mmol) sodium were suspended in 150 mL of dry, boiling toluene. 16 g (100 mmol) diethylmalonate was added dropwise. After all the sodium had reacted, 10 g of the chloromethylated polystyrene (41 mmol of CH_2Cl) was added and the mixture was refluxed for 6 h. After cooling, the resin was filtered, washed with methanol and acetone, and then treated with acetone in a Soxhlet apparatus. The product was dried at 80 °C in vacuo. The polymer contains 7.1 mmol oxygen per g polymer (11.32 wt. % O) which corresponds to 1.78 mmol malonic ester groups per g polymer. The Cl content is 0.84 mmol per g polymer (2.99 wt. % Cl). IR (KBr): 1736 cm^{-1} (v C=O ester).

Polymers **20** *containing covalently bound moieties of amide groups:* 2 mmol of 3,6-diaza-1,8-octanediamine (Aldrich) and 0.96 g (5 mmol) triiso-propanolamine were dissolved in 30 mL of pyridine. 100 mg of the malonic ester polymer (0.178 mmol of malonic ester groups) were added and the mixture was refluxed for 24 h. After the mixture had cooled, the resin was filtered, washed with methanol and water and then treated with acetone in a Soxhlet apparatus for 2 h. The products were dried at 80 °C in vacuo. According to N analysis, the polymers **20** contain 1.71 mmol malonic amide groups per g polymer. IR (KBr): 1670 cm^{-1} (v C=O amide), weaker absorption 1736 cm^{-1} (v C=O ester).

Metal ion uptake: The measurements of the transition metal uptake by the amine- or amide-modified polymers were carried out in a 1×1 cm quartz cell at 25 °C. Typically, 3 mL of a solution (5×10^{-2} mol L^{-1}) of the transition metal chloride and 10 mg of amine- or amide-modified polymers were used. The uptake was measured at the absorption maximum of the transition metal ion in aqueous solution: Cu(II) 812 nm, Ni(II) 392 nm, Co(II) 509 nm. For **20** the following amounts of metal ions in mmol g^{-1} were found by elemental analysis: Cu(II) 80, Ni(II) 50, Co(II) 38 (for other polymer bound amines and amides see [61]).

Measurement for regeneration: 10 mg of the metal-containing polymer were suspended in 10 mL of diluted HCl (5%) and stirred for 24 h. Then the polymer was filtered and washed with 200 mL of distilled water. Diluted NaOH was added to adjust to a pH value of about 6. Then water was added to a total volume of 250 mL. The metal content was measured complexometrically. For metal ion-loaded **20** ~80% were found in solution.

Experiment 5-8: Binding of a Mo-Complex at Silica, Compound 25 (Section 5.1.2) [75]

Synthesis of 25: A mixture of 3 g silica and 25 g octadecyldimethyl[3-(trimethoxysilyl)propyl]ammonium chloride in 50 mL methanol and 5 mL toluene was heated at 70 °C with stirring for 24 h. After separation from the mixture, the modified silica was washed with methanol and dried at 70 °C under vacuum. The modified silica was added to a hydrobromic acid solution of molybdenum trioxide. The mixture was stirred and the precipitate was filtered, washed with methanol and dried under vacuum. The halogen content was obtained by the Volhard method. Molybdenum was analyzed by the EDTA method. The resulting polymer-bound catalyst is used in oxidation reactions.

Experiment 5-9: Phenylphosphonic Acid Resins 41 (Section 5.1.4) [114]

Synthesis of polymers 41: Poly(styrene-*co*-divinylbenzene) (2 wt. %) was used as the starting polymer. 17 g of this copolymer is placed in a round-bottomed flask together with 128 mL of PCl$_3$ and allowed to swell for 30 min. Fresh, powdered aluminum chloride is added (20.2 g) and the entire mixture is shaken at 20 °C for 24 h. After that time the beads are filtered off, washed briefly with dioxane and dropped into a 1-L beaker containing ice and sodium chloride solution. After washing with distilled water, the beads are left overnight in ca. 3 M NaOH solution followed by standard conditioning. A polymer-bound phosphinic acid is obtained by this procedure.

1 g of the polymer obtained, containing 4.05 mmol of phosphinic acid groups per gram is placed in a 100 mL Erlenmeyer flask together with 20 g of dry 1,2-dichloroethane. After 2 h of preswelling, 2.20 g (17 mmol; 4.2 × excess in respect to phosphinic groups) of diisopropylethylamine is added followed by an equimolar amount of trimethylchlorosilane. The entire mixture is shaken

under Ar for 24 h. After that time 9.0 mmol of appropriate electrophile, for example methyl chloroformate, is added and shaking under Ar is continued for 48 h. Workup consists of rinsing the polymer with 1,2-dichloroethane, dioxane and 1 M HCl in which the polymer is kept overnight, followed by passing distilled water through the polymer.

Hydrolysis of the resulting resins containing methyl and ethyl ester functionalities is done by refluxing the polymer in 3 M NaOH solution for 4 h. After the reaction, the polymer is washed with distilled water and conditioned with 1 M HCl, water, 1 M NaOH, water, 1 M HCl and finally with water. The phosphonic acid content is 3.7 mmol g^{-1}.

Determination of acid capacity: Acid capacity is measured by immersing a known amount of centrifuged polymer in the protonated form in 100 mL of 0.1 M NaOH solution for 24 h and titrating the resultant solution with 0.1 M HCl. pK_a of the resins is measured using ca. 0.1 g (dry weight) of polymer equilibrated with 45 mL of 1 M KCl. To this stirred (200 rev. min^{-1}) suspension 0.15 mL of 0.1 M KOH is injected every 90 min using a Kühn-and-Bayer liquid processor. The pH is recorded every 15 min. An acid capacity of 6 mmol g^{-1} was determined.

Metal-ion uptake: In order to determine the performance of the resins as cation exchangers a resin equivalent to 0.1 meq of acid capacity is shaken with 20 mL of the appropriate metal 10^{-4} N ion solution in nitric acid for 24 h and the metal concentration is determined using the atomic absorption method with wavelength set at 213.9, 228.8, 232.9 and 324.8 nm for Zn(II), Cd(II), Ni(II) and Cu(II), respectively. The affinity of the resins towards Eu(III) is determined by contacting a resin equivalent to 0.1 mmol of phosphorus with 10 mL of 3·10^{-4} N europium nitrate solutions in nitric acid. The final metal concentration is measured using emission atomic spectroscopy with a wavelength of 459.4 nm. For the result see [114].

Experiment 5-10: Covalent Binding of a 5,10,15,20-Tetraphenylporphyrin Derivative at Polystyrene for Polymer 53 and Photoredox Activity (Section 5.2.1) [132–135]

Chemicals: Metal-free 5,10,15,20-tetrakis-(4-pyridyl)porphyrin and metal-free 5,10,15,20-tetrakis-(4-aminophenyl)-porphyrin (Porphyrin Products, fax 001-801-753-6731, e-mail: porphyrin@cachenet.com; Porphyrin Systems, fax 0049-451-502-1130, e-mail: contact@porphyrin-systems.de); linear poly(chloromethylstyrene) containing 6.56 mmol Cl per g polymer with $\overline{M}_n$= 50,000 mol g^{-1} as a 3:2 mixture of 3- and 4-isomers (Aldrich); cross-linked chloromethylated styrene/divinylbenzene (2%) copolymer containing 1 mmol Cl per g polymer (Aldrich).

Equipment: Slide projector (or another light source with a quartz halogen lamp emitting visible light); quartz or glass cell 1 × 1 cm; IR-, UV-cut-off filters if available or 1 cm water filter for cutting IR light, no UV filter necessary if a glass cell is used; if available, equipment to measure the light intensity e.g. with a bolometer; UV-vis spectrometer (Fig. 5-8).

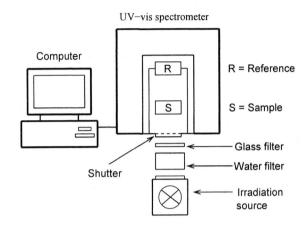

UV–vis spectrometer

Computer

R = Reference

S = Sample

Glass filter

Water filter

Shutter

Irradiation source

Figure 5-8. Measuring equipment for photo-redox activities.

*5,10,15,20-Tetrakis-(N-methyl-4-pyridylphenyl)-porphyrin zinc(II) complex tetraiodide or -chloride **47** (R = –C₅H₄N⁺CH₃; M = Zn):* 200 mg (0.32 mol) metal-free tetra(4-pyridyl)porphyrin was dissolved in 20 mL DMF at 120 °C. 350 mg (1.6 mmol) zinc acetate dihydrate was added in four portions over 5 h at 120 °C. The zinc porphyrin was precipitated with saturated aqueous KCl, filtered and intensively washed with water and dried in vacuo over P_4O_{10} at 60 °C. Methylation of the tetrapyridylporphyrin was achieved by treating the product with 20 mL methyl iodide at RT. The methylated porphyrin was filtered and dissolved in water. After filtration the solution was freeze-dried to obtain the tetraiodide salt. Iodide can be exchanged against chloride with a strong basic chloride-loaded ion exchanger from aqueous solution. Yield 88 mg (30%). UV-vis in water at λ: 620, 570, 520, 425 nm.

*5,10,15,20-Tetrakis-(4-aminophenyl)-porphyrin zinc(II) complex **47** (R = C₆H₄-NH₂; M = Zn):* 200 mg (0.3 mmol) of the metal-free tetraaminophenyl-porphyrin was dissolved in 20 mL DMF. 132 mg (0.6 mmol) zinc acetate dihydrate was added, and the mixture was heated for 5 h at 100 °C. The porphyrin was precipitated with saturated aqueous KCl solution, filtered, intensively washed with water and dried in vacuo over P_4O_{10} at 100 °C. UV-vis in DMF at λ (ε in L mol^{-1} cm^{-1}): 614, 566, 524, 436 (125,000) nm.

*Example of the synthesis of the positively charged linear polystyrene **53** containing a covalently bound porphyrin:* 6 mg (0.0083 mmol) of 5,10,15,20-

tetrakis-(4-aminophenyl)-porphyrin zinc(II) complex was dissolved in 100 mL of DMF and heated under inert gas at 80 °C. 3.34 g (33 mmol) triethylamine and a solution of 0.5 g (3.3 mmol) of the linear poly(chloromethylstyrene) in 50 mL of DMF was added dropwise with vigorous stirring. After 5 h at 80 °C the solution was cooled, the polymer precipitated with diethyl ether and filtered. The green-colored polymer **53** was dissolved in water, dialyzed against water and isolated by freeze-drying. Alternatively, the polymer can be dissolved in a small amount of DMF, precipitated with ether or methanol and treated in a Soxhlet apparatus with methanol. Yield around 0.45 g. The amount of polymer-bound porphyrin was determined in DMF taking the absorption at $\lambda = 434$ nm with $\varepsilon = 125,000$ L mol^{-1} cm^{-1} into account. A loading of 0.0079 mmol of porphyrin per g polymer was determined. UV-vis in DMF at λ: 610, 564, 436 nm.

Example of the synthesis of a positively charged cross-linked polystyrene containing covalently bound porphyrin: 18 mg (0.0244 mmol) of the tetra-aminophenylporphyrin zinc(II) complex was dissolved in 50 mL DMF. 1 g of the cross-linked chloromethylated polystyrene and then after 30 min 0.5 mL of triethylamine were added. The mixture was heated while slowly stirring for 24 h at 80 °C. The isolated polymer was treated with DMF. The amount of polymer-bonded porphyrin was determined from these solutions taking the absorption at $\lambda = 434$ nm with the absorption coefficient $\varepsilon = 125,000$ L mol^{-1} cm^{-1} into account. The polymers were further treated with acetone and water and then dried. The polystyrene contains 0.0076 mmol of porphyrin per g of polymer.

Photoredox measurements: Measurements were carried out in a quartz or glass cell (1 × 1 cm) at RT under argon. Typically, a solution (3 mL) containing sensitizers (10^{-6} mol L^{-1}), 2-mercaptoethanol as donor (0.54 mol L^{-1}) and methylviologen (7.5 10^{-4} mol L^{-1}) in water was used. The cell was placed into the light beam path of a UV-vis spectrophotometer with water as reference. Perpendicularly to the analytical light beam path, the quartz cell was irradiated with visible light (400–700 nm) by opening a shutter. A slide projector (between 100 and 250 W, 12–24 V) is used as a light source, and the light intensity in mW cm^{-2} can be determined e.g. with a bolometer. Water, IR- and UV-cut-off filters are used. Irradiation and registration of the spectra were performed automatically by linking to a computer. The concentration of $MV^{+\cdot}$ formed was calculated from the increased absorbance at 610 nm, ε being 13,700 L mol^{-1} cm^{-1}. The time dependence of the formation of $MV^{+\cdot}$ can be determined from the irradiation time. In addition, the concentration of sensitizer, acceptor and donor can be varied. For comparison with the polymer-bound porphyrins, the low molecular weight 5,10,15,20-tetrakis-(*N*-methyl-4-pyridylphenyl)porphyrin zinc complex is employed. The results of the measurements are described in [134,135].

Experiment 5-11: Covalent Binding of Tetracarboxyphthalocyanine 48 (R = –COOH; M = Co(II)) at Modified Silica for Compound 57 and Catalytic Oxidation of Thiols (Section 5.2.1) [137–139]

Synthesis of phthalocyanine-2,9,16,23-tetracarboxylic acid cobalt(II) complex 48 (R = –COOH; M = Co(II)): 0.26 g (2 mmol) water-free cobalt chloride, 1.54 g (8 mmol) 1,2,4-benzenetricarboxylic acid anhydride, 4.8 g (80 mmol) urea and around 20 mg ammonium molybdate are intensively mixed and heated under inert gas in a 100-mL glass vessel at 270 °C for 1 h. The blue-colored reaction mixture was pulverized, stirred in 400 mL 6 M aqueous HCl, filtered, intensively washed with 6 M HCl and then water and acetone. The product was treated in a Soxhlet apparatus with acetone and dried. The phthalocyanine tetracarboxylic acid amide obtained was saponified to the tetracarboxylic acid as follows. 1 g of the tetracarboxylic acid amide was heated in a mixture of 6 g KCl and 15 mL 2 M aqueous KOH at 90 °C until the ammonia evolution became weak. The filtered product was dissolved in a fairly small amount of 2 M NaOH and after filtration from byproducts separated with 2 M HCl and isolated by centrifugation. Purification was achieved by dissolving in 0.05 M NaOH, separation with 2 M HCl and centrifugation until the filtrate is free of chloride ions. Yield from 1 g tetracarboxylic acid amide 0.56 g (38%). IR (KBr): 1698, 1373, 1326, 1150, 1091, 743 cm^{-1}.

Synthesis of the phthalocyanine-2,9,16,23-tetracarboxylic acid tetrachloride cobalt complex 48 (R = –COCl; M = Co(II)): 0.5 g of the tetracarboxylic acid was added to 1.5 mL thionyl chloride containing 2 drops of pyridine. The mixture was heated for 20 h under reflux at 80 °C. After centrifugation the blue-colored acid chloride was dried at 100 °C in vacuo over P$_4$O$_{10}$. Yield 0.48 g (88%). IR (KBr): 1657, 1522, 1323, 1091, 749, 719 cm^{-1}.

Synthesis of modified silica 56: Two different types of low surface area silica were employed. The silica Fractosil 2500 (grain size = 80 μm) from Merck AG has a macroporous structure with relatively large pores (>250 nm) and a surface area (BET) of 8.3 m^2 g^{-1}. The monosphere silica from Merck exhibits a spherical non-porous structure with a surface area (BET) of only 1.7 m^2 g^{-1}. The silica species were functionalized with 3-aminopropyl groups (~0.5 mmol per g SiO$_2$) on the surface as follows. The silica was dried in vacuum (<1 Pa at 200 °C). A total of 20 g of silica was charged in a 250-mL round-bottomed flask with 150 mL dried *m*-xylene. 2.2 g (10 mmol) 3-aminopropyltriethoxysilane was added, and the mixture was refluxed for 8 h. The filtered product was washed thoroughly with acetone. The solid 56 was dried under vacuum at 80 °C.

Synthesis of the covalently bound 57: 0.164 g (0.2 mmol) of 48 (R = –COCl; M = Co(II)) was dissolved in 25 mL DMF with a trace of pyridine. To this solution 2 g of **56** was added. The mixture was stirred at 100 °C for 20 h, filtered, washed with DMF and acetone and air-dried. The blue solid was stirred 8 h in 100 mL 0.01 M sodium hydroxide to dissolve unreacted phthalocyanine derivative and to saponify residual carboxylic chloride groups of bound phthalocyanine. The solid was separated, washed with 0.01 M sodium hydroxide and with water until the filtrates were colourless. The air-dried product was extracted with pyridine for 8 h to remove residual unreacted phthalocyanine. The final product **57** was dried under vacuum at 80 °C over P_4O_{10}. The degree of loading of the silica with the cobalt phthalocyanine was determined as follows. 200 mg of a sample was treated with 2 mL 65% HNO_3 and 1 mL H_2SO_4. The mixture was heated in a Teflon autoclave for 1 h at 150 °C. After cooling, 5 mL of 30% NaOH was added. The solid was removed and the liquid transferred into a 10-mL volumetric flask which was filled with distilled H_2O. The cobalt content was measured with an AAS-Spectrometer. A cobalt or cobalt phthalocyanine content of ~1.3 µmol per g carrier was deter-mined for the silica used.

Experiment 5-12: Coordinative Binding of Phthalocyanine 48 (R = – H; M = Co(II)) at Modified Silica for Compound 62 and Catalytic Oxidation of Thiols (Section 5.2.2) [153,154]

Chemicals: Phthalocyanine cobalt(II) complex (Aldrich).

Equipment: Thermostated glass apparatus consisting of a 50-mL gas buret and a 100-mL reaction vessel to measure the oxygen consumption during catalytic oxidations.

*Synthesis of modified silica **61**:* Different types of silica dried at 200 °C under vacuum were placed in a 250-mL flask with 150 mL dried *m*-xylene. 2 g (10 mmol) 3-chloropropyltrimethoxysilane and 1.4 g (20 mmol) imidazole were added, and the mixture was refluxed for 8 h. An excess of imidazole was used to bind the hydrochloric acid formed. The filtered product was washed with 0.01 M NaOH to remove hydrochloric acid and unbonded imidazol. Afterwards the solids were washed thoroughly with water and acetone. The solids were dried under vacuum at 80 °C.

*Synthesis of coordinatively bound **62**:* 12 mg (20 µmol) cobalt(II) phthalocyanine was dissolved in 25 mL dry DMF. To this solution 2 g of the modified silica were added. The blue suspension was stirred at 25 °C for 4 h. The solids were separated and washed twice with 10 mL DMF. Afterwards the solids were washed thoroughly with acetone and dried under vacuum at 80 °C over P_4O_{10}. The resulting light blue products were found after cobalt analysis to contain the chelate in the range of 4.5 µmol per g carrier. For the coordinative binding of different kind of cobalt(II) phthalocyanines, naphthalocyanines or tetraphenylporphyrins, **47**, **48**, **49** either unsubstituted or substituted by nitro, phenoxy, alkoxy or nitrile groups can be employed [153]. The catalytic activity for the oxidation of 2-mercaptoethanol was measured (see Section 5.4.5). Turnover frequencies (TON) of ~400 min^{-1} were determined. But with other porphyrin-type compounds TON up to ~2500 min^{-1} were obtained [153].

Catalytic oxidation of 2-mercaptoethanol: A thermostated glass apparatus consisting of a 50-mL gas buret and 100-mL reaction vessel which are filled with oxygen were used. The catalyst, containing 0.5 µmol of phthalocyanine, was added to the reaction vessel, and 10 mL of a buffer solution pH 9.0 was added. After rinsing the apparatus for 10 min with dioxygen, 500 µL (7.1 mmol) 2-mercaptoethanol was added. The apparatus was closed immediately, and the dioxygen consumption was measured volumetrically at 25 °C. During the measurements the solution was stirred with a magnetic bar (1000 rpm). The mole ratio mercaptoethanol/phthalocyanine was ~14,000. During the measurements no leaching of the phthalocyanines from the silica into solution was observed. The oxygen consumption over time was measured and plotted as described in [137]. Turnover frequencies (TON) in min^{-1} were taken from linear slopes of the conversion *versus* time plots. For the phthalocyanine on Fractosil 2500 a TON of 850 min^{-1} and on monosphere silicon of 1000 min^{-1} were determined. For comparison, the low molecular weight cobalt phthalocyanine tetracarboxylic acid exhibits a TON of around 850 min^{-1}. But the heterogeneous catalyst can be re-used. Results of the measurements are described in [137,139].

Experiment 5-13: Ionic Binding of the Tetraanionic Phthalocyanine 48 (R = –SO₃H; M = Al(OH)) at a Cationic Cross-Linked Polystyrene for Compound 66 and Photooxidation of Phenols (Section 5.2.3) [167,168]

Chemicals: Amberlite® IRA 400 (Fluka).

Equipment: Thermostated glass apparatus consisting of a 50-mL gas buret and a 100-mL reaction vessel to measure the oxygen consumption during photooxidation; 250 W slide projector with a quartz halogen lamp.

Synthesis of phthalocyanine-2,9,16,23-tetrasulfonic acid aluminum(III) complex 48 (R = –SO₃;, M = Al(OH)): 10.73 g (40 mmol) 4-sulfophthalic acid monosodium salt, 1.28 g (24 mmol) ammonium chloride, 24.02 g (400 mmol) urea and 0.185 g (0.15 mmol) ammonium molybdate (all chemicals dried) were intensively mixed and placed in a 250 mL glass vessel under dry inert gas. 2 g (15 mmol) water-free aluminum trichloride (purity 99.99%) was added under inert gas. The mixture was heated at first at 140 °C and then within 30 min with stirring to 190 °C, followed by heating at 210 °C for 24 h under inert gas. The blue-colored reaction product was pulverized, treated for 24 h with 1 M aqueous

HCl saturated with NaCl and isolated by centrifugation. Additional products can be obtained by reducing the volumes of the aqueous HCl/NaCl. The product was dissolved in a small amount of water, filtered and freeze-dried. The residue was treated for 72 h in a Soxhlet apparatus with ethanol (96%) to extract sodium chloride. Yield 2.65 g (27%) of the tetrasodium salt. IR (KBr): 3168, 1630, 1427, 1362, 1189, 1032, 794 cm^{-1}. UV-vis in water (containing 0.1 M cetyltrimethylammonium chloride) at λ (ε/L mol^{-1} cm^{-1}): 674 (132,000), 606, 343 nm.

Synthesis of the ionically bound 66: As an example we describe the synthesis of polymers containing 4 µmol g^{-1} photosensitizer. Commercial Amberlite® IRA 400 (Fluka) was dried at 25 °C in vacuo for 24 h, then 2.5 g was suspended in 100 mL distilled water. **48** (R = –SO$_3$H; M = Al(OH)) (10 µmol) was dissolved in 40 mL water and was added dropwise during 30 min with gentle stirring. The mixture was heated for 3 h at 60 °C. The solution became colorless and the Amberlite® intensively colored. The polymer was isolated by filtration, washed with water and then ethanol. No desorption of the dyes could be observed under the stated reaction conditions. After drying overnight in vacuo 2.5 g of the polymer **66** was obtained. Due to quantitative binding the polymer contains 4 µmol phthalocyanine per g polymer.

Photooxidation of Phenol: The photooxidation of phenols is described in Sections 13.4 and 13.5.7).

5.5   References

1. A.D. Pomogailo, I.E. Uflyand, *Macromolecular Metal Chelates*, Nauka, Moscow, 1991.
2. A.D. Pomogailo, Polymeric Immobilized Metallocomplex Catalysts, Nauka, Moscow 1988.
3. E.A. Bekturov, S.E. Kudaibergenov, *Catalysis by Polymers*, Hüthig und Wepf Verlag, Heidelberg, 1996.
4. D.C. Sherrington, *Pure Appl. Chem.* **1988**, *60*, 401.
5. F. Ciardelli, E. Tsuchida, D. Wöhrle, *Macromolecule–Metal Complexes*, Springer-Verlag, Berlin, 1996.
6. L.V. Miroshnik, A.M. Dubina, V.N. Tolmachev, *Koord. Khim.* **1980**, *6*, 870. J.W. Rosthauser, A. Winston, *Macromolecules* **1981**, *14*, 538.
7. E.A. Bekturov, *Macromol. Symp.* **2000**, *156*, 231.
8. E. Kalalova, O. Populova, S. Stokrova, P. Stopka, *Collect. Czech. Chem. Commun.*, **1983**, *48*, 2021.
9. A.A. Bol'shov, A.D. Pomogailo, I.D. Leonov, in: *Kompleksnaya Pererabotka Mangyshlakskoi Nefti (Complex Processing of Mangyshlak Petroleum)*, Nauka, Alma-Ata, 1972, Vol. 4, p. 110.
10. A. Warshawsky, A. Deshe, G. Rossey, A. Patchornik, *React. Polym.* **1984**, *2*, 301.
11. P.S. Drago, E.D. Nyberg, A.G. El A'mma, *Inorg. Chem.* **1981**, *20*, 2461.

12. E. Tsuchida, Y. Karino, H. Nishide, Y. Kurimura, *Makromol. Chem.* **1974**, *175*, 161.
13. G. Wulff, W. Vesper, R. Crobe-Einsler, A. Sarhan, *Makromol. Chem.* **1977**, *178*, 2799. G. Wulff, *Angew. Chem.* **1995**, *107*, 1958.
14. K. Mosbach, *Trends in Biochem. Sci.* **1994**, *19*, 19. M.J. Whitcombe, M.E. Rodriguez, P. Villar, E.N. Vulfson, *J. Am. Chem. Soc.* **1995**, *117*, 7105.
15. A.A. Efendiev, D.D. Orudgev, V.A. Kabanov, *Dokl. Akad. Nauk SSSR* **1980**, *255*, 1393.
16. A.A. Efendiev, *Macromol. Symp.* **1994**, *80*, 289; ibid, **1998**, *131*, 29.
17. U. Braun, W. Kuchen, *Chem.-Ztg.* **1984**, *107*, 255.
18. H. Chen, M.M. Olmstead, R.L. Albright, J. Devenyi, R.H. Fish, *Angew. Chem.* **1997**, *109*, 624.
19. X. Zeng, G.M. Murray, *Sep. Sci. Technol.* **1996**, *31*, 2403.
20. R. Kunin, *Ion Exchange Resins*, Wiley and Sons, New York, 2nd Edn, 1958.
21. D.C. Sherrington, P. Hodge (Eds.), *Synthesis and Separations Using Functional Polymers*, Wiley and Sons, New York, 1988.
22. S.D. Alexandratos, D.W. Crick, *Ind. Eng. Chem. Res.* **1966**, *35*, 635.
23. D.C. Szlag, N.J. Wolf, *Clean Products and Processes* **1999**, *1*, 117.
24. R.A. Beauvais, S.D. Alexandratos, *React. Funct. Polym.* **1998**, *36*, 113.
25. G.M. Ritcey, A.W. Ashbrook, *Solvent Extraction: Principles and Application to Process Metallogy*, Part 1, Elsevier, Amsterdam, 1984.
26. B.L. Rivas, S.A. Pooley, M. Luna, K.E. Geckeler, *J. Appl. Polym. Sci.* **2001**, *82*, 22.
27. E. Ranucci, A. Sandgren, N. Andronova, A.-C. Albertsson, *J. Appl. Polym. Sci.* **2001**, *82*, 1971.
28. C. Palezzi, F. Pinna, G. Strukul, *J. Mol. Catal. A: Chem.* **2000**, *151*, 245.
29. C. Wang, T. Takayama, S. Nakamura, *J. Polym. Sci. A: Polym. Chem.* **1997**, *35*, 3561.
30. B.L. Rivas, G.V. Seguel, *Polym. Bull.* **1998**, *40*, 431.
31. Y. Dan, Q. Wang, *Polym. Intern.* **2001**, *50*, 1109.
32. B.L. Rivas, H.A. Maturana, I.M. Peric, S. Villegas, *Polym. Bull.* **1999**, *43*, 277.
33. V. Montembault, V. Folliot, J.-C. Soutif, J.-C. Brosse, *React. Polym.* **1994**, *22*, 81.
34. C.-C. Wang, C.-Y. Chang, C.-Y. Chen, *Macromol. Chem. Phys.* **2001**, *202*, 882.
35. Y. Okamoto, *Macromol. Chem., Macromol. Symp.* **1992**, *59*, 83. Y. Okamoto, J. Kido, in: *Macromolecular Complexes*, E. Tsuchida (Ed.), VCH Publishers, New York, 1991, p. 93.
36. B.M. Culbertson, D. Xie, A. Thakur, *Macromol. Symp.* **1998**, *131*, 11.
37. M.J. Tiera, V.A. De Oliveira, H.D. Burrows, M. Miguel, M.G. Neumann, *Colloid Polym. Sci.* **1998**, *276*, 206.
38. K. Inui, T. Noguchi, T. Miyata, T. Uragami, *J. Appl. Polym. Sci.* **1999**, *71*, 233.
39. B.L. Rivas, G.V. Seguel, C. Ancatripai, *Polym. Bull.* **2000**, *44*, 445.
40. C. Fenn-Barrabaß, A. Pohlmeier, W. Knoche, H.D. Narres, M.J. Schwager, *Colloid Polym. Sci.* **1998**, *276*, 627.
41. B.L. Rivas, G.V. Seguel, *Polym. Bull.* **2001**, *46*, 271.
42. G.S. Vinodkumar, B. Mathew, *Eur. Polym. J.* **1998**, *34*, 1185.
43. K.J. Kamble, D.N.J. Patkar, *Polym. Sci., Polym. Chem.* **1995**, *33*, 629.

44. O. Sanli, G. Asman, *J. Appl. Polym. Sci.* **2000**, *77*, 1096.
45. K. Burger, *Carbohydrate Res.* **2001**, *332*, 197.
46. W.H. Meyer, *Adv. Mater.* **1998**, *10*, 439.
47. S. Takeoka, K. Horiuchi, S. Yamagata, E. Tsuchida, *Macromolecules* **1991**, *24*, 2003.
48. R.R.M. Hikmet, I. Michels, *Adv. Mater.* **2001**, *13*, 338.
49. C.K. Chiang, G.T. Davies, *Macromolecules* **1985**, *18*, 825.
50. J.R.R. Giles, M.P. Greenhall, *Polym. Commun.* **1986**, *27*, 360.
51. N. Oyama, O. Hatozaki, *Macromol. Symp.* **2000**, *156*, 171.
52. R. Sinta, B. Lamb, J. Smid, *Macromolecules* **1983**, *16*, 1382.
53. Q. Zhao, R.A.J. Bartsch, *Appl. Polym. Sci.* **1995**, *57*, 1465.
54. K. Yamashita, K. Kurita, *React. Funct. Polym.* **1996**, *31*, 47.
55. G. Uysal, S. Memon, M. Yilmaz, *React. Funct. Polym.* **2001**, *47*, 165.
56. N. Dumont, A. Favre-Requillon, *Sep. Sci. Technol.* **1996**, *31*, 1001.
57. J.R. Patel, D.H. Sutaria, M.N. Patel, *High Perform. Polym.* **1994**, *6*, 123.
58. U.K. Samal, P.L. Nayak, S.J. Lenka, *Appl. Polym. Sci.* **1993**, *47*, 1315; ibid. **1993**, *47*, 367.
59. T. Kaliyappan, S.C. Murugavel, P. Kannan, *Polym. Intern.* **1998**, *45*, 278; *Eur. Polym. J.* **1997**, *33*, 59.
60. C.S. Harris, D.F. Shriver, M.A. Ratner, *Macromolecules* **1986**, *19*, 987.
61. V. Nicolaus, D. Wöhrle, *Angew. Makromol. Chem.* **1992**, *198*, 179.
62. M. Chanda, G.L. Rempel, *React. Polym.* **1993**, *19*, 213.
63. U. Schilde, H. Kraudelt, E. Uhlemann, *React. Polym.* **1994**, *22*, 101.
64. B.L. Rivaso, E.D. Pereira, *Polym. Bull.* **2000**, *45*, 69.
65. N. Farell, in: *Transition Metal Complexes as Drugs and Chemotherapeutic Agents,* W.O. Foye (Ed.), Kluwer Academics, Dordrecht, 1989.
66. W.R. Waud, in *Cancer Chemotherapeutic Agents,* W.O. Foye (Ed.), American Chemical Society, Washington, DC, 1995, p.121.
67. E.W. Neuse, B.B. Patel, C.W.N. Mbonyana, *J. Inorg. Organomet. Polym.* **1991**, *1*, 147.
68. E.W. Neuse, G. Caldwell, A.G. Perlwitz, *J. Inorg. Organomet. Polym.* **1995**, *5*, 195.
69. E.W.Neuse, G. Caldwell, *J. Inorg. Organomet. Polym.* **1997**, *7*, 163.
70. G. Caldwell, E.W. Neuse, C.E.J. van Rensburg, *J. Inorg. Organomet. Polym.* **1997**, *7*, 217.
71. V.V. Khutoryansky, P. Kujawa, J.M. Rosiak, *Macromol. Chem. Phys.* **2001**, *202*, 1089.
72. W.Li, M. Coughlin, R.L. Albright, R.H. Fish, *React. Funct. Polym.* **1995**, *28*, 89.
73. R. Antony, G.L. Tembe, M. Ravindranathan, R.H. Ram, *J. Mol. Catal. A: Chem.* **2001**, *171*, 159.
74. T. Hirao, M. Higuchi, S. Yamaguchi, *Macromol. Symp.* **1998**, *131*, 59.
75. Y. Kurusu, *J. Inorg. Organomet. Polym.* **2000**, *10*, 127.
76. M. Kaneko, A. Yamada, *Adv. Polym. Sci.* **1984**, *55*, 1.
77. M. Kaneko, A. Yamada, in: *Metal Containing Polymer Systems*, J.E. Sheats, C.E. Carraher, C.U. Pittman (Eds.), Plenum Publishing Corporation, New York, 1985, p. 289.

78. E.A. Bekturov, *Macromol. Symp.* **2000**, *156*, 231.
79. R. Antony, G.L. Tembe, M. Ravindranathan, R.N. Ram, *Eur. Polym. J.* **2000**, *36*, 1579.
80. H.J. Hoorn, P. de Joode, W.L. Driesen, J. Reedijk, *Recl. Trav. Chim. Pays-Bas* **1996**, *115*, 191.
81. A.M.J. Jorna, A.E.M. Boelrijk, H.J. Horn, J. Reedijk, *React. Funct. Polym.* **1996**, *29*, 101.
82. N. Moriguchi, T. Tsugaru, S. Amiya, *J. Mol. Struct.* **1999**, *477*, 181; ibid. **2000**, *523*, 93; ibid. **2001**, *562*, 205; ibid. **2001**, *562*, 215.
83. M. Kaneko, S. Nemoto, A. Yamada, Y. Kurimura, *Inorg. Chim. Acta* **1980**, *44*, L289.
84. Y. Kurimura, N. Shinozaki, K. Shigehara, E. Tsuchida, M. Kaneko, A. Yamada, *Bull. Chem. Soc. Jpn.* **1982**, *55*, 380.
85. G.G. Talanova, L. Zhong, O.V. Kravchenko, K.B. Yatsimirskii, R.A. Bartsch, *J. Appl. Polym. Sci.* **1999**, *74*, 849; ibid. **2001**, *80*, 849.
86. Y. Zhang, C.B. Murphy, W.E. Jones, *Macromolecules* **2002**, *35*, 630.
87. J.-M. Kim, B.-O. Chong, K.H. Park, *Polym. Bull.* **2000**, *44*, 79.
88. T. Miyajima, H. Nishimura, H. Kodama, S. Ishiguro, *React. Funct. Polym.* **1998**, *38*, 183.
89. P.M. van Berkel, W.L. Driessen, J. Reedijk, D.C. Sherrington, *Eur. Polym. J.* **1997**, *33*, 129; *React. Funct. Polym.* **1997**, *32*, 139; *Eur. Polym. J.* **1997**, *33*, 303; *React. Funct. Polym.* **1995**, *28*, 39.
90. H. Kise, H. Sato, *Makromol. Chem.* **1985**, *186*, 2449.
91. A. Boudakgi, J. Jezierska, B.N. Kolarz, *Macromol. Chem., Macromol. Symp.* **1992**, *59*, 343.
92. N. Kabay, A. Kataki, T. Sugo, H.J. Egawa, *Appl. Polym. Sci.* **1993**, *49*, 599.
93. L. Weiping, F. Ruowen, L. Yun, Z. Hammin, *React. Polym.* **1994**, *22*, 1.
94. Y.K. Argrawall, K.V. Rao, *React. Funct. Polym.* **1996**, *31*, 225.
95. D.E. Bergbreiter, N. Koshti, J.G. Frauchina, J.D. Frels, *Angew. Chem.* **2002**, *112*, 1082.
96. M.J. Haron, W.M.Z.W. Yunus, M.Z. Desa, A. Kassim, *Talanta* **1994**, *41*, 805.
97. G.N. Rao, S. Srivastava, S.K. Srivastava, M. Singh, *Talanta* **1996**, *43*, 1821.
98. S. Thamizharasi, A. Reddy, S. Balasubramanian, *Eur. Poly. J.* **1998**, *34*, 503.
99. M.K. Dalal, M.J. Upadhyay, R.N. Ram, *J. Mol. Catal. A: Chem.* **1999**, *142*, 325; ibid. **2000**, *159*, 285.
100. A.Z. El–Sonbati, A.A. El–Bindary, N.A. El–Deeb, *React. Funct. Polym.* **2002**, *50*, 131.
101. S. Thamizharasi, A. Reddy, S. Balasubramanian, *Eur. Polym. J.* **1998**, *34*, 1605.
102. U.K. Jain, A. Handa, S.S. Pait, P. Shrivastav, Y. K. Argrawal, *Anal. Chim. Acta* **2001**, *429*, 237
103. N. Manolova, M. Ignatova, L. Rashkov, *Eur. Polym. J.* **1998**, *34*, 1133.
104. M. Kan, *React. Funct. Polym.* **1996**, *31*, 207.
105. S.J. Choi, K.E. Geckeler, *Polym. Intern.* **2000**, *49*, 1519.
106. B.L. Rivas, S.A. Pooley, M. Soto, K.E. Geckeler, *J. Appl. Polym. Sci.* **1999**, *72*, 741.
107. I.E. Uflyand, A.D. Pomogailo, N.D. Golubeva, V.N. Sheinker, *Proc. XVII Europ.*

Congr. Mol. Spectros., Madrid, 1985, p. 152.

108. A.D. Pomogailo, I.E. Uflyand, N.D. Golubeva, *Kinetics and Catalysis,* **1985,** *26,* 1104.

109. I.E. Uflyand, A.D. Pomogailo, M.O. Gorbumova, A.G. Starikov, V.N. Sheinker, *Kinetics and Catalysis* **1987,** *28,* 613; I.E. Uflyand, A.S. Kuzharov, M.O. Gorbumova, V.N. Sheinker, A.D. Pomogailo, *React. Polym.* **1989,** *11,* 221.

110. K. Yamashita, K. Kurita, K. Ohara, *React. Funct. Polym.* **1996,** *31,* 47.

111. T.F. Baumann, J.G. Reynolds, G.A. Fox, *React. Funct. Polym.* **2000,** *44,* 111.

112. S.-P. Huang, K.J. Franz, E.H. Arnold, J. Dvenyi, R.H. Fish, *Polyhedron* **1996,** *15,* 4241.

113. J.M. Sanchez, M. Hidalgo, V. Salvado, *React. Funct. Polym.* **2001,** *46,* 283.

114. A.W. Trochimczuk, *React. Funct. Polym.* **2000,** *44,* 9; ibid. **2001,** *48,* 141.

115. J.M. Sanchez, M. Hidalgo, M. Valiente, V. Salvado, *J. Polym. Sci. A: Polym. Chem.* **2000,** *38,* 269.

116. D. Wöhrle, P. Buttner, *Polym. Bull.* **1985,** *13,* 57.

117. H. Aeissen, D. Wöhrle, *Makromol. Chem.* **1981,** *182,* 2961.

118. R.S. Paredes, N.S. Valera, L.F. Lindoy, *Austr. J. Chem.* **1986,** *39,* 1081.

119. D.C. Sherrington, J.K. Karjalainen, L. Canali, H. Deleuze, *Macromol. Symp.* **2000,** *156,* 125.

120. M.D. Angelino, P.E. Laibinis, *J. Polym. Sci. A: Chem.* **1999,** *37,* 3888.

121. F. Minutolo, D. Pini, A. Petri, P. Salvadori, *Tetrahedron Asymm.* **1996,** *7,* 2293.

122. P.K. Dhal, B.B. De, S. Sivaram, *J. Mol. Catal. A: Chem.* **2001,** *177,* 71.

123. A. Syamal, M.M. Singh, D. Kumar, *React. Funct. Polym.* **1999,** *39,* 27.

124. E. Tsuchida, H. Nishide, *Adv. Polym. Sci.* **1977,** *24,* 1.

125. D. Wöhrle, *Adv. Polym. Sci.* **1983,** *50,* 45.

126. D. Wöhrle, in: *Phthalocyanines — Properties and Applications*, C.C. Leznoff, A.B.P. Lever (Eds.), VCH, New York, 1989, p. 55.

127. Yu.S. Avlasevich, O.G. Kulinkovich, V.N. Knyukshto, K.N. Solov'ev *J. Appl. Spectroscopy* **1999,** *66,* 597; ibid. **2000,** *67,* 663.

128. Yu.S. Avlasevich, T.A. Chevtchouk, V.N. Knyukshto, O.G. Kulinkovich, K.N. Solov'ev, *J. Porphyins Phthalocyanines* **2000,** *4,* 579.

129. V.N. Knyukshto, Yu.S. Avlasevich, O.G. Kulinkovich, K.N. Solov`ev, *J. Fluorescence* **1999,** *9,* 371.

130. R. Zhou, L. Tang, K.E. Geckeler, W. Göpel, *Makromol. Chem. Phys.* **1994,** *195,* 2409.

131. M. Sanchez, N. Chap, J.-B. Cazaux, B. Meunier, *Eur. J. Inorg. Chem.* **2001,** *1775.*

132. D. Wöhrle, G. Krawczyk, *Makromol. Chem.* **1986,** *187,* 2535.

133. D. Wöhrle, G. Krawczyk, M. Paliuras, *Makromol. Chem.* **1988,** *189,* 1001, 1013.

134. D. Wöhrle, J. Gitzel, G. Krawczyk, E. Tsuchida, H. Ohno, I. Okura, T. Nishisaka, *J. Macromol. Sci.* **1988,** *A25,* 1227.

135. D. Wöhrle, M. Paliuras, I. Okura, *Makromol. Chem.* **1991,** *192,* 819.

136. J. Gitzel, H. Ohno, E. Tsuchida, D. Wöhrle, *Polymer* **1986,** *27,* 1781.

137. T Buck, D. Wöhrle, G. Schulz-Ekloff, A. Andreev, *J. Mol. Catal.* **1991,** *70,* 259.

138. H. Fischer, G. Schulz-Ekloff, T. Buck, D. Wöhrle, *Erdöl, Erdgas, Kohle* **1994,** *110,* 128.

139. G. Schneider, D. Wöhrle, W Spiller, J. Stark, G. Schulz-Ekloff, *Photochem.*

Photobiol. **1994**, *60*, 333.

140. D.W.J. McLallien, P.L. Burn, N.L. Anderson, *J. Chem. Soc., Perkin Trans.* **1997**, *1*, 2581.

141. J.C. Biazotto, H.C. Sacco, K.J. Ciaffi, C.R. Neri, A.G. Ferreira, Y. Iamamoto, O.A. Serra, *J. Non-Cryst. Solids* **1999**, *247*, 134; ibid. **2000**, *273*, 186.

142. F.G. Doro, J.R.L. Smith, A.G. Ferreira, M.D. Assis, *J. Mol. Catal. A: Chem.* **2000**, *164*, 97.

143. Y. Rohlfing, D. Wöhrle, M. Wark, G. Schulz-Ekloff, J. Rathousky, A. Zukal, *Studies in Surface Science and Catalysis* **2000**, *129*, 295; Y. Rohlfing, *Master Thesis*, Universität Bremen **1999**.

144. E. Tsuchida, *J. Inorg. Nucl. Chem.* **1978**, *40*, 1241.

145. E. Tsuchida, K. Honda, H. Sata, *Makromol. Chem.* **1975**, *176*, 2251.

146. H Nishide, E. Tsuchida, *Biopolymers* **1978**, *17*, 191.

147. E. Tsuchida, *Artificial Red Cells*, John Wiley & Sons, Chichester, 1995.

148. R.I. Kureshy, N.H. Khan, S.H.R. Abi, P. Iyer, *React. Funct. Polym.* **1997**, *34*, 153.

149. M. Devenney, J. Grimshaw, J. Trocha-Grimshaw, *J. Chem. Soc., Perkin Trans.* **1998**, *2*, 997.

150. S.E.J. Bell, M. Devenney, J. Grimshaw, S. Hara, J.M. Rice, J. Trocha-Grimshaw, *J. Chem. Soc., Faraday Trans.* **1998**, *94*, 2955.

151. T. Yoshida, K. Kamato, M. Tsukamoto, T. Iida, D. Schlettwein, D. Wöhrle, M Kaneko,. *J. Electroanal. Chem.* **1995**, *385*, 209.

152. F. Zhao, J. Zhang, T. Abe, D. Wöhrle, M. Kaneko, *J. Mol. Catal. A: Chem.* **1999**, *145*, 245.

153. T. Buck, H. Bohlen, D. Wöhrle, G. Schulz-Ekloff, A. Andreev, *J. Mol. Catal.* **1993**, *80*, 253.

154. H. Fischer, G. Schulz-Ekloff, T Buck, D. Wöhrle, *Erdöl, Erdgas, Kohle* **1994**, *10*, 128.

155. V. Huc, F. Armand, J.P. Bourgoin, S. Palacin, *Langmuir* **2001**, *17*, 1928.

156. R. Purello, E. Bellacchio, S. Gurrieri, R. Lauceri, A. Raudino, L.M. Scolaro, A.M. Santoro, *J. Phys. Chem. B* **1998**, *102*, 8852.

157. R. Lauceri, T. Campagno, A. Raudino, R. Purello, *Inorg. Chim. Acta* **2001**, *317*, 282.

158. R. Purrello, A. Raudino, L.M. Scolaro, A. Loisi, E. Bellacchio, R. Lauceri, *J. Phys. Chem. B* **2000**, *104*, 10900.

159. M.V. Vinodu, M. Padmanabhan, *J. Polym. Sci. A: Polymer Chem.* **2001**, *39*, 326.

160. T. Mathew, M. Padmanabhan, *J. Appl. Polym. Sci.* **1996**, *59*, 23.

161. T. Yamaue, T. Kawai, M. Onoda, K. Yoshino, *J. Appl. Phys.* **1999**, *85*, 1626.

162. A.B. Solovieva, S.L. Kotova, S.A. Zaviyalov, T.N. Rmyantseva, N.N. Glagolev, S.F. Timashev, *Russ. J. Phys. Chem. Suppl. 1* **2000**, *74*, 92.

163. J. van Welzen, A.M. van Herk, A.L. German, *Makromol. Chem.* **1987**, *188*, 1923.

164. A.M. van Herk, A.H.J. Tullemanns, J. van Welzen, A.L. German, *J. Mol. Catal.* **1988**, *44*, 269.

165. M. Hassanein, W.T. Ford, *Macromolecules* **1988**, *21*, 526.

166. W. Spiller, D Wöhrle, G. Schulz-Ekloff, W.T. Ford, G. Schneider, J. Stark, *J. Photochem. Photobiol. A: Chem.* **1996**, *95*, 161.

167. R. Gerdes, O. Bartels, G. Schneider, D. Wöhrle, *Intern. J. Photoenergy* **1999**, *41*.

168. R. Gerdes, O. Bartels, G. Schneider, G. Schulz-Ekloff, *Polym. Adv. Technol.* **2001**, *12*, 152.
169. H.C. Sacco, Y. Iamamoto, J.R.L. Smith, *J. Chem. Soc., Perkin Trans.* **2001**, *2*, 181.
170. F. Bedioui, Y. Bouher, C. Sorel, J. Devynck, *Electrochim. Acta* **1993**, *38*, 2485.
171. M.-H. Liu, Y.O. Su, *J. Chin. Chem. Soc.* **1999**, *46*, 115.
172. W. Paik, I.-H. Yeo, J. Suh, Y. Kim, E. Song, *Electrochim. Acta* **2000**, *45*, 3833.
173. J.R. Reynolds, M. Pyo, Y.-J. Quiu, *J. Electrochem. Soc.* **1994**, *141*, 35.
174. R. Cabala, J. Skarda, K. Potje-Kamloth, *Phys. Chem. Chem. Phys.* **2000**, *2*, 3283.
175. V.C. Nguyen, K. Potje-Kamloth, *Thin Solid Films* **1999**, *338*, 142; *J. Phys. D: Appl. Phys.* **2000**, *32*, 2230.
176. B. Fabre, S. Barlet, R. Cespuglio, G. Bidan, *J. Electroanal. Chem.* **1999**, *426*, 75.
177. H.-L. Krauss, J. Nockel, *Z. Naturforsch. B* **1965**, *20*, 630.
178. M.E. Vol'pin, Y.N. Novikov, *J. Am. Chem. Soc.* **1975**, *97*, 3366.
179. T.R. van den Ancker, C.L. Raston, *J. Organomet. Chem.* **1998**, *550*, 283.
180. E.S. Chandrasekaran, R.H. Grubbs, *J. Organomet. Chem.* **1976**, *120*, 49; *J. Am. Chem. Soc.* **1977**, *99*, 4517.
181. K.E. Branham, J.W. Mays, *Appl Organomet. Chem.* **1997**, *11*, 213.
182. M. Altmann, V. Enkelmann, G. Lieser, U.H.F. Bunz, *Adv. Mater.* **1995**, *7*, 726.

6 Metal Complexes as Part of Linear or Cross-linked Macromolecules via the Ligand

Dieter Wöhrle

In type II macromolecules the ligand of a metal complex is part of a macromolecular chain or network (Section 1.2.1). Most described examples consist of monocyclic or cyclic polymeric ligands with ligand atoms L that are able to bind a metal ion. The complex binding abilities of the polymer ligands are comparable to the analogous low molecular weight ligands, and therefore it is essential to know the complexing behavior of metal ions with the low molecular analogues.

Two principal routes are used to synthesize the polymeric metal complexes described in Sections 6.1 and 6.2 (Fig. 6-1):

- Route A: Reaction of a bi- or higher-functional ligand or metal complex with itself or with another similar comonomer, followed by metallation. Both possibilities have advantages. If a substituted low molecular weight metal complex is employed in the polyreactions, the known coordination geometry is retained and is also present in the polymer metal complex. On the other hand it is analytically easier to characterize the more soluble polymeric ligand before introducing a metal ion.

- Route B: Several examples are also known where a ligand/chelate precursor is reacted with a polymer ligand or in another case in the presence of a metal ion as template directly under construction of a polymeric metal complex. Difficulties can arise from side-reactions which mean that other reaction products can also be formed. The disadvantage is that such side-products are also covalently incorporated into the polymer, and no purification is possible. Therefore detailed design of the synthesis and then careful analysis is necessary.

In addition, this chapter also considers dendrimers in which the ligand of a metal complex is part of the dendritic structure.

The incorporation of a rigid metal complex in a polymer chain reduces the solubility and processibility as is known for aromatic polyamides or polyesters. Polymeric metal complexes with aliphatic alkylene moieties between the chelate units or bulky groups as substituents are easier to handle. Cross-linked polymeric metal complexes are, of course, more difficult to analyze.

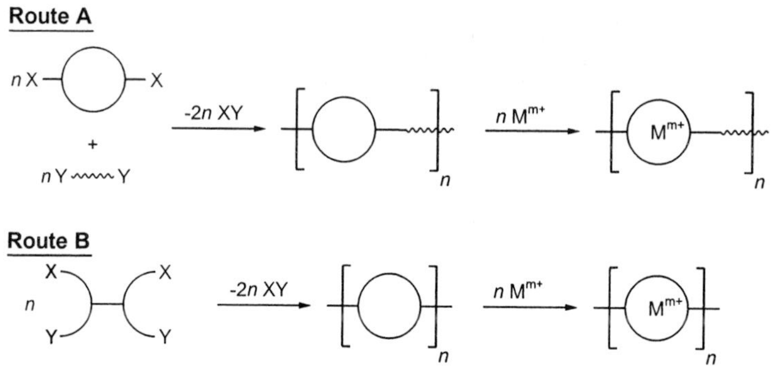

Route A

Route B

Figure 6-1. Two routes for the preparation of metal complexes as part of a macromolecule via the ligand.

Electropolymerization of suitable substituted metal complexes (Section 6.3), dendrimers (Section 6.4) and hydrogen-bonded networks (Section 6.5) are also dealt with in this chapter.

6.1 Polymeric Metal Complexes with Noncyclic Organic Ligands

One group of characteristic examples of noncyclic polymeric metal complexes has Schiff bases as part of a polymer chain ([1,2] and references cited therein). As shown by the general formula **1**, in addition to variation of the metal ion, these polymers allow modification of the polymeric chain R, the substituents R″ and the bridging group R′. Older papers described brown to black-colored polymeric Schiff-base complexes that were often insoluble because no solubilizing groups had been introduced. These polymers were investigated for adsorption of solvent molecules and gas separation [3,4], and as complexing agents towards toxic metals [5].

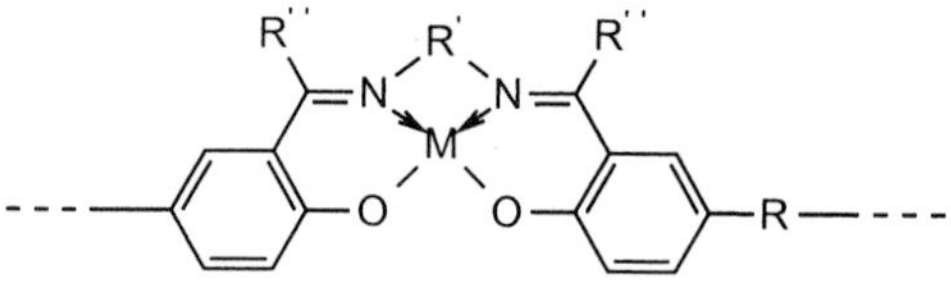

1

One possibility for the synthesis of such polymers is the reaction of a bifunctional Schiff-base ligand or chelate with another bifunctional monomer [1]. The advantage is that side-reactions, which may occur during ligand formation, are avoided because the ligand moiety is employed in the polymerization reaction. An example of the reaction of a substituted low molecular weight metal complex is the reaction of a dihydroxy-substituted Schiff-base complex with dicarboxylic acid dichlorides (Eq. 6-1) [6,7].

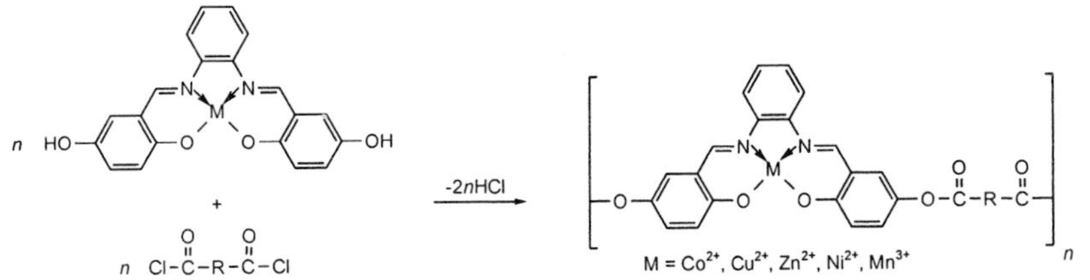

$$(6\text{-}1)$$

A series of Schiff-base Cu(II)-complexes have been synthesized by a transesterification reaction of random liquid crystal terpolymers with a functionalized tetradentate low molecular weight Cu(II)-complex [8]. Between 5 and 20 mol% of the organometallic unit was incorporated without disrupting the liquid crystallinity. The aim of the research was to obtain new magnetoactive organic systems which combine the anisotropic paramagnetic susceptibility of metal entities and the cooperative reorientation of liquid crystals in external fields (for a review see [9]).

A second method describes the initial preparation of the polymer ligand, which is then converted into the polychelate by reaction with a transition metal ion [1,3–6,10–12]. An advantage of this procedure is that – in contrast to the polychelates – the polyligands are soluble in polar organic solvents. This method, which leads to quite pure polychelates because of the   moderate reaction conditions, first uses the preparation of alkali metal salts of polymeric Schiff-base ligands followed by metallization with several transition metal ions or alkaline earth ions [11]. One example is the reaction of the polymeric ligand derived from 4,4′-dihydroxy-3,3′-diacetylbiphenyl and 4,4′-diaminodiphenyl-methane with Mn(II), Co(II), Cu(II), Ni(II) and Zn(II) to yield **2** [13].

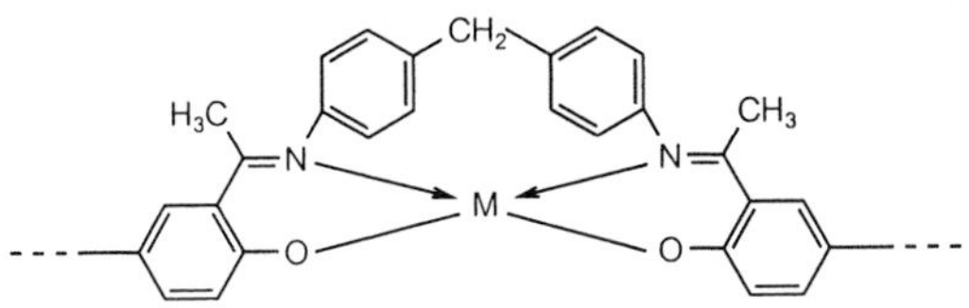

2

A third approach for the synthesis of polymeric Schiff-base chelates is to prepare at first *in situ* a precursor chelate of the diamine and a M(II), or an *o*-hydroxyaldehyde and a M(II), and to add then the second component *o*-hydroxyaldehyde, or a diamine [1,14,15], respectively.

Because of the numerous interesting properties of low molecular weight tetradentate Schiff-base chelates related to catalysis and electrochemistry [16,17], some more recent papers describe the synthesis of more-soluble polymers. A polymer ligand **3** with reasonable solubility contains dialkoxy-phenylene bridges between the salen units [18]. They are synthesized via the reaction of the corresponding bis-hydroxyaldehydes with diamines in a CHCl$_3$/toluene mixture in the presence of acetic acid as catalyst (Eq. 6-2). The yellow solids possess a molecular weight $M_n < 30$ kDa ($P_n < 35$).

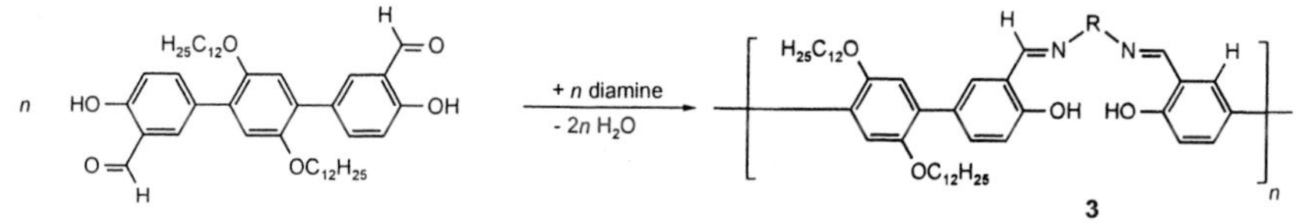

$$(6-2)$$

On the other side, a totally insoluble cross-linked Schiff-base complex was prepared as follows. The glycidyl ether of bisphenol A was reacted with a Schiff-base chelate (prepared from 2,4-dihydroxybenzaldehyde and 1,3-diaminopropane followed by metallation) in a molar ratio of, for example, 6:1 at 150–200 °C to obtain the polymer **4** as shown in Eq. 6-3 (see Experiment 6-1, Section 6.6) [19]. Polymeric copper complexes in particular exhibit good thermal stabilities and good mechanical properties such as tensile strength.

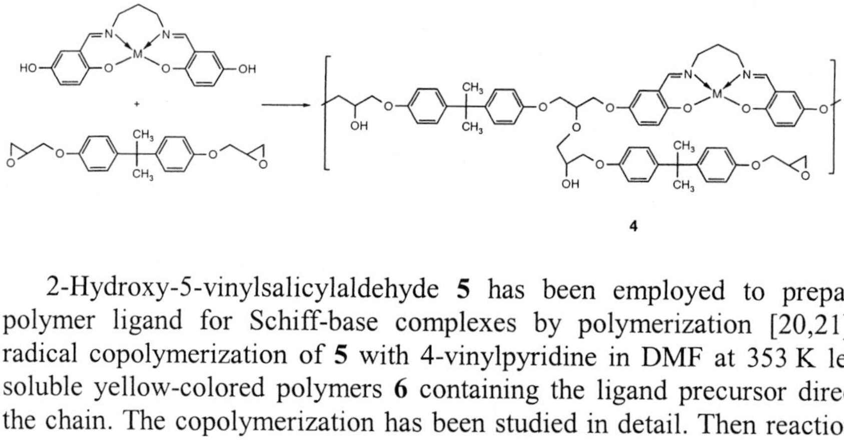

$$(6-3)$$

2-Hydroxy-5-vinylsalicylaldehyde **5** has been employed to prepare the polymer ligand for Schiff-base complexes by polymerization [20,21]. The radical copolymerization of **5** with 4-vinylpyridine in DMF at 353 K leads to soluble yellow-colored polymers **6** containing the ligand precursor directly in the chain. The copolymerization has been studied in detail. Then reaction with 1,2-diaminoethane resulted in the now insoluble cross-linked polymer ligands **7** followed by the introduction of Co(II) to the polymer chelates **8** (Eq. 6-4) [21]. Owing to the presence of the donor pyridine in the comonomer unit, the polymeric Co(II) chelate absorbs oxygen reversibly with the formation of a 1:1 superperoxo complex (molar ratio O$_2$:Co = 1). In contrast, the low molecular weight Co(II) salen takes oxygen up with the formation of a μ–peroxo complex (molar ratio O$_2$:Co = 0.5) [21].

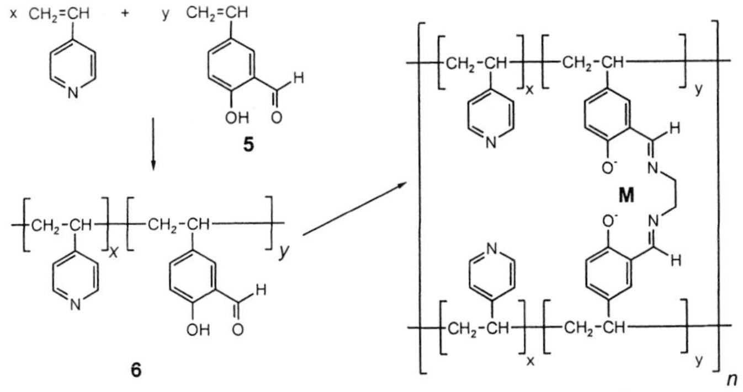

$$(6-4)$$

7: M = 2H$^+$
8: M = Co^{2+}

In addition, the salicylaldehyde derivative **5** has been reacted with 1,2-diaminoethane to divinylsalenes which were afterwards copolymerized with styrene to cross-linked polymers in which Co(II) was introduced [21]. Also divinyl-substituted Mn(III) salen complexes have been successfully copolymerized [22,23]. The aim of this interesting research is the synthesis of heterogeneous catalysts for the asymmetric epoxidation of olefins using chiral Mn(III)-salen complexes [24].

Poly(salicylaldoxime-3,5-diylmethylene) **9** containing chelate-forming groups in the main chain was obtained by polycondensation reaction of salicylaldoxime with formaldehyde in ethanol/water at pH 9 (see Experiment 6-2, Section 6.6) [25]. The polymer exhibits high selectivity towards Cu(II) ions compared to other metal ions. A similar chelating ion-exchange 1-hydroxy-2-naphthaldoxime-formaldehyde polymer **10** has been prepared [26]. The polymer behaves as a selective chelating ion exchanger for some metal ions. A poly(vinylchloride)-based membrane shows Nernstian behavior for Ni(II) and a sensor-based response time of 10 s (for a more detailed description of ion exchangers see Section 5.2).

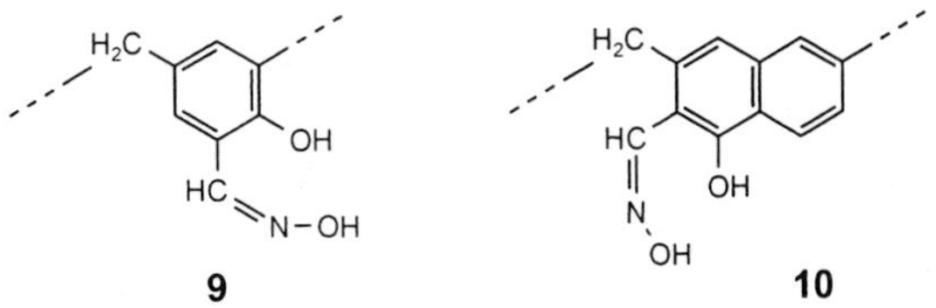

9   10

Thermotropic liquid crystalline polymers containing β-diketonato groups **11** give inter-chain complexation of the diketonato groups with Cu(II) and Ni(II) [27]. ESR indicates that square-planar coordination is present in the case of Cu(II), orientated parallel to the fiber axis. Too high a metal content suppresses the mesogenic properties owing to the insolubility and reduced mobility of the mesogenic units. A polymer **14** having β-β-triketo units in the main chain has been obtained by polycondensation of the ester **13** with the acetylaceto compound **12** in ethylene glycol dimethyl ether (DME) (Eq. 6-5) [28]. The polymer with $M_n = 4670$ is very soluble in various organic solvents. Addition of Cu(II) produced quantitative inter-chain complexes. Copper complexation was carried out at hydroxyl-functionalized liquid crystalline homo- and copolyazomethines **15** [9,29,30]. Flexible –CH$_2$–CH$_2$– groups R give rise to intra-chain coordination whereas rigid aromatic groups R give rise to inter-chain coordination with cross-linking. A nematic mesophase was only observed for polymers with a metal content lower than 30% (molar proportion of Cu(II) with respect to repeat units). Also soluble intra-chain metal complexes of poly(1,2-phenylenedithiocarbamate) [31] and insoluble inter-chain metal complexes of poly(2-hydroxy-4-methoxyacetophenone thiourea) [32] are described.

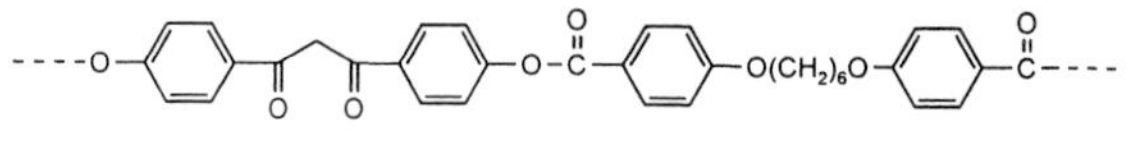

11

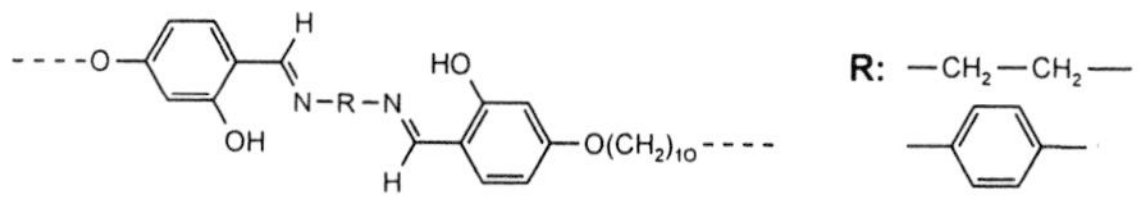

15

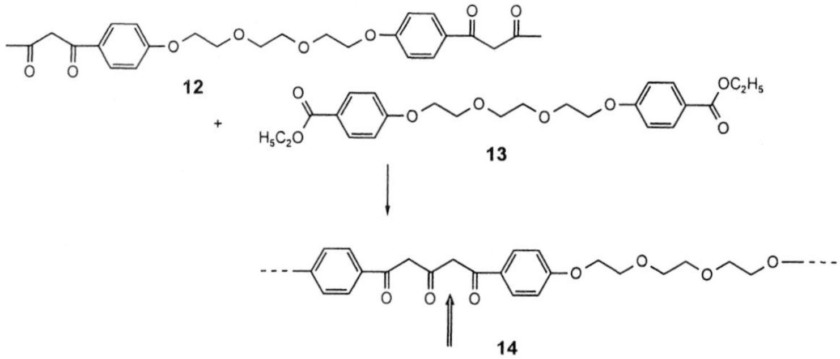

12

13

14

(6-5)

Polymers **16** have been prepared by the polycondensation reaction of 4,4′-bis-(hydroxyethoxy)biphenyl with 4,4′-dibenzo-18-crown-6-ketoalkylene esters [33]. The reactions were carried out with an equimolar mixture of the starting materials at 150–180 °C in the presence of tetraisopropyl orthotitanate. The polyesters **16** possess a degree of polymerization of 10–25. Films can be formed by casting from solution. Their complexing properties with alkali metal chlorides in methanol/chloroform were investigated. The stability constants exhibit the odd–even effect depending on the number of methylene groups in the spacer. The stability constants in solution are lower for polymers than for the corresponding monomers.

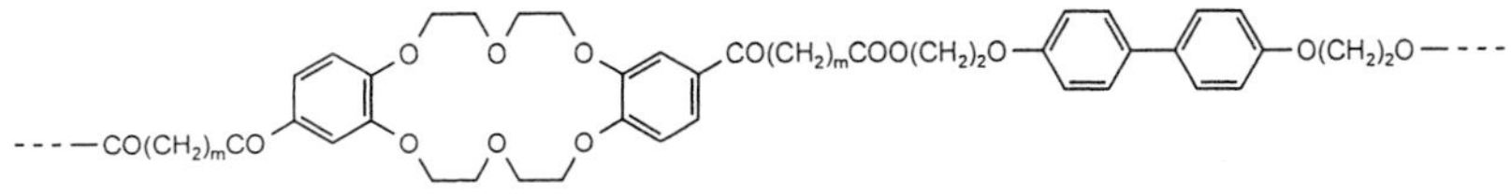

16

Poly(bithiophenes) **17** in which bithiophene units are incorporated directly into macrocyclic binding sites exhibit an ionochromic response with a twisting mechanism (Eq. 6-6) and the binding properties of the analogous crown ether systems [34]. **17** sodium gave a large blue shift of 91 nm at λ_{max} compared to **17**, which is explained by the additive effect of destroying conjugation at many points along a highly conjugated system.

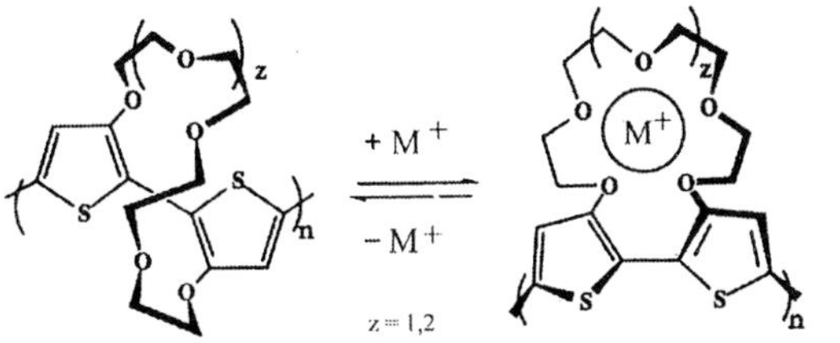

$$(6\text{-}6)$$

17 (planar)   **17M** (twisted)

Oligopyridines such as 2,2'-bipyridyl have been used extensively as fundamental metal-binding sites. These metal complexes have a rich and well-characterized photophysical redox chemistry, energy-transfer processes and are also interesting as sensors, in non-linear optics and in electroluminescence [35–40] (For a very recent review see: U.R. Schubert, C. Eschbaumer, *Angew. Chem. Int. Ed.* **2002**, *41*, 2892). Various 4,4'-disubstituted2,2'-bipyridyl containing aromatic polyamides **18** have been reacted with ruthenium(II) ions to yield high-performance [tris(polymer)ruthenium(II) complex] polymers as shown in **19** for polymers with *p*-phenylene units (Eq. 6-7) [41]. These polymers are soluble in polar solvents such as DMF. Their formation was monitored by optical spectroscopy. The three-dimensional polymer complexes absorb at $\lambda = 620$ nm. Similar homo- and copolyesters containing 3,3'-bipyridyl units form ionic complexes with Fe(II) and Cu(II) [9,42]. ESR measurements indicated an octahedral coordination geometry for the copper(II)-modified polymers. Polymers bearing a low metal content exhibit smectic or nematic behavior which is suppressed by increasing metal content due to intra-chain metal cross-links. Polycondensation of pyromellitic dianhydride and dipyridylamines leads to ligands which eventually yield colored insoluble intra-chain complexes with Fe(III), Cr(III), Co(II), etc. [43]. Block junctions of linear poly(styrene)-poly(ε-caprolacton) **20** were decorated with 2,2'-bipyridine chromophores [44]. Reaction with Ru(bpy)$_2$Cl$_2$ yields luminescent intra-chain

PS–Ru(bpy)$_2$bpy–PCL complexes whereas with Fe(II) inter-chain complexes are obtained.

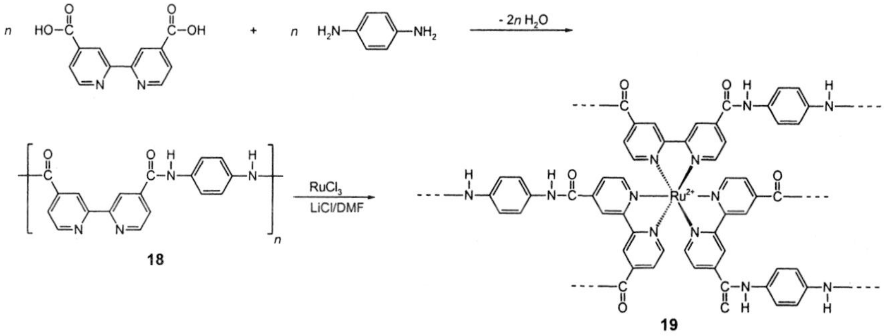

(6-7)

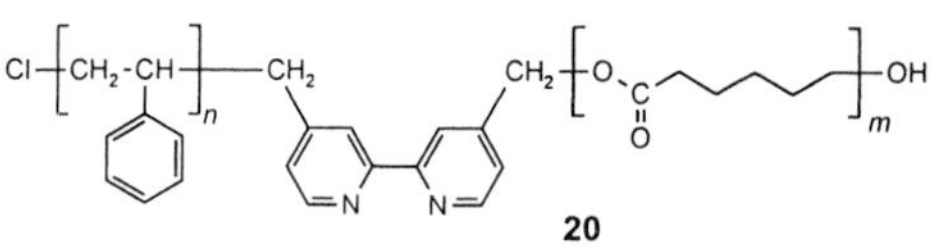

Various conjugated polymers composed of 9,9-dioctylfluorene and 2,2'-bipyridine, which are alternatively linked by C–C single bonds, vinylene bonds or ethynylene bonds, have recently been described [45]. For example, the ethynylene-bridged polymer **21** is obtained in high yield as a yellow solid with good solubility by Heck reaction of the monomers shown in Eq. 6-8. Optical absorption and photoluminescence spectra of the different polymers were measured. The transition metal ion-sensing properties were measured in solution by changes in the absorption spectra and fluorescence quenching. The results reveal clear guidance for molecular design toward a polymer-based chemosensor using 2,2'-bipyridyl as the recognition site: A C–C single bond link between conjugated units (compared with vinylene and ethynylene links) and lower backbone rigidity gave higher response sensitivity.

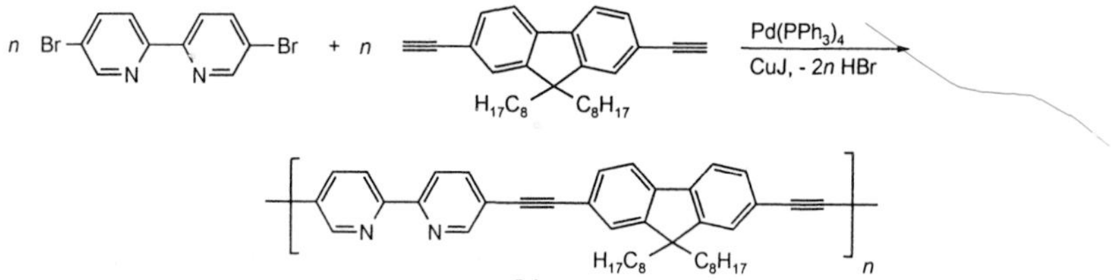

(6-8)

A series of soluble polyarylene ethynylene π-conjugated polymers with M_n values of 10–15 kDa containing (5,5'-diethynyl-2,2'-bipyridine)Re^I (CO)$_3$Cl as part of the polymer backbone have been synthesized by Pd-mediated coupling chemistry [46]. The intensity and lifetime of fluorescence is reduced as the mole fraction of Re in the polymer increases. A π-conjugated polymer containing a dithiafulvalene and a bipyridyl unit was obtained by cycloaddition polymerization of aldothioketone derived from 5,5'-diethynyl-2,2'-bipyridine [47]. Addition of Ru(bpy)$_3$Cl$_2$ resulted in an intrachain polymer Ru complex **22** partly soluble in DMSO. Cyclic voltammetry indicates electronic interaction between the donor dithiafulvalene and the Ru complex unit. Finally, soluble interchain polythiophene Ru(bipy)$_3$$^{n+}$ complexes (an example is shown in **23**) have been mentioned [48]. The polymer complexes display the high-conducting properties of the polymer and the redox properties of isolated metal centers as shown. Solutions of the red–orange complex in acetonitrile polymerize electrochemically with cycling between –0.5 and 1.2 V vs. Ag/Ag$^+$.

Recently a series of Ru(II) chelating poly(heteroaryleneethynylene)s **25** with high molecular weight have been synthesized by Pd-catalyzed cross-coupling of diethynylbenzene and dibromopyridine Ru(II) chelates **24** (Eq. 6-9) [49]. These interesting results show that well-defined Ru complexes can also be employed in polyreactions. Good solubilities in organic solvents are given by suitable substituents: R$_1$ (long-chain alkoxy) and R$_2$ (*tert*-butyl). UV-vis spectra reveal electronic delocalization between the backbone and the Ru(II) chelate. The films were found to have good photoconducting properties.

$$(6\text{-}9)$$

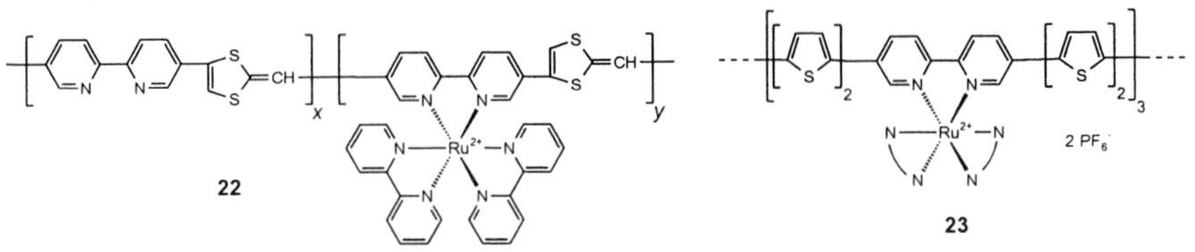

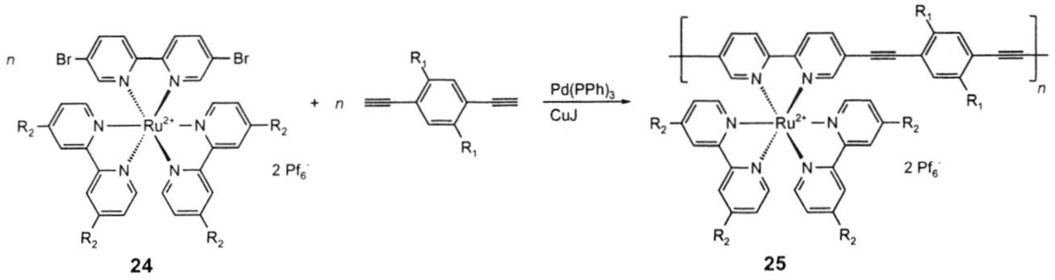

Readily soluble poly(2,9-*o*-phenanthroline-(2',5'-dihexyl)-4,4''-*p*-ter-phenylene)s with a degree of polymerization P_n up to 39 have been prepared via a Pd-catalyzed polycondensation reaction of 2,9-bis-(*p*-bromophenyl)-*o*-phenanthroline and 2,5-dihexylbenzene-1,4-diboronic acid [50]. Some cyclic oligomers could be separated from the polymer. Diamagnetic transition metal complexes of well-defined constitution as shown in **26** were obtained with Cu(I) and Ag(I) without branching and cross-linking.

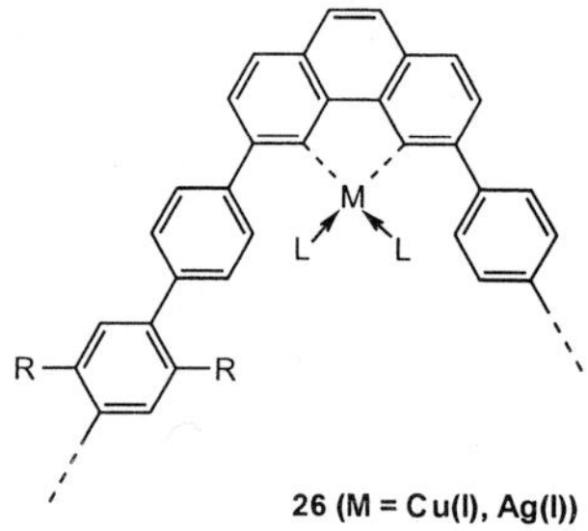

26 (M = Cu(I), Ag(I))

Novel organometallic poly(arylene)s with 1,3-type (cyclobutadiene)cobalt moieties **27** in the main chain are prepared by Ni(0)-mediated dehalogenative polycondensation of monomers having (cyclobutadiene)cobalt moieties [51]. The polymers with M_n of ~20 kDa exhibit thermotropic liquid crystallinity. The synthesis of a variety of homometallic and heterometallic oligomers and polymers such as **28** is possible via nucleophilic aromatic substitution reactions of dichlorophenylene–Fe$^+$Cp and dihydroxyphenylene–Ru$^+$Cp complexes in DMF in the presence of K_2CO_3 [52].

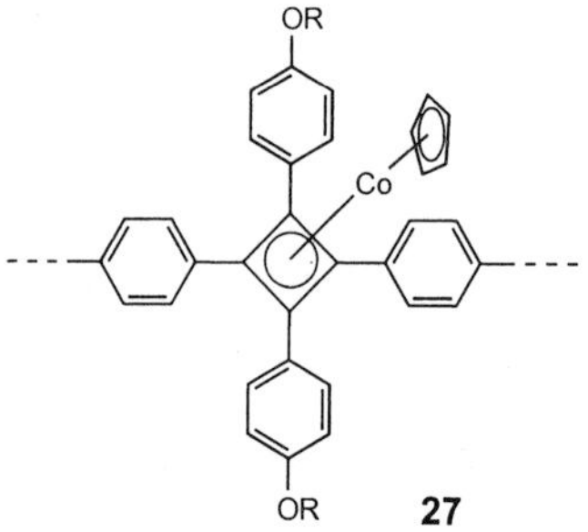

6.2 Polymeric Metal Complexes with Cyclic Organic Ligands

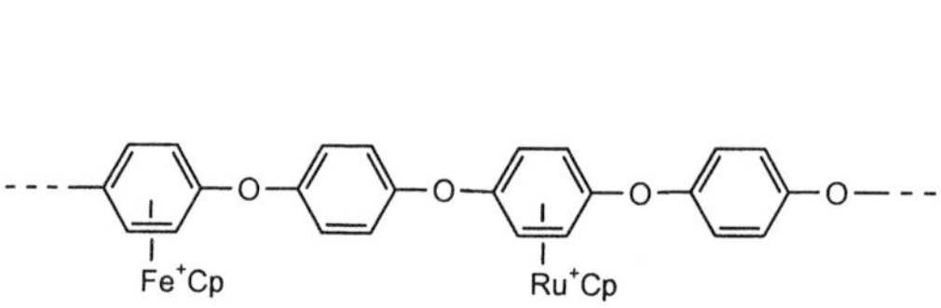

Porphyrins and phthalocyanines are representative examples of cyclic ligands and their metal complexes. They are prepared either by polycyclization reactions of bifunctional/higher-functional ligand precursors or polyreactions of bifunctional/higher-functional porphyrins and phthalocyanines.

6.2.1 Cyclization of Bifunctional and Higher-Functional Ligand/Chelate Precursors

The well-known Rothemund reaction for the synthesis of 5,10,15,20-tetrasubstituted porphyrins **29** from aromatic or heteroaromatic aldehydes and pyrrole has the disadvantage of producing a large quantity of side products such as polypyrrole, linear condensation products and chlorines. Therefore, no structurally uniform polymeric porphyrins are expected if terephthalaldehyde is reacted with pyrrole in boiling propionic acid. Polymers from this reaction have been described [53]. The authors cleaned the product from polypyrrole and low molecular weight porphyrins by washing with methanol and chloroform, and introduced different metal ions by treatment with metal salts in an ethanol/water mixture. But the polymers which exhibited specific conductivities of $\sim 10^{-8}$ S cm^{-1} (up to $\sim 10^{-3}$ S cm^{-1} by doping with I$_2$) were analyzed only by IR-spectra.

The synthesis of polymeric phthalocyanines was investigated more intensively. Low molecular weight phthalocyanines **30** are prepared by cyclotetramerization of unsubstituted or substituted phthalic acid derivatives such as phthalonitriles or phthalic acid anhydride. For a polycyclotetramerization (for reviews see [54–56]) bifunctional monomers based on tetracarbonitriles such as 1,2,4,5-tetracyanobenzene for polymers **31** [57–66], various oxy-, arylenedioxy- or alkylenedioxy-bridged diphthalonitriles for polymers **32** [67–74] and other nitriles [75–79] or tetracarboxylic acid derivatives like 1,2,4,5-benzenetetracarboxylic acid dianhydride [80–85] are employed mainly in the presence of metal salts or metals in bulk at higher temperatures (Eqs. 6-10, 6-11). Because each phthalocyanine group in polymers **31**, **32** is tetra-substituted, there will be a statistical mixture of four structural isomer units of D_{2h}, C_{4h}, C_{2v} and C_s symmetry [86]. Equations for calculating the vast number of isomers for this type of polymers have been derived [87]. This is not the case for polymers **31**. Also polymers **33** prepared from bis(phthalonitrile) linked via crown ether moieties with various metal ions in the core of the phthalocyanine ligand exhibit good uniformity [78,79]. A network polymer with crown ether moieties removes various metal ions (Na(I), K(I), Rb(I), Cs(I), Ag(I)) from solution.

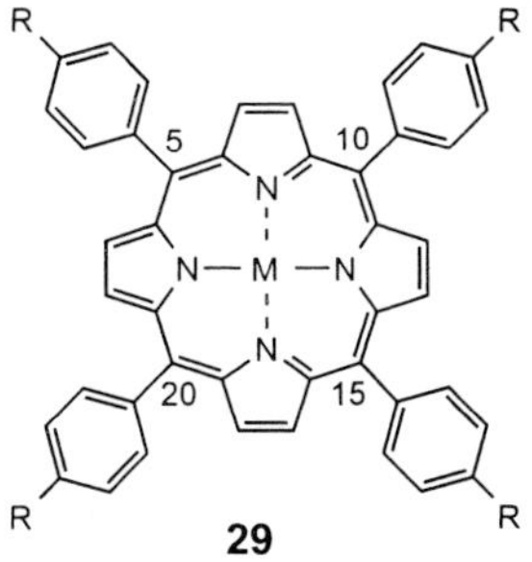

29

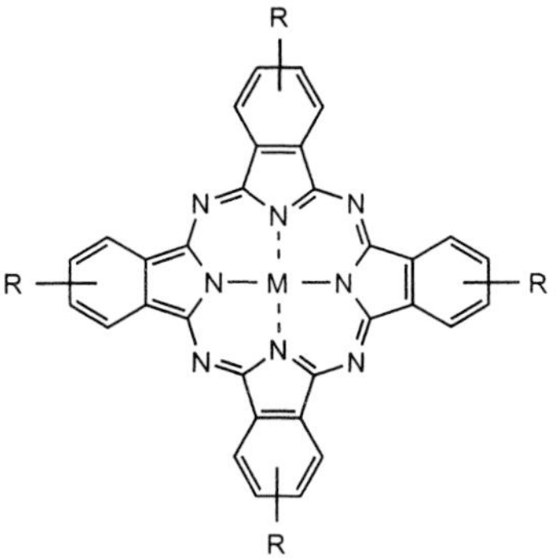

30

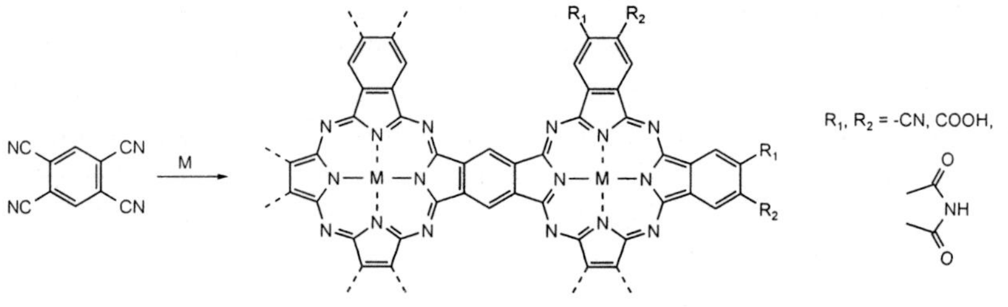

$$R_1, R_2 = -CN, COOH,$$

(6-10)

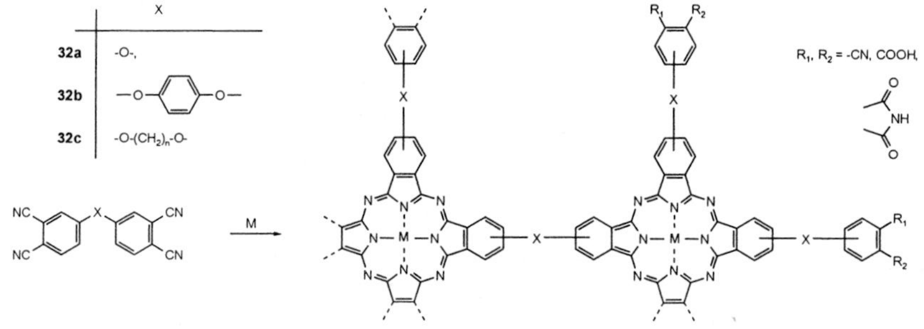

31

	X
32a	-O-,
32b	—O—⟨⟩—O—
32c	-O-(CH$_2$)$_n$-O-

$$R_1, R_2 = -CN, COOH,$$

(6-11)

32a-c

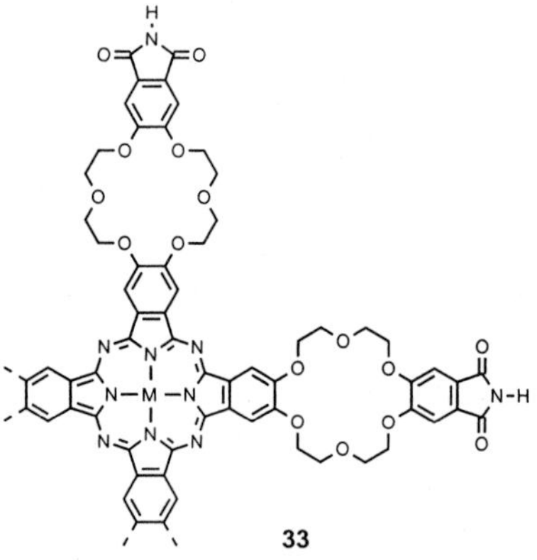

33

Low molecular weight phthalocyanines can be purified by zone sublimation and liquid chromatography, depending on whether they are unsubstituted or contain suitable substituents. Common instrumental techniques are employed for their analytical characterization. In contrast, polymeric phthalocyanines are insoluble in organic solvents (sometimes only partially soluble in concentrated sulfuric acid) and are involatile. Therefore, purification from unreacted monomers, octacyanophthalocyanine **34** as an intermediate product, metal salts and perhaps some by-products are only possible by treatment with organic solvents (e.g. in a Soxhlet apparatus) or dilute acids. During the polycyclotetramerization of tetracarbonitriles the formation of the by-products polyisoindolenine **35** and polytriazine **36** can occur. They are covalently incorporated as co-units into the polymers and cannot be separated from the phthalocyanine structural elements [55,56,58]. Therefore it was necessary to find a way of preparing structurally uniform polymeric phthalocyanines.

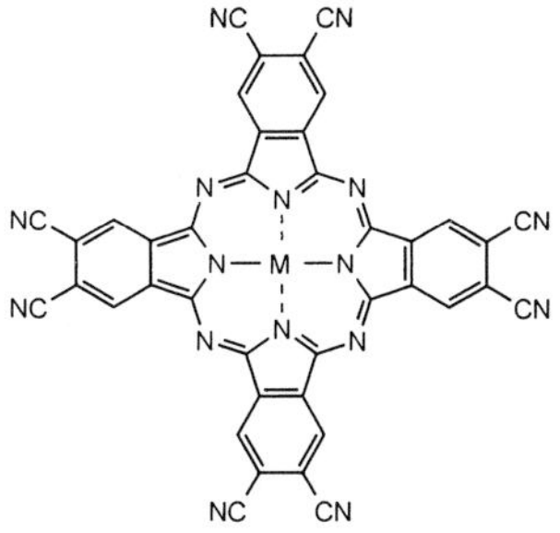

For complete characterization of the polymers the following points must be considered: structural uniformity, the nature of end groups, metal content and degree of polymerization (molecular weight). Only a very few reports contain adequate statements on these points and describe the preparation of properly structurally uniform polymers [58,67,68]. It should be pointed out that the preparation of structurally uniform polymeric Pc is restricted to well-defined reaction conditions. For various investigations, such as electrical, photo-electrical, catalytic and photocatalytic properties, thin films on flat surfaces (e.g. glass, Ti, ITO, KCl) or coatings on particles (e.g. SiO_2, TiO_2, Al_2O_3) are necessary. Because polymeric phthalocyanines are insoluble and involatile, special techniques must be employed. These include the reaction of films or coatings of metals or metal salts and gaseous tetracarbonitriles with flat surfaces [88–94] or powdered particles [95,96].

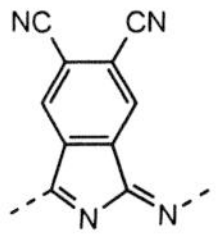

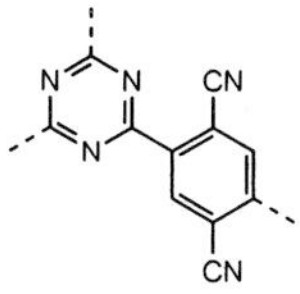

Four methods employing tetracarbonitriles as bifunctional monomers for the formation of polymeric phthalocyanines **31** and **32** have been described (see Experiment 6-3, Section 6.6 for different preparations):

- Method A: Powdered polymeric CuPc **31** by the reaction of tetracyanobenzene with $CuCl_2$ in bulk [58].
- Method B1: Powdered polymeric ZnPc **31** and **32** by solution reaction of tetracarbonitriles in pentanol-1 in the presence of Li-pentanolate [97].
- Method B2: Powdered copolymer by the solution reaction of tetracyanobenzene and the bridged tetracarbonitrile in pentanol-1 in the presence of Li-pentanolate [97].
- Method C: Thin films of polymeric CuPc **31** on glass by the reaction of a Cu-film with tetracyanobenzene [89].

- Method D: Coatings of polymeric CoPc **31** on SiO_2 by the reaction of $Co_2(CO)_8$ adsorbed on SiO_2 with tetracyanobenzene [96].

The polycyclotetramerizations yielding polymeric phthalocyanines occur according to the stoichiometric equations shown for a two-electron reduction and the formation of the dianion of the structural phthalocyanine unit:

- Method A: $2n(C_{10}H_2N_4) + nCuCl_2 \rightarrow [(C_{20}H_4N_8)^{2-} Cu^{2+}]_n + nCl_2$
- Method B: $2n(C_{10}H_2N_4) + \text{alcohol} + 2Li^+ \rightarrow [(C_{20}H_4N_8)^{2-} Li^{2+}]_n + \text{oxidized alcohol}$
- Method C: $2n(C_{10}H_2N_4) + nCu^0 \rightarrow [(C_{20}H_4N_8)^{2-}Cu^{2+}]_n$
- Method D: $2n(C_{10}H_2N_4) + 0.5n\,Co_2(CO)_8 \rightarrow [(C_{20}H_4N_8)^{2-}Co^{2+}]_n + 4CO$

For method A the theoretical stoichiometric molar ratio tetracarbonitrile: $CuCl_2$ of 2:1 was employed because elemental analysis showed the maximum metallation [58]. The results of several experiments showed that the solution method B1 needs an excess of Li-pentanolate, probably because the alcoholate anion is consumed by oxidation [97]. In the case of methods C and D a higher amount of nitrile was used than is shown in the equations in order to leave no unreacted metal on the surfaces [89,96]. The excess of unreacted tetracarbonitrile could be removed by washing with acetone. Method B takes place under mild reaction conditions and is also useful to prepare copolymers [97].

Only a small amount of information is available for the mechanism of the formation of the polymer structure. It is suggested that four molecules of tetracyanobenzene ainitially react to form **34** which reacts further to the polymer **31** [55,56,58]. In addition, it has been shown that the reaction of Cu-films with gaseous tetracyanobenzene at $T < 350°C$ produces films of **34**, whereas at $T > 350°C$ films of the polymer **31** are obtained [89]. The mechanism of film growth using method C was discussed in [88,89]. After formation of the first few layers of polymeric phthalocyanines, copper atoms diffuse from the copper film to the growing polymer film surface in order to react with the tetracarbonitrile, at first to **34** and then to oligomeric and polymeric phthalocyanines (Fig. 6-2). Depending on the thickness of the deposited Cu-film (~1.5 to 20 nm) adhering films of **31** with thicknesses of ~46 to 230 nm were obtained. In every case an average value of ~25 was determined for the ratio of the thicknesses of the polymer film and the copper film. The films exhibit good electrical conductivities.

Figure 6-2. Film growth of polymeric phthalocyanines from tetracarbonitrile on metal plates.

For the preparation of coatings of phthalocyanines on SiO_2 or TiO_2 using method D two routes were used (Fig. 6-3): route a – after adsorption of metal carbonyls at T_1 (40–60°C) their decomposition to metal at T_2 (130–320°C) and subsequent reaction with the nitrile at T_3 (200–350°C); route b – direct reaction of the adsorbed metal carbonyl at T_4 (180–250°C with the nitrile) [96]. The loading of polymeric phthalocyanines **31** on quartz particles of $\approx$ 2 wt.% was calculated from the amount of metal carbonyl employed and by parallel experiments on the reaction of phthalonitrile with $Co_2(CO)_8$.

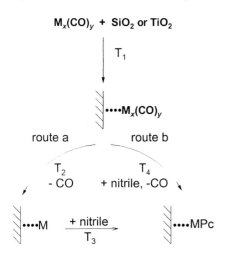

$M_x(CO)_y$ + SiO_2 or TiO_2

T_1

••••$M_x(CO)_y$

route a   route b

T_2 T_4
- CO   + nitrile, -CO

••••M $\dfrac{+\ nitrile}{T_3}$ ••••MPc

Figure 6-3. Coatings of phthalocyanines on SiO_2 and TiO_2 from nitriles.

UV-vis spectra allow us at first to determine the presence of phthalocyanine units in the polymer through the Q-band transition at ~700 nm (spectra taken in reflection) or ~740 nm (spectra taken in concentrated H_2SO_4). As in low molecular weight phthalocyanines **30** the ratio of the intensities in the UV (Soret-band transition):Vis (Q-band transition) = <1. For the polymers this intensity ratio indicates the absence of polyisoindolenine **35** co-units. It should be pointed out that for the bulk reaction of method A and the gas-phase reaction of method C only the reactions of tetracyanobenzene with Cu or $CuCl_2$ result in structurally uniform polymers, with an intensity ratio in the UV-vis spectra of $I_{UV}:I_{Vis}$ = <1 [58]. Using Mg, Al, V, Cr, Mn, Fe, Co, Ni and Zn or their salts, the ratios are in most cases >> 1 showing structurally non-uniform polymers. This is also indicated by the IR spectra. If bridged diphthalonitriles are used in the reaction with metals or metal salts, only Cu or $CuCl_2$ lead to structurally uniform polymeric phthalocyanines **32** [67,68]. In contrast, methods B and D allow the synthesis of structurally uniform polymers with other metal ions than Cu. This was recently shown for method D employing $Co_2(CO)_8$, $W(CO)_8$, $M(CO)_8$ or $Cr(CO)_8$ in the reaction with tetracyanobenzene [96]. For polymer **31** electronic transitions up to $\approx$ 1400 nm in the NIR indicate conjugation in the polymer plane [97].

The nature of the end groups depends on the reaction conditions used. After methods A and C, the presence of nitrile groups is shown by only weak absorption in the IR spectra at ~2220 cm^{-1} [58]. After method B the nucleophilic attack of the alcoholate anion at the carbon atom of the nitrile group results in the initial formation of alkoxyisoindolenine groups, and these react further to phthalocyanine units. The workup procedure leads, by hydrolysis of the isoindolenine groups, to imide groups which are clearly seen in the IR spectra at ~1770 and 1720 cm^{-1} [97]. In the case of method D partial saponification of nitrile groups leads to both residual nitrile groups and to imide groups [96].

The insolubility of the polymeric phthalocyanines in organic solvents makes molecular weight determinations difficult. Additionally, different annelations of phthalocyanine units in a polymer must be taken into account: a two-dimensional plane structure, a one-dimensional ladder structure, and different arrangements of phthalocyanine rings between these two cases. Therefore several isomers exist for a polymer molecule with a definite molecular weight. One possibility is the determination of the degree of polymerization (number of connected phthalocyanines) by exact end-group determinations: quantitative IR spectroscopy comparing the intensity of the nitrile end groups to another ring vibration. This method was applied in practice in a few cases [58,67,68,82] and evaluated theoretically [98,99]. For example, after quantitative nitrile group determination the polymeric Cu-phthalocyanine 31 prepared after method A consists most probably as a nanomer. Careful determinations of the maximum average values of the molecular weight by the IR method shows that the values increase with increasing flexibility between the two reactive sides of a tetracarbonitrile or the bridging group X: for 31 > 4000, for 32 with X = –O– > 12000, for 32 with X = –O–arylene–O– > 6000, for 32 with X = –O–alkylene–O– up to infinity.

Polymeric phthalocyanines such as 31 exhibit good thermal stability under inert gas up to ~500 °C and under oxidative conditions up to ~350 °C [59]. One way to enhance the electrical conductivity is by enlarging the conjugated π-electron system in the two-dimensional plane of the polymeric phthalocyanines. Compressed powders of polymer 31 exhibit intrinsic conductivities in the order of 10^{-3}–10^{-7} S cm^{-1} [55,56,100,101]. The conductivity increases to 10^{-2}–10^{-3} S cm^{-1} for thin films of 31 prepared by the reaction of a Cu-film with tetracyanobenzene as described before [89].

The Faradaic activity of thin films of polymeric phthalocyanines 31 (M = Cu(II)) on titanium foils in electrochemical cells has been investigated [94]. Electrodes with the polymer film dipped into an aqueous solution of K$_3$Fe(CN)$_6$/K$_4$Fe(CN)$_6$ exhibit a high Faradaic activity and reversibility comparable to a bare platinum electrode. The electrically conductive polymer allows an efficient electron transfer to redox couples in solution. Thin films of bridged polymeric phthalocyanines (M = Cu(II)) **32** have been prepared on Ti-plates by reaction of a thin Cu-film with gaseous diphthalonitriles [93]. In contact with an aqueous electrolyte containing the redox couple $K_3Fe(CN)_6/K_4Fe(CN)_6$ the electrodes show photoelectrochemical behavior. Depending on the kind of bridging group in **32**, p- or n-type conductivity can be obtained. Therefore either cathodic photoreductions or anodic photooxidations of species in solution were observed under irradiation.

Cobalt complexes of polymeric phthalocyanines have been employed in aqueous alkaline solution as heterogenous catalysts in the oxidation of thiols to disulfides (MEROX, mercaptan oxidation process in the petroleum industry, Eq. 6-12, see Sections 5.2 and 5.4, Experiments 5-11 and 5-12). The catalytic activities of the polymer **31** (M = Co(II)) are higher than those of dissolved low molecular weight phthalocyanines, and both complexes exhibit better activities on charcoal than on SiO_2 as carrier. This is the result of better electrical contact between different reaction centers, which facilitates a multi-electron process in the oxidation of R–S⁻ to R–S–S–R and reduction of O_2 to H_2O [95]. Another advantage of the heterogeneous catalysts in comparison to the dissolved low molecular weight phthalocyanines is their easy re-use.

$$4R-S^- + O_2 + H_2O \rightarrow 2R-S-S-R + 4OH^- \qquad (6-12)$$

6.2.2   Polyreactions of Bifunctional or Higher-Functional Porphyrins and Phthalocyanines

Considerable attention had been focused recently on the synthesis of molecular rods based on porphyrins and on their potential applications as molecular-scale electronics, optical devices or sensors and for solar energy conversion. Monodisperse rods of definite length (dimers, trimers, tetramers, etc.) have been extensively reviewed in *The Porphyrin Handbook* [102,103]. Porphyrins are connected via single bonds, triple bonds, vinylenephenylene bridges or fused coplanar aromatic bridges. Because these well-defined oligomers like dimers and trimers are described in detail in these reviews (see also [104]), in the following we shall mainly describe polymers obtained by connecting porphyrin rings.

Polynuclear porphyrins directly connected via a single bond between porphyrin moieties are prepared by a *meso–meso* coupling reaction of a 5,15-diarylporphyrin such as **37** (M = e.g. Zn(II), Eq. 6-13) [104–107]. *Meso–meso* coupling of porphyrins with two free reactive sides to polymer **38** can be effected by treatment with an Ag(I) salt or by electrochemical oxidation with high regioselectivity. In the case of **37** with R = –C(CH₃)₃ the reaction product consists of 44% starting material and coupling products **38**: $n = 2$, 20%; $n = 3$, 2.9%; $n = 4$, 0.9 %.... $n = 8$, 0.05 %; and higher oligomers. Polymers **38** with n = 2–8 were separated by GPC-HPLC. In the UV-vis spectra splitting of the

Soret bands by exiton coupling is seen [104,108]. As the number of porphyrins increases, the low-energy Soret band is shifted to longer wavelengths, while the high-energy Soret band remains, so that the splitting energy increases progressively. The repeating individual chromophores have some electronic interactions in the ground state. The extended linear conformation precludes the formation of any stacked energy sink, and the electronic interactions are considerably enhanced in the excited state.

$$(6\text{-}13)$$

R = -C(CH$_3$)$_3$, -OC$_8$H$_{17}$, etc.

37   38

The synthesis was recently extended from grid porphyrin arrays to soluble windmill arrays [109]. Flanking with a peripheral Ni β-octaalkylporphyrin was achieved by Ag(I)-promoted coupling reactions of 1,4-phenylene-bridged linear porphyrin arrays, which are composed of a central Zn β-free porphyrin. Up to 48 porphyrin rings were connected. Efficient singlet energy transfer from the peripheral porphyrins to the central Zn-porphyrins was found.

Linear butadiyne-linked polymeric porphyrins **39, 40** and an arylacetylene-linked polymer **41** were prepared by Glaser–Hay or Heck–Sonogashira coupling of Zn-complexes of 5,15-diethynylporphyrins, in most cases with unreactive substituents in the 10,20-positions to give solubility of the resulting polymers [110–112]. Oligomers of **40** from dimer to 13-mer ($M_n = 53$ kDa) were detected by MALDI-TOF MS [110]. Polymer **41** was obtained by Pd-catalysed Heck–Sonogashira coupling of a 5,15-diethynyl porphyrin with 1,4-diiodophenylene [112]. A series of conjugated Zn-porphyrin oligomers linked by linear butadiyne bridges were converted into ladder-structured complexes by self-organization with the bidendate ligand 1,4-diazabicyclo[2.2.0]octane [113]. UV-vis spectra of the linear butadiyne-linked polymeric porphyrins show that the Q-band is significantly red-shifted and intensified compared with the Q-band of a simple porphyrin, indicating that there is substantial delocalization across the butadiyne bridges [110,114]. The solid-state absorption spectrum of **39** consists of two bands at 500 and 860 nm and is similar to that of the polymer in solution. The use of degenerate four-wave mixing at 1064 nm established that the magnitudes of the tensor for the polymer **39** is comparable with existing materials [110,114]. The microscopic polarizability per macrocycle of the polymer is around three orders of magnitude greater than the monomer which indicates inter-macrocycle conjugation in the nonlinearity. Photoexcitation indicates that the polarizability increases with increasing length of the oligomers [115].

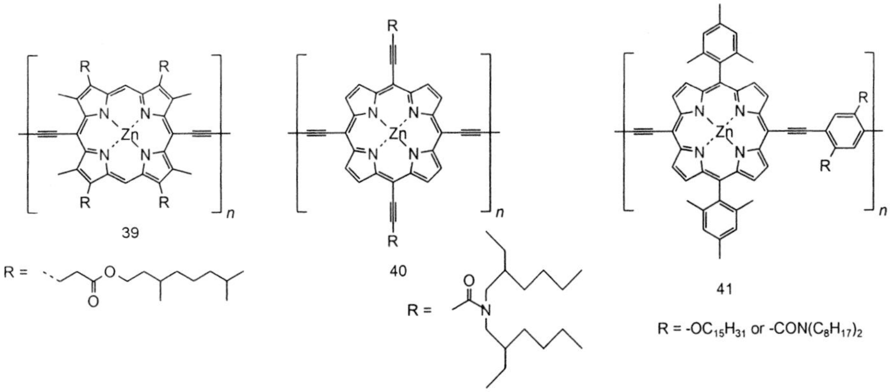

39

R =

40

R =

41

R = -OC$_{15}$H$_{31}$ or -CON(C$_8$H$_{17}$)$_2$

Zn-porphyrins with two 4-bromophenyl- or two 5-bromo-2-thenyl units at the 5,15 positions have been converted by dehalogenative polycondensations to polymeric porphyrin zinc-complexes [111,116]. For example, polycondensation of the monomer **42** in DMF in the presence of bis(1,5-cyclooctadiene)nickel(0) (Ni(Cod)$_2$) and triphenylphosphine gave the polymer **43** in 73% yield (Eq. 6-14) [116]. The monomer **42** and the polymer **43** show the Soret band (416 nm) and the Q-bands (546 and 584 nm) at the same positions. Thin films of **43** are electrochemically active in both the oxidative and reductive regions. The 4-phenylene-bridged polymers possess a molecular weight M_n of 38 kD [111]. 1,4-Bis(chloromethyl)-2,5-(dihexyloxy)benzene has been reacted in different molar ratios with Al(III)- and Zn(II)-complexes of 5,15-bis(mesityl)-10,20-bis(4-(chloromethyl)phenyl)porphyrin to polymers containing vinylene-phenylene and porphyrin units in the main chain [117]. The conjugated polymers contained between 1.7 and 7.0 wt.% porphyrin moieties. The electrical conductivities of iodine-doped films are in the order of 1–3 S cm^{-1} for pure poly(vinylenephenylene) and also for porphyrin-containing polymers.

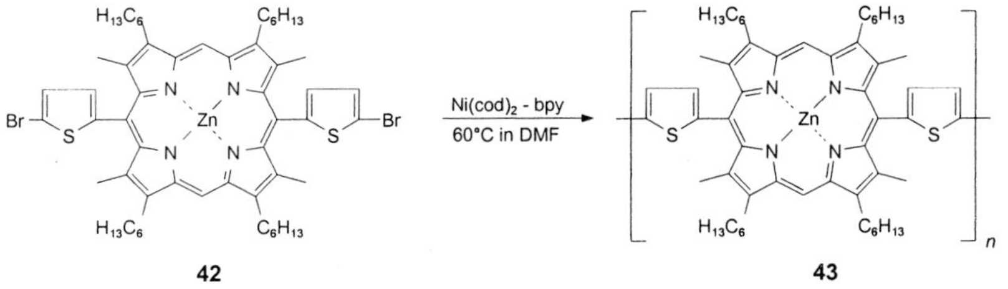

42   **43**   (6-14)

A Heck coupling reaction of 1,4-diiodo-2,5-dialkoxybenzene with the bis(vinylphenyl)-substituted porphyrin **44** led to soluble and processable polymers **45** with direct linkages between vinylenephenylene units (Eq. 6-15) [118]. After GPC polymers with reasonable molecular weights $\overline{M}_w \sim 10^4$ mol g^{-1} were obtained. Good photoconductivities and quantum yields for photocharge generation (e.g. 2.8 %) were observed (applied field ~620 kV cm^{-1}). Because of steric hindrance the porphyrin and vinylenephenylene groups are out of plane and every porphyrin behaves comparably to a monomeric porphyrin. Some reports describe the electronic structures of two-dimensional porphyrin polymers linked by triple-bonds or phenylene units [119] and consisting of fused rigid porphyrin-oligomer wires [120]. Indeed, a completely fused triporphyrin exhibits an extremely long wavelength absorption at 1251 nm [121].

$$(6\text{-}15)$$

The interfacial polycondensation technique, in which reactive comonomers are dissolved in separate immiscible solutions and are thereby constrained to react only at the interface between two solutions, has been used to synthesize chemically asymmetrical polymeric porphyrin (M = Zn, Cu, Ni, Pd or 2H) films [122–124]. Tetrakis(4-aminophenyl)-, tetrakis(4-hydroxyphenyl)porphyrins **(29**, R = –NH$_2$, –OH) or aliphatic amines in one solvent were reacted with tetrakis(4-chlorocarboxyphenyl)porphyrins **(29** (R= –CO–Cl) or aliphatic diacylchlorides, respectively, in the other solvent (see Experiment 6-4, Section 6.6). Figure 6-4 shows schematically the formation of asymmetric polyamide porphyrin films. The dependence of the film growth on monomer concentration and time has been studied in detail. Typical film thicknesses are in the range of 0.1–10 µm. The unique chemical asymmetry is shown by distinctive differences in the concentration and type of functional groups present. The photoactivities of the polymeric porphyrin films were measured in dry sandwich cells.

In contrast to porphyrins, the synthesis of monodisperse and polymeric linear rods of phthalocyanines has not been much developed in practice. The difficulty lies in the synthesis in good yields of unsymmetrically substituted phthalocyanines bearing two reactive substituents in positions 2 and 9 of the phthalocyanine **30**. An ambitious synthesis seeks to prepare linear, conjugated ladder polymers, in which the phthalocyanine units are linked by an oligoacene bridge [125–127]. The polymers are to be assembled using a Diels–Alder reaction between bis(diene) and bis(dienophile) phthalocyanine monomers followed by aromatization of six-membered rings in the linking group. Some relevant preparations have been achieved including the synthesis of a number of oligomeric structures based on phthalocyanines and hemiporphyrazines.

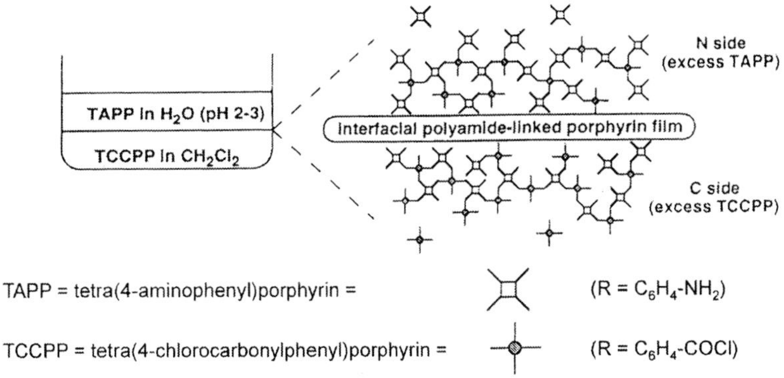
N side
(excess TAPP)
TAPP in H_2O (pH 2-3)
TCCPP in CH_2Cl_2
interfacial polyamide-linked porphyrin film
C side
(excess TCCPP)
TAPP = tetra(4-aminophenyl)porphyrin =
($R = C_6H_4\text{-}NH_2$)
TCCPP = tetra(4-chlorocarbonylphenyl)porphyrin =
($R = C_6H_4\text{-}COCl$)

Figure 6-4. Illustration of the interfacial polycondensation process between differently substituted porphyrins.

A linear phthalocyanine containing polyesters **46** has been described [128,129]. Recently, a linear polymeric phthalocyanine **47** prepared in a multistep synthesis with an iptycene architecture was briefly described [130]. At first two differently substituted 1,3-dihydro-1,3-diiminoisoindolenines were statistically reacted. In the mixture of phthalocyanines obtained, the ethylenedioxythiophene part is introduced and the bis(ethyleneoxythiophene) oppositely substituted phthalocyanine was isolated by flash chromatography. Afterwards the thiophene parts in the phthalocyanine were electrochemically polymerized to obtain **47**.

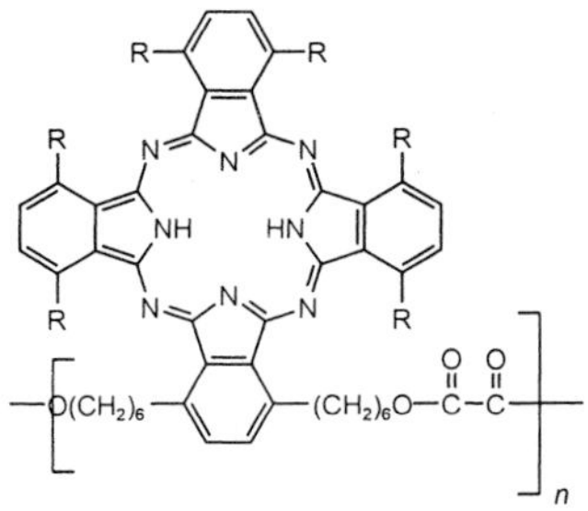

46 **R = alkyl**

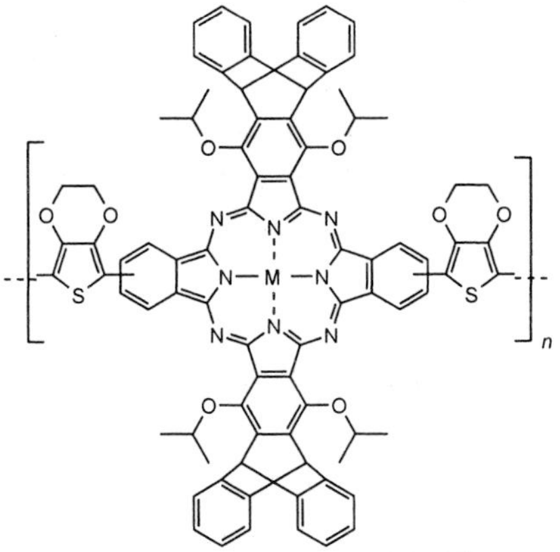

47 **(M = Ni)**

A series of bisfunctional Fe(III)- and Co(II)-phthalocyanines and their polymers have been synthesized as models for catalase, peroxidase, oxidase and oxygenase ([131] and references cited therein). Copolyesters **48** containing Fe(III)- and Co(II)-phthalocyanines were obtained by polycondensation of phthalocyanine dicarboxylic acid dichlorides with terephthalic acid dichloride and aliphatic diols. Green or blue colored fibers could be obtained by melt spinning of the copolyesters containing less than 1 mol% of the metal complex [132].

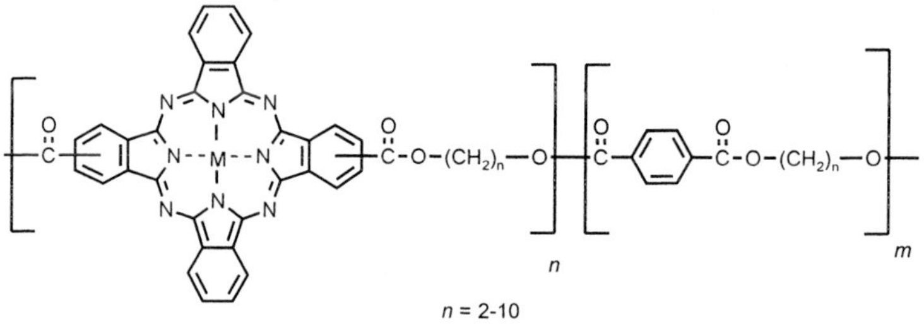

n = 2-10

48

Tetrafunctional phthalocyanines like the tetraamines and tetracarboxylic acid derivatives of **30** (R = –NH$_2$, –COOH) have also been used as starting materials in polycondensations. Owing to the high functionality (>2), network formation easily occurs, and insoluble polymers are obtained that are difficult to characterize for degree of network formation, molecular weight, end groups, etc. Therefore in many cases no detailed structure determination was carried out. The polymers were mainly investigated for their thermal stability. The main results have been reviewed [54–56], and only an overview of the strategy for the preparation of such polymers is given below.

Metal–2,9,16,23-tetraaminophthalocyanines have been employed in the synthesis of polyimides and as curing agents for epoxy resins [133–135] Variables such as molar concentrations of the reagents, solvents and temperature were investigated to improve the conditions of the polycondensation. Solutions of the polyamide-acid copolymers can be used to fabricate films or fibers. The polyimide copolymers obtained in the second step of the reactions are insoluble. Excellent thermal stabilities up to 500 °C in air and 600 °C under vacuum were reported.

6.3　Electropolymerization of Metal Complexes

One fruitful strategy for obtaining conducting polymeric metal complexes and models of enzymes and other catalytically active sites has been the electro-assisted design of modified electrode surfaces by metal complex structures. In general, highly cross-linked networks are obtained in which metal complexes are building blocks via their ligands. The oxidative electropolymerization of pyrroles, thiophenes, anilines, phenols, etc. to highly conducting polymers is well established. In order to achieve rapid electron transfer in a conducting polymer matrix to model enzyme and catalytically active sites at the electrode surface, suitable substituted metal complexes and their ligands, such as porphyrins, phthalocyanines, Schiff-base complexes, ruthenium pyridyl complexes, etc., were employed as monomers for electropolymerization.

The following general statements are drawn from several publications:

- All electrochemically stable metal complexes that are substituted by pyrrole, thiophene, aromatic amino, aromatic phenol and some other groups such as vinyl, are suitable for anodic electropolymerizations from solution (organic solvent or water) in the presence of a conducting salt (counter ions in the oxidized films).
- The formation of polymers on a conducting electrode surface (ITO, Pt, carbon) occurs irreversibly after the electrochemical oxidation of the above-mentioned groups at around 0.7–1.1 V vs. SCE. Redox-active metals in the core of the ligand such as Fe, Co are electrochemically oxidized reversibly before oxidation of the electropolymerizable group. Oxidation of the ligand is sometime also observed.
- Electropolymerizations are carried out either potentiodynamically (by cyclic scanning of a potential range) or potentiostatically (at a constant anodic potential). Electropolymerization is indicated by an increase in the growth of polymer film thickness up to several μm, an increase in the anodic current and also by monitoring the total charge passed during the electrooxidative (or reductive) polymerization process.
- The polymers obtained are in general cross-linked and therefore insoluble. Therefore an exact structure determination is not possible. The metal complex is not destroyed during the electropolymerization, as can be seen from UV-vis spectra.
- The polymer films remain electrochemically active when they are employed again in a fresh electrolyte.
- The morphology (controlled, for example, by AFM) and the physical parameters (e.g. conductivity) depend largely upon the electrochemical polymerization conditions: potentiostatic/potentiodynamic, solvent, conducting salt. Therefore the film properties such as conductivity and electrocatalysis must also be carefully optimized.

Figure 6-5 lists most of the pyrrole-, thiophene-, amino- and hydroxy-substituted porphyrins employed for electropolymerization and the references. Monomers **49a – k** [136–147] are based on synthetic porphyrins and **50a, b** on deuteroporphyrin [148–150]. Surprisingly, the electrooxidation of metal complexes of protoporphyrin-IX dimethyl ester, possibly via the vinyl groups, leads to the deposition of electroactive porphyrin films on the electrode surface [151–153].

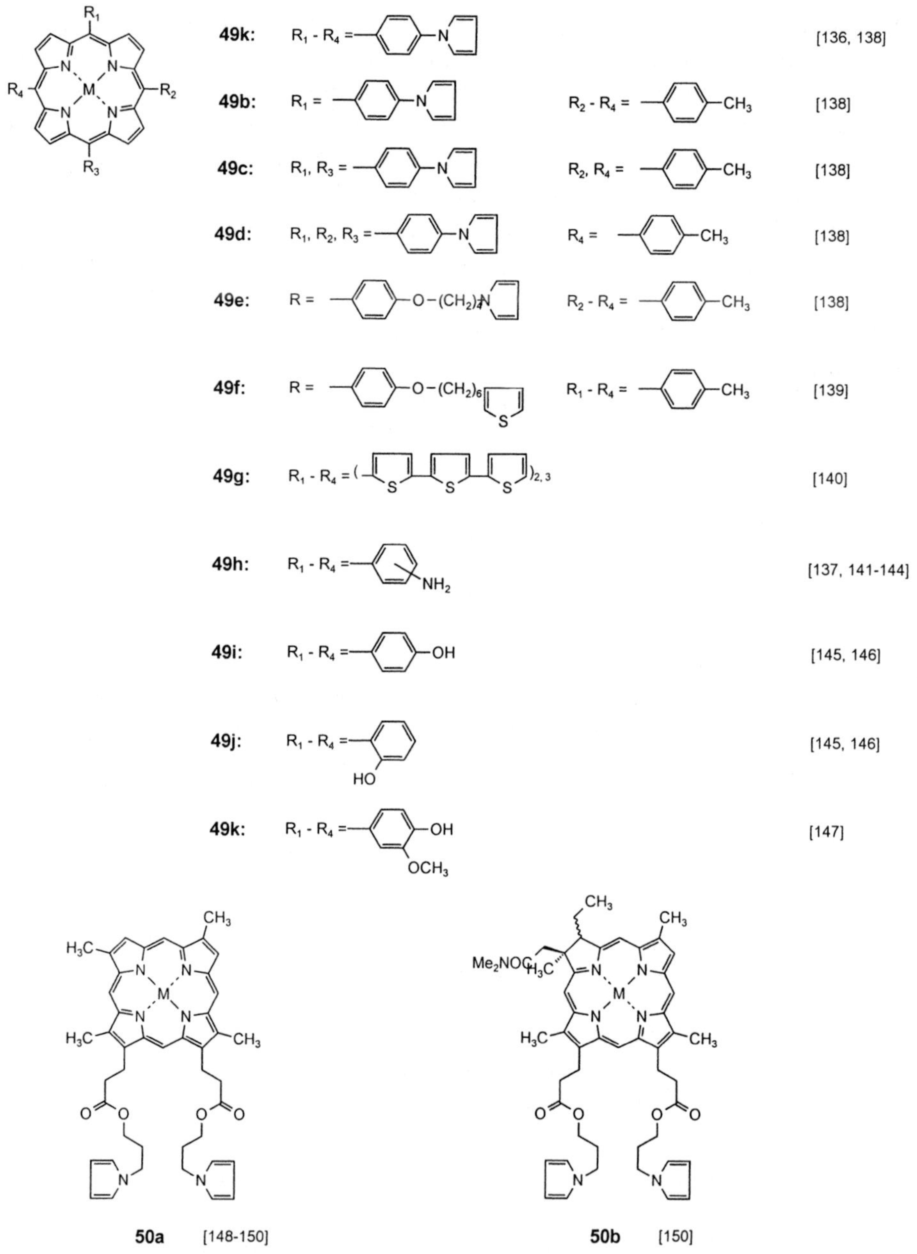

49k:	$R_1 - R_4 =$		[136, 138]
49b:	$R_1 =$	$R_2 - R_4 =$ —CH$_3$	[138]
49c:	$R_1, R_3 =$	$R_2, R_4 =$ —CH$_3$	[138]
49d:	$R_1, R_2, R_3 =$	$R_4 =$ —CH$_3$	[138]
49e:	$R =$ —O–(CH$_2$)$_3$	$R_2 - R_4 =$ —CH$_3$	[138]
49f:	$R =$ —O–(CH$_2$)$_6$	$R_1 - R_4 =$ —CH$_3$	[139]
49g:	$R_1 - R_4 =$ ( )$_{2,3}$		[140]
49h:	$R_1 - R_4 =$ NH$_2$		[137, 141-144]
49i:	$R_1 - R_4 =$ —OH		[145, 146]
49j:	$R_1 - R_4 =$ HO		[145, 146]
49k:	$R_1 - R_4 =$ —OH OCH$_3$		[147]

50a [148-150]

50b [150]

Figure 6-5. Examples of porphyrins employed for electropolymerization.

Only sample results of the electropolymerization of these monomers are mentioned (for an experiment on the electropolymerization of **49h** see Section 6.6). Electrochemical polymerization of pyrrole-substituted porphyrins is achieved by cyclovoltammetry in dichloromethane or acetonitrile solutions containing the monomer and the supporting electrolyte. Figure 6-6 shows the cyclic voltammograms of the manganese porphyrin **49e** [138]. A peak corresponding to Mn(III)/Mn(II) occurs at $E = -0.32$ V vs. SCE, and two irreversible oxidation steps are observed at $E_{p1} = 0.75$ and $E_{p2} = 1.05$ V vs. SCE. The pyrrole group is oxidized irreversibly at around 1.0–1.3 V vs. SCE. Thus, repeated scanning of the potential over the range – 0.8 V to +1.3 V results in the formation of a polyporphyrin film at the electrode surface. For **50a** (M = Cu) repeatedly scanning the potential over a range from –0.4 to 0.75 V vs. Ag/AgCl results in continuous growth of the cyclovoltammetric peaks, which shows the formation of the electrogenerated polymeric film on the electrode surface (Fig. 6-7) [149]. This increase corresponds to (i) the increase in the one-electron oxidation wave of **50a** and (ii) the appearance of an increase of the reversible oxidation/rereduction wave of the polypyrrolic backbone.

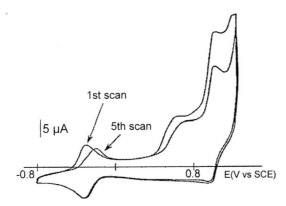

Figure 6-6. Electropolymerization of a 1 mM solution of the porphyrin **49e** at a vitreous carbon disc electrode (area 0.071 cm^2) in acetonitrile and 0.1 M tetrabutylammonium tetrafluoroborate (scan rate 100 mV s^{-1}). Only 1st and 5th successive runs shown.

In the UV-vis absorption spectra of polyporphyrins a shift of some absorption bands is seen. UV-vis spectra in transmission of the polymer films show broadening of the Soret and Q-bands. Spectroelectrochemical characterization of films exhibits reversible color change with reversible charge/discharge behavior within a controlled potential range. The polymer films can be seen as a large number of stacked monolayers of redox centers. Charge transfer can occur by electron self-exchange reactions between neighboring redox active molecules resulting in electron hopping and a diffusion-limited current.

Electroassisted biominetic oxidations are of special interest using electropolymerized metalloporphyrin films with molecular oxygen or oxygen donors such as PhIO, H_2O_2 or $KHSO_5$. Oxidation with oxygen is achieved by its reduction with two electrons from the porphyrinic catalytic site. This leads to the formation of high-valent metal-oxo species, formally a $[Fe(V)=O]^+$ or $[Mn(V)=O]^+$ complex. This species is highly reactive and responsible for hydrocarbon oxidation to alcohols, ketones and epoxides. Some results for the electroassisted oxidation of hydrocarbons by the Mn(III) polymer films with molecular oxygen are described in [138]. Co-complexes of electropolymerized amino- and pyrrole-substituted porphyrins have been investigated for O_2 reduction [149,154,155].

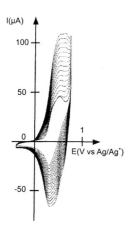

Figure 6-7. Electropolymerization of a 1 mM solution 50a (M = Cu) at a glassy carbon electrode (diameter 5 mm) in acetonitrile and 0.1 M tetrabutylammonium perchlorate.

Electrodes modified with polymer porphyrins can act as sensors or detectors for different compounds. For films of **50a** (M = Mn(III)) dipped in an organic solvent an excellent biominetic activation of oxygen in the presence of 1-methylimidazole and benzoic anhydride was observed in the cathodic regime [149]. Electrodes covered with **50a** (M = Fe(III)) also exhibit excellent recognition properties towards cyanide. The determination of phenols under anodic potential conditions has been intensively investigated with films of protoporphyrin–IX dimethyl ester [151–153]. With M = Ni a detection limit of 13 ng mL^{-1} was found for 4-nitrophenol [151]. Polymer porphyrin films based on 5,15-bis(2-aminophenyl)porphyrin have been investigated for selective binding of various inorganic anions [143]. Zinc 5,10,15-trimesityl-20-(4-ethynylphenyl)-porphin was covalently bound to a polypyrrole surface modified with 4-iodobenzenethiol to get the modified surface shown in Fig. 6-8 [156].

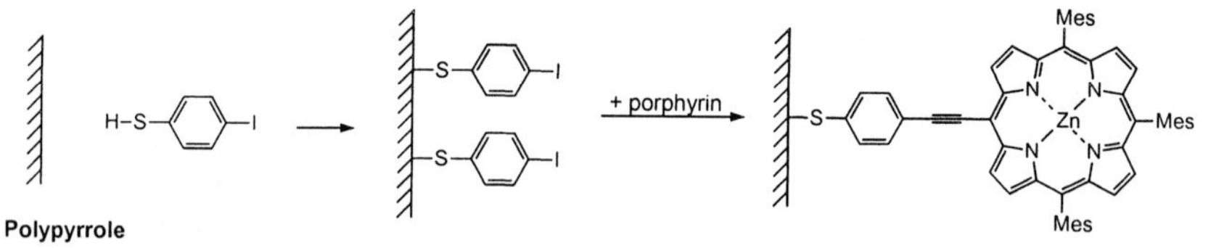

Figure 6-8. Binding of a porphyrin at a polypyrrole surface.

Also some phthalocyanine derivatives such as 2,9,16,23-tetraaminophthalo-cyanine **51a** (see Experiment 6-5, Section 6.6) [157–164], 2,9,16,23-tetrakis(2-

hydroxyphenoxy)phthalocyanine **51b** [165] and 2,9,16,23-tetrakis-(pyrrole-1-yl-alkoxy)-phthalocyanines **51c** [166] have been employed for oxidative electropolymerization (see Experiment 6-5, Section 6.6).

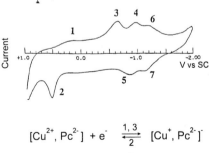

51a: R = —NH₂

51b: R =

51c: R = —O—(CH₂)ₙ-N

n = 2, 3

After the pioneering work reviewed in [157] a recent publication studied again in detail the oxidative electropolymerization of **51a** (M = Cu) on ITO and glassy carbon in DMSO with sodium perchlorate as supporting electrolyte [164]. Figure 6-9 shows the cyclic voltammogram obtained after 10 cycles. Three redox couples are identified: one corresponds to the central metal ion, and the others correspond to redox processes involving the ligand. The proposed assignments are summarized in the equations shown in Fig. 6-9. During further cycles the peaks shift as discussed [164]. In Fig. 6-10 repeated cyclovoltammograms are shown. Peak 1 decreases, and peaks 2 and 3 increase. The increase of peak 3 is due to increasing film thickness and charge/discharge of the film. Spectroelectrochemical studies of the color change under charge/discharge were also made. For the anodic electropolymerization of **51a** a mechanism analogous to that for aniline was discussed [164].

$$[Cu^{2+}, Pc^{2-}] + e^- \; \underset{2}{\overset{1,3}{\rightleftharpoons}} \; [Cu^+, Pc^{2-}]^-$$

$$[Cu^+, Pc^{2-}]^- + e^- \; \underset{5}{\overset{4}{\rightleftharpoons}} \; [Cu^+, Pc^{3-}]^{2-}$$

$$[Cu^+, Pc^{3-}]^{2-} + e^- \; \underset{7}{\overset{6}{\rightleftharpoons}} \; [Cu^+, Pc^{4-}]^{3-}$$

Figure 6-9. Above: Cyclic voltammogram of the 10th cycle during the electropolymerization of 51a (M= Cu) in DMSO (concentrations: 10^{-3} M 51a, 0.1 M sodium perchlorate, scan rate 50 mV s^{-1}). Below: Interpretation of the peak assignments.

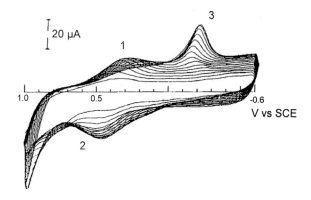

Figure 6-10. Repeated cyclic voltammogram for the first 15 cycles during the electropolymerization of 51a (M= Cu) in DMSO (concentrations: 10^{-3} M 51a, 0.1 M sodium perchlorate; scan rate 50 mV s^{-1}). Peak assignments s. Figure 6–9.

Electrochemical polymerizations of **51c** ($n = 2, 3$; M = 2H, Zn, Ni, Co) at ITO working electrodes have been carried out in dichloromethane/tetrabutylammonium hexafluorophosphate under potentiodynamic and potentiostatic conditions containing 10^{-3} mol L^{-1} monomer and 0.1 mol L^{-1} supporting electrolyte [166].

Polymer films of **51a** (M = Co, Ni, Lu) on ITO have been investigated under different potentials for their electrochromic behavior [162,166,167]. The polymer from **51a** (M = Ni(II)) exhibit an electrical conductivity $\sigma \sim 10^4$ S cm^{-1} and an activation energy $E_a \sim 0.2$ eV [157].

Polymeric phthalocyanine films prepared from **51a** have been investigated as electrocatalysts, in biological applications and as amperometric biosensors [157]. The electropolymerized films exhibited better activities than analogous low molecular weight phthalocyanine films. Some examples are given:

- Electrocatalytic oxidation of oxalic acid by a polymer from **51a** (M = Co) [168].
- Electrocatalytic oxidation of hydrazine by polymers from **51a** (M = Co, Fe) [169–171].
- Electrocatalytic oxidation of various thiols by a polymer from **51a** (M = Co) [172,173].
- Electrocatalytic oxidation of sulfide by a polymer from **51a** (M = Co) [174]
- Electrocatalytic reduction of oxygen by a polymer from **51a** (M = Fe) [154,175].
- Electrocatalytic oxidation of reduced nicotinamide-adenine glycolhydrolase by polymers from **51a** (M = Co, Ni, Zn) [161].
- Electrochemistry of redox proteins by a polymer from **51a** (M = Co) [157].
- Amperometric glucose sensor by polymers from **51a** (M = Cu, Co) [176].
- Ethanol sensor based on alcohol dehydrogenase by a polymer from **51a** (M = Ni) [157].
- Electrocatalytic determination of *tert.*-butylhydroxyanisol by a polymer from

51a (M = Ni) [177].
- Potentiometric pH sensor in the pH range 1–13 by polymers from **51a** (M = Ni, Cu) [178].

Besides porphyrins and phthalocyanines, which have already been described in more detail, several other suitable substituted metal complexes have been employed for electropolymerizations. Only a few examples are given below (for review see [179]). Oxidative electropolymerization of chelates of salicylaldimes have been successfully carried out [180–182]. Cu(II) and Ni(II) complexes of the thiophene-containing Schiff-base complex **52** have been oxidatively polymerized at +0.9 V vs. Ag/AgCl in CH_2Cl_2 [182]. Surprisingly, the polymerization occurs at the benzene rings – marked by arrows in **52** – and not at the annelated thiophene ring. The terthiophene chelate **53** shows polymerization via the thiophenes as marked by arrows. The electrochromic properties of films as conducting electrodes have been demonstrated [182].

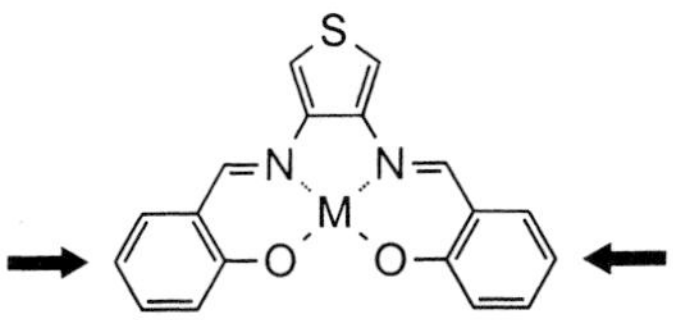

52 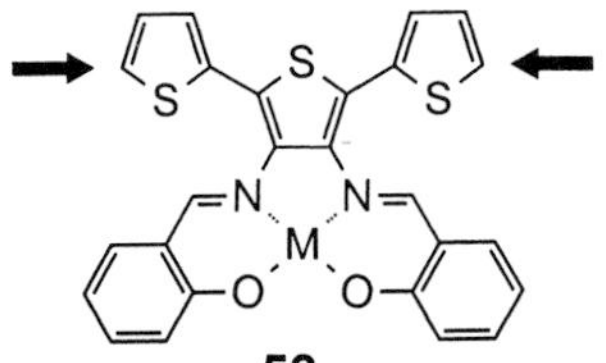   **53**

Oxidative electropolymerization of chelates modified by functionalized pyrroles has been described in [180]. Pyrrole-containing Ru(II) 2,2'-bipyridyl complexes **54** have been deposited electrochemically in CH_3CN to more than 10,000 monolayers. Analogously, several amino-substituted Ru(II) complexes containing 4-aminopyridine **55**, 3-aminopyridine **56** or 5-amino-1,10-phenanthroline **57** have been successfully employed for oxidative electropolymerization [183]. The polymers form stable, electrochemically active films on Pt, SnO_2 or vitreous carbon, but the films are not stable upon repeated oxidative cycling in the presence of a trace of H_2O. In contrast, electropolymerization of $[Rh(III)(L)_2Cl_2]^+$ complexes (L = pyrrole-substituted 2,2'-bipyridine or 1,10-phenanthroline) on carbon yields active electrode materials for the electrocatalytic hydrogenation of organic compounds in aqueous solution [184,185].

Enantioselective hydrogenation of prochiral aromatic ketones has been successfully achieved employing a chiral electrode covered with a polymer obtained by oxidative electropolymerization of a complex based on the 2,2'-bipyridyl **58** (composition $[Rh(133)_2Cl_2]^+$) [186].

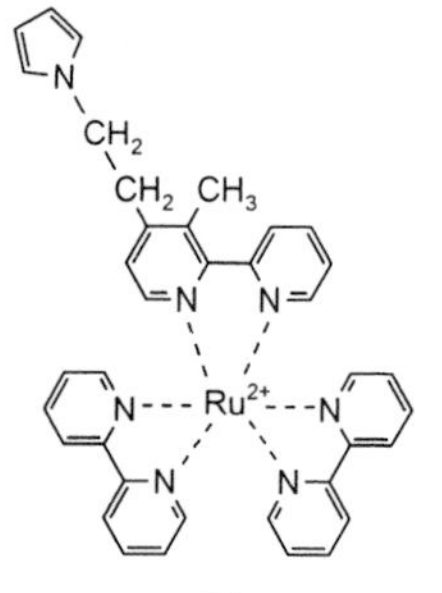

54

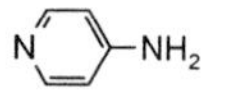

55 (4-pyNH₂)

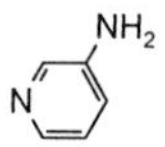

56 (3-pyNH₂)

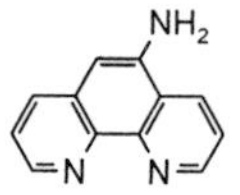

57 (5-phenNH₂)

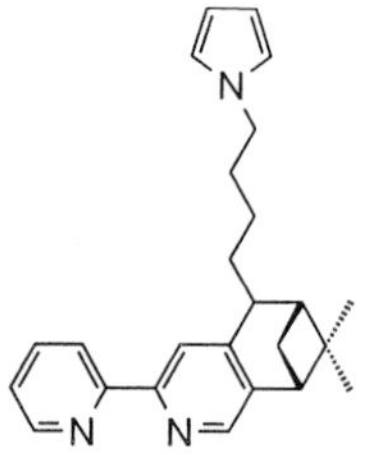

58

A new advance in preparing polymers containing metals by electropolymerization was shown by the synthesis of conjugated conducting polymetallorotanes [187–189]. The monomers consists of two building blocks: a macrocyclic phenanthroline and a bis(bithiophene)-bipyridine [187,188] or bis(thiophene)phenanthroline [189]. These two compounds were complexed by transition metal ions like Cu⁺ or Zn²⁺ [187,188]. Electropolymerization was achieved by anodic electropolymerization on Pt or ITO electrodes by continuously cycling a solution of these metal complexes between ~0.2 and ~1.2 V vs Ag/AgCl. Red-colored films of the polymers (one example is shown in **59**) were obtained.

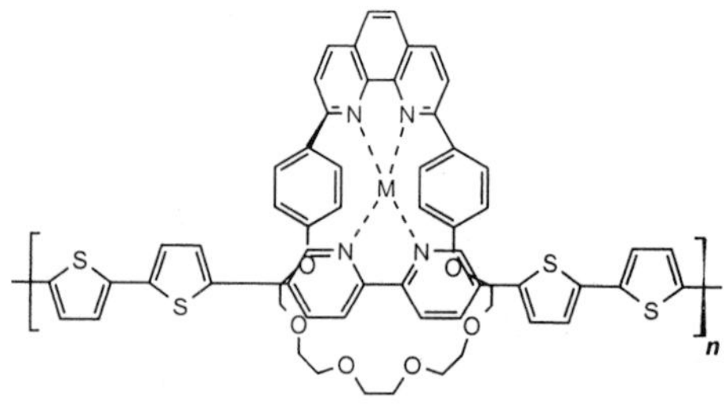

59

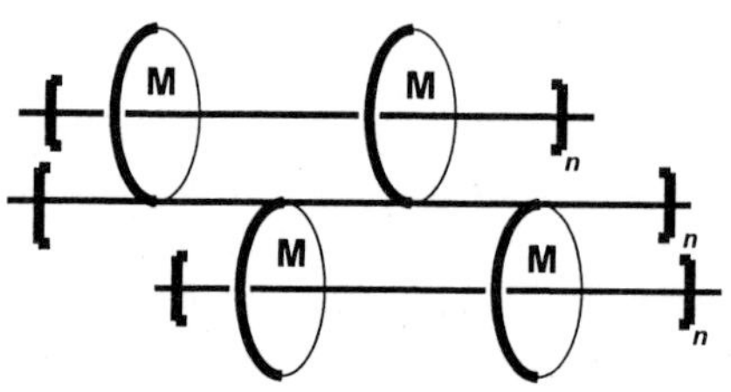

60

The reversibility of the demetallation–remetallation processes was investigated for possible use in metal sensors [187,189]. The polymer films exhibit two discrete redox processes which are associated with two conductivity maxima (σ up to $3.5 \cdot 10^{-2}\,\mathrm{S\,cm^{-1}}$) [187]. In addition, three-strand ladder polymers of comparable structure as schematically shown in **60** were synthesized with conductivities up to 60 S cm^{-1} [188].

6.4 Dendrimers Constructed via the Ligand of a Metal Complex

Dendrimers are macromolecules of uniform mass that contain a core, successive layers of branched repeat units and peripheral groups [190,191]. Because of their precision molecular architecture and three-dimensional structure, dendrimers are important artificial materials. Functionalities can be introduced in the molecular core, the repeat or branching unit and in the peripheral units of the dendrimer. These functionalities can be active centers with photophysical, photochemical, electrochemical, catalytical, sensoric, etc. properties. The properties of such moieties are altered by the homogeneous surrounding of the dendritic structure, which can lead to materials with enhanced properties. Metallodendrimers are an increasingly represented class of molecules in the area of dendrimer chemistry (for reviews see [192,193]). The use of metals in dendrimer structures as active parts has resulted in molecules with potentially useful physical properties. In the most interesting cases, these properties are not found in the free metal-bearing subunit of the dendrimer or in analogous non-metallated dendrimers. We shall briefly describe synthetic strategies for obtaining dendrimers. In the divergent strategy, increasing generations of a dendrimer are prepared step by step starting from an active core. In contrast, the often advantageous convergent strategy involves adding previously prepared dendrons to an active core molecule.

Macrocyclic metal complexes are the principal materials incorporated in the dendrimer structure (for reviews see [192,193]). Examples are: metallo dendrimers with a porphyrin core [194–204], with a phthalocyanine core [205–209], with crown ethers in the repeat unit [210–212] or with a cyclic amine core [213].

The syntheses of some dendrimers containing porphyrin-type compounds are shown. Dendrimer **61** is prepared by the reaction of Zn tetrakis(3,5-dihydroxyphenyl)porphyrin with poly(benzylether)-dendrons (Eq. 6-16) [194]. Structurally comparable dendrimer porphyrins bearing negatively or positively charged groups at the peripheral units and therefore being water soluble have been described [195]. A large multiporphyrin array **62** consists of four dendritic wedges of a zinc porphyrin heptamer which were anchored to a focal free-base porphyrin [196]. Finally, the dendritic Zn-phthalocyanine **63** has been reported. This was prepared by the reaction of octakis[3,5(dicarboxy)-phenoxy] ZnPc with triply branched carboxylethylesters containing amino groups [205]. In another approach dendron-substituted phthalonitriles were reacted to phthalocyanines [207,208].

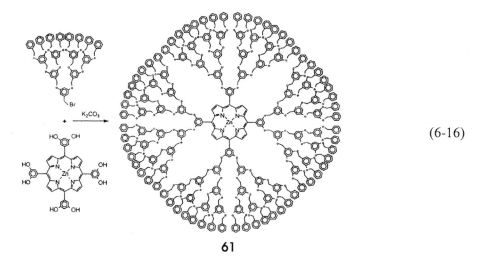

(6-16)

61

It is important now to discuss how the properties of metallodrimers are altered in comparison to the analogous low molecular weight metal complexes. The encapsulation of a metal complex in the core of a dendrimer as an artificial system is comparable to a biologically active system, such as cytochrome P450 (electron transfer), hemo/myoglobin (gas transport) or the antenna/reaction center of photosynthesis (photo-induced energy, electron transfer), in which the active center is surrounded by a protein. For a dendrimer with a Zn-porphyrin core and peripheral anionic carboxylic acid groups the Soret bands in the electronic spectra are blue-shifted with higher and higher generations [214]. The fluorescence of a dendritic porphyrin is quenched more effectively by vitamin K_3 with a dendrimer of the fourth than with a dendrimer of the fifth generation. The larger dendrimer can more easily take up the small quencher molecule.

A self-organizing system of self-organizing dendrimers with oppositely charged peripheral groups and porphyrin and Zn-porphyrin cores exhibits efficient fluorescence quenching [195]. In a recent interesting report the dendritic multiporphyrin array 62 was investigated as a mimic of the bacterial light-harvesting antenna complex [196]. The dendritic wedges with the zinc porphyrins act as energy-donating units and the free-base porphyrin in the core as electron acceptors for efficient energy funneling of visible light photons. When a solution of the low molecular weight analogous Zn-porphyrin was excited at $\lambda = 544$ nm the fluorescence from the Zn-porphyrin at $\lambda = 589$ and 637 nm was detected (Fig. 6-11A). In contrast, excitation of 62 at 544 nm resulted in an emission at $\lambda = 658$ nm which is characteristic for the free base porphyrin (Fig. 6-11B).

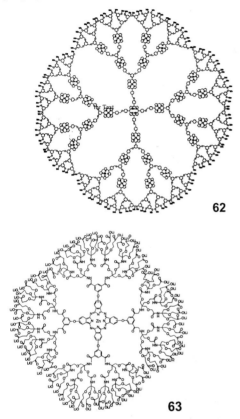

62

63

This demonstrates an efficient energy transfer from the excited singlet state of the dendritic Zn-porphyrin to the local free-base porphyrin. Time-resolved fluorescence spectra confirm this statement.

Investigation of the electrochemical behavior of dendrimers with Zn- or Fe-porphyrin cores have shown easier oxidation and more difficult reduction, which is explained by a shielding of the porphyrin core and reduced contact with the external solvent [199,200]. With increasing size of the dendrimer electron transfer to the central porphyrin core is more and more reduced, which is also explained by shielding of the active part [194].

The O_2-binding of dendrimers with a Fe(II)-porphyrin core has been studied as a model for hemo/myoglobin [203]. Interestingly, the lifetime of the oxygen adduct increases with increasing shielding of the Fe-porphyrin (compare Fig. 2-8 in Section 2.3.1). In the case of a fifth generation dendrimer, 95% of the O_2-adduct still exists after storage for two months. Finally, a suggestion has been made that porphyrin-type core dendrimers might be used as catalysts in the epoxidation of olefins [202] or oxidation of thiols [206].

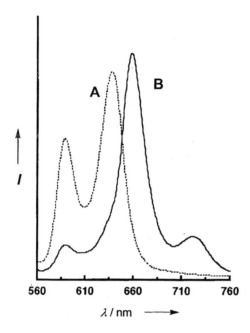

Figure 6-11. Fluorescence spectra of the dendrimer **62** (B) and an analogous low molecular weight Zn-porphyrin (A) (see text).

6.5   Hydrogen-Bonded Networks via the Ligand

The flat, rigid and thermally stable metalloporphyrin framework has turned out to be an extremely versatile and particularly attractive building block for the designed tailoring of porous solids [215]. One possibility is the non-covalent supramolecular synthesis of new molecular solids with structural and functional similarity to inorganic molecular sieves. Besides coordination polymers via the metal, interporphyrin hydrogen bonding via substitutents at the ligand has been studied. Covalent/coordinative interactions including the participation of the metal of a complex in the formation of a framework is discussed in Sections 7.2 and 7.3.

The aim to crystal-engineer cross-linked open networks of multiporphyrin aggregates with high structural integrity is a promising perspective for new materials with interesting properties. Surprisingly, cooperative hydrogen bonding can provide enough mechanical strength to sustain a large amount of empty space in organic crystals [216–218]. Spontaneous self-assembly of supramolecular aggregates by suitably substituted large porphyrins can result in porous networks with large void volumes.

Tetra(4-amidophenyl)porphyrin **29** (R = –CONH$_2$) prepared by slow cooling of a solution of the porphyrin in DMSO in the presence of a dimethylsulfoxide template forms networks that do not interpenetrate. This solid has interporphyrin cavities of about $0.6–0.7 \times 1.0$ nm^2 [219]. The Zn complex of tetra(4-carboxyphenyl)porphyrin **29** (R = –COOH) reveals different networks. With Na or K 18-crown-6 chlorides as templates non-interpenetrating networks with voids of 0.85×1.1 nm^2 and 0.3×0.65 nm^2 have been prepared by dissolving the porphyrin **29** (R = –COOH) and crown components in hot methanol in the presence of NaCl or KCl. Slow evaporation of these solutions led to single crystals of the composite materials (Fig. 6-12) [220]. The [18-crown-6 Na or K]$^+$ fits within the width of the large interporphyrin cavity and turns out to be an excellent template for this mode of layered porphyrin–crown ether assembly (for more details see [220]). To prevent interpenetration it is essential to use sizable solubilizing and templating agents with a low tendency to interact with the Zn-porphyrin lattice. This was also achieved with tetra(aminophenylporphyrin) as templating agent to nucleate an open lattice structure [221].

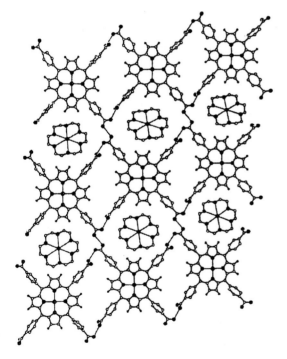

Figure 6-12. View of a molecular network of **29** (R = –COOH; M = Zn) with [18-crown-6 Na]$^+$ in the cavities.

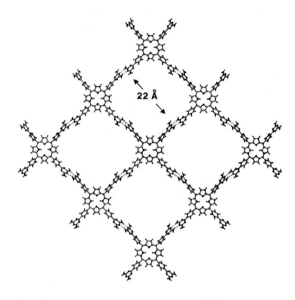

Figure 6-13. Network of large **29** (R = 3′,5′-diaminotriazine; M = Zn) building blocks sustained by multiple hydrogen bonds.

Tetra[4-(3′,5′-diaminotriazino)phenyl]porphyrin **29** (R = 3′,5′-diaminotriazine; M = Zn) has also been used as a molecular building block, and crystals were obtained in a mixture of DMF and THF [222]. The diaminotriazine functions had higher hydrogen-bonding capacity, spatial flexibility and directionality features than the carboxylic acid groups. The uniquely structured solid consists of flat multi–porphyrin networks with large voids of 2.2×2.2 nm^2 (Fig. 6-13). Finally a 5,15-bis(imidazol-4-yl)-10,20-bis(4-dodecyloxyphenyl)-porphyrin assembly through imidazol-imidazole-hydrogen bonding has been reported [223].

6.6 Experimental

Experiment 6-1:   Polymeric Schiff-base Complex 4 from Bisphenol A and a Dihydroxy-Substituted Schiff-base Complex (Section 6.1) [19]

Synthesis of the low molecular weight dihydroxy-substituted Schiff base Cu(II)-complex (see Eq. 6-3): A methanolic solution (10 mL) of 1,3-diaminopropane (0.5 mL, 10.65 mmol) was added dropwise to a methanolic solution (100 mL) of 2,4-dihydroxybenzaldehyde (2.94 g, 21.28 mmol) at 0 °C, and the mixture was stirred for 20 min. The color of the mixture gradually changed to yellow. An aqueous solution (40 mL) of copper(II) acetate monohydrate (2.13 g, 10.64 mmol) was then added dropwise at 0 °C, and the mixture was stirred for

20 min. Upon adding a solution of 2 mol L^{-1} sodium hydroxide, the metal complex precipitated and was subsequently isolated by filtration and dried under vacuum. The copper complex was obtained as a green solid (3.87 g, 87%) by recrystallization from hot methanol. The brown-colored Co(II)- and the green-colored Ni(II)-complexes were prepared in the same manner.

Synthesis of the polymeric Schiff base complexes **4** *(see Eq. 6-3)*: A mixture of the diglycidyl ether of bisphenol A, a metal complex and tetrabutylammonium hydroxide was degassed under vacuum and then cast into a mold and cured by heating in a hot-air oven. The completeness of curing was confirmed by the disappearance of the epoxy group absorption at 917 cm^{-1} in IR spectra. With a molar ratio of metal complex:bisphenol A derivative:tetra-butylammonium hydroxide = 1:6:0.2 the curing was carried out at 160 °C for 4 h.

Experiment 6-2:   Metal Complexes of Poly(salicylaldoxime-3,5diyl-methylene) 9 (Section 6.1) [25]

Synthesis of poly(salicylaldoxime-3,5-diylmethylene **9**: The polymer was prepared by dissolving 13.7 g (0.1 mol) of salicylaldoxime and 16.2 g (0.2 mol) of 37% aq. formaldehyde in 65 mL of 50% aqueous ethanol and enough NaOH to raise the pH to about 9. The mixture was refluxed for 9 h after which the reaction mixture was evaporated and the resulting resin was cured at 120 °C for 24 h. The resin was then washed with aqueous ethanol and water until neutral. After Soxhlet extraction by benzene for 24 h, the brown resin was dried at 90 °C for 24 h in vacuum, crushed and sieved (35–60 mesh).

Metal-ion uptake in a static system: A dry 0.5 g sample of the polymer was suspended in 70 mL of 0.25 M sodium acetate. With 0.10 mol L^{-1} hydrochloric acid the solution adjusted to the required pH (5–8) and left to equilibrate. The mixture was charged with 5 ml aqueous metal ion solution containing 40.0 mg of the metal ions and shaken at 25 °C. After being shaken for a definite contact time, the mixture was filtered and the amount of the metal ions remaining in the filtrate was determined by atomic absorption spectroscopy or by chelatometric titration with EDTA. A series of experiments were undertaken in which the pH was varied between 5 and 8 at a fixed contact time of 3 h. Experiments were also carried out in distilled deionized water (in the absence of acetate buffer) in which the exposure time was varied from 1 to 24 h (for details see [25]).

Metal-ion uptake in a dynamic system: Sorption experiments in a dynamic system were carried out using a glass column (length: 10 cm, internal diameter: 1.2 cm) filled with 1.0 g resin. The resin was washed with distilled–deionized water (10 bed volumes), then a solution containing 40 ppm Cu^{2+} was passed through at a flow rate of 3 mL min^{-1}. For determination of the breakthrough curve, the eluate from the column was collected in 50 mL fractions and the copper cation concentration was determined by atomic absorption spectroscopy.

Experiment 6-3: Polymeric Phthalocyanines 31, 32 (Section 6.2.1) [58,89,96,97]

Equipment: Heating oven to 500 °C, thick-walled glass ampoules of ~25 mL.

Synthesis of 1,2,4,5-tetracyanobenzene (after a modified procedure described in [224]): This tetracarbonitrile is prepared via a three-step procedure. Benzene-1,2,4,5-tetracarboxylic acid is first converted into its diimide. 54.5 g (0.25 mol) of the acid and 50 g (0.83 mol) urea were heated in 250 mL 1,2-dichlorobenzene under reflux at 130 °C for 4 h and then at 160 °C for 2 h. The benzene-1,2,4,5-tetracarboxylic acid diimide obtained was directly employed for the synthesis of the tetraamide by dissolving it in 370 mL 33% aqueous NH₄OH. After stirring overnight, the product was filtered, washed with water and dried. Yield of white-colored benzene-1,2,4,5-tetracarboxylic acid tetraamide 50 g (80%) (mp 285 °C with decomposition). For the preparation of 1,2,4,5-tetracyanobenzene 50 g (0.2 mol) of the tetraamide was suspended in 300 mL dry DMF. Distilled thionyl chloride was added dropwise with stirring in two portions: at first 45 mL and then after 2 h a further 30 mL in such a way that the temperature does not exceed 50 °C. The mixture was stirred overnight at 40–50 °C, filtered and poured onto 1 kg of ice. The yellow precipitate of impure 1,2,4,5-tetracyanobenzene was filtered, washed until neutral with cold water and dried in vacuo over P_4O_{10}. The tetracarbonitrile was purified either by recrystallization from toluene or by sublimation. Yield of white-colored 1,2,4,5-tetracyanobenzene ~25 g (~70%). mp 266 °C. Elemental analysis: Calculated C 67.40%, H 1.12%, N 31.46%; found C 67.11%, H 1.24%, N 31.98%. IR (KBr): 2225, no C=O between 1800 and 1650, intensive C≡N at 2225 cm⁻¹.

Synthesis of 4,4'-(1,4-phenylenedioxy)diphthalonitrile (for a description of the preparation of different oxy-, arylenedioxy- and alkylenedioxy-bridged diphthalonitriles see [67,68]): 11.01 g (100 mmol) hydroquinone was dissolved in 100 mL dry DMSO under nitrogen. After addition of 15.85 g (60 mmol) fuse-dried K_2CO_3 the solution was stirred for 15 min. 19.05 g (110 mmol) 4-nitrophthalonitrile was added, and the solution was stirred for 20 h at RT. The reaction product was precipitated by pouring the solution into 1 L cold 0.5 wt.-% aqueous HCl. The isolated product was dissolved in 250 mL acetone and 5 g activated charcoal was added. After filtration the tetracarbonitrile was precipitated with distilled water, isolated and dried in vacuo over O_4O_{10}. Yield of ~90% (mp 251 °C). IR (KBr): 2225, 1245 cm⁻¹. MS (70 eV): 362 (100%, M⁺).

Method A for the preparation of powdered 31 (M = Cu(II)) [58]: 890.8 mg (5 mmol) 1,2,4,5-tetracyanobenzene was mixed with 336 mg (2.5 mmol) dry $CuCl_2$. The mixture was transferred to a glass ampoule. All these operations were carried out under dry argon conditions in a glove box. After evacuating and flushing with dry argon three times, the tube was sealed under vacuum. The ampoule was placed in a 400 °C hot-air oven, and heated for 4 h. After cooling, the ampoule was opened cautiously (excess pressure by Cl_2!). The polymer partially appeared as a film on the glass wall. The isolated polymer was weighed and purified by treating in a Soxhlet apparatus with methylene chloride (~2 h), acetone (~3 h) and DMF (~4 h). Further workup was carried out under stirring with 1 M nitric acid at room temperature, followed by washing with water and acetone, and drying for 24 h at 200 °C in vacuo. Yield 0.90 g (85%). Elemental analysis: Calculated for **31** with an infinite two-dimensional plane structure containing 2 H_2O per $Cu((C_{20}H_8CuN_8O_2)_n)$: C 52.69%, H 1.77%, Cu 13.94%, N 24.59%; calculated for **31** with an infinite one-dimensional chain structure containing 2 H_2O per Cu $(C_{30}H_{10}CuN_{12}O_2)_n$: C 56.82%, H 1.59%, Cu 10.02%, N 26.52%; calculated for **31** as polymer with nine phthalocyanine units containing 2 H_2O per Cu $(C_{240}H_{84}Cu_9N_{96}O_{18})$: C 55.73%, H 1.64%, Cu 11.06%, N 26.00%; found C 55.1%, H 1.79%, Cu 10.1%, N 25.3%.

*Method B1 for the preparation of powdered **31** and **32b** (M = Zn(II)) [97]:* To obtain **31** 250 mg (1.4 mmol) 1,2,4,5-tetracyanobenzene was suspended in 20 mL pentanol-1 (distilled and dried over molecular sieve) under dry inert conditions; to obtain **32b** 1.4 mmol 4,4'-(1,4-phenylenedioxy)diphthalonitrile was employed. Then 50 mg metallic lithium (7.1 mmol) was added. The mixture was stirred and heated in a dry inert gas atmosphere under reflux. 30 min later the precipitated polymer **31** was separated by filtering and washed with ethanol. In the case of polymer **32b** a reaction time of two weeks is necessary to produce an insoluble polymer. Without further purification the raw products were metallized with, for example, a 10-fold excess of zinc acetate in DMF for 24 h at 100 °C. The polymers were purified by subsequent treatment in a Soxhlet apparatus with CH_2Cl_2 followed by DMF. Yield **31** 337 mg (57%). Yield **32b** >90%.

Method B2 for the preparation of a powdered copolymer [97]: Corresponding to a molar ratio of 1:1, 351 mg (1 mmol) tetracyanobenzene and 149 mg (1 mmol) of the diphthalonitrile were suspended in dry pentanol-1. After the addition of 50 mg metallic lithium the reaction mixture was stirred and heated under reflux in an inert atmosphere. The precipitated copolymer was separated after 30 min and washed with ethanol. Metallization and purification were carried out as described above. Yield 452 mg (80%).

*Method C for the preparation of a thin film of **31** and copper on quartz glass [89]:* Quartz plates (size 2 cm × 3 cm) were pre-treated by washing with detergent, cleaning with methanol in an ultrasonic bath and additionally by glowing in a non-luminous flame. The quartz substrate was settled in the sample holder inside the sputtering chamber of a UI_{VAC} RF sputtering plant. The radio frequency sputtering was carried out under $5 \cdot 10^{-3}$ Torr Ar pressure. The presputtering time was 5 min in every case. For example a Cu-film of 9.8 nm was sputtered. Alternatively, the copper can be vapor-deposited. The Cu-covered plate was placed into a glass ampoule containing 6 mg tetracyanobenzene. The ampoule was sealed under vacuum as soon as possible in order to avoid oxidation of the copper layer. The pressure in the tube was ~$2 \cdot 10^{-5}$ Torr. Then the ampoule was heated in a 400 °C air oven for 4 h. The layer thickness of the film of **31** (M = Cu(II)) determined by a Sloan Dektak (mechanical stylus method) was 228 nm.

Method D for the preparation of a coating of 31 and Co$_2$(CO)$_8$ on SiO$_2$ [96]: Quartz powder (surface area 350 m^2 g^{-1}) was first dried for 3 h at 300 °C under vacuum. Then under dry inert gas atmosphere, 89 mg (0.25 mmol) Co$_2$(CO)$_8$ was added to 1.5 g of dried quartz powder in a 25 mL glass vessel. The glass was shaken for 3 h at 40 °C to allow a homogeneous distribution of the metal carbonyl on the carrier. 178 mg (1 mmol) of tetracyanobenzene was added to the glass vessel. The glass vessel was sealed under vacuum and heated with shaking for 3 h at 120 °C and then for 3 h at 180 °C in a hot oven. The dark blue-black powder was treated for 8 h in a Soxhlet apparatus with acetone.

Experiment 6-4: Interfacial Polycondensation to Polymeric Porphyrins (Section 6.2.2, Fig. 6-4) [123]

Chemicals: Porphyrin derivatives were purchased as described in Section 5.4.10. **29** (R = COCl) is prepared by refluxing **29** (R = COOH) in distilled thionyl chloride followed by complete removal of the thionyl chloride and vacuum drying. Metal-containing porphyrins (M = Cu(II), Zn(II)) are prepared by refluxing the metal-free porphyrin with the corresponding metal acetate in methanol or methanol/DMSO solution. Both metal-free and metal-containing porphyrins can be used for the film preparation.

Preparation of thin films by interfacial polycondensation: A typical procedure for preparation of a free-floating interfacial film was as follows. A 1 mM solution of **29** (R = COCl) in chloroform was added to a Petri dish, and an equal volume of 1 mM **29** (R = –OH) in pH 11 carbonate buffer was carefully layered on top. After 1–3 h reaction time, the aqueous layer was removed by pipet until the interfacial film breaks and folds over the retreating aqueous buffer solutions. The wet film is then deposited onto an appropriate substrate for subsequent analyses, wetted with 2-propanol, and allowed to air-dry overnight. The same technique provided polyporphyrin films made from **29** (R = COCl) and aliphatic polyamines such as ethylenediamine, diethylene-triamine and poly(ethyleneimine), although significantly shorter reaction times (several minutes) are sufficient and substantially thicker films result.

Experiment 6-5: Electropolymerization of Amino-Substituted Porphyrins and Phthalocyanines (Section 6.3) [225]

Equipment for the electropolymerization: A 5 mL glass cell with gas inlet closed with a silicone stopper containing an ITO working elecrode, an Ag wire as internal reference electrode and a platinum wire counter electrode; for electropolymerization a potentiostat connected with a function generator and a X–Y recorder or computer (see Fig. 6-14).

Preparation of 5,10,15,20-tetrakis(4-aminophenyl)porphyrin zinc(II) complex 49h: The preparation is described in Section 5.4, Experiment 5-10.

Preparation of 2,9,16,23-tetranitrophthalocyanine nickel(II) complex: 1.26 g (4.8 mmol) nickel sulfate hexahydrate, 3.7 g (17.5 mmol) 4-nitrophthalic acid, 0.45 g (8.4 mmol) ammonium chloride and 6.0 g (0.1 mol) urea were mixed. After addition of 2.5 mL nitrobenzene the mixture was slowly heated to 185 °C and maintained at this temperature for 4.5 h.

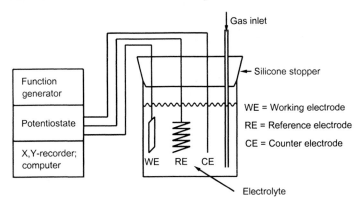

Figure 6-14. Measuring equipment for the electropolymerization.

The blue-colored reaction product was washed thoroughly with ethanol (to remove nitrobenzene), boiled for 5 min with 1 M HCl saturated with NaCl and filtered after coiling. Then the residue was treated with 50 mL 1 M NaOH containing 20 g NaCl at 90 °C until no more ammonia was evolved. The product was treated again after filtration with 1 M HCl and 1 M NaOH, and then thoroughly washed with water. Finally, the tetranitrophthalocyanine was treated for 24 h with methanol, then for 24 h with acetone in a Soxhlet apparatus and dried. Yield 3.2 g (97%). IR (in KBr): 1612, 1521, 1338, 825, 753 cm^{-1}. UV-vis in H$_2$SO$_4$ at λ: 766, 633, 307 nm.

Preparation of 2,9,16,23-tetraaminophthalocyanine nickel(II) complex 51a: 3.3 g (4.4 mmol) tetranitrophthalocyanine and 9.6 g (40 mmol) Na$_2$S·9H$_2$O in 100 mL water were stirred at 50 °C for 5 h. After filtration and washing with water the residue was twice treated first with 250 mL 1 M HCl and then with 170 mL 1 M NaOH. The tetraaminophthalocyanine was thoroughly washed with methanol and dried in vacuo over P_4O_{10}. Yield 2.56 g (90%). IR (in KBr): 3350, 1611, 1528, 853, 808 cm^{-1}. UV-vis in DMF at λ: 718, 645, 430, 312 nm.

Electropolymerization: A 1 × 1 cm ITO glass plate (purified with soap solution, water, ethanol) was used as the working electrode. A Pt wire counter electrode was purified with H_2SO_4/H_2O_2, water and ethanol. The Ag wire as reference electrode was polished and then cleaned with water and ethanol. The solution for the electropolymerization of the phthalocyanine derivative contains 6.30 mg (10^{-5} mol) **51a** (M = Ni) and 342 mg (10^{-3} mol) tetrabutylammonium perchlorate in 10 mL dry DMF, and for the electropolymerization of the porphyrin derivative **49h** (M = Zn) 7.4 mg (10^{-5} mol) **49h** and 342 mg (10^{-3} mol) tetrabutylammonium perchlorate in 10 mL dry methylene chloride. 4.5 mL of one of the solutions was filled into the glass cell under a stream of nitrogen. The ITO electrode was connected to a copper holder. The potential was scanned continuously between –0.4 V and +0.8 V for **51a** or 0.0 and 0.9 V for **49h**, respectively, vs. SCE at a rate of 10 mV/s. For calibration of the reference electrode, the porphyrin derivatives were replaced by ferrocene (E° = 0.4 V vs. SCE) in the same electrolyte.

6.7 References

1. D. Wöhrle, *Adv. Polym. Sci.* **1983**, *50*, 45.
2. A.K. Dey, *J. Indian Chem. Soc.* **1986**, 63, 357.
3. W. Sawodny, R. Meyer, R. Grünes, *J. Chem. Res. (S)*, **1983**, 6.
4. W. Sawodny, R. Grünes, H. Reitzle, *Angew. Chem. Int. Ed. Engl.*, **1982**, *21*, 775.
5. F.A. Bottino, P. Finocchiaro, E. Libertini, A. Mamo, A. Recca, *Polymer Commun.* **1983**, *63*.
6. M. Spiratos, G.I. Rusu, A. Airinei, A. Ciobanu, *Angew. Makromol. Chem.* **1982**, *107*, 33.
7. M. Macru. S. Lizarescu, G.E. Grigoriu, *Polym. Bull.* **1986**, *16*, 103.
8. J.S. Moore, S.I. Stupp, *Polymer Bull.* **1988**, *19*, 251; *Macromolecules* **1988**, *21*, 1217, 1222, 1228.
9. L. Oriol, J.L. Serrano, *Adv. Mater.* **1995**, *7*, 348.
10. R.D. Patel, *Makromol. Chem.* **1986**, *187*, 1871.
11. R. Grünes, W. Sawodny, *Inorg. Chim. Acta* **1983**, *70*, 247.
12. M.N. Patel, S.H. Patel, M.S. Setly, *Angew. Makromol. Chem.* **1981**, *67*, 69.
13. M.N. Patel, B.N. Jani, *J. Macromol. Sci.-Chem. A.* **1985**, *22*, 1517.
14. G. Manecke, R. Wille, *Makromol. Chem.* **1970**, *137*, 61; ibid. **1972**, *160*, 111.
15. M.N. Patel, S.H. Patel, Synth. React. Inorg. Met.-Org. Chem. **1983**, 13, 133.
16. E. Eichhorn, A. Ricker, B. Speiser, H. Stahl, *Inorg. Chem.* **1997**, *36*, 3307.
17. R.D. Archer, *Coord. Chem. Rev.* **1993**, *128*, 49.
18. D.H. Kim, H.I. Cho, T. Zhung, L.-M. Do, K.-M. Bark, G.C. Shin, S.C. Shin, *Eur.*

Polym. J. **2002**, *38*, 133.

19. N. Chantarasiro, T. Tuntulani, P. Tongraung, *Eur. Polym. J.* **2000**, *36*, 2000.
20. D. Wöhrle, H. Bohlen, G. Meyer, *Polym. Bull.* **1984**, *11*, 151.
21. D. Wöhrle, H. Bohlen, J.K. Blum, *Makromol. Chem.* **1986**, *187*, 2081.
22. B.B. De, B.B. Lohray, S. Sivaram, P.K. Dhal, *Tetrahedron Asymm.* **1995**, *6*, 2105; *J. Polym. Sci.: Polym. Chem.* **1997**, *35*, 1809.
23. F. Minutolo, D. Pini, A. Petri, P. Salvadori, *Tetrahedron Asymm.* **1996**, *7*, 2293.
24. P.K. Dhal, B.B. De, S. Sivaram, *J. Mol. Catal. A: Chem.* **2001**, *177*, 71.
25. K.A.K. Ebraheem, S.T. Hamdi, *React. Funct. Polym.* **1997**, *34*, 5.
26. G.N. Rao, S. Srivastava, S.K. Srivastava, M. Singh, *Talanta* **1996**, *43*, 1821.
27. K. Hanabusa, Y. Tanimura, T. Suzuki, T. Koyama, H. Shirai, *Makromol. Chem.* **1991**, *192*, 233; ibid. **1992**, *193*, 2149.
28. K. Naka, A. Azuma, Y. Chujo, *Polym. Bull.* **1998**, *40*, 701.
29. L. Oriol, F.J. Alonso, J.I. Martinez, M. Pinol, J.L. Serrano, *Macromolecules* **1994**, *27*, 1869.
30. P.J. Alonso, J.I. Martinez, L. Oriol, M. Pinol, J.L. Serrano, *J. Macromol. Sci. Chem.* **1990**, *A27*, 1379.
31. A. El–Shekeil, H. Al–Maydama, A. Al–Karbooly, *J. Inorg. Organomet. Polym.* **1998**, *8*, 167.
32. M.M.Patel, H.B. Pancholi, *J. Inorg. Organomet. Polym.* **1996**, *6*, 51.
33. A.V. Ovchinnikov, V.V. Zuev, *Eur. Polym. J.* **1998**, *34*, 1549.
34. T. M. Swager, M.J. Marsella, *Adv. Mater.* **1994**, *6*, 595; *J. Am. Chem. Soc.* **1993**, *115*, 12214.
35. K. Kalyanasundaram, *Coord. Chem. Rev.* **1982**, *46*, 159.
36. A. Juris, F. Balzani, *Coord. Chem. Rev.* **1988**, *84*, 85.
37. J.-P. Sauvage, J.-P. Collin, *Chem. Rev.* **1994**, *94*, 993.
38. Q. Wang, L. Yu, *J. Am. Chem. Soc.* **2000**, *122*, 11806.
39. S. Yu, S. Hou, W.K. Chan, *Macromolecules* **2000**, *33*, 3259.
40. K.D. Ley, K.S. Schanze, *Coord. Chem. Rev.* **1998**, *171*, 287.
41. O.N. Petzold, I.I. Harruna, *Inorg. Organomet. Polym.* **2000**, *10*, 231, 249.
42. K. Hanabusa, J. Higashi, T. Kojama, H. Shirai, N. Hojo, A. Kurose, *Makromol. Chem.* **1989**, *190*, 1.
43. A. Majumdar, M. Biswas, *Polym. Bull.* **1991**, *26*, 145.
44. A.P. Smith, C.L. Fraser, *Macromolecules* **2002**, *35*, 594.
45. R. Liu, W.-L. Yu, H. Pei, S.-Y. Liu, Y.-H.Lai, W. Huang, *Macromolecules* **2001**, *34*, 7932.
46. K.D. Ley, K.S. Schanze, *Coord. Chem. Rev.* **1998**, *171*, 287.
47. K. Naka, T. Uemura, Y. Chujo, *J. Polym. Sci. A: Chem.* **2001**, *39*, 4083.
48. S. Zhu, T.M. Swager, *Adv. Mater.* **1996**, *8*, 497.
49. T. Pautzsch, E. Klemm, *Macromolecules*, **in press**.
50. U. Velten, M. Rehahn, *Macromol. Chem. Phys.* **1998**, *199*, 127.
51. Y. Sawada, I. Tomita, I.L. Rozhanskii, T. Endo, *J. Inorg. Organomet. Polym.* **2000**, *10*, 221.
52. C.R. de Denus, L.M. Hoffa, E.K. Todd, A.S. Abd-El-Aziz, *J. Inorg. Organomet. Polym.* **2000**, *10*, 189.
53. Y.-P. Wang, Y.-H. Lei, R.-M. Wang, *Indian J. Chem.* **1998**, *36b*, 1054.

54. D. Wöhrle, *Adv. Polym. Sci.* **1983**, *50*, 45.
55. D. Wöhrle, in: *Phthalocyanines – Properties and Applications*, C.C. Leznoff, A.B.P. Lever (Eds.), VCH, New York, **1989**, 55ff.
56. D. Wöhrle, Macromol. Rapid Commun. **2001**, 22, 68.
57. A.A. Berlin, A.I. Sherle, *Inorg. Macromol. Rev.* **1971**, *1*, 235.
58. D. Wöhrle, U. Marose, R. Knoop, *Makromol. Chem.* **1985**, *186*, 2209.
59. D. Wöhrle, B. Schulte, *Makromol. Chem.* **1985**, *186*, 2229.
60. C.J. Norrel, H.A. Pohl, M. Thomas, K.D. Berlin, *J. Polym. Sci., Polym. Phys. Ed.* **1974**, *12*, 913.
61. T. Hara, Y. Ohkatsu, T. Osa, *Bull. Chem. Soc. Jpn.* **1975**, *48*, 85.
62. R. Bannehr, G. Meyer, D. Wöhrle, *Polym. Bull.* **1980**, *2*, 841.
63. D.R. Boston, J.C. Bailar, *Inorg. Chem.* **1972**, *11*, 1578.
64. J.W.-P. Lin, L.P. Dudek, *J. Polym. Sci., Polym. Chem. Ed.* **1985**, *23*, 1579, 1589.
65. A.I. Sherle, V.V. Promyslova, N.I. Shapiro, V.P. Epshtein, A.A. Berlin, *Vysokomol. Soedin Ser. A* **1980**, *22*, 1258. M.Ya. Kushnerev, A.I. Sherle, *Vysokomol. Soedin Ser. A* **1981**, *23*, 1187.
66. J. Bellido, J. Cardoso, T. Akachi, *Makromol. Chem.* **1981**, *182*, 713.
67. D. Wöhrle, B. Schulte, *Makromol. Chem.* **1988**, *189*, 1167.
68. D. Wöhrle, B. Schulte, *Makromol. Chem.* **1988**, *189*, 1229.
69. A.W. Snow, J.R. Griffith, N.P. Marullo, *Macromolecules* **1984**, *17*, 1614.
70. R.Y. Ting, T.M. Keller, N.P. Marullo, P. Peyser, T.R. Price, C.F. Poranski Jr., *Polym. Prepr. (Am. Chem. Soc., Div. Polym. Chem.)* **1981**, *22(1)*, 50.
71. N.P. Marullo, A.W. Snow, *ACS Symp. Ser.* **1982**, *195*, 325.
72. J.A. Hinkley, *J. Appl. Polym. Sci.* **1994**, *29*, 3339; *ACS Symp. Ser.* **1985**, *282*, 43.
73. T.M. Keller, J. Polym. Sci., Polym. Lett. Ed. **1986**, 24, 211.
74. T.M. Keller, J. Polym. Sci., Polym. Lett. Ed. **1987**, 25, 2569.
75. G. Manecke, D. Wöhrle, *Makromol. Chem.* **1967**, *116*, 36; ibid. **1968**, *120*, 176.
76. R. Bannehr, G. Meyer, D. Wöhrle, N.I. Jaeger, *Makromol. Chem.* **1981**, *182*, 2633.
77. A.G. Gürek, Ö. Bekaroglu, *J. Porphyrins Phthalocyanines* **1997**, *1*, 67; ibid **1997**, *1*, 227.
78. V. Ahsen, E. Yilmazer, Ö. Bekaroglu, *Makromol. Chem.* **1988**, *189*, 2533.
79. E. Musluoglu, Z.Z. Özturk, V. Ahsen, A. Gül, Ö. Bekaroglu, *J. Chem. Res. (S)* **1993**, 6.
80. C.S. Marvel, J.H. Rassweiler, *J. Am. Chem. Soc.* **1958**, *80*, 1197.
81. A.S. Akopov, T.N. Lomova, B.D. Berezin, *Izv. Vyssh. Uchebn. Zaved., Khim. Khim. Tekhnol.* **1976**, *19*, 1177; ibid, **1978**, *21*, 663.
82. D. Wöhrle, E. Preußner, *Makromol. Chem.* **1985**, *186*, 2189.
83. W.C. Drinkard, J.C. Bailar, *J. Am. Chem. Soc.* **1959**, *81*, 4795.
84. J. Blomquist, L.C. Moberg, L.Y. Johansson, R. Larsson, *J. Inorg. Nucl. Chem.* **1981**, *43*, 2287.
85. L. Kreja, A. Plewka, *Electrochim. Acta* **1980**, *25*, 1283; *Angew. Makromol. Chem.* **1982**, *102*, 45.
86. M. Hanack, D.Y. Meng, A. Beck, *J. Chem. Soc., Chem. Commun.* **1993**, 58.
87. G. Knothe, Makromol. Chem., Theory Simul. **1993**, 2, 503.
88. D. Wöhrle, B. Schumann, V. Schmidt, N.I. Jaeger, *Makromol. Chem., Macromol. Symp.* **1997**, *8*, 195.

89. D. Wöhrle, V. Schmidt, B. Schumann, A. Yamada, K. Shigehara, *Ber. Bunsenges. Phys. Chem.* **1987**, *91*, 975.
90. M. Yudaska, K. Nakanishi, T. Hara, M. Tanaka, S. Kurita, *Synth. Met.* **1987**, *19*, 775.
91. M. Ashida, Y. Ueda, H. Yanagi, K. Sayo, *J. Polym. Sci., Polym. Chem. Ed.* **1989**, *27*, 883.
92. H. Yanagi, K. Tsukatani, H. Yamaguchi, M. Ashida, D. Schlettwein, D. Wöhrle, *J. Electrochem. Soc.* **1993**, *140*, 1942.
93. H. Yanagi, M. Wada, Y. Ueda, M. Ashida, D. Wöhrle, *Makromol. Chem.* **1992**, *193*, 1903.
94. D. Wöhrle, R. Bannehr, B. Schumann, G. Meyer, N.I. Jaeger, *J. Mol. Catal.* **1983**, *21*, 255.
95. D. Wöhrle, T. Buck, U. Hündorf, G. Schulz-Ekloff, A. Andreev, *Makromol. Chem.* **1989**, *190*, 961.
96. D. Wöhrle, O.N. Suvorova, N. Trombach, E.A. Schupack, R. Gerdes, N.L. Baziakina, O. Bartels, A.P. Matveev, G. Schnurpfeil, *J. Porphyrins Phthalocyanines* **2001**, *5*, 381.
97. R. Benters, *Master Thesis*, Universität Bremen 1999.
98. G. Knothe, D. Wöhrle, *Makromol. Chem.* **1989**, *190*, 1573.
99. G. Knothe, Makromol. Chem., Theory Simul. **1992**, 1, 187; J. Inorg. Organomet. Polym. **1994**, 4, 325.
100. D. Djurado, S. Tadlaoui, A. Hamwi, J.C. Cousseins, *Synth. Met.* **1991**, *41-43*, 2595.
101. M.S. Liao, K.T. Kuo, *Polymer J.* **1993**, *25*, 947.
102. J.-H. Chou, H.S. Nalwa, M.E. Kosal, N.A. Rakow, K.S. Suslick, *The Porphyrin Handbook,* K.M. Kadish, K.M. Smith, R. Guilard (Eds.), Academic Press, San Diego, 2000, Vol. 6, pp. 43.
103. J.-C. Chambron, V. Heitz, J.-P. Sauvage, in: *The Porphyrin Handbook,* K.M. Kadish, K.M. Smith, R. Guilard (Eds.), Academic Press, San Diego, 2000, Vol. 6, p. 1.
104. N. Aratani, A. Osuka, Y.H. Kim, D.H. Jeong, D. Kim, *Angew. Chem. Int. Ed. Engl.* **2000**, *39*, 1458.
105. R.G. Khoury, L. Jaquinod, K.M. Smith, *Chem. Commun.* **1997**, 1057.
106. N. Yoshida, N. Aratani, A. Osuka, *Chem. Commun.* **2000**, 197.
107. T. Okawa, Y. Nishimoto, N. Yoshida, N. Ono, A. Osuka, *Chem. Commun.* **1998**, 337.
108. Y.H. Kim, D.H. Jeong, D. Kim, S.C. Jeoung, H.S. Cho, S.K. Kim, N. Aratani, A. Osuka, *J. Am. Chem. Soc.* **2001**, *123*, 76.
109. A. Nakano, T. Yamazuki, Y. Nishimura, I. Yamazaki, A. Osuka, *Chem. Eur. J.* **2000**, *6*, 3254.
110. T.E.O. Screen, K.B. Lawton, G.S. Wilson, N. Dolney, R. Ispasaia, H.L. Anderson, *J. Mater. Chem.* **2001**, *11*, 312.
111. T. Yamamoto, N. Fukushima, H. Nakajima, T. Maryama, I. Yamaguchi, *Macromolecules* **2000**, *33*, 5988.
112. B. Jiang, S.-W. Yang, D.C. Barbini, W.E. Jones, *Chem. Commun.* **1998**, 213.
113. P.N. Taylor, H.L. Anderson, *J. Am. Chem. Soc.* **1999**, *121*, 11538.

114. S.M. Kuebler, R.G. Deuning, H.L. Anderson, *J. Am. Chem. Soc.* **2000**, *122*, 339.
115. J.J. Piet, P.N. Taylor, B.R. Wegewijs, H.L. Anderson, A. Osuka, J.M. Warman, *J. Phys. Chem. B* **2001**, *105*, 97.
116. T. Yamamoto, H. Nakajima, N. Hayashida, K. Shiraishi, H. Kokubo, *Polym. Adv. Technol.* **2000**, *11*, 658.
117. B. Jiang, W.E. Jones, *Macromolecules* **1997**, *30*, 5575.
118. Z.N. Bao, Y.M. Chen, L.P. Yu, *Macromolecules* **1993**, *26*, 5281; ibid. **1994**, *27*, 4629.
119. K. Tanaka, N. Kosai, H. Maryama, H. Kobayashi, *Synth. Met.* **1998**, *92*, 253.
120. J.R. Reimers, T.A. Lu, M.J. Crossley, N.S. Hush, *Nanotechnology* **1996**, *7*, 424.
121. A. Tsuda, H. Furata, A. Osuka, *Angew. Chem. Int. Ed. Engl.* **2000**, *39*, 2549.
122. W.Li, C.C. Wamser, *Langmuir* **1995**, *11*, 4061.
123. C.C. Wamser, *J. Am. Chem. Soc.* **1989**, *111*, 8485.
124. C.C. Wamser, *Mol. Cryst. Liq. Cryst.* **1991**, *194*, 65.
125. M. Hanack, P. Stihler, *Eur. J. Org. Chem.* **2000**, *2*, 203.
126. B. Hauschel, R. Jung, M. Hanack, *Eur. J. Inorg. Chem.* **1999**, *1*, 1434.
127. M. Hanack, P. Stihler, *Macromol. Symp.* **1998**, *131*, 49.
128. G.C. Bryant, M.J. Cook, T.G. Ryan, A.J. Thorne, *Tetrahedron* **1996**, *52*, 809.
129. M.J. Cook, *Adv. Mater.* **1995**, *7*, 877.
130. R.P. Kingsborough, T.M. Swager, *Angew. Chem.* **2000**, *112*, 3019.
131. H. Shirai, *J. Porphyrins Phthalocyanines* **1998**, *2*, 31.
132. H. Shirai, K. Hanabusa, M. Kitamura, E. Masuda, N. Hojo, *Makromol. Chem.* **1984**, *185*, 2537.
133. B.N. Achar, G.M. Fohlen, J.A. Parker, J. Keshavaya, *Polyhedron* **1987**, *6*, 1463.
134. B.N. Achar, G.M. Fohlen, J.A. Parker, *J. Polym. Sci., Polym. Chem. Ed.* **1982**, *20*, 2773, 2781.
135. B.N. Achar, G.M. Fohlen, J.A. Parker, *J. Polym. Sci., Polym. Chem. Ed.* **1983**, *21*, 1025, 3063; **1984**, *22*, 319; **1985**, *23*, 801.
136. F. Bedioui, J. Devynck, C. Bied-Charreton, *Acc. Chem. Res.* **1995**, *28*, 30.
137. A. Bettelheim, B.A. White, S.A. Raybuck, R.W. Murray, *Inorg. Chem.* **1987**, *26*, 1009.
138. F. Bedioui, J. Devynck, C. Bied-Charreton, *J. Mol. Catal. A: Chem.* **1996**, *113*, 3.
139. B. Ballarin, R. Seeber, L. Tassi,; D. Tonelli, *Synth. Met.* **2000**, *114*, 279.
140. H. Maruyama, H. Segawa, S. Sotoda, T. Sata, N. Kosai, S. Sagisaka, T. Shimidzu, K. Tanaka, *Synth. Met.* **1998**, *96*, 141.
141. K.A. Pressprich, S.G. Maybary, R.E. Thomas, R.W. Linton, E.A. Irene, R.W. Murray, *J. Phys. Chem.* **1989**, *93*, 5568.
142. J. Hayon, A. Raveh, A. Bettelheim, *J. Electroanal. Chem.* **1993**, *359*, 209.
143. R. Volf, T.V. Shishkanova, P. Matejka, M. Hamplova, V. Kral, *Anal. Chim. Acta* **1999**, *381*, 197.
144. S.-S. Huang, H. Tang, B.-F. Li, *Microchim. Acta* **1998**, *128*, 37.
145. T.L. Blair, J.R. Allen, S. Daunert, L.G. Bachas, *Anal. Chem.* **1993**, *65*, 2155.
146. T.J. Savenije, R.B.M. Koehorst, T.J. Schaafsma, *J. Phys. Chem. B.* **1997**, *101*, 720.
147. T. Malinski, A. Ciszweski, J. Bennet, J.R. Fish, *J. Electrochem. Soc.* **1991**, *138*, 2008.
148. S. Cosnier, A. Walter, F.-P. Montforts, *J. Porphyrins Phthalocyanines* **1998**, *2*, 39.

149. S. Cosnier, C. Gondran, R. Wessel, F.-P. Montforts, M. Wedel, *J. Electroanal. Chem.* **2000**, *488*, 83.
150. M. Wedel, A. Walter,; F.-P. Montforts, *Eur. J. Org. Chem.* **2001**, *9*, 1681.
151. V.C. Dall'Orto, C. Danilowicz, S. Sobral, A. LoBalbo, I. Rezzano, *Anal. Chim. Acta* **1996**, *336*, 195.
152. V.C. Dall'Orto, C. Danilowicz, J. Hurst, A. LoBalbo, I. Rezzano, *Electroanalysis* **1998**, *10*, 127.
153. R. Paolesse, C.D. Natale, V.C. Dall'Orto, I. Rezzano, M. Mascini, A. D'Amico, *Thin Solid Films* **1999**, *354*, 245.
154. O.E. Mouahid, A. Rokotondrainibe, P. Crouigneau, J.M. Leger, C. Lamy, *J. Electroanal. Chem.* **1998**, *455*, 209: **1997**, *426*, 117.
155. F. Vilchez-Aguado, S. Guitirrez-Granados, *New J. Chem.* **1997**, *21*, 1009.
156. T.W. Hanks, B. Bergman, P. Dillon, *Synth. Met.* **2001**, *121*, 1431.
157. T.F. Guarr, in: *Handbook of Organic Conducting Molecules and Polymers*; H.S. Nalwa, (Ed.) John Wiley & Sons: Chichester, 1997, Vol. 2, p. 461.
158. H. Li, T.F. Guarr, *J. Chem. Soc. Chem. Commun.* **1989**, 832.
159. E.M. Baum, H. Li, T.F. Guarr, J.D. Robertson, *Nucl. Inst. Meth.* **1991**, *356/57*, 761.
160. H. Li, T.F. Guarr, *Synth. Met.* **1990**, *38*, 243.
161. F. Xu, H. Li, S.J. Cross, T.F. Guarr, *J. Electroanal. Chem.* **1994**, *368*, 221.
162. D.J. Moore, T.F. Guarr, *J. Electroanal. Chem.* **1991**, *314*, 213.
163. F. Xu, H. Li, Q. Peng, T.F. Guarr, *Synth. Met.* **1993**, *55–57*, 1668.
164. K.L. Brown, H.A. Mottola, *Langmuir*, **1998**, *14*, 3411.
165. M. Kimura, T. Horai, K. Hanabusa, H. Shirai, *Chem. Lett.* **1997**, *7*, 653.
166. N. Trombach, O. Hild, D. Schlettwein, D. Wöhrle, *J. Mater. Chem.*, **2002**, *12*, 879.
167. H.F. Li, T.F. Guarr, *J. Electroanal. Chem.* **1991**, *297*, 169.
168. H.F. Li, T.F. Guarr, *J. Electroanal. Chem.* **1991**, *317*, 189.
169. Q. Peng, T.F. Guarr, *Electrochim. Acta* **1994**, *39*, 2629.
170. E. Trollund, P. Ardiles, M.J. Aguirra, S.R. Biaggio, R.C. Rocha-Filho, *Polyhedron*, **2000**, *19*, 2303.
171. P. Ardiles, E. Trollund, M. Isaacs, F. Armijo, J.C. Canales, M.J. Aguirre, M.J. Canales, *J. Mol. Catal. A: Chem.* **2001**, *165*, 169.
172. K. Qi, R.P. Baldwin, H.F. Li, T.F. Guarr, *Electroanalysis* **1991**, *3*, 119.
173. P. Ardiles, E. Trollund, M. Isaacs, F. Armijo, M.J. Aguirre, *J. Coord. Chem.* in press.
174. Y.-H. Tse, P. Janda, A.B.P. Lever, *Anal. Chem.* **1995**, *67*, 981.
175. G. Ramirez, E. Trollund, M. Isaacs, F. Armijo, J. Zagal, J. Costamagna, M.J. Aguirre, *Electroanalysis* in press.
176. Z. Sun, H. Tachikawa, *Anal. Chem.* **1992**, *64*, 1112.
177. M.A. Ruiz, M.G. Blasquez, J.M. Pingarron, *Anal. Chim. Acta* **1995**, *305*, 49.
178. T.-F. Kang, Z.-Y. Xie, H. Tang, G.-L. Shen, R.-Q. Yu, *Talanta*, **1997**, *45*, 291.
179. A. Deronzier, J.-C. Moutet, *Acc. Chem. Res.* **1989**, *22*, 249.
180. K.A. Goldsby, J.K. Blaho, L.A. Hoferkamp, *Polyhedron*, **1989**, *8*, 113.
181. L.A. Hoferkamp, K.A. Goldsby, *Chem. Mater.* **1989**, *1*, 348.
182. J.L. Reddinger, J.R. Reynolds, *Macromolecules* **1997**, *30*, 673; *Synthetic Metals*, **1997**, *84*, 225.
183. C.D. Ellis, L.D. Margerum, R.W. Murray, T.J. Meyer, *Inorg. Chem.* **1983**, *22*,

1283.

184. I.M.F. De Oliviera, J.-C. Moutet, *J. Electroanal. Chem.* **1993**, *361*, 203.
185. S. Chardon–Noblat, I.M.F. De Oliviera, J.-C. Moutet, S. Tingry, *J. Mol. Catal.: Chem.* **1995**, *99*, 13.
186. J.-C. Moutet, C. Duboc-Toia, S. Menage, S. Tingry, *Adv. Mater.* **1998**, *10*, 665.
187. S.S. Zhu, T.M. Swager, *J. Am. Chem. Soc.* **1997**, *119*, 12568.
188. J. Buey, T.M. Swager, *Angew. Chem.* **2000**, *112*, 622.
189. P.L. Vidal, M. Billon, B. Divisia-Blohorn, G. Bidan, J.M. Kern, J.P. Sauvage, *Chem. Commun.* **1998**, *629*.
190. G.R. Newcombe, C.N. Moorefield, F. Vögtle, *Dendric Molecules: Concepts, Syntheses and Perspectives*, VCH, Weinheim, 1996.
191. A.J. Berresheim, M. Müller, K. Müllen, *Chem. Rev.* **1999**, *99*, 1747; A. Archut, F. Vögtle, *Chem. Soc. Rev.* **1999**, *27*, 237; M. Fischer, F. Vögtle, *Angw. Chem.* **1999**, *11*, 934.
192. C. Gorman, *Adv. Mater.* **1998**, *10*, 295.
193. S. Hecht, J.M.J. Frechet, *Angew. Chem.* **2001**, *113*, 77.
194. K.W. Pollack, F.M. Sanford, J.M.C. Frechet, *J. Mater. Chem.* **1998**, *8*, 519.
195. N. Tomioka, D. Takusa, T. Takahashi, T. Aida, *Angew. Chem.* **1998**, *110*, 1611.
196. Y. Tomoyose, D.-L. Jiang, R.-H. Jiu, T. Aida, *Macromolecules* **1996**, *29*, 5236.
197. M.-S. Choi, T. Aida, T. Yamazaki, I. Yamazaki, *Angew. Chem.* **2001**, *113*, 3294.
198. D.K. Smith, F. Diederich, *Chem. Eur. J.* **1998**, *4*, 1353.
199. P.J. Danliker, F. Diederich, A. Zingg, J.-P. Gisselbrecht, *Helv. Chim. Acta* **1997**, *80*, 1773.
200. P. Weyermann, J.-P. Gisselbrecht, C. Boudon, F. Diederich, *Angew. Chem.* **1999**, *111*, 3400.
201. K.W. Pollak, J.W. Leon, J.M.J. Frecht, *J. Chem. Mater.* **1998**, *10*, 30.
202. P. Bhyrappa, J.K. Young, J.S. Morre, K.S. Suslick, *J. Am. Chem. Soc.* **1996**, *118*, 5708; *J. Mol. Catal. A* **1996**, *113*, 109.
203. D.-L. Jiang, T. Aida, *Chem. Commun.* **1996**, 1523.
204. D.-L. Jiang, T. Aida, *J. Am. Chem. Soc.* **1998**, *120*, 10895.
205. M. Kimura, K. Nakada, Y. Yamaguchi, K. Hanabusa, H. Shirai, N. Kobayashi, *Chem. Commun.* **1997**, 1215.
206. M. Kimura, Y. Sugihara, T. Muto, K. Hanabusa, H. Shirai, N. Kobayashi, *Chem. Eur. J.* **1999**, *5*, 3495.
207. N.B. McKeown, *Adv. Mater.* **1999**, *11*, 67.
208. M. Brewis, G.J. Clarkson, A.M. Holder, N.B. McKeown, *Chem. Commun.* **1998**, 969.
209. G.J. Clarkson, N.B. McKeown, K.E. Treacher, *J. Chem. Soc., Perkin Trans. 1* **1995**, 1817.
210. T. Nagasaki, O. Kimura, M. Ukon, S. Arimori, *J. Chem. Soc., Perkin Trans. 1* **1994**, 75.
211. C. Galliot, D. Prevote, A.-M. Caminade, J.-P. Majoral, *J. Am. Chem. Soc.* **1995**, *117*, 5470.
212. V. Percec, G. Johansson, G. Unga, J. Zhou, *J. Am. Chem. Soc.* **1996**, *118*, 9855.
213. M. Enomoto, T. Aida, *J. Am. Chem. Soc.* **1999**, *121*, 874.
214. R. Sadamoto, N. Tomioka, T. Aida, *J. Am. Chem. Soc.* **1996**, *118*, 3978; R.H. Jin, T.

Aida, S. Inoue, *Chem. Commun.* **1993**, 1261.
215. K.M. Kadish, K.M. Smith, R. Guilard (Eds.), *The Porphyrin Handbook*, Academic Press, Orlando, FL, 2000, Vol. 6.
216. I. Goldberg, *Chem. Eur. J.* **2000**, *6*, 3863.
217. M.J. Zaworotko, *Nature* **1997**, *386*, 220.
218. H.J. Choi, T.S. Lee, M.P. Suh, *Angew. Chem.* **1999**, *111*, 1490.
219. R. Krishna Kumar, S. Balasubramanian, I. Goldberg, *Chem., Commun.* **1998**, 1435.
220. Y. Diskin-Posner, R. Krishna Kumar, I. Goldberg, *New J. Chem.* **1999**, *23*, 885.
221. Y. Diskin-Posner, I. Goldberg, *Chem. Commun.* **1999**, 1961.
222. S. Dahal, I. Goldberg, *J. Phys. Org. Chem.* **2000**, *13*, 382.
223. N. Nagato, S. Kugimiya, Y. Kobuke, *Chem. Commun.* **2000**, 1389.
224. A. Epstein, B.S. Wildi, *J. Phys. Chem.* **1960**, *32*, 324.
225. O. Hild, *Masters Thesis*, Universität Bremen, 1999; *Doctoral Thesis*, Universität Bremen, 2003.

7 Metals or Metal Complexes as Part of Linear or Crosslinked Macromolecules via the Metal

Dieter Wöhrle

A great variety of possibilities are available when a metal is part of a linear or crosslinked macromolecule (Section 1.2.1, Fig. 7-1). One possibility is the covalent incorporation of a metal into a homochain (metal–metal connections) or a heterochain (metal–heteroatom connections). Coordinative bonds between a metal ion and another coordinating donor group can lead to linear chains or to supramolecular organizations when coordination polymers are formed. π-Bonds between π-rich aromatic compounds and metal ions can result in polymeric metallocenes. Covalent and coordinative bonds are realized in cofacially stacked metal complexes. In addition, dendrimers containing the interaction of metals and another group in different bonds are treated in this chapter.

Often these polymers are insoluble, although in several cases this can be overcome by adding bulky substituents. Several coordination polymers are stable only in the solid crystalline state. Properties so far described are electrical conductivity, photoactivity, non-linear optical behavior, liquid crystallinity and the formation of porous networks.

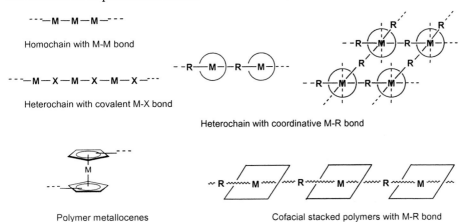

Figure 7-1. Schematic representations of metals and metal complexes as part of a macromolecule via the metal.

7.1   Homochain Polymers with Covalent Metal–Metal Bonds

Examples of polymers of the type $-(M)_n-$ of structure **1** containing a bond in the main chain are known with semimetals or metals such as B, Si, Ge, Sn, As, Sb, Th and Po (for polysilanes see Section 1.5) [1,2]. Polygermane homopolymers of the composition $-[(C_6H_{13})_2Ge]_n$ (molecular weight $M_n \sim 10^4$ Da) and copolymers of the composition $\{[(C_6H_{13})_2Ge]_n[(C_6H_{13})_2Si]_n\}$ ($M_n \sim 10^5$) have been prepared by condensation of the corresponding dichlorogermanes and dichlorosilanes with sodium metal in toluene [3]. Thin films of the polymers absorbing at $\lambda \sim 320$ nm were prepared by spin-coating from isooctane solution. They are thermally stable and resistant against oxidation, and they show large third-order non-linear optical susceptibilities. The formation of extended two-dimensional structures with three-fold backbone binding between sp^3-hybridized atoms has been reported for silicon in the form of polysilyne $-(SiH)_n-$ [4] based on a topochemical transformation of Zintlphase $CaSi_2$ [5] which already exhibits the two-dimensional structure of the Si-backbone product. Recently, a crystalline layered polygermyne $-(GeH)_n-$ **1** (part of the unit cell is shown) has been prepared by treating Zintlphase $CaGe_2$ on Ge substrates with aqueous HCl (Eq. 7-1) [6]. This polymer forms stacked hydrogen-terminated Ge layers with a layer distance of 0.565 nm (polysilyne 0.51 nm). The yellow-colored polygermyne shows fluorescence at $\lambda = 920$ nm.

$$nCaGe_2 + 2n\ HCl \rightarrow (GeH)_{2n} + nCaCl_2 \qquad (7-1)$$

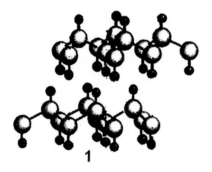

1

One-dimensional metals of d^8-complexes with a quadric-planar surrounding are good electrical conductors (For a very recent review on polymers with transition metal chains see: J.K. Bera, K.R. Dunbar, *Angew. Chem.* **2002**, *114*, 4633). Partial doping of tetracyanoplatinate (Kroogmann salt) with columnar stacks results in metallic-like conductivity [7]. A linear chain polymer containing Rh–Rh bonds has been obtained by galvanostatic reduction of

$[Rh_2(CH_3CN)_{10}](BF_4)_4$. The polymer $\{[Rh(CH_3CN)_4](BF_4)_{1.5}\}_n$ exhibits Rh–Rh distances of 0.28442 and 0.29277 nm with Rh in the oxidation state 1.6 [8]. Calculations on the band structure were reported.

The combination of d^{10} metal centers such as Au(I) with Tl(I) has resulted in stable adducts by metal–metal interactions. Examples are polymeric chain compounds $[AuTl(CH_2PS(C_6H_5)_2]^-{}_n$ **2** [9] and $[TlOP(C_6H_5)_2 \cdot Au(C_6F_5)_2]_n$ **3** [10]. The anionic ruthenium cluster $[N(PPh_3)_2]_2[Ru_6C(CO)_{16}]$ has been reacted with AgNO$_3$ in THF and an acetone-soluble red-colored oligomer $[N(PPh_3)_2]$-$[AgRu_6C(CO)_{16}]$ isolated [11]. The Ru clusters are connected via Ag ions, and the linear chains are separated by the bulky $[N(PPh_3)_2]^+$.

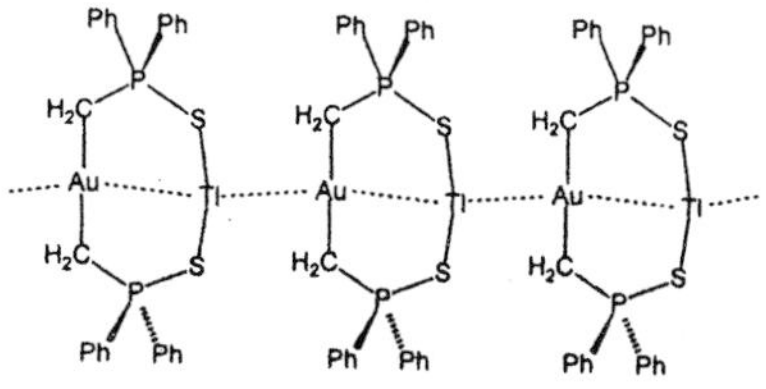

7.2 Heterochain Polymers with Covalent Bonds between Metals and Another Element

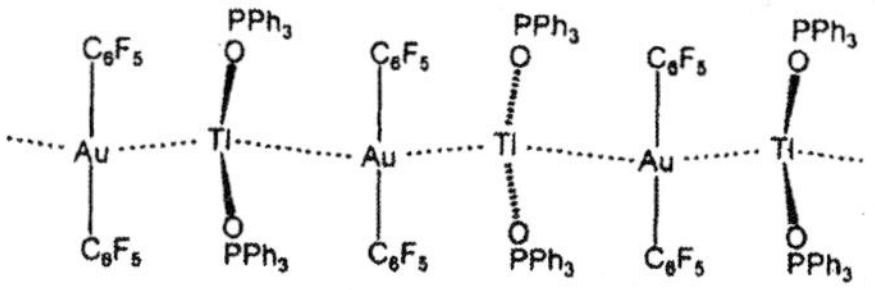

Heterochain polymers of the type $-(M-X)_n-$ contain polar M–X bonds (for reviews see [1,2,12–15]). Such polymers are often prepared by polycondensation of a bifunctional metal halide (M = B, Si, Ge, Sn, Pb, Sb, Ni, Pd, Pt, Ti, Hf) with a bifunctional Lewis base such as a diol, diamine, dihydrazine, dihydrazide, dioxime, diamideoxime, dithiol, diacetylene (Eq. 7-2). Another possibilitiy is the polyaddition of a bifunctional metal hydride to bifunctional alkenes (Eq. 7-3). Mn and mg containing poly(p–xylylenes) of the composition $-(-CH_2-C_6H_4-CH_2-M-)_n-$ were produced by solid–state UV–photopolymerization of a cocondensated mixture of p–xylylene with Mn or mg at 80 K [16]. Prolonged storage of the polymers at ambient temperature under vacuum led to gradual decomposition.

$$n\text{hal}M(R)_x\text{hal} + n\text{H–L–R'–L–H} \rightarrow -[-M(R)_x-L-R'-L]_n- + 2n\text{ HCl} \qquad (7\text{-}2)$$

$$n\text{HM}(R)_x\text{H} + n\text{H}_2\text{C=CH–R'–CH=CH}_2 \rightarrow$$
$$-[-M(R)_x-CH_2-CH_2-R'-CH_2-CH_2-]_n- \qquad (7\text{-}3)$$

Several copper-catalyzed synthetic methods have been developed for the synthesis of metal-poly(yne)s: dehydrohalogenation of α,ω-bisethynyl compounds [17,18], oxidative coupling of metal-terminated oligoethynyl compounds [19] and alkynyl–ligand exchange [20]. One example of the synthesis of a film-forming Pt-containing (also Pd, Ni) polymer **4** following the dehydrohalogenation route is shown in Eq. 7-4 (see Experiment 7-1, Section 7.7) [17,21,22]. These polymers exhibit molecular weights of around 10^5 Da. High molecular weight Ni- and Pt-containing polymers **5** have been obtained in good yields via the alkynyl–ligand exchange approach with α,ω-diethynyl compounds as shown in Eq. 7-5 [20]. Well-defined polymers **6** with molecular weights of around 10^5 Da could be prepared via oxidative coupling of *trans*-$PtCl_2(PR_3)_2$ complexes with metal-terminated oligoethynyl compounds (Eq. 7-6) [23]. More recently there have been reports of synthesized polymers **7** with 2,3-diphenylthieno[3,4]pyrazino building blocks [24] and chiral poly(yne)s **8** containing 1,1'-bi-2-naphthol existing in a helical conformation [25]. Other examples are polyarylene cobalt-cyclopentadienylenes prepared by the reaction of diacetylenes with $CpCo[P(C_6H_5)_2]$ [26].

Moreover, the synthesis and electronic structure of rigid-rod octahedral Fe, Ru and Os σ-acetylide complexes have been described [27–29].

$$n \; Cl\text{-}Pt\text{-}Cl \;+\; n \; HC\equiv C\text{-}C\equiv C\text{-}Pt\text{-}C\equiv C\text{-}C\equiv CH \xrightarrow[(H_5C_2)_2NH]{CuX} \left[Pt\text{-}C\equiv C\text{-}C\equiv C\right]_{2n} \;+\; 4n \; HCl \qquad (7\text{-}4)$$

with $P(C_4H_9)_3$ phosphine ligands; polymer **4**

$$nHC\equiv C\text{-}Y\text{-}C\equiv CH \;+\; n \; HC\equiv C\text{-}Ni\text{-}C\equiv CH \longrightarrow \left[Ni\text{-}C\equiv C\text{-}Y\text{-}C\equiv C\right]_{2n} \;+\; 2n \; HC\equiv CH \qquad (7\text{-}5)$$

polymer **5**

$$n \; (H_3C)_3Sn\text{-}C\equiv C\text{-}\langle\text{aryl}\rangle\text{-}C\equiv C\text{-}Sn(CH_3)_3 \;+\; n \; Cl\text{-}Pt\text{-}Cl \longrightarrow \left[Pt\text{-}C\equiv C\text{-}\langle\text{aryl}\rangle\text{-}C\equiv C\right]_n \;+\; 2n \; ClSn(CH_3)_3 \qquad (7\text{-}6)$$

polymer **6**

Solution properties indicate that the metal-poly(yne)s exist in a rod-like structure. They display lyotropic nematic mesophases [30] and form crystallites with a diameter of up to 50 nm [31]. Absorption and luminescence spectra of the polymers show that π-electron conjugation is expanded over the whole polymer chain, and third-order non-linear optical properties are exhibited. For polymer **7** good photocurrents were found in sandwich-diodes.

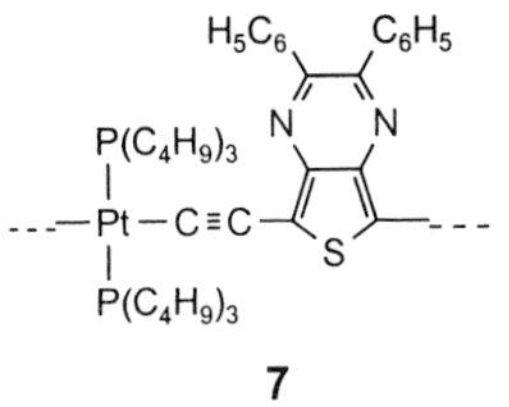

An oligomer **9** was prepared with the aim of obtaining a corresponding polymer with metalla-dinitrogen functionality [32]. The alkylation of [WCl$_4$(dimethoxyethane)] using MgBr(mesitylene) occurs with fast absorption of N$_2$ loading to the green solid W-dimer, which undergoes stepwise reduction to the violet oligomer **9** (Eq. 7-7). This oligomer or a possible corresponding polymer can be seen as chains with perpendicular π-systems made of alternating W and two N atoms and carrying two electrons for each N atom as well as the d-electrons of tungsten:

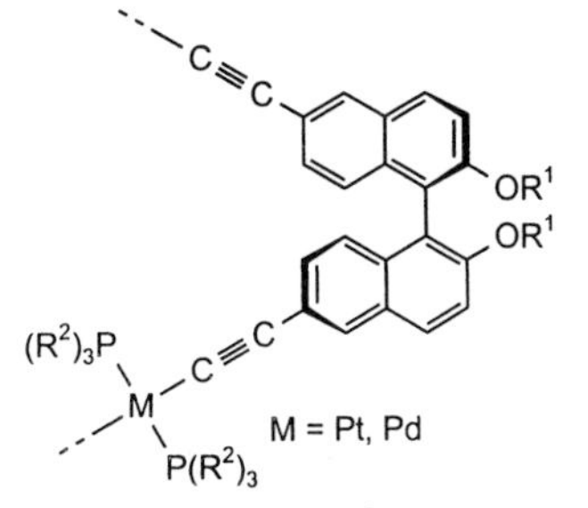

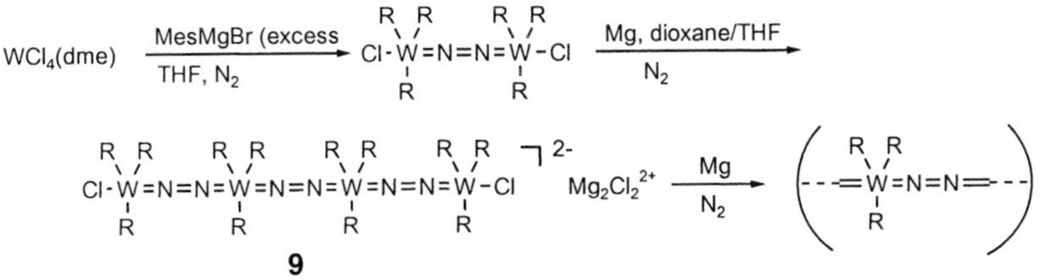

$$(7\text{-}7)$$

One-dimensional chains of heterometallic polymers with covalent metal–metal bonds are interesting as molecular conductors (see references cited in [33]). A complex system of four different elements surrounded by insulating organic materials is described in [33]. An alloy K$_6$Ag$_2$Sn$_2$Te$_9$ was treated with 1,2-diaminoethane, and a saturated aqueous solution of tetraethylammonium iodide was added to the resulting solution. One-dimensional chains of composition (Et$_4$N)$_4$[Au(Ag$_{1-x}$)Au$_x$Sn$_2$Te$_9$] (x = 0.32) **10** were obtained. Band structures are discussed and a band gap of 0.45 eV was found.

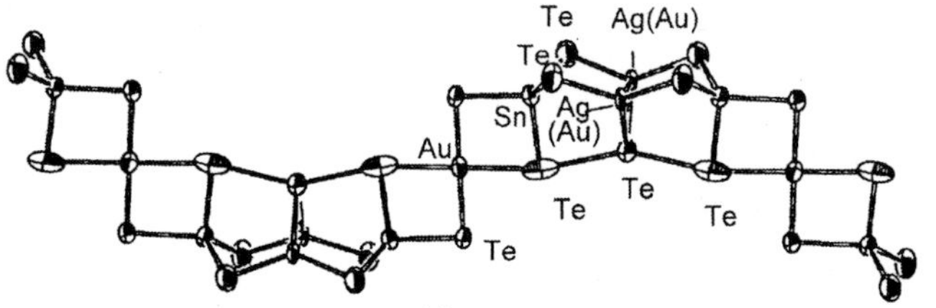

10

A golden-coloured polymeric organometallic oxide of formula $\{H_{0.5}[(CH_3)_{0.92}ReO_3]\}\infty$ **11** has been prepared by heating methylthioxorhenium in water [34]. The structure was described in terms of double layers with corner-sharing CH_3ReO_5 octahedra (A,A′) containing intercalated water molecules (B) in a AA′BB′... layer sequence. Suggestions of other inorganic macromolecules are given: LiB_x (linear, unbranched borynide chains in a lithium matrix) [35], two-dimensional layers of self-organized 1,2-bis(chloromercurio)-tetrafluorobenzene [36], framework-structured Sb(III)-phosphate [37], graphite-structured $[(Me_3Sn)_3O]Cl$ [38], three-dimensional structured $RbCuSb_2SI_4$ [39], layer-structured $[Cu_4(OH)_4][Re_4(Te)_4(CN)_{12}]$ [40], linear polymeric Mo/Ag/S-complexes [41a] (all these references contain additional literature to comparable macromolecules). The reaction of P_2-ligand complexes of the composition $[Cp_2Mo_2(CO)_4-P_2]$ (Cp = cyclopentadienyl, P = dipenylphosphanylmethyl) with $AgNO_3$ or CuBr leads to one-dimensional red-colored crystalline polymers bridged by Ag(I) or $Cu(I)_2Br_2$, respectively [41b].

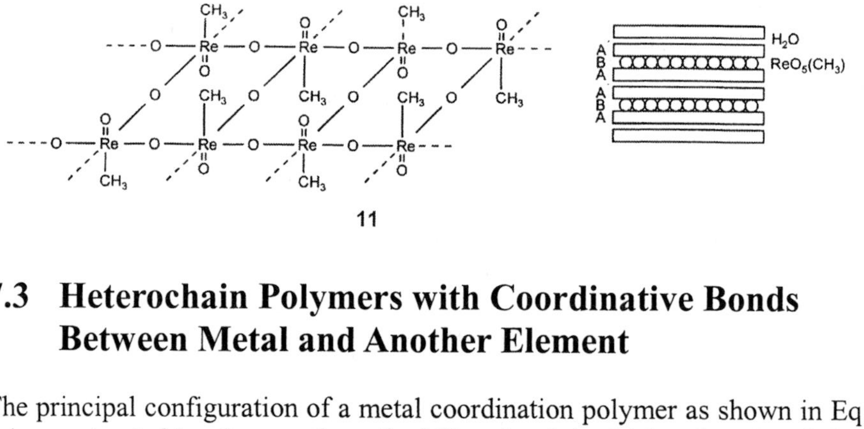

11

7.3 Heterochain Polymers with Coordinative Bonds Between Metal and Another Element

The principal configuration of a metal coordination polymer as shown in Eq. 7-8 is constructed by the reaction of a bifunctional (or higher-functional) ligand with a metal ion having different coordination geometries. Besides chain-forming coordination polymers special attention is given in this section to the supramolecular organization of these polymers.

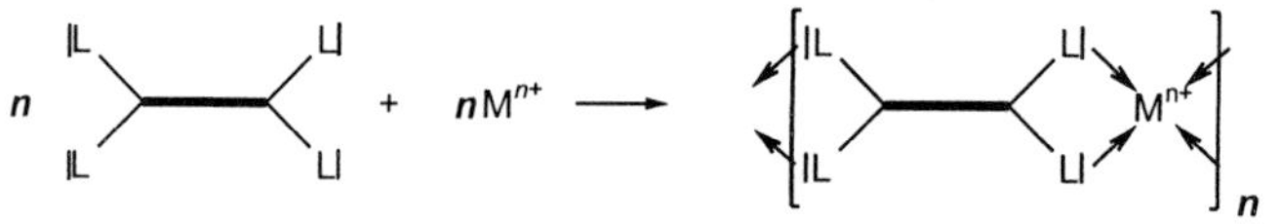

$$(7-8)$$

7.3.1 Chain-Forming Coordination Polymers

Several examples of organic coordination polymers described some decades ago are reviewed in [1,43,44]. Chelating groups, for example, are =O, =S, =NR, $-NR_2$, $-O^-$, $-S^-$, $-NR^-$. Most of these polymers are insoluble, and it is difficult to characterize them analytically in detail. Many of them exhibit high electrical conductivities as compressed powders. Some of the examples reviewed in the above-mentioned literature are:

- Reaction of dicarboxylic acids with $SnCl_2$ or uranyl salts [44,45].
- Polymers from transition metal ions with aromatic bis(*o*-hydroxy acids) [46], dihydroxyquinones [47–49], dihydroxyquinoxalines or -quinolines [50–53].
- Coordination polymers of transition metal ions with different tetrathiolates such as tetrathiooxalate [54], tetrathiosquarate [55], tetrathiofulvalene tetrathiolate [56,57], benzene- or naphthalenetetrathiolates [58,59]. Such polymers exhibit electrical conductivities up to $30\ S\ cm^{-1}$ (for the investigation of the electronic structure see [60]).

More recently, the detailed structures of some coordination polymers have been successfully investigated by careful synthesis. Solubility problems can be overcome by using small tetrahedral centers (Be, B, C, etc.) [61]. One example of a readily soluble polymer **12** is obtained by ring-opening polymerization of $(CH_2)_x$-linked beryllium β-diketones (Eq. 7-9) [62]. Strong solvent interactions with metal-ion centers [61] also enhance the solubility. One example is $[UO_2(OOC–CH=CH–COO)(DMSO)_2]$ **13** with M_n of 10^4 Da (R.D. Archer et al. in [14]).

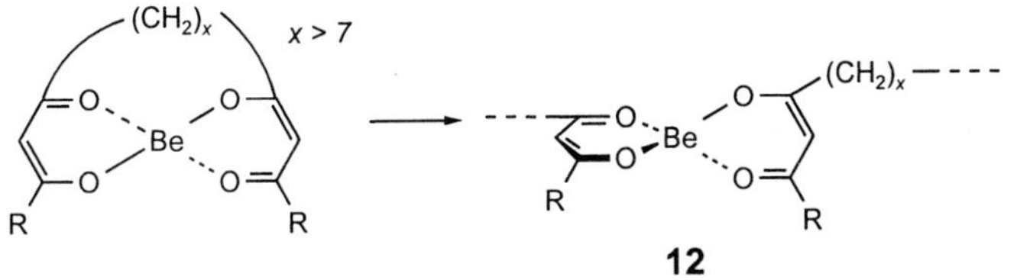

$$(7\text{-}9)$$

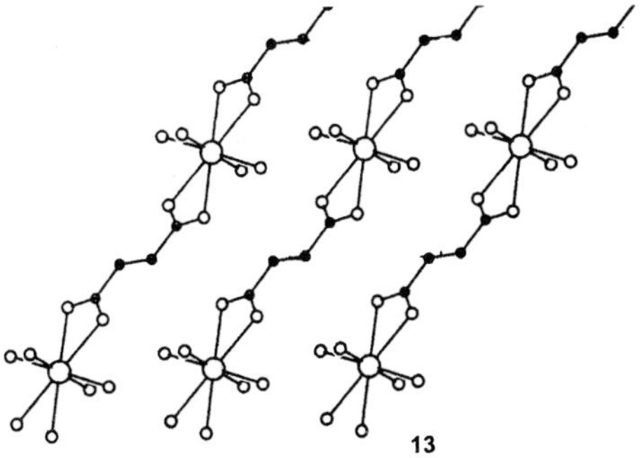

12

13

Another concept described in [61] points out that eight-coordinate centers tend to be non-rigid and thus give good solubility to coordination polymers. In the polycondensation of $Ce(OR)_4$ or $Ce(acetylacetonate)_4$ with bis(dihydroxy Schiff bases) **14** in hot DMSO to the polymer **15**, the polar solvent solubilizes the bridging ligand, the metal monomer and the growing polymer chain (Eq. 7-10) [61]. Zr(IV) coordination polymers have also been described [63]. Analogous reactions with trivalent lanthanide species and sodium salts of bridging Schiff-base ligands **14** in DMSO form anionic lanthanide polyelectrolytes **15** (Y(III), La(III), Gd(III), Lu(III) instead of Ce(IV)) (Eq. 7-10) [61,64]. By end-capping reactions and nmR of Ce(IV) polymers an average number of repeating units of 36.8 and M_n of 28,700 Da were found [61]. For the Ce(IV) and Zr(IV) coordination polymers specific conductivities of $\sim 10^{-7}$ S cm^{-1} in the undoped and of $\sim 10^{-3}$ in the iodine-doped state were found [63]. The Eu(III) polyelectrolytes with M_n up to 10^4 Da are highly luminescent [64]. The synthesis of **15** with M = Y(III) or Eu(III) instead of Ce(IV) is described in Experiment 7-2, Section 7.7. The polymers are soluble in DMSO and are polyelectrolytes. The luminescence properties of mixed polymeric Y/Eu complexes are enhanced over the polymeric Eu complexes.

$$(7\text{-}10)$$

14   **15**

Treatment of 1,3-bis(4-pyridyl)propane with perchlorates of Co(II) or Ni(II) gives one-dimensional coordination polymers [65]. The reaction of 3-phenyl-4-hydroxycyclobut-3-ene-1,2-dione with M(NO$_3$)$_2$ ·x H$_2$O (M = Mn, Co, Ni, Zn) in acetonitrile solution forms linear polymers [Mn(μ-C$_6$H$_5$C$_4$O$_3$)-(C$_6$H$_5$C$_4$O$_3$)(H$_2$O)$_3$]$_n$ **16** [66]. The Mn(II)-coordination polymer contains octahedral Mn centers that are bridged μ-1,3 by a phenylsquarate ligand to two neighboring metal atoms and further coordinated to a pendent phenylsquarate ligand and three water molecules. One-dimensional metal chains containing bridging Cs(I) and diketo-localized Ni(II) have been obtained by the reaction of Ni(II) acetate with a coronate ligand in the presence of Cs salts in methanol [67]. Hydroquinone can self-assemble into one-, two- and three-dimensional polymers by suitable interaction with transition metal ions [68]. The key step is the initial coordination of the transition metal fragment Mn(CO)$_3$$^+$ to hydroquinone. Different polymers have different geometrical requirements which depend on subsequent proton loss and σ-complexation of the oxygen atoms to metal ions. One example is shown in **17**. Coordination polymers with good solubilities, consisting of Ru(II) β-diketone units linked by butadiyne bridges, have been synthesized by CuCl-catalyzed polymerization of ethynyl-2,4-pentanedione Ru-complexes [69]. The polymers exhibit a molecular weight

M_n of 5400 Da (around 11 monomer units).

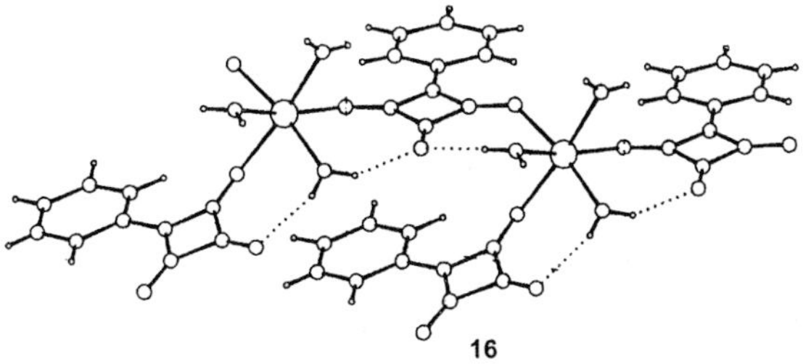

16

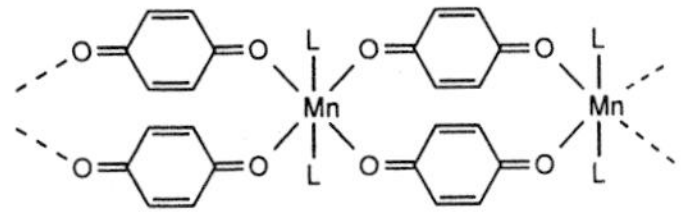

17

Ultrathin films in the nanometer range of poly(metal tetrathiooxalates) (M = Cu(II), Ni(II), Pd(II), Co(II)) have been prepared by sequential adsorption of divalent metal cations and tetrathiooxalate dianions on charged substrates (quartz slides with an upper layer of poly(allylamine hydrochloride) as shown in Eq. 7-11 [70]. Non-modified ITO glass or ZnSe substrates have also been also used. The modified quartz slide was dipped alternately into aqueous solutions of the tetrathiooxalate and a metal salt, and cleaned in between. The dark-brownish layers show a very broad absorption extending from the UV to the NIR. The electrical conductivities are in the order of $\sim 10^{-3}$ S cm^{-1}.

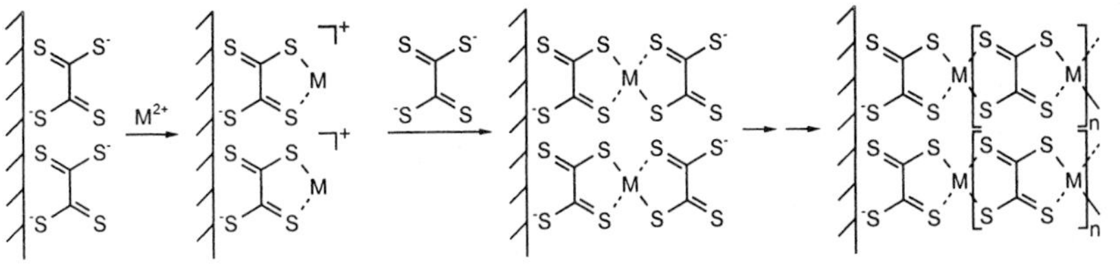

$$(7\text{-}11)$$

Red crystals consisting of a three-dimensional network were obtained by reacting 1,3-dithiole-2-thione-4,5-dithiolate **18** with Ag(I) [71]. The network consists of [Ag$_4$(thiol ligand)$_3$] repeating units. The four Ag(I) ions are in a distorted tetrahedral arrangement, and the tetrahedral frame is held together by the three thiol ligands. A Cu(I) complex of 4,5-ethylenedithio-1,3dithiole-2-thione **19** contains Cu$_4$I$_4$ clusters bridged by the bidentate ligand to form one-dimensional chains linked by S$\cdots$S contacts [72]. The interaction of Cu(I) with tetrakis(alkylthio)tetrathiafulvalenes **20** leads to the formation of 1:1 metal–ligand-complexes, in which the Cu(I) ions are tetrahedrally coordinated by four sulfur atoms to form coordination polymer chains **21** [73]. The iodine-doped complexes exhibit electrical conductivities of $\sim 10^{-5}$ S cm^{-1}. Reaction of dithiooxamide **22** with Fe(II), Co(II), Ni(II), Cu(II) or Zn(II) produces 1:1 brown to black coordination polymers (see Experiment 7-3, Section 7.7) [74]. Electrical conductivities were measured, and the different temperature dependences discussed.

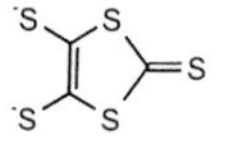

18 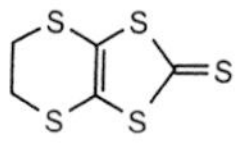   **19** 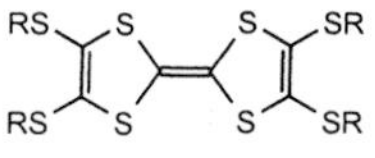   **20** 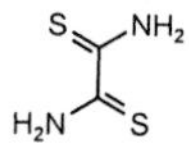   **22**

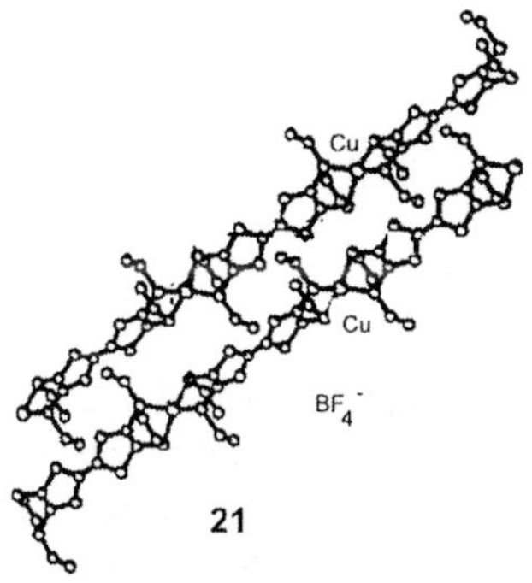

21

Nitriles are well known as ligands for transition metal ions. The interaction with bi- or higher-functional nitriles can lead to polymers. Yellowish-orange crystals were obtained by addition of succinonitrile to a water/acetone solution of $CoCl_2$ or $NiCl_2$ [75]. They consisted of a chain structured polymer as shown in **23** with the nitrile group trans oriented. A preparative method is given in Experiment 7-4, Section 7.7. Pt cyano-bridged supramolecular complexes have been reviewed [76]. For the preparation of a polymer the trinuclear complex **24** was first prepared from $[Fe(II)(CN)_6]^{4-}$ and $[Pt(II)(NH_3)_4]^{2+}$ in aqueous solution. Polynuclear complexes are formed by electrochemical treatment of the trinuclear complex above its redox potential in the presence of $[Pt(NH_3)_4]^{2+}$. The polymer complex precipitates on the working electrode as a thin layer. Besides one-dimensional polymers, higher-dimensional polymers have also been obtained, depending on the conditions. Because these mixed-valence polymers absorb in the visible region of light (there is an intensive metal–metal charge-transfer band at 350–450 nm), they are interesting for photo-induced electron transfer. The polymers are capable of oxidizing chloride to the energy-rich chlorine. Cyano cuprates are obtained by the reaction of Cu(I)CN with organo-Li compounds in THF [77]. In the solid state a polymer chain **25** exists with alternating cuprate $[Ar_2Cu]^-$ anions and $[LiCNLi]^+$ cations (with two stabilizing THF molecules).

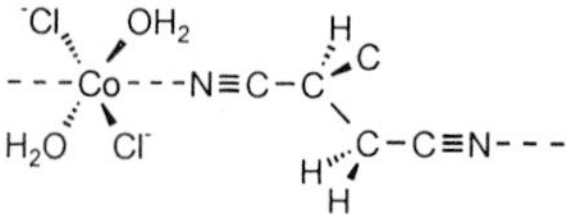

23 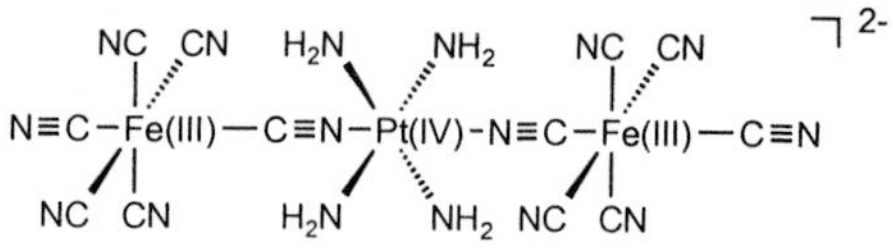   **24**

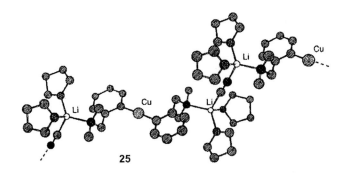

25

Thermodynamically very stable fibrous yellow-colored coordination polymers of Cu(I) and Ag(I) have been obtained by ligand-exchange reactions of acetonitrile complexes of e.g. Cu(I) against phenanthroline monomers in a solvent mixture of acetonitrile and tetrachloroethane (Eq. 7-12) [78]. *n*-Hexyl side-chains are necessary for good solubility of the polymer in less polar solvents. Because the coordination sphere of each metal center is fully saturated by exactly two chelating moieties of the ligand monomers and all carbon atom are ortho to the phenanthroline-N-atoms substituted by bulky groups (requiring pseudotetrahedral coordination of Cu(I)), branching and cross-linking are avoided. Intramolecular ring closure is also avoided by an intrinsically rigid *p*-terphenylene bridge. Highly soluble Ru(II) coordination polymers **27** have been prepared by reaction of the metal monomer $[Ru(R_2bpy)Cl_3]_x$ with a tetrapyridophenazine [79,80]. The polymers with M_w up to $4.7 \cdot 10^4$ Da (~43 units) are considered as ribbon-like polyelectrolytes with a coiled shape. Liquid crystalline coordination polymers with a comparable structure based on oligopyridines have been briefly reviewed recently [81]. Soluble π-conjugated coordination polymers with a ruthenium(II)-bipyridyl complex in the main chain have been prepared from a soluble metal complex monomer and a bridging ligand via coordination [82]. Coordination polymers with M_n up to 9500 were obtained by refluxing an ethanol–water suspension of [4,4'-dinonyl-2,2'-bipyridyl]Ru(III) with the bridging ligand 2,3-bis(2'-pyridyl)pyrazine.

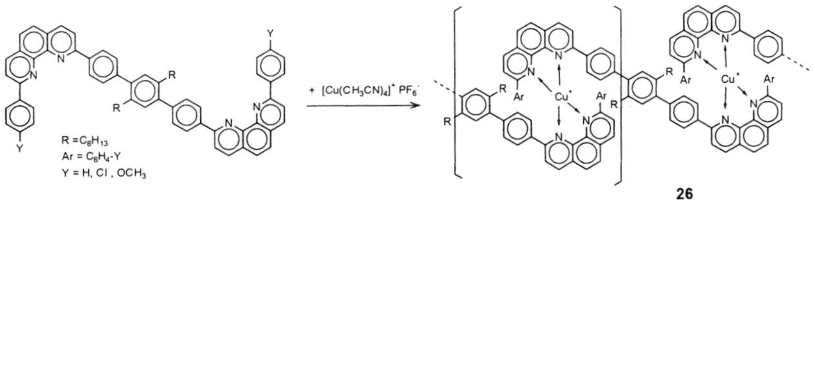

$$R = C_6H_{13}$$
$$Ar = C_6H_4\text{-}Y$$
$$Y = H, Cl, OCH_3$$

$$+ \ [Cu(CH_3CN)_4]^+ \ PF_6^-$$

(7-12)

26

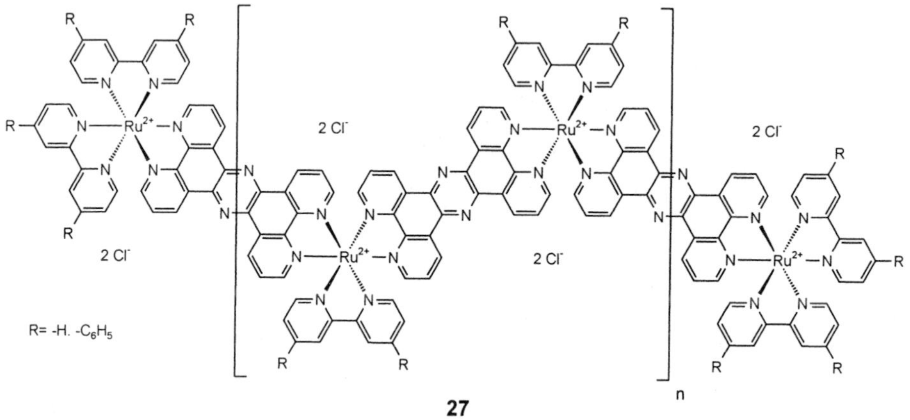

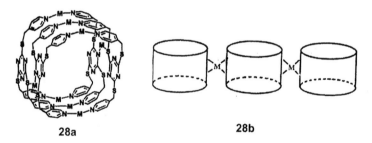

27

Organo-copper compounds are important for the C–C link. In the solid state such copper compounds can exist as linear polymers. An example is the cyanocuprat. 2,4,6-Tris[(4-pyridyl)-methylsulfanyl]-1,3,5-triazine with Ag⁺ forms a one-dimensional chain polymer with nanometer tubes **28a** (M = Ag(I)) [83]. These tubes are connected via Ag–N and Ag–S bonds to form linear chains **28b** (see Experiment 7-5, Section 7.7). Other comparable polymers are mentioned in [83].

28a　　　　　　**28b**

One-dimensional coordination polymers based on Ag(I) complexes with aromatic nitrogen-donor ligands have recently been reviewed in [84]. The main focus was on coordination polymers **29** based on Ag(I) cations with Lipyridyl ligands (X = linear, angular or a flexible linker) with various different geometries. The process of coordination polymerization is totally reversible because of the high lability of the Ag–donor atom bond. For example, in the case of Ag(I) complexes with N,N'-bis(2-pyridylmethyl)piperazine a one-dimensional polymer **30** with a 1:1 composition was obtained.

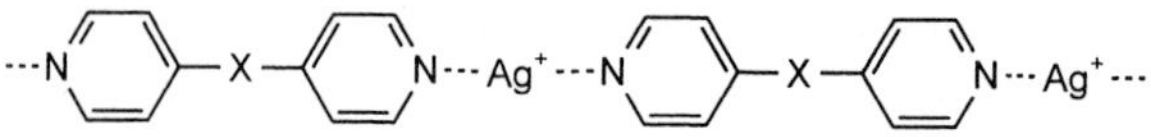

29

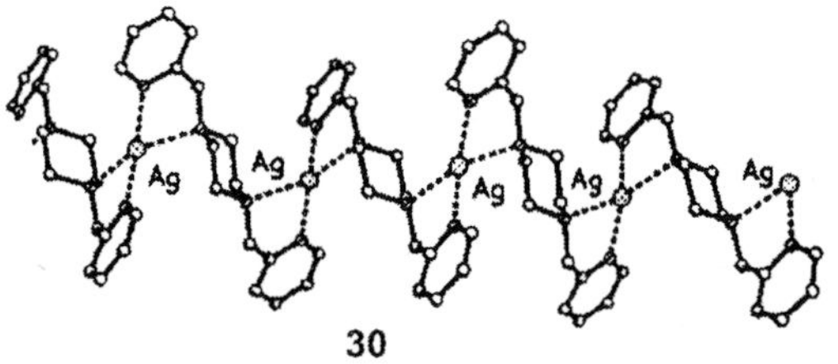

30

A few other examples of chain-forming coordination polymers are mentioned. A polymer **31** was obtained under inert conditions by the reaction of 4,4′,5,5′-tetramethyl-2,2′-bisphosphinine with naphthalene/Na in dimethoxyethane [85]. The bisphosphinine is reduced two-fold. In the deep purple oxygen-sensitive crystals the chain is held together by electrostatic interactions through alternation between three and one Na(I) cations. The reaction of pentaphosphaferrocene with CuX (X = Cl, Br, I) was used to prepare yellow-brown crystals of one- and two-dimensional polymers in which the phospha-ferrocene units are connected by CuX units [86]. One-, two- and three-dimensional "tubes" were prepared by employing dimetal building units [87]. For example, a one-dimensional tube was obtained from a Rh_2 compound with malonic acid and linkers such as *trans*-1,2-bis(4-pyridyl)ethylene.

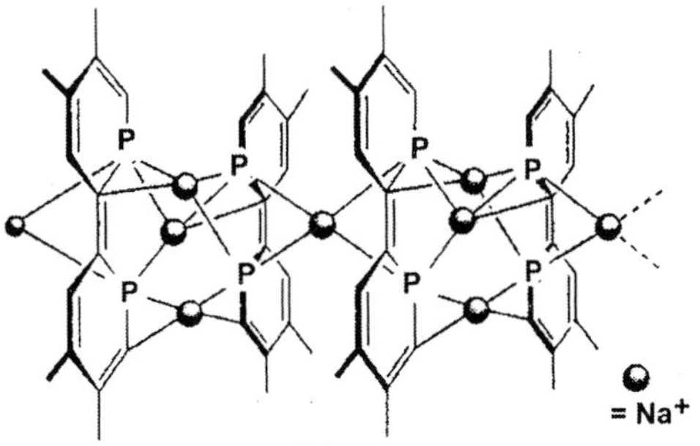

= Na⁺

31

7.3.2 Supramolecular Organization of Coordination Polymers

Numerous papers have been recently published on the crystal engineering of supramolecular solids with coordinative bonds in two- and three-dimensional network structures. The literature mentioned below contains selected results as examples. Further references are given therein. Tetrahedrally, trigonally or octahedrally acting metal-ion centers (e.g. Zn(II), Cd(II), Ag(I), Cu(I)) coordinate with ligands containing N-donor (amines, N-heterocycles), O-donor (carboxylic acids), cyano or thioether ligands. The interaction can be seen also as a linking of Lewis-acid metal centers with polyfunctional Lewis bases. The synthesis of new microporous materials is of especial interest, as are their reversible inclusion of guest molecules and their uses as catalysts or magnetic materials.

The supramolecular organization of coordination polymers is classified into non-penetrating and penetrating networks. The coordination numbers and geometries of the metal centers and the functionalities of ligands allow us in principle to predetermine the lattice structure. But it is difficult to predict the formation of non-penetrating or penetrating networks. Often guest molecules such as solvents are included in the lattice. Thermal stabilities are low compared to inorganic networks such as zeolites. An excess of a strong monofunctional Lewis base destroys the structure of the coordination polymers by coordinating to the metal ion.

7.3.2.1   Non-Interpenetrating Coordination Polymers

Several compounds with a simple formula e.g. AuI, $PdCl_2$, MoI_3, AuCN are in fact cross-linked owing to coordination of halide or pseudohalide bridges [88]. Cyano-bridged one- to three-dimensional coordination polymers based on $[M(CN)_6]^{3-}$ (M = Fe, Cr, Mn, etc.) have attracted great attention because of their rich structures and magnetic behavior [89]. Prussian Blue, $Fe_4[Fe(CN)_6]_3 \cdot H_2O$, consists of a three-dimensional network and is obtained by mixing dilute equimolar solutions of $K_4[Fe(CN)_6]$ and $FeCl_3$. The product is a colloid with an average diameter of 23 nm and a molecular weight of $\sim 7 \cdot 10^6$ Da [90]. This polymer is interesting as a photosensitizing device, rechargeable battery material, memory device and for electrochromic displays. The complex KV(II)[Cr(III)(CN)$_6$] is a molecular magnet up to 103 °C [91]. The three-dimensional polymers $[SmFe(CN)_6] \cdot 4H_2O$ and $[TbCr(CN)_6] \cdot 4H_2O$ exhibit long-range ferromagnetic ordering below 3.5 K and the highest known Curie temperature (T_C= 11.7 K) for 4f-3d molecule-based magnets [92]. Reaction of $(C_6H_5)_4P[Ru(III)(acetylacetone)_2(CN)_2]$ with Mn^{2+} produces a novel cyano-bridged polymer of the composition $[Mn(II)[Ru(III)(acetylacetone)_2(CN)_2]_2$. The polymer has a diamond-like structure and exhibits ferromagnetic ordering below T_c = 3.6 K [93]. Unusual magnetic properties were also measured for two-dimensional nets prepared from $K_3[Fe(CN)_6]$ and 2,2'-bipyrimidine with $Nd(NO_3)_3$ [94]. Alternating fused rows or rhombus-like $Fe_2Nd_2(CN)_4$ rings and six-sided $Fe_4Nd_4(CN)_4$ rings form the net, and the bipyrimidine coordinates to the Nd ions in a chelating fashion. Mesoscopic layers of a cyano-bridged Cu/Ni-coordination compound were obtained by first coating quartz or membrane filters with an aqueous layer of the detergent **32** [95]. This film was dipped in an aqueous solution of $K_2[Ni(CN)_4]$ and then in an aqueous solution of $Cu(NO_3)_2$. Thus a two-dimensional layer of $Cu[Ni(CN)_4]$ was formed between the layers of **32** (Fig. 7-2). From $[Re_6S_8(CN)_6]^{4+}$ clusters and $[Co_2(H_2O)_4]^{4+}$ complexes, coordination polymers have been formed which exhibit different colors with different solvents owing to the porosity of the lattice [96]. Three-fold symmetric phenylacetylene nitrile silver salts show a honeycomb topology [97].

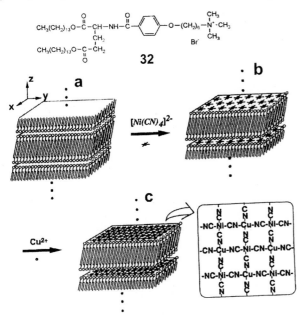

32

Figure 7-2. Schematic view of the step-wise synthesis of an aggregate of Cu[Ni(CN)$_4$];
a: layers of **32**. b: film of **32** with K$_2$[Ni(CN)$_4$]. c: film of **32** with Cu[Ni(CN)$_4$].

Moving on now to consider two- and three-dimensional networks based on bifunctional space ligands, several basic structures as shown in Fig. 7-3 were assembled a few years ago [98,99]: a) diamond analogues [100], b) honeycomb analogues [101], c) square-lattice analogues [102], d) ladder-and-stair analogues [98,103], e) brick analogues [104], f) octahedral structures [105,106], g) helical structures [107] (for other examples see [108] and references cited therein). These coordination polymers are held together by "coordinative covalent" bonds. Preparations are often relatively easy. In a one-pot synthesis a metal salt is mixed in solution with the bifunctional ligand and the coordination polymer is obtained by slow crystallization. The molar ratio of the two compounds is very important. The host lattice often includes solvent molecules. A few examples, including the recently described coordination polymers, are given below.

Mixing a solution of Co(NO$_3$)$_2$·6H$_2$O with 4,4′-bipyridine (molar ratio 1:1.5) yields after few hours red crystals of **33** [98]. In air the crystals are stable for only a few hours. X-ray analysis shows that **33** consists of a molecular ladder structure (Fig. 7-4). The 4,4′-bipyridine ligands coordinate to Co^{2+} and form the side part and the rung of a ladder in the direction of the *a*-axis by self-organization.

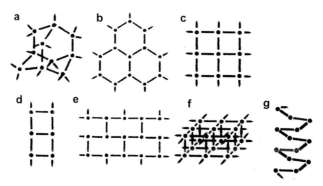

Figure 7-3. Schematic representation of simple network structures (metal units as points, ligand units as small rods).

Infinite ladders, which include solvent molecules, are shifted by a rung distance against each other and stacked like a stair in the direction of the b-axis. A two-dimensional square lattice of **34** is obtained from Cd^{2+} and 4,4'-bipyridine (see Experiment 7-6, Section 7.7) [102,109]. Aromatic guest molecules can be incorporated, and the polymer catalyze the cyanosilylation of aldehydes [102]. The structure of [Ag(4,4'-bipyridine)NO$_3$] consists of linear silver-ligand chains which are crosslinked by an Ag–Ag interaction [110]. Infinite channels (2.3 × 0.6 nm) are formed which can reversibly incorporate PF_6^-, MoO_4^{2-}, BF_4^- and SO_4^{2-} ions. Pyrazine has also been used as a building block [111].

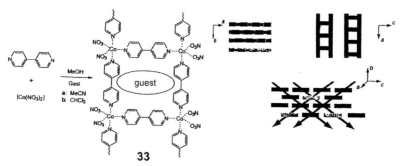

Figure 7-4. Synthesis and molecular structure of the coordination polymer **33** from Co(NO$_3$)$_2$ and 4,4'-bipyridine.

Inserting a biphenylene unit (0.85 nm longer than **33**) in the 4,4'-bipyridine ligand between the two bipyridyls gave rise to coordination polymers with an extraordinarily large square grid, with dimensions of about 2 × 2 nm [112]. The guest o-xylene (used as solvent) was included and occupies 58% of the crystal volume. The large square cavities are packed and create big rectangular channels. Square-grid coordination polymers from a bipyridine derivative with dimensions as large as 2.5 × 2.5 nm have also been described [113].

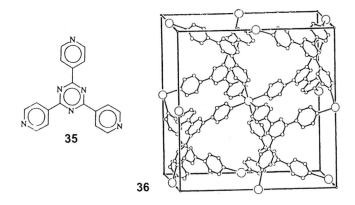

34

Reacting 2,4,6-tri(4-pyridyl)-1,3,5-triazine **35** with HgClO$_4$ in tetrachloro-ethane yields three-dimensional structures **36** containing solvent molecules (Fig. 7-5) [114]. The Hg centers form a slightly distorted octahedral geometry. An infinite network **38** was obtained from hexakis(imidazol-1-ylmethyl)benzene **37** as ligand interacting with CdF$_2$ (Fig. 7-6) [115] (for other imidazol derivatives as ligand see [116]).

35

36

Figure 7-5. Unit cell of the coordination polymer **36** prepared from HgClO$_4$ and the ligand **35** (Hg are large circles, C and N atoms are small circles).

Helical coordination polymers containing large chiral cavities or channels are interesting for stereospecific syntheses, separation of enantiomers and stereoselective catalysis. Crystals of **39** are obtained from a solution of nickel(II) acetate and benzoic acid in methanol which is covered with a solution of 4,4'-bipyridine in the presence of benzene or nitrobenzene [107]. Each helix turn contains three complex units and the chain consists of Ni(II) (with binding of benzoate) and the bipyridine. Because each helix is shifted against the next, large cavities are formed containing, for example, benzene or nitrobenzene.

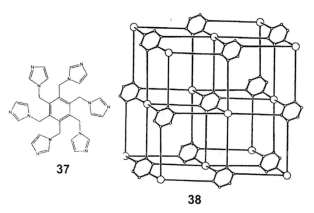

37

38

Figure 7-6. Infinite three-dimensional network of the coordination polymer **38** prepared from CdF$_2$ and **37** (Cd as large circles, imidazole bridges as lines).

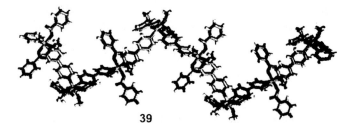

39

A column layer-structured coordination polymer has been obtained from Cu^{2+} by reaction with pyrazine-2,3-dicarboxylate and pyrazine [117]. The polymer consists of two-dimensional layers of Cu-pyrazinedicarboxylate which are connected by columns of pyrazine. After desorption of water at 100 °C, methane could be adsorbed reversibly. Single-, double- and three-dimensional structures of rare-earth metal coordination polymers have been formed by hydrothermal synthesis through the reaction of a rare-earth metal(III) nitrate (M = La, Ce, Eu) with 3,5-pyrazoledicarboxylic acid in water for 3 days at 150 °C in a Teflon autoclave [118]. The three-dimensional framework contains nine-coordinated lanthanide metal centers. The hydrothermal reaction of nickel chloride with 1,2,4,5-benzenetetracarboxylic acid anhydride and 4,4'-bipyridine in aqueous solution at 170 °C led to the formation of green crystals of **40** consisting of a three-dimensional neutral structure (see Experiment 7-7, Section 7.7) [119]. Each Ni(II) has an octahedral environment: three oxygen atoms from two of the tetracarboxylic acid ligands, two nitrogen atoms from the bipyridine and one aqua ligand. Each tetracarboxylic acid ligand links four symmetry-related Ni(II) into an infinite two-dimensional framework [Ni$_2$(tetracarboxylic acid)(H$_2$O)$_2$]$_n$ (Fig. 7-7a). Adjacent lamellas are interlinked by the bipyridine to form a three-dimensional network with vacancy dimensions of 0.455 ×

0.946 nm filled with water (Fig. 7-7b). The network is stable up to ~200 °C. Comparable structures were obtained, for example, by the reaction of Co^{2+}, benzenetetracarboxylic acid and 2,2'-bipyridyl [120]. Ferromagnetic interactions were measured between the Ni(II) ions. The reaction of Cu(2-methylpyrazine-5-carboxylic acid)$_2$ with either CdCl$_2$, CuCl or CdBr$_2$ at 130 °C in water yields one-, two- or three-dimensional coordination polymers stable up to ~240 °C [121].

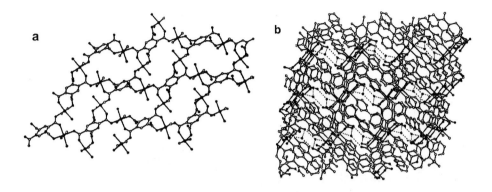

Figure 7-7. Crystal structure of **40**, a: two-dimensional lamella of Ni(II) and tetra-carboxylic acid. b: three-dimensional structure with vacancies occupied by water molecules.

Polymeric structures based on networks of *homoleptic* eight-coordinate La(III) centers linked through 2,2'-bipyridine-*N,N'*-dioxide have been prepared, for example, by the reaction of La(CF$_3$SO$_3$)$_3$ and 2,2'- and 4,4'-bipyridine-*N,N'*-dioxide in methanol at 20 °C [122]. The reaction of FeCl$_2$, 1,10-phenanthroline and benzene-1,2,4,5-tetracarboxylate in water at 160 °C in a Teflon reactor produces a two-dimensional polymeric structure of [Fe(phenanthroline)-(benzenecarboxylate)]$_n$ [123]. Multi-dimensional metalorganic coordination polymers with metal–aromatic interaction which exhibit strong photoluminescent behavior have been prepared by different Ag(I)-naphthalene-carboxylates with hexamethylenetetramine in CH$_2$Cl$_2$ [124].

Two-dimensional networks have been formed by four square carboxylate-bridged di-metal(II) building units using the reaction of Zn(NO$_3$)$_2$ or Cu(NO$_3$)$_2$ with benzene-1,3-dicarboxylate in methanol or DMF in the presence of pyridine [125]. Hydrothermal treatment of Ni(NO$_3$)$_2$ with disodium fumarate in water at 170 °C leads to a complex [Ni$_2$(fumarate)$_2$·H$_2$O] with a three-dimensional polymeric structure formed by chains of fused Ni(II) octahedral units [126]. Nickel succinate presents a remarkable three-dimensional Ni–O–Ni framework with porosity and high thermal stability (~400 °C) [127]. A two-dimensional metalorganic coordination polymer **41** has been obtained by the reaction of $Zn(ClO_4)_2$ with ethyl *S*-lactate (see Experiment 7-8, Section 7.7) [128]. The Zn(II) is located in the center of a plane formed by four oxygen atoms from two lactic acid ligands (Fig. 7-8). Each lactate ligand offers one of two carboxylate O atoms and one hydroxyl O. The other carboxylic O takes part in coordination to the neighbouring Zn(II). Additionally, networks based on binaphthylene-dicarboxylic acids with Mn(II), Co(II), Ni(II) [129], dirhodium tetra(trifluoro-acetate) [130] and coordination compounds of 1,4-dihydroxybenzoquinones [131] have been reported. Two-dimensional magnetic materials based on nitroxides are prepared from nitronyl-nitroxide ligands by their interaction with Mn(II) [132].

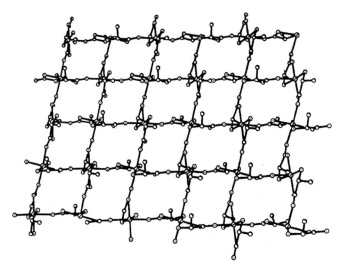

Figure 7-8. View of the two-dimensional structure of the coordination polymer **41**.

The reaction of the Ni(II)-complex **42** of bis(2-pyridylcarbonyl)amine with Fe(II) perchlorate in methanol results in dark-purple hexagonal crystals of **43** [133]. Figure 7-9 shows the stepwise growth to a graphite-like polymer coordination complex with large cavities. Two- and three-dimensional polyrotaxane coordination polymers have been described recently [134]. The first step results in the pseudorotaxane **46** by threading the cucurbituric **44** with the bipyridyl derivative **45** (Fig. 7-10). In the second step the polyrotaxane **47** is obtained by reaction of **46** with $Cu(NO_3)_2$ in the presence of oxalate ions. In **47** compound **44** is threaded on a two-dimensional polymer network. The stacking of the layers at a distance of 1.29 nm results in one-dimensional channels.

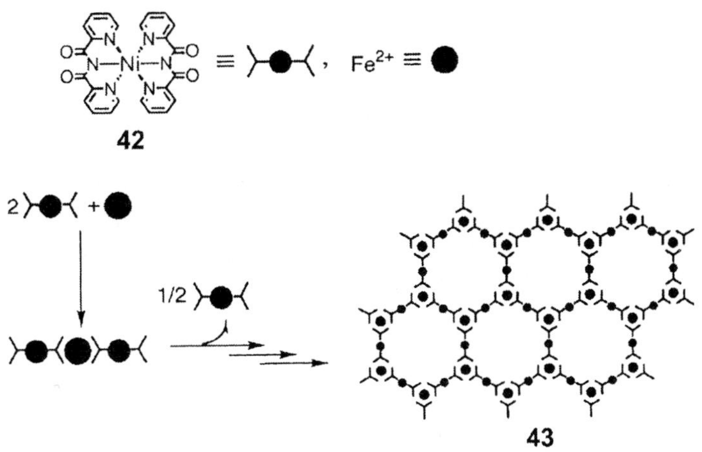

Figure 7-9. Schematic representation of the stepwise growth of the Ni-complex **42** with Fe(II) perchlorate to the coordination polymer **43**.

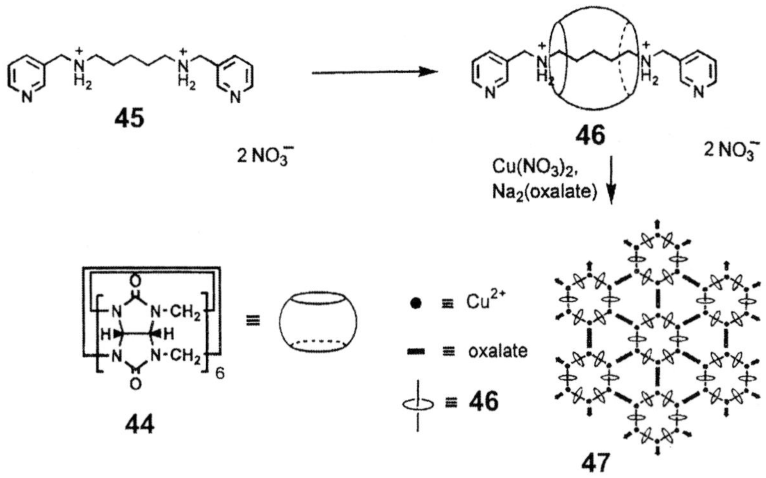

Figure 7-10. Synthetic steps for the preparation of the two-dimensional polyrotaxane **47**.

Another concept is based on neutral, polyfunctional Lewis-acid spacers for the construction of stair and ladder structures [103]. Ortho-phenylene(indium-bromide) **48** exists as a THF stabilized dimer. The p-orbitals at indium are orientated perpendicular to the diindacyclus. Reaction of **48** with pyrazine in a molar ratio of 1:1 or 1:2 in THF results in the stair-structured molecule **49** whereas in a molar ratio of 1:4 the ladder-structured **50** (both contain additional THF) is obtained (Fig. 7-11). The polymers are insoluble in nonpolar solvents but can be dissolved in coordinating solvents, destroying the polymer structure.

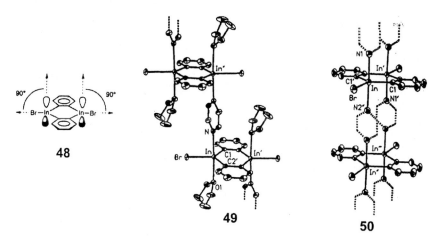

Figure 7-11. Molecular structure of ortho-phenylene/indium-bromide **48** and its coordination polymers **49**, **50** by reaction with pyrazine.

In nature, porphyrins participate in electron transfer, energy transfer and redox-catalysis. Therefore these macrocycles are also interesting as building blocks in coordination polymers. Tetrasubstituted porphyrins that are substituted by groups capable of coordinating to metal ions are employed for the preparation of porphyrin coordination polymers. Mixing $Hg(ClO_4)_2$ in methanol with 5,10,15,20-tetrapyridylporphyrin in 1,1,2,2-tetrachloroethane yields crystals of a coordination polymer [135]. The Hg centers form by coordinating to the pyridyl groups in an infinite net. Tetrachloroethane and ClO_4^- are encapsulated in the cavities. Networks **51** with large cavities are obtained when 5,10,15,20-tetrapyridylporphyrin is reacted with Co^{2+} or Mn^{2+} in aqueous solution under conditions of hydrothermal synthesis in a Teflon autoclave at 200 °C for 2 days (see Experiment 7-9, Section 7.7) [136]. Hexagonal channels with a size of 1.3 × 1.6 nm are formed (Fig. 7-12). The network is stable up to 400 °C. This would allow the use of these networks as hosts for other molecules analogous to inorganic molecular sieves.

5,10,15,20-Tetracarboxytetraphenylporphyrin has been employed as another ligand [137]. The Zn-complex of this porphyrin was treated with 4,4'-bipyridine in a mixture of methanol and ethyl benzoate to give crystals of **52**. Figure 7-13 shows interlinked arrays of the porphyrin and bipyridyl. The structure represents an open three-dimensional network in which the individual metallo-porphyrin units are cross-linked both axially and equatorially.

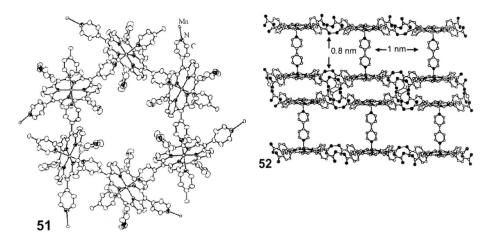

Figure 7-12. View of one cavity of the coordination polymer **51** prepared from Mn(II) tetrapyridylporphyrin.

Figure 7-13. Part of the crystal structure of the coordination polymer **52** from Zn(II) tetracarboxyphenylporphyrin and 4,4'-bipyridine.

7.3.2.2 Interpenetrating Coordination Polymers

As mentioned before, coordination polymers can form porous structures (with channels and/or cavities) in which solvent molecules are included. In several cases interpenetrating structures are obtained in which cavities belonging to one lattice framework are occupied by one or more independent lattice frameworks. These interpenetrating structures can be separated only by splitting bonds. Some relationships to catenanes, rotaxanes and molecular knots can be considered. A classification of interpenetrating coordination polymers with a catenated structure is given in Fig. 7-14: a, b) linear structures, c) interpenetration of ladders, d) inclined interpenetration of ladders, e) interpenetration of undulating layers, f) interpenetration of multiple layers [138]. A recent review describes in detail penetrating networks [139]. A very few examples are given below.

The ligand 1,4-bis(imidazol-1-ylmethyl)benzene **53** interacts with Ag^+ to form a polymer **54** of the composition $[Ag_2(\mathbf{53})_3(NO_3)_2]_\infty$ [139]. Two one-dimensional chains penetrate in a polycatenane-analogous structure (Fig. 7-15). However, in contrast, the reaction of **53** with Zn^{2+} yields a polymer **55** of the composition $[Zn(\mathbf{53})_2](NO_3)_2 \cdot 4.5\ H_2O$ [140]. This polymer consists of two independent two-dimensional networks lying parallel. Zinc is tetracoordinated (Fig. 7-16).

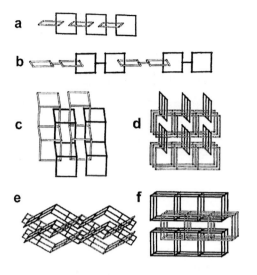

Figure 7-14. Classification of interpenetrating coordination polymers; a and b: one-dimensional linear structures; c and d: two-dimensional interpenetration of ladders; e and f: three-dimensional interpenetration of undulating and multiple layers.

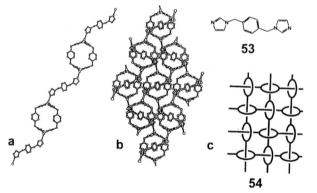

Figure 7-15. Coordination polymer **54** prepared from the ligand **53** and AgNO₃; a: one single chain; b: polyrotaxane analogous layer of different chains; c: schematic representation of the structure.

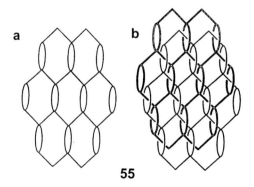

Figure 7-16. Schematic representation of the coordination polymer **55** prepared from the ligand **53** and Zn(NO₃)₂. Zn are at the centers of four-fold knots; the ligand **53** represents the connections between them; a: part of the network; b: interpenetrating network.

An example of a three-dimensional penetrating network is the solvated

An example of a three-dimensional penetrating network is the solvated $[(ZnCl_2)_3(35)_2]$ **56** [139]. The triazine molecules are part of threefold connecting knots (Fig. 7-17). The zinc atoms have a nearly tetrahedral coordination geometry with two N atoms of **35** and two chlorine ligands. A distorted network is obtained. The self-assembly of $CuSO_4$ in water and 1,3-bis(4-pyridyl)propane in ethanol results in **57** which presents a three-dimensional architecture sustained by two different types of coordination polymer: one-dimensional ribbons of rings and two-dimensional layers (Fig. 7-18) [138].

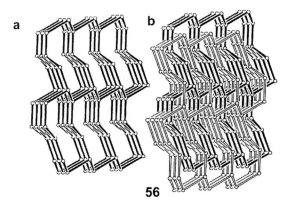

56

Figure 7-17. Schematic representation of the coordination polymer **56** prepared from the ligand **35** and $ZnCl_2$; a: part of the network; b: interpenetrating network.

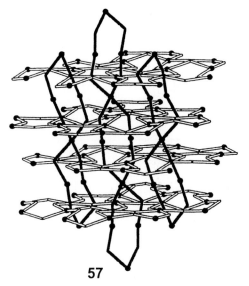

57

Figure 7-18. Schematic representation of the coordination polymer **57** (simplified by showing only the central methylene atoms).

7.4   Metallocenes as Part of a Polymer Chain

The arrangement of bifunctional π-electron-rich charged aromatics connected by π-bonds in the main chain as shown in the general formula **58** has been described in some reviews [141–145]. Polymers containing directly linked metallocenes are obtained by, for example, step-growth polycondensation of 1,1-dilithiumferrocenes with 1,1-diiodoferrocenes (M <10,000 Da) [146]. Poly-(1,1-ruthenocenylenes) have also been described [146]. Face-to-face polydecker sandwich complexes via a naphthalene spacer **59** have been prepared from a monomer dianion by reaction with a transition metal ion [144,147]. Polydecker sandwich complexes containing 2,3-dihydro-1,3-diborolo ligands and Rh^{2+} are described in [145]. The Barat complex [Cp_3Ba]⁻ in the solid state consists of a linear coordination polymer **60** in which Ba^{2+} ions are surrounded in a tetrahedral coordination by four Cp anions (Fig. 7-19) [148]. The polymer **60** is prepared by reaction of a Wittig reagent with CpH to form [Cp]⁻[R_4P]⁺ which reacts in situ to [Cp_3Ba]⁻[Bu_4P]⁺ (Eq. 7-13). The reaction of the bimetallic tetracarboxylate $Rh_2(OOC-nC_7H_{15})_4$ with 1,1'-bis(4-pyridylethynyl)ferrocene in CH_2Cl_2 yields an orange-colored oligomer [$Rh_2(OOC-nC_7H_{15})_4$ (pyridylferrocene)]$_n$ [149].

$$Bu_3P{=}CHCH_2CH_2CH_3 + CpH + Cp_2Ba \rightarrow [Cp_3Ba]^-[Bu_4P]^+ \qquad (7\text{-}13)$$

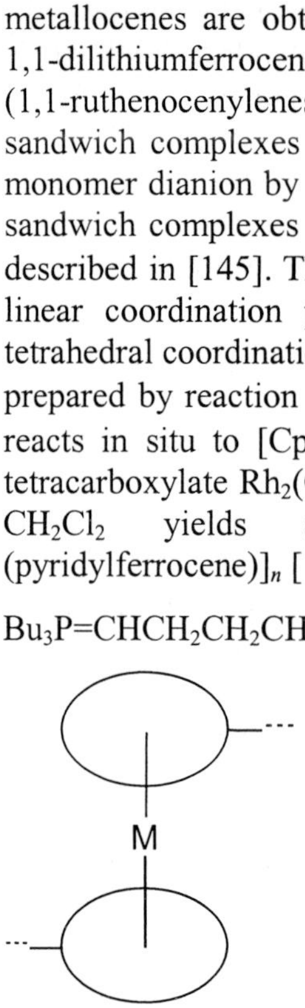

58 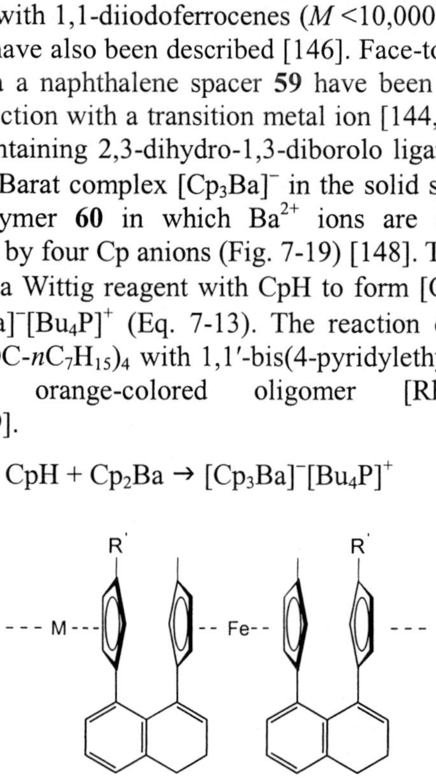   **59**

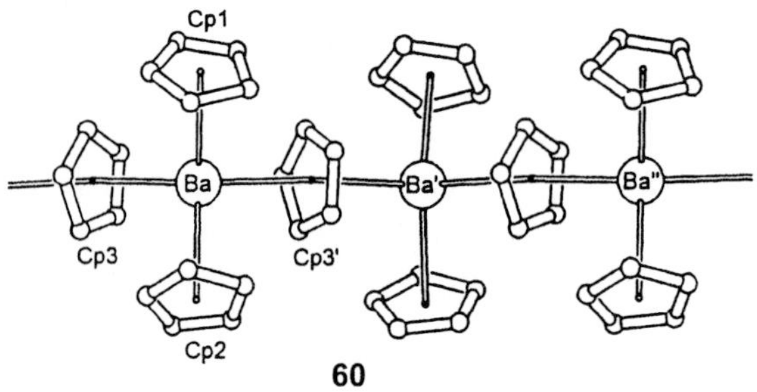

60

Figure 7-19. Part of the structure of the linear [Cp_3Ba]⁻ chain of the polymer **60**.

Several papers have reported on polyesters, polyamides and polyurethanes containing metallocenylenes [146,150] (see references cited in [151]) and polyferrocenylenes bridged by $-(CH_2)_n-$, $-(SiR_2)-$, $-(GeR_2)-$ [142,146,152]. The synthesis of poly(1,1'-ferrocenylene-p-oligophenylenes) such as **61** with a degree of polycondensation of ~55 was realized by a Pd-catalyzed polycondensation of 1,1'-bis(p-bromophenyl)ferrocene with 2,5-dialkylbenzene-1,4-diboronic acids [153]. Also polyurethanes, polyureas, polyethers, polyvinyls and several copolymers, with direct metal–metal bonds along the backbone, have been synthesized [154]. One sample for a Mo–Mo bond stabilized by CO groups is the polymer **62** from an aliphatic dichloride and the corresponding diamino-Mn compound [151,155]. In solution and in the solid state these polymers exhibit photodecomposition under photolysis of the Mo–Mo bond.

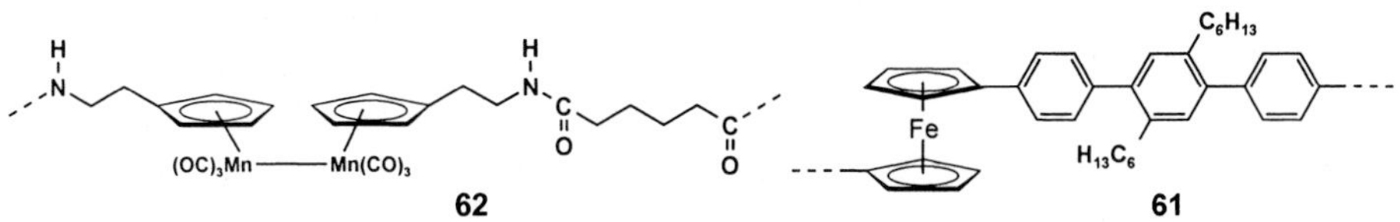

Interesting polymers are obtained by the ring-opening polymerization of various metallocenophanes **63** to high molecular weight poly(metallocenes) **64** ($M_w = 10^5-10^6$, $M_n >10^5$ Da) (Eq. 7-14) [141,142,150,156–163].Between the metallocenes, the polymers contain the following bridging blocks X in **64**: – (CR_2–CR_2)– [141,163], –(SiR_2)– [141,157], –(GeR_2)– [141], –(PR)– [141], – (SnR_2)– [158], –(S)– [141], –[B(N($SiMe_3$)$_2$)$_2$)]– [159]. The ring opening can be achieved thermally, via anionic or cationic initiation or by the use of transition metal catalysts (see Experiment 7-10, Section 7.7). Different copolymers have also been prepared [157,160,161,164]. Examples of a di-block- and a tri-block-copolymer are shown in **65, 66**. These poly(metallocenes) are interesting materials for coating dielectrics [162] and as precursors for magnetic ceramics [165]. The block-copolymers have been found to undergo microphase separation and in a suitable solvent to form interesting morphologies based on isolated or collections of spherical and cylindrical micelles/nanotubes [160,161,164,165]. It is pointed out in [165] that such superlattices may be used for the fabrication of magnetic multilayers for data storage, patterned redox-active thin-film electronic devices and designer ceramic architectures.

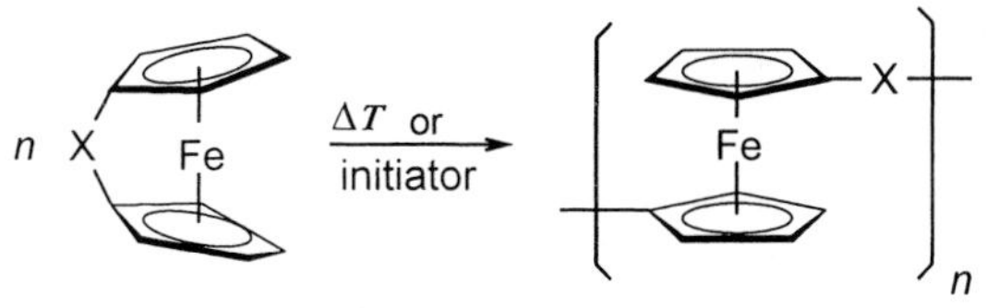

(7-14)

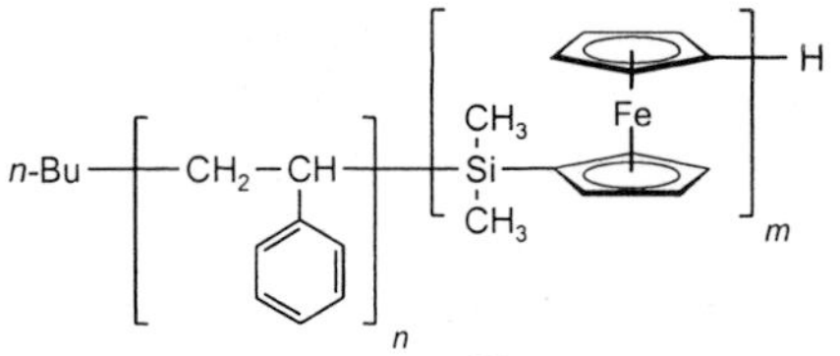

65

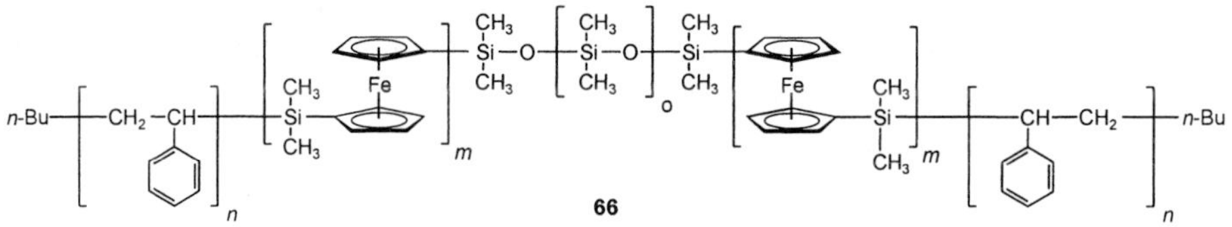

66

7.5   Cofacially Stacked Polymeric Metal Complexes

Stacked arrangements **67** of macrocyclic metal complexes – porphyrins, phthalocyanines, naphthalocyanines (see formulae **47**, **48** and **49** in Chapter 5) and others – in a cofacial orientation are obtained by connecting the metal ions in the core with different bivalent atoms or groups. This subject has been intensively reviewed [166–169] and described in detail by Hanack and Dini [170], so only a brief description is given here. These materials have interesting properties, including electrical conduction, photoconduction and non-linear optical and electroluminescent behavior. To prepare thin-film devices, for example by coating or Langmuir–Blodgett techniques, the solubility of these polymers is improved by introducing bulky groups at the macrocycle. The degree of polymerization of macrocycles can be up to around 200. A mathematical model for determining the number of isomers in tetrasubstituted cofacially stacked polymeric phthalocyanines has been developed [171]. The polymers can be subdivided into three main groups as described below.

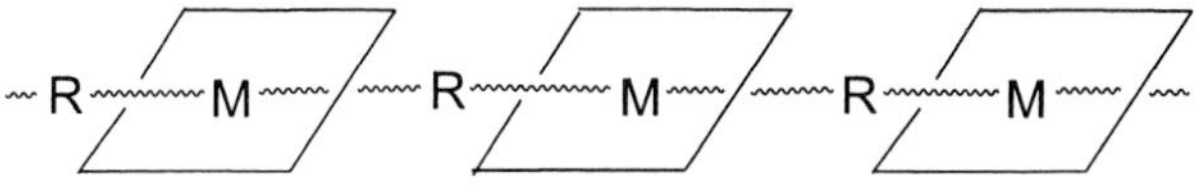

67

a. Covalent/covalent connections
    Covalent/covalent bonds in μ-oxo-bridged polymers [PcM(IV)O]$_n$ (Pc = phthalocyanine) (Fig. 7-20) with tetravalent Si, Ge or Sn have been obtained by polycondensation of dihydroxy-metal macrocycles such as the porphyrins, phthalocyanines and naphthalocyanines (M = Si(OH)$_2$, Ge(OH)$_2$, Sn(OH)$_2$) in a high boiling solvent (the HO groups can be activated by better leaving groups such as CF$_3$COO$^-$ or under catalysis by FeCl$_3$) or in the solid state (under

topochemical control) [172]. The interplanar distances are: 333, 353 and 382 pm for $[PcSiO]_n$, $[PcGeO]_n$ and $[PcSnO]_n$, respectively. Vapor deposition of the $Si(OH)_2Pc$ monomer on the surface can result in oligomers with the stack axis perpendicular and polymers with the stack axis parallel to the substrate surface [173]. In addition to oxygen, bridging groups were covalently included between the macrocyclic ring systems (Fig. 7-20). These include bivalent alcohols, phenols [174], carboxylic acids [174,175], aliphatic and aromatic diamines [176], ethynylene [177], diethynyl-*p*-phenylene [178] and oligosiloxanes (see experiments 7−11 and 7−12, chapter 7.7) [179−181].

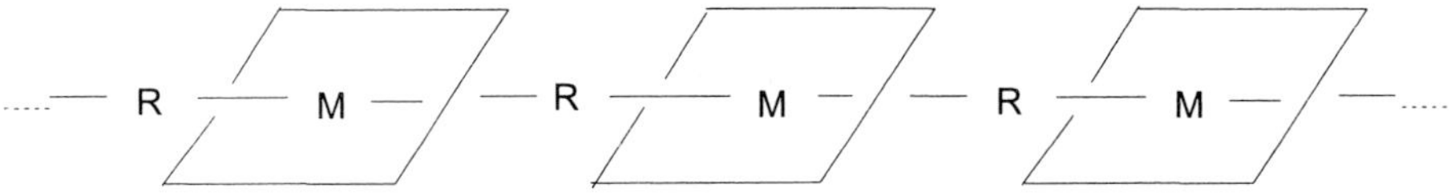

$M = Si(IV), Ge(IV), Sn(IV)$

$R = -O-, -O-R^1-O-, -OOC-R^2-COO-, -HN-R^2-NH-,$

$-C \equiv C-, \quad -C \equiv C-C_6H_4-C \equiv C-, \quad -(O-SiR^4{}_3)_n-O- \ (n = 2, 4, 6)$

Figure 7-20. Examples of cofacial covalent/covalent connections.

Phthalocyanines with M = $Si(OH)_2$ or $Ge(OH)_2$ were covalently incorporated into polyesters during the polycondensation of terephthalic acid dimethylester and ethylene glycol [182]. With only 10^{-4} molar amounts of dye, the polyesters are intensely blue-colored. Good solubility can be achieved in water by axial substitution at the central metal or with hydrophilic polymers. Phthalocyanines with the tetravalent M = $SiCl_2$ were reacted with the sodium salt of methoxypoly(oxyethylene) (M_w = 5000 Da) to give the blue-colored polymer **68** which is soluble in water and some organic solvents [183]. The reaction of phthalocyanines with the trivalent M = AlCl with poly-(oxyethylene) or poly(vinylalcohol) also led to water-soluble polymers having covalent bonds of the polymers at the Al(III) [184]. These water-soluble materials have been tested in the photodynamic therapy of cancer.

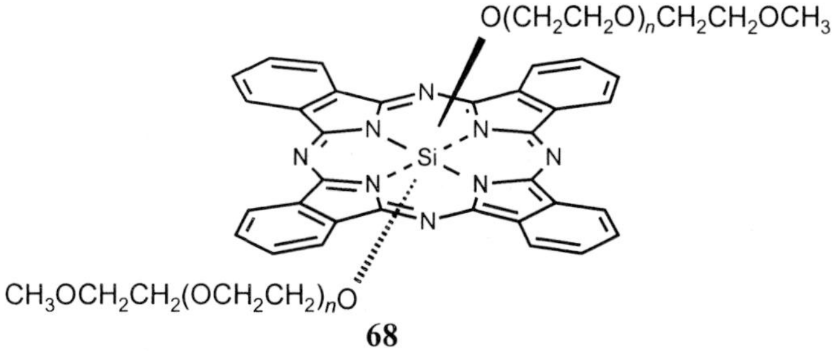

$O(CH_2CH_2O)_nCH_2CH_2OCH_3$

$CH_3OCH_2CH_2(OCH_2CH_2)_nO$

68

b. Covalent/coordinative connections

Covalent/coordinative bonds between macrocycles in polymers $[PcM(III)L]_n$ are possible with trivalent Al, Ga, Cr and F as a bridging group by vapor deposition, or Co(III), Fe(III) and –CN, –SCN or –N$_3$ as a bridging group, for example by Na-ligand splitting from Na(metal(III)-macrocycle)(ligand)$_2$ compounds (Figure 7-21).

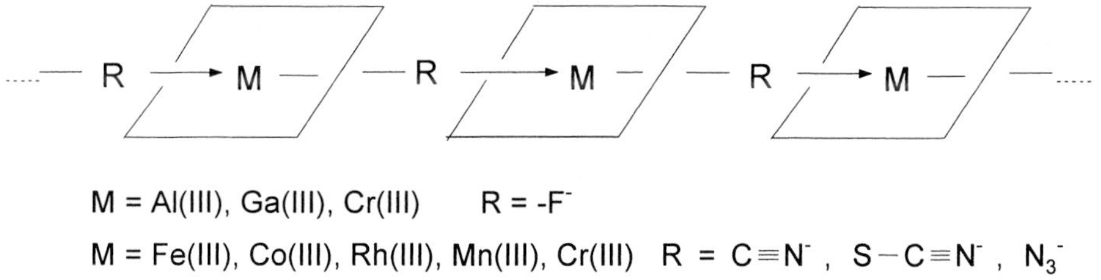

M = Al(III), Ga(III), Cr(III)   R = -F$^-$

M = Fe(III), Co(III), Rh(III), Mn(III), Cr(III)   R = C≡N$^-$, S–C≡N$^-$, N$_3^-$

Figure 7-21. Examples of cofacial covalent/coordinative connections.

We mention a few examples. The stacked fluorine-bridged complexes $[PcMF]_n$ (M = Al(III), Ga(III)) have been prepared by treatment of the hydroxylate PcMOH with aqueous HF or as needle-like crystals by sublimation under vacuum at ~500 °C [185,186]. The separation between PcAl units is 400 pm. Polymers with the general formula $[PcM(CN)]_n$ (M = Cr, Mn, Fe, Co, Rh) can be prepared using PcM(II) as starting material through several steps including the oxidation from PcM(II) to PcM(III) [187,188]. For thiocyanate-bridged $[PcM(SCN)]_n$ (M = Cr, Mn, Fe, Co) several routes have been described [189]. One possibility is the treatment of $K^+[PcM(Cl_2)]^-$ with KSCN to $K^+[PcM(SCN)_2]^-$ followed by removal of KSCN in water.

c. Coordinative/coordinative connections

Coordinative/coordinative bonds are obtained with metal ions in the macrocycles which are able to react in solution with stochiometric amounts of bifunctional donors to give polymers (Fig. 7–22).

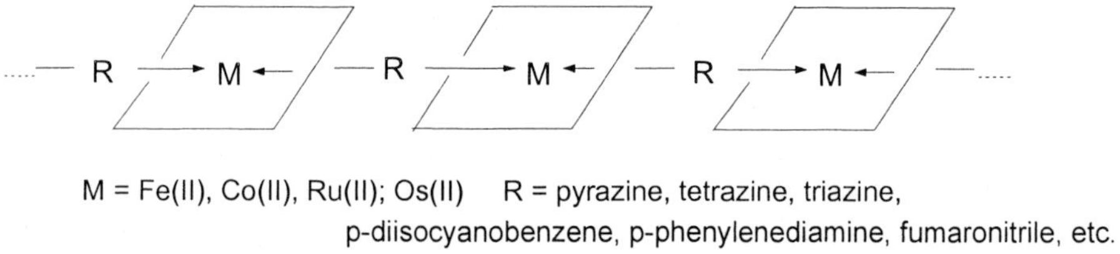

M = Fe(II), Co(II), Ru(II); Os(II)   R = pyrazine, tetrazine, triazine, p-diisocyanobenzene, p-phenylenediamine, fumaronitrile, etc.

Figure 7-22. Examples of cofacial coordinative/coordinative interactions.

These polymers have been reviewed in detail recently [170], so only a few examples are given. Pyrazine-bridged polymers $[PcM(pyr)]_n$ (M = Fe(II), Ru(II), Co(II), Os(II)) have been prepared, for example in the case of Fe, by heating PcFe with an excess of pyrazine in chlorobenzene [190] or, in the case of Ru, by heating the adduct PcRu(pyz)$_2$ at 300 °C [191]. The polymers [PcM(tetrazine)]$_n$ (M = Ru(II), Os(II), Fe(II)) are obtained by heterogeneous reaction of the PcM(II) in a slight excess of tetrazine at in chlorobenzene (experiment 7–13, chapter 7.7) [166,192,193].

The cofacially stacked polymeric phthalocyanines can exhibit high electrical conductivities in either the undoped or the doped state (Table 7-1) [166,168,170]. For comparison, low molecular weight phthalocyanines MPc or naphthalocyanines MNc (M = Fe, Cu, Co, Ni, Zn) exhibit values between σ = 10^{-8} and 10^{-15} S cm^{-1}. For undoped polymers, Table 7-1 shows that polymerization through cyano- and tetrazine-bridging is particularly effective in increasing the conductivity. This effect can be ascribed to the existence of additional interactions (π-orbital interactions through the bonds) [170,194].

Table 7-1. Conductivity values of some polymeric cofacially stacked phthalocyanines (Pc) and naphthalocyanines (Nc).

Undoped samples		Doped samples			
Compound	σ (S cm^{-1})	Compound	Dopant (moles mol Pc)	I$_2$ per	σ (S cm^{-1})
[2,3NcFe(tetrazine)]$_n$	$3 \cdot 10^{-1}$	[PcAlF]$_n$	3.3		5.1
[2,3NcCo(CN)]$_n$	$1 \cdot 10^{-1}$	[PcFe(pyz)]$_n$	2.6		0.2
[PcFe(tetrazine)]$_n$	$2 \cdot 10^{-2}$	[PcCo(CN)]$_n$	1.6		0.6
[PcFe(CN)]$_n$	$6 \cdot 10^{-3}$	[PcSiO]$_n$	1.6		1.4
[PcFe(pyz)]$_n$	$1 \cdot 10^{-6}$				
[PcSiO]$_n$	$1 \cdot 10^{-6}$				

7.6 Metallodendrimers

Metallodendrimers are an interesting class of molecules in the area of dendrimer chemistry. They combine dendritic structures with the specific activity of metal complex centers. Metal coordination has facilitated the synthesis of a number of dendritic, supramolecular structures. Metals have been incorporated in all of the topologically different parts of dendrimers: in the repeat or branching unit, in the molecular core and in the peripheral units. Because this field of metallodendrimers has been reviewed recently [195–197], only a few examples are given below. Other supramolecular organizations such as catenanes and rotaxanes have been mentioned previously in this chapter.

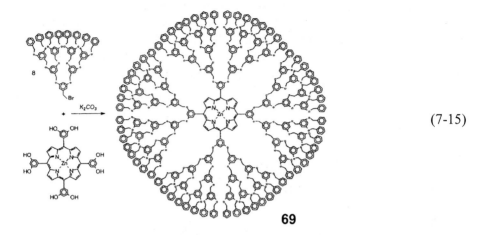

(7-15)

69

The synthesis of a dendrimer **69** of the fourth generation with a substituted tetraphenylporphyrin in the cores is shown in Eq. 7-15 [198]. The convergent strategy is more useful than the divergent one as it avoids several preparative steps. Structurally comparable dendrimer porphyrins bearing negatively or positively charged groups at the peripheral units making them water soluble have been described [199]. A silicon(IV) phthalocyanine **70**, substituted at the axial positions with dendritic groups, shows no aggregation by phthalocyanine–phthalocyanine interactions in the solid state [200,201]. Dendrimers built around [Ru(bpy)$_3$]$^{2+}$ (bpy = 2,2′-bipyridine) have been investigated for photoactive and redox-active properties [202].

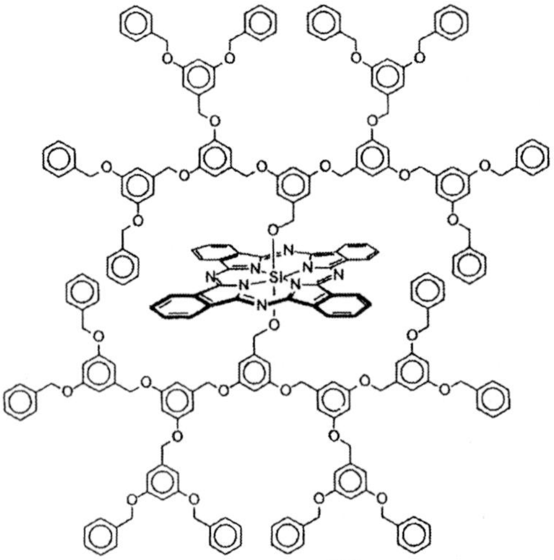

70

One example of a dendrimer substituted at the peripheral units with a metal complex is shown in Eq. 7-16. **71** is obtained by treating the fourth generation of the poly(propylene imine) dendrimer (DAB-dendr-$(NH_2)_x$) ($x = 32$) with chlorocarbonylferrocene and the PF_6^- salt of chlorocarbonylcobaltocenium [202]. Statistical substitution by the two different metallocenes is observed.

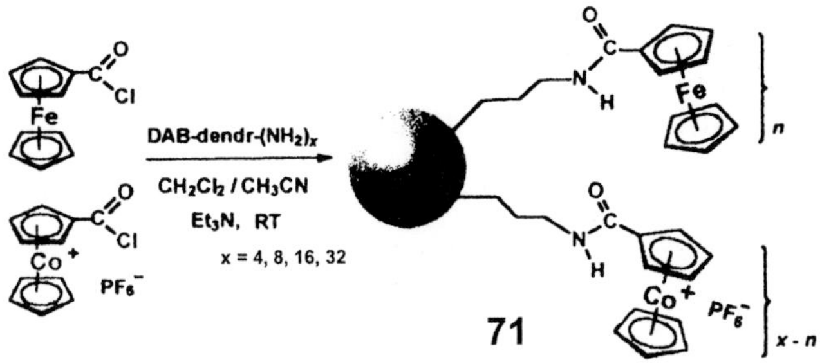

(7-16)

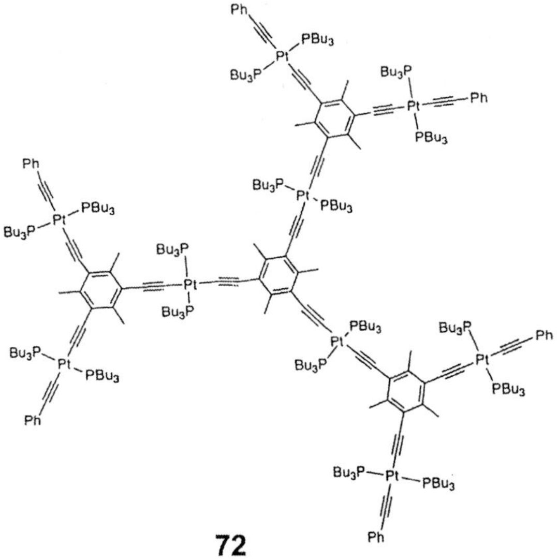

71

72

Different types of metallodendrimers with metal compounds at each repeating unit throughout the dendrimer have been reviewed [195,197]. Because of their high concentration of metal centers, they can be considered as nanoparticle equivalents. One example is the platinum acetylide dendrimer **72** obtained by a stoichiometric transmetallation to form acetylides [204]. This metallodendrimer has a precedent in earlier syntheses of linear Pt-acetylide coordination polymers. Several papers describe the synthesis of various Ru pyridyl complexes as repeating units for photochemical investigations and non-linear optics [197,205–208].

The properties of metallodendrimers with a view to their potential applications have been reviewed in [195–197]: sensors, binding of small molecules, catalysis, reactive centers, molecular antennas by visible light irradiation.

7.7 Experimental

Experiment 7-1:   Synthesis of the Platinum Poly(yne) 4 (Section 7.2) [21]

Preparation of the polymer 4: Copper(I) iodide (7 mg, 0.037 mmol) was added under argon or nitrogen to a stirred oxygenated solution of *trans*-bis-[butadiinyl]-bis-[tributylphosphan]platinum (1.38 g, 2.0 mmol) and bis-[tributylphosphan]-dichloroplatinum (1.34 g, 2.0 mmol) in 50 mL diethylamine. The solution was gentle refluxed with stirring for 24 h and then evaporated to dryness under reduced pressure. The residue was dissolved in methylene chloride, and the solution was filtered by alumina-column chromatography to separate the polymer from the copper salt. After evaporation the pale-yellow product was purified by repeated precipitation from methylene chloride into methanol. Then a benzene solution was frozen and then freeze-dried under reduced pressure. Yield 2.5 g (96%). Molecular weight M_w ~100,000 Da. IR (KBr): 1999 (C≡C) cm^{-1}.

Experiment 7-2:   Synthesis of Luminescent Schiff-Base Electrolytes 15 (Section 7.3.1) [64a,64d]

We describe the preparation of **15** with M = Y, Eu and X= –CH$_2$– according to Eq. 7-10. Reference [64d] also contains the synthesis of Ce-containing polymers.

Preparation of N,N',N'',N'''-tetrasalicylidene-3,3',4,4'-tetraaminodiphenyl-methane (14, X = –CH$_2$–) [64d]: 3,3',4,4'-Tetraaminodiphenylmethane (0.838 g, 3.67 mmol) was added to neat salicylaldehyde (10 mL, 95.5 mmol) at 75 °C. After the mixture was heated at this temperature for 2 h, 30 mL of methanol was introduced to the yellow reaction solution and the resulting mixture was heated under reflux conditions for 30 min. The system was then chilled to below 0 °C and filtered to isolate the crude Schiff-base ligand. Purification was accomplished by dissolution of the ligand in 25 mL of CH$_2$Cl$_2$ and precipitation with 100 mL of petroleum ether. Yield: 2.07 g (87% or 43% overall, based on the original diamine). Mp = 195 °C. ^{1}H-NMR (DMSO–d_6): 4.10 (s, 2); 6.91–7.68 (m, 22); 8.92 (s, 4); 12.89–13.04 (m, <4) ppm.

Preparation of the polyelectrolyte [(15 Y)⁻Na⁺]ₙ (X = –CH₂–) [64b]: **14** (0.9050 g, 1.405 mmol) was dissolved in 45 mL of DMSO with 0.7153 g (1.405 mmol) of Y(NO₃)₃·3DMSO at 75 °C under N₂. NaOH (0.2698 g, 6.744 mmol) in 10 mL of methanol was added to the solution. The resulting solution was allowed to react at 75 °C under N₂ for 16 h. The solution was filtered and then slowly a mixed solvent (ether:toluene 3:1) was added with stirring until the solution became turbid; then another 20 mL of the mixed solvent was added to the solution. After several minutes, an oil sank to the bottom of the flask to form a film. The solution was decanted from the oil, and the oil film was transformed to a precipitate by washing with methanol. The decanted solution was treated with another 30 mL of the mixed solvent and another product was separated as above. The product was dried in vacuo at 100 °C. Yield ~80%. M_n = 20,000 Da.

Preparation of the polyelectrolyte [(15 Y_xEu_{1-x})⁻Na⁺]ₙ (X = –CH₂–) [64b]: **14** (500 g, 0.776 mmol) was dissolved in 25 mL of DMSO with a total of 0.776 mmol of Eu(NO₃)₃·3DMSO and Y(NO₃)₃·3DMSO at 75 °C under N₂. The mole ratio of Y(III) to Eu(III) was varied from 1:9 to 6:1 as well as 0:1 and 1:0 (see [64b]). NaOH (0.149 g, 3.73 mmol) in 5 mL of methanol was added to each solution, which was then allowed to react for 16 h at 75 °C under N₂ with stirring. Spontaneous precipitation occurred during the reaction in some trials. The products were obtained by adding 100 mL of a mixed precipitant (1:1 (v/v) methanol and water containing 5 g of NaNO₃) to each solution. The products were filtered, washed with distilled water and dried in vacuo at 100 °C. Yield: > 97%. M_w ~10,000 Da.

Experiment 7-3: Synthesis of Polydithiooxamide Metal Complexes from 22 (Section 7.3.1) [74]

Preparation of polymers from dithiooxamide 22: In a three-necked 100 mL flask, equipped with a reflux condenser and a glass inlet tube, a mixture of cobalt chloride hexahydrate (2.38 g, 0.01 mol) in 20 mL ethanol and dithiooxamide **22** (1.2 g, 0.01 mol) in 20 mL ethanol and 5 mL DMSO were added and stirred at 70–80 °C, under a thin stream of nitrogen gas, for 24 h. At the end of the reaction, the mixture was poured into 300 mL of distilled water. The dark powdery precipitated polymer complex was filtered by suction in a Buchner funnel, washed thoroughly five times with hot distilled water, twice with diethyl ether and then dried under vacuum at 40 °C for 24 h. The light-brown polymeric cobalt complexes were obtained in a yield of 87%. IR (KBr): ~1300–1390 (C–N), ~1080–1170 (C=S) cm⁻¹.

Experiment 7-4:   Synthesis of a Polymeric Cobalt(II) Complex 23 with Bridging Succinonitrile (Section 7.3.1) [75]

Preparation of the complex $[Co(NC-CH_2-CH_2-CN)Cl_2(H_2O)_2]_n \cdot 2nH_2O$ 23: 0.49 mL (27 mmol) H_2O was added to 0.58 g (4.5 mmol) $CoCl_2$ and the mixture was then dissolved in 10 mL of dry acetone giving a clear royal-blue solution. 1.20 g (15 mmol) of succinonitrile was added while the reaction mixture was stirred frequently. The reaction was carried out while stirring gently. After about 30 min, a yellow-orange precipitate appeared, which was removed after a total reaction time of 24 h, when the solution had become significantly lighter in color. The crystals obtained were washed with a few milliliters of dry acetone, dried and stored in a refrigerator. For long-term storage, keeping it cooled and under argon is recommended. The polymer 23 is obtained in a yield of 80%. IR spectra are discussed in [75].

Experiment 7-5:   Synthesis of the Silver(I) Coordination Polymer 28b with Nanometer Tubes (Section 7.3.1) [83]

Preparation of the coordination polymer 28b: At first a solution of 2,4,6-tris[(4-pyridyl)-methylsulfanyl]-1,3,5-triazine (1.5 mmol) was prepared by heating a solution of 4-picolylchloride hydrochloride (4.5 mmol), Sodium trithiocyanorate (1.5 mmol) and sodium methylate (4.5 mmol) in 25 mL methanol at 50 °C. To this solution $AgNO_3$ (3.0 mmol) dissolved in 15 mL DMF was added. After 2 h stirring $AgClO_4$ (1.0 mmol) dissolved in 10 mL DMF was added. After 2 h stirring a colorless solution was obtained and filtered. Diethylether was diffused into this solution to obtain colorless needles of 28b·2H$_2$O in a yield of 0.76 g (61%). IR (KBr): 3450 (m), 2929 (w), 2906 (w), 1676 (s), 1616 (s), 1479 (s), 1429 (m), 1261 (s), 1090 (s), 802 (s), 704 (m), 621 (m) cm^{-1}.

Experiment 7-6:   Synthesis of the Two-Dimensional Coordination Polymer 34 from Cadmium(II) and 4,4'-Bipyridine (Section 7.3.2.1) [109]

Preparation of $[Cd(4,4'-bpy)_2(H_2O)_2(NO_3)_2 \cdot 4H_2O]_n$ 34: A methanolic solution (10 mL) of 4,4'-bipyridine (0.156 g, 10 mmol) was added dropwise to a stirred aqueous solution (5 mL) of $Cd(NO_3)_2 \cdot 4H_2O$ (0.308 g, 1.0 mmol) at 50 °C over 15 min. An ethanolic solution (10 mL) of 4,4'-bipyridine (0.156 g, 1.0 mmol) was then added. The resulting colorless solution was allowed to stand in air at RT for three weeks, yielding colorless block-like crystals (64% yield based on bipyridyl). IR (KBr): 3409 (m), 3107 (w), 3058 (w), 1609 (s), 1539 (w), 1489 (s), 1384 (vs), 1278 (s), 1222 (m), 1075 (m), 1039 (m), 1011 (w), 983 (w), 807 (m) cm^{-1}.

Experiment 7-7: Synthesis of the Three-Dimensional Coordination Polymer 40 from Nickel(II), Benzenetetracarboxylic Acid and 4,4′-Bipyridine (Section 7.3.2.1) [119]

Preparation of the coordination polymer 40. A mixture of NiCl$_2$·6H$_2$O (0.12 g, 0.50 mmol), 1,2,4,5-benzenetetracarboxylic anhydride (0.10 g, 0.46 mmol), 4,4′-bipyridine·2H$_2$O (0.10 g, 0.52 mmol) and NH$_3$·H$_2$O (0.10 mL, 1.3 mmol) in H$_2$O (18 mL) was heated at 170 °C for six days under autogeneous pressure in a sealed 30 mL Teflon-lined stainless-steel vessel. Green crystals were isolated after the reaction solution was cooled gradually, and washed with water and ethanol. Yield 87%. IR (KBr pellet): 1645 (m), 1614 (m), 1576 (s), 1537 (s), 1493 (m), 1471 (m), 1417 (s), 1392 (s), 1367 (s), 1325 (m), 1308 (m), 1261 (s), 1221 (s), 1140 (m), 1130 (m), 1111 (m), 1078 (s), 1020 (m), 914 (m), 885 (m), 862 (s), 818 (s), 804 (s) cm^{-1}.

Experiment 7-8: Synthesis of the Two-Dimensional Chiral Coordination Polymer 41 from Zinc(II) and Ethyl *S*-Lactate (Section 7.3.2.1) [128]

Preparation of the coordination polymer 41: A heavy-walled Pyrex tube containing a mixture of Zn(ClO$_4$)$_2$·6H$_2$O (0.372 g, 1 mmol), ethyl *S*-lactate (2 mL), pyridylacrylic acid (0.149 g, 1 mmol), H$_2$O (0.2 mL), pyridine (0.1 mL) and ethanol (0.1 mL) was frozen in liquid N$_2$, sealed under vacuum and placed inside f an oven at 75 °C. The colorless block crystals were harvested after heating for 16 h. Yield 29.0% (0.0705 g) based on Zn(ClO$_4$)$_2$·6H$_2$O. IR (KBr): 3000 (m), 1600 (sh, s), 1460 (w), 1415 (m), 1365 (m), 1310 (m), 1285 (w), 1100 (s), 1090 (m), 1050 (m), 890 (w), 850 (m) cm^{-1}.

Experiment 7-9: Synthesis of the Tetrapyridylporphyrin Coordination Network 51 (Section 7.3.2.1) [136]

Preparation of the coordination network 51: A mixture of 0.2384 g CoCl$_2$·2H$_2$O, 0.1243 g metal-free 5,10,15,20–tetrapyridylporphyrin, aqueous CsOH solution (0.3 mL, 50 wt. %), 4 mL CH$_3$COOH and 4 mL water was heated in a Teflon-coated 23 mL steel autoclave for 48 h at 200 °C. Then the autoclave was cooled at 9 K h^{-1} to 70 °C. The purple-colored crystals were filtered, washed with methanol and obtained in a yield of 85% (0.142 g based on the porphyrin). X–ray diffraction showed that the material exists in only one phase.

Experiment 7-10: Synthesis of the Poly(ferrocenylsilane) 64 (X = –Si(CH₃)₂) (Section 7.4) [157]

Preparation of 1,1'-dilithioferrocene–TMEDA [209]: The whole procedure is carried out under an inert, dry atmosphere with freshly dried solvents stored under inert gas using standard Schlenk line techniques. *n*–Butyllithium (160 mL, 2.3 M) in *n*-hexane), was added to 30 g (0.16 mol) of ferrocene (recrystallized from hexane), followed by 30 mL (0.20 mol) of *N.N,N',N'*-tetramethylethylenediamine (TMEDA, distilled from BaO under argon). The reaction was allowed to proceed overnight, ca. 16 h. The product was filtered off, washed with 3 × 100 mL of hexane, and dried under vacuum. This gave 44.9 g (92% yield) of an orange pyrophoric powder, sensitive to moisture. The compound was recrystallized from THF at –40 °C.

Preparation of [1]Silaferrocenophane 63 [157a]: The preparation was carried out under inert gas using the Schlenk technique or in a glove-box. Over a period of 5 min, 4.6 mL of Me₂SiCl₂ (38 mmol) was added dropwise to a suspension of 10.0 g of dilithioferrocene–TMEDA (36.4 mmol) in 500 mL of diethyl ether at –30 °C during which the reaction mixture changed from orange-yellow to red. The reaction mixture was then slowly left to warm to 20 °C over 2 h and filtered, and solvent and excess Me₂SiCl₂ were removed under vacuum. The crude product was then sublimed at room temperature onto a cold probe. Thus 6.98 g (79%) of red crystalline [1]ferrocenophane was obtained. ¹H NMR (CDCl₃): 0.51 (s, 6H, Me), 4.08 (t, 4H, Cp), 4.48 (t, 4H, Cp) ppm.

Thermal ring-opening polymerization of 63 [157b]: A sample of the monomer 63 (450 mg, 1.51 mmol) was flame-sealed in an evacuated (0.01 mm Hg) Pyrex tube and heated at 150 °C for 3 h. The tube contents were dissolved in THF, filtered, and precipitated into methanol. The fibrous orange solid of 64 was filtered off, washed with methanol (3 × 5 mL), and dried under vacuum. Yield 340 mg (75%). $M_w = 1.2 \cdot 10^5$ Da.

Transition metal-catalyzed ring-opening polymerization of 63 [157b]: A solution of the monomer 63 (350 mg, 1.17 mmol) in toluene (5 mL) was stirred for 12 h in the presence of PtCl₂ (1 mol %) as catalyst. The precipitated solid was separated by filtration, washed with THF (3 × 5 mL) until the washings were colorless, and then washed with methanol (2 × 5 mL). The powdery orange solid of 64 was then dried under vacuum. Yield 310 mg (87%).

Anionic ring-opening polymerization of 63 initiated by n-*BuLi [157a]:* The reaction was carried out by addition of 34.4 μL of a 0.18 M solution of *n*-BuLi in hexane to a solution of 150 mg (0.5 mmol) of **63** in 2 mL of THF (molar ratio **63**: BuLi = 100). After 60 min the reaction was terminated by addition of few drops of H_2O. The polymer **64** was isolated by precipitation into hexane and dried under vacuum. Yield 62%. M_w =22,000 Da.

Experiment 7-11: Synthesis of Poly−[phthalocyaninatosilicium-(IV)−μ−(oxo)] 68 (M= Si(IV), R= −O−) (Chapter 7.5) [174,210)

Dihydroxy−phthalocyaninato-silicium(IV) (200 mg) (Aldrich) was heated in vacuo (~10^{-3} torr) for 12 h at 440 °C. The polymer was obtained in a yield of 190 mg (~100 %). If better soluble phthalocyanines such as octaalkoxy-substituted derivatives are used, the reaction can be carried out in boiling toluene under catalysis of $FeCl_3$ [211].

Experiment 7-12: Synthesis of poly-[phthalocyaninatogermanium-(IV)-μ-(1,4-dioxyphenylene)] 68 (M= Ge(IV), R= -O-C₆H₄-O) (Chapter 7.5) [174,212]

Dihydroxy-phthalocyaninato-germanium(IV) (0.62 g, 1 mmol) and hydroqui-none (110 mg, 1 mmol) were heated in 10 mL 1-chloronaphthalene in inert gas under reflux for 5 h. The suspension was filtered and the residue was treated intensively with ethanol in a Soxhlet apparatus. Yield 0.38 g (55 %).

Experiment 7-13: Synthesis of poly-[phthalocyaninatoeisen(II)-μ-(1,2,4,5-tetrazine)] 71 (M= Fe(II), R= tetrazine) (Chapter 7.5) [213]

1,2,4,5-Tetrazine (0.41 g, 5 mmol) was dissolved in 70 mL chlorobenzene, and β-FePc (chapter 5, **48** with M= Fe(II) and R= -H) (1.14 g, 2 mmol was added. The suspension was stirred at 110 – 120 °C in the dark for 48 h. The dark-blue precipitate was filtered, washed with ethanol and dried in vacuo at 60 °C. Yield 1.36 g (97 %).

7.8   References

1. D. Wöhrle, Polymers with Metals in the Backbone, in: *Handbook of Polymer Synthesis*, Vol. B, H. Kricheldorf (Ed.), Marcel Dekker, New York, 1992.
2. I. Manners, Angew. Chem. Int. Ed. Engl. **1996**, 35, 1602.
3. T. Kodaira, A. Watanabe, O. Ito, K. Mochida, *Adv. Mater.* **1995**, 7, 917.
4. G. Schott, *Z. Chem. (Leipzig)* **1992**, 6/7, 194.

5. H. Schäfer, B. Eisenmann, W. Müller, *Angew. Chem. Int. Ed. Engl.* **1973**, *12*, 694.
6. G. Vogg, M.S. Brandt, M. Stutzmann, *Adv. Mater.* **2000**, *12*, 1278.
7. K. Kroogman, *Angew. Chem.* **1969**, *81*, 10.
8. G.M. Finnis, E. Canadell, C. Compana, K.R. Dunbar, *Angew. Chem.* **1996**, *108*, 2946.
9. S. Wang, G. Garzon, C. King, *Inorg. Chem.* **1989**, *28*, 4623.
10. O. Crespo, E.J. Fernandez, P.G. Jones, *Chem. Commun.* **1998**, 2233.
11. J.S. Miller (Ed.), *Extended Linear Chain Compounds*, Plenum Press, New York, 1982, Vol. 1.
12. C.E. Carraher, J.E. Sheats, C.U. Pittman (Eds.), *Organometallic Polymers*, Academic Press, New York, 1978.
13. C.E. Carraher, J.E. Sheats, C.U. Pittman (Eds.), *Advances in Organometallic and Inorganic Polymer Science*, Marcel Dekker, New York, 1982.
14. J.E. Sheats, C.E. Carraher, C.U. Pittman (Eds.), *Metal-Containing Polymeric Systems*, Plenum Press, New York, 1985.
15. L. Oriol, J.L. Serrano, *Adv. Mater.* **1995**, *7*, 348.
16. L. Alexandrova, D. Likhatchev, S. Muhl, G. Gerasimov, *J. Inorg. Organomet. Polym.* **1998**, *8*, 157.
17. N. Hagihara, K. Sonogashira, S. Takahashi, *Adv. Polym. Sci.* **1981**, *41*, 149.
18. A.S. Hay, *J. Polym. Sci. A-1* **1969**, *7*, 1625.
19. S. Takahashi, J. Polym. Sci., Polym. Chem. Ed. **1980**, 18, 661.
20. K. Sonogashira, *J. Organomet. Chem.* **1980**, *188*, 1237.
21. K. Sonogashira, S. Takahashi, N. Hagihara, *Macromolecules* **1977**, *10*, 879.
22. K. Sonogashira, Y. Fujikura, *J. Organomet. Chem.* **1978**, *145*, 101.
23. S.J. Davies, B.F.G. Johnson, M.S. Khan, J. Lewis, *Chem. Commun.* **1991**, 187.
24. M. Younus, A. Köhler, S. Cron, *Angew. Chem.* **1998**, *110*, 1998.
25. K. Onitsuka, Y. Harada, F. Takei, S. Takahashi, *Chem. Commun.* **1998**, 643.
26. H. Nishihara, H. Shimura, A. Ohkubo, N. Matsuda, K. Aramaki, *Adv. Mater.* **1993**, *5*, 752.
27. Z. Atherton, C.W. Faulkner, S.L. Ingham, *J. Organomet. Chem.* **1993**, *462*, 265.
28. C.W. Faulkner, S.L. Ingham, M.S. Khan, *J. Organomet. Chem.* **1994**, *482*, 139.
29. M.S. Khan, A.K. Kakkar, R.H. Friend, *J. Organomet. Chem.* **1994**, *472*, 247.
30. A. Abe, S. Tabata, N. Kimura, *Polym. J.* **1991**, *23*, 69; *Macromolecules* **1991**, *24*, 6238.
31. A.E. Dray, R. Rachel, R.H. Friend, *Macromolecules* **1992**, *25*, 3473.
32. E. Solari, J. Hesschenbrouck, R. Scopelliti, C. Floriani, N. Re, *Angew. Chem.* **2001**, *113*, 958.
33. S.S. Dhingra, G.R. Kowach, R.K. Kremer, *Angew. Chem.* **1997**, *109*, 1127.
34. M.R. Mattner, W.A. Herrmann, R. Berger, C. Gerber, J.K. Gimzewski, *Adv. Mater.* **1996**, *8*, 654.
35. M. Wörle, R. Nesper, *Angew. Chem.* **2000**, *112*, 2439.
36. M. Tschinkl, A. Schier, J. Riede, F.P. Gabbai, *Angew. Chem.* **1999**, *111*, 3769.
37. B.A. Adair, G.D. de Delgado, A.K. Cheetham, *Angew. Chem.* **2000**, *112*, 761.
38. B. Räke, P. Müller, H.W. Roesky, I. Uson, *Angew. Chem.* **1999**, *111*, 2069.
39. J.A. Hanko, M.G. Kanatzidis, *Angew. Chem.* **1998**, *110*, 354.
40. Y.V. Mironov, A.V. Virovets, S.B. Artmekina, V.E. Feodorov, *Angew. Chem.* **1998**,

110, 2656.

41. a) Y. Heng, W. Zhang, X. Wu, T. Sheng, Q. Wang, P. Lin, *Angew. Chem.* **1998**, *110*, 2662. b) J. Bai, E. Leiner, M. Scheer, *Angew. Chem.* **2002**, *114*, 820.

42. F. Ciardelli, E. Tsuchida, D. Wöhrle (Eds.), *Macromolecule Metal Complexes*, Springer, Berlin, 1996.

43. A.K. Dey, *J. Indian Chem. Soc.* **1986**, *63*, 557.

44. T.A. Ibidapo, *Adv. Polym. Sci.* **1990**, *30*, 1151.

45. C.E. Carraher, J.A. Schroeder, *J. Polym. Sci., Polym. Lett. Ed.* **1975**, *13*, 215.

46. P.C. Srivastava, K.B. Pandeya, H.I. Nigram, *J. Inorg. Nucl. Chem.* **1973**, *35*, 3613.

47. J.T. Wrobleski, D.B. Brown, *Inorg. Chem.* **1979**, *18*, 498; 2738.

48. C.R. Theocharis, *Chem. Commun.* **1987**, 80.

49. T.R. Rao, P.R Rao, P. Lingaiah, L.S. Deshmukh, *Angew. Makromol. Chem.* **1991**, *191*, 177.

50. S.N. Poddar, N. Saha, *Indian J. Appl. Chem.* **1970**, *33*, 244.

51. V. Banerjie, *Polymer Bull.* **1979**, *1*, 685.

52. V. Banerjie, Makromol. Chem. Rapid Commun. **1980**, 1, 41.

53. V.D. Kumari, M. Khave, K.N. Munishi, *Indian J. Chem.* **1985**, *24A*, 72.

54. J.R. Reynolds, C.P. Lillya, J.C.W. Chien, *Macromolecules* **1987**, *20*, 1184.

55. R.R. Schumater, E.M. Engler, *J. Am. Chem. Soc.* **1977**, *99*, 5521.

56. N.M. Rivera, E.M. Engler, R.R. Schumater, *Chem. Commun.* **1979**, 184.

57. J. Ribas, P.C. Cassoux, *R. Seances Acad. Sci.* **1981**, *293*, 665.

58. C.W. Dirk, E.A. Mintz, K.F. Schoch, T.J. Marks, in: *Advances in Organometallic and Inorganic Polymer Science*, C.E. Carraher, J.E. Sheats, C.U. Pittmann (Eds.), Marcel Dekker, New York, 1986, p. 275.

59. B.K. Teo, F. Wudl, J.J. Hauser, A. Krüger, *J. Am. Chem. Soc.* **1977**, *99*, 4862.

60. M.C. Böhm, *Phys. Stat. Sol. (B)* **1984**, *121*, 255.

61. R. D. Archer, *Inorganic and Organometallic Polymers*, Wiley-VCH, New York, 2001.

62. R.W. Kluiber, J.W. Lewis, *J. Am. Chem. Soc.* **1960**, *82*, 5777.

63. J.A. Cronin, S.M. Palmer, R.D. Archer, *Inorg. Chim. Acta* **1996**, *251*, 81.

64. a) H. Chen, R.D. Archer, *Macromolecules* **1995**, *28*, 1609. b) ibid. **1996**, *29*, 1957. c) R.D. Archer, H. Chen, L.C. Thompson, *Inorg. Chem.* **1998**, *37*, 2089. d) H. Chen, J.A. Cronin, R.D. Archer, *Inorg. Chem.* **1995**, *34*, 2306.

65. M.J. Plater, M.R.J. Foremann, T. Gelbrich, M.B. Hursthouse, *Inorg. Chim. Acta* **2001**, *318*, 171.

66. A. Williams, B.D. Alleynes, L.A. Hall, *Inorg. Chim. Acta* **2001**, *313*, 56.

67. R.W. Saalfrank, H. Maid, N. Mocren, F. Hampel, *Angew. Chem.* **2002**, *114*, 323.

68. M. Oh, G.B. Carpenter, D.A. Sweigart, *Angew. Chem.* **2001**, *113*, 3291.

69. Y. Hashino, Y. Hagihara, *Inorg. Chim. Acta* **1999**, *292*, 64.

70. M. Pyrasch, D. Amirbeyki, B. Tieke, *Adv. Mater.* **2001**, *13*, 1188.

71. T. Sheng, X. Wu, Q. Wang, X. Gao, P. Liu, *Polyhedron* **1998**, *17*, 4519.

72. J. Dai, M. Munakata, L. Wu, Y. Suenaga, *Inorg. Chim. Acta* **1997**, *258*, 65.

73. M. Yamamoto, X. Gan, T. Kuroda-Sowa, M. Maekawa, Y. Suenaga, M. Munakata, *Inorg. Chim. Acta* **1997**, *261*, 169.

74. A. El-Shekeil, M.A. Khalid, H. Al-Maydama, A. Al-Karbody, *Europ. Polym. J.* **2001**, *37*, 575.

75. O.I. Fengler, A. Ruoff, *Spectrochim. Acta Part A* **2002**, *58*, 567.
76. C.C. Shang, B. Pfennig, A.B. Bocarsly, *Coord. Chem. Rev.* **2000**, *208*, 33.
77. N. Krause, *Angew. Chem.* **1999**, *111*, 83; C.M.P. Kronenberg, J. Jastrzebski, A.L. Spek, G. van Koten, *J. Am. Chem. Soc.* **1998**, *120*, 9688.
78. U. Velten, B. Lahn, M. Rehahn, *Macromol. Chem. Phys.* **1997**, *198*, 2789.
79. R. Knapp, A. Schott, M. Rehahn, *Macromolecules* **1996**, *29*, 478.
80. S. Kelch, M. Rehahn, *Macromolecules* **1997**, *30*, 6185.
81. C. Tschierske, *Angew. Chem.* **2000**, *112*, 2547
82. G. Konishi, K. Naka, Y. Chujo, *J. Inorg. Organomet. Polym.* **1999**, *9*, 179.
83. M. Hong, Y. Zhao, *Angew. Chem.* **2000**, *112*, 2586.
84. A.N. Khlobystov, A.J. Blake, N.R. Champness, D.A. Lemenovskii, A.G. Majouga, N.V. Zyk, M. Schröder, *Coord. Chem. Rev.* **2001**, *222*, 155.
85. R. Rsoa, N. Mezailles, L. Ricard, F. Mathey, P. Le Floch, *Angew. Chem.* **2001**, *113*, 4608.
86. J. Bai, A.V. Virovets, M. Scheer, *Angew. Chem.* **2002**, *114*, 1809.
87. F.A. Cotton, C. Lin, C.A. Murrillo, *Acc. Chem. Res.* **2001**, *34*, 759.
88. N.H. Ray, *Inorganic Polymers*, Academic Press, New York, 1978; F.G.A. Stone, W.A.G. Graham (Eds.), *Inorganic Polymers* Academic Press, New York, 1962.
89. M. Ohba, N. Usuki, N. Fukita, H. Okawa, *Angew. Chem.* **1999**, *111*, 1909; K.V. Langenberg, S.R. Batten, K.J. Berry, D.C.R. Hockless, B. Moubaraki, K.S. Murray, *Inorg. Chem.* **1997**, *36*, 5006; S.M. Holmes, G.S. Girolami, *J. Am. Chem. Soc.* **1999**, *121*, 5593.
90. J.W. Rosthauser, A. Winston, *Macromolecules* **1981**, *14*, 538; H.J. Buser, *Inorg. Chem.* **1997**, *16*, 2704; M. Kaneko, X.-H. Hou, A. Yamada, *Bull. Chem. Soc. Jpn.* **1984**, *57*, 156; M. Kaneko, *J. Macromol. Sci. Chem.* **1987**, *A24*, 357.
91. S.M. Holmes, G.S. Girolami, *J. Am. Chem. Soc.* **1999**, *121*, 5593.
92. F. Hulliger, M. Ladolt, H. Vetsch, *J. Solid State Chem.* **1976**, *18*, 283.
93. W.-F. Yeung, W.-L. Man, W.-T. Wong, T.-C. Lau, S. Gao, *Angew. Chem.* **2001**, *113*, 3121.
94. B. Q. Ma, S. Gao, G. Su, G.X. Xu, *Angew. Chem.* **2001**, *113*, 448.
95. N. Kimizuka, T. Handa, I. Ichinose, T. Kunitake, *Angew. Chem.* **1994**, *106*, 2576.
96. L.G. Beauvais, M.P. Shores, J.R. Long, *J. Am. Chem. Soc.* **2000**, *122*, 2763.
97. Z. Xu, S. Lee, Y.-H. Kiang, A.B. Mallik, N. Tsomaia, K.T. Mueller, *Adv. Mater.* **2001**, *13*, 637.
98. P. Losier, M.J. Zaworotko, *Angew. Chem.* **1996**, *108*, 2957.
99. M.J. Zaworotko, *Angew. Chem.* **1998**, *110*, 1269.
100. L.R. MacGillivray, S. Subramanian, M.J. Zaworotko, *Chem. Commun.* **1994**, 1325; M.J. Zaworotko, *Chem. Soc. Rev.* **1994**, 283.
101. G.B. Gardner, D. Venkataraman, J.S. Moore, S. Lee, *Nature* **1993**, *374*, 792.
102. M. Fujita, Y.J. Kwon, S. Washizu, K. Ogura, *J. Am. Chem. Soc.* **1994**, *116*, 1151.
103. F.P. Gabbai, A. Schier, J. Riede, *Angew. Chem.* **1998**, *110*, 646.
104. M. Fujita, *J. Am. Chem. Soc.* **1995**, *117*, 7287.
105. S. Subramanian, M.J. Zaworotko, *Angew. Chem.* **1995**, *107*, 2295.
106. M.R. Mattner, W.A. Herrmann, *Adv. Mater.* **1996**, *8*, 654.
107. B. Kumar, C. Seward, M.J. Zaworotko, *Angew. Chem.* **1999**, *111*, 584.
108. C. Janiak, *Angew. Chem.* **1997**, *109*, 1499.

109. M.L. Toang, S.L. Zheng, H.-M. Chen, *Polyhedron* **2000**, *19*, 1809.
110. B.F. Hoskins, R. Robson, *J. Am. Chem. Soc.* **1990**, *112*, 1546.
111. L. Carlucci, G. Ciani, *Angew. Chem.* **2002**, *114*, 1987.
112. K. Biradha, Y. Hongo, M. Fujita, *Angew. Chem.* **2000**, *112*, 4001.
113. N.G. Pschirer, U.H.F. Bunz, H.C. zur Loye, *Angew. Chem.* **2002**, *114*, 603.
114. S.R. Batten, B.F. Hoskins, R. Robson, *Angew. Chem.* **1995**, *107*, 884.
115. B.F. Hoskins, R. Robson, D.A. Slizys, *Angew. Chem.* **1997**, *109*, 2861.
116. L.P. Wu, T. Kamikawa, M. Munakata, *Inorg. Chim Acta* **1997**, *256*, 155; N. Masciocchi, G.A. Ardizzoia, *Inorg. Chem.* **2001**, *40*, 6983.
117. M. Kondo, T. Okubo, S. Kitagawa, *Angew. Chem.* **1999**, *111*, 190.
118. L. Pan, X. Huang, Y. Wu, N. Zheng, *Angew. Chem.* **2000**, *112*, 537.
119. C.D. Wu, C.-Z. Lu, S.-F. Lu, H.-H. Huang, J.S. Huang, *Inorg. Chem. Commun.* **2002**, *5*, 171.
120. J.M. Plater, M.R. Foreman, R.A. Howie, J.M.S. Skakle, A.M.Z. Slawin, *Inorg. Chim. Acta* **2001**, *315*, 126.
121. D.M. Ciurtin, M.D. Smith, H.-C. zur Loye, *Inorg. Chim. Acta* **2001**, *324*, 46.
122. D.-L. Long, A.J. Blake, N.R. Champness, C. Wilson, M. Schröder, *Angew. Chem.* **2001**, *113*, 2510.
123. D.-Q. Chu, J.-Q. Xu, *Eur. J. Inorg. Chem.* **2001**, 1135.
124. S.-L. Zhong, M.-L. Tong, X.-M. Chen, *Organometallics* **2001**, *20*, 5319.
125. S.A. Bourne, J. Lu, A. Mondal, B. Moulton, M.J. Zaworotko, *Angew. Chem.* **2001**, *113*, 2169.
126. S. Konar, F. Lloret, N.R. Chaudhuri, *Angew. Chem.* **2002**, *114*, 1631.
127. P.M. Forster, A.K. Cheetham, *Angew. Chem.* **2002**, *114*, 475.
128. Z.-F. Chen, J. Zhang, R.-G. Xion, X.-Z. You, *Inorg. Chem. Commun.* **2000**, *3*, 493.
129. Y. Cui, O.R. Evans, H.L. Ngo, P.S. White, W. Lin, *Angew. Chem.* **2002**, *114*, 1207.
130. F.A. Cotton, E.V. Dikarev, M. A. Petrukhina, *Angew. Chem.* **2001**, *113*, 1569.
131. S. Kitagawa, S. Kawata, *Coord. Chem. Rev.* **2002**, *224*, 11.
132. K. Fegy, D. Lueneau, T. Ohm, C. Paulsen, P. Rey, *Angew. Chem.* **1998**, *110*, 1331.
133. A. Kamiyama, T. Nogushi, T. Kajiwara, T. Ito, *Angew. Chem.* **2000**, *112*, 3260.
134. E. Lee, J. Heo, K. Kim, *Angew. Chem.* **2000**, *112*, 2811; E. Lee, J. Kim, J. Heo, D. Whang, K. Kim, *Angew. Chem.* **2001**, *113*, 413.
135. S.R. Batten, B. Hoskins, R. Robson, *Angew. Chem.* **1995**, *107*, 884.
136. K.-J. Lin, *Angew. Chem.* **1999**, *111*, 2894.
137. Y. Diskin-Posner, S. Dahal, I. Goldberg, *Angew. Chem.* **2000**, *112*, 1344.
138. L. Carlucci, G. Ciani, M. Moret, D.M. Proserpio, S. Rizatto, *Angew. Chem.* **2000**, *112*, 1566.
139. S.R. Batten, R. Robson, *Angew. Chem.* **1998**, *110*, 1558.
140. B.F. Hoskins, R. Robson, D.A. Slizys, *J. Am. Chem. Soc.* **1997**, *119*, 2952; *Angew. Chem.* **1997**, *109*, 2430.
141. I. Manners, *Angew. Chem.* **1996**, *108*, 1713.
142. I. Manners, *Adv. Mater.* **1994**, *6*, 68.
143. I. Manners, *Can. J. Chem.* **1998**, *76*, 371.
144. M. Rosenblum, *Adv. Mater.* **1994**, *6*, 159.
145. W. Siebert, *Russ. Chem. Rev.* **1991**, *60*, 784.
146. E.W. Neuse, H. Rosenberg. In *Metallocene Polymers*, Marcel Dekker, New York

1970. E.W. Neuse, *J. Macromol. Sci.-Chem.* **1981**, *A16*, 3.

147. H.M. Nugent, M. Rosenblum, P. Klemarczyk, *J. Am. Chem. Soc.* **1993**, *115*, 3848.
148. S. Harder, *Angew. Chem.* **1998**, *110*, 1357.
149. W.-M. Xue, F.E. Kühn, *Eur. J. Inorg. Chem.* **2001**, 2041.
150. O. Nuyken, V. Burkhardt, T. Pöhlmann, M. Heberhold, *Makromol. Chem., Macromol. Symp.* **1991**, *44*, 195; R.D. Miller, *Angew. Chem. Adv. Mater.* **1989**, *101*, 1773.
151. G.F. Nieckarz, D.R. Tyler, *Inorg. Chim. Acta* **1996**, *242*, 303.
152. M. Hnyene, A. Yassar, M. Escorne, A. Percheron-Guegan, F. Garnio, *Adv. Mater.* **1994**, *6*, 564.
153. R. Knapp, U. Velten, M. Rehahn, *Polym.* **1998**, *23*, 5827.
154. S.C. Tenhaeff, D.R. Tyler, *Organometallics* **1991**, *10*, 473; ibid. **1991**, *10*, 1116; ibid **1992**, *11*, 1466.
155. G.F. Nieckarz, J.J. Litty, D.R. Tyler, *J. Organomet. Chem.* **1998**, *554*, 19.
156. I. Manners, *Polyhedron* **1996**, *15*, 4311.
157. a) Y. Ni, R. Rulkens, I. Manners, *J. Am. Chem. Soc.* **1996**, *118*, 4102. b) P. Gomez-Elipe, R. Resendes, P.M. MacDonald, I. Manners, *J. Am. Chem. Soc.* **1998**, *120*, 8348.
158. R. Rulkens, L. Lough, I. Manners, *Angew. Chem. Int. Ed. Engl.* **1996**, *35*, 1805.
159. H. Braunschweig, R. Dirk, M. Müller, P. Nguven, R. Resendes, D.P. Gates, I. Manners, *Angew. Chem.* **1997**, *109*, 2433.
160. R. Resendes, J.A. Massay, H. Dorn, K.N. Power, M.A. Winnik, I. Manners, *Angew. Chem.* **1999**, *111*, 2738.
161. J. Raez, R. Barjovanu, J.A. Massay, M.A. Winnik, I. Manners, *Angew. Chem.* **2000**, *112*, 4020.
162. R. Resendes, I. Manners, *Adv. Mater.* **2000**, *12*, 327.
163. J.M. Nelson, I. Manners, *Chem. Eur. J.* **1997**, *3*, 573.
164. J.A. Massay, K.N. Power, A. Mitchell, M.A. Winnik, I. Manners, *Adv. Mater.* **1998**, *10*, 1559.
165. M.J. MacLachlan, I. Manners, G.A. Ozin, *Adv. Mater.* **2000**, *12*, 675.
166. M. Hanack, M. Lang, *Adv. Mater.* **1994**, *6*, 819; ibid, *Chemtracts - Org. Chem.* **1995**, *8*, 131.
167. C.W. Dirk, E.A. Mintz, K.F. Schoch, T.J. Marks, in: *Advances in Organometallic and Inorganic Polymer Science*, C.E. Carraher, J.E. Sheats, C.U Pittmann, (Eds.), Marcel Dekker, New York, 1986, p. 275.
168. H. Schultz, H. Lehmann, M. Rein, M. Hanack, *Structure and Bonding* **1991**, *74*, 41.
169. T.J. Marks, *Angew. Chem.* **1990**, *102*, 886.
170. M. Hanack, D. Dini, in: *The Porphyrin Handbook*, K.M. Kadish, K.M. Smith, R. Guilard (Eds.), Academic Press, San Diego, 2003; Vol. 15.
171. G. Knothe, *Makromol. Chem., Theory Simul.* **1993**, 2, 917.
172. K.G. Beltsios, S.H. Carr, *J. Polym. Sci., Part C, Polym. Lett.* **1989**, 27, 355.
173. T. Kobayashi, C. Furakawa, T. Ogawa, S. Isoda, *J. Porphyrins Phthalocyanines* **1997**, *1*, 297.
174. G. Meyer, D. Wöhrle, *Makromol. Chem.* **1974**, *175*, 715; ibid. **1974**, *175*, 728.
175. U. Lauter, M. Schulze, G. Wegner, *Macromol. Rapid. Commun.* **1995**, *16*, 239.

176. H.Z. Chen, M. Wang, S.L. Wang, *J. Polym. Sci,. A, Polym. Chem.* **1997**, *35*, 91, 959; *J. Appl. Polym. Sci.* **1997**, *64*, 1769.
177. M. Hanack, K. Mitulla, G. Pawlowski, L.R. Subramanian, *J. Organomet. Chem.* **1981**, *204*, 315; *Chem. Ber.* **1982**, *115*, 2836,
178. I.W. Shim, W.M. Risem, *J. Organomet. Chem.* **1984**, *260*, 171.
179. T. Sauer, W. Caseri, G. Wegner, *Mol. Cryst. Liq. Cryst.* **1990**, *183*, 387.
180. K.J. Wynne, J. Inorg. Organomet. Polym. **1992**, 2, 79.
181. N. Sasa, K. Okada, K. Nakamura, S. Okada, *J. Mol. Struct.* **1998**, *446*, 163.
182. G. Meyer, P. Plieninger, D. Wöhrle, *Angew. Makromol. Chem.* **1978**, *72*, 173.
183. S. Müller, D. Wöhrle, M. Shopova, B. Roeder, C. von Schönermark, *SPIE. The International Society for Optical Engineering*, **1994**, *2325*, 339.
184. N. Brasseur, R. Ouellot, C.L. Madeleine, J.E. van Lier, *British J. Canc.* **1999**, *80*, 1533.
185. R.S. Nohr, P.M. Kuznesof, K.J. Wynne, M.E. Kenney, P.G. Siebenmann, *J. Am. Chem. Soc.* **1981**, *103*, 4371.
186. P.M. Kuznesof, K.J. Wynne, R.S. Nohr, M.E. Kenney, *Chem. Commun.* **1980**, 121.
187. M. Hanack, *Israel J. Chem.* **1985**, *25*, 205.
188. J. Metz, M. Hanack, *J. Am. Chem. Soc.* **1983**, *105*, 828.
189. M. Hanack, C. Hedtmann-Rein, A. Datz, U. Keppeler, X. Münz, *Synth. Met.* **1987**, *19*, 787.
190. O. Schneider, M. Hanack, *Chem. Ber.* **1983**, *116*, 2088.
191. W. Kobel, M. Hanack, *Inorg. Chem.* **1986**, *25*, 103.
192. M. Hanack, S. Deger, A. Lange, *Coord. Chem. Rev.* **1988**, *83*, 115.
193. U. Kepler, S. Deger, A. Lange, M. Hanack, *Angew. Chem.* **1987**, *99*, 349.
194. W.J. Pietro, T.J. Marks, M. A. Ratner, *J. Am. Chem. Soc.* **1985**, *107*, 5387.
195. C. Gorman, *Adv. Mater.* **1998**, *10*, 295.
196. S. Hecht, J.M.J. Frechet, *Angew. Chem.* **2001**, *113*, 77.
197. F.J. Stoddart, T. Welton, *Polyhedron* **1999**, *18*, 3575.
198. K.W. Pollak, E.M. Sanford, J.M.J. Frechet, *J. Mater. Chem.* **1998**, *8*, 519.
199. N. Tomioka, D. Takasu, T. Takahashi, T. Aida, *Angew. Chem.* **1998**, *110*, 1611.
200. M. Brewis, G.J. Clarkson, V. Goddard, M. Helliwell, A.M. Holder, N.B. McKeown, *Angew. Chem., Int. Ed. Engl.* **1998**, *37*, 1092; N.B. McKeown, *Adv. Mater.* **1999**, *11*, 67.
201. M. Kimura, Y. Sugihara, T. Muto, K. Hanabusa, H. Shirai, N. Kobayashi, *Chem. Eur. J.* **1999**, *5*, 3495.
202. F. Vögtle, M. Plevoets, L. De Cola, V. Balzami, *J. Am. Chem. Soc.* **1999**, *121*, 6290.
203. C.M. Casado, B. Gonzales, I. Cuadrado, B. Alonso, M. Moran, J. Losada, *Angew. Chem.* **2000**, *112*, 2219.
204. N. Ohshiro, F. Takei, K. Onitsuka, S. Takahashi, *Chem. Lett.* **1996**, 871.
205. V. Balzani, S. Campagna, G. Denti, A. Juris, S. Serroni, M. Venturi, *Acc. Chem. Res.* **1998**, *31*, 26.
206. S. Campagna, G. Denti, S. Serroni, A. Juris, M. Venturi, V. Balzani, *Chem. Eur. J.* **1995**, *1*, 211.
207. J. Issberger, F. Vögtle, L. De Cola, V. Balzani, *Chem. Eur. J.* **1997**, *3*, 706.
208. H. Le Bozec, T. Le Bouder, O. Maury, A. Bondon, I. Ledoux, *Adv. Mater.* **2001**,

13, 1677.

209. M.S. Wrighton, M.C. Palazotto, A.B. Bocarsly, *J. Am. Chem. Soc.* **1978**, *100*, 7264.
210. M. Hanack, J. Metz, G. Pawlowski, *Chem. Ber.* **1982**, *115*, 2836.
211. E. Orthmann, G. Wegner, *Angew. Chem.* **1986**, *98*, 1114.
212. K. Fischer, M. Hanack, *Chem. Ber.* **1983**, *116*, 1860.
213. O. Schneider, M. Hanack, *Angew. Chem.* **1983**, *95*, 804.

8 Metal Complexes or Clusters Physically Embedded in Macromolecules

Anatolii Pomogailo and Dieter Wöhrle

8.1 Introduction

Metal complexes, metal clusters and other nanosized particles can be embedded in organic or inorganic macromolecules without "strong" interactions to the carrier matrix. In the case of metal complexes an organic polymer can function as a "solid" solvent. Preparation usually involves mixing the polymer and metal complex, using a variety of techniques. The metal complex is either dissolved in a monomolecular state or is part of the solid material as small particle crystallites. Host–guest interactions and the polarity of the macromolecule influence such properties as binding of small molecules, electron transfer, photoinduced electron transfer, etc.

On the other hand in the case of nanoparticles of metals, their oxides, etc., the macromolecules provide both ligand protection and stabilization of the nanosized particles. Different techniques such as microencapsulation, mechanochemical dispersion or formation/structural organization of nanoparticles in macromolecules are used for preparing such systems. The catalytic activities of these materials, in which the nanoparticles form the active part, are particularly studied.

8.2 Metal Complexes Embedded Physically in Macromolecules

This section concentrates on metal complexes dissolved or dispersed in macromolecules. In principle, any metal complex (or metal salt) can be incorporated into an organic or inorganic macromolecule, either as separate molecules or aggregated, to produce a solid composite material. Because several hundred papers have been published in this field, the following survey considers only few examples. Chapters 9 through 14, which consider properties, contain further examples. Biological systems, in which macromolecular proteins combine with metal complexes, are described in Chapter 2. Incorporating metal complexes into organic polymers offers the ability to prepare more flexible films with higher mechanical stability. In addition, polar macromolecules enhance the activity in devices in which metal complexes are the active parts. Van der Waals/hydrophobic interactions, or some electrostatic and/or coordinative interactions between the macromolecules and the metal compound cannot be ruled out.

Films have been prepared from solutions containing the polymer or a sol–gel precursor and a metal complex homogeneously dissolved, or from the dissolved polymer and dispersed metal complex on suitable carriers (Pt, Au, carbon in the form of graphite or glassy carbon) and inorganic semiconductors (ITO, SnO_2, Si, GaAs). The layer thicknesses varied a great deal, extending from approximately, 50 nm to a few micrometers. The number of active centers of a metal complex in the films coated on a carrier is on the order of 10^{-10}–10^{-6} mol cm^{-2}. The apparent concentration of a metal complex in a macromolecule is often as high as 0.1–5 M. The different methods, partly reviewed in [1], are:

- Casting from solution. The solution is spread onto a carrier and the solvent is evaporated carefully.
- Spin coating. A small amount of solution is dropped onto a spinning carrier. The thickness of the resulting film depends on the rotation speed, the evaporation rate of the solvent and the initial viscosity (concentration of polymer and metal complex) of the solution.
- Dip coating. A carrier is dipped into a solution and then dried. Higher concentrations and longer soaking times yield thicker coatings.
- Coating and adsorption process. First, a layer of the polymer is coated onto a carrier from solution. Then, in a second step, this coated material is placed in contact with an aqueous solution containing the metal complex for times which vary from a few minutes to several hours.
- Inclusion during formation of macromolecules. The metal complex is added during the formation of inorganic molecular sieves and is thus incorporated.

In order to obtain both a homogeneous distribution of the metal complex and the formation of a smooth film, the experimental conditions must be optimized carefully. Low concentrations of a metal complex (<1–10 wt%) in a polymer can result in mononuclear distribution. The polymer appears in this case to function as a solvent that minimizes complex—complex interaction by separation. Depending on the solubility of the metal complex in the solvent used for casting, higher concentrations (>1–10 wt%) can result in aggregation or microcrystal formation.

8.2.1   Incorporation in Organic Polymers

Examples of the incorporation of metal complexes and their properties are described in Chapter 13 (for experimental methods see Sections 13.5 and 14.5):

- Ru-red in Nafion membranes for water oxidation to O_2 (Section 13.2.3.1).
- Phthalocyanines **1** and porphyrins **2** in Nafion films for proton reduction to H_2 (Section 13.2.3.2).
- Phthalocyanines **1** in poly(vinylpyridine) films for carbon dioxide reduction to CO (Section 13.2.3.3).
- Tris(2,2′–bipyridine)ruthenium **3** in cellulose and polysiloxane films for photoinduced electron transfer (Section 13.2.3.4).
- Ru-red and **1–3** in polymer films for charge transport (Sections 14.1–14.3).

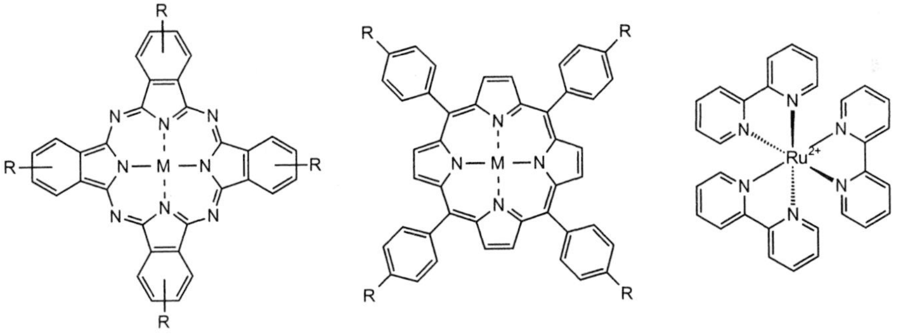

1   2   3

The mechanism of charge transport in polymer films is described in Sections 13.2.3.4 and 14.1.2, but some additional discussion is given here. The mechanism of charge transport in the above-mentioned polymer films of poly(vinylpyridine) and Nafion containing monomolecular dissolved Zn or Co phthalocyanines **1** and porphyrins **2** has been investigated in detail using potential step chronoamperospectrometry [2–6]. The results show again how greatly the polymer matrix and the type of included macrocyclic compounds influence its properties. When the redox molecule is bound to the framework by a chemical interaction such as chemical binding or coordination, its mobility in the matrix will be seriously restricted, so that charge propagation occurs via a hopping mechanism. When the redox molecule is bound only by a physical interaction, such as electrostatic and hydrophobic interactions, its mobility becomes less restricted, and charge propagation in the matrix can occur via a charge-hopping and/or a diffusion mechanism [4]. For zinc phthalocyanine in Nafion, an equilibrium between the monomer and dimer of the complex is also important [6]. Charge transport takes place by physical displacement of the ZnPc monomer and not by charge hopping, and the contribution of the dimer to the charge transport is negligible.

Tris(2,2′-bipyridine)ruthenium(II) **3** has been dispersed in a Nafion membrane coated on an ITO electrode, and oxidation and re-reduction of $Ru(bpy)_3^{2+}/Ru(bpy)_3^{3+}$ studied by cyclic voltammetry. A critical value of approximately 1.3 nm was found and calculated for the distance between the

Ru-complexes for efficient charge hopping between redox centers. The maximum of a Poisson distribution of distances is ≥ 0.1 mol L^{-1} of $Ru(bpy)_3^{2+}$ [7,8].

Intermolecular interactions of metal complexes can be identified by photophysical and photochemical measurements. Some metal complexes, such as $Ru(bpy)_3^{2+}$ **3** or $Ru(II)$ tris(4,7-diphenyl-1,10-phenanthroline)$^{2+}$, exhibit a high quantum yield and long lifetime of excited states, which is quenched by acceptors such as oxygen (energy transfer via formation of singlet oxygen) or MV^{2+} (electron transfer and formation of the methylviologen cation radical) [9–11]. Inclusion in polymers can enhance the lifetimes because of the microenvironmental effect of the surrounding polymer and reduced bimolecular annihilation processes [10].

Thin films consisting of a mixture of different metal phthalocyanines **1** (R = –H) with polymers were prepared by drop coating or spin coating on conducting ITO or gold (see Experiment 8-1, Section 8.4) [12–15]. Polymers of different polarity and coordination ability such as polystyrene, poly(acrylo-nitrile), poly(1-vinylcarbazole), poly(4-vinylpyridine) or poly(vinylidene fluor-ide) were used. In order to achieve conducting pathways between the porphyrin-type compounds in the film, approximately equal weights of the polymer and the phthalocyanine were dissolved in a polar solvent such as N,N-dimethylacetamide and drop coated on a carrier. Homogeneously colored, pin-hole-free films containing clusters of phthalocyanines were obtained. The electronic absorption spectra of films containing Zn(II) phthalocyanine **1** (R = –H) show significant differences from those obtained from solutions or from films prepared by vapor deposition (Fig. 8-1). Dissolved phthalocyanines exhibit the Q-band transition at ~670 nm, but layers with different polycrystalline structures give rise to different electronic spectra. The α- and β-modifications are characterized by a slip stack orientation of adjacent phthalocyanine rings, giving distinguishable bands in the visible region [16]. For example, the spectrum of **1** (R = –H; M = Zn(II)) in Fig. 8-1 shows that particles embedded in polymers exhibit a different polymorphic arrangement of molecules. The amount of splitting indicates the strength of the intermolecular electronic interaction, which is thus most intense in β-crystals, intermediate in α-crystals, and weak in the zinc complex dispersed in poly(vinylidene fluoride).

We now consider some results regarding the electrochemical and photoelectrochemical reduction of oxygen [13–15]. Thin films of different porphyrin-type compounds **2** within a polymer matrix (1:1 by weight) were prepared by a drop coating technique on ITO. This process produces smooth homogenously colored films. Vapor deposited films were also prepared. Photoelectrochemical measurements were conducted in 0.5 M KNO_3 (pH 5) during front side illumination through the electrolyte at a dioxygen concentration of $1.3 \cdot 10^{-3}$ mol L^{-1}, corresponding to a pure O_2 atmosphere (760 torr) at room temperature (see Experiment 8-1, Section 8.4).

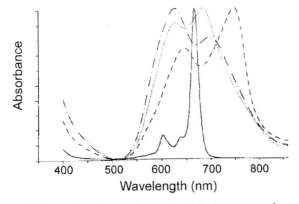

Figure 8-1. Visible absorption spectra of **1** (M = Zn; R = –H). –·–, α- and ---, β-modification in film prepared by vapor deposition; ···, film prepared from mixture (1:1) with poly(vinylidene fluoride); —, solution in *N,N*-dimethylacetamide.

When the electrode potential is scanned negatively in the dark in the presence of O_2 using films of **1** (M = Zn, R = –H), no cathodic current corresponding to reduction of dioxygen is observed. Significant irreversible reduction of dioxygen only occurs in the presence of light, which is characteristic of p-type conducting phthalocyanines. A cathodic photocurrent density as high as 0.1 mA cm^{-2} (conversion efficiency at 622.5 nm ~2%) was measured at –120 mV vs. NHE, for cast films of **1** in poly(vinylidene fluoride) (Table 8-1). The process is highly specific for dioxygen. Such reactions are of interest for measuring either oxygen concentration or light intensity [14]. Other acceptors, e.g. potassium ferricyanide and *p*-benzoquinone show only small differences between the cathodic current in the dark and under illumination. The highest photocurrent densities are measured at a weight ratio of the components between 0.5 and 1. Better intermolecular contact between metal complex particles leads to increased electrical conductivity in the bulk. The influence of the polymer environment on the photoelectrochemical behavior is significant (Table 8-1). The vapor-deposited film of **1** consists of the α-modification (Fig. 8-1), which exhibits higher dark and lower photoconductivity than other modifications. In cast films of other polymers and **1**, the phthalocyanine behaves differently. As the polarity of the polymer increases, the proportion of the α-modification decreases and that of another modification increases. This tendency runs parallel to an enhanced cathodic photocurrent (Table 8-1). The low activity in the presence of poly(2-vinylpyridine) may be due to the extremely high Lewis basicity of the polymer, which results in strong donation from the pyridine to the zinc central metal ion. This seems to have a strong influence on the phthalocyanine p-semiconducting properties. Poly(vinylidene fluoride) produces a high cathodic photocurrent. Because of the large dipole moment of 2.1 D of this polymer, a sizable electric field of 10^6 V cm^{-1} exists in the microenvironment of a monomer unit. This could enhance electron–hole dissociation and transport compared to recombination, and thus contributes to an increase in the photocurrent. Cast films of **1** with M = Co(II) or Fe(II) (metal ions with open d-shell electron configurations) exhibit no cathodic photocurrents in the presence of oxygen-saturated electrolytes (Table 8-1). The Co(II) and Fe(II) complexes are easily oxidized to Co(III) and Fe(III) complexes, and they can form localized ionized states in the bulk of the film. These states reduce the mobility of the charge carriers by acting as trapping states, thus decreasing the photoconductivity. However, the Fe and Co chelates are well known as active electrocatalysts for dioxygen reduction in fuel cells. Thin films containing Zn and Co 5,10,15,20-tetraphenylporpyhrins **2** (R = –H) are less active electrochemically and photoelectrochemically. In general, porphyrins show lower conductivity and photoconductivity than phthalocyanines because of less favorable intermolecular interactions.

Polymer films containing phthalocyanines have also been electrochemically investigated for their electrochromic reductions and reoxidations [12,17–18].

Table 8-1. Dioxygen reduction at various electrodes of **1**(R = –H) and **2** (R = –H). Measurements in 0.5 M KNO_3. Current densities during cyclic voltammetry at an electrode potential of –120 mV vs. NHE.

Chelate	Polymer	Current density, $\mu A\ cm^{-2}$		i_I/i_D
		Dark (i_D)	Irradiation (i_I)	
1(M = Zn)[a]	-	10	46	4.6
1(M = Zn)[b]	poly(4-vinylpyridine)	5	7	1.4
1(M = Zn)[b]	poly(styrene)	5	12	2.4
1(M = Zn)[b]	Nafion	9	30	3.3
1(M = Zn)[b]	poly(acrylonitrile)	4	14	3.5
1(M = Zn)[b]	poly(vinylchloride)	4	25	6.3
1(M = Zn)[b]	poly(n-vinylcarbazol)	5	35	7.0
1(M = Zn)[b]	poly(vinylidene fluoride)	6	100	16.7
1(M = Co)[b]	poly(vinylidene fluoride)	240	240	1
2(M = Zn)[a]	–	7	17	2.4
2(M = Co)[a]	–	6	10	1.7

[a] Vapor-deposited film, thickness 50 nm. [b] Cast films in mixture with polymers (1:1 by weight), thickness 250 nm.

It should be pointed out that a specific modification of Ti(O)Pc **1** (R = –H) in organic polymers is the most widely used photoconductor in bilayer photoreceptor copiers and laser printers [19]. A thin film containing ethylene-bridged silicon phthalocyanine polymer dispersed in poly(vinylidene fluoride) (1:4 by weight) obtained by casting from a DMF slurry on quartz was orientated in an electrical field [20]. A double-layered receptor device for measuring the photoconductivity was also prepared. Organic photovoltaic cells consisting of poly(N-vinylcarbazole) containing 5 wt% Zn-complex of 5,10,15,20-tetra(o-nitro)phenylporphyrin **2** (R = –NO₂) as active layers (prepared by spin coating from solution onto ITO) and tris(8-hydroxyquinolinato)Al exhibit an overall conversion yield of only ~0.03% [21]. The Pt-complex of phthalocyanine **1** (M = Pt; R = –H) and poly-bisphenol-A-carbonate were dispersed (1:1 by weight) in chloroform [22]. Thin transparent films were obtained by spin coating, and the electrochemical and spectroelectrochemical properties of the composite films were investigated. The material has a conductivity of 0.5–1 S cm⁻¹ and a positive thermoelectric power. Electroluminescent polymer blends of various host materials (derivatives of poly(vinylenephenylene), poly(methyloxystyrene), poly(N-vinylcarbazole)) with 0.1 wt% of the Pt complex of octaethylporphyrin have been investigated for photoinduced energy and charge transfer [23].

Anionic   metal   complexes,   such   as   tetrasulfophthalocyanine   or disulfoferrocene, can be incorporated from solution during the oxidative electrochemical polymerization of pyrrole [24–27]. After re-reduction of the polypyrrole film the phthalocyanine derivative remains in the film, and the observed electrochemical conductivity is explained by ion transport of small electrolyte cations [26].

Co(II) and Fe(II) Schiff bases and porphyrin complexes have been included by casting in various copolymers containing vinylpyridine or N-vinylimidazole units. Details on the preparation of thin films and their use in O₂-transporting membranes are given in Chapter 9. The use of Pt and Pd porphyrins in polymer films as optical sensors is described in Section 10.1.

The excitation and emission spectra of lanthanide metal ions, such as Tb(III) or Eu(III) in polycarboxylates (ionic interaction), have been investigated in detail [28,29]. The purpose was to study the structure of ionomers in solution and as solids. The linear increase of the luminescence intensity with increasing concentrations of lanthanides in a polymer film indicates a mononuclear distribution at lower concentrations. In a copolymer of styrene and acrylic acid a decrease of the luminescence intensity at > 4–6 wt%, and in poly(acrylic acid) or poly(styrene-co-maleic acid) at > 15 wt%, indicates the formation of ionic aggregates [30].

Luminescence behavior and electron transfer of various transition metal complexes with 4,4′-bipyridyl, EDTA or 1,2-diaminoethane in the interaction with polyelectrolytes have been studied in solution in detail [31,32]. The systems were prepared by mixing solutions of the polymers (in excess) and the metal complexes. The measurements showed a monomolecular distribution of the metal complexes and allowed the study of the pH-dependent conformational transition of the polymer chains as shown in Fig. 8-2. Polystyrene was doped with small amounts of porphyrins **2** (M = 2H; R = –C₆H₅ and –C₆F₅) by casting a film from toluene solution [33]. The fine-structured fluorescence and excitation spectra were investigated. Alcoholic solutions of Nafion containing the metal-free or Zn complex of **2** ($R = –C_6H_5$) were cast on glass slides [34]. A 1-μm thick film was obtained with a porphyrin concentration of 0.05 mol L^{-1} of a porphyrin. Emission and quenching of the photoexcited state was studied.

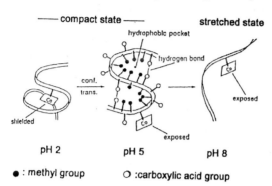

Figure 8-2. Schematic representation of conformational transitions of Co(III) (ethylenediamine)$_2$ at poly(methacrylic acid) at different pH values.

   Transition metal salts are stabilized by interaction with part of a polymer chain. Examples are: polystyrene $AlCuCl_4$-Co complexes [35,36], CuCl complexes at polystyrene modified with amino groups [35], $PdCl_2$ and $RhCl$ complexes at poly(styrene-co-butadiene) [37,38], $PdCl_2$ at poly(acrylonitrile) [39] and $RhH(CO)(PPh_3)_3$, $PtCl_2$-$SnCl_2$, $RuCl_3$-$CoCl_2$, $Rh_2(CO)_4Cl_2$ at different polymer phosphine ligands [40], $PdCl_2$ at different organic polymers (poly(benzimidazole), cyanomethylated cross-linked polystyrene, cross-linked poly(acrylonitrile)) [41] and carboxylated Co(II) and Fe(III) phthalocyanines in rayon fibres [42]. These complexes have been investigated mainly as catalysts in different reactions. Polymer (e.g. poly(ethylene oxide)/inorganic salts) complexes have been actively investigated for solid-state ionic conductivity to develop materials for commercial applications (batteries, electrochromic devices, moisture or gas sensors) [43–47]. Films are prepared easily by casting a solution of the polymer and metal salt followed by drying.

8.2.2   Incorporation in Inorganic Macromolecules

Xerogel silica was prepared from tetraethoxysilane in the presence of different porphyrins and a template (e.g. pyridine) in an alcohol/water mixture at between 25 and 120 °C [48]. The following metal-free or Fe complexes of porphyrins were employed: **2** with pentafluorophenyl, 2.6-dichlorophenyl-3-sulfonic acid, 4-carboxyphenyl or 4-methylpyridinium instead of $–C_6H_4R$. A mixture of 3 mL tetraethoxysilane and 3 mg of the porphyrin was used. It was found that the structure and morphology of the inorganic/organic hybrid materials depended on the kind of porphyrin and the xerogel preparation conditions. Side selection spectra and vibronic analysis were carried out for metal-free **2** ($–C_6H_5$, $–C_6F_5$

instead of $-C_6H_4R$) in xerogel silica [48].

The Fe complex of **1** (R = $-SO_3H$) has been incorporated into sol–gel silica materials by acidic or basic hydrolysis of tetramethoxysilane or methyltrimethoxysilane [49]. For example, 24 mg of the phthalocyanine per milliliter water was added to a mixture of 7.38 mL tetramethoxysilane and 1.75 mL water. The Fe complex was incorporated in an aggregated form. The composite was tested as a catalyst in oxidation reactions.

Ni, V(O), Pd and Pt complexes of etioporphyrin have been adsorbed on SiO_2 with loadings between 0.5 and 5.0 wt.% [50]. SiO_2 was added to a solution of the porphyrins in $CHCl_3$ and the solvent was then evaporated. Catalytic properties for hydrocarbon cracking and hydrogenation were studied.

Various metal complexes such as metal phthalocyanines, metal salenes or Ru pyridyl complexes have been incorporated in molecular sieves such as cavity-structured zeolites (faujasites, supercages with 1.3-nm diameter), channel-structured aluminium phosphates ($AlPO_4$-5, channel diameter 0.73 nm) and channel-structured silicates MCM-41 (channel diameter 3.2 nm) [51–53]. Different strategies were applied for the inclusion of the phthalocyanines. For example, whereas the zeolite-encaged phthalocyanines (**1** R = –H; M = Co(II), Ru(II), etc.) are synthesized by the reaction of a transition metal ion-exchanged zeolite with phthalonitrile in a closed-bomb vessel [54], in the cases of $AlPO_4$-5 and MCM-41 substituted derivatives of phthalocyanines were added to the mixture during the hydrothermal synthesis of the molecular sieve [55,56].

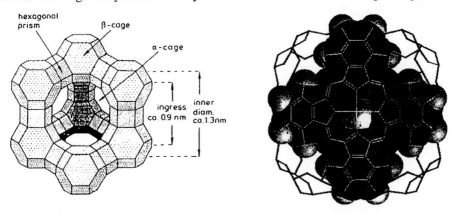

Figure 8-3. Molecular sieve zeolite faujasite. Left: Structure of the zeolite with supercages of 1.3-nm diameter. Right: Model for the incorporated metal phthalocyanine **1** (R = –H) with diameter 1.2 nm.

The cavity size of the faujasite zeolite agrees well with the diameter of the phthalocyanine (Fig. 8-3) [54]. Phthalocyanines included in MCM-41 are found in the columnar-orientated detergent (used as templates for synthesis of the molecular sieve) of the channels (Fig. 8-4) [56]. Loadings of $\sim 10^{-5}$ mol per g ram of molecular sieve were achieved. The materials are intensely green-colored and contain the phthalocyanines in the monomeric state as shown in the UV-Vis reflectance spectra by the Q-band transition at ~680 nm. One example of the inclusion of the phthalocyanine **1** (R = –O–CH$_2$–CH$_2$–N$^+$(CH$_3$)$_3$; M = Zn(II)) is shown in Fig. 8-4. Aggregation for the inclusion **1** (M = Co, Al(OH); R = –COOH, –SO$_3$H) is described in Experiment 8-2, Section 8.4. The molecular sieve-encapsulated phthalocyanines have been investigated as catalysts [54], in photophysical hole burning optical information storage [57] and in photoinduced charge separation [51].

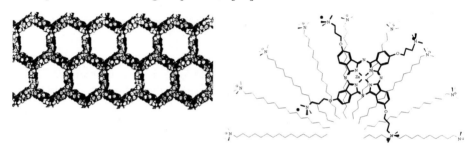

Figure 8-4. Molecular sieve silicate MCM–41. Left: Structure of hexagonal pores with 3.1-nm diameter. Right: Model of the incorporation of **1** (R = –O–CH$_2$–CH$_2$–N$^+$(CH$_3$)$_3$; M = Zn(II) in the columnar orientated detergent cetyltrimethylammonium chloride surrounded by the MCM–41 channels.

8.3 Metal-Cluster and Nanosized Particles Embedded in Polymers

The study of ligand-protected metal nanoparticles and their oxides is the most intensively developed field of the physical chemistry of ultradispersed materials. Without solving the serious problem of the structural design of such macrocomplexes it is difficult to find and optimize their fields of application. Macromolecules (including natural ones) have long been used as stabilizers of nanosized particles. The steric stabilization of metal particles by polymers can be considered as shielding by a protective colloid. Such a stabilization occurs because the spatial dimensions of even low molecular weight polymers are comparable to or greater than the range of London's attraction forces. Attempts to estimate the stabilizing ability of polymers quantitatively go back to Faraday's time [58]. The necessity to enhance the stability and regulate the reverse transition of such systems draws attention to their morphology, structure and architecture.

8.3.1   The General Characteristics of Metallopolymer Systems

Nanos means a dwarf in Greek. One nanometer (nm) equals 10^{-9} m. Materials whose individual crystallites or phases making up their structural base (small forms of a substance) are below 100 nm in size in at least one dimension are called nanostructures. Although the limit is conditional and is accepted for convenience, this size marks the beginning of the regions in which disordered structures are becoming significant. It correlates with the characteristic size of a variety of phenomena (see below). Because different metals and their oxides have different limiting characteristic values for various physical properties, the 100-nm limiting value is inevitably something of a convention.

A protective polymer interacts with nanoparticles in two principal different ways. These are physical (determined by van der Waals forces, dipole interactions or weak hydrogen bonds) or chemical adsorption. The noncovalent interaction of nanoparticles with a macromolecule is very weak (about 10^{-4} J m^{-2}). For chemisorption the effectiveness of such an interaction is determined by the number of polar groups of the adsorbed polymer per unit of surface, independently of the shape of macromolecule (unfolded or globular). It is important that these functional groups should react strongly with the surface atoms of nanoparticles, for example playing the role of electron donors [59]. The effectiveness of polymers considerable increases if they include the centers of a specific interaction which acts simultaneously on the mechanism of both charge and steric stabilization. Polymers are used not only as stabilizers but often as ingredients. In the latter case such a system can be considered as a metallopolymer nanocomposite. Although linear homopolymers are convenient models for studying theoretical aspects of the stabilization of metal nanoparticles, copolymers are more often used for these purposes. The stabilizing effect is determined not only by the nature of functional groups but even by the composition and the unit distribution in the chain of a copolymer (statistical, alternate, block, grafted ones). We shall consider the main methods of synthesis of metallopolymers in which metalloparticles are physically encapsulated in polymers, although there is no doubt that these processes are always accompanied by a chemical binding.

8.3.2   Microencapsulation of Metallocomplexes and Nanoparticles by Polymers

The term "microencapsulating" implies obtaining nanoparticles of a compound in protecting covers which consist of film-forming polymeric materials [60]. The encapsulated substance forming the nucleus of the microcapsule may be a metal complex or a metal nanoparticle. The polymeric cover separates the nanoparticles from each other and from the environment and serves also as a stabilizer. The cover can be made from natural polymers, for instance proteins and polysugars, or from synthetic polymers (of addition or polycondensation types), either pre-prepared or obtained from monomers with subsequent polymerization.

For example, efficient catalysts for α-olefin polymerization can be obtained [61] by radical homo- and co-polymerization of suitable monomers in solutions or suspensions of catalyst components MX_n ($TiCl_4$, VCl_4, $ZrCl_4$, $Ti(OR)_4$, $TiCl_3$, $VO(OR)_3$, etc.; Eq. (8-1); Experiment 8-3, Section 8.4). In principle, MX_n itself can be an initiator of polymerization.

$$Monomer(s) + MX_n + Solvent \xrightarrow{Initiation} MX_n \text{ (in solution of polymer)} \longrightarrow$$

$$\xrightarrow[\text{of solvent}]{\text{Removing}} MX_n \text{ (in polymer)}$$

(8-1)

The forming macromolecules (e.g. PS) incorporate MX_n into microcapsules. A macroporous structure with included MX_n forms when solvent is removed.

The shape of the microcapsules is usually that of the encapsulated substance and the cover thickness depends on the microencapsulation conditions and is varied according to the intended uses of the products. It may be possible to process the resulting composition directly into goods (e.g. by pressing), or the polymeric cover may enable better combination of the nanoparticles with other polymers.

Methods of coacervation, physical adsorption, precipitation by a non-solvent or solvent evaporation are used for microencapsulation. A developed surface may be prepared by removing solvent from a vitrified polymer by sublimation at low temperature under vacuum.

Extrusion (in which particles of the compound to be encapsulated are covered by pressing them through the film-forming substance), spray-coating in a quasi-liquid layer, vapor condensation and polymerization on the surface of nanoparticles have also found great application. For example, hydrosols of Fe_2O_3 (or TiO_2) with an average particle size of 3–5 nm adsorb on latex granules obtained by terpolymerization of styrene, butadiene and acrylic acid and having negative charges on their surface (–COO^- on the polymer and SO_3^{2-} on the emulsifying agent; Fig. 8-5). Comonomers of the latex can be added to such a system and copolymerization carried out [62,63] resulting in three-layered nanocomposites with Fe_2O_3 or TiO_2 in the middle layer.

There are numerous methods of microencapsulation. The principal fields of application of microencapsulated clusters and nanosized particles are mainly given in patent literature. Examples are inorganic pigments, including iron, aluminium and silicon oxides, in PVA (poly(vinyl alcohol)) matrixes, titanium dioxide in PAC and its salts, and isolating and waterproof coatings for $LiAlH_4$ and lithium from poly(acrylates). Iron and cobalt nanoparticles for magnetic tapes are microencapsulated in PMMA. Photoemulsions with microcapsules of silver bromide are obtained by irradiating dispersions of AgBr particles in suitable monomers. Photoconductive nanocomposites are obtained similarly using CdS, and microcapsules with aluminum particles and air in one polymeric cover are used in detonating compositions, and so on.

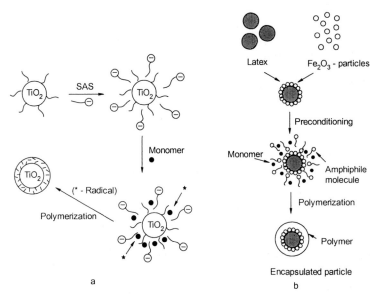

Figure 8-5. A scheme of the emulsion polymerization on the surface of nanoparticles (a) and their encapsulation (b).

8.3.3   Mechanochemical Dispersion of Nanoparticles with Polymers

Nanoparticles of metals may be co-dispersed with a polymer matrix in different ways: in high-speed planetary mills, disk integrators or vibrators of various types including ultrasonic. Magnetodynamic dispersers are used for the dispersion of magnetic particles in polymers (specifically, γ-Fe_2O_3).

The variation of the Helmholtz energy during dispersion of a homogeneous metal with volume V to particles with radius r can be expressed in Eq. (8-2) where σ is the increase in the system's free energy due to the formation of a unit of new interface surface.

$$\Delta F(r) = \sigma V/r \qquad\qquad (8\text{-}2)$$

The only characteristic example is shown here [64]: co-breakage (dispersion) of solid halides $TiCl_3$, VCl_3, etc. in matrices of common PE, PP, and those (PE-L) which are modified with functional groups L by graft polymerization of 4-vinylpyridine or MMA [65]. The presence of PE does not greatly affect the crystallite size during the initial grinding period. Evidently, an agglomerate of $TiCl_3$ is covered by PE, which, however, neither enters between crystallites nor prevents their reaggregation. Dispersion of MCl$_3$ with polymers having functional groups occurs by another mechanism – the particles' size (the zone of coherent distribution, ZCD) depends upon the dispersion time and the L/M ratio (Table 8-2; Experiment 8-4, Section 8.4).

Table 8-2. The change of crystallite sizes during grinding.

Sample	The ZCD sizes on 300 and 003 axes, nm					
	Grinding time, h					
	0	0.3	1	3	5	10
TiCl$_3$/PE	47	14	13	14.5	12.5	13
	25	18	17	19	22	20
TiCl$_3$/PE-gr-Poly4-VPy	47	32	20	10	~5	–
	25	10	8	~5	–	–
TiCl$_3$/PP	47	–	25	–	–	–
	25	–	5	–	–	–
TiCl$_3$/PP-gr-PMMA	47	–	25	–	–	–
	25	–	18	–	–	–
VCl$_3$/PP	300	–	45	–	–	–
VCl$_3$/PP-gr-PAN	300	–	72	–	–	–

This is not only associated with physical processes but with the chemical interaction of MCl$_3$ with polymer functional groups of the corresponding macrocomplexes, which occurs simultaneously with the decrease in crystallite salt size (Eq. (8-3)). Moreover the previously unknown structure δ-VCl$_3$ forms as a result of grinding VCl$_3$ in a planetary mill [66].

$$PE\text{-}L + (MCl_3)_n \rightarrow [(MCl_3)_n \leftrightarrow (MCl_3)_m] + MCl_3 \cdot L\text{-}PE \ (m < n) \qquad (8-3)$$

The formation of macrocomplexes was observed [67] during the mechano-chemical dispersion of AlCl$_3$·6H$_2$O with styrene and divinylbenzene copolymers modified by iminodiacetic groups, the co-dispersion of salts with PVC and in other circumstances.

A modern method of mechanochemical grinding is ultrasonic (US) dispersion of metal suspensions using the stabilizing properties of water-soluble polymers. The addition of surfactants provides an adsorptional decrease in the durability of the dispersing substance (Rebinder's effect) and stabilization of the resulting suspensions. It seems promising to use metallomonomers, especially sodium acrylate, as surfactants, which are able to polymerize under the conditions of high shifting stress and contact forces present at the dispersion of metals in polymer–monomer solutions [68].

8.3.4   Spray-Coating of Polymers by Atomic Metals

Various methods of spray-coating polymers by metalloparticles can be applied in physical methods of incorporating metallocomplexes, clusters and nanoparticles into polymers. In recent years these have been intensively developed, mainly for vacuum metallization of polymeric materials in electronics (the manufacture of communicational plates, plane cables and other units for microelectronic equipment). Investigations of processes occurring on polymer surfaces during spray-coating by atomic metals are also of fundamental importance since they allow us to understand the nature of the interactions and adhesion in the interface surface formed [69].

Polymer zones in contact with metal ions or nanoparticles can become negatively charged by the injection into the polymers of photoelectrons, originating from clusters, with subsequent entrapping by charge traps [70]. Theoretically, an antiprocess is also possible when an outflow of electrons occurs from a neighboring polymer zone into positively charged clusters or nanoparticles. In both cases, such polymer zones should get some charge (positive or negative) corresponding to other participants in the polymer surface. In recent years such trends have been investigated very intensively for many reasons. We only mention the main methods of atomic metal spray-coating without describing them in detail: cryochemical synthesis (the spray-coating of thin polymer films cooled to low temperatures), thermal evaporation (in which the flow of spray-coating metal is generated by sublimation or by thermal evaporation of block metal) and spray-coating by metal under low-temperature polymerization in the plasma of a glow discharge, etc.

8.3.5   Formation of Nanocomposites in Polymer Solutions

This process, which involves a reduction of transition metal compounds in polymer solutions, is a complicated multistage way for the formation of cluster and nanosized particles in polymers. As a rule, such materials are resistant in storage in dry air, whereas the reduced products of base metals are quickly oxidized to oxides while drying in air or on contact with moisture. A part of the deposited substance, as the first stage of its investigation, is detected by chemical analysis or by the increase in weight (Eq. (8-4)), where m_b is the mass of polymer with metallic particles and m_p is the mass of the parent polymer, which can vary in percentage terms from a fraction to hundreds, depending on the initial concentrations.

$$\Delta m = (m_b - m_p)/m_p \tag{8-4}$$

The second stage of investigation is the determination of the metallocomposite's phase composition, the characteristics of the obtained layer (thickness, compactness) and the sizes of particles and crystallites. The typical scheme of film formation by reduction of palladium compounds in 10% polyamidoimine solutions (in THF or N-methylpyrrolidone) with $NaBH_4$ is as follows [71]. The first stage involves mixing components to attain MX_n uniform distribution in solution, probably obtaining chelate structures (Fig. 8-6, points).

The second stage is the formation of a film from the metallopolymer solution by a glaze method and slow (10–15 h) evacuation of the solvent to a residual percentage of 5–10%. The resulting metastable membrane film is freed from parent salt by washing with methanol or water. Finally, the third stage, reduction, is very quick: the formed particles are linked to polymer chains (Fig. 8-6, bold points).

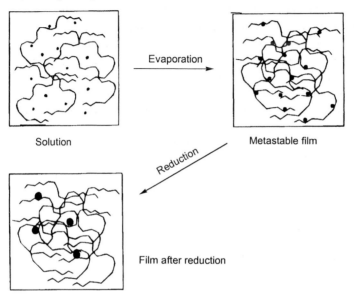

Figure 8-6. The main stages of formation of nanoparticles in solution.

Depending on the conditions of preparation, Pd particle diameter is 1–3 nm (~36–960 atoms). This indicates that favorable conditions occur for the simultaneous formation of a great number of nanoparticle nuclei by the chemisorption of reduced ions by protective polymers. Thus, the reduced metal is mainly consumed in forming nuclei, and only a very small part is involved in their growth. The result is the formation of quite small nanoparticles, which is compounded by their high aggregation stability provided by the protective role of the polymer. The so-called theory of matrix isolation was developed to describe nanoparticle formation in polymer solutions, and adequately describes the dependence of their sizes on the reaction conditions and the nature of the stabilizing matrix [72].

8.3.6   Structural Organization of Nanoparticles in Block Copolymers

This type of metallopolymer is interesting because it includes block copolymers that provide a ligand-controlled synthesis of nanoparticles. Many block-copolymers in organic solutions, e.g. poly(styrene-4-vinylpyridine), poly-(styrene-acrylic acid) etc., exist as inverted micelles: they are separated (segregated) into microphases with spherical, cylindrical and lamellar morphology [73]. In such microphase-separated structures, a polar component acts as a nucleus, which is located in the PS coating. Salts of many metals, which are insoluble in pure toluene, are soluble in toluene micelles of such copolymers. Metal ions are bound to functional groups of the micelle nucleus by covalent or ionic bonds. Such bonding (loading) can attain relatively high values (e.g. in the best variants – 1 g metal salts per g polar comonomer of the block-copolymer). Almost calibrated nanosized particles are generated under reduction, and these are simultaneously stabilized by non-polar blocks of the amphiphilic block-copolymers (corona) [74]. In addition, a correlation is observed between the radii of the ionic nuclei of the inverted micelles and the sizes of the nanoparticles generated.

The principle a: and a concrete scheme b: of nanocomposite preparation by such a method is shown in Fig. 8-7.

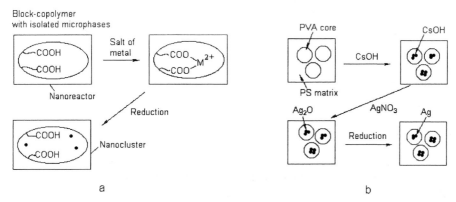

Figure 8-7. a: A principal scheme of block-copolymer's synthesis and b: a scheme of the incorporation of Ag nanoparticle into PS-block-poly(vinylalcohol) (PVA).

The structure of such metallopolymers can be regulated by varying the ratio of sizes of block-copolymer (general molecular weight, composition) and micelle parameters (size of nucleus/cover, form, polydispersion), and this enables us to use block-copolymers as nanoreactors or templates. The ratio between the molecular weights of the first (polar) and the second (hydrocarbon) components (M_1 and M_2) is important to provide steric stabilization: $M_1 = nM_2$ (where n is the number of the second component chains). As a rule, M_1 is 10^4–10^5, and $M_2 = 10^3$–10^4.

It is important that a nucleus can undergo cross-linking in the case of styrene and 2-vinylpyridine block-copolymers (e.g. by 1,4-diiodobutane) [75]. The monodispersed microspheres formed have a lamellar morphology with a "hairy-ball" structure. The thickness of the quaternized polyvinylpyridine nucleus swollen in water is 35 nm, and the microdomain distance of the lamellar phase is 65 nm. Such a structure of the polymer phase allows localization of nanoparticles (diameter 10–20 nm) in the block-copolymer nucleus only.

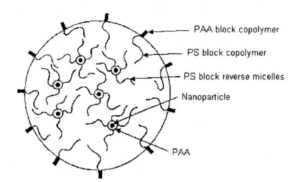

Figure 8-8. A scheme of large micelles containing nanoparticles.

The structural organization of such systems can be more complicated: for example, micelles of large compounds. These are microsized spheres of high polydispersion, consisting of ensembles of inverted micelles stabilized by a thin layer of hydrophobic chains (Fig. 8-8) [76]. At 12% of polar components in a copolymer less-complicated micelle-type compounds are generated with an average diameter of 64 nm, including 58 inverted micelles and 87 stabilizing chains with an average surface area of 148 nm^2 per stabilizing chain. The shape of the nanoparticles generated: e.g. from the quadric cube to the triangular tetrahedron in the case of platinum can be controlled.

8.3.7   Metallopolymer Formation in Heterogeneous Systems

In the case of insoluble polymers, the processes of metallopolymer formation are more complicated because of the diffusion of reduced ions in a polymer matrix. This mechanism has been observed in detail during metalloparticle formation in films of vitreous or crystalline polymers, previously deformed in adsorption-active liquid media (this method was generalized in review [77]). After such a treatment, a porous structure appears in a matrix of, for example, isotropic isotactic polypropene by a mechanism of delocalized crazing (the pore volume is < 45%, diameter 3–6 nm). The system of interpenetrating open pores penetrable by reagents is divided by zones of non-orientated block-polymer, in which metal salts are injected. Nanoparticles of reduced metal are localized in pores through a counterflow diffusion. The metal particles are obtained by consecutive steps: the penetration of metal ions and reducing agent into the polymeric matrix, the deeper diffusion of the reagents and, finally, the chemical reaction itself.

The size of the nanoparticles generated depends more on the reaction conditions and the structural parameters of the porous polymer, and less on the nature of the metal. An increased metal content in the polymer is attained mainly by an increase in the particle size rather than in the number of particles.

The architecture of such nanocomposites and the topochemistry of the metal layer over a transverse section of the polymer matrix are defined by a reaction zone width, which depends on the ratio between a diffusion coefficient D and a chemical reaction rate constant k. At $D<<k$ the rate of metal particle deposition is limited by the diffusion rate, and the reaction zone width is minimal (Fig. 8-9).

Figure 8-9. Model schemes of the cross-sections of nanoparticles.

At $D>>k$, the reaction zone extends across the whole transverse section of the polymer film. Therefore, regulation of the ratio between various parameters (viscosity of solution, temperature, concentration of reagents, etc.), makes it possible to obtain materials with different model schemes. It is possible to obtain particles of different chemical content by the reduction of metal ions depending on the nature of the polymer matrix. For example, in a swelling matrix (PVS, cellulose, etc), copper oxide is formed by reduction of Cu^{2+}, while copper(0) forms mainly in a porous matrix (PE, PTFE).

An original way of forming metallopolymers is the rapid thermal decomposition of precursors (often metal carbonyls) in a polymer melt-solution [78]. In contrast to solution, the short-range order of the parent polymer is conserved in fusion, and its vacancies become available for localizing generated metalloparticles. Primarily, these are included into interspherulite disorder, into looser zones of polymeric matrix, into space between lamellas and in spherulite centers. A strong interaction between nanoparticles and polymeric chains is observed in such materials. This impedes a segmental movement of the amorphous phase because of the decrease in free volume and the generation of a firm structural lattice: cross-linking of chains produces a more rigid matrix. It has been shown for Co-containing particles that the structure includes a metallic nucleus with a small oxide cover (3–10 nm) isolated in a polymer matrix.

Probably, materials based on graft polymers should be related to the same type of metallopolymers. The properties of such "double-layered" macroporous materials (particularly block-copolymers) depends on the localization of a graft layer as a thin film on the support polymer surface (PE, PP PTFE, PS are best) or using highly dispersed powders. For this purpose a graft of small amounts of functionalized monomer (acrylic acid, allyl alcohol, methylmethacrylate, vinylpyridine, etc.) is important. The optimal thickness of a graft layer is usually not more than 10–30 nm [79]. The scheme of the principal conversions in a graft layer is shown in Fig. 8-10 using highly dispersed PE powder (particle size 100–200 μm) as an example, whose surface was modified by gas-phase graft polymerization of acrylic acid (PE-gr-PAA). For polymer substrates as films, this method can be efficient for the metallization of plastics, and also for the surface of synthetic fibers (see Experiment 8-5, Section 8.4).

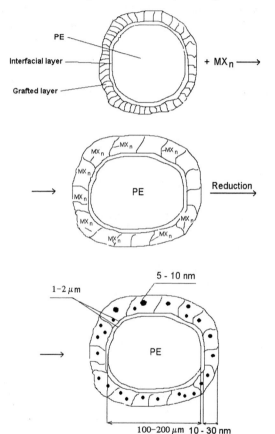

Figure 8-10. The principal scheme of nanoparticle formation in a grafted layer.

It should be noted that in a graft layer with crazing polymers, at least when copper nanoparticles are being generated, the reaction is observed to be reversible: owing to the formation of a galvanic pair Cu^0/Cu^{2+}, nanoparticles with sizes less than 3 nm are oxidized in time (a few hours) and converted into mononuclear complexes with migration of copper ions: the specific metallic luster of the polymeric particles disappears.

The products prepared by reduction of metal ions in nanoporous polymers as nanoreactors, e.g. ion-exchange resins, are also heterogeneous materials. The pores perform as transporting arteries for the infiltration of nanosized particles or their precursors into the surface layer of the polymer. They are classified into three types depending on pore size: micropores ($r < 1.5$ nm), mesopores ($r = 1.5$–30 nm) and macropores ($r = 30$–6000 nm). Moreover, pores are divided into closed (isolated from each other and having no entry to the surface) and open (channels connected with each other and the surface). As a rule, polymers include pores of different types and sizes. Though there has been no systematic research into metalloparticle generation in polymer pores, this field should be developed intensively, for example, for the synthesis of heterogeneous catalysts. It should be noted that compounds of Rh, Ru and Pt fixed by ion-exchange on the perfluorinated resin Nafion, are reduced in a flow of hydrogen at 473 K, and metallic clusters with sizes 2.88, 3.3 and 3.4 nm, respectively, have been obtained [80]. The same situation is considered for the stabilization of Raney nickel or other metals.

8.3.8 Metallopolymers Obtained by Polymerization (Polycondensation) of their Stabilizing Cover

A search for self-regulating systems, in which synthesis of a polymer matrix, nucleation and growth of metal particles occur simultaneously, may provide the best solution of the problem of nanoparticle stabilization and structural organization in polymers. Therefore, we need to elaborate methods of preparation of metallopolymers with the "microencapsulated particle in a polymeric cover" architecture obtained *in situ*. This is realized by the generation of cluster dispersions in polymer matrices when, as a result, the growth of metalloparticles is limited. There are many ways to achieve this: polymerization of vinyl monomers after mechanical metal dispersion (see Section 8.3.3), when freshly prepared metal surfaces perform as initiators; injection of metalloorganic compounds decomposing at temperatures close to those of polymerization in polymer systems; γ-co-irradiation of a precursor and monomer at room temperature; polymerization of metal-containing monomers (Chapter 4), etc. The main stages of such methods are: dissolution of a metal salt and initiator in the monomer (for example, methylmethacrylate, MMA), block-polymerization at some temperature and, finally, formation of the metal composite, often by post-heating of the products obtained (Eq. (8-5)).

$$\text{MMA} + \text{MX}_n \xrightarrow{\text{Solution}} \text{M}^{n+}/\text{MMA} \xrightarrow{\text{Initiation}}$$

$$\xrightarrow{} \text{M}^{n+}/\text{PMMA} \xrightarrow{\text{Heating}} \text{M}_n/\text{PMMA} \qquad (8\text{-}5)$$

$$\text{Solid solution} \qquad\qquad \text{Solid solution}$$

In such nanocomposites, the polymer cover thickness is regulated by the concentration ratio and polymerization conditions. The main difficulties in this method (especially in solidifying epoxy, formic and other resins) are the need to provide sedimentation stability in monomer–cluster particle systems and also to accommodate the internal stresses that appear in such systems.

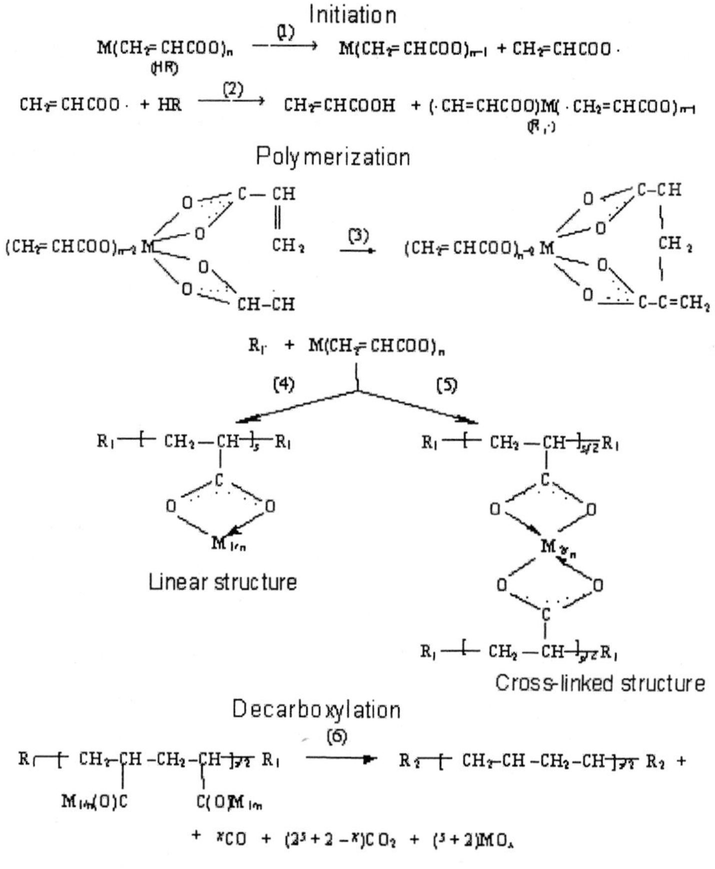

Initiation

$$M(CH_2=CHCOO)_n \xrightarrow{(1)} M(CH_2=CHCOO)_{n-1} + CH_2=CHCOO \cdot$$
(HR)

$$CH_2=CHCOO \cdot + HR \xrightarrow{(2)} CH_2=CHCOOH + (\cdot CH=CHCOO)M(\cdot CH_2=CHCOO)_{n-1}$$
(R_r)

Polymerization

$$R_r + M(CH_2=CHCOO)_n$$

(4)   (5)

(8-6)

Linear structure

Cross-linked structure

Decarboxylation

(6)

$$R_1 \left[CH_2\text{-}CH\text{-}CH_2\text{-}CH \right]_{y/2} R_1 \longrightarrow R_2 \left[CH_2\text{-}CH\text{-}CH_2\text{-}CH \right]_{y/2} R_2 +$$
$M_{l+y}(O)C$ $C(O)M_{lm}$

$$+ \quad ^xCO + (2^3 + 2 - ^x)CO_2 + (3 + 2)MO_x$$

An interesting type of metallopolymer preparation by this method is the solid-phase polymerization of metal-containing monomers (Chapter 4) accompanied by thermal conversion of generated metallopolymers (Experiment 8-6, Section 8.4) [81], for example transition metal salts and unsaturated carboxylic acids (acrylates, maleates, etc.). Their thermal transformation proceeds via three main consecutive temperature-determined macrostages. Firstly, dehydration (desolvation) of parent monomers (303–473 K), secondly, solid-phase homo- and copolymerization of the dehydrated monomer (473–523 K) and, finally, decarboxylation of the prepared polymer to a metal-containing phase and the non-oxygenic polymeric matrix stabilizing it (at $T > 523$ K). The principal scheme of such a process is shown in Eq. (8-6) where MO_x is a general expression for metal nanoparticles at $x = 0$ or for its oxide, at $x > 0$).

Control of the thermolysis temperature at the final stage, together with kinetic observations (amount and composition of gas evolution), permit regulation of the morphology, topography and dispersion of the products obtained. Nearly spherical metal-containing particles with high charge density are observed, existing individually as well as in aggregates of 3–10 particles, allocated rather uniformly through a matrix with a lower charge density. They are characterized by comparatively narrow size distribution, with an average diameter of 4–9 nm and a distance of 8–10 nm between particles. Great progress is expected in the field of controlled thermolysis of similar bimetallic monomers, e.g. co-crystals of different metal metallomonomers [82].

8.3.9 Architecture of Metallopolymers with Dendrimers

The metallopolymers considered above are used relatively often (see, for example, [59]). New types of nanocomposites, with an architecture arising from the unusual structure of highly branched polymers, have been intensively investigated. Their place in a macromolecular system is shown in Fig. 8-11.

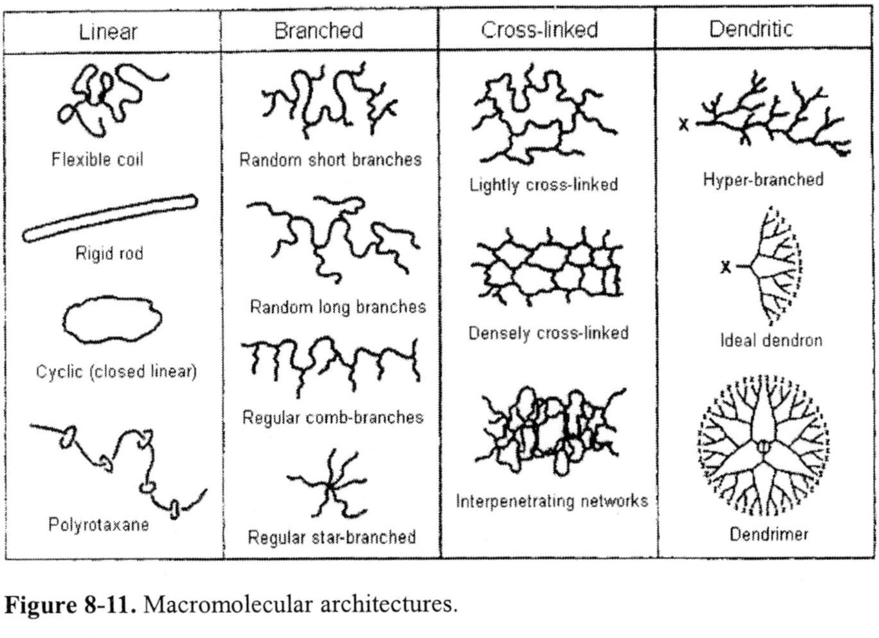

Figure 8-11. Macromolecular architectures.

Special attention is focused on dendrimers (see Sections 6-4 and 7-6), a new class of regular polymers characterized by a tree-type structure, generating from one center, by a great number of branching centers and by the absence of closed rings [83]. Sometimes they are called cascade polymers or polymers with controlled molecular architecture [84]. Poly(amidoamine) or poly(propylene-amine) with a diaminobutane nucleus are commonly used for this purpose. In such compounds, a special role is played by terminal fragments, whose number increases exponentially with the number of generations. Organometallic dendrimer derivatives are also known [85], as well as their numerous macromolecular metal complexes (e.g. ruthenium, palladium, platinum, etc.) [86]. Some dendrimers contain quite large spaces, so that they can be used as "molecular containers". Thus, they serve as a basis for construction of dendrimer-template nanocomposites [87]. The fourth-generation dendrimer of poly(amidoamide) and its copper-containing metallocomplex are shown in Fig. 8-12.

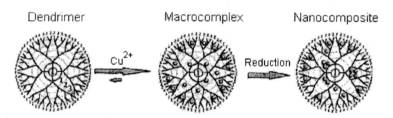

Figure 8-12. Construction of dendrimer nanocomposites on the base of poly(amidoamine) and Cu^{2+}.

In the first stage, Cu^{2+} forms a macrocomplex with poly(amidoamide), and after reduction by hydrazine hydrate nanoparticles of Cu^0 with size 2–4 nm are obtained; the dendrimer's diameter is 4.5–6 nm. Nanoparticles of copper, synthesized in a dendrimeric nanoreactor, are stable for more than three months at room temperature in an oxygen-free atmosphere. The dynamics of Cu^0 nanoparticle formation in a dendrimer of polypropylenamine (Am_n) with a diaminobutane nucleus (DAB) as a function of generation has been studied by various physicochemical methods [88], including EXAFS. In fact, the DAB-Am_n-Cu_x^{2+} ($n = 4$, 8, 16, 32, 64; $x = n/2$) complex forms in the first stage. The geometry of the Cu^{2+} tripropylendiamine complex with terminal groups in any dendrimer generation repeats a tetragonal pyramid with three nitrogen and two (including axial) oxygen atoms. DAB-Am_n-$Cu^0_{cluster}$ forms if the complex is reduced with $NaBH_4$. The size of the clusters formed dimishes as the generation number of the dendrimer increases, and the cluster size is a function of a n/x ratio. Such clusters are monodispersed with very small sizes, e.g. for $n = 64$, $x = 16$ $r_{cluster} = 0.80 \pm 0.16$ nm.

Dendrimers functionalized by specific groups, e.g. thiol groups for gold nanoparticles [89], are also promising for the preparation of dendron-stabilized nanoparticles with a narrow size distribution. In this case, the dendrimer plays a dual role: as a stabilizing agent and as a permanent selective ligand, forming a cover around Au nanoparticles. The scheme of dendron-stabilized nanoparticles (generations 1–3) is shown in Fig. 8-13.

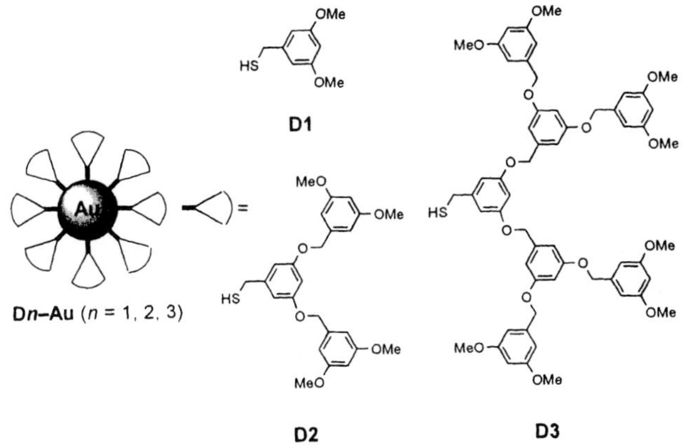

Figure 8-13. Dendron-stabilized Au nanoparticles.

Thus, particles with a very narrow distribution were observed – in the case of a second generation dendrimer, nanoparticles with a monodisperse nucleus of 2.4 ± 0.2 nm. It is important that not only individual dendrimers can be used for metallopolymer preparation, but also their dispersed mixtures with polymer matrices, which form new types of polymer–inorganic nanomaterials. For example, the highest poly(amidoamide) generations in water were put in swollen polymeric patterns of poly(2-hydroxyethylmethacrylate); Cu^{2+}, Au^{2+} or Pt^{4+} ions of a complex bound to dendrimer [90] were added to such a composition. Reduction of the metal ions resulted in new types of inorganic hybrid materials.

It should be noted that dendrimers are objects of supramolecular chemistry, and the microstructure of metal complexes based on such dendrimers is defined by the presence of van der Waals, weak hydrogen or hydrophilic–hydrophobic interactions. Their conversions have a template-directed self-organizing character similar to that in biological processes. The strategy of assembling large self-organized nanoparticles by consecutive precipitations of negatively charged macromolecules on positively charged metallocomplexes was discussed in [91]. Positively charged PS (stabilized by cetyltrimethylammonium) with a diameter of 73 ± 14 nm was used as a template, on which a composite, each layer of which contains 7600 metal ions, was obtained by repeated adsorption of polystyrene sulfonate and Fe^{2+} complex with terpyridine (Fig. 8-14) followed by centrifugation and washing of polyelectrolyte. Undoubtedly, such methods, based on a layer-by-layer technique, should be developed intensively.

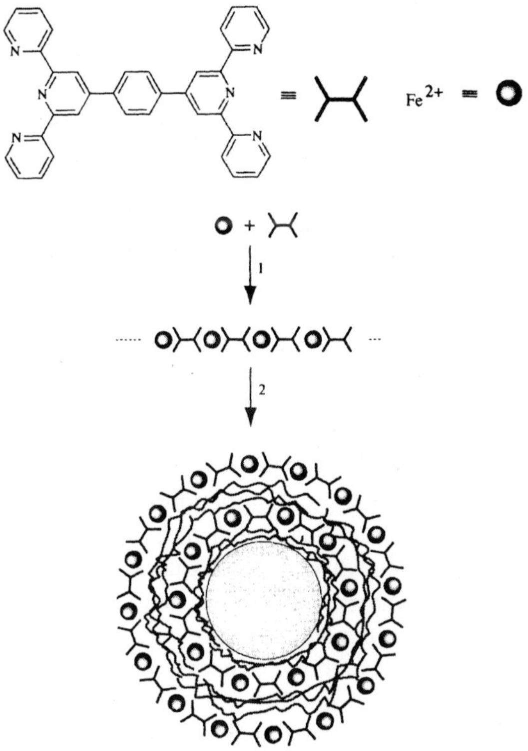

Figure 8-14. Template
assembling of
nanocomposites.

8.3.10 Self-organized Hybrid Metallopolymers-Langmuir–Blodgett Films

Another type of promising material with a unique architecture involves metal-containing nanoparticles in Langmuir–Blodgett (LB) films. They are two-dimensional (2D) generations, whose elements' sizes are not more than 2–10 nm [92]. Although this field is only just beginning to develop, methodological approaches have already been investigated. The LB technology has been modified to allow the introduction of different sensor groups or their precursors [93], and nanoparticles of metallocomplexes, into a self-organized polymolecular layer. Many limits of such technology may be overcome by using polymeric films. How the composition and structure of polymers and copolymers and the conditions influence the formation, properties and structural organization of such film monolayers has been analyzed in a review [94].

Electrostatic interactions between charged metalloparticles dispersed in the subphase and charged surface monolayers, i.e. between cationic polyelectrolyte and anionic nanoparticles, are used for the preparation of regular multilayer assemblies in Langmuir–Blodgett films. This method can be demonstrated [95] using the generation of TiO$_2$ self-organized layers as an example. Cationic TiO$_2$ particles obtained by acidic hydrolysis of TiCl$_4$ (average size ~4 nm), organize into layer-by-layer structures on the surface of a cationic polymer – poly(sodium 4-styrene sulfonate) (PSS) or poly(allylamine hydrochloride) [96]. Optically transparent films, organized at a molecular level, with a thickness up to 120 layers (60 bi-layers, thickness of each estimated to be 3.6 nm) gather on a substrate surface (metal, silicon, polymer) (Fig. 8-15). This strategy of material design for metal-film electronics will perhaps allow production of various combinations with a semiconductor structure of metal-dielectric and nanosized centers: *p-n, p-n-p, n-p-n*, etc.

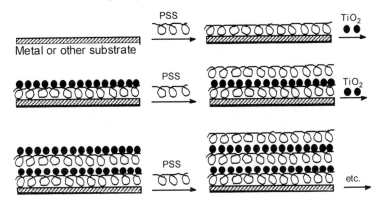

Figure 8-15. Schematic illustration for preparing multilayer TiO$_2$/polymer films.

Great attention has been paid to the construction of Langmuir–Blodgett films to be used in various sensors, e.g. metals, bound to imidazole- [97], azobenzene-containing [98], polyaniline [99] and other films.

8.3.11 Design of Polymer–Inorganic Nanomaterials Based on Incorporated Metalloparticles

This class of the newest nanocomposites possesses synergy between the properties of the parent components. The peculiarity of their architectural arrangement means that the organic phase is able to catch nanoparticles in a special "trap" or a polymeric network. However, many natural silicates (so-called smectite) including, for example, hectorite and montmorillonite (with the mica structure), consist of interchanged layers of cations and negatively charged silicates layers. Such *host* layers, with regulated systems of percolation pores and channels, form inclusion compounds with *guests* including monomer molecules.

The metallopolymer structures obtained by sol–gel synthesis (by controlled hydrolysis of Si, Ti, Zr or Al alkoxy derivatives, etc.) in the presence of organic polymers (most frequently poly(n-butylacrylate), poly(vinylpyrrolidone), poly-(methylacrylamide), etc. [59]) are determined by interphase interactions between the inorganic and polymeric components. Such materials are divided into two main types. The first are amorphous inorganic nanocomposites formed *in situ* by hydrolysis–condensation reactions in polymer solutions. An alternative method is intrusion of the polymer (or its precursor) into oxogel by mixing or by impregnating polymers into the pores of an oxide xerogel. Uncontrolled phase separation can take place if this method is used. The second type is nanocomposites obtained by sol–gel synthesis and having intercalation properties, in which organic molecules or polymers are included as guests. One characteristic example is V_2O_5, in whose interplanar distance (1.322 nm in this case) polyethylenoxide is incorporated from aqueous solution [100].

A more conventional method involves incorporation of monomers such as aniline, pyrrole, thiophene, etc. into the gel followed by their post-intercalation oxidizing polymerization [101]. Hybrid reticular nanocomposites, in which inorganic and polymeric components are bound by strong covalent or ionic chemical bonds, are of particular importance. Such materials are prepared by the formation of secondary networks after the preliminary synthesis of primary ones which, in addition, should be functionalized by an appropriate method. Other methods have also been developed, based on the simultaneous generation of two different networks (including interpenetrating ones) from molecular precursors having simultaneous organic and inorganic functionalities and reacting by different mechanisms (e.g. polyaddition, polycondensation, hydrolysis–condensation and so on). Such methods are effective in sol–gel technology in template synthesis. This last implies processes of product preparation by a method of assembling from simple components occurring in conditions of strict stereochemical reagent orientation in small reaction volumes which give rise to a higher convergence of reacting molecules than reactions in solution or in the solid state. Template methods of nanoparticle preparation in micro- and nanoporous membranes of Al_2O_3, nanotubes, fibers, etc. provide examples [102].

It is important to notice that oxidizing polymerizations of aniline, pyrrole or thiophene intercalated to layered aluminosilicates, V_2O_5, MoO_3, and some metal halides (e.g. FeOCl, α-$RuCl_3$ [103]) give highly oriented layers of guest–host type and lead to the generation of polyconjugated anisotropic polymers (Fig. 8-16). Finally, intercalation systems can be divided by their architecture and pore properties into firm pores with fixed volume, isolated parallel lattice channels and connected network channels. The concentration and spatial arrangement of guests is dictated by the topology and chemical nature of the host inner surface. Selective intercalation behavior (of molecular-sieve type) is observed in such matrices. Another type is a small host lattice (with layered or chained structure, e.g. metallophosphates) providing "flexible" pores whose dimensionality can adapt to that of the guest. Of course, intercalation chemistry is defined by pore character as well as by the nature of guest and host (mainly their semiconducting, acid–base and exchange properties).

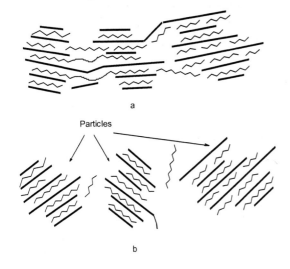

a

Particles

b

Figure 8-16. Structure of the nanocomposites obtained by intercalation polymerization of pyrrole in V_2O_5 *in situ* as a film (a) and a powder (b).

8.3.12 Conclusion

This chapter has analyzed the principal methods, structural organization and architecture of a wide range of metallopolymers obtained by incorporation of metal particles into polymers, as well as promising new ones. The multiplicity of architectural structures in such self-organized systems is a reflection of the infinite variety of natural objects coupled with the synthetic possibilities of chemistry. Even the term "supramolecular architecture" has appeared [104]. Almost all architectural forms (including those obtained by sol–gel methods in thin-layered films) are used in nature for generation of metal particles in biopolymers and their analogues. Primarily, this relates to supramolecular polyfunctional systems such as enzymes, liposomes and cells. The processes of nanocomposite preparation are very similar to biomineralization, biosorption etc.

The primary problem in synthetic and natural nanocomposites is the establishment of a correlation between the molecular structure and the properties of the materials obtained, which indeed is a part of the fundamental content–structure–property problem. It seems to us that one of the ways of solving this problem is to investigate the catalytic properties of polymer-immobilized metal complexes, clusters and nanoparticles in different organic synthesis reactions as a function of their molecular organization (see Chapters 11 and 12).

8.4 Experimental

Experiment 8-1:   Dispersion of Porphyrin-Type Compounds in Different Organic Polymers and Photoelectrochemical Oxygen Reduction (Section 8.2.1) [14,15]

Chemicals: Zinc phthalocyanine **1** (R = –H) (Aldrich) zinc 5,10,15,20-tetraphenylporphyrin (R = –H) (Aldrich); poly(vinylidene fluoride) (Aldrich); ITO (indium tin oxyde).

Equipment for photoelectrochemical measurements: 10 mL gas-tight glass cell with gas inlet, the appropriate working electrode, a saturated calomel reference electrode (SCE) and a platinum counter electrode; for measurements a potentiostat connected with a function generator and an X-Y recorder or computer; for irradiation a 250 W slide projector with a quartz-halogen lamp; water filter for cutting IR light.

Preparation of polymer films with an incorporated porphyrin-type compound: Conducting glass ITO was cut into plates of 1×1.5 cm^2 and cleaned with distilled water and methanol. Solutions of a porphyrin-type compound (0.10 g; in the case of **1** dispersed or in the case of **2** dissolved) and a polymer (0.10 g L^{-1}) in *N,N*-dimethylacetamide served as coating solutions. For electrode preparation, of the solution (10–70 μL) was dropped on an ITO plate. The solvent was removed in vacuo ($\sim 10^{-3}$ mbar) at 70 °C using a glass tube oven. Film thicknesses of 200–400 nm with roughness of 40–80 nm were obtained. No pinholes or cracks could be detected using an optical microscope at a magnification of 1200 ×.

Photoelectrochemical measurements: After the appropriate film preparation a square of 1 cm^2 electroactive area was contacted by a glass-covered copper wire using conducting adhesive Rex bond T–700 (Muromachi Kagaku Kogyo Co. Ltd.). This connection and the non-active electrode area were sealed by Ciba–Geigy Araldite Rapid epoxy resin. The experiments were performed in a gas-tight 10 mL glass cell using the appropriate working electrode, a saturated calomel reference electrode (SCE) and a platinum counter electrode. Cyclic voltammetry was performed at 20 mV s^{-1} using a potentiostat/galvanostat connected to a voltage scan generator and to an X-Y recorder. The film was illuminated from the electrolyte side if not mentioned otherwise. The 0.5 M KNO$_3$ electrolyte was 1.3 mM in O$_2$ (saturated solution). Current densities were calculated from the geometrical surface area of the electrode.

Experiment 8-2: Incorporation of Phthalocyanines into the Molecular Sieve Si-MCM-41 during Hydrothermal Synthesis (Section 8.2.2) [56]

Chemicals: Phthalocyanine-2,9,16,23-tetracarboxylic acid cobalt(II) complex **1** (R = –COOH) (see Experiment 5-11, Section 5.4), phthalocyanine-2,9,16,23-tetrasulfonic acid aluminium(III) complex **1** (R = –SO$_3$H) (see Experiment 5-13, Section 5.4); Ludox As 40 (Aldrich, 40 wt.% colloidal silica in water); tetraethylammonium hydroxide (TEAOH); polypropylene vessel; cetyl-trimethylammonium chloride (CTAC); ethanol.

Synthesis of the incorporated phthalocyanine: Ludox As 40 (92.5 g), tetraethylammonium hydroxide (92.5 g of a 20 wt.% solution in water) and de-ionized water (75 g) were placed in a polypropylene vessel and stirred at room temperature for 22 h. A mixture of CTAC (5 g of a 25 wt.% solution in water) and one of the phthalocyanines (~300 mg, ~0.25 mol) was added slowly (5 mL min^{-1}) to gel (10.4 g) with stirring. This mixture was stirred for 15 min and aged overnight at room temperature. The composition of the final gel mixture was 9.61% SiO$_2$, 4.8% TEAOH, 77.47% H$_2$O and 8.12% CTAC. The mixture was reacted without stirring for 40 h at 100 °C in a polypropylene autoclave. The resulting green product was recovered by filtration and washed intensively with water and hot ethanol. The MCM-41 structure was proven by X-ray diffraction. The content of the phthalocyanine in the molecular sieve is ~10^{-5} mol g^{-1}.

Experiment 8-3: Microcapsulation of TiCl$_3$ in Polystyrene (Section 8.3.2) [61]

Method A: Refined and freshly distilled styrene (5 g), TiCl$_4$ (0.14 g) and dinitrile azoisobutyric acid (0.05 g)were loaded into a glass ampoule under an inert atmosphere, and the ampoule was sealed. Polymerization was carried out at 70 °C for 2 h. The vessel was shaken until thickening occurred(for first 2 h). The product obtained was ground to yield TiCl$_3$ encapsulated in polystyrene (8.4 mg of Ti g^{-1}).

Method B: Polystyrene (2.5 g) was dissolved in refined and freshly distilled benzene (50 mL). To this solution Ti(OC$_4$H$_9$)$_4$ (1.0 g) was added, and the mixture was stirred until a homogeneous solution formed. Then it was added dropwise via a thin capillary into a Dewar vessel with liquid nitrogen. The granules obtained were put in a glass ampoule provided with an outlet for evacuation. Then micellar benzene was sublimed for 5 h at the temperature of liquid nitrogen. After that the temperature was slowly raised over 3 h to room temperature at which a final sublimation was carried out. The product obtained was ground to yield Ti(OC$_4$H$_9$)$_4$ encapsulated in polystyrene (10 mg of Ti g^{-1}).

Experiment 8-4: Mechanochemical Grinding (Section 8.3.3) [64]

PE, PP (S_{sp} ~ $4\,m^2\,g^{-1}$) or PE-gr-poly(4-vinylpyridine) (grafted degree 7.5 wt.%) are put into an agate planetary mill "Fritisch pulverisette" with working volume 0.4 L (sphere volume 0.08 L, speed of rotation 20 TN s^{-1}) and software tools. Loading of $TiCl_3$ or VCl_3 into the mill and sampling are carried out in a pressure-tight box under a helium atmosphere. After the appropriate time of grinding the mill was put in the box again, opened and sampling for analysis (X-ray, IR etc.) was carried out. The products obtained are used as catalysts for polymerization of ethylene or stereospecific polymerization of propylene.

Experiment 8-5: Incorporation of Clusters and Nanoparticles in a Surface Layer of Polymer (Section 8.3.7) [105].

A suspension of PE-gr-poly(acrylic) acid (grafted degree 5.0 wt.%) in ethanol was put into a temperature-controlled glass reactor at 50 °C. A solution of $Cu(OCOCH_3)_2$ in ethanol was added to the suspension and stirred for 1 h. Then the suspension was filtered, washed with ethanol and dried. The product obtained was reduced by $NaBH_4$. The content of Cu and the size of nanoparticles formed were analysed.

Experiment 8-6: Thermolysis of Metallomonomers (Section 8.3.8) [81].

Thermolysis of metal acrylates ($MAcr_n$) was carried out in an isothermal regime in a nonisothermal reactor (at the experimental temperature T_{exp}); the heated volume of the compound amounted to 0.03 V (where V is the reaction vessel volume; the basic part of the vessel was at room temperature (T_r). $MAcr_n$ was studied at T_{exp}= 350–390 °C ($CoAcr_2$), 300–360 °C ($NiAcr_2$), 200–370 °C ($FeAcr_3$) and 330–370 °C (Fe_2NiAcr_8). The kinetics of gas evolution were recorded using a membrane zero-manometer. The mass loss of the sample and the total amounts of gaseous products at T_r were determined at the end of the experiments. The low-temperature fractionation of gaseous products was carried out at 77 K. The metal content in metallopolymer products was analyzed, and the size of the particles was determined by transmission electron microscopy.

8.5 References

1. M. Kaneko, D. Wöhrle, *Adv. Polym. Sci.* **1988**, *84*, 141.
2. J. Zhang, F. Zhao, M. Kaneko, *J. Porphyrins Phthalocyanines* **2000**, *4*, 65.
3. F. Zhao, J. Zhang, M. Kaneko, *J. Porphyrins Phthalocyanines* **2000**, *4*, 158.
4. F. Zhao, J. Zhang, T. Abe, M. Kaneko, *J. Porphyrins Phthalocyanines* **1999**, *3*, 238.

5. J. Zhang, T. Abe, M. Kaneko, *J. Porphyrins Phthalocyanines* **1998**, *2*, 93.
6. M. Yagi, H. Fukiya, M. Kaneko, *J. Electroanal. Chem.* **2000**, *481*, 69.
7. R.-J. Lin, T. Onikubo, M. Kaneko, *J. Electroanal. Chem.* **1993**, *348,* 189.
8. J. Zhang, M. Yagi, M. Kaneko, *Macromol. Symp.* **1996**, *105*, 59; *J. Electroanal. Chem.* **1996**, *412,* 159.
9. J.N. Demas, B.A. DeGraff, *Makromol. Chem., Macromol. Symp.* **1992**, *59*, 35; *J. Macromol. Sci. Chem.* **1988**, A 25, 1189.
10. M. Kaneko, A. Yamada, *Adv. Polym. Sci.* **1984**, *55*, 1. M. Kaneko, A. Yamada, in: *Metal Containing Polymer Systems*, J.E. Sheats, C.E. Carraher, C.U. Pittman (Eds.), Plenum Press, New York, 1985, p. 249.
11. M. Kaneko, S. Hayakawa, *J. Macromol. Sci. Chem.* **1988**, *A 25*, 1255.
12. D. Wöhrle, H. Kaune, B. Schumann, N.I. Jaeger, *Makromol. Chem.* **1986**, *187*, 2947.
13. M. Kaneko, D. Wöhrle, D. Schlettwein, V. Schmidt, *Makromol. Chem.* **1988**, *189*, 2419.
14. D. Schlettwein, M. Kaneko, A. Yamada, D. Wöhrle, N.I. Jaeger, *J. Phys. Chem.* **1991**, *95*, 1748.
15. D. Schlettwein, N.I. Jaeger, D. Wöhrle, *Makromol. Chem., Macromol. Symp.* **1992**, *59,* 267.
16. M.J. Stillman, T. Nyokong, in *Phthalocyanines – Properties and Applications*, C.C. Leznoff, A.B.P. Lever (Eds.), VCH Publishers, New York, 1989, p. 133.
17. D.Wöhrle, D. Schlettwein, M. Kirschenmann, M. Kaneko, A. Yamada, *J. Macromol. Sci. Chem.* **1990**, *A27* 1239.
18. D. Schlettwein, D. Wöhrle, N.I. Jaeger, *Ber. Bunsenges. Phys. Chem.* **1991**, *95*, 1526.
19. K.-Y. Law, *Chem. Rev.* **1993**, *93*, 449.
20. H.-Z. Chen, M. Wang, S.-L. Yang, *Thin Solid Films* **1999**, *357*, 208.
21. H.-S. Kim, C.-H. Kim, C.-S. Ha, J.-K. Lee, *Synth. Met.* **2001**, *117*, 289.
22. I.L. Kogan, K. Yakushi, *Electrochim. Acta.* **1998**, *43*, 2053; *J. Mater. Chem.* **1997**, *7*, 2231.
23. V. Cleave, R.H. Friend, N. Tessler, *Adv. Mater.* **2001**, *13*, 44.
24. T. Skotheim, M. Velazquez, C.A. Linhous, *J. Chem. Soc., Chem. Commun.* **1985**, 612.
25. R.A. Bull, F.R. Fan, A.J. Bard, *J. Electrochem. Soc.* **1984**, *131*, 687.
26. C.S. Choi, H. Tachikawa, *NTIS Chem.* **1990**, *90*, 20; AD-A 217 677.
27. D.J. Walton, C.E. Hall, *Synth. Met.* **1991**, *45*, 363.
28. Y. Okamoto, *Makromol. Chem., Macromol. Symp.* **1992**, *59*, 83.
29. Y. Okamoto, J. Kido, in: *Macromolecular Complexes*, E. Tsuchida (Ed.), VCH Publishers, New York, 1991, p. 143.
30. Y. Okamoto, J. Kido, H.G. Brittain, S. Paoletti, *J. Macromol. Sci. Chem.* **1988**, *A25,* 1383.
31. Y. Kurimura, Y. Sairenchi, S. Nakayama, *Makromol. Chem., Macromol. Symp.* **1992**, *59*, 199.
32. Y. Kurimura, in: *Macromolecular Complexes*, E. Tsuchida (Ed.), VCH Publishers, New York, 1991, p. 93.
33. S.M. Arabei, S.G. Kulikov, A.V. Veret-Lemarinier, J.P. Galoup, *Chem. Phys.* **1997**,

    163, 177.
34. F. Zhao, J. Zhang, M. Kaneko, *J. Photochem. Photobiol. A: Chem.* **1998**, *119*, 53.
35. H. Hirai, *J. Macromol. Sci. Chem.* **1990**, A27, 1293.
36. N. Toshima, K. Kanaka, M. Komiyama, H. Hirai, *J. Macromol. Sci. Chem.* **1988**, *A25*, 1349.
37. L.M. Bronshtein, I.E. Larikova, P.M. Valetsky, *Polym. Sci. USSR,* **1987**, *29*, 2653.
38. E.S. Mirzoeva, L.M. Bronshtein, P.M. Valetsky, *Polym. Sci. USSR* **1989**, *31*, 2898.
39. C.-G. Jia, F.-Y. Jin, H.-Q. Pan, M.-Y. Hung, Y.-Y. Jiang, *Macromol. Chem. Phys.* **1984**, *195*, 751.
40. C. Pan, H. Zong, *Macromol. Symp.* **1994**, *80*, 265.
41. D.C. Sherrington, H.-G. Tang, *Macromol. Symp.* **1994**, *80*, 193.
42. H. Shirai, K. Hanabusa, T. Koyama, H. Tsuiki, E. Masuda, *Makromol. Chem., Macromol. Symp.* **1992**, *59*, 155.
43. F. Ciardelli, E. Tsuchida, D. Wöhrle, *Macromolecule–Metal Complexes*, Springer, Berlin, 1996.
44. K. Ono, *High Polym. Jpn.* **1985**, *34*, 766.
45. K. Orihara, H. Yonekura, *J. Macromol. Sci. Chem.* **1990**, *A27*, 1217.
46. P. Hegenmüller, W.V. Gool (Eds.), *Solid Electrolytes*, Plenum Press, New York 1980.
47. B. Bogdanov, C. Uzov, M. Michaelov, *Acta Polym.* **1992**, *43*, 202.
48. H.C. Sacco, K.J. Ciuffi, A. Serra, Y. Iamamoto, *J. Non-Cryst. Solids* **2001**, *284*, 174.
49. A.B. Sorokin, P. Buisson, A.C. Pierre, *Microporous Mesoporous Mater.* **2001**, *46*, 87.
50. I.T. Caga, I.D. Carnell, J.M. Winterbottom, *J. Chem. Technol. Biotechnol.* **2001**, *76*, 179.
51. G. Albert, T. Bein (Eds.), *Comprehensive Supramolecular Chemistry,* 1996, Vol. 7, Solid State Supramolecular Chemistry: Two- and Three-Dimensional Inorganic Networks, Pergamon, Oxford.
52. D. Wöhrle, G. Schulz-Ekloff, *Adv. Mater.* **1994**, *6*, 875.
53. G. Schulz–Ekloff, D. Wöhrle, R.A. Schoonheydt, *Microporous Mesoporous Mater.* **2002**, *51*, 91.
54. K.J. Balkus, in *Phthalocyanines – Properties and Applications*, C.C. Leznoff, A.B.P. Lever (Eds.), VCH Publishers, New York, 1996, Vol. 4, p. 287.
55. D. Wöhrle, A.K. Sobbi, O. Franke, G. Schulz-Ekloff, *Zeolites* **1995**, *15*, 450.
56. M. Ganschow, D. Wöhrle, G. Schulz-Ekloff, *J. Porphyrins Phthalocyanines* **1999**, *3*, 299.
57. M. Ehrl, F.W. Deeg, C. Bräuchle, O. Franke, A. Sobbi, G. Schulz-Ekloff, D. Wöhrle, *J. Phys. Chem.* **1994**, *98*, 47.
58. M. Faraday, Philos. Trans. R. Soc. London 1857, 147, 145.
59. A.D. Pomogailo, A.S. Rozenberg, I.E. Uflyand, *Metal Nanoparticles in Polymers*, Khimiya, Moscow, 2000.
60. V.D. Solodovnik, *Microcapsulation,* Khimiya, Moscow, 1980.
61. A.D. Pomogailo, Pat. USSR 925965, 1982.
62. A.B.R. Mayer, J.E. Mark, *J. Polym. Sci. B: Polym. Phys.* **1997**, *35*, 1207.
63. H. Du, P. Zhang, F.Liu, S. Kan, D. Wang, T. Li, X. Tang, *Polym. Intern.* **1997**, *43*,

274.

64. S.L. Saratovskikh, A.D. Pomogailo, O.N. Babkina, F.S. Dyachkovskii, *Kinetika i Kataliz* **1984**, *25*, 464.

65. D.A. Kritskaya, A.D. Pomogailo, A.N. Ponomarev, F.S. Dyachkovskii, *J. Appl. Polym. Sci.* **1980**, *25*, 349; *J. Polym. Sci.: Polym. Symp.* **1980**, *68*, 23.

66. D. Siew Hew Sam, P. Courtine, J.C. Jannel, *Macromol. Chem. Rapid Commun.* **1986**, *6*, 631; *Europ. Polym. J.* **1986**, *22*, 89.

67. N. Toshima, T. Teranishi, H. Asanuma, Y. Saito, *Chem. Lett.* **1990**, 819.

68. P.N. Logvinenko, T.V. Dmitrieva, *Kolloidny Zh.* **1990**, *52*, 1067.

69. E. Sacher, J.J. Pireaux, S.P. Kowalczyk (Eds.), *Metallization of Polymers. ACS Symp. Ser. No. 440*, ACS, Washington DC, 1990.

70. H.M. Meyer, S.G. Anderson, L.J. Atanasoska, J.H. Weaver, *J. Vac. Sci. Technol. A* **1988**, *6*, 30.

71. L. Tröger, H. Hünnefeld, S. Nunes, M. Oehring, D. Fritsch, *J. Phys. Chem. B* **1997**, *101*, 1279.

72. O.E. Litmanovich, I.M. Papisov, *Vysokomolekul. Soedin.* **1999**, *41*, 1824; O.E. Litmanovich, E.A. Eliseeva, A.A. Litmanovich, I.M. Papisov, *Vysokomolekul. Soedin. A* **2001**, *43*, 1315.

73. E. Helfand, Z.R. Wasserman, *Macromolecules* **1976**, *9*, 879; ibid, **1978**, *11*, 960; ibid, **1980**, *13*, 994.

74. M. Moffitt, H. Vali, A. Eisenberg, *Chem. Mater.* **1998**, *10*, 102; R.T. Clay, R.E. Cohen, *Supramolecular Science* **1995**, *2*, 183.

75. R. Saito, S. Okamura, K. Ishizu, *Polymer* **1992**, *33*, 1099; ibid, **1995**, *36*, 4515; ibid, **1996**, *37*, 5255.

76. L. Zhang, A. Eisenberg, *Science* **1995**, *268*, 1728; *J. Am. Chem. Soc.* **1996**, *118*, 3168.

77. A.L. Volynskii, E.S. Trofimchuk, N.I. Nikonorova, N.F. Bakeev, *Zh. Obchsh. Khimii* **2002**, *72*, 575.

78. S.P. Gubin, I.D. Kosobudskii, *Usp. Khimii* **1983**, *52*, 1350; G.Yu. Yurkov, S.P. Gubin, D.A. Pankratov, Yu. A. Koshkarov, A.V. Kozynkin, Yu.I. Spichkin, T.I. Nedoseikina, I.V. Pirog, V.G. Vlasenko, *Neorg. Materialy* **2002**, *38*, 186.

79. A.D. Pomogailo, *Polymer-Immobilized Catalysts*, Nauka, Moscow, 1988.

80. G.D. Chryssicos, V.D. Mattera, Jr., A.T. Tsatsas, W.N. Risen, Jr., *J. Catal.* **1985**, *93*, 430.

81. A.S. Rozenberg, G.I. Dzhardimalieva, A.D. Pomogailo, *Polym. Adv. Technol.* **1998**, *9*, 527.

82. A.D. Pomogailo, A.S. Rozenberg, G.I. Dzhardimalieva, M. Leonowicz, *Adv. Mat. Sci.,* **2001**, *1*, 19.

83. D.A. Tomalia, P.R. Dvornic, in: *The Polymeric Materials Encyclopedia*, CRC Press, Boca Raton, FL, 1996, Vol. 3, p. 1814.

84. K.L. Wooley, C.J. Hawker, J.M. Frechet, *J. Chem. Soc., Perkin Trans., I* **1991**, 1059.

85. M.A. Hearshaw, J.R. Moss, *Chem. Commun.* **1999**, 1.

86. F. Zeng, S.C. Zimmerman, *Chem. Rev.* **1997**, *97*, 1681.

87. L. Balogh, D.A. Tomalia, *J. Am. Chem. Soc.* **1998**, *120*, 7355; M. Zhao, L. Sun, R.M. Crooks, *J. Am. Chem. Soc.* **1998**, *120*, 4877.

88. P.N. Floriano, C.O. Noble, J.M. Schoonmaker, E.D. Poliakoff, R.L. McCarley, *J. Am. Chem. Soc.* **2001**, *123*, 10,545.
89. M.-K.Kim, Y.-M. Jeon, W.S. Jeon, H.-J. Kim, S.G. Hong, C.G. Park, K. Kim, *Chem. Commun.* **2001**, 667.
90. F. Gröhn, G. Kim, B.J. Bauer, E.J. Amis, *Macromolecules* **2001**, *34*, 2179.
91. D.G. Kurth, F. Caruso, G. Schüler, *Chem. Commun.* **1999**, 1579.
92. A.D. Pomogailo, *Usp. Khimii* **2000**, *69*, 60; ibid, **1997**, *66*, 750.
93. G. Roberts (Ed.), *Langmuir–Blodgett Films*, Plenum Press, NewYork, 1990.
94. V.V. Arslanov, *Usp. Khimii* **1994**, *63*, 3.
95. H. Shin, R.J. Collins, M.R. DeGuire, A.N. Heuer, C.N. Sukenik, *J. Mater. Res.* **1995**, *10*, 692.
96. Y. Liu, A. Wang, R. Claus., *J. Phys. Chem. B* **1997**, *101*, 1385.
97. H. Jeong, B.-I. Lee, W.J. Cho, C.-S. Ha, *Polymer* **2000**, *41*, 5525.
98. Riul, Jr., D.S. Dos Santos, Jr., K. Wohnrath, *Langmuir* **2002**, *18,* 239.
99. M. Ferreira, K. Wohnrath, R.M. Torresi, *Langmuir* **2002**, *18*, 540.
100. E. Ruiz-Hitzky, P. Aranda, B. Casal, *J. Mater. Chem.* **1992**, *2*, 581.
101. M.G. Kanatzidis, C.-G. Wu, H.O. Marcy, C.R. Kannewurf, *J. Am. Chem. Soc.* **1989**, *111*, 4139; *Chem. Mater.* **1990**, *2*, 221.
102. C.J. Lakshmi, B.B. Patrissi, C.R. Martin, *Chem. Mater.* **1997**, *9*, 2544.
103. L.Wang, M. Rocci-Lane, P. Brazis, C.R. Kannewurf, Y.-I. Kim, W. Lee, J.-H. Choy, M.G. Kanatzidis, *J. Am. Chem. Soc.* **2000**, *122*, 6629.
104. T. Bein (Ed.), Supramolecular Architecture: Synthetic Control in Thin Films and Solids. ACS Symposium Series No. 499, ACS, Washington, DC, 1992.
105. A.D. Pomogailo, N.D. Golubeva, *J. Inorg. Organometal. Polym.* **2001**, *11*, 67.

C PROPERTIES

9 Binding of Small Molecules

Jongok Won, Yong Soo Kang and Hiroyuki Nishide

9.1 Introduction

One of the typical chemical characteristics of metal ions and/or their complexes is their ability to bind specifically and reversibly with small molecules, such as alkene, carbon monoxide, molecular nitrogen, and molecular oxygen [1–3]. For instance, ethylene coordinatively binds with silver or copper ions, and molecular oxygen with metalloporphyrins. When such metal ions or metalloporphyrins are molecularly bound in macromolecules to form metal complexes, the reversible binding of small molecules with the metal complexes in macromolecules offers extensive potential applications, such as separation membranes, absorbents, and sensors.

A macromolecule provides a suitable functional group for either complexation with metal ions of metal salts or attaching metal complexes into the macromolecular matrix through a coordination bond; typical examples of the former are silver-polymer complexes, while cobalt porphyrins incorporated on a polymer backbone through coordination bonding are examples of the latter.

Since the structure and properties of metal complexes in macromolecules are affected by the surrounding macromolecules [4–5], metal complexes in macromolecules exhibit characteristic behavior in binding reactions with small or gaseous molecules as well as unique chemical reactivities and physicochemical properties. When a small molecule contacts a metal complex in a macromolecular matrix, it may dissolve in the macromolecule as a result of coordinative binding with the metal complex if there is a specific interaction between the small molecule and the metal complex. In many cases, such as a silver-polymer complex with olefin or a metalloporphyrin with molecular oxygen, the binding is reversible and specific, resulting in an enhanced total solubility and transport. Since the binding of small molecules to metal complexes is a simple phenomenon or elementary step, with no further reaction involved, monitoring the adduct formation and transport of small gaseous molecules in macromolecules containing a metal complex is a simple but powerful tool for characterizing the reactivity and functionality of metal complexes in macromolecules.

In organometallic chemistry, complexes between metal salts and olefins, such as Zeise's salt, have been known since the early nineteenth century. Silver or copper ions are known to form a reversible adduct with olefins [6–7]. The bond between the olefin molecule and the metal ion is regarded as a dative σ bond to a suitable hybrid orbital of the metal (*s* or *sp* in silver ion). This bond is analogous to the dative carbon–metal σ bond between carbon monoxide and transition metals and metal–carbon (antibonding orbital) π bond by back-donation (Fig. 9^{-1}) [6–7].

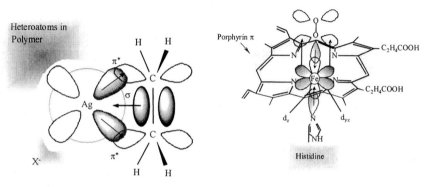

Figure 9-1. Bonding between silver ion and olefin.

Figure 9-2. Binding of molecular oxygen in hemoglobin/myoglobin.

The reversible binding behavior between silver ions and olefins can be utilized to separate unsaturated hydrocarbons (olefins) from saturated ones (paraffins) [8]. The extent and reversibility of the bonding between silver ions and olefins vary depending on the structure of the olefin and the type of macromolecule. Therefore, the binding ability and its reversibility for specific applications can be tailored using macromolecular ligands. The roles of the macromolecule (polymer) surrounding the central silver ion are as follows: first, tuning the strength of the binding between the silver ion and the olefin, second, providing a hydrophobic environment through the formation of a micro-domain, third, prolonging the lifetime of the metal complexes by immobilizing the silver ions, and fourth, enabling the fabrication of various forms including films, beads, and so on.

A similar binding of small molecules is also observed in iron and cobalt porphyrins [2–3,9–10]. Iron(II), d^6, and cobalt(II), d^7, are not particularly soft metal cations, yet the "softening" (symbiotic) action of the tetrapyrrole ring system in metalloporphyrins permits molecular oxygen binding. The porphyrins contain a highly resonant planar ring system in which four nitrogen atoms are fixed in the center at a spacing that is ideal for their unshared electrons to bond with a metal ion. The iron porphyrin group has attracted the most attention because it plays a central biological role in the binding of molecular oxygen, O_2.

In a living body, oxygen is transported or stored with the help of hemoglobin or myoglobin. Iron(II) ions have a coordination number of six. When a ferrous ion forms a chelate with a porphyrin, two of its coordination sites are still unfilled. Therefore, one of the functions of the polypeptide chain of hemoglobin and myoglobin is to provide a histidyl residue as a position where one of the nitrogen atoms links to the fifth coordination position of the iron ion, as shown in Fig. 9-2.

The sixth position of the iron is left open and now has the ability to bind molecular oxygen reversibly. Molecular oxygen is far from a typical ligand, but it may be considered a soft ligand with some π-bonding capacity. Details are given in Section 2.3.1, *Oxygen Transport Hemes*. The oxygen-binding reaction is rapid and reversible, as presented in Fig. 9-2. However, irreversible oxidation of the iron(II) may occur to remove its oxygen binding capability (see Fig. 2-8 of Chapter 2, and Section 9.2.4). The effects of synthetic polymer ligands on the reversible oxygen binding are described in this chapter, by comparing those of the polypeptide ligand of hemoglobin [4,11].

In this chapter, the specific and reversible bindings of small molecules with metal complexes in macromolecules are described using the examples of silver-polymer complexes and metalloporphyrins coordinatively bound in macromolecules. Next, their structures are characterized, followed by potential applications such as gas-separation membranes and gas-carrying materials.

9.2 Binding of Small Molecules to Metal Complexes in Macromolecules

9.2.1 Formation of Silver-Polymer Complexes

Polymer complexes are composed of a metal complex bound to a chain of a linear macromolecule via a Type I coordinative bond. (see Section 1.2.1, *Classification by Kind of Metal Complex/Metal Binding*). The formation process of metal-polymer complexes can be schematically written as in Eq. (9-1), where M and X are a metal ion and its counter anion, respectively.

$$\text{(9--1)}$$

A complexation process in the solid state involves the dissociation of metal salts into ions (losing lattice energy) and the coordination of the cation to a heteroatom (or polar group) of a macromolecule (gaining coordination bonding or complexation energy). It is known that the lower the lattice energy, the easier it is to make metal-polymer complexes. Thus, the lattice energy is one important parameter in forming metal-polymer complexes. The complexation energy is associated with the coordinative interaction between the metal cation and the polar groups in the polymer backbone. For example, polar oxygen or nitrogen atoms coordinate with transition metal cations, resulting in the homogeneous complexation of the metal cations into a polar macromolecular matrix. Specifically, oxygen atoms from an ether or amide group in poly(ethylene oxide) (PEO), poly(2-ethyl-2-oxazoline) (POZ), or poly(N-vinylpyrrolidone) (PVP) coordinate with silver cations to form silver-polymer complexes [12–13]. The structures of the repeat unit of the polymers are presented in Fig. 9-3. The significance of silver-polymer complexes was discussed above in the Introduction.

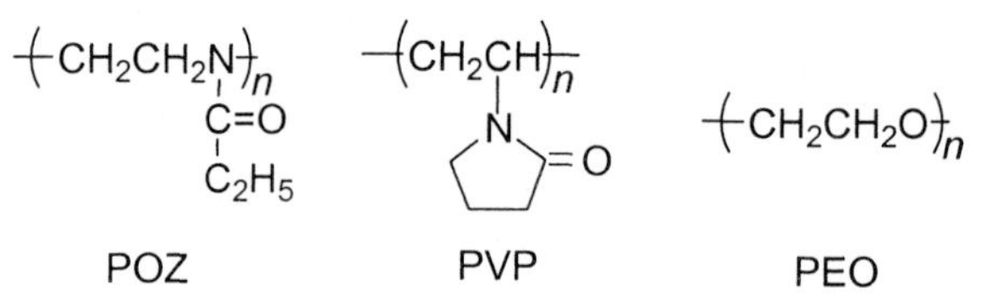

POZ   PVP   PEO

Figure 9-3. Structure of the polymers POZ, PVP, PEO.

When silver ions are coordinated by heteroatoms in a polymer matrix to form polymer complexes, their structure and properties will be changed, mainly because of the coordinative interaction between the silver ions and polar groups in the polymers. This change in structure can be seen clearly by the variation of the intersegmental distance and glass transition temperature (T_g) of the silver-polymer complex. The intersegmental distance and d-spacing have been determined using wide angle X-ray scattering (WAXS) based on Bragg's law. In a WAXS study, the first broad maximum peak for $AgCF_3SO_3$-POZ complexes was assigned to the interchain distance and decreased with the silver concentration at low silver concentrations; the silver ions become transient crosslinks and pull the polymer chains by coordinative interaction, thereby decreasing the interchain distance.

The T_g of the silver-POZ complex initially increases with the silver concentration, reaches a broad maximum, then slightly decreases again. The increase in T_g with the complexation of the silver ions is primarily attributable to both the reduced chain mobility in the transient crosslinks of the polymer segments due to the silver cations and the dangling of silver cations on the polymer chain [14].

Gas permeability is also one of the parameters used to detect a change in the structure of a polymer. Since paraffins, such as propane, do not have any specific interaction with silver ions and thus only permeate via normal sorption and diffusion transport, the permeability behavior of propane is mostly determined by the microstructure and chain mobility of a polymer complex. As such, the propane permeabilities of POZ films decrease rapidly with $AgCF_3SO_3$ concentration, which is consistent with the results of the d-spacings and glass transition temperatures.

Metal ion complexation affects the structure and physicochemical properties of polymer ligands, and this is generally observed in metal-polymer complexes, for example, Li-polymer complexes (electrolytes).

9.2.2 Olefin Binding and Reversibility

When small molecules, such as olefins, come into contact with a metal complex in macromolecules, such as silver-polymer complexes, a significant amount of olefins are absorbed by the silver-polymer complexes. The propylene solubilities in Ag-POZ complex films ([Ag$^+$] in AgBF$_4$: [C=O] in POZ = 1:1, ca. 80 wt.% of AgBF$_4$ in the polymer matrix) were plotted against the propylene pressure, as shown in Fig. 9-4 [15]. The propylene solubility in pure POZ was very small. However, the propylene solubilities in a 1:1 AgBF$_4$-POZ complex (190 and 240 cm^3 (STP)of propylene per 1 cm^3 of silver-polymer complex at 50 and 190 kPa, respectively) significantly increased ~100 times compared to those in pure POZ because of the chemically specific binding of propylene into silver ions.

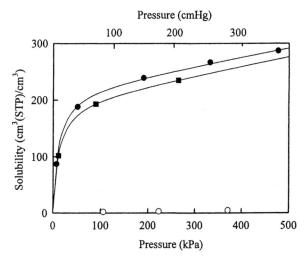

Figure 9-4. Solubility of propylene in solid silver-polymer complex films at 25 °C [15]. [Ag$^+$]:[C=O] = 1:1. ●:AgBF$_4$-POZ; ■: AgBF$_4$-PVP; ○:pure POZ. The propylene solubility in pure PVP film was immeasurably small and was not plotted in the figure.

Propylene dissolves in two modes, a physical dissolution mode in the matrix and a binding mode resulting from a reversible chemical reaction with metal complexes. Henry's law commonly expresses the physical dissolution mode for small molecules, while the Langmuir model adequately describes the reversible binding mode, as shown in Eq. (9-2). Mathematically, a Langmuir adsorption isotherm for a small molecule in a porous media is identical to the expression of the olefin concentration bound to the metal complex.

$$Ag^+ (\,polymer\,) + olefin \rightleftharpoons olefin{-}\,Ag^+ (\,polymer\,) \qquad (9\text{-}2)$$

$$K = \frac{[\text{Ag} - \text{olefin}]}{[\text{Ag}^+][\text{olefin}]} \qquad (9\text{-}3)$$

The total solubility is the sum of the contributions of both the dissolution mode based on Henry's law and the binding mode using the Langmuir model: the dual mode sorption [11]. At equilibrium, the total solubility of olefin is shown by Eq. (9-4), where C is the total concentration of the olefin gas absorbed in the sample, p is the applied olefin pressure, k_D is the solubility coefficient of the olefin gas for Henry's law mode, K is the olefin binding equilibrium constant of Eq. (9-2), as defined in Eq. (9-3), and C_c is the saturated amount of the olefin gas bound to the silver complex.

$$C = k_D p + \frac{C_c K p}{1 + K p} \qquad (9\text{-}4)$$

The propylene solubilities can be analyzed using the dual sorption model of Eq. (9-4) and the parameters are listed in Table 9-1. The k_D value is very small and represents a small physical dissolution term, which only contributes 7% to the total solubility at 100 kPa, whereas the binding Langmuir term provides a major contribution represented by a large C_c value of 220 $(\text{cm}^3(\text{STP})/\text{cm}^3)$ and a large K value of 0.12 (1/cmHg) in comparison with 0.06 for oxygen binding to the picketfence cobalt porphyrin-OPy membrane (see Fig. 9-10), representing the strong affinity of propylene for a silver complex. These results indicate that the total solubility mostly depends on the binding of propylene by a reversible chemical reaction between the propylene and the silver complex in the polymer.

Table 9-1. Binding equilibrium constant (K), solubility coefficient (k_D), and saturated amount of the gas bound to the silver complex (C_c) of the solid silver-POZ complex for propylene [16].

	AgBF$_4$-POZ	AgCF$_3$SO$_3$-POZ
k_D (cm^3(STP)/ cm^3 cmHg)	0.20	0.20
K (1/cmHg)	0.12	0.04
C_c (cm^3(STP)/cm^3)	222	131

The olefin solubility in a silver-PEO complex has been also reported. AgBF$_4$-PEO absorbed 45 cm^3 (STP) of propylene per 1 g of silver-PEO complex, at 30 °C and 93 kPa [17]. The relationship between the olefin solubility and the structure can be understood by an *ab initio* calculation based on the density functional theory of the model system. The theoretical calculation shows that the bond length between the silver ion and the closest anion atom in the AgBF$_4$-PEO film changes from 2.309 to 2.506 Å with the addition of ethylene, and the free energy for the formation of an ethylene adduct with AgBF$_4$ is favorable for an ethylene-silver complex in PEO [15, 18–19] (Fig. 9-5).

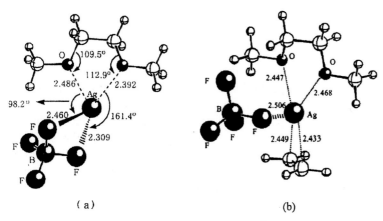

(a)   (b)

Figure 9-5. Structures of complexes of a: AgBF$_4^{-1}$,2-dimethoxy ethane complex, b: AgBF$_4^{-1}$,2-dimethoxy ethane-ethylene adduct where 1,2-dimethoxy ethane is a model compound of poly(ethylene oxide) [15]. The silver ion is coordinated by two oxygen atoms from 1,2-dimethoxy ethane and two F atoms from the anion to make silver-polymer complexes, when AgBF$_4$ is complexed with 1,2-dimethoxy ethane. One of the two F atoms bound to the silver ion is replaced by an ethylene molecule, when one ethylene molecule approaches the complex in an ethylene environment.

The reversible olefin binding to silver ions in solid polymer complexes has been investigated using FTIR and UV-Vis spectroscopies (see Experiment 5^{-1}, Section 9.5) [20–22]. The series of IR spectra in Fig. 9-6 demonstrate the reversible olefin coordination to the silver-poly(vinylmethylketone) (PVMK) complex [21]. The free C=O stretching peak at 1709 cm^{-1} shifts to a lower frequency owing to the silver complex formation, as shown in Fig. 9-6a, demonstrating the coordination of carbonyl groups to silver ions. A silver-PVMK complex film ([Ag$^+$] in AgBF$_4$:[C=O] in PVMK = 1:2; 58 wt.% of AgBF$_4$ in the polymer matrix) only showed a coordinated C=O peak at 1681 cm^{-1}, implying that all the carbonyl oxygen is coordinated to silver ions. When the film was exposed to 207 kPa of propylene, then purged with nitrogen gas to sweep up any excess propylene, a new peak appeared at 1586 cm^{-1} (Fig. 9-6c). This new peak represents the C=C stretching vibration of the coordinated propylene (cf. $v_{C=C}$ of free 1-hexene = 1640 cm^{-1}). It is interesting to note that the peak at 1586 cm^{-1} remains even after out-gassing at 10^{-5} torr and room temperature for 4 h. However, the exposure of the propylene coordinated complex to 207 kPa of 1,3-butadiene and subsequent purging with nitrogen produced a new peak at 1558 cm^{-1} with the concomitant disappearance of the peak for the coordinated propylene at 1586 cm^{-1}. The peak at 1558 cm^{-1} represents the C=C stretching frequency of the coordinated 1,3-butadiene (Fig. 9-6d). The peak at 1558 cm^{-1} was again replaced by a peak at 1586 cm^{-1} with the re-introduction of propylene into the silver-PVMK complex previously coordinated with 1,3-butadiene.

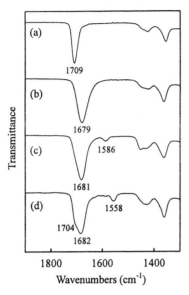

Figure 9-6. IR spectra for the solid-state interactions of the silver-PVMK complex and olefins: (a) PVMK, (b) AgBF$_4$-PVMK ([Ag$^+$]:[C=O] = 1:2), (c) propylene-coordinated film and (d) 1,3-butadiene-coordinated film [21].

This reversible olefin coordination can also be observed based on spectral changes in the UV absorption spectra [21]. A strong and broad absorption band at 220–230 nm for the propylene coordinated with the AgBF$_4$-PVMK complex ([Ag$^+$]:[C=O] = 1:2) disappeared and a new band appeared at around 258 nm for the coordinated 1,3-butadiene when the propylene coordinated film was exposed to a 1,3-butadiene atmosphere followed by a nitrogen gas purge. The absorption band of the coordinated propylene at 220–230 nm reappeared when propylene was introduced into the cells containing the 1,3-butadiene coordinated film.

Because the concentration of the metal complex moiety in the polymer domain of a solid-state film is higher than that of a homogeneously solubilized complex in solution, even a simple spectrophotometer is effective for observing the gaseous molecule-binding reactions and ligand-bonding characteristics of complexes.

Both IR and UV spectroscopy indicate that coordinated propylene is labile enough to be easily replaced by other olefins. The rapid exchange reaction seems to be a determining factor for olefin transport and a driving force for olefin-facilitated transport.

9.2.3 Formation of Metalloporphyrin-Polymer Complexes

Polymer complexes of metalloporphyrins are formed by mixing solutions of a polymer ligand with a metalloporphyrin, as schematically written in Eq. (9-5), where M, L, and the square of dashed lines are a metal ion, such as iron or cobalt ion, a nitrogenous residue of a polymer ligand, and a porphyrin planar ring, respectively.

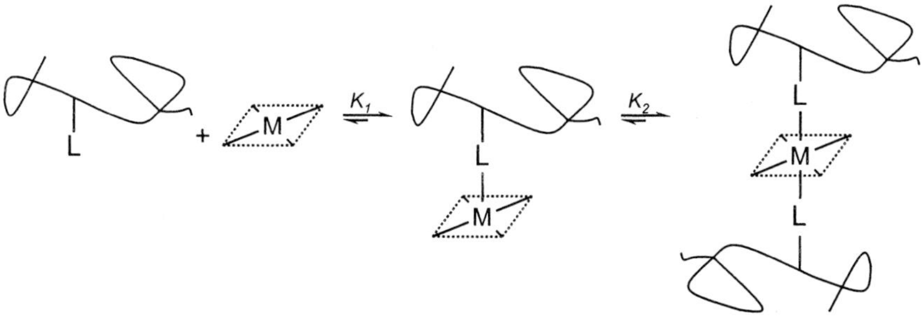

$$(9-5)$$

Typical examples of metalloporphyrins and polymer ligands are presented in Figs. 9-7 and 9-8, respectively. Other metalloporphyrin-related complexes include metallophthalocyanines and cobalt Schiff-base chelates: the latter (Co-salen) is shown at the bottom of Fig. 9-7. A polymer ligand is made to coordinate to a vacant site or the fifth coordination position of the complex. The ligand coordination in the pendant-type metalloporphyrin allows the formation of an oxygen adduct, as represented in Fig. 9-2. However, when the polymer ligand further coordinates to the sixth coordination position of the metalloporphyrin, as shown in Eq. (9-5), the metalloporphyrin loses its oxygen-binding ability. Consequently, the structure and hence the functions of a metalloporphyrin are significantly affected by the attachment of a polymer ligand.

For reversible oxygen binding, the oxidation potential of a metalloporphyrin must be within such a range that a certain amount of electron density is located on the bound oxygen molecule, yet not so much that irreversible oxidation of the metal accompanying the formation of a superoxide anion ($O_2^{-\bullet}$) occurs [10, 23–24] (Fig. 9-2). Iron(II) and cobalt(II) ions often satisfy this first requisite. An aromatic and nitrogenous ligand, such as imidazole and pyridine, contributes to this requisite, from the viewpoint of an axial ligand of a metalloporphyrin. The coordination number of one or two, that is, the molar ratio of the coordinated axial ligand to the metal ion of porphyrin (Eq. (9-5)), and the formation constants of metalloporphyrin-polymer complexes have been well studied by spectroscopic titration [4].

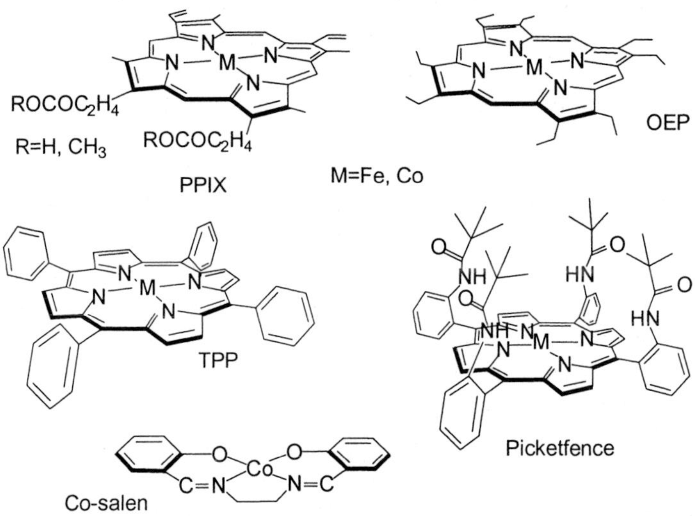

Figure 9-7. Examples of metalloporphyrins.

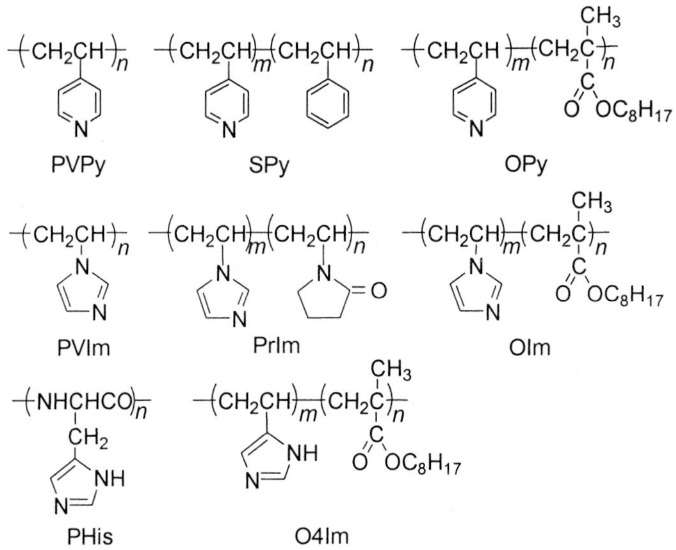

Figure 9-8. Examples of polymer ligands.

In solutions, the coordination number is the one that preferentially forms five-coordinate metalloporphyrin complexes with a K_1 value of $10–10^3$ M^{-1} for almost all imidazole- and pyridine-based ligands. In the case of cobalt porphyrin, the K_2 value for a six-coordinate complex is very small <1 M^{-1} [25], and the cobalt porphyrin scarcely forms any six-coordinate porphyrin complex. But, the K_2 value for a six-coordinate iron porphyrin has a comparable value to that K_1 for the five-coordinate. For example, when casting or evaporating a solution to yield a solid sample of metalloporphyrin-polymer complex, the solution becomes highly concentrated and the formation of six-coordinate metallo(especially iron) porphyrin complexes, which is not active in oxygen binding, cannot be ignored.

ESR spectroscopy is effective for determining such coordination structures for cobalt(II) and iron(III) porphyrins [25]. For example, the ESR parameters $g_\perp = 2.3$ and $g_{//} = 2.0$: Both signals are split into hyperfine lines by the ^{59}Co nucleus, and several of the parallel signals are further split into equal intensity super-hyperfine lines. The latter are derived from the nitrogen nucleus of the nitrogenous ligand at the fifth coordination site of a cobalt porphyrin.

The effects of polymer ligands in the complexation with a metalloporphyrin have been compared with those of the corresponding monomeric ligands. The formation constants K_1 of metalloporphyrin-polymer complexes are $10-10^2$ times larger than those of monomeric complexes in a solution [4]. This phenomenon appears to be general for metal-polymer complexes, and can be explained by assuming that the concentration of the ligand is locally higher in the domain of a polymer ligand. This stability in the complexation provides an advantage for polymer complexes so that a minimal excess supply of a polymer ligand to a metalloporphyrin establishes the formation of a complex in both solution and solid states.

Apart from the coordinative bond, an electrostatic or hydrophobic interaction between a metalloporphyrin and a polymer ligand can also play a role in complexation, especially in an aqueous solution [4]. It is well known that metalloporphyrins easily aggregate in a solution (and even in the solid state) based on a stacking interaction. The dispersion effects of polymers on porphyrin aggregation have been reported. For example, aggregated iron protoporphyrin IX (PPIX, Fig. 9-7, M = Fe; R = –H) is effectively dispersed by a copolymer of vinylimidazole and vinylpyrrolidone (PrIm in Fig. 9-8), which is a water-soluble but hydrophobic polymer ligand. Iron porphyrin is dispersed by the hydrophobic effect of a polymer ligand and/or tightly coordinated on a polymer ligand matrix to prevent aggregation. The secondary structure effect of a polymer ligand has been also studied; poly(L-histidine) (PHis in Fig. 9-8) has an α-helical conformation that yields a six-coordinate iron-PPIX complex with large K_2 values.

Casting the complex solution of a metalloporphyrin and a polymer ligand such as poly(vinylimidazole-co-alkylmethacrylate) (OIm in Fig. 9-8), for example, on a Teflon plate and drying, yielded both a homogenously dispersed metalloporphyrin in a solid state and a mechanically tough membrane [26]. The T_g of the membrane was increased, for example, from –1 °C for the OIm itself to 6 °C and the membrane became brittle after the incorporation of 40 wt.% cobalt porphyrin; this was explained by the stiffening effect of the incorporated planar and rigid porphyrin molecules on the segmental motion of the polymer chains. The chemical structure of a polymer ligand has to be carefully designed to obtain a solid membrane of metalloporphyrin-polymer complexes that is feasibly applicable to gas permeation experiments.

The membrane was homogenously and deeply red in color with a metalloporphyrin. X-ray diffractometry of the membrane provides no significant pattern, indicating an amorphous state for the polymer in the membrane and denying the aggregation of the metalloporphyrin. The homogenous complexation in the membrane was also supported by a homogenous metal-ion distribution in the membrane, as determined by an X-ray microanalysis. After slow and careful solvent evaporation (such as under an inert atmosphere), UV-Vis and ESR spectroscopies on the membrane supported a homogeneous metalloporphyrin complexation having the same coordination structure even in a solid or solvent-free state over the entire membrane specimen. Accordingly, another advantage of polymer complexes is that a large amount of a gaseous-molecule carrier (metalloporphyrin) can be incorporated into the solid matrix.

9.2.4 Oxygen-Binding and Reversibility

A five-coordinate metalloporphyrin binds molecular oxygen from air selectively, rapidly, and reversibly (Eq. (9-6));

$$M(II)PL + O_2 \rightleftharpoons LPM-O_2 \tag{9-6}$$

$$LPM-O_2 + M(II)PL \longrightarrow LPM-O_2-MPL \longrightarrow PM(III)-O-M(III)P \tag{9-7}$$

$$LPM-O_2 + H^+(H_2O) \longrightarrow M(III)PL + HO_2^\bullet \tag{9-8}$$

$$LPM-O_2 \longrightarrow L + PM-O_2 \longrightarrow M(III)PL + O_2^{-\bullet} \tag{9-9}$$

In Eqs. (9-6), (9-7), (9-8), and (9-9), M(II)P and M(III)P represent metalloporphyrins composed of metal(II) ions and metal(III) ions, respectively. In addition to a five-coordinate metalloporphyrin formation and its appropriate oxidation potential (as mentioned above), the following requisites for reversible oxygen-binding have been elucidated to suppress the side-reactions of an oxygen adduct or irreversible oxidation of a metalloporphyrin, as represented by Eqs. (9-7) through (9-9) [10,23].

There is a strong driving force towards the irreversible formation of a stable μ-oxo-M(III)P dimer, as represented by Eq. (9-7). An oxygen-bound complex, or oxygen adduct, rapidly reacts with another deoxy M(II)PL, forming a binuclear dioxygen-bridged complex. This binuclear complex is irreversibly converted to an oxo-bridged M(III) dimer. In other words, the first problem of reversible oxygen-binding is how to inhibit dimerization.

A lot of work has been concerned with overcoming this problem, and two approaches have been successful. The first is an elegant steric approach, where a planar porphyrin is sterically substituted in a fashion that inhibits dimerization. A typical example is a picketfence metalloporphyrin or *meso*-α, α,α,α-tetrakis(*o*-pivalamidophenyl)porphyrinatometal (picketfence in Fig. 9-7), which has four pivalamido groups on one side of the porphyrin plane that provide a cavity for protecting the bound oxygen from dimerization [9]. Another realistic approach attaches the metalloporphyrins to a rigid polymer chain to prevent the two metalloporphyrins from reacting with each other and dimerizing. By these means (or a combination of the first and second approach), reversible oxygen-binding to metalloporphyrins and cobalt-Schiff-base complexes has been achieved under limited conditions.

The next requisite is that the proton-driven oxidation (see Eq. (9-8)) has to be retarded with a hydrophobic environment. The protein of hemoglobin embeds iron PPIX and protects it from oxidation; that is, the hydrophobic domain of the globular protein excludes water molecules and suppresses proton-driven oxidation (see Section 2.3.1).

Another requisite is that the oxidation induced by loss of a ligand (see Eq. (9-9)) has to be retarded with a stable five-coordinate complexation. Although a metalloporphyrin complex with an axial ligand (L in Eq. (9-9)) is thermodynamically stable, with a large K_1 as described in the previous section, the metalloporphyrin is still labile in the ligand complexation. A momentary dissociation of the axial ligand in the oxygen-bound complex causes an electron transfer from the metal(II) ion to the bound oxygen molecule to yield M(III)P and a superoxide anion. A stable metalloporphyrin-polymer complex with a large K_1 value has been characterized in comparison with monomeric complexes. For these reasons, metalloporphyrin-polymer complexes are expected to become efficient oxygen carriers with a sufficient operational lifetime.

For example, the operational lifetime of a picketfence cobalt porphyrin-polymer complex in a solid state is much longer than that of monomeric complexes and the complex in a solution, and it maintains its oxygen-binding ability for over a month at room temperature under an air atmosphere. For a cobalt(II) porphyrin bound to a polymer ligand in a solid state, irreversible oxidation via Eq. (9-7) is inhibited because it is molecularly dispersed and immobilized on the polymer chain. The hydrophobic environment around a cobalt porphyrin excludes water vapor, thereby suppressing the proton-driven irreversible oxidation in Eq. (9-8). The static bonding of an axial ligand with a cobalt porphyrin in a solid state retards the process of Eq. (9-9). The cooperation of these effects strikingly inhibits irreversible oxidation and prolongs the lifetime of cobalt porphyrin in a solid polymer membrane. This enhanced stability of the carriers provides a great advantage for solid membranes of polymer complexes.

Oxygen-binding is monitored by a reversible change in the UV-Vis absorption spectrum of an aprotic solvent solution of a metalloporphyrin-polymer complex (for an experiment see Section 9.5.3). For example, a dichloromethane solution of a cobalt octaethylporphyrin (OEP in Fig. 9-7) complex of OIm was cooled to −15 °C [27]. The absorption of the red-colored CoOEP-OIm solution (inset in Fig. 9-9) changed from the spectrum (dark red; λ_{max} = 392 and 550 nm) ascribed to a deoxy or five-coordinate cobalt porphyrin under a nitrogen atmosphere to the spectrum (bright red; λ_{max} = 413, 528 and 561 nm) ascribed to an oxycobalt porphyrin or O_2:Co = 1:1 adduct immediately after exposure to oxygen. The deoxy–oxy spectral change was reversible in response to the oxygen partial pressure with isosbestic points at 400, 519, 537, and 558 nm. For a cobalt protoporphrin IX dimethylester complex of poly(4-vinylpyridine-*co*-styrene) (PPIX, R = −CH₃ and SPy in Figs. 9-7 and 9-8, respectively) deoxy: λ_{max} = 404 and 555 nm, oxy: λ_{max} = 420, 543, and 575 nm, isosbestic points = 412, 548, and 565 nm. Such a reversible spectral change in response to the oxygen partial pressure with isosbestic points is a crucial piece of evidence that a metalloporphyrin acts as an effective oxygen carrier from an equilibrium perspective. When a transparent (but red-colored) membrane could be prepared by selecting the polymer ligand, the oxygen-binding was also confirmed and measured using visible absorption spectroscopy.

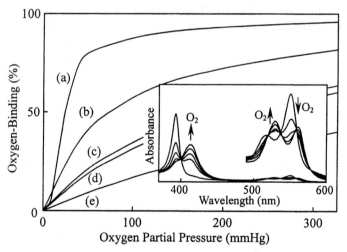

Figure 9-9. Oxygen-binding equilibrium curves for (a) cell-free hemoglobin in pH 7.4 aqueous solution at 37 °C, (b) Co picketfence-OIm in the membrane state at 25 °C, (c) Co picketfence-OIm in toluene solution at 25 °C, (d) CoOEP-OIm in CH₂Cl₂ at −15 °C, and (e) Co picketfence-OPy in toluene solution at 25 °C. Inset: UV-Vis absorption spectral change of the CoOEP-OIm complex in CH₂Cl₂ at −15 °C in response to oxygen partial pressure from 0 to 760 mmHg.

Oxygen-binding equilibrium curves were drawn from the UV-Vis absorption spectral changes, as shown in Fig. 9-9. The equilibrium curves obeyed Langmuir isotherms to provide the oxygen-binding equilibrium constant K or -binding affinity p_{50} (oxygen partial pressure at which half of the metalloporphyrins bind with oxygen), as defined by Eq. (9-10). The curves in Fig. 9-9 also show that the oxygen-binding affinity depends both on the species of metalloporphyrin and polymer ligand and on the solution (solvent species) or solid state (and also on the temperature, as described later). In other words, the oxygen-binding affinity can be controlled with the chemical structures of the metalloporphyrin and polymer ligand and tuned to be adequate for the separation of oxygen from air.

$$LPCo + O_2 \underset{k_{off}}{\overset{k_{on}}{\rightleftarrows}} LPCo \cdot O_2 \qquad K(= 1/p_{50}) = k_{on} / k_{off} \qquad (9\text{-}10)$$

In addition to UV-Vis absorption spectroscopy, oxygen-binding to a metalloporphyrin complex fixed in a solid or solvent-free membrane and film has also been confirmed in an *in situ* membrane state with general spectroscopies. For example, by attaching a membrane fragment to the cell window of a simple IR spectrometer, strong IR absorption at 1150 cm^{-1} for $^{16}O_2$ and 1060 cm^{-1} for $^{18}O_2$ attributed to an end-on-type coordination of molecular oxygen to the metal ion appeared with an increase in the oxygen partial pressure (inset in Fig. 9-10) [28].

The amount of oxygen solubility is greatly enhanced by the incorporation of a metalloporphyrin complex in a polymer, which was quantitatively measured using a microvolumetric measurement, as described in Section 9.2.2. (Fig. 9-10) [29]. For example, the membrane of an OPy complex containing 30 wt.% picketfence cobalt porphyrin absorbed ca. 7 cm^3 (STP) oxygen gas/g polymer, which is more than 500 times greater than that of physically dissolved nitrogen. This extraordinarily large amount of dissolved oxygen in the polymer membrane arises from chemically specific and reversible oxygen-binding to the cobalt porphyrin complex in the polymer. As such, one of the advantages of solid polymer complexes is that a large amount of cobalt porphyrin can be incorporated into the membrane through the coordination of a nitrogenous residue of the polymer.

Figure 9-10 shows the effect of the cobalt porphyrin concentration on the oxygen solubility in the OPy sample, indicating that the amount of dissolved oxygen corresponds to the cobalt porphyrin concentration in the membrane. Figure 9-10 also reveals that the oxygen absorption behavior responds to the atmospheric oxygen pressure and is analogous to the olefin absorption in silver-polymer complexes. The oxygen absorption curves (Fig. 9-10) correspond to a Langmuir isotherm for the chemical dissolution in the complex and Henry's law for the physical absorption, to give K, k_D, and C_c (see Section 9.2.2 and Eq. (9-4)). The curves drawn using the dual mode sorption were in good agreement with the experimental results. The C_c value increased whereas the K value was independent of the cobalt porphyrin concentration in the sample. This data agreed with that determined spectroscopically.

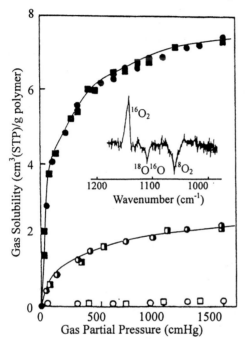

Figure 9-10. Oxygen absorption into the cobalt porphyrin-polymer membranes at 25 °C. ●, ■: absorbed oxygen volume per membrane unit weight (OPy membrane containing 30 wt.% of picketfence cobalt porphyrin), ○,□: absorbed nitrogen volume per membrane unit weight, and ◐,◧ absorbed oxygen volume (7 wt.% of cobalt porphyrin). Inset: Differential IR absorption of the CoPicketfence-OIm membrane under $^{16}O_2$ and $^{18}O_2$ atmosphere.

It is known that metal-coordinated gaseous molecules are photodissociated under flash irradiation, and that their rapid re-binding reactions can be analyzed, based on the oxygen-binding with a metalloporphyrin (Eq. (9-11)) [11, 28]. The photodissociation and binding of gaseous molecules in a solid metal-polymer membrane were observed by improving pulse and laser flash spectroscopic techniques (for details see Experiment 5-3, Section 9.5). The oxygen-binding rate constant k_{on} and releasing rate constant k_{off} in Eq. (9-10) (see also Eq. (9-12)) were estimated by second-order kinetics, and listed with the oxygen-binding (equilibrium) affinity values in Table 9-2.

$$LPCo–O_2 \xrightarrow{h\nu} LPCo + O_2 \xrightarrow{k_{app}} LPCo–O_2 \qquad (9-11)$$

$$k_{app} = k_{on}[O_2] + k_{off} \qquad (9-12)$$

Table 9-2 first shows that the oxygen-binding affinity of a metalloporphyrin complex can be controlled by variation of the metal species, porphyrin structure, and axial ligand species. The oxygen affinity of an iron porphyrin is around twice that of the corresponding cobalt porphyrin, while an iron porphyrin suffers irreversible oxidation owing to its lower oxidation potential, and its lifetime is much shorter than that of a cobalt porphyrin (a cobalt porphyrin is superior to an iron porphyrin as an oxygen carrier, except that its toxicity inhibits its medical usage). A linear relation between the affinity and the basicity of the axial ligand has been summarized for a cobalt porphyrin complexed with various nitrogenous and axial ligands [28,33]: with this relationship we can explain the affinity of OIm > OPy in Table 9-2. It can also be noted in Table 9-2 that such higher oxygen affinities (smaller p_{50} or larger K in Eq. (9-10) are caused by smaller oxygen-releasing rate constants (k_{off}, that is, $K = k_{on} / k_{off}$).

A simple and planar cobalt porphyrin, such as CoTPP and CoOEP, with a (polymer-) ligand formed an oxygen adduct in a cool solution (but did not at room temperature). When linear logarithmic p_{50} and k_{on} versus $1/T$ plots (van't Hoff and Arrhenius plots) were measured, their extrapolations to room temperature produced p_{50}, k_{on}, and k_{off} values at 25 °C for CoTPP and CoOEP complexes [27]. The p_{50} value of 10^{2-3} cmHg indicates a very weak oxygen-binding affinity for these cobalt porphyrin complexes at 25 °C (ca. 1/30 of the Co picketfence's affinity), but suggests that an oxygen adduct can be observed for planar cobalt porphyrins under a high oxygen pressure, for example, 10 atm. The extrapolated k_{off} value of the order of 10^8 s^{-1} for the planar cobalt porphyrin complex was more than 200 times larger than that for the picketfence cobalt porphyrin, and this significantly large k_{off} is the origin of the very low oxygen-binding affinity of planar cobalt porphyrin complexes (or the reason why an oxygen adduct could only be observed at a low temperature).

Table 9-2. Oxygen-binding affinity (p_{50}) and binidng rate constants (k_{on} and k_{off}) for the metalloporphyrin-polymer complexes at 25 °C.

Metalloporphyrin	Axial ligand	State	p_{50} cm Hg	$10^{-7} k_{on}$ M^{-1} s^{-1}	$10^{-4} k_{off}$ s^{-1}	Ref.
Hemoglobin[a]	–	soln	1.4	1.7[b]	0.0013[b]	30
CoTPP	OIm	soln[c]	790[d]	130[d]	17,000[d]	4,27
CoPicketfence	OIm	soln[c]	25	84	73	4,9
CoOEP	OIm	membr	980[d]	5.0[d]	760[d]	4,27
CoPicketfence	OIm	membr	7.6	1.7	1.2	11,26
CoPicketfence	OPy	membr	16	1.9	2.6	31
FePicketfence	OIm	membr	3.2	0.47	0.13	32

[a] Cell-free-hemoglobin in pH 7.4 aqueous solution; [b] 37 °C; [c] Soln = CH$_2$Cl$_2$ solution, [d] Extrapolated values from van't Hoff and Arrhenius plots of p_{50}, k_{on}, and k_{off}, respectively, over the temperature range −35 °C to −15 °C.

Table 9-2 also indicated that the k_{on} and k_{off} values of the complexes in the solid membranes are smaller than those of the complexes in a solution, but still of the order of 10^7 and 10^3, respectively. This means that cobalt porphyrins are still kinetically active for oxygen-binding after fixing as a dry solid polymer complex. As such, from a kinetic point of view, the oxygen-binding reaction in cobalt porphyrins exhibits great promise for the facilitated transport of oxygen through a membrane.

9.3   Small Molecule Transport Through Membranes of Metal Complexes in Macromolecules

9.3.1   Membrane Gas Separation

Membrane separation of gaseous small molecules through dense (non-porous) polymeric membranes occurs because of differences in solubility and diffusivity, while membrane performance is characterized by permeability and selectivity. The permeability of component i, P_i, is defined as the product of the diffusion and solubility coefficients (D_i and S_i, respectively) (Eq. (9-13)).

$$P_i = D_i S_i \qquad (9\text{-}13)$$

The selective property of a membrane towards a mixture is generally expressed by the ideal separation factor or selectivity. The ideal separation factor α_{ij} is defined as the ratio of two pure gaseous molecule permeabilities (Eq. (9-14)).

$$\alpha_{ij} = P_i/P_j \qquad (9\text{-}14)$$

In membrane separation processes for two components, the feed stream is a mixture, and as such, the ideal separation factor is often unsuitable for describing the real separation performance. Therefore, the selectivity A_{ij} is defined differently, where x_i and y_i are the concentrations of component i for the feed and permeate streams, respectively (Eq. (9-15)).

$$A_{ij} = (y_i/y_j)/(x_i/x_j) \qquad (9\text{-}15)$$

A_{ij} depends on the membrane material properties and the experimental conditions, whereas α_{ij} only depends on the membrane material properties. Thus, the permeability and ideal separation factor are mostly used to characterize the transport properties of membrane materials. Note that A_{ij} becomes α_{ij} when the permeate side is held in a vacuum for pure gas experiments.

A composite membrane, composed of a thin (~1 μm) top layer of metal-polymer complex on an asymmetric porous support, is mainly used for practical applications to provide a high separation performance and mechanical strength. Membrane separation usually occurs on the top dense layer of metal-polymer complexes, while the asymmetric porous support provides mechanical strength. Accordingly, the thickness of the top layer is important in determining the separation properties and should be measured precisely. However, since a precise measurement of the top layer is often extremely difficult, the permeability is hard to measure. Consequently, the thickness normalized permeability, hereafter called the permeance Q_i, is frequently used and defined as $Q_i = P_i/L$, where L is the effective top layer thickness of the membrane. The permeance Q_i depends on the membrane material properties as well as the membrane structure. As a result, the permeation property of a membrane can be characterized by either its permeability or permeance.

To achieve a high separation performance, the permeability and selectivity should both be high. However, a trade-off behavior between the permeability and selectivity is commonly observed in polymeric membranes [34]: the permeability is high when the selectivity is low, and vice versa. One effective way to overcome such trade-off behavior is to employ the concept of *facilitated transport* [35]. In facilitated transport membranes, carriers are incorporated into the membrane. The carrier is any compound, often a transition metal ion or complex, that binds *reversibly* with a specific gaseous molecule. The reversible binding for a specific gaseous molecule induces additional mass transport, referred to as "carrier-mediated" transport. Thus, the total mass transport is the summation of the carrier-mediated transport and Fickian transport, facilitating the permeation of the specific gaseous molecule, thereby resulting in a high permeability and high selectivity. The characteristics of facilitated or carrier-mediated transport result from the occurrence of a reversible binding reaction in combination with a Fickian diffusion process. The choice of the carrier (or metal complex) is a key factor in facilitated transport. For example, a high selectivity is obtained if the carrier's binding reaction is reversible and kinetically active to a particular small molecule and the rates of the binding and releasing reactions of the small molecule with the carrier are very high.

In facilitated transport membranes, the carriers can be dissolved in a liquid solvent or fixed to a solid polymer matrix. When the carriers are dissolved in a liquid medium, the carriers at the upstream interface react with a specific small molecule to form an adduct, which then moves across the membrane and releases the small molecule from the adduct due to a releasing reaction, as shown in Fig. 9-11a. As such, the separation property depends on the binding and releasing reaction rates as well as the mobility of the adduct in the liquid medium.

A liquid-free, solid-state membrane containing fixed carriers can also provide high permeability as well as high selectivity. Figure 9-11b schematically represents facilitated transport membranes with fixed carriers. The fixed carriers at the upstream interface bind with a specific small molecule to form an adduct. The small molecule bound to the carrier is then released in the matrix or transferred to other carriers in response to a concentration gradient. The binding and releasing reactions occur repeatedly across the membrane. At the membrane-downstream interface, the carrier adduct releases the small molecule to the downstream side, thereby resulting in facilitated transport. Thus, the small molecule's solubility, and binding–releasing reaction rates between the carrier (the metal complex) and the small molecule are major parameters in determining the transport properties. Fixed carrier membranes in a solid state provide the following advantages over liquid membranes: (1) higher stability of the carrier, (2) higher mechanical strength, (3) no liquid loss, (4) possibility of incorporating a larger amount of the carrier in the matrix, and (5) flexibility as regards controlling the membrane thickness to enhance the flux.

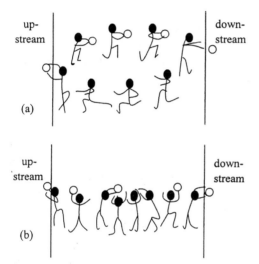

Figure 9-11. Facilitated transport in (a) liquid, and (b) solid state.

Since the transport of a small molecule through a fixed carrier membrane depends largely on the kinetics, reversibility and affinity of the small molecule to the metal complex carrier, the transport properties of a small molecule are a good index for the activity of a metal complex in a macromolecule. Therefore, the following sections give a brief review of facilitated olefin and oxygen transport.

9.3.2   Facilitated Olefin Transport

The separation of olefin/paraffin gas mixtures is one of the most energy-intensive processes in the petrochemicals industry, because it is mainly performed by cryogenic distillations. Membrane processes using the concept of facilitated transport have been considered as an intriguing alternative to cryogenic distillation, as they can simultaneously improve both permeability and selectivity. Silver ions incorporated in liquid membranes act as olefin carriers for facilitated olefin transport, resulting in a high separation performance for propylene/propane or ethylene/ethane mixtures. However, such liquid membranes suffer from the disadvantage that liquid must be added to the feed stream to activate the silver salt (solvate the silver ions) and then removed from the permeate stream [36–39]. Therefore, a facilitated transport membrane in a solid state is more desirable.

To make facilitated transport membranes in a solid state, silver salts, such as $AgBF_4$, are complexed with polar polymers, PEO [40], POZ, and PVP [13] (see Fig. 9-3), as described in Experiment 5-2, Section 5.5. Figure 9-12 shows the propylene permeance through $AgBF_4$-POZ and $AgBF_4$-PVP membranes with an increasing amount of silver and constant feed pressure at 414 kPa [12]. The propylene permeance through the membranes shows a similar trend with an increasing silver salt content, and no significant increase in the propylene permeance is observed until the mole fraction of silver to carbonyl reaches > 0.25, from which point the propylene permeance increases almost linearly with the silver concentration. This linear increase in the propylene permeance is consistent with theoretical predictions based on the concentration fluctuation theory, as described in Section 9.3.3. While no significant dependence on either polymer matrix is evident under these experimental conditions, the anions of the silver salt have a significant effect on the facilitated transport [12].

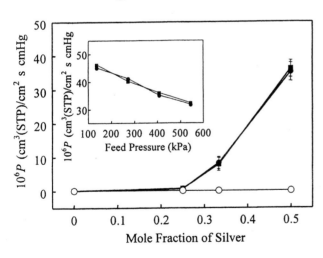

Figure 9-12. Propylene permeance and propylene/propane selectivity with increasing carrier concentration [12]. ●: $AgBF_4$-POZ, ■: $AgBF_4$-PVP and ○: propane.

The mole fraction of silver ions to carbonyl oxygen is fixed at 0.5 in the silver-POZ (or -PVP) complexes, that is, $[Ag^+]:[C=O] = 1:1$ (~80 wt.% of $AgBF_4$). The permeance of pure propylene through the membranes is then plotted as a function of the feed pressure, as shown in the inset in Fig. 9-12. The polymer membranes containing $AgBF_4$ in POZ or PVP exhibited a propylene permeance of 4.5×10^{-5} cm³(STP) cm^{-2} sec^{-1} cmHg^{-1} at 140 kPa, whereas the propylene permeance through the pure POZ or PVP membrane was only $\sim 5.0 \times 10^{-8}$ cm^3(STP) cm^{-2} sec^{-1} cmHg^{-1}. As such, the propylene permeance with the silver-polymer complex membranes is more than three orders of magnitude higher than that with the pure POZ or PVP, clearly demonstrating the facilitated transport of propylene due to the presence of silver ions. Conversely, the propane permeance decreased from 5.0×10^{-8} to 3.0×10^{-9} cm^3(STP) cm^{-2} sec^{-1} cmHg^{-1} with the silver concentration. Thus, the ideal separation factor of propylene over propane is approximately 15,000. As a result, a propylene concentration higher than 98% is observed in the downstream side after only a single pass of the silver-polymer complex membranes when 50% propylene is used as the feed stream. For further details, refer to Experiment 9-2, Section 9.5.

Rubbery polyether-based polymer membranes containing silver ions also exhibit a good separation performance for an olefin/paraffin gas mixture. The pure gas permeation through membranes consisting of PEO and AgBF$_4$ is presented in Table 9-3 [40]. The ethylene/ethane and propylene/propane selectivities of the pure PEO membrane were only 1.2 and 2.5, respectively. The PEO membrane containing 33 wt.% AgBF$_4$ ([Ag$^+$]:[-O-] = 1:8) also had poor separation properties. However, when the silver concentration reached [Ag]:[-O-] = 1:1, the pure ethylene and propylene permeances are 50–100 times higher than those of the pure PEO membrane. As such, these results have direct implications for facilitated olefin transport in solid-state polymer membranes.

Table 9-3. Effect of silver concentration on olefin/paraffin separation in AgBF$_4$-PEO membrane at 689.1 kPa feed pressure and 23 °C [40].

AgBF$_4$ (wt.%)	[Ag]:[-O-]	Pure gas permeance (10^{-6} cm^3(STP) cm^{-2} sec^{-1} cmHg^{-1})		Pure gas selectivity	
		C$_2$H$_4$	C$_3$H$_6$	C$_2$H$_4$/C$_2$H$_6$	C$_3$H$_6$/C$_3$H$_8$
0	no Ag	0.55	0.89	1.2	2.5
33	1:8	0.18	0.22	1.8	2.0
50	1:4	1.6	2.6	>160	>260
67	1:2	10	10	>1000	>1000
80	1:1	55	48	>5500	>4000

Membrane performance also depends on the thickness of the selective layers of a silver-polymer complex, that is, the thinner the selective layer, the better the transport performance of the membrane. A thickness of several micrometers for the selective layer was the limitation in conventional methods used to prepare composite membranes. Based on exciting developments in the field of nanostructure science and technology, there have been several attempts to reduce the thickness of the selective layer to a nanometer length scale [41–43]. Schematic methods for preparing nano-thin selective layer membranes are generalized in Fig. 9-13, based on the use of nanometer-sized dendrimers, star polymers, and the preparation of polymer brushes. A several nanometer-thin selective layer in a facilitated transport membrane has been found to produce an improved performance. Although the range and nature of functionalities that can be accessed through nanostructuring is just beginning to unfold, its tremendous potential for revolutionizing the way in which materials and products are created is already clear, and it will have a great impact on future scientific and commercial applications.

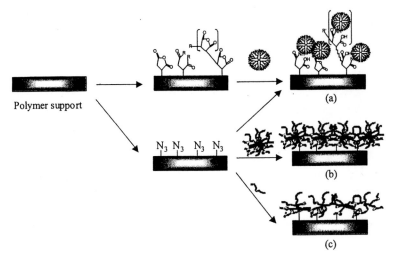

Figure 9-13. Schematic methods for the preparation of nano-thin selective layer membranes.

9.3.3   Facilitated Oxygen Transport

The facilitated transport of molecular oxygen through a solid polymer complex membrane was first achieved by incorporating a picketfence cobalt porphyrin, characterized with both rapid kinetics and a strong affinity for oxygen-binding, as a chemically selective oxygen-binding site [5,11]. The oxygen permeability coefficient P_{O2} for the membrane is shown in Fig. 9-14 [5,44]. P_{O2} is larger than P_{N2} and steeply increases with a decrease in the oxygen upstream pressure (p_{O2}). Conversely, P_{N2} is small and independent of the nitrogen upstream pressure (p_{N2}), because the fixed cobalt porphyrin does not interact with nitrogen. P_{O2} is also small and independent of p_{O2} for the control membrane composed of an inert Co(III) complex, which does not interact with oxygen. As such, the active cobalt porphyrin fixed in the polymer membrane facilitates oxygen transport and enhances the oxygen permeation, represented by the shaded area in Fig. 9-14. P_{O2} also increases with the cobalt porphyrin concentration in the polymer membrane. The oxygen/nitrogen selectivities (P_{O2}/P_{N2}) were 3.2, 4.4, and 7.5 for membranes containing 0, 2.5, and 4.5 wt.% picketfence cobalt porphyrin, respectively. These results indicate a high selectivity and high permeability for a solid-state polymer membrane containing an oxygen carrier.

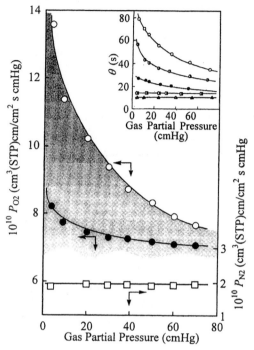

Figure 9-14. Oxygen and nitrogen permeability coefficients (P_{O2} and P_{N2}) for the cobalt porphyrin-polymer complex membrane. The cobalt porphyrin concentration in the OIm membrane: O: 4.5 wt.%; ●: 2.5 wt.%; and □: nitrogen for 4.5 wt.% at 30 °C. Inset: Effect of gas partial pressure on induction period (θ) for the permeation through the membrane of 4.5 wt.% cobalt porphyrin. oxygen at O: 20 °C; ◐: 25 °C, and ●: 30 °C; nitrogen at □ 25 °C; and oxygen at △: 25 °C for the inert cobalt(III) porphyrin membrane.

The time course of the permeation of gaseous molecules through the membranes exhibited an induction period followed by permeation with a constant slope (steady state). The induction period (θ) for oxygen permeation was longer than θ for nitrogen and was also prolonged with a decrease in p_{O2} (inset in Fig. 9-14) [44]. In contrast, θ for nitrogen permeation was short and independent of p_{N2}. This behavior indicates that oxygen clearly interacts with the cobalt porphyrin in the polymer membrane and that its diffusivity in the membrane is reduced by repeated oxygen binding and releasing on the fixed cobalt porphyrin. θ_{O2} was prolonged depending on the cobalt porphyrin concentration in the membrane. θ_{O2} and the p_{O2}-dependence of θ_{O2} were enhanced at lower temperatures, because the oxygen-binding rate constants decreased and the oxygen affinity K increased with a decreasing temperature. As such, θ is the best parameter to distinguish whether or not the metal complex fixed in the polymer membrane contributes to the transport of a gaseous molecule through the membrane.

Figure 9-11b represents the facilitated transport of oxygen via a fixed carrier (cobalt porphyrin) in a solid membrane. The fixed carrier picks up oxygen specifically from air at the upstream interface. This specific and reversible oxygen-binding reaction establishes a steep gradient in the oxygen concentration across the membrane from the upstream side to the downstream side, in response to which the oxygen taken up in the membrane is transferred via the fixed carriers to the downstream side (if the passage of the oxygen by the fixed carrier is efficient and rapid). That is, the driving force behind the facilitated oxygen transport is the concentration gradient of oxygen in the membrane or the selectively enhanced solubility of oxygen, as demonstrated, for example, by the results in Fig. 9-10.

The above results and consideration suggest that a dual-mode transport model, derived from the dual-mode absorption model (Eq. (9-4)), is applicable to an analysis of the oxygen transport in cobalt porphyrin-polymer membranes. Figure 9-15 is a schematic representation of the oxygen permeability in a cobalt porphyrin-polymer membrane governed by two modes [11]. That is, the oxygen permeation is equal to the sum of the first term, which represents the physical permeation mode, and the second term, which represents the carrier-mediated mode. For the physical mode (upper permeation route in Fig. 9-15), oxygen physically dissolves in the polymer membrane according to Henry's law, and the dissolved oxygen then diffuses physically. For the carrier-mediated mode (the lower permeation route in Fig. 9-15), oxygen is specifically and chemically taken up by the cobalt porphyrin fixed in the membrane and hops from one cobalt porphyrin to another by repeating a binding and releasing reaction to and from the cobalt porphyrin. The oxygen transport is thus accelerated by this carrier-mediated mode in addition to the physical mode.

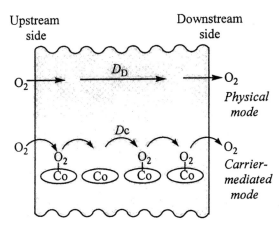

Figure 9-15. Dual-mode oxygen transport in the solid-state polymer membrane containing cobalt porphyrin as a fixed carrier of oxygen.

The dual-mode-transport model is mathematically described in Eq. (9-16).

$$P_{O2} = k_D D_D + \frac{D_c C_c K}{1 + K p_{O2}} \qquad (9\text{-}16)$$

Here, D_D is the physical diffusion coefficient of oxygen in the membrane, and D_C is the postulated diffusion coefficient of oxygen that hops via the fixed cobalt porphyrin carriers. k_D and C_C are measured by an absorption experiment (for example, Fig. 9-10). C_C corresponds to the concentration of active cobalt porphyrin incorporated in the membrane and K is the oxygen-binding equilibrium constant of the cobalt porphyrin fixed in the membrane, which is spectroscopically measured (for example, as given in Table 9-2).

The effect of p_{O2} on P_{O2} in Fig. 9-15 was analyzed using Eq. (9-16), that is, P_{O2} was plotted versus $1(1 + Kp_{O2})$ [5,44]. The plots showed a linear relationship to yield the diffusion coefficients, $D_D = 10^{-6} \text{ cm}^2 \text{ s}^{-1}$ and $D_C = 10^{-8} \text{ cm}^2 \text{ s}^{-1}$. The D_D value agreed with the previously reported diffusion coefficient of oxygen in an amorphous polymer with a similar T_g. D_C, representing the oxygen diffusion via the fixed cobalt porphyrin carriers, is smaller than D_D, because the chemical binding reaction of oxygen with the cobalt porphyrin suppresses the oxygen diffusivity. D_C increased with C_C or the active cobalt porphyrin concentration in the membrane. D_C has been found to be inversely proportional to the average distance between the cobalt porphyrin sites [45], thereby supporting the postulation that D_C represents the oxygen diffusivity hopping via the fixed cobalt porphyrin carriers, as illustrated in Fig. 9-15. An increase in the cobalt porphyrin concentration in the membrane, that is C_C, resulted in an apparent increase in D_C. The multiplication of the increases in both C_C and D_C significantly enhances the second (chemical) term in Eq. (9-16). As a result, both P_{O2} and the oxygen/nitrogen permselectivity increase with the active cobalt porphyrin concentration in the membranes, which can be clearly ascribed to the cobalt porphyrin-mediated transport or facilitated oxygen transport by the fixed cobalt porphyrin carriers.

A previous study of the relation between the oxygen-hopping diffusivity, D_C, and the oxygen-binding reactivity, k_{on} and k_{off}, using a series of cobalt porphyrin species as the fixed carriers, elucidated that D_C is related to both k_{on} and k_{off} and that k_{off}, in particular, reflects D_C [46]. A polymer membrane containing a cobalt porphyrin with a larger rate constant for oxygen release leads to a larger diffusion constant of oxygen, yielding a more efficient oxygen transport or higher oxygen selectivity. Applying this indication, OIm complex membranes containing a large amount of simple and planar cobalt porphyrins (40 wt.%) (TPP and OEP in Fig. 9-7) were prepared with a sub-micron thickness [47]. The CoTPP and CoOEP complexes in the membranes were characterized by a low oxygen affinity but extraordinarily large oxygen-releasing rate constants (see Table 9-2). P_{O2} was of the order of 10^{-9} to $10^{-10} \text{ cm}^3 \text{(STP) cm cm}^{-2} \text{ sec}^{-1} \text{ cmHg}^{-1}$ and much higher than P_{N2}, while P_{N2} was independent of the upstream nitrogen pressure. P_{O2} steeply increased with a decrease in Δp_{O2} [= upstream oxygen pressure – downstream pressure (76 cmHg)]. These results indicate that the oxygen-binding and -releasing kinetics for the carrier CoTPP complex and its amount predominantly affect the facilitated oxygen transport. The P_{O2} for the control and inactive Co(III)TPP membranes was not enhanced at a low p_{O2}, while P_{O2} for the active membranes significantly increased to give a selectivity $(P_{O2}/P_{N2}) > 120$ and facilitation factor $= 30$ at $\Delta p = 1$ cmHg. A downstream oxygen concentration of $>90\%$ was obtained after just a single pass of the membrane when compressed air ($\Delta p = 1$ cmHg) was applied to the upstream side.

The importance of the reaction kinetics between the small molecule and the carrier and the carrier concentration was also analyzed using the concentration fluctuation theory [48,49]. In developing the concentration fluctuation theory, it was assumed that the linear concentration profile across the membrane fluctuates slightly owing to the reversible chemical reaction between the carrier and the small molecule, as shown Fig. 9-16. When an oxygen molecule binds with a carrier to form a carrier-oxygen adduct, the oxygen concentration near the carrier is slightly decreased instantaneously. When the carrier-oxygen adduct releases an oxygen molecule into the matrix, the oxygen concentration is slightly increased instantaneously. These continuous binding and releasing reactions induce a small concentration fluctuation from the average concentration. According to Cahn's theory, this fluctuating concentration profile results in an increased chemical potential, thereby increasing driving force for mass transport as described in Eq. (9-17), where $\overline{P_f}$ and $\overline{P}$ are the permeabilities of the facilitated transport membrane and membrane matrix, respectively, p_0 and p_d are the applied pressure and pressure fluctuation due to the reversible reaction, respectively, k_2 and K are the molecule-releasing reaction rate and equilibrium constants of the small molecule-binding reaction, respectively, L is the membrane thickness, and C_B^0 is the carrier concentration [48,49].

$$\frac{\overline{P_f}}{\overline{P}} = 1 + (\frac{p_d}{p_0}) \frac{2\pi k_2 L^2 C_B^0}{\overline{P}} \frac{\ln(1+Kp_0)}{p_0} \qquad (9\text{-}17)$$

This model suggests that $\overline{P_f}/\overline{P}$ is linearly proportional to p_d, k_2, C_B^0, and L^2 and logarithmically proportional to K. Furthermore, it increases with a decreasing p_0, which is an intrinsic advantage of facilitated transport membranes, as observed experimentally. The linear increase in $\overline{P_f}/\overline{P}$ relative to C_B^0 and k_2 represents the importance of both the carrier concentration and the releasing-reaction rate between the carrier and the small molecule. Note that p_d can be empirically obtained from the slope of a plot of $\overline{P_f}/\overline{P}-1$ versus $(1/p_0)(2\pi k_2 L^2 C_B^0)(\ln(1+Kp_0)/p_0)$ [48,49].

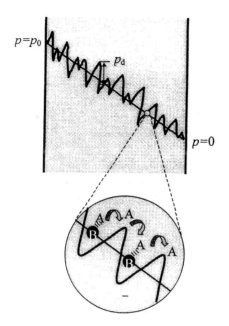

Figure 9-16. Concentration or pressure fluctuation profile in a fixed carrier membrane.

Oxygen and nitrogen are the top two commodity chemicals. Oxygen is used for combustion, chemical oxidation, respiratory therapy, refreshment, aquaculture, and wastewater treatment, while nitrogen is used for purging gases in explosion protection, food storage, and the manufacture of fine electronics and pharmaceuticals. While air is separated to produce oxygen and nitrogen using mainly energy-intensive cryogenic processes, gas-separation membranes are very attractive as they are a simple compact device with low initial capital and running costs. To surpass the competitive cryogenic process, the requisites for the next generation of air-separation membranes are a high permeability and high selectivity (for example, oxygen permeability of more than 10^{-8} cm^3(STP) cm cm^{-2} sec^{-1} cmHg^{-1} and oxygen/nitrogen selectivity of more than 10). As such, facilitated transport membranes containing oxygen-binding cobalt porphyrins, as discussed in this section, are expected to be candidates for such next generation air-separation membranes. Since air is free and abundant and has the same composition everywhere, its efficient usage is an important issue for near-future technology.

9.3.4   Other Small-Molecule Transport

Small-molecule transport other than olefin and molecular oxygen facilitated transport through a metal-polymer complex containing carriers (silver ions and metalloporphyrins, respectively) has also received attention, such as nitrogen transport by a cyclopentadienylmanganese and benzenechromium complex attached to a polymer [50,51], and CO_2 and H_2S transport based on ion-exchange membranes containing a diamine complexing agent [52,53]. The facilitated transport of SO_2 through supported liquid membranes has also been observed in membranes containing water, aqueous neutral salt solutions, and aqueous sodium hydroxide solutions as the carrier [54].

9.4 Other Applications of Metal Complexes in Macro-molecules with Small Molecule-Binding Abilities

The main application of the specific and reversible binding ability of small molecules to metal complexes in macromolecules is in gas separation membranes. Another application of the binding between gaseous small molecules and a metal-polymer complex is in absorbents. Powdered silver-polymer complexes absorb olefins selectively under atmospheric pressure (or at ambient temperature) by the formation of an adduct, then releases the absorbed gas under reduced pressure (or at a high temperature) using a pressure (or thermal) swing method. Since most metal complexes are easily deactivated by a reaction with other components in a gas mixture, especially water, the polymers in metal-polymer complexes play an important role in protecting and supporting the metal complexes. When silver-polymer complexes were prepared from a mixture of polystyrene, silver(I) chloride and aluminum chloride in the matrix, the absorption of ethylene (contained in water) under 1 atm at room temperature by the silver-polymer complex was very rapid. The equilibrium molar ratio of absorbed ethylene to the admitted silver ion was 1.01 after 2 h [55]. The solid polymer complex repeatedly exhibited an absorption of ethylene without any apparent deterioration, even in the presence of water.

Copper-polymer complexes also retain an important position as regards binding with ethylene. When a solid polymer complex was prepared from polystyrene beads, copper(I) chloride, and aluminum chloride, absorption at room temperature by the polymer complex was rapid, and the equilibrium molar ratio of absorbed ethylene to the admitted copper(I) chloride was 1.40. The amount of ethylene absorbed by 1 g of the polymer complex solid was 89 cm^3(STP) [56].

A porous polymer complex containing 30% picketfence cobalt porphyrin absorbed ~4 cm^3(STP) of oxygen per gram of polymer, which is more than 500 times that of physically dissolved nitrogen, and provided oxygen-enriched air (oxygen concentration in the product flow: 45%) when using a pressure swing method [26]. In addition, polymer complexes with a cobalt Schiff base (Co-salen in Fig. 9-7) have been tested as a sensor and oxygen absorbent [57].

Several optical sensors have already been introduced that selectively respond to oxygen in the atmosphere or water. A calorimetric oxygen sensor has been tested using a solid polymer membrane or coating involving a picketfence cobalt porphyrin. The polymer cobalt porphyrin membranes are characterized by a deep-red color, and the oxygen-adduct formation is accompanied by a color change or sharp deoxy–oxy spectral change with the isosbestic points (for example, see inset in Fig. 9-9). When the oxygen response was monitored at the oxy absorption maximum at 547 nm, the 90% response time for a 20-μm-thick fluorinated OIm-picketfence cobalt porphyrin membrane was 5 s and the operational lifetime was longer than three months [58]. Membranes of picketfence cobalt porphyrins complexed with three kinds of polymer ligands, OPy, OIm, and O4Im, were characterized by a different oxygen affinity, which increased by a factor of 10. The three cobalt porphyrin complex membranes displayed a selective and continuous response over a very wide range of oxygen partial pressures from 0.01 to 76 cmHg [31].

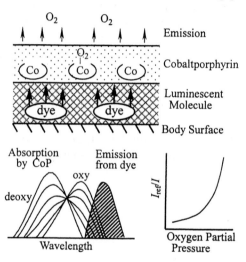

Figure 9-17. A novel oxygen-sensing coating based on combination of a luminescent dye molecule and an oxygen-binding cobalt porphyrin. Lower left: Overlapping of absorption by the cobalt porphyrin with emission from the dye molecule. Lower right: Stern-Volmer plots (I_{ref}/I: normalized luminescence intensity).

The most frequently used optical oxygen sensors are related to the luminescence quenching of organic dyes by molecular oxygen. A novel oxygen sensing coating has been reported based on a combination of the luminescence quenching by oxygen and oxygen-binding of cobalt porphyrins [59,60]. As illustrated in Fig. 9-17, the body surface of a model is first coated with a polymer layer containing luminescent dye molecules, such as pyrene. The emission from the dye molecules is then decreased by quenching with oxygen which obeys the Stern–Volmer equation (for plots of the relative luminescence intensity versus the oxygen partial pressure, see Chapter 10). The dye molecule layer is further coated with a polymer-picketfence cobalt porphyrin layer. The UV-Vis absorption of the cobalt porphyrin is red-shifted in response to the oxygen-adduct formation or oxygen pressure, which overlaps and reduces the emission from the dye molecules. The sum of the two emissions decreases along with an increase in the oxygen pressure, yielding a non-Stern–Volmer type response or high sensitivity to the oxygen pressure. This polymer cobalt porphyrin coating is currently being tested as an oxygen pressure (that is atmospheric total pressure)-sensitive paint for fluctuating wind-tunnel experiments related to aircraft and automobile development.

The reversible oxygen-binding property of a picketfence cobalt porphyrin allows the efficient accumulation of oxygen from an aqueous solution, which contributes to an increase in the diffusion-limited current for the electroreduction of oxygen, as illustrated in Fig. 9-18 [61]. When a cobalt porphyrin-OIm complex is coated on a glassy carbon surface, the advantages of the polymer complex include nonvolatility, insolubility in the aqueous phase, and a long operational lifetime for the cobalt porphyrin. The reduction current of oxygen for the cobalt porphyrin membrane was significantly larger than that for the control membrane without any cobalt porphyrins and remarkably increased with the cobalt porphyrin concentration. For example, the reduction current of oxygen for a 50 mM cobalt porphyrin membrane was around 10 times higher than that for the control membrane (Fig. 9-18). The reduction current obeyed a Langmuir-type model, while that of the control membrane followed a Henry-type model. The results of rotating platinum ring–glassy carbon disk voltammetry suggested that over 90% of the oxygen was reduced by four electrons to hydroxyl, regardless of the basicity of the aqueous solution. These results demonstrate that the cobalt porphyrin complex accumulates oxygen from the aqueous phase, then supplies it to the electrode surface, which could be practically applied to air-assisted batteries and fuel cells.

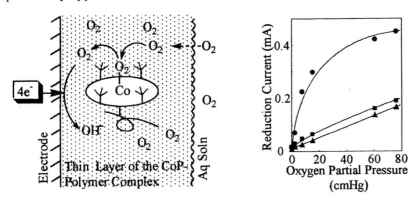

Figure 9-18. Schematic representation of the electroreduction of oxygen accumulated in the cobalt porphyrin membrane at an electrode. Reduction current of oxygen for the electrode modified with ●: 50 mM, ■: 5 mM cobalt porphyrin, and ▲: without cobalt porphyrin.

The oxygen-binding ability of a variety of cobalt(II) ion complexes in solution has also been well studied. An important class of such complexes is represented by the cobalt(II) amine complexes in an aqueous solution. Their oxygen adducts have been characterized as μ-peroxo cobalt(III) complexes (or $[O_2]:[Co] = 1:2$ type or sandwich type), as shown in Eq. (9-18), using the example of the cobalt-hexammine complex.

$$2Co(II)(NH_3)_6 + O_2 \rightleftharpoons (NH_3)_5Co(III) -O_2^{2-} Co(III)(NH_3)_5 + 2NH_3 \qquad (9-18)$$

Oxygen adducts are readily formed by the air bubbling of a solution of cobalt amine complex. Similar μ-peroxo cobalt(III) complexes are formed by the reaction of oxygen with a variety of cobalt(II) chelates. The amine ligands in oxygen-binding complexes also have been extended to polyamines, such as poly(ethyleneimine) (details are given in Experiment 9-4, Section 9.5).

In the blood stream of mammals, serum albumin (molecular weight 66,500 Da) is the major plasma protein and serves two main major roles: maintaining the colloid osmotic pressure, and transporting many helpful materials around the body. Employing these fundamental functions, serum albumin has been used as a plasma expander and a therapeutic drug carrier in the circulatory system. From the viewpoint of clinical application, providing serum albumin with an oxygen-carrying capability is of significant interest [62,63]. Accordingly, an iron TPP derivative with a covalently linked imidazole ligand, for example, 2-[8-{N-(2-methylimidazolyl)}octanoyloxymethyl]-5,10,15,20-tetrakis(α,α,α,α-o-pivaloylamino)phenylporphinatoiron, has been synthesized, and a small amount of its ethanol solution was incubated with an aqueous solution of recombinant human serum albumin. As a result, the iron porphyrin was included into the albumin and the colorless albumin solution became a deep-red-colored solution. The iron porphyrin included in the albumin and solubilized in the aqueous solution can bind and release oxygen reversibly under physiological conditions (in aqueous media, pH 7.3, 37 °C), like hemoglobin and myoglobin. When a maximum of eight iron porphyrin molecules were incorporated into the albumin host (Fig. 9-19), the oxygen transport amount is superior to that of hemoglobin (4 iron porphyrins involved in the protein with a molecular weight of 64,500 Da, see Chapter 2). Spectroscopies supporting the iron porphyrin incorporation do not reveal any changes in the albumin's higher structures, and the iron porphyrins are surrounded by a hydrophobic environment. The iron porphyrin-albumin solution exhibits a good compatibility with whole blood. Consequently, this entirely synthetic iron porphyrin-recombinant albumin is currently being studied as a new type of blood substitute [30,63].

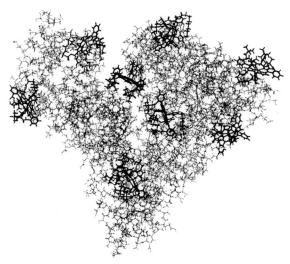

Figure 9-19. Three-dimensional structure of the serum albumin including iron porphyrins as an oxygen-carrying blood substitute.

Selective and reversible binding of small molecules to the metal complexes in macromolecules is now providing potential applications of metal-polymer complexes as gas transporting materials, gas separation membranes, and gas sensors. This chapter has indicated that the small-molecule-binding kinetics and equilibria are strongly affected by the surrounding macromolecular environment and that the metal-polymer complexes as a material display a performance different from that of the simple mixture of a transition metal ion or complex with a polymer. The small-molecule binding to a metal ion is the simplest and most elementary reaction; for example, the first step of the catalytic actions described in the following chapters, 11 and 12. The features of metal-polymer complexes described by the binding properties of small molecules are also appropriate for these properties. While this chapter focused on the metal-polymer complexes having a homogeneous structure, the specific advantages of metal-polymer complexes are quite variable depending on the selection of the metal ion and polymer species and tailor-made structural variety. Expected examples are an alignment of metal moieties along a polymer matrix, a sequential array of different metal complexes, an interfacial assembly of metal complexes on a nanometer-sized asymmetric membrane, and so on. They could yield unique binding performances such as an integrated small-molecule-binding process, a cascade-like flux of gaseous molecules, and the formation of a gaseous molecule transport channel. The metal-polymer complexes with such designed structures may open new systems for the superefficient utilization of gaseous small molecules.

9.5   Experimental

Experiment 9-1:   Spectroscopic and Sorption Experiments with Silver-Polymer Complexes (Section 9.2.2) [20–22]

Authors: H. Lee, J. Won, Y. S. Kang

Silver polymer complexes were made by dissolving a silver salt, silver triflate ($AgCF_3SO_3$) or silver tetrafluoroborate ($AgBF_4$), in a polar polymer, POZ, PVP, or PVMK. The POZ and PVMK were purchased from Aldrich Chemical Co, the PVP from Polysciences Inc. $AgBF_4$ (98%) and the $AgCF_3SO_3$ (99+%) from Aldrich Chemical Co.

For the spectroscopic and sorption experiments, the silver-polymer complex solution (20 wt.% in water) was cast on a glass plate and dried under a nitrogen atmosphere. The cast films were finally dried overnight in a vacuum at room temperature to remove any residual water. After drying completely, the films were lifted from the glass plate and FTIR measurements performed. The FTIR spectra for the solid-state interactions of the silver-PVMK complex and olefin are shown in Fig. 9-6.

For the absorption experiments, a film was placed in a sample chamber, which had been evacuated for more than 12 h. Pure propylene gas was then introduced into the chamber and allowed to equilibrate. The solubility was obtained by measuring the difference between the initial and final pressures. Once the chamber pressure remained constant, additional gas was introduced and again allowed to equilibrate. In this incremental manner, the propylene solubility as a function of the gas pressure was observed. The solubility of propylene in the solid silver-polymer complex films is shown in Fig. 9-4.

Experiment 9-2:   Membrane Performance of Silver-Polymer Complexes (Section 9.3.2) [12]

Authors: J.H. Kim, Y.S. Kang, J. Won

Silver-polymer complexes were prepared by dissolving a silver salt, $AgCF_3SO_3$ or $AgBF_4$, in a polar polymer, POZ or PVP. For the permeation experiments, an RK Control Coater was used to coat the silver-polymer complex solution (20 wt.% in water) onto a microporous membrane support to give composite membranes. The composite membranes were first dried at 40 °C for a day in a light-protected convection oven under nitrogen, then subsequently in a vacuum oven for a day at room temperature. The membrane samples were then cut into

2 cm × 2 cm sheets and mounted in a permeation test cell. The permeation properties of the pure and dry gases were measured at feed pressures between 140 and 550 kPa at 23 °C. The volumetric gas-flow rates were determined using a soap-bubble flow-meter, set up as shown in Fig. 9-20. A cell containing a homogeneous membrane with a known thickness was pressurized with a chosen gas. The extent of gas permeation through the membrane was then measured. The gas permeability was determined from the steady-state flow based on the known membrane thickness [35]. The propylene permeance and propylene/propane selectivity with an increasing silver concentration are shown in Fig. 9-12.

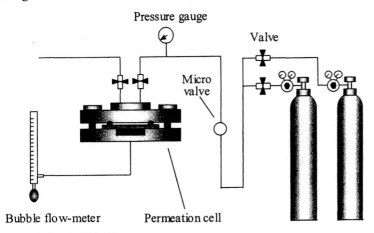

Figure 9-20. Bubble flow-meter set up.

Experiment 9-3:   Measurement of Oxygen-Binding Rate Constants for Solid State Cobalt Porphyrin-Polymer Complexes (Section 9.2.4) [27]

Authors: H. Shinohara and H. Nishide

Cobalt 2,3,7,8,12,13,17,18-octaethylporphyrin (CoOEP in Fig. 9-7) was purchased from Aldrich Chemical Co. The OIm (Fig. 9-8) was prepared by the radical polymerization of octylmethacrylate and 1-vinylimidazole with azobisisobutyronitrile, and used as the polymer ligand (content of vinylimidazolyl residue = 30 mol.% ; $[\eta] = 0.64$ g dL^{-1}, toluene, 30 °C, by elemental analysis and viscometric measurement, respectively). Distilled dichloromethane solutions of CoOEP and OIm were mixed together to obtain a CoOEP-OIm complex with mole ratio of [CoOEP]:[imidazolyl residue of OIm] = 1:10 ([CoOEP] = 0.02 mmol L^{-1}). A red transparent membrane was obtained by casting the solution on a glass plate under a nitrogen atmosphere. The membrane was dried in vacuo overnight to completely remove the solvent.

The flash photolysis measurements were carried out with a pulse laser flash spectrophotometer equipped with a kinetic data processor (UNISOKU FR-2000). The laser flash was applied perpendicular to the light path of the spectrophotometer, while the polymer membrane was placed at the crossing of the laser flash and the light path at 45° to both in a four-window quartz cell with a septum rubber cap. For the low-temperature measurements, the dry gas was charged in a spectrophotometer to avoid frosting on the cell windows. The rapid absorption change was recorded with a contact-type photomultiplier to cancel any noise caused by scattered light. An example of the recombination–time curve of oxygen on the CoOEP-OIm membrane at −15 °C is given in Fig. 9-21, which shows that the reaction was rapidly completed within a microsecond, despite the low temperature. Changing the monitoring wavelength from 400 to 420 nm allowed the differential spectrum to be measured before and after the flash photolysis. The negative extreme in the differential spectrum, 413 nm, was selected as the monitoring wavelength. This wavelength agreed with the absorption maximum of the oxygen adduct in the UV region.

Plots of the apparent oxygen-binding rate constant k_{app} estimated from the time course of the absorbance change versus the oxygen partial pressure are shown in the inset in Fig. 9-18. The linear relation indicates second-order kinetics for the oxygen binding to the cobalt porphyrin fixed in the polymer membrane. The oxygen-binding and oxygen-releasing rate constants, k_{on} and k_{off}, were calculated from the slope and intercept (as listed in Table 9-2).

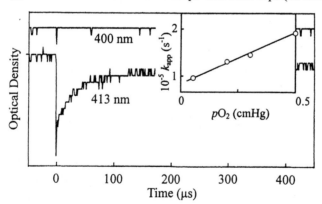

Figure 9-21. Recombination reaction of the photodissociated oxygen to the CoOEP complex of OIm in the solid membrane state. Spectral change after laser flash irradiation at −15 °C. Inset: plots of apparent oxygen-binding rate constant (k_{app}) versus oxygen partial pressure (pO_2).

Experiment 9-4:   Oxygen Uptake by Cobalt Poly(ethyleneimine) Complexes and its Electroreduction at an Electrode (Section 9.4) [26,64]

Authors: B. Shentu and H. Nishide

A cobaltpoly(ethyleneimine) (PEI) complex forms a μ-peroxo-type oxygen adduct in an aqueous solution, as represented by Eq. (9-18). The oxygen-binding (Eq. (9-19); N = the amine group of PEI) can be examined by very simple volumetry and the electrochemical reduction of oxygen at an electrode.

$$-(NHCH_2CH_2)_n-$$

PEI

$$\left[\begin{array}{c} -CH_2CH_2-N-CH_2CH_2NH- \\ | \\ CH_2 \\ | \\ -CH_2CHO- \end{array} \right]_n$$

Crosslinked PEI

$$2Co(II)N_6 + O_2 \rightleftharpoons N_5Co(III)\text{-}O_2^{2-}\text{-}Co(III)N_5 + 2N$$

$$\xrightarrow{10H^+} 2Co(II) + 10NH^+ + O_2 \qquad (9\text{-}19)$$

Oxygen uptake by an Aqueous Co-PEI Complex Solution: PEI ($M_n = 6 \times 10^4$ Da, 50 wt.% solution in H_2O) was purchased from Aldrich Chemical Co. An aqueous solution of PEI (40 mL; 1.5 unit mol) was charged into one of two 100-mL cylinders, then an aqueous solution of cobalt(II) chloride (8 mL; 0.5 mol) was charged into the other cylinder. The $CoCl_2$ solution was transferred to the PEI cylinder through a three-way stopcock connecting both cylinders, then the mixed solution was returned to the first cylinder and the process repeated several times. The mixed solution turned pink, indicating the formation of a cobalt complex. Air (exactly 100 mL) was supplied to one empty cylinder, then this air was supplied to the cylinder that contained the cobalt complex solution. The pink solution immediately turned brown, then the solution was repeatedly transferred from one cylinder to another to complete the oxygen-binding reaction. Next, the solution was kept in one cylinder and the air was removed to another, where the piston of the gas cylinder was kept open to the atmosphere. The volume content of the cylinder was recorded for calculating the volume of gas remaining, or the oxygen volume absorbed by the solution ($100 - 80 = 20$ cm³(STP)). The gas was transferred to a tube and recognized as nitrogen (for example, by extinguishing a flaming taper).

An acetic acid solution (1N, 5 mL) was introduced into the empty cylinder, and the solutions were transferred several times from one cylinder to another.

The gas that evolved from the mixed solution was attributed to the decomposition of the oxygen-adduct by the excess acid (Eq. (9-19)). The volume content of the gas cylinder was recorded (20 cm^3(STP)), and the evolved gas collected by a tube for a burn test, which supported the expectation that the evolved gas was oxygen.

Oxygen Electroreduction on Electrode Modified with the Co-PEI Complex: Electrochemical measurements were carried out in a conventional three-compartment glass cell at room temperature. The working electrode was either an MnO$_2$/C paste electrode or a glassy carbon disk electrode. The MnO$_2$/C electrode was provided by Matsushita Electric Co., and the electrode assembly consisted of a carbon paste pellet containing MnO$_2$ particles as the catalyst, which was embedded in a nickel-mesh electrode, adhered on one side to a poly(tetrafluoroethylene) plate to provide structural integrity with the other side exposed to the electrolyte solution. The apparent size of the electrode assembly was 3×3 mm^2. The glassy carbon disk electrode was polished before each experiment with a 0.05 μm alumina paste. The auxiliary electrode (a coiled platinum wire) was separated from the working solution by a fine-porosity frit. The reference electrode was a commercial saturated calomel electrode, which was placed in the male cell compartment. A dual potentiogalvanostat and universal programmer were employed with an X-Y recorder to obtain the current decay curves in potential step experiments and cyclic voltammograms. The electrolyte solution was an acetate buffer (pH = 5.6). The concentration of oxygen in the electrolyte solution was controlled by saturating the solution with a composition-defined mixed gas of oxygen and nitrogen at 25 °C and 1 atm, to form the oxygen adduct.

Reduction current of oxygen (i) relative to time (t) plots were recorded in the prepared solutions by dissolving PEI and CoCl$_2$ in the pH 5.6 acetate buffer. A larger current for the reduction of oxygen was obtained in the Co-PEI solution, and the time-independent parameter $it^{1/2}$ increased with the concentration of Co-PEI. The physical dissolution of oxygen in the buffer solutions without the Co-PEI complex obeyed a Henry-type model, where $it^{1/2}$ was several times smaller and linearly proportional to the oxygen concentration. Conversely, $it^{1/2}$ variation in the presence of the Co-PEI complex conformed to a Langmuir isotherm for dual-mode dissolution, which steeply increased at low concentrations of oxygen and was predominant over the physical flux, and finally became constant at higher concentrations at a partial oxygen pressure of 15 cmHg. The flux curve of the enriched oxygen corresponded to the oxygen-binding equilibrium curve of the Co-PEI complex.

A cobalt complex with a cross-linked PEI can provide an insoluble and swellable oxygen-enriching membrane that can modify an electrode surface even in pure electrolyte solutions. Poly(epichlorohydrin) ($M_n = \sim7 \times 10^5$ Da) was obtained from Acros Chemical Co. and purified by precipitation from a solution of benzene to methanol. A solution of polyepichlorohydrin (0.03 g) and PEI (0.03 g) in DMF (6 mL) was heated to 60 °C slowly enough to prevent gelation. After stirring for 1 h, the hot solution (200 µL) was transferred onto the surface of a glassy carbon disk electrode (area = 0.28 cm^2). The solvent was allowed to evaporate at 100 °C for 2 h. The thickness of the resulting polymer film was approximately 70 µm based on the density of the polymer ($\sim$1 g cm^{-3}). A cobalt ion was incorporated into the polymer film by immersing the coated electrode in an aqueous saturated solution of $CoCl_2$ for 24 h to give the desired cobalt(II) complex film.

In a cyclic voltammogram recorded with a glassy carbon disk electrode modified by the crosslinked Co-PEI complex film, the uncatalyzed reduction of oxygen was observed at potentials lower than –0.3 V. As such, a significant increase in the current magnitude was noted for the electrode surface modified with the thin Co-PEI membrane under an air atmosphere due to the oxygen enrichment from the aqueous phase.

Accordingly, a Co-PEI complex that can reversibly bind and release oxygen according to the concentration of oxygen can be regarded as an oxygen enricher in aqueous solutions, resulting in the enhancement of the diffusion-limited reduction current of oxygen at the electrode. Since the polymer complex-modified electrode has no leachable components, this makes it suitable for use in oxygen batteries to enhance the oxygen reduction current.

9.6   References

1. E. Martell, M. Calvin, *Chemistry of the Metal Chelate Compounds*, Prentice-Hall, New York, 1952.
2. R.D. Jones, D.A. Summerville, F. Basolo, *Chem. Rev.* **1979**, *79*, 139.
3. J.E. Huheey, *Inorganic Chemistry: Principles of Structure and Reactivity*, Harper International SI Edition, Cambridge, 1983.
4. E. Tsuchida, H. Nishide, *Adv. Polymer Sci.* **1997**, *24*, 1.
5. H. Nishide, E. Tsuchida, in: *Macromolecule Metal Complexes*, F. Ciardelli, E. Tsuchida, D. Wöhrle (Eds.), Springer-Verlag, Berlin, 1996, Chapter 4.2.
6. F.R. Hartley, *Chem. Rev.* **1973**, *73*, 163.
7. C.D. Beverwijk, G.J.M. van der Kerk, A.J. Leusink, J.G. Noltes, *Organomet. Chem. Rev. A.* **1970**, *5*, 215.
8. D.J. Safarik, R.B. Eldridge, *Ind. Eng. Chem. Res.* **1998**, *37*, 2571.
9. J.P. Collman, *Acc. Chem. Res.* **1977**, *10*, 265.
10. E. Tsuchida, H. Nishide, *Topics Curr. Chem.* **1986**, *132*, 64.

11. H. Nishide, E. Tsuchida, in: *Polymer for Gas Separation*, N. Toshima (Ed.), VCH Publishers, New York, 1992, Chapter 6.
12. Y. Yoon, J. Won, Y. S. Kang, *Macromolecules* **2000**, *33*, 3185.
13. S.U. Hong, T.H. Jin, J. Won, Y.S. Kang, *Adv. Mater.* **2000**, *12*, 968.
14. S. Choi, J.H. Kim, Y.S. Kang, *Macromolecules* **2001**, *34*, 9087.
15. S.U. Hong, C.K. Kim, Y.S. Kang, *Macromolecules* **2000**, *33*, 7918.
16. S.U. Hong, J.Y. Kim, Y.S. Kang, *Polym. Adv. Technol.* **2001**, *12*, 177.
17. S. Sunderrajan, B.D. Freeman, C.K. Hall, *Ind. Eng. Chem. Res.* **1999**, *38*, 4051.
18. C.K. Kim, H.S. Kim, J. Won, Y.S. Kang, C.K. Kim, C.K. Kim, *J. Phys. Chem. A* **2001**, *105*, 9024.
19. C.K. Kim, J. Won, H.S. Kim, Y.S. Kang, H.G. Li, C.K. Kim, *J. Comput. Chem.* **2001**, *22*, 827.
20. J.H. Jin, S.U. Hong, J. Won, Y.S. Kang, *Macromolecules* **2000**, *33*, 4932.
21. H.S. Kim, J.H. Ryu, H. Kim, B.S. Ahn, Y.S. Kang, *Chem. Commun.* **2000**, 1261.
22. J.H. Ryu, H. Lee, Y.J. Kim, Y.S. Kang, H.S. Kim, *Chem. Eur. J.* **2001**, *7*, 1525.
23. M. Momenteau, C.A. Reed, *Chem. Review.* **1994**, *94*, 659.
24. K.M. Kadish, K.M. Smith, R. Guilard (Eds.), *The Porphyrin Handbook*, Academic Press, Oxford, Vol. 4, 1996.
25. D. Dolphin (Ed.), *The Porphyrins*, Academic Press, New York, Vol. V, 1978.
26. H. Nishide, X.S. Chen, E. Tsuchida, *Functional Monomers and Polymer*, K. Takemoto, R.M. Ottenbrite, M. Kamachi (Eds.), Marcel Dekker, New York, 1997, Chapter 6.
27. H. Shinohara, T. Arai, H. Nishide, *Macromol. Symp.* **2002**, *186*, 135.
28. H. Nishide, E. Tsuchida, in: *Macromolecular Complexes*, E. Tsuchida, (Ed.), VCH Publishers, New York, 1991, Chapter 6.
29. H. Nishide, H. Kawakami, S. Toda, E. Tsuchida, Y. Kamiya, *Macromolecules* **1991**, *24*, 5851.
30. E. Tsuchida (Ed.), *Blood Substitutes*, Elsevier, Amsterdam, 1998.
31. Y. Suzuki, H. Nishide, E. Tsuchida, *Macromolecules* **2000**, *33*, 2530.
32. M. Ohyanagi, H. Nishide, K. Suenaga, E. Tsuchida, *Macromolecules* **1998**, *21*, 1590.
33. E. Tsuchida, H. Nishide, M. Ohyanagi, O. Okada, *J. Phys. Chem.* **1988**, *92*, 6461.
34. L.M. Robeson, *J. Membr. Sci.* **1991**, *62*, 165.
35. M. Mulder, *Basic Principles of Membrane Technology*, 2nd Edn. Kluwer Academic Publishers, Dordrecht, 1996.
36. J.S. Yang, G.H. Hsiue, *J. Membr. Sci.* **1998**, *138*, 203.
37. W.S. Ho, D.C. Dalrymple, *J. Membr. Sci.* **1994**, *91*,13.
38. T. Yamaguchi, C. Baertsch, C.A. Koval, R.D. Noble, C.N. Bowman, *J. Membr. Sci.* **1996**, *117*, 151.
39. S. Bai, S. Sridhar, A. A. Khan, *J. Membr. Sci.* **1998**, *147*, 131.
40. Pinnau, L.G. Toy, *J. Membr. Sci.* **2001**, *184*, 39.
41. B.J. Cha, Y.S. Kang, J. Won, *Macromolecules* **2001**, *34*, 6631.
42. Y.S. Park, J. Won, Y.S. Kang, *Langmuir* **2000**, *16*, 9662.
43. Y.S. Park, Y.S. Kang, J. Won, *J. Appl. Polym. Sci.* **2002**, *83*, 2369.
44. H. Nishide, M. Ohyanagi, O. Okada, E. Tsuchida, *Macromolecules* **1987**, *20*, 417
45. H. Nishide, H. Kawakami, T. Suzuki, Y. Azechi, Y. Soejima, E. Tsuchida,

Macromolecules **1991**, *24*, 6306.

46. H. Nishide, T. Suzuki, H. Kawakami, E. Tsuchida, *J. Phys. Chem.* **1994**, *98*, 5084.
47. H. Shinohara, T. Arai, H. Araihara, H. Nishide, *Int. Cong. Membr. & Proc. Supp. Abstr.* **2002**, 265.
48. Y.S. Kang, J.-M. Hong, J. Jang and U.Y. Kim, *J. Membr. Sci.* **1996**, *109*,149.
49. J.-M. Hong, Y.S. Kang, J. Jang and U.Y. Kim, *J. Membr. Sci.* **1996**, *109*, 159.
50. H. Nishide, H. Kawakami, Y. Kurimura, E. Tsuchida, *J. Am. Chem. Soc.* **1989**, *111*, 7175.
51. Y. Kurimura, F. Ohta, J. Gohta, H. Nishide, E. Tsuchida, *Macromol. Chem.* **1982**, *183*, 2889.
52. R. Quinn, D.V. Laciak, *J. Membr. Sci.* **1997**, *131*, 49.
53. J.J. Pellegrino, M. Ko, R. Nassimbene, M. Einert, *Gas Separation Technology*, E.F. Vansant, R. Dewolfs, (Eds.), Process Technology Proceedings, 8, Elsevier, Amsterdam, 1990.
54. M. Teramoto, Q. Huang, T. Maki, H. Matsuyama, *Separation and Purification Technology* **1999**, *16*, 109.
55. H. Hirai, S. Hara, M. Kamiyama, *Angew. Makromol. Chem. Lett.* **1985**, *130*, 207.
56. H. Hirai, K. Kurima, K. Wada, M. Komiyama, *Chem. Lett.* **1985**, 1513.
57. D. Wöhrle, H. Bohlen, *Makromol. Chem.* **1986**, *187*, 2081.
58. S. Rösli, E. Pretsch, W.E. Morf, E. Tsuchida, H. Nishide, *Anal. Chim. Acta* **1997**, *338*, 119.
59. Y. Amao, I. Okura, H. Shinohara, H. Nishide, *Polymer J.* **2002**, *34*, 411.
60. K. Asai, Y. Egami, T. Sakai. H. Nishide, *J. Thermophys. Heat Trans.*, **2002**, *16*, 109.
61. B. Shentu, K. Oyaizu, H. Nishide, *Chem. Lett.* **2002**, 712.
62. E. Tsuchida, T. Komatsu, K. Hamamatsu, Y. Matsukawa, J. Wu, *Bioconjugate Chem.* **1999**, *10*, 797.
63. T. Komkatsu, Y. Matsukawa, E. Tsuchida, *Bioconjugate Chem.* **2002**, *13*, 397.
64. K. Oyaizu, H. Nakano, B. Shentu, H. Nishide, *J. Mater. Chem.* **2002**, 3162.

10 Physical and Optical Sensors

10.1 Optical Sensors

Ichiro Okura and Yutaka Amao

10.1.1 Principle of Optical Gas Sensors

Recently, optical gas sensors based on the luminescence or absorption changes of dye compounds have been paid much attention. Optical sensors for CO_2, NO_2, SO_2 and oxygen have been developed using various dye compounds. Among various gases, oxygen is an immensely important chemical species, essential for life. The determination of oxygen levels is required in many different fields. In environmental analysis, for example, oxygen measurement provides an indispensable guide to the overall condition of the ecology, and it is routine practice to monitor oxygen levels continuously in the atmosphere and in water. In the medical field, oxygen levels in exhaled air or in the blood of a patient are key physiological parameters for judging health. Such parameters should ideally be monitored continuously. The measurement of oxygen levels is also essential in industries which utilize metabolizing organisms: yeast for brewing and bread making, and the plants and microbes that are used in modern biotechnology, such as those producing antibiotics and anticancer drugs. In this section, the background to oxygen measurements is described and the work to develop new optical oxygen sensors which utilize the luminescence of metal complexes is discussed.

Several methods for oxygen detection have been reported so far, some based on titration [1], amperometry [2], chemiluminescence [3] or thermo-luminescence [4]. Winkler titration has been employed for the measurement of oxygen for many years and is considered, to some extent, to be the standard method [1]. However, the time-consuming and cumbersome nature of the titration has hindered its application to process monitoring. Clark-type electrodes provided a breakthrough technique for the measurement of oxygen and have become the conventional method. The Clark cell is robust and quite reliable. However, it is bulky and not readily miniaturized without great expense. Since the cells are based on reduction of oxygen at the cathode and the diffusion-limited passage of oxygen through the film, any factor that can change the diffusion resistance, such as fouling of the film or a change in flow conditions in the testing fluid, can cause measurement error [5,6]. They also suffer from electrical interference and since they consume oxygen, can easily generate misleading data. Thus, there is constant interest in new, superior techniques for oxygen detection, and optical oxygen sensors represent one of the new hopeful possibilities in this area. Optical oxygen sensors should be inexpensive, easily miniaturized and simple to use. They should not suffer from electrical interference or consume oxygen. Many of the chemical sensors developed to measure oxygen employ luminescence as the principle of measurement. Optical sensors based on photoluminescence quenching usually measure the luminescence intensity [7,8] or lifetime [9] of an organic dye immobilized in a polymer film, which varies with the concentration of the analyte. However, in the development of the optical oxygen sensor, the detection method as well as the choice of dye and matrix material and the film preparation method are crucial factors [10].

Recent developments in optical oxygen sensors have been reviewed [11]. In 1985, a new optical method based on phosphorescence lifetime quenching was introduced for the measurement of oxygen in biological systems [12]. This technique seemed to be very attractive because it did not suffer many of the practical limitations of the fluorescence intensity quenching methods used by previous workers [13,14]. Also, the relatively long phosphorescence lifetime of the materials used allowed the use of less sophisticated measurement instrumentation. The possibility of using this technique to develop a new gaseous oxygen sensor based on film immobilization of organic dyes, especially the room-temperature phosphorescent metalloporphyrins, has been explored. Most optical oxygen sensors respond specifically and reversibly to molecular oxygen by a change in the intensity of luminescence emitted from a probe molecule.

As shown in the Jablonski diagram (Fig. 10-1) luminescence (spontaneous emission of light) is produced in certain molecules following excitation by light from their ground state to higher energy states. When the excited molecule returns to the ground state, it emits light (fluorescence or phosphorescence) at a longer wavelength than the original stimulating light (Stokes shift). Fluorescence is a short-lived emission from the singlet state with the electrons spin-paired. As the transition from the singlet state to the ground singlet state is spin-allowed (i.e. with no change in multiplicity), it has a high probability of occurrence and the decay time of fluorescence is usually short (10^{-9} to 10^{-7} s; i.e. the same order as the lifetime of an excited singlet state). On the other hand, phosphorescence involves a change in electron spin, and the transition from the lowest excited triplet state to the singlet ground state occurs at a low probability and phosphorescence is a long-lived emission (about 10^{-5} to 10 s) [15,16]. Owing to the relative energies of the excited states, phosphorescence occurs at longer wavelengths than fluorescence. Given the fact that molecules in the triplet state have long lifetimes, they are particularly susceptible to interactions with other molecules. Oxygen quenching is diffusion-limited and can be described by the Stern–Volmer relationship (Eq. (10-1)) [17]. This is valid for both phosphorescence and fluorescence lifetimes and intensities, where, I_0 and I are the luminescence intensities, and τ_0 and τ are lifetimes in the absence of oxygen and in the presence of oxygen (concentration $[O_2]$), respectively.

$$I_0 / I = \tau_0 / \tau = 1 + k_q \tau_0 [O_2] \qquad (10\text{-}1)$$

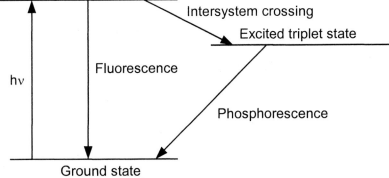

Figure 10-1. Jablonski diagram.

The quenching rate constant, k_q, is dependent on the physical constants of the system and is described by a modified Smoluchowski equation (Eq. (10-2))[18–20], where, N is Avogadro's number, p is a factor related to the probability of each collision causing quenching and to the radius of interaction between the donor and oxygen. D_A and D_B are the diffusion coefficients for the donor and acceptor, respectively. In general, under defined experimental conditions, k_q can be considered constant. Using Eq. (10–1), it is possible to make intensity or lifetime measurements of phosphorescence (or fluorescence) then relate this measurement to the oxygen concentration.

$$k_q = 4\pi N p (D_A + D_B) \cdot 10^3 \qquad (10\text{-}2)$$

Kautsky discovered that the phosphorescence [21] and the fluorescence [22] of surface-adsorbed dyes such as trypaflavin, benzoflavin, safranin, chlorophyll and porphyrins were sensitively quenched by molecular oxygen, which, in its electronic ground state, is a triplet molecule. This led to the design of what now appears to have been the first optical sensor for oxygen. Depending on the measurement of either phosphorescence or fluorescence, extremely high sensitivities (typical detection limit in phosphometry 0.067 Pa oxygen) are achieved [23]. Zhakharov and Grishaeva [24] have utilized this effect to devise an opto-sensor for low oxygen levels as they occur in wastewater. Detection limits are reported to be as low as 0.5 mg oxygen per liter. The fast growing interest in optical sensors has led to a variety of other sensor types in the past. Bergman [25] and others [26–30] have used polycyclic aromatic hydrocarbons (PAHs) which are found to be efficiently quenched by oxygen in the 0–40 kPa range. The PAHs are either dissolved in a polymer [27–30], soaked into porous glass [25] or covalently immobilized on a glass support [26]. Peterson et al., by combining Kautsky's adsorption technique with the sensitivity of the PAHs, developed a fiberoptic sensor for oxygen based on the oxygen-sensitive fluorescence of Amberlite-adsorbed perylene dibutyrate [13].

Changes in luminescence intensity can therefore be used to monitor oxygen [14]. It has also been shown that the fluorescent dye pyrene butyric acid (PBA) might be used to determine oxygen in biological microenvironments. Subsequently, PBA has been used by several workers to measure oxygen in tissue [31,32]. PBA is also used in a gaseous oxygen sensor [28], with a thin layer of the dye (10-μm thick) trapped behind a 6-μm thick gas-permeable Teflon film. For *in vivo* oxygen concentration measurement, a catheter tip sensor that can be used in tissue or introduced into blood vessels has been developed [5]. The fluorescence dye, perylene dibutyrate, used in this system is held behind a 25-μm thick porous polypropene gas-permeable film. The oxygen concentration is calculated using an equation derived from the Stern–Volmer relationship, including a correction for the non-linearity of the response.

Recently, phosphorescence quenching-based sensors have increasingly become the focus of attention as they have several advantages over fluorescence-based ones. The longer excited state lifetimes of phosphorescent indicators give rise to high quenching efficiency by oxygen. In addition, the long excitation and emission wavelengths are more compatible with the available optical monitoring technology. Historically, the first optical sensors for oxygen described in the literature are based on room-temperature phosphorescence quenching of immobilized dyes, though these dyes tended to be photolabile [33]. Indeed, this tends to be a general problem associated with excited triplet state molecules and is one of the major drawbacks associated with their use. As molecules in the triplet state have long lifetimes, they should in principle form the basis of an extremely sensitive oxygen sensor. Until recently, however, there have been few molecules that exhibit a strong phosphorescence yield at room temperature. Recently, several probes that are suitable for room temperature phosphorescence applications have been presented. More recent room-temperature phosphorescence quenching-based sensors have utilized metal chelates such as tetrakis(pyrophosphito)diplatinate(II) [34] and 8-hydroxy-7-iodo-5-quinolinesulfuric acid (ferron) chelates [35], the organic dyes camphorquinone [36] and erythrosin B [37] and metalloporphyrins [38]. Optical oxygen sensors using metalloporphyrins, pyrene and fullerene are described in the following sections.

10.1.2 Luminescence Quenching-based Oxygen Sensors

Recently, a variety of devices and sensors based on photo-luminescent quenching of organic dyes for measurement of oxygen concentration on solid surfaces have been developed. Many optical oxygen sensors are composed of organic dyes, such as polycyclic aromatic hydrocarbons (pyrene and its derivatives, quinoline and phenanthrene) [39–45] transition metal complexes (ruthenium [46–51], osmium [52] or rhenium-polypyridine complexes [53]), and metalloporphyrins [54–56], immobilized in an oxygen-permeable polymer (e.g. silicone polymer or polystyrene). Platinum porphyrins and organic dyes show only weak prompt fluorescence and phosphorescence yields Φ_P that vary over the range of 10^{-4} to 1 at room temperature. Phosphorescence lifetimes of platinum porphyrins are long ($\tau \sim 3.0$ ms) [57]. Platinum octaethylporphyrin (PtOEP) displays a particularly strong room temperature phosphorescence with a high quantum yield ($\Phi_P < 0.5$) and a long lifetime ($\tau = \sim 100$ μs) [57]. Some optical oxygen sensors based on phosphorescence quenching by oxygen of PtOEP immobilized in a polymer film have been developed [58,59] (for a film preparation see Experiment 10-1, Section 10.1.5). Since organic dyes interact with polymer molecules, the optical sensing ability strongly depends on the properties of the polymer matrices. Oxygen-permeable polymers with low diffusion barriers for oxygen are desirable. The oxygen permeability of a fluoropolymer film is generally large [60,61]. As oxygen affinity is induced by the large electronegativity of fluorine, the oxygen permeability of a fluoropolymer film will be large. Thus, a fluoropolymer is suitable for the above-stated requirements. It may be possible to develop a highly sensitive oxygen sensor using a fluoropolymer as a polymer matrix.

In this section, we describe the optical oxygen-sensing properties of platinum porphyrin in new fluoropolymer films (poly(isobutylmethacrylate-co-2,2,2-trifluoroethyl-methacrylate) (poly-IBM$_m$-co-TFEM$_n$) [62], poly(styrene-co-2,2,2-trifluoroethylmethacrylate) (poly-Sty$_m$-co-TFEM$_n$) [63] and poly(styrene-co-pentafluorostyrene) (polySty$_m$-co-PFS$_n$) [64] (chemical structures are shown in Fig. 10-2; for the synthesis of the copolymers see Experiment 10-2, Sections 10.1.5).

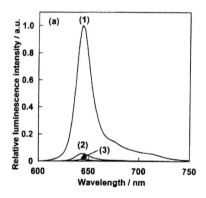

Poly-IBM$_m$-co-TFEM$_n$ Poly-Sty$_m$-co-TFEM$_n$ Poly-Sty$_m$-co-PFS$_n$

Figure 10-2. Chemical structures of new fluoropolymers.

PtOEP immobilized in fluoropolymer films showed luminescence at 645 nm when excited at the wavelength attributed to the Q-band (535.0 nm), as shown in Fig. 10-3. Parts (a), (b) and (c) show PtOEP immobilized in poly-IBM$_1$-coTFEM$_{3.28}$, poly-Sty$_1$-co-TFEM$_{0.62}$ poly-Sty$_1$-co-PFS$_{0.21}$, respectively. The luminescence spectra were obtained under (1) argon-saturated, (2) air--saturated, and (3) oxygen-saturated conditions. In all cases, the luminescence intensity of the film depended on the oxygen concentration as shown in Fig. 10-3.

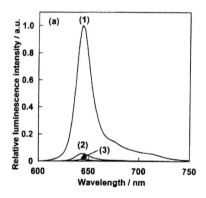

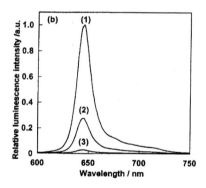

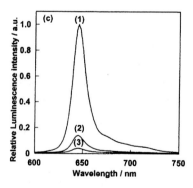

Figure 10-3. Luminescence spectra of PtOEP immobilized in poly-IBM$_1$-co-TFEM$_{3.28}$ (a), poly- Sty$_1$-co-TFEM$_{0.62}$ film (b) and poly-Sty$_1$-co-PFS$_{0.21}$ film (c). (1) 100% argon; (2) 20% oxygen; and (3) 100% oxygen. Excitation wavelength was 535.0 nm.

The luminescence intensity of PtOEP in the fluoropolymer film decreased with increase in oxygen concentration in all cases. These results indicate that the luminescence of PtOEP in a fluoropolymer film was quenched by oxygen. And PtOEP immobilized in a fluoropolymer film can be used as a material for an oxygen sensor. In general, a sensor with I_0/I_{100} of more than 3.0 is a suitable oxygen sensor [65]. All of the I_0/I_{100} values of poly-IBM$_m$-co-TFEM$_n$ films were more than 70.0. The I_0/I_{100} values of all of the poly-Sty$_m$-co-TFEM$_n$ films were more than 20.0. The I_0/I_{100} values of all of the poly-Sty$_m$-co-PFS$_n$ films were more than 24.0. These results indicate that PtOEP immobilized in a fluoropolymer film is a highly sensitive oxygen sensor. In the cases of poly-IBM$_m$-co-TFEM$_n$ and poly-Sty$_m$-co-TFEM$_n$ films, the I_0/I_{100} values increased with increase in TFEM units in fluoropolymers.

The oxygen-sensing properties of PtOEP immobilized in a poly-Sty$_m$-co-PFS$_n$ film is characterized by the Stern–Volmer quenching constant, K_{SV}, obtained from Eq. (10-3), where I_0 and I are the luminescence intensities in the absence and presence of oxygen, respectively. K_{SV} was obtained from a linear plot of $(I_0/I) - 1$ versus $[O_2]$.

$$(I_0/I) - 1 = K_{SV}[O_2] \tag{10-3}$$

Figure 10-4 shows Stern–Volmer plots of PtOEP immobilized in poly-IBM$_m$-co-TFEM$_n$ films. For PtOEP, the plots exhibit considerable linearity at oxygen concentrations in the range of 0–100%. The K_{SV} values for all of the poly-IBM$_m$-co-TFEM$_n$ films ($>0.70\%^{-1}$) were larger than those for the poly-IBM film ($K_{SV} = 0.66\%^{-1}$). The K_{SV} value increased with increase in the TFEM unit in a poly-IBM$_m$-co-TFEM$_n$. These results indicate that PtOEP immobilized in a poly-IBM$_m$-co-TFEM$_n$ film is a reliable optical oxygen sensor.

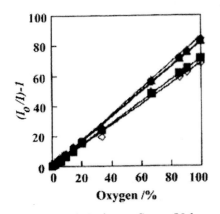

Figure 10-4. Stern–Volmer plots for PtOEP immobilized in poly-IBM$_m$-co-TFEM$_n$ films. ■: $n = 0.034$; ●: $n = 0.48$; ▲: $n = 1.25$; ◆: $n = 1.92$; □: $n = 3.28$; ◇: poly-IBM film.

Figure 10-5 shows Stern–Volmer plots of PtOEP immobilized in poly-Sty$_m$-co-TFEM$_n$ films. Plots of PtOEP immobilized in poly-Sty$_m$-co-TFEM$_n$ ($n > 0.27$)

films exhibit considerable linearity at oxygen concentrations in the range of 0–100%. The Stern–Volmer constant, K_{SV}, increased with increase in TFEM units in poly-Sty_m-co-$TFEM_n$. Plots of PtOEP immobilized in polystyrene (PS) and poly-Sty_1-co-$TFEM_{0.069}$ film, on the other hand, exhibit considerable linearity at lower oxygen concentrations, although the curvature decreases at higher oxygen concentrations. At higher concentrations, Stern–Volmer plots of a sensor based on luminescence quenching is non-linear because of the simultaneous presence of static and dynamic quenching.

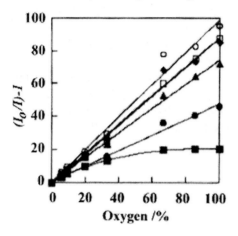

Figure 10-5. Stern–Volmer plot for PtOEP immobilized in poly-Sty_m-co-$TFEM_n$ films. ■: $n = 0.069$; ●: $n = 0.27$; ▲: $n = 0.41$; ◆: $n = 0.62$; □: $n = 1.00$; ○: $n = 2.50$; △: PS film.

Our group previously reported that PtOEP immobilized in a PS film has some different oxygen-accessible sites. As there are two oxygen-accessible sites in the sensing film, one site with good accessibility and the other site with poor accessibility, the intensities from the PtOEP film are quenched by oxygen according to the Stern–Volmer equation at lower oxygen concentrations (~0–20%). Demas et al. also reported a multi-site model with an oxygen-accessible site and an oxygen inaccessible site [66]. In this model, the sensor molecule can exist in two or more sites that each has its own characteristic quenching constant. The Stern–Volmer plot becomes as shown in Eq. (10-4), where f_n is the fractional contribution to the site with good accessibility for oxygen or the site with poor accessibility for oxygen, and K_{SVn} is the quenching constant for each site.

$$I_0/I = [\Sigma(f_n/(1+K_{SVn}[O_2]))]^{-1}, \tag{10-4}$$

In Fig. 10-5(a), for PS and poly-Sty_1-co-$TFEM_{0.069}$ films, the solid line is the best fit using the above equation ($n = 2$). Thus, there are two oxygen-accessible sites for a sensing film: a site with good accessibility ($K_{SV1} = 0.10\%^{-1}$ and $f_1 = 0.980$ for the PS film and $K_{SV1} = 0.50\%^{-1}$ and $f_1 = 0.986$ for the poly-

Sty$_1$-co-TFEM$_{0.069}$ film) and a site with poor accessibility ($K_{SV2} = 0.051\%^{-1}$ and $f_2 = 0.020$ for the PS film and $K_{SV2} = 0.049\%^{-1}$ and $f_2 = 0.014$ for the poly-Sty$_1$-co-TFEM$_{0.069}$ film). The value of K_{SV2} is very low and makes little contribution compared with K_{SV1}. The values of K_{SV} for poly-Sty$_m$-co-TFEM$_n$ ($n > 0.41$) films (more than $0.75\%^{-1}$) were larger than those for the PS film ($K_{SV1} = 0.10\%^{-1}$). These results indicate that PtOEP immobilized in a poly-Sty$_m$-co-TFEM$_n$ film is a reliable optical oxygen sensor. The K_{SV} value increased with increase in the TFEM units in poly-Sty$_m$-co-TFEM$_n$.

Figure 10-6 shows Stern–Volmer plots of PtOEP immobilized in poly-Sty$_m$-co-PFS$_n$ films. The plots of PtOEP immobilized in poly-Sty$_m$-co-PFS$_n$ films exhibit considerable linearity at oxygen concentrations ranging from 0 to 100%. The K_{SV} values of all of the poly-Sty$_m$-co-PFS$_n$ films (more than $0.23\%^{-1}$) were larger than that of the PS film ($K_{SV1} = 0.10\%^{-1}$). These results indicate that PtOEP immobilized in a poly-Sty$_m$-co-PFS$_n$ film is a useful optical oxygen sensor at oxygen concentrations ranging from 0 to 100%. K_{SV} was not affected by the composition ratio of Sty and PFS unit in poly-Sty$_m$-co-PFS$_n$.

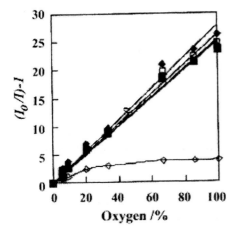

Figure 10-6. Stern–Volmer plots for PtOEP immobilized in poly-Sty$_m$-co-PFS$_n$ film. ■: $n = 0.089$; ●: $n = 0.21$; ▲: $n = 1.11$; ◆: $n = 1.33$; □: $n = 1.67$; ◇: PS film.

The oxygen-sensing properties (I_0/I_{100} and K_{SV} values) of PtOEP immobilized in fluoropolymer, polystyrene and poly-IBM films are given in Table 10-1. The values of I_0/I_{100} and K_{SV} were large for all fluoropolymer films. In the cases of poly-IBM$_m$-co-TFEM$_n$ and poly-Sty$_m$-co-TFEM$_n$ films, the I_0/I_{100} and K_{SV} values increased with increase in the TFEM units in the fluoropolymer. On the other hand, I_0/I_{100} and K_{SV} were not affected by the composition ratio of Sty and PFS unit in poly-Sty$_m$-co-PFS$_n$. For all fluoropolymers, the interfacial intermolecular force between the gaseous phase and the surface of the polymer film was lower than that in the case of the poly-IBM or polystyrene film, and oxygen affinity was induced by the large electronegativity of fluorine. Thus, the large degree of oxygen permeability of synthesized fluoropolymers is induced by low surface free energy and the large electronegativity of fluorine compared with those of poly-IBM or polystyrene. Hence, highly sensitive optical sensors have been developed using fluoropolymers, which have large degrees of oxygen permeability as polymer matrices.

Table 10-1. I_0 / I_{100} and K_{SV} values of PtOEP immobilized in polymer films.

Sensing film	I_0 / I_{100}	$K_{SV} / \%^{-1}$
polystyrene	4.5	0.10
poly-IBM	67.3	0.66
poly-IBM$_1$-co-TFEM$_{0.034}$	70.8	0.69
poly-IBM$_1$-co-TFEM$_{0.48}$	82.3	0.81
poly-IBM$_1$-co-TFEM$_{1.25}$	82.4	0.81
poly-IBM$_1$-co-TFEM$_{1.92}$	83.5	0.83
poly-IBM$_1$-co-TFEM$_{3.28}$	85.4	0.84
poly-Sty$_1$-co-TFEM$_{0.069}$	20.6	0.50
poly-Sty$_1$-co-TFEM$_{0.27}$	45.8	0.50
poly-Sty$_1$-co-TFEM$_{0.41}$	72.3	0.75
poly-Sty$_1$-co-TFEM$_{0.62}$	84.8	0.88
poly-Sty$_1$-co-TFEM$_{1.00}$	87.6	0.87
poly-Sty$_1$-co-TFEM$_{2.50}$	94.8	0.90
poly-Sty$_1$-co-PFS$_{0.089}$	24.5	0.235
poly-Sty$_1$-co -PFS$_{0.21}$	24.9	0.249
poly-Sty$_1$-co -PFS$_{1.11}$	24.6	0.236
poly-Sty$_1$-co-PFS$_{1.33}$	27.2	0.262
poly-Sty$_1$-co-PFS$_{1.67}$	25.6	0.256

10.1.3 Optical Oxygen-Sensing Systems Based on Triplet–Triplet Absorption of Organic Dye

Optical oxygen-sensing systems based on the phosphorescence intensities of various platinum porphyrin-doped polymers or sol–gel glass have been described above. Since most organic compounds, however, have no phosphorescence at room temperature, the number of oxygen-quenchable compounds is extremely limited. Triplet–triplet (T_1-T_n) absorption decay on a laser flash photolysis allows us to measure the photoexcited triplet lifetime of such non-phosphorescent compounds [67–69]. A quenching analysis of the photoexcited triplet state of the compound is based on the decrease in lifetime in the presence of oxygen. If the quenching of the photoexcited triplet state is entirely diffusional, the lifetime is related to the oxygen concentration (or pressure) by the following Stern–Volmer Eq. (10-5), where τ is the lifetime of the photoexcited triplet state, the subscript "0" denotes the value in the absence of oxygen, and K_{SV} is the Stern–Volmer quenching constant.

$$\tau_0/\tau = 1 + K_{SV}[O_2],\tag{10-5}$$

The above method extends the number of indicators available for oxygen sensor. In this section, we describe some examples to use the new systems using $T_1\text{-}T_n$ absorption quenching for oxygen concentration measurement, such as zinc tetraphenylporphyrin (ZnTPP) [67] and fullerene C_{60} immobilized in a polystyrene (PS) film [68].

Upon absorption of light, a molecule is raised from the singlet ground state to an excited singlet state which may be converted to a triplet manifold by intersystem crossing. By internal conversion the lowest triplet state (T_1) can be produced in a relatively high concentration. Therefore, it is not difficult to measure the absorption spectra, which are due to the spin-allowed transitions from T_1 to the higher triplet states T_n. As T_1-to-S_0 transitions are forbidden, the lifetimes of the triplet states are several orders of magnitude longer than those of the singlet excited states e.g. fluorescence lifetimes of polycyclic aromatic molecules are less than 100 ns, while the lifetimes of the triplet states are as long as several seconds, particularly in rigid media. In fluid solutions, the triplet lifetimes are between several tens of nanoseconds and hundreds of milliseconds, depending on the experimental conditions.

The principle of laser flash photolysis and its oxygen-sensing application is shown in Fig. 10-7 [69]. In laser flash photolysis techniques, a strong xenon or laser lamp is used to generate T_1 states via S_0-to-S_1 absorption, and this is followed by intersystem crossing from S_1 to T_1. The concentrations of species in the triplet state are determined via the transient T_1-to-T_2 absorption, which is monitored by means of a second light source. This method is particularly useful when the lifetimes of the triplet state are short (in the range of nanoseconds).

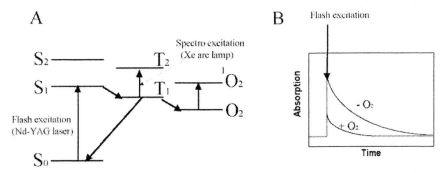

Figure 10-7. Schematic diagrams of laser flash photolysis (A) and time profiles of sensing dye with and without oxygen (B).

A typical sensing system for measuring oxygen is shown in Fig. 10-8. Laser flash photolysis is carried out using an OPO-Nd-YAG laser with a pulse width of 10 ns at room temperature. A xenon arc lamp as a monitoring light beam is coupled to one end of an optical fiber. The light reflected by the oxygen-sensing film is transmitted by the same fiber to a monochromator and photomultiplier. Different oxygen standards (in the range 0–100%) in a gas stream are produced by controlling the flow rates of oxygen and argon gases entering a mixing chamber. The total pressure is maintained at 760 Torr. A mixture of 30 wt.% polymer and zinc porphyrin in THF was cast onto 1.4×5.0-cm non-luminescent glass slides. The zinc porphyrin concentration in the membrane was approximately 2.9×10^{-5} mol dm^{-3}. The membranes were dried at room temperature and stored in the dark prior to use. The thicknesses of the films, measured by a micron-sensitive caliper, were between 50.0 and 120 µm.

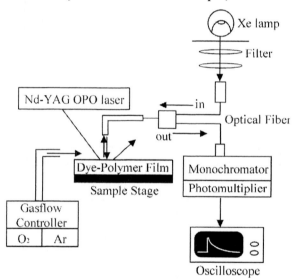

Figure 10-8. Oxygen-sensing system for measuring of T_1-T_n absorption of dye immobilized in polymer film.

Figure 10-9 shows the time profiles of the triplet state of ZnTPP (monitored at 470 nm) under argon (1) and oxygen (2) after laser flash photolysis. The decays in the absence and the presence of oxygen obey first-order kinetics. The fast decay under oxygen, compared to that under argon, indicates that the excited triplet state of ZnTPP is quenched by oxygen. This result indicates that ZnTPP in a PS film can be used as a material for an oxygen sensor. Figure 10-10 shows the relationship between the photoexcited triplet lifetime of ZnTPP and the oxygen concentration. The triplet lifetimes of ZnTPP are calculated from the first-order plot. The lifetime decreases with increase in oxygen concentration. Figure 10-11 shows a Stern–Volmer plot. The Stern–Volmer plot of ZnTPP exhibits considerable linearity ($r = 0.988$). The sensing properties are greatly affected by the thickness of ZnTPP in the PS film. In general, a thinner film needs less time for the oxygen inside the film to reach equilibrium with the outside environment. Figure 10-12 shows the relationship between the Stern–Volmer constant and film thickness. Film thickness seems to have little effect on the Stern–Volmer constant (K_{sv}).

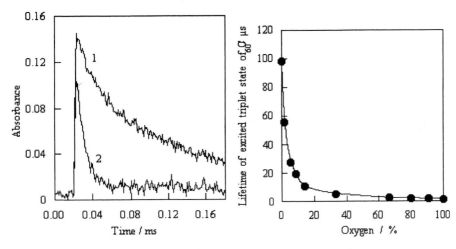

Figure 10-9. Decay curve of the photo-excited triplet state of ZnTPP in PS film under 0% (1) and 100% (2) oxygen.

Figure 10-10. Lifetime change of the photoexcited triplet state of ZnTPP in PS film under various oxygen concentrations.

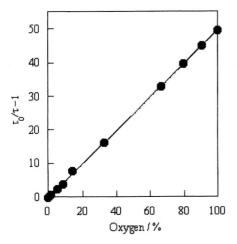

Figure 10-11. Stern–Volmer plot for ZnTPP in PS film.

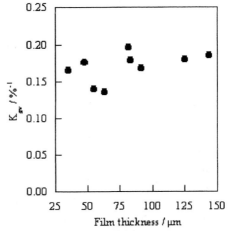

Figure 10-12. Effect of film thickness on Stern–Volmer coefficient in ZnTPP in PS film.

Another candidate probe for an oxygen sensor is fullerene. Fullerene possesses useful electronic and photochemical properties [70–81]. The quantum yield of the intercrossing system from the photoexcited singlet state to the photoexcited triplet state of fullerene is estimated to be 1.0, and quenching efficiency of the triplet state by oxygen is about 100%. Thus, fullerene is an attractive compound for optical oxygen-sensing using T_1-T_n absorption change. Figure 10-13 shows the transient absorption spectra of C_{60} in a PS film excited at 532 nm. The band at 750 nm is attributed to the T_1-T_n absorption. Figure 10-14 shows typical decays of the triplet state of C_{60} (monitored at 750 nm) under argon (1) and oxygen (2). The decays were followed by first-order kinetics in the absence and presence of oxygen. The decay of absorption of the photoexcited triplet state of C_{60} under oxygen was fast compared to that under argon, indicating that effective quenching of the excited triplet state of C_{60} by oxygen occurred. This result shows C_{60} is also a useful oxygen-sensing material.

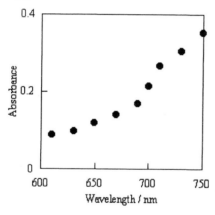

Figure 10-13. Transient absorption spectrum of the photoexcited triplet state of C_{60} in PS film.

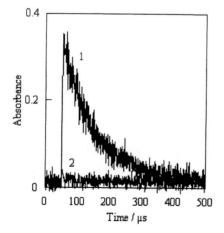

Figure 10-14. Decay curve of the photoexcited triplet state of C_{60} in PS film under 0% (1) and 100% (2) oxygen.

Figure 10-15 shows the relationship between the photoexcited triplet lifetime of C_{60} and oxygen concentration. The triplet lifetimes of C_{60} were calculated from the first-order plot. The lifetime decreases with increase in oxygen concentration. Figure 10-16 shows a Stern–Volmer plot of C_{60} in a PS film. The Stern–Volmer plot of C_{60} exhibits considerable linearity ($r^2 = 0.999$); K_{sv} is estimated to be 0.50%$^{-1}$. The limit of oxygen concentration detection was about 0.3%.

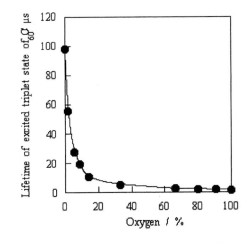

Figure 10-15. Lifetime change of the photoexcited triplet state of C_{60} in PS film under various oxygen concentrations.

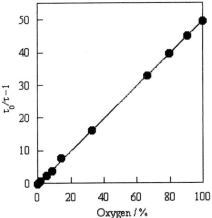

Figure 10-16. Stern–Volmer plot for C_{60} in PS film.

The photostability of the sensing film is an important factor for applications to optical sensors. The fact that no spectrum change was observed after laser irradiation indicates that C_{60} in a PS film has good photostability under laser irradiation.

10.1.4 Optical Oxygen Sensors Using an Ultra-Thin Film of a Macromolecule Dye

Many optical oxygen sensors are composed of organic dyes, such as polycyclic aromatic hydrocarbons (pyrene and its derivatives, quinoline and phenanthrene) and organometallic compounds, immobilized in an oxygen-permeable polymer (e.g. silicone polymer and polystyrene). Since organic dyes directly interact with polymer molecules, the properties of sensing films depend strongly on the properties of the polymer matrices. To overcome these problems, chemisorption films and Langmuir–Blodgett (LB) films have been used. Much interest has been shown in these techniques as possible effective methods for designing a solid surface with well-defined composition, structure and thickness for interfacial optical studies (for the preparation of a macromolecular dye layer on an alumina plate see Experiment 10-3, Section 10.1.5) [82–85]. Chemisorption films are formed using spontaneous binding between a thiol group and a metal surface (Au, Ag or Pt) or between a carboxyl group and a metal oxide surface (e.g. Al_2O_3, Fe_2O_3 or TiO_2). As the chemisorption film technique is very convenient, this film is widely used for optical and optoelectronic devices. Among the various available chemisorption film techniques, the use of a compound with a carboxyl functional group is the most widely used technique for preparing a chemisorption film of organic compounds on a metal aluminum oxide. As the sensing dyes are arranged on the solid surface directly by using this technique, the chemisorption film may possess a low diffusion barrier for oxygen. Thus, a highly sensitive device for oxygen-sensing can be obtained by using a chemisorption film. In this section the fabrication of TCPP metal complexes (platinum complex, PtTCPP; palladium complex, PdTCPP; metal-free complex, H_2TCPP) and pyrene derivatives as luminescence probe for oxygen-sensing, chemisorption films onto alumina plate and those optical oxygen-sensing properties were described (experiment 10−3, chapter 10.1.5).

TCPP metal complex chemisorption films were prepared as follows. An alumina plate was immersed in 0.1 mmol dm^{-3} of a TCPP metal complex in ethanol solution at room temperature for 24 h. The plate was then washed with water and ethanol several times. The TCPP metal complex that had physically adsorbed onto the alumina was removed by ultrasonication, and then the film was dried under vacuum overnight.

TCPP metal complex chemisorption films showed strong phosphorescence when excited at wavelengths attributed to Q bands (538 nm for PtTCPP, 523 nm for PdTCPP and 548 nm for H_2TCPP), as shown in Fig. 10-17. The phosphorescence spectra shapes of PtTCPP (a), PdTCPP (b) and H_2TCPP (c) chemisorption films were almost the same as those in the ethanol solution (maxima of phosphorescence spectra at 665 nm for PtTCPP, 701 nm for

PdTCPP and 653 nm for H_2TCPP films, and 663 nm for PtTCPP, 700 nm for PdTCPP and 651 nm for H_2TCPP in ethanol solution). For all TCPP metal complexes, the phosphorescence intensity of the film depended on the oxygen concentration. The intensity decreased with increase in oxygen concentration as shown in Fig. 10-18. This result indicates that the phosphorescence of TCPP metal complex films was quenched by oxygen. The I_0/I_{100} values of PtTCPP, PdTCPP and H_2TCPP films were estimated to be 16.5, 17.7 and 2.58, respectively, indicating that PtTCPP and PdTCPP chemisorption films are highly sensitive devices for oxygen.

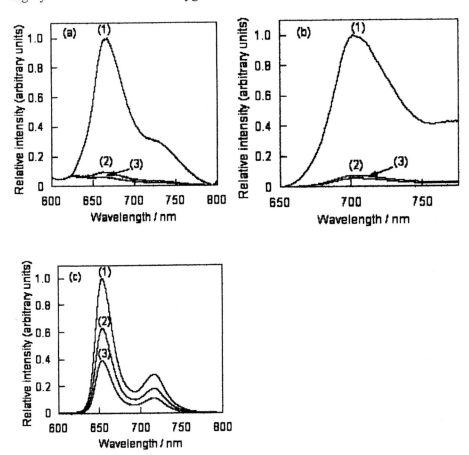

Figure 10-17. Phosphorescence spectra change of TCPP metal complex films under various oxygen concentrations excited at 538 nm for PtTCPP (a), 523 nm for PdTCPP (b) and 548 nm for H_2TCPP (c). (1) argon-saturated, (2) air-saturated and (3) oxygen-saturated conditions.

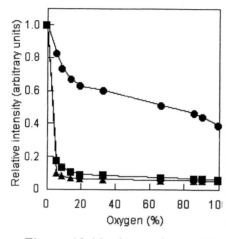

Figure 10-18. Relative phosphorescence intensity changes of TCPP metal complex films under various oxygen concentrations. Excitation and emission wavelength were 538 and 665 nm for PtTCPP (■), 523 and 701 nm for PdTCPP (▲) and 548 and 653 nm for H$_2$TCPP (●), respectively.

Figure 10-19 shows Stern–Volmer plots of the relationships between phosphorescence intensities of TCPP metal complex chemisorption films and oxygen concentration. For all the films, the plots exhibit considerable linearity at lower oxygen concentrations. At higher oxygen concentrations, on the other hand, Stern–Volmer plots of all films are nonlinear because of the simultaneous presence of static and dynamic quenching. Demas et al. reported a multi-site model in which the oxygen-sensing film has some differently oxygen-accessible sites [66]. According to this model, each site has its own individual characteristic quenching constant. Since the observed phosphorescence intensity is the sum of emissions from different oxygen-accessible sites with their own characteristic quenching constants, the Stern–Volmer relationship is given by Eq. (10-5).

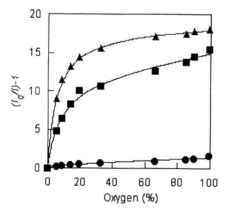

Figure 10-19. Stern–Volmer plot for TCPP metal complex films. PtTCPP (■), PdTCPP (▲) and H$_2$TCPP (●), respectively. The solid lines are the best-fit curve using Eq. 10-4 ($n = 2$).

For all films, the best-fit curve was obtained when n was equal to 2 in Eq. (10-4). In Fig. 10-19, the solid line is the best fit using the above equation

($n = 2$). The correlation factor of the plots, r^2, estimated to be 0.992 for PtTCPP, 0.999 for PdTCPP and 0.999 for H$_2$TCPP by the least-squares method, indicated that the TCPP metal complex thin-film sensors are calibrated by the modified Stern–Volmer equation (Eq. (10-4)). Thus, there are two oxygen-accessible sites in the sensing film; one is a site with good accessibility for oxygen ($K_{sv1} = 1.64\%^{-1}$ and $f_1 = 0.924$ for PtTCPP, $K_{sv1} = 4.56\%^{-1}$ and $f_1 = 0.945$ for PdTCPP, $K_{sv1} = 0.358\%^{-1}$ and $f_1 = 0.405$ for H$_2$TCPP) giving rise to dynamic quenching, and the other is a site with poor accessibility for oxygen ($K_{sv2} = 0.0030\%^{-1}$ and $f_2 = 0.076$ for PtTCPP, $K_{sv2} = 0.0019\%^{-1}$ and $f_2 = 0.055$ for PdTCPP, $K_{sv2} = 0.0079\%^{-1}$ and $f_2 = 0.595$ for H$_2$TCPP) giving rise to static quenching. In the cases of PtTCPP and PdTCPP films, K_{sv2} is very low ($0.0030\%^{-1}$ for PtOEP and $0.00190\%^{-1}$ for PdTCPP) and its contribution is low ($f_2 = 0.076$ for PtTCPP and $f_2 = 0.055$ for PdTCPP) compared with K_{sv1}. As the diffusion rate of oxygen in the film is rapid, the phosphorescence of PtTCPP or PdTCPP is fully quenched by oxygen. Thus, the site with poor accessibility for oxygen is attributed to the static quenching site in the film (emission from background scattering, etc). For H$_2$TCPP, K_{sv2} is also very low compared with K_{sv1}. However, the contribution of K_{sv2} (f_2 value) is larger than that of the f_1 value. The phosphorescence of H$_2$TCPP is also fully quenched by oxygen. However, H$_2$TCPP displays fluorescence and phosphorescence in the same wavelength region. Since no quenching of the fluorescence of H$_2$TCPP by oxygen occurred, the component of K_{sv2} includes the fluorescence of H$_2$TCPP, the emission from background scattering, and so on.

Among polycyclic aromatic hydrocarbons, pyrene derivatives display strong fluorescence with high quantum yields and long lifetimes. Among pyrene derivatives, pyrene-1-butyric acid (PBA) is suitable as an optical oxygen sensor using chemisorption films because of the formation of a stable film on alumina and the carboxyl group of PBA. In this section, we describe the fabrication of PBA as a fluorescence probe for oxygen-sensing, the fabrication of a chemisorption film on an alumina plate and the optical oxygen-sensing properties of a PBA film.

The PBA chemisorption film shows fluorescence at 376, 396 and 474 nm when excited at 365 nm, as shown in Fig. 10-20. The fluorescence spectra shape of the PBA film was almost the same as that in an ethanol solution (0.1 mmol dm^{-3} PBA). Emissions of PBA monomers (376 and 396 nm) and excimers between PBA molecules (474 nm) were observed. Figure 10-20 shows that the fluorescence intensity of the film at 376, 396 and 474 nm depended on the oxygen concentration. Thus, the fluorescence of a PBA film is quenched by oxygen and this film is useful as an optical oxygen sensor. The intensity decreased with increase in oxygen concentration as shown in Fig. 10-21. This result indicates that the fluorescence of the PBA film was quenched by oxygen.

The I_0/I_{100} value of the PBA film was estimated to be 6.14. These results indicate that a PBA film is a highly sensitive device for sensing oxygen.

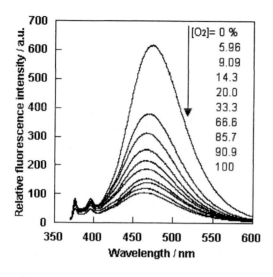

Figure 10-20. Fluorescence spectrum change of PBA chemisorption film under various oxygen concentrations excited at 365 nm.

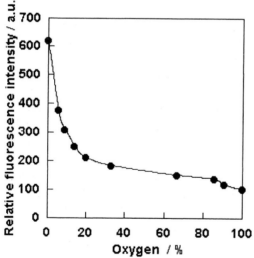

Figure 10-21. The relative fluorescence intensity change of PBA chemisorption film under various oxygen concentrations. Excitation and emission wavelengths were 365 and 474 nm, respectively.

Figure 10-22 shows a Stern–Volmer plot of the relationship between the fluorescence intensities of a PBA film at 474 nm, attributed to PBA excimer emission, and the oxygen concentration. At higher PBA concentrations, the emission of the PBA excimer is sensitized by energy transfer from the PBA monomer.

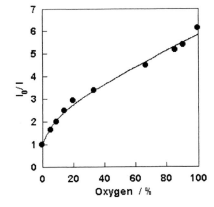

Figure 10-22. Stern–Volmer plot for a PBA chemisorption film. The solid line is a fitting curve using Eq. 10–4. Excitation and emission wavelength were 365 and 474 nm, respectively.

The fluorescence of the PBA monomer observed at 376 and 396 nm was also quenched by oxygen. There are two processes of quenching PBA monomer emission: one is direct quenching by oxygen and the other is quenching by energy transfer to the PBA excimer. However, a Stern–Volmer plot of the relationship between the fluorescence intensities of PBA film at 376 nm, attributed to PBA monomer emission, and the oxygen concentration exhibits considerable linearity, and the Stern–Volmer quenching constant is estimated to be $0.0070\%^{-1}$. Thus, the fluorescence quenching process of the PBA monomer only is due to direct quenching by oxygen. On the other hand, a Stern–Volmer plot between the fluorescence intensities of PBA film at 474 nm, attributed to the PBA excimer emission, and the oxygen concentration exhibits considerable linearity at lower oxygen concentrations ($< 40\%$). At higher oxygen concentrations, on the other hand, a Stern–Volmer plot of the film is nonlinear because of the simultaneous presence of some differently oxygen-accessible sites. The best-fit curve was obtained when n was equal to 2 in Eq. (10-4), as shown in Fig. 10-22. The correlation factor of the plots, r^2, was estimated to be 0.998 by the least-squares method, indicating that the PBA film sensor is also calibrated by the modified Stern–Volmer equation (Eq. (10-4)). Thus, there are two oxygen-accessible sites in a sensing film: a site with good accessibility for oxygen ($K_{sv1} = 0.21\%^{-1}$, $f_1 = 0.820$) and a site with poor accessibility for oxygen ($K_{sv2} = 0.0065\%^{-1}$, $f_2 = 0.180$). These results indicate that the PBA chemisorption film is a highly sensitive device for oxygen compared to the other oxygen sensor using pyrene. An important factor for application of a PBA film as an optical oxygen-sensing material is its photostability. To characterize the photostability of a PBA film, the reflectance spectrum of the film was measured after continuous irradiation using a 150 W tungsten lamp on the film for 24 h. Only a small spectrum change was observed, indicating that the PBA film is stable against irradiation.

10.1.5 Experimental

Experiment 10.1-1: PtOEP Immobilized in a Polymer Film (Section 10.1.2) [58,59]

Preparation of a polymer film with PtOEP: PtOEP immobilized in a polymer film was formed by casting a mixture (0.1 mL) of 10 wt.% polymer film and PtOEP in toluene onto non-luminescent glass slides 1.4×5.0 cm in size. The PtOEP concentration in the film was $\sim 2.9 \times 10^{-5}$ mol dm^{-3}. The films were dried at room temperature and stored in the dark prior to use. The thickness of each film was determined by using a micron-sensitive calliper. The thicknesses of the prepared films were 50–80 μm. For ZnTPP, ZnTFPP and C$_{60}$ immobilized in PS film, the films are prepared by the same method.

Experiment 10.1-2: Synthesis of Fluoropolymers (Section 10.1.2) [62–64]

Synthesis of poly(styrene-co-trifluoroethylmethacrylate) (poly-Sty$_n$-co-TFEM$_m$): Poly-Sty$_n$-co-TFEM$_m$ was synthesized by the following method. Sty, TFEM and AIBN were dissolved in toluene. The reaction mixture was heated at 80 °C for 5 h under a nitrogen atmosphere. After the mixture was cooled to room temperature, the polymer was precipitated with methanol. The solid was collected by filtration, washed with methanol to remove unreacted monomer and finally dried in vacuum. To obtain samples with different molar ratios of Sty and TFEM, the initial concentrations of Sty and TFEM were changed. The composition ratio of Sty and TFEM unit was determined using the molar absorption coefficient of PS at 280 nm. The molecular weight was determined using gel permeation chromatography (column: Plegel 5-μm MIXED-D, Polymer Laboratories; detector: Shimadzu, RID-10A; eluent, THF). The system was calibrated using polystyrene standards. The number-average molecular weight (M_n), the weight-average molecular weight (M_w) and the polydispersity (M_w/M_n) of synthesized polymers are as follows: poly-Sty$_1$-co-TFEM$_{0.069}$: $M_n = 8391$, $M_w = 10,551$ and $M_w/M_n = 1.26$. Poly-Sty$_1$-co-TFEM$_{0.27}$: $M_n = 8403$, $M_w = 10,571$ and $M_w/M_n = 1.25$. Poly-Sty$_1$-co-TFEM$_{0.41}$: $M_n = 8496$, $M_w = 12,165$ and $M_w/M_n = 1.43$. Poly-Sty$_1$-co-TFEM$_{0.62}$: $M_n = 14,194$, $M_w = 18,450$ and $M_w/M_n = 1.30$. Poly-Sty$_1$-co-TFEM$_{1.00}$: $M_n = 9570$, $M_w = 12,726$ and $M_w/M_n = 1.33$. Poly-Sty$_1$-co-TFEM$_{2.50}$: $M_n = 9700$, $M_w = 13,017$ and $M_w/M_n = 1.34$.

Synthesis of poly(styrene-co-pentafluorostyrene) (poly-Sty$_n$-co-PFS$_m$): Poly-Sty$_n$co-PFS$_m$ was synthesized by the following method. Sty, PFS and AIBN were dissolved in toluene. The reaction mixture was heated at 80 °C for 5 h under a nitrogen atmosphere. After the mixture was cooled to room temperature, the polymer was precipitated with methanol. The solid was collected by filtration, washed with methanol to remove unreacted monomer and finally dried in vacuum. To obtain samples with different molar ratios of Sty and PFS, the initial concentrations of Sty and PFS were changed. The composition ratio of Sty and PFS unit was determined using the molar absorption coefficient of PS at 280 nm. The molecular weight was determined using gel permeation chromatography (column: Plegel 5-μm MIXED-D, Polymer Laboratories; detector: Shimadzu, RID-10A; eluent, THF). The system was calibrated using polystyrene standards. Poly-Sty$_1$-co-PFS$_{0.089}$: $M_n = 6532$, $M_w = 10,451$ and $M_w/M_n = 1.60$. Poly-Sty$_1$-co-PFS$_{0.21}$: $M_n = 7718$, $M_w = 13,458$ and $M_w/M_n = 1.74$. Poly-Sty$_1$-co-PFS$_{1.11}$: $M_n = 7859$, $M_w = 11,621$ and $M_w/M_n = 1.48$. Poly-Sty$_1$-co-PFS$_{1.33}$: $M_n = 6378$, $M_w = 10,971$ and $M_w/M_n = 1.72$. Poly-Sty$_1$-co-PFS$_{1.60}$: $M_n = 4078$, $M_w = 6952$ and $M_w/M_n = 1.70$.

Synthesis of poly(isobutylmethacrylate-co-trifluoroethylmethacrylate) (poly-IBM$_n$-co-TFEM$_m$): IBM, TFEM and AIBN were dissolved in toluene and the reaction mixture was heated at 80 °C for 5 h under an argon atmosphere. After the mixture was cooled to room temperature, the polymer was precipitated with methanol. The solid was collected by filtration, washed with methanol to remove unreacted monomer and finally dried in vacuum. To obtain samples with different molar ratios of IBM and TFEM, the initial concentrations of IBM and TFEM were changed. The ratio of IBM and TFEM units was determined by measuring the specific extinction coefficient of poly-IBM at 270 nm ($\varepsilon_{IBM} = 0.0065$ dm^3 cm^{-1} g^{-1}). The molecular weight was determined using gel permeation chromatography (column, Plegel 5-μm MIXED-D, Polymer Laboratories(Amherst, MA, USA); detector: Shimadzu (Tokyo, Japan), RID-10A; eluent, tetrahydrofuran). The system was calibrated using polystyrene standards. Poly-IBM$_1$-co-TFEM$_{0.034}$: $M_n = 24,911$, $M_w = 26,670$ and $M_w/M_n = 1.07$. poly-IBM$_1$-co-TFEM$_{0.48}$: $M_n = 14,425$, $M_w = 15,637$ and $M_w/M_n = 1.08$. poly-IBM$_1$-co-TFEM$_{1.25}$: $M_n = 10,846$, $M_w = 11,527$ and $M_w/M_n = 1.06$. poly-IBM$_1$-co-TFEM$_{1.92}$: $M_n = 12,234$, $M_w = 13,569$ and $M_w/M_n = 1.11$.

Experiment 10.1-3: Macromolecular Dye Chemisorption Layer (Section 10.1.4) [82–85].

Preparation of a macromolecular dye layer on an alumina plate: An alumina plate was prepared by electrically oxidizing the surface of aluminum plate. The aluminum plate (1.2 × 4 cm) was washed with NaOH aqueous solution for 2 min and then was electrically oxidized in 1.0 mol dm^{-3} H$_2$SO$_4$ solution for 30 min. After oxidation, the plate was washed with H$_3$PO$_4$ solution for 10 min. The alumina plates prepared by anodic oxidation were dried in vacuum at 80 °C for 5 h and stored in vacuum prior to use. The dye chemisorption films were prepared as follows. An alumina plate was dipped into dye compounds (0.1 mmol dm^{-3}) in ethanol solution at room temperature for 24 h. After dipping, the plate was washed with water and ethanol several times. The dye compound physically adsorbed onto the alumina was removed by ultrasonication and then the film was dried under vacuum overnight.

10.2 Electrochemical Sensors Based on Self-Assembled Monolayers (SAMs) of Metallomacrocycles

Takeo Ohsaka and C. Retna Raj

10.2.1 Introduction

By definition a chemical sensor is a device that transforms chemical information into an analytically useful signal [86]. It consists of two components, a receptor (a chemical recognition system) and a physico-chemical transducer. The transducer is a detector or electrode that transfers the signal from the output domain of the recognition system to the electrical domain. An electrochemical sensor is a chemical sensor with an electrochemical transducer. There are many varieties of electrochemical sensors and they can be divided into voltammetric, amperometric, potentiometric and conductometric sensors depending on the primary signal used to monitor the analyte concentration [87–90]. A schematic illustration of an electrochemical sensor is given in Fig. 10-23. In electrochemical biosensors a biological recognition system is integrated with an electrochemical transducer. A successful marriage between the biological recognition system, for instance enzymes, and the transducer would result in a third-generation biosensor.

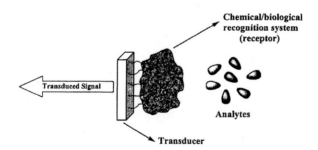

Figure 10-23. Schematic illustration of an electrochemical sensor.

The self-assembling of organo-sulfur compounds on a coinage-metal surface provides a tunable platform for the fabrication of sensor devices. The self-assembled monolayer (SAM) of alkanethiols is one of the versatile interfaces for accomplishing the functionalization and patterning needed for the construction of a biosensor. The flexibility in controlling the thickness of the film and the possibility of introducing functional groups, which in turn control the nature of the interface, make the SAMs very popular in the field of electrochemical sensors. The redox-inactive SAMs of thiols and disulfides with functional groups can be used for the immobilization of biological recognition systems, enzymes, by either covalent or noncovalent (electrostatic, hydrophobic or hydrophilic) interaction [91]. By this means, electrical communication between the electrode and the redox enzyme is easily achieved. A SAM with ionic terminal groups can be utilized as an efficient interface for the measurement of neurochemicals, pH, metal ions, etc [92]. SAMs with redox-active terminal groups find application in the electrocatalytic sensing of biologically important molecules such as NADH and also in constructing mediated biosensors, where the electron transfer between the electrode and enzyme is achieved by the use of surface-confined redox mediators. In this chapter we describe the possible utilization of the SAMs of a macrocyclic nickel(II) complex (MNC) and a metalloenzyme (superoxide dismutase, SOD) for sensing applications.

10.2.2 A Voltammetric Sensor Based on Self-Assembly of a Macrocyclic Nickel(II) Complex

10.2.2.1 Nickel in Biology

Studies on the role of nickel in biological processes have received much attention because of the involvement of nickel in the active site of many hydrogenases [93]. The most noteworthy characteristics of the nickel sites in the hydrogenases are the existence of stable tervalent nickel and the low Ni(III/II) redox potential (~ -150 to -400 mV vs. NHE) [94]. The redox potential of the Ni(III/II) couple in the enzyme is much lower than the potentials of commonly observed Ni(III/II) couples in synthetic complexes [95]. Different approaches have been made to increase the stability of Ni(III) and thus lower the redox potential of the Ni(III/II) redox couple in synthetic complexes [96]. Studies on the stabilization of the Ni(III) state in synthetic complexes could afford information regarding the stabilization of Ni(III) in the enzymes. The nickel redox centers of the enzymes are considered to be the binding site for the catalytic cycle in the biological process [93].

10.2.2.2 Electrochemistry of MNC

The macrocyclic nickel(II) complex, MNC (Fig. 10-24; Experiments 10-4 and 10-5, Section 10.2.4) was shown to self-assemble on a gold (Au) electrode via Au–S bond formation to yield a redox-active monolayer of nickel(II) complex [97]. MNC showed a symmetrical reversible redox wave (Fig. 10-25) corresponding to the Ni(III/II) redox couple at 0.82 V vs. Ag/AgCl in 0.1 M $NaNO_3$ (pH 2).

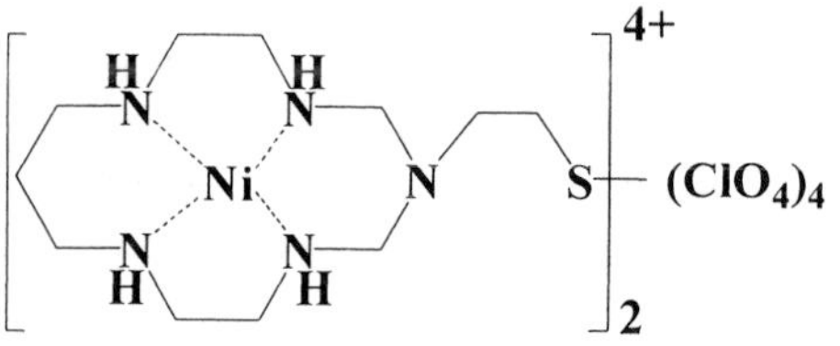

Figure 10-24. Structure of the macrocyclic nickel(II) complex (MNC).

All the electrochemical characteristics of MNC are consistent with those expected for a surface-confined redox species. For instance, the peak shapes are independent of scan rate (ν), the ratio of i_p^a/i_p^c at a given ν is unity, and the peak height varies linearly with ν in the range of 0.02–1 V s^{-1}. The single sharp redox wave for the self-assembly of MNC, which possesses two identical electroactive nickel(II) metal centers, establishes that the formation of the monolayer bound via sulfur to the Au electrode proceeds from disulfide, whereby the produced monolayer is uniform [98]. The surface coverage of the nickel redox centers, 1.44×10^{-10} mol cm^{-2}, was obtained by graphical integration of the anodic wave, which corresponds to a square close-packed monolayer (Fig. 10-26) of nickel redox centers on the Au electrode surface [97]. The SAM of MNC on the electrode may be regarded as a kind of macromolecular assembly held together by van der Waals or coordinative bonds. The macrocyclic ring of MNC is oriented perpendicularly to the plane of the electrode surface and one molecule occupies an area of 118 Å^2. The close packing of nickel(II) redox centers also appears to be reasonable in terms of the size of the perchlorate counter anions, which must be immobilized along with the nickel(II) redox centers in the SAM. The effective radius of an alkyl chain (2.49 Å) in Fig. 10-26 was calculated from the experimental value of an area occupied by a molecule of docosanethiol in the hexagonal close-packing arrangement at the Au electrode surface of 21.4 Å^2 [99]. Considering the relevant dimensions in Fig. 10-26, the free space available within four adjacent hydrocarbon chains is calculated to exceed a square of 8.5 Å $\times$ 8.7 Å.

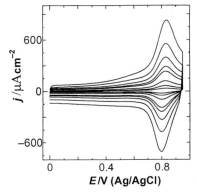

Figure 10-25. Cyclic voltammograms obtained for the SAM of MNC in 0.1 M NaNO$_3$ aqueous solution. Scan rates: 50, 100, 200, 400, 500, 700, 1000 and 1500 mV s^{-1}.

The radius of a hydrated perchlorate anion is 2.24 Å [100], and thus the two perchlorate counter anions of the nickel(II) redox center can reside conveniently in the SAM itself. Note that the coordination of axial ligands produces only small changes in the molecular volume and dimensions; hence the Ni(III/II) redox reaction, where the Ni (III) ion is of octahedral geometry with axially coordinated supporting electrolyte anions/solvent molecules, would not affect the compactness and order of the SAM.

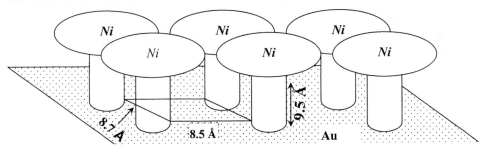

Figure 10-26. Geometry of the proposed square close packing of nickel(II) redox centers on the Au surface.

It is interesting to note that the formal potential ($E^{0'}$) of SAM of MNC depends on the anions present in the supporting electrolyte and is independent of solution pH. The $E^{0'}$ value of MNC in aqueous 0.1 M NaNO$_3$ is 0.82 V vs. Ag/AgCl, which shifted to a less positive potential when the supporting electrolyte was changed to sulfate or phosphate. The change of $E^{0'}$ value was reversible when the MNC electrode was exchanged between aqueous supporting electrolyte solutions. These phenomena could be explained by the fact that the nickel (III) complexes of tetraazamacrocycles exhibit *trans*-octahedral geometry with the axial positions of the octahedral complex occupied by supporting electrolyte anions [101].

Table 10-2. Formal potential of MNC in the presence of different supporting electrolytes[a].

Supporting electrolyte	$E^{0'}$ V vs. Ag/AgCl
NaNO$_3$	0.820
p-Toluene sulfonic acid	0.815
CF$_3$COOH	0.695
Na$_2$SO$_4$	0.530
Na$_2$HPO$_4$	0.370
NH$_4$PF$_6$	0.850

[a] Cyclic voltammograms were recorded in an aqueous solution 0.1 M in supporting electrolyte.

The axial coordination of supporting electrolyte anion at the nickel redox center increases the electron density and hence will favor the oxidation of Ni(II). This increase in the electron density at the metal center favors the stabilization of tervalent nickel and shifts the formal potential to the less positive side. The $E^{0'}$ value varies with the identity of the supporting electrolyte anion irrespective of the magnitude of its charge (Table 10-2). This unique reversible binding of anions with the nickel(III) redox center, would in fact make the SAM of MNC an electrochemical sensor for detecting electroinactive anions. This will be discussed in the following section.

10.2.2.3 Anion Recognition

The recognition of electroinactive anions is a topic with enormous application value because they play numerous indispensable roles in biological and chemical processes. The study of macrocyclic metal complexes, in which a change in electrochemical behavior can be effectively used to monitor interactions with guest species, is an increasingly important area of molecular recognition. In particular the concept of anion coordination at the metal center would, in principle, be an excellent approach to the sensing of electroinactive anions. We have initiated a research project aimed at the design and construction of an innovative electrochemical sensory reagent for anions, based on macrocyclic nickel(II) complexes, as the redox behaviors of these metal complexes are expected to be dependent on the anions [102]. In this section the potential utilization of the SAM of MNC is described.

As described in the previous section, the electrochemical characteristics of the self-assembly of MNC is dependent on the anion present in the supporting electrolyte and is independent of the solution pH, which makes the self-assembly of MNC an excellent and versatile electrochemical anion sensor. The addition of millimolar amounts of different kinds of anions was found to induce different magnitudes of negative shift in the $E^{0'}$ of the SAM of MNC (Table 10-3) [103].

Table 10-3. Voltammetric response for the binding of anions with the SAM of MNC[a].

Anions	$E^{0'}/$ V vs. Ag/AgCl
None	0.82
Cl⁻	0.80
SO_4^{2-}	0.71
$H_2PO_4^-$	0.73
AMP	0.73
ADP	0.64
ATP	0.58

[a] Formal potential of the metal complex was recorded in an aqueous solution of 0.1 M NaNO₃ at a scan rate of 100 mV s⁻¹. The sodium salt of the respective anions was added at 1 mM concentration.

The observed shift in the $E^{0'}$ of the SAM of MNC was found to be reversible when transferring the electrode between different electrolyte solutions. Furthermore, the self-assembly of MNC could sense the biologically important adenosine mono-, di- and tri-phosphates, AMP, ADP and ATP and the results are given in Table 10-3. The observed results highlight the fact that among the different anions tested, the shift in the $E^{0'}$ of MNC is highest for ATP, and among the electroinactive anions the response factor is highest for SO_4^{2-} and $H_2PO_4^-$ anions. The self-assembly is very stable and it could be used repeatedly for sensing application.

10.2.2.4 Voltammetric H₂O₂ Sensor

The measurement of H₂O₂ is of great importance as this compound plays an important role in the assay of different substrates of oxidoreductases such as ethanol, lactate, glucose, etc. and in industrial food laboratories.

An electrochemical sensor based on the SAM of MNC can be constructed, based on its electrocatalytic property towards the oxidation of H₂O₂. The electrocatalytic activity of the self-assembly of MNC towards H₂O₂ in an aqueous solution of 0.1 M NaNO₃ (pH 2.7) is illustrated in Fig. 10-27 [104]. As described earlier, the SAM of MNC shows a symmetrical redox wave at 0.82 V vs. Ag/AgCl in the absence of H₂O₂. In the presence of H₂O₂ an enormous increase in the anodic peak current (I_p^a) is observed with the disappearance of the cathodic peak. The I_p^a for the electrocatalytic oxidation of H₂O₂ does not increase linearly with the square root of scan rate, $v^{1/2}$, showing that the electron transfer is not diffusion controlled but catalytic-reaction controlled. The anodic peak potential (E_p^a), of the electrocatalytic oxidation of H₂O₂ (0.87 V vs. Ag/AgCl) is found to be more positive than that of MNC (0.82 V), indicating that the electroproduced Ni(III) complex is involved as an intermediate. Moreover, the electrocatalytic property of

MNC depends on the concentration of the supporting electrolyte anion. For instance, the I_p^a for the electrocatalytic oxidation of H_2O_2 is relatively large at low concentrations of electrolyte, showing that the supporting electrolyte NO_3^- anions significantly take part in the oxidation process.

As can be seen from Fig. 10-27 the CVs of MNC shows an unusual sharp "inverted V" shape in both anodic and cathodic scans. Generally, the cyclic voltammograms for an electrocatalytic oxidation mediated by a surface immobilized species show an anodic peak in the anodic scan without any anodic current in the reverse scan.

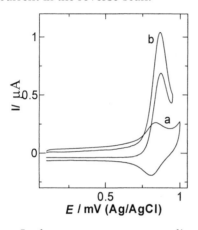

Figure 10-27. Cyclic voltammograms of MNC in an aqueous solution of 0.1 M $NaNO_3$ (pH 2.7) in the absence (a) and presence (b) of 1 mM H_2O_2. Scan rate 100 mV s^{-1}.

In the present case, an anodic peak is observed for the electocatalytic oxidation of H_2O_2 during anodic as well as cathodic scans, showing that the electrocatalysis at the SAM of MNC is a potential-dependent process but not a conventional redox state-dependent process. Furthermore, in the hydrodynamic voltammetric studies, the SAM of MNC shows an "inverted V" shape cyclic voltammogram in the presence of H_2O_2 instead of an S-shaped curve with a plateau current, which is purely controlled by the mass transport of the substrate molecules [105]. These observations establish that the electrocatalytic activity of the MNC towards the oxidation of H_2O_2 is dependent on the applied potential, and the transport of the substrate molecules to the electrode surface is not the cause of the observed unique voltammetric behavior. Thus the electrocatalytic oxidation of H_2O_2 at the SAM of MNC follows an inner-sphere electron-transfer mechanism as illustrated in Figure 10-28. Since the Ni(III) complexes are of octahedral geometry with axially coordinated solvent molecules/anions [101], the electroproduced Ni(III) complex of SAM is considered to undergo an equilibrium with the nitrate-coordinated intermediate X. Since the electrocatalytic oxidation of H_2O_2 by the SAM of MNC is an inner-sphere electron-transfer process, the intermediate Y is formed in the presence of H_2O_2 by the substitution of NO_3^- anion by H_2O_2 in the intermediate X or directly from the uncoordinated Ni(III) complex. Intramolecular electron transfer from the hydroperoxide anion to the Ni(III) metal center leads to the formation of Z. The electrocatalytic behavior of MNC towards the oxidation of H_2O_2 at different concentrations of NO_3^- can be explained by considering Fig. 10-28. High concentrations of $NaNO_3$ lead to the formation of more and more nitrate-coordinated intermediate X and to stabilization of X over intermediate Y. The stabilization of X and the shift in the equilibrium between "X" and "Y" toward "X" could slow down the electrocatalytic oxidation of H_2O_2 by the SAM of MNC. At suitably low concentration of $NaNO_3$ (10 mM), the intermediates X and Y seem to be not stabilized so that all the H_2O_2 molecules diffusing to the monolayer-solution interface quickly undergo electrocatalytic oxidation.

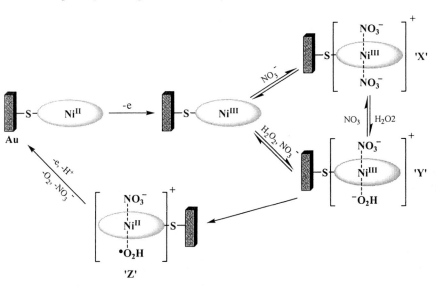

Figure 10-28. Reaction mechanism for the electrocatalytic oxidation of H_2O_2 by the self-assembly of MNC in aqueous solution of $NaNO_3$.

10.2.2.5 Voltammetric NADH Sensor

The electrochemical sensing of NADH is of great interest in the development of a dehydrogenase-based amperometric biosensor owing to the ubiquitous use of NADH as a cofactor for over 300 enzymes and in the fine chemicals industry using NAD^+-dependent biocatalysts [106]. The oxidation of NADH at bare and modified electrodes has been well studied and the oxidation process is dependent on the nature of the electrode used. The direct electrochemical oxidation of NADH at the bare electrode, irrespective of its nature, requires a high overpotential, despite the formal potential of the NAD^+/NADH redox couple at pH 7, which is reported to be

–0.56 V (SCE) [107]. The high overpotential observed in the direct oxidation is believed to be due to the very high potential of the $NADH^{+}/NADH$ redox couple, and fouling of the electrode surface by the adsorption of oxidized product [108,109]. The high overpotential for the oxidation of NADH can be considerably decreased through the use of pretreated electrodes or redox mediators either in solution or immobilized on the electrode surface, which can catalyze the oxidation process [109–115]. Here we describe the potential application of the self-assembly of MNC for the sensing of NADH.

The electrocatalytic oxidation of NADH in 0.1 M $NaNO_3$ by the self-assembly of MNC is illustrated in Fig. 10-29. A dramatic enhancement of the anodic peak current consistent with a very strong electrocatalytic effect is observed in the presence of NADH (Fig. 10-29b). Also no cathodic peak current is observed for the SAM of MNC in the presence of NADH, which indicates that the electroproduced Ni(III) complex is effectively consumed in the catalysis.

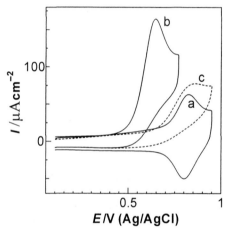

Figure 10-29. Cyclic voltammograms of MNC (a,b) and bare Au (c) in the absence (a) and presence (b,c) of 500 µmol L^{-1} of NADH in 0.1 mol L^{-1} $NaNO_3$. Scan rate 100 mV s^{-1}.

At the self-assembly of MNC a 210 mV decrease in the overpotential with respect to the bare electrode for the NADH oxidation was observed. The anodic peak potential (E_p^a) for the electrocatalytic oxidation of NADH is 0.64 V vs. Ag/AgCl at a potential scan rate of 100 mV s^{-1} and shifts to more positive potential region as the scan rate is increased. At the bare Au electrode the oxidation of NADH occurs at 0.85 V vs. Ag/AgCl. The plot of peak current against square root of scan rate is linear, indicating the electron transfer is diffusion-controlled. The E_p^a for the electrocatalytic oxidation of NADH is 180 mV less positive than that of the MNC itself. Although such a negative shift is expected whenever the electroproduced Ni(III) is rapidly consumed in the electrocatalysis, it is quite surprising to observe such a large negative shift for the catalysis at the self-assembly of MNC. The apparent catalytic rate constant for the oxidation of NADH at the SAM of MNC was calculated using the Koutecky–Levich equation and was found to be 2×10^8 mL mol^{-1} s^{-1}. This value is relatively higher than that reported in the literature and is explained by considering an inner-sphere electron transfer mechanism for the oxidation of NADH.

The electrocatalytic oxidation of NADH has also been carried out in the presence of other supporting electrolytes such as Na$_2$SO$_4$ and phosphate buffer solutions (Table 10-4). Although $E^{0'}$ of the Ni(III/II) surface redox wave shifts toward the less positive potential when the supporting electrolyte is changed from NaNO$_3$ to Na$_2$SO$_4$, $E_p{}^a$ for the oxidation of NADH becomes more positive.

Table 10-4. Formal potentials of SAM of MNC and anodic peak potentials, current densities and apparent rate constants (k_{cat}) for the electrocatalytic oxidation of NADH at SAM of **1** in different supporting electrolytes[a].

Supporting electrolyte	$E^{0'b}$ V (Ag/AgCl)	$E_p{}^{ab}$ V (Ag/AgCl)	Current density,[b] μA cm^{-2}	k_{cat} cm^3 mol^{-1} s^{-1}
0.1 M NaNO$_3$	0.820	0.640	150	2.0×10^8
0.1 M NaNO$_3$+ 2 mM NaH$_2$PO$_4$	0.707	0.691	133	6.2×10^7
0.1 M NaNO$_3$ + 0.7 mM ATP	0.638	0.730	135	5.3×10^7
0.033 M Na$_2$SO$_4$	0.530	0.700	130	7.0×10^7
0.1 M phosphate buffer	0.393	-	-	-

[a] [NADH] : 0.5 mM; ionic strength : 0.10. [b] obtained at the scan rate of 100 mV s^{-1}.

The catalytic current obtained in Na$_2$SO$_4$ was relatively low when compared with that in NaNO$_3$ (Fig. 10-30A and Table 10-4). The decrease in the catalytic current in Na$_2$SO$_4$ could be due to the binding of SO$_4{}^{2-}$ anions at the Ni(III) redox center. In the case of 0.1 M phosphate buffer, the monolayer-modified electrode scarcely catalyzes the oxidation of NADH (Fig. 10.30B) and it could be due to the very strong binding of the phosphate anions at the Ni(III) redox center. The apparent rate constant for the catalytic oxidation of NADH in Na$_2$SO$_4$ was found to be relatively low (7.0×10^7 mL mol^{-1} s^{-1}) when compared with that obtained in NaNO$_3$ (the ionic strength of both solution is 0.10). Similarly the rate constants obtained in NaNO$_3$ containing other anions are relatively low when compared to that obtained in NaNO$_3$ alone (Table 10-4). As the NADH molecule contains two phosphate groups, the interaction of NADH through the phosphate groups at the Ni(III) redox center is expected. However, the competitive binding of other supporting electrolyte anions at the Ni(III) redox center could have hindered the interaction with NADH and therefore the electrocatalytic rate constant is decreased.

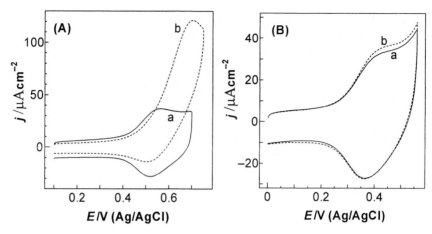

Figure 10-30. Cyclic voltammograms of MNC in the absence (a) and presence (b) of NADH (500 µM) in (A) aqueous solution of 0.033 M Na$_2$SO$_4$ and (B) 0.1 M phosphate buffer. Scan rate 100 mV s^{-1}.

Since the supporting electrolyte anions have the key role in the electrocatalysis, the following reaction sequences can be proposed for the observed electrocatalysis at the monolayer-modified electrode. Initially the Ni(II) is oxidized to Ni(III) and the NADH molecule interacts through its phosphate groups with Ni(III) to form an intermediate. Most probably the intermediate thus formed undergoes an intramolecular electron transfer followed by a very fast deprotonation of the cation radical (NADH$^{+\cdot}$) to produce the radical (NADH$^{\cdot}$). The next electron-transfer step to produce NAD$^+$ could occur either at the same nickel center or at the adjacent nickel center. Since the heterogeneous electron transfer reaction of SAM of MNC is very fast (1.3 × 10^3 s^{-1}), the second electron transfer most likely occurs at the same nickel center. Thus electrocatalysis might occur through a unique interaction of NADH with the Ni(III) redox center. The electrocatalytic rate for the oxidation of NADH at the monolayer-modified electrode at different concentrations of NADH was found to decrease with increasing concentration of NADH, which supports the idea of interaction between the NADH and the Ni(III) redox center. A similar decrease in the rate constant and the formation of a charge-transfer complex between NADH and the catalyst have been reported for the oxidation of NADH at electrodes modified with phenoxazine and phenothiazine dyes [112].

10.2.2.6 Voltammetric pH Sensor

The fabrication of a miniaturized pH sensor for the measurement of pH in the chemical and biological systems has received considerable interest. pH-sensitive glass electrodes and other fiber-optic and electrochemical pH sensors were employed [116–118]. Here we describe a proof-of-concept result demonstrating a new approach based on an electrode derivatized with two individually addressable redox components, a two-terminal voltammetric sensor with internal reference for the measurement of solution pH. One of the electrode-bound components is insensitive to the solution pH and therefore can serve as a reference molecule whereas the other is sensitive to the pH of the medium and would serve as the indicator. The measurement is accomplished by knowing the difference between the peak potentials (ΔE_p) of the reference and indicator molecules, and the magnitude of ΔE_p can be related to the solution pH. Rubinstein first introduced this concept using a polymer-bound pH-sensitive quinone derivative and a pH-independent redox species Ru(bpy)$_3^{2+}$ [119]. Later Wrighton et al. [120] and Willner et al. [121] extended that idea using a different reference molecule, ferrocene. In all these reports the quinone redox couple served as the pH indicator. The accuracy of pH measurement was compromised by the very broad peaks of both reference and indicator molecules and the surface-confined quinone system is not ideal because of the observed large peak separation between the anodic and cathodic peaks. Moreover, as the reduced form of quinone is known to react with oxygen, the dissolved oxygen interferes with its voltammetric measurement.

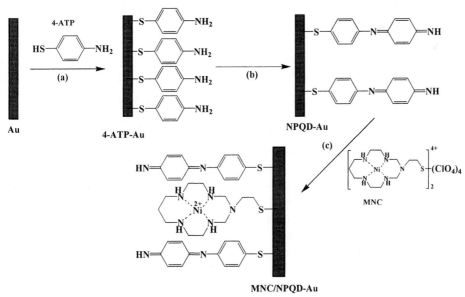

Figure 10-31. Schematic illustration of the construction of a pH sensor based on the SAMs of NPQD and MNC on a Au electrode. (a) 10 mM 4-ATP in methanol for 1 h. (b) potential cycling between −0.2 and 0.7 V vs. Ag/AgCl in 0.1 M phosphate buffer (pH 7.2). (c) 0.6 mM MNC in methanol for 12 h.

In our approach, an electrochemical interface was created by modifying a Au electrode with macrocyclic nickel(II) complex (MNC) and a diimine (NPQD) by a self-assembling technique. The former serves as a reference whereas the latter serves an indicator molecule [122]. The advantage of the present system is that the dissolved oxygen does not influence the redox potential of either reference or indicator molecules and therefore dissolved oxygen need not be removed before measurement. Figure 10-31 illustrates the fabrication of the bifunctional pH sensor. The self-assembly of NPQD was electrochemically created on the electrode surface [123] and it shows a redox response at 0.23 V vs. Ag/AgCl in neutral pH (Fig. 10-32A, see Experiment 10-6, Section 10.2.4).

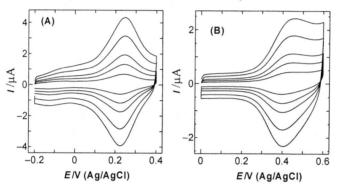

Figure 10-32. Cyclic voltammograms of surface-confined (A) NPQD and (B) MNC in 0.1 M phosphate buffer at different scan rates. Scan rate: 100, 200, 300, 500 and 700 mV s^{-1}.

The redox peak shifts by -56 ± 2 mV when the solution pH increases by one unit, indicating that the redox reaction involves two protons and two electrons. As shown in Fig. 10-32B the self-assembled MNC shows a reversible redox peak characteristic of a surface-confined Ni(II)/Ni(III) redox couple with $E^0 = 0.39 \pm 0.01$ V vs. Ag/AgCl in 0.1 M phosphate buffer (pH 7.2) and the formal potential is insensitive to the solution pH. The surface coverages of NPQD and MNC were estimated to be $(1.3 \pm 0.2) \times 10^{-10}$ and $(1.4 \pm 0.2) \times 10^{-10}$ mol cm^{-2}, respectively.

The electrochemical response obtained for the bifunctional electrode is shown in Fig. 10-33. The voltammogram consists of two couples of redox waves corresponding to the electrochemical processes of NPQD and MNC units. A linear dependence of peak currents as a function of sweep rate was observed for both waves, implying that the redox species are confined on the electrode surface. Here it is considered that the adsorption of MNC occurs at the gold electrode through the defect sites of the NPQD monolayer as shown in Fig. 10-31. At a given pH the formal potential of NPQD at the MNC/NPQD-Au electrode remained the same as at the NPQD-Au electrode. Protons do not take part in the redox reaction of MNC and therefore it is insensitive to the solution pH (Fig. 10-34).

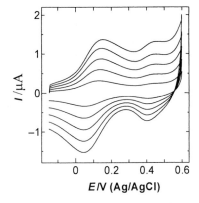

Figure 10-33. Voltammetric response of the bifunctional electrode in 0.1 M Na₂HPO₄ (pH 9.2) at different scan rates. Scan rate: 100, 200, 300, 400 and 500 mV s⁻¹.

On the other hand, as the redox reaction of NPQD involves protons, it is sensitive to the solution pH. The cyclic voltammetric response of the bifunctional electrode at different pH values is shown in Fig. 10-35. The redox potential of MNC is unaltered at all pH values, whereas the redox response of NPQD is sensitive to the solution pH in accordance with Fig. 10-34. For instance, the formal potential of NPQD was negatively shifted by 112 ± 2 mV as the pH was changed from 6.6 to 8.5. The formal potentials of both MNC and NPQD are insensitive to dissolved oxygen, which was demonstrated by measuring their values in the presence and absence of oxygen.

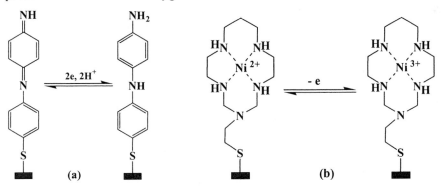

Figure 10-34. Redox reactions of the SAMs of (a) NPQD and (b) MNC on a Au electrode.

The use of a microelectrode assembly with MNC and NPQD was also tested for the measurement of solution pH in the same way as the macro electrode. A very similar result was obtained in each case. The advantages of the present system rest on the fact that, unlike the previous systems, dissolved oxygen does not influence the redox processes of either species and therefore the electrolyte solutions need not be deaerated before the measurements. Furthermore, unlike the quinone system, which shows a large ΔE_p value, the present NPQD system is reversible at all the pH values examined. This electrode can be effectively used to measure the pH values of physiological solutions.

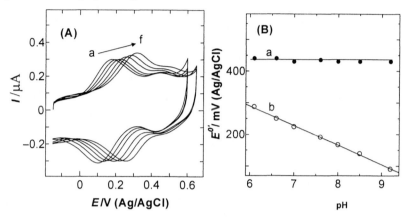

Figure 10-35. (A) Cyclic voltammograms obtained for the bifunctional self-assembly at different pH values. (a) 8.5, (b) 8.04, (c) 7.63, (d) 7.05, (e) 6.6 and (f) 6.08. Scan rate 100 mV s^{-1}. (B) Plots of formal potentials of MNC (a) and NPQD (b) against the solution pH.

10.2.3 Electrochemical Superoxide Biosensor Based on Self-Assembly of Metalloenzymes

The superoxide anion (O_2^-), the primary species of the so-called reactive oxygen species (ROS), is generated as a reduced intermediate of molecular oxygen in significant quantities in a variety of biological systems. Under normal physiological conditions, O_2^- undergoes disproportionation by noncatalytic and enzymatic reactions, resulting in a rather low and undetectable endogenous physiological concentration. However, increases in the activity of O_2^- occur in response to traumatic brain injury ischemia-reperfusion and hypoxia, and O_2^- may be involved in the etiology of aging, cancer, and progressive neurodegenerative diseases such as Parkinson's disease. Thus O_2^- is of great importance in the determination of cellular damage, and thus, a specific and sensitive method for durable and reliable measurement of O_2^- is essential.

Many attempts have been made to develop a sensor for the detection of O_2^-, including spectrophotometric and electrochemical techniques [124]. Electrochemical methods have drawn extensive attention because they have many advantages such as direct detection, high sensitivity, measurement *in vivo*, and so on. Third-generation biosensors are proposed as the most attractive in biosensor development, and much effort has been paid to the design of a third-generation biosensor for O_2^- based on the reaction of O_2^- with cyt *c* [125]. The inherent property of cyt c as a peroxidase to reduce H_2O_2 [126] makes the sensor suffer from interference from H_2O_2 either in the XOD/xanthine O_2^- generating system or endogenously coexisting in biological systems. This greatly limited the application of cyt c-based O_2^- sensors for detection of O_2^- in biological systems.

Superoxide dismutases (SODs) are ubiquitous metalloenzymes in oxygen-tolerant organisms and are regarded as "natural macromolecular proteins". They efficiently protect the organism from the toxic effect of the superoxide ion by efficiently catalyzing its dismutation to O_2 and H_2O_2 via a cyclic oxidation–reduction electron-transfer mechanism. The fabrication of a third-generation electrochemical superoxide biosensor based on the direct electrochemical redox reaction of SOD would be very promising, as SOD has been well addressed for the dismutation of O_2^- to O_2 and H_2O_2 with strong activity and great specificity [127]. However, the direct electron transfer between the enzyme and the electrode is very slow and it is rather difficult to use such strategy. Therefore the construction of a third-generation superoxide sensor based on the redox electrochemistry of SOD would be successful only if a direct electron transfer is achieved. One of the fascinating approaches that accelerate the electron transfer between electrode and metalloenzyme is the use of promoters. We have explored the possible application of self-assembled monolayers of cysteine (Cys) as a promoter to study the direct electron transfer of SOD. Surprisingly, the direct electron transfer is successfully achieved and it opens the gate for the fabrication of third-generation superoxide sensors (see Experiment 10-7, Section 10.2.4) [128]. Here an important issue needs to be addressed before the fabrication of the sensor device: the dismutation of O_2^- catalyzed by SOD produces H_2O_2 and this would be detected, usually via its oxidation, for the measurement of O_2^-. The oxidation of H_2O_2 requires a high potential and it causes the simultaneous oxidation of some coexisting electroactive species in biological samples, e.g. ascorbic acid and uric acid. More importantly, H_2O_2 is a metabolite of the degradation of O_2^- as well as a product from the enzymatic reaction of endogenous oxidases such as monoamine oxidase and L-amino acid oxidase. Thus interference from the coexisting electroactive species and endogenous H_2O_2 greatly limits the practical applications of the sensors. Therefore a bidirectional sensor based on the direct electron transfer of SOD, which can detect O_2^-, would be of great value.

10.2.3.1 Direct Electron Transfer of SOD at Cys/Au Electrode

The voltammetric response of SOD at bare and Cys/Au electrodes is depicted in Fig. 10-36. At the Cys/Au electrode SOD shows a quasi-reversible redox peak whereas no electrochemical response was obtained at the bare Au electrode. This suggests that direct electron transfer between SOD and the Au electrode does not occur at the bare electrode but the Cys/Au monolayer on the electrode surface significantly promotes the electron transfer. The formal potential, $E^{0'}$, of SOD was estimated to be 0.065 V (Ag/AgCl), which is slightly smaller than the values observed in the literature [129,130]. The redox peak was found to increase linearly with the concentration of SOD in solution up to 0.56 mM. The redox response of SOD is very stable, and the peak current and peak potential remained the same during the repeated use of the same electrode. When the Cys/Au electrode was transferred to the supporting electrolyte containing no SOD after being used in measuring a SOD solution, it showed a very similar redox response as in Fig. 10-36a, indicating that the SOD is adsorbed on the monolayer assembly. This was further demonstrated by the linear increase of peak current against the scan rate. The peak current is proportional to the scan rate rather than the square root of scan rate, indicating that the electron-transfer process is adsorption controlled.

In order to understand the electrode reaction associated with the redox peak of SOD at the Cys/Au electrode, the Cu- and Zn-free derivatives, E_2Zn_2SOD and Cu_2E_2SOD, of SOD were prepared by removing Cu(II) or Zn(II) from the native SOD and reconstituting Cu_2Zn_2SOD from E_2Zn_2SOD and Cu^{2+}. The absorption spectrum of the reconstituted SOD is in good agreement with that of the native SOD with an absorption maximum of 680 nm [131].

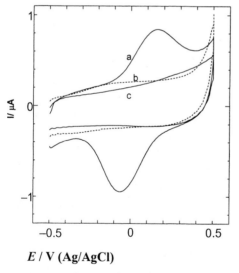

Figure 10-36. Cyclic voltammograms obtained at a Cys/Au (a,b) and a bare Au (c) electrode in 25 mM phosphate buffer in the presence (a,c) and absence (b) of 0.56 mM SOD. Sweep rate 100 mV s^{-1}.

Figure 10-37 shows the voltammetric response of native and reconstituted SODs at the Cys/Au electrode. The E_2Zn_2SOD does not shows any redox response in the potential window of –0.5 to 0.5 V vs. Ag/AgCl. On the other hand, as can be seen from the Fig. 10-37 the Cu_2E_2SOD shows a reversible redox peak, which is very similar to that of the native SOD. This implies that the observed direct electron transfer of SOD at the Cys-modifed electrode is associated with the redox reaction of the copper moiety (not of the zinc moiety). It further demonstrates that the copper moiety is the active center not only for superoxide ion dismutation but also for the direct electron transfer of SOD promoted by the SAM of Cys, which forms a strong basis for the development of SOD-based amperometric biosensors for superoxide ion. This will be discussed in the following section.

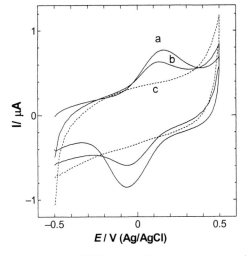

Figure 10-37. Cyclic voltammograms obtained at a Cys/Au electrode in 25 mM phosphate buffer containing (a) 0.56 mM SOD, (b) 0.56 mM Cu_2E_2SOD and (c) 0.56 mM E_2Zn_2SOD at 100 mV s^{-1}.

Bovine SOD is a dimer made up of two identical subunits (151 amino acid residues, MW: 16,000 Da). It has a net negative charge at pH 7.0 (pI = 4.9). The pK_a values of the –COO$^-$ and –NH$_2$ groups of Cys are 1.71 and 10.78, respectively, and thus the Cys immobilized on the gold electrode via S-Au bonding can be expected to behave as a zwitterion at pH 7.0. Therefore, the orientation of SOD on the gold electrode through the SAM of Cys and the resulting facilitation of its electron transfer may be considered to be not due simply to an electrostatic interaction between the SOD molecule and these functional groups, but also to a unique interaction on the molecular level. For example, the –NH$_2$ and –COO$^-$ groups of Cys are considered to function cooperatively to bind SOD; hydrogen bonding between the –NH$_2$ group and the –COO$^-$ group of Thr 135 occurs and the –COO$^-$ group interacts favorably with the positively charged –NH$^+_3$ of the amino acid residues (e.g. Lys 134 and Arg 141) surrounding the Cu^{2+} site [132], as schematically illustrated in Fig. 10-38. The SOD immobilized on the gold electrode via the Cys monolayer was found to possess an inherent activity for the dismutation of superoxide ion, and it is considered that the SOD is oriented on the gold electrode via the Cys monolayer so as to allow rapid electron transfer to and from the Cu^{2+} active site [128].

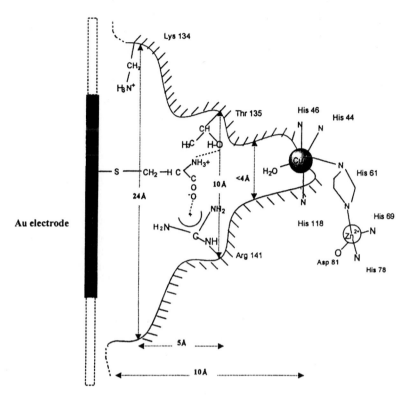

Figure 10-38. Schematic representation of the active-site channel above the Cu(II) site of SOD for facilitated electron transfer.

10.2.3.2 SOD-Based Third-Generation Biosensor for Superoxide Ion

As described above, the self-assembly of Cys promotes the electron transfer between SOD and the Au electrode. This Cys-promoted rapid and direct electron transfer of SOD and its relevance to the redox reaction of the copper complex moiety in SOD formed a strong basis for the development of a SOD-based third-generation biosensor for O_2^- because the copper complex moiety has been well documented as the active site for the catalytic dismutation of O_2^- [128]. Figure 10-39 shows CVs at the bare Au, Cys/Au and SOD/Cys/ Au electrodes in PBS (O_2 saturated) containing 0.002 unit of XOD and 50 μM xanthine, i.e. in the presence of O_2^-. Both cathodic and anodic peak currents corresponding to the redox reaction of the SOD confined on the electrode are significantly increased, compared with those in the absence of O_2^- (Fig. 10-36a) [133–135]. Such a redox response was not observed at the bare Au and Cys/Au electrodes in the presence of O_2^-. The large anodic and cathodic currents starting from about 600 and –300 mV vs. Ag/AgCl were due to the oxidation of uric acid and the reduction of H_2O_2 and O_2, respectively. H_2O_2 and uric acid are coproduced in the xanthine/XOD based O_2^- generating system. The observed increase in the anodic and cathodic current response of SOD/Cys/Au in the presence of O_2^- can be considered to result from the oxidation and reduction of O_2^-, respectively, which are effectively mediated by the SOD confined on the electrode. That is, the reaction mechanism indicated in Eq. (10-6) can explain the enhanced oxidation current observed in the anodic scan.

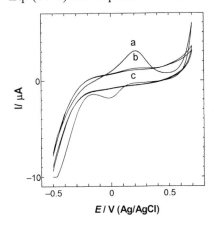

Figure 10-39. Cyclic voltammograms obtained at (a) SOD/Cys/Au, (b) bare Au, and (c) Cys/Au electrodes in O_2 saturated phosphate buffer containing 0.002 unit of XOD and 50 μM xanthine. Sweep rate 100 mV s^{-1}.

SOD (Cu(I)) - e ⟶ SOD (Cu(II))

(10-6)

SOD (Cu(II)) + O_2^- ⟶ SOD (Cu(I)) + O_2

Similarly, the reduction current is enhanced in the cathodic scan according to Eq. (10-7).

SOD (Cu(II)) + e ⟶ SOD (Cu(I))

(10-7)

SOD (Cu(I)) + O_2^- $\xrightarrow{\text{H}^+}$ SOD (Cu(II)) + H_2O_2

Such a bidirectional electromediation (electrocatalysis) by SOD/Cys/Au is essentially based on the inherent specificity of SOD for the dismutation of O_2^-; i.e. SOD catalyzes both the reduction of O_2^- to H_2O_2 and the oxidation to O_2 via a redox cycle of its Cu complex moiety as well as the direct electron transfer of SOD we obtained at Cys/Au. Thus, this coupling between the electrode and enzyme reactions of SOD could facilitate the development of the third-generation biosensor for O_2^-.

By taking into account the possibility of both oxidation and reduction of O_2^- at SOD/Cys/Au, the electrode was polarized at 300 or -200 mV vs. Ag/AgCl for the measurement of O_2^-. Figure 10-40 shows the typical current–time response of SOD/Cys/Au at the applied potential of 300 and -200 mV vs. Ag/AgCl on addition of xanthine and SOD to the O_2-saturated PBS containing 0.002 units of XOD. After a stable background current was obtained under the applied potential conditions, xanthine (50 nmol L^{-1}) was pipetted into the PBS to generate O_2^-. The introduction of xanthine into the electrolyte solution produced a rapid and obvious increase in the anodic current (Fig. 10-40A). To verify that the observed anodic current is attributable to the oxidation of O_2^- rather than to the coproducts of the XOD/xanthine-based O_2^- generating reaction, i.e. uric acid and H_2O_2, 6 μmol L^{-1} SOD was added to the solution since SOD, a selective scavenger of O_2^-, can specifically dismutate O_2^-.

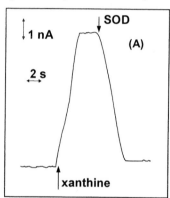
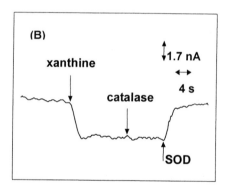

Figure 10-40. (A) Current–time response obtained at a SOD/Cys/Au electrode in O_2-saturated phosphate buffer containing 0.002 unit of XOD upon the addition of 50 nmol L^{-1} xanthine and subsequent addition of 6 μmol L^{-1} SOD. Electrode was polarized at 300 mV. (B) Current–time response of the SOD/Cys/Au electrode O_2-saturated phosphate buffer containing 0.002 unit of XOD upon the addition of 30 nmol xanthine and subsequent addition of 590 units of catalase and 6 μM SOD. The electrode was polarized at -200 mV.

This addition caused the anodic current to decrease by >95% within 6 s, indicating that only O_2^- generated by the XOD/xanthine system is concerned in the anodic current response measured at SOD/Cys/Au. To examine the direct oxidation of O_2^- at SOD/Cys/Au without the above-mentioned redox mediation via SOD, the response of apo-SOD/Cys/Au toward O_2^- was also measured. A much smaller response was obtained at apo-SOD/Cys/Au, indicating that the oxidation of O_2^- at SOD/Cys/Au was based on the SOD enzyme amplification. The minor current response observed at apo-SOD/Cys/Au may be considered to be due to the direct oxidation of O_2^-, independent of the SOD catalytic dismutation, probably because of the small size of O_2^- and its permeation through the apo-SOD/Cys layer and pinholes in the Cys monolayer. As shown in Fig. 10-40B the addition of xanthine to the solution resulted in an obvious increase in the cathodic current when the electrode is polarized at –200 mV, but such a cathodic current was not observed at the bare Au and Cys/Au electrodes. As can be seen from Fig. 10-40, the introduction to the solution of catalase, which has a high catalytic activity toward the dismutation of H_2O_2, did not result in any change in the current response, suggesting that the observed cathodic current is not due to the reduction of H_2O_2 coproduced in the XOD/xanthine-based O_2^- generating reaction. On the contrary, the addition of SOD caused the cathodic current to decrease to the background level. Thus, the observed cathodic response is ascribable to the reduction of O_2^- mediated by the SOD confined on the electrode. The two pathways (i.e. anodic and cathodic processes) of the response of the SOD/Cys/Au to O_2^- through the recognition and amplification by SOD can be schematically illustrated in Fig. 10-41 [133–135].

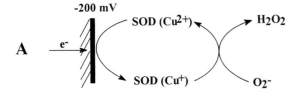

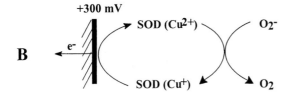

Figure 10-41. Amperometric response of the bidirectional sensor, SOD/Cys/Au towards O_2^- at (a) –200 mV and (b) 300 mV vs. Ag/AgCl.

Well-defined steady-state current responses were obtained at both 300 and –200 mV vs. Ag/AgCl during the successive concentration changes of O_2^-, and the currents increased stepwise with successive additions of xanthine. The generated O_2^- can undergo spontaneous dismutation into O_2 and H_2O_2 under the experimental conditions. The counterbalance between the generation of O_2^- and its degradation results in a constant equilibrium concentration of O_2^-. The steady-state currents at 300 and –200 mV vs. Ag/AgCl were proportional to the rate of O_2^- generation in the examined range of 13–130 nm min^{-1}. The sensitivity of SOD/Cys/Au was found to be 24 and 22 nA cm^{-2} (nM min^{-1})$^{-1}$ at 300 and –200 mV, respectively. The detection limit was evaluated based on a signal-to-noise ratio of 3:1 and calculated to be 5 nm at 300 mV and 6 nm at −200 mV, respectively. The response time of the sensor was measured as the time to reach 95% of the maximum change in response to a step injection of xanthine and found to be less than 6 s. For the stability test, the anodic and cathodic responses for O_2^- generated by the XOD/xanthine system were recorded four times each day and the current responses were found to be constant for at least one week.

10.2.4 Experimental

Experiment 10.2-1: Synthesis of macrocyclic nickel(II) complex (MNC) (Section 10.2.2.2) [136,137]

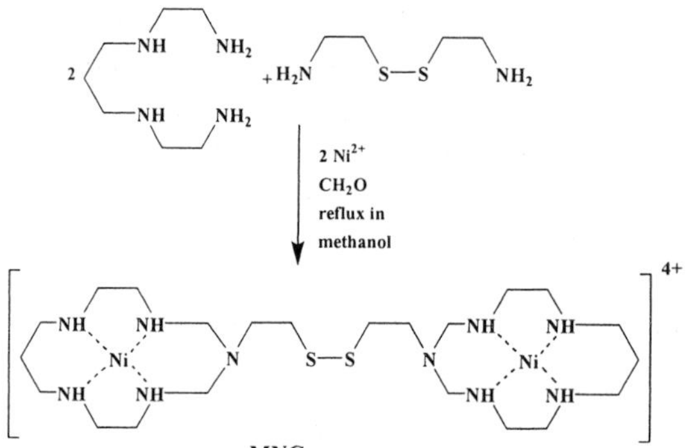

MNC

The macrocyclic nickel(II) complex (MNC), dinickel(II) (2,2′-bis(1,3,5,8,12-pentaazacyclotetradec-3-yl)-ethyl disulfide) perchlorate was synthesized using the following procedure. A sample of NiCl.6H$_2$O (4.8 g) was dissolved in methanol (180 mL) to which N,N-bis(2-aminoethyl)-1-3-propanediamine (3.6 g) was added slowly with stirring. To the resulting solution mixture, formaldehyde (3.4 mL of a 36% solution, 0.04 mol) and 2-aminoethyl disulfide dihydrochloride (2.3 g, 0.01 mol) were added in small portions, and the resulting mixture was heated at reflux for 30 h until a dark reddish-brown solution resulted. The solution was cooled to room temperature and filtered. Water (90 mL) was added to the solution, and the volume was reduced on a rotary evaporator to 70 mL. Excess sodium perchlorate or perchloric acid was added to the filtrate, and the formed pale yellow crystals were recrystallized from hot water. Yield = 20%; Elemental analysis: Calcd. For $C_{22}H_{56}N_{10}Cl_5NaNi_2O_{22}S_2$: C, 22.12; H, 4.69; N, 11.73; Cl, 14.485. Found: C, 22.32; H, 4.64; N, 11.73; Cl, 15.13. IR (Nujol mull): 3195 cm^{-1} (N-H$_{str}$). ^{13}C-NMR (CD$_3$NO$_2$): 67.6 ppm (N-C-N); 35.9 and 41.6 ppm (N–C–C–S); 27.8 and 38.9 ppm (N–C–C–C–N; 48.9 and 50.9 ppm (N–C–C–N). MS (FAB): 937 [M-ClO$_4$]$^+$; 837 [M-2ClO$_4$]$^+$.

Experiment 10.2-2: Self-assembly of MNC on gold electrode (Section 10.2.2.2) [97,104]

The self-assembly of MNC was made by immersing a clean Au electrode into a methanol solution of MNC (0.5 mmol L^{-1}). The prepared MNC-Au electrode was washed with copious amounts of methanol and water before it was subjected to experiments.

Experiment 10.2-3: Self-assembly of *N*-phenylquinone diimine (NPQD) (Section 10.2.2.6) [122,123]

The self-assembly of NPQD was electrochemically created on the Au electrode by the following procedure. First the Au electrode was modified with the self-assembly of 4-aminothiophenol (4-ATP) and the 4-ATP-Au electrode was subjected to potential cycling between –0.2 and 0.7 V vs. Ag/AgCl in 0.1 mol L^{-1} phosphate buffer at a scan rate of 100 mV s^{-1} (5–7 cycles). The prepared self-assembly was characterized by electrochemical and *in situ* FTIR techniques. The mixed self-assembly of MNC and NPQD was fabricated by immersing the NPQD-Au electrode into a methanol solution of 0.6 mmol L^{-1} MNC for 12 h.

Experiment 10.2-4: Preparation of SOD immobilized Au electrode (Section 10.2.3) [134,135]

The self-assembly of cysteine (Cys) was formed by soaking the clean Au electrode in an aqueous solution of 1 mmol L^{-1} Cys for 10 min and rinsing with copious amounts of water to remove the nonchemisorbed Cys. SOD was immobilized on the SOD/Au electrode by soaking it in 25 mmol L^{-1} phosphate buffer containing 0.56 mM SOD or apo-SOD. The prepared SOD electrode was rinsed with water and kept at 4 °C while not being used.

10.3 References

1. D.A. Skoog, D.M. West, F.J. Holler, *Fundamentals of Analytical Chemistry*, Saunders, Philadelphia, 5th edn., 1988, p. 344.
2. L.C. Clark, US Pat., 2 913 386 (1959).
3. T. M. Freeman, W.R. Seitz, *Anal. Chem.* **1981**, *53*, 98.
4. H.D. Hendricks, US Pat., 3 709 663 (1973).
5. X.M. Li, H.Y. Wong, in: Transient D.O. Measurement Using a Computerized Membrane Electrode. Horizons of Biochemical Engineering, S. Aiba (Ed.), , University of Tokyo Press, Tokyo, 1987, p. 213.
6. X.M. Li, H.Y. Wong, US Pat., 4 921 582 (1990).
7. G.J. Mohr, O.S. Wolfbeis, *Anal. Chim. Acta* **1995**, *316*, 239.
8. A.V. Vaughan, M.G. Baron, R. Narayanaswamy, *Anal. Comm.* **1996**, *33*, 393.
9. H.N. McMurray, P. Douglas, C. Busa, M.S. Garley, *J. Photochem. Photobiol. A:*

Chem. **1994**, *80*, 283.

10. S.M. Ramasamy, R.J. Hurtubise, *Anal. Chim. Acta.* **1983**, *152*, 83.
11. M.R. Surgi, in: *Applied Biosensors*, D.L. Wise (Ed.), Butterworth, Guildford, 1989, Ch. 9.
12. J.M. Vanderkooi, D.F. Wilson, *Oxygen transport to tissue VIII.* I.S. Longmuir (Ed.), Plenum, New York, 1986, p. 189.
13. J.I. Peterson, R.V. Fitzgerald, D.K. Buckhold, *Anal. Chem.* **1984**, *56*, 62.
14. J.L. Gehrich, D.W. Lubber, N. Opitz, D.R. Hansmann, W.W. Miller, J.K. Tusa, M. Yafuso, *IEEE Trans. Biomed. Eng.* **1986**, BME-33, 117.
15. G.N. Lewis, M. Kasha, *J. Am. Chem. Soc.* **1944**, *66*, 2100.
16. P.F. Lott, R.J. Hurtubise, *J. Chem. Edu.* **1974**, *51*, A315.
17. S. Fischkoff, J.M. Vanderkooi, *J. Gen. Physiol.* **1975**, *65*, 663.
18. M. Smoluchowski, *Z. Phys. Chem.* **1917**, *92*, 129.
19. W.K. Subczynski, J.S. Hyde, *Biophys. J.* **1984**, *45*, 743.
20. J.M. Vanderkooi, G. Maniara, T.J. Green, D.F. Wilson, *J. Biol. Chem.* **1987**, *262*, 5476.
21. H. Kautsky, A. Hirsch, *Z. Anorg. Allg. Chem.* **1935**, *222*, 126.
22. H. Kautsky, G.O. Mler, *Z. Naturforsch.* **1947**, *2A*, 167.
23. H. Kautsky, A. Hirsch, F. Davidsher, *Ber. Dtsch. Chem. Ges.* **1932**, *65*, 1762.
24. I.A. Zakharov, T.I. Grishaeva, *Zhur. Prikl. Specktrosk.* **1982**, *36*, 980; Engl. Ed. p. 697.
25. Bergmann, *Nature* **1968**, *218*, 396.
26. O.S. Wolfbeis, H. Offenbacher, H. Kroneis, H. Marsoner, *Mikrochim Acta.* **1984**, *I*, 153.
27. H.W. Kroneis, H.J. Marsoner, *Sens. Actuators* **1983**, *4*, 587.
28. D. Lber, N. Opitz, *Sens. Actuators,* **1983**, *4*, 641.
29. M.E. Cox, D. Dunn, *Appl. Optics* **1985**, *24*, 2114.
30. O.S. Wolfbeis, H.E. Posch, H. Kroneis, *Anal. Chem.* **1985**, *57*, 2556.
31. I.S. Longmuir, J.A. Knopp, *J. Appl. Physiol.* **1976**, *41*, 598.
32. M.H. Mitnick, F.F. Jis, *J. Appl. Physiol.* **1976**, *41*, 593.
33. H. Kautsky, *Trans. Faraday Soc.* **1939**, *35*, 216.
34. X.M. Li, K.Y. Wong, *Anal. Chim. Acta* **1992**, *262*, 27.
35. Y.M. Liu, R. Pereiro-Garcia, M.J. Valencia-Gonzalez, M.E. Diaz-Garcia, A. Sanz-Medel, *Anal. Chem.* **1994**, *66*, 836.
36. J.M. Charlesworth, *Sens. Actuators B* **1994**, *22*, 1.
37. N. Velasco-Garcia, R. Pereiro-Garcia, M. Diaz-Garcia, *Spectrochim. Acta* **1995**, *51A*, 895.
38. D.M. Papkowsky, G.V. Ponomarev, W. Trettnak, P. O'Leary, *Anal. Chem.* **1995**, *67*, 4112.
39. T. Ishiji, M. Kaneko, *Analyst* **1995**, *120*, 1633.
40. Sharma, O.S. Wolfbeis, *Appl. Spect.* **1988**, *42*, 1009.
41. E.D. Lee, T.C. Werner, R. Seitz, *Anal. Chem.* **1987**, *59*, 279.
42. S.M. Ramasamy, R.J. Hurubise, *Anal. Chim. Acta* **1983**, *152*, 83.
43. H.W. Kroneis, H.J. Marsoner, *Sens. Actuators.* **1983**, *4*, 587.
44. W. Xu, R. Schmidt, M. Whaley, J.N. Demas, B.A. DeGraff, E.K. Karikari, B.L. Farmer, *Anal. Chem.* **1995**, *67*, 3172.

45. J. Olmsted, *Chem. Phys. Lett.* **1974**, *26*, 33.
46. P. Hartmann, M.J.P. Leiner, M.E. Lippitsch, *Anal. Chem.* **1995**, *67*, 88.
47. M.G. Sasso, F.H. Quina, E.J.H. Bechera, *Anal. Biochem.* **1986**, *156*, 239.
48. E. Singer, G.L. Duveneck, M. Ehrat, M. Widmer, *Sens. Actuators A* **1994**, *41–42*, 542.
49. E.R. Carraway, J.N. Demas, B.A. DeGraff, J.R. Bacon, *Anal. Chem.* **1991**, *63*, 332.
50. J. R. Bacon, J. N. Demas, *Anal. Chem.* **1987**, *59*, 2780.
51. X.M. Li, H.Y. Wong, *Anal. Chim. Acta* **1992**, *262*, 27.
52. W.Y. Xu, K.A. Kneas, J.N. Demas, B.A. DeGraff, *Anal. Chem.* **1996**, *68*, 2605.
53. L. Sacksteder, J.N. Demas, B.A. DeGraff, *Anal. Chem.* **1993**, *65*, 3480.
54. D.B. Papkovsky, G.V. Ponomarev, W. Trettnak, P. O'Leary, *Anal. Chem.* **1995**, *67*, 4112.
55. J. Vanderkooi, G. Maniara, J. Green, D.F. Wilson, *J. Biol. Chem.* **1987**, *262*, 5476.
56. Mills, A. Lepre, *Anal. Chem.* **1997**, *69*, 4653.
57. K. Kalyanasundaram, *Photochemistry of Polypyridine and Porphyrin Complexes*, Academic Press, New York, 1992, p. 500.
58. S-K. Lee, I. Okura, *Anal. Sci.* **1997**, *13*, 535.
59. S-K. Lee, I. Okura, *Anal. Commun.* **1997**, *34*, 185.
60. N. Yi-Yan, R.M. Felder, W.J. Koros, *J. Appl. Polym. Sci.* **1980**, *25*, 1755.
61. A.G. Pittman, *Fluoropolymers*, Wiley-Interscience, New York, 1972, p. 446.
62. Y. Amao, K. Asai, T. Miyashita, I. Okura, *Polymer J.* **1999**, *31*, 1267.
63. Y. Amao, K. Asai, T. Miyashita ,I. Okura, *Chem. Lett.* **1999**, 1031.
64. Y. Amao, K. Asai, T. Miyashita, I. Okura, *Anal. Commun.* **1999**, *36*, 367.
65. B.D. MacCraith, C.M. McDonagh, G. O'Keeffe, E.T. Keyes, J.G. Vos, B. O'Kelly, J.F. McGilp, *Analyst* **1993**, *118*, 385.
66. J.N. Demas, B.A. DeGraff, W. Xu, *Anal. Chem.* **1995**, *67*, 1377.
67. T. Furuto, S-K. Lee, K. Asai, I. Okura, *Chem. Lett.* **1998**, 61.
68. Y. Amao, K. Asai, I. Okura, *Chem. Lett.* **1998**, 95.
69. T. Furuto, S.-K. Lee, Y. Amao, K. Asai, I. Okura, *J. Photochem. Photobiol. A: Chem.* **2000**, *132*, 81.
70. Hirsch, *The Chemistry of the Fullerenes*, Georg Thieme Verlag, Stuttgart, 1994.
71. J.W. Arbogast, C.S. Foote, M. Kao, *J. Am. Chem. Soc.* **1992**, *114*, 2277.
72. J.W. Arbogast, A.O. Darmanyan, C.S. Foote, Y. Rubin, F.N. Diederich, M.M. Alvarez, S.J. Anz, R.L. Whetten, *J. Phys. Chem.* **1991**, *95*, 11.
73. J.W. Arbogast, C.S. Foote, *J. Am. Chem. Soc.* **1991**, *113*, 8886.
74. R.R. Hung, J.J. Grabowski, *J. Phys. Chem.* **1991**, *95*, 6073.
75. M. Terazima, N. Hirota, H. Shinohara, Y. Saito, *J. Phys. Chem.* **1991**, *95*, 9080.
76. N.M. Dimitrijevic, P.V. Kamat, *J. Phys. Chem.* **1992**, *96*, 4811.
77. C. Taliani, G. Ruani, R. Zamboni, R. Danieli, S. Rossini, V.N. Denisov, V.M. Burlakov, F. Negri, G. Orlandi, F. Zerbetto, *J. Chem. Soc., Chem. Commun.* **1993**, 220.
78. F. Diederich, C. Thilgen, *Science* **1996**, *271*, 317.
79. S.I. Khan, A.M. Oliver, M.N. Paddon-Row, Y. Rubin, *J. Am. Chem. Soc.* **1993**, *115*, 4919.
80. R.M. Williams, J.M. Zwier, J.W. Verhoeven, *J. Am. Chem. Soc.* **1995**, *117*, 4093.

81. H. Imahori, K. Hagiwara, T. Akiyama, S. Taniguchi, T. Okada, and Y. Sakata, *Chem. Lett.* **1995**, 265.
82. R.G. Nuzzo, F.A. Fusco, and D.L. Allara, *J. Am. Chem. Soc.* **1987**, *109*, 2358.
83. M.D. Porter, T.B. Bright, D.L. Allara, and C.E.D. Chidsey, *J. Am. Chem. Soc.* **1987**, *109*, 3559.
84. P.E. Laibinis, G.M. Whitesides, *J. Am. Chem. Soc.* **1992**, *114*, 1990.
85. Ulman, An Introduction to Ultrathin Organic Films From Langmuir–Blodgett to Self-Assembly, Academic Press, San Diego, CA, 1990.
86. D.R. Thevenot, K. Toth, R.A. Durst, G.S. Wilson, *Pure Appl. Chem.* **1999**, *71*, 2333.
87. Brajter-Toth, J.Q. Chambers, *Electroanalytical Methods for Biological Materials*, Marcel Dekker, New York, 2002.
88. J. Janata, M. Josowicz, P. Vanysek, D.M. DeVaney, *Anal. Chem.* **1998**, *70*, R179.
89. P. Pantano, W.G. Kuhr, *Electroanal.* **1995**, *7*, 405.
90. P. Buhlmann, *Chem. Sens.* **1998**, *13*, 93.
91. Willner, E. Katz, *Angew. Chem. Int. Ed.* **2000**, *39*, 1180.
92. D.A. Mandler, I. Turyan, *Electroanalysis* **1996**, *8*, 207.
93. V.M. Fernandez, K. Schinder, in: *Bioinorganic Chemistry of Nickel*, J.R. Lancaster, Jr. (Ed.), VCH, Florida, 1988
94. H.-J. Kruger, R.H. Holm, *J. Am. Chem. Soc.* **1990**, *112*, 2955.
95. K. Nag, A. Chakravorty, *Coord. Chem. Rev.* **1980**, *33*, 87.
96. H.-J. Kruger, G. Peng, R.H. Holm, *Inorg. Chem.* **1991**, *30*, 734.
97. K.V. Gobi, T. Okajima, K. Tokuda, T. Ohsaka, *Langmuir* **1998**, *14*, 1108.
98. H.A. Biebuyck, G.M. Whitesides, *Langmuir* **1993**, *9*, 1766.
99. L. Strong, G.M. Whitesides, Langmuir **1988**, 4, 546.
100. D.M.P. Mingos, A.L. Rohl, *Inorg. Chem.* **1991**, *30*, 3769.
101. R.I. Haines, A. McAuley, *Coord. Chem. Rev.* **1981**, *39*, 77.
102. P.V. Bernhardt, G.A. Lawrance, *Coord. Chem. Rev.* **1990**, *104*, 297.
103. K.V.Gobi, T. Ohsaka, *J. Electroanal. Chem.* **2000**, *485*, 61.
104. K.V. Gobi, F. Kitamura, K. Tokuda, T. Ohsaka, *J. Phys. Chem. B* **1999**, *103*, 83.
105. A.J. Bard, L.R. Faulkner, *Electrochemical Methods – Fundamentals and Applications*, John Wiley, New York, 1980.
106. A.P.F. Turner, I. Karube, *Biosensor-Fundamentals and Applications*, Oxford University Press, Oxford, 1987.
107. W.M. Clark, *Oxidation–Reduction Potential of Organic Systems*, Robert E. Krieger Publishing Company, Huntington, 1972.
108. W.J. Blaedal, R.A. Jenkins, *Anal. Chem.* **1975**, *47*, 1337.
109. L. Gorton, E. Dominguez, Bioelectrochemistry, in: *Encyclopedia of Electro-chemistry*, G.S. Wilson (Ed.), Vol. 9, Wiley-VCH, Toronto, 2002.
110. C.R. Raj, T. Ohsaka, *Electrochem. Commun.* **2001**, *3*, 633.
111. C.R. Raj, K.V. Gobi, T. Ohsaka, *Bioelectrochem.* **2000**, *51*, 181.
112. L. Gorton, J. Chem. Soc., Faraday Trans. I **1986**, 82, 1245.
113. M.J. Lobo, A.J. Miranda, P. Tunon, *Electroanalysis* **1997**, *9*, 191.
114. T. Ohsaka, K. Tanaka, K. Tokuda, *Chem. Commun.* **1993**, 222.
115. C.R. Raj, T. Ohsaka, *Bioelectrochem.* **2001**, *53*, 251.
116. G. Eisenman (Ed.), *Glass Electrodes for Hydrogen and Other Cations*, Marcel

Dekker, New York, 1967.

117. D.O. Wipf, F. Ge, T.W. Spaine, J.E. Baur, *Anal. Chem.* **2000**, *72*, 4921.
118. S.A. Grant, K. Bettencourt, P. Krulevitch, J. Hamilton, R. Glass, *Sens. Actuators: B. Chem.* **2001**, *72*, 174.
119. Rubinstein, *Anal. Chem.* **1984**, *56*, 1135.
120. J.J. Hickman, D. Ofer, P.E. Laibinis, G.M. Whitesides, M.S. Wrighton, *Science* **1991**, *252*, 688.
121. M. Lahav, E. Katz, I. Willner, *Electroanalysis* **1998**, *10*, 1159.
122. C. R. Raj, T. Okajima, T. Ohsaka, *Electrochem. Commun.* **2002**, *4*, 330.
123. C. R. Raj, F. Kitamura, T. Ohsaka, *Langmuir* **2001**, *17*, 7378.
124. M. Pontie, F. Bedioui, *Analysis* **1999**, *27*, 564.
125. F. Lisdat, B. Ge, E. Ehrentreich-Forster, R. Reszka, F. Scheller, *Anal. Chem.* **1999**, *71*, 1359.
126. K. V. Gobi, F. Mizutani, *J. Electroanal. Chem.* **2000**, *484*, 172.
127. V. Lovich, A. Scheeline, *Anal. Chem.* **1997**, *69*, 454.
128. Y. Tian, M. Shioda, S. Kashara, T. Okajima, L. Mao, T. Hisabori, T. Ohsaka, *Biochim. Biophys. Acta* **2002**, *1569*, 151.
129. S. Descroix, F. Bedioui, *Electroanalysis* **2001**, *13*, 524.
130. X. Wu, X. Meng, Z. Wang, Z. Zhang, *Chem. Lett.* **1999**, 1271.
131. M.W. Pantoliano, P.J. McDonnell, J.S. Valentine, *J. Am. Chem. Soc.* **1979**, *102*, 6454.
132. J.A. Tainer, E.D. Getzoff, J.S. Richardson, D.C. Richardson, *Nature* **1983**, *306*, 284.
133. T. Ohsaka, Y. Tian, L. Mao, T. Okajima, *Anal. Sci.* **2001**, *17*, 1379.
134. Y. Tian, L. Mao, T. Okajima, T. Ohsaka, *Anal. Chem.* **2002**, *74*, 2428.
135. T. Ohsaka, Y. Tian, M. Shioda, S. Kasahara, T. Okajima, *Chem. Commun.* **2002**, 990.
136. K.V. Gobi, K. Tokuda, T. Ohsaka, *Electrochim. Acta* **1998**, *43*, 1013.
137. K.V. Gobi, T. Ohsaka, *Electrochim. Acta* **1998**, *44*, 269.

11 Catalysis by Soluble Macromolecular Metal Complexes

Edward A. Karakhanov and Anton L. Maximov

The application of catalytic systems based on macromolecular metal complexes is one of the attractive lines of development of metal complex catalysis [1–7]. The use of macromolecular fragments in a metal complex catalyst enables one to substantially change the microenvironment of the catalytic site and, thereby, the catalytic properties of the metal complex. The main role in such a change (as, for example, in enzymes) is played by the submolecular structures formed by macromolecular metal complexes. These structures can selectively bind the substrate, alter the geometry and the energy of the transition state and cause mutual activation of the participants in the catalytic reaction [1].

Increasing attention in catalysis is now being paid to the concept of attaching metal complexes to soluble polymers [8–15]. Such soluble macromolecular metal complex catalysts retain the homogeneous character of their low molecular analogues and have minimal diffusion limitation. The substrate binding selectivity and the rate of the process in the presence of such catalysts are generally governed by the polarity, the conformation of the polymer fragment and the presence of groups capable of reacting with the substrate and the reaction product. For some polymers the possible change in this conformation, depending on the conditions (solvent, pH, temperature), can alter the degree and the character of the binding, and hence, the selectivity of the process [14–16]. Different structures of soluble polymer support give three different types of soluble polymer catalyst based on linear polymers (polyethers and polyethylene) containing a terminal metal complex, linear polymers (polyacids and polyamides, linear polystyrene), containing metal complexes in the polymer chain and dendrimers (Fig. 11-1).

One of the promising trends in catalyst design is the use of water-soluble macromolecular metal complex catalysts. Catalysis in aqueous–organic two-phase systems is of considerable academic and industrial interest [17] The design of water-soluble complex catalysts is the object of much concentrated attention. The most obvious advantage of aqueous biphasic reactions is that the catalyst can be easily separated from organic products and reused. The Ruhrchemie-Rhone-Poulenc process for rhodium-catalyzed hydroformylation of propene using water-soluble trisulfonated triphenylphosphine (TPPTS) as a ligand has demonstrated the industrial significance of the concept of reaction in aqueous two-phase systems.

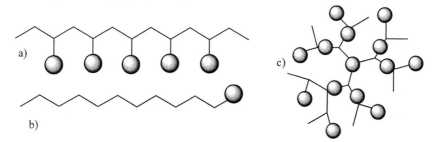

Figure 11-1. Different types of soluble macromolecular complex catalysts: a: metal complex groups in the chain of a polymer; b: metal complex groups in the end of a linear polymer; c: dendrimer.

Another promising approach in designing an active and selective metal complex catalytic system is the use of compounds with cavity-containing binding blocks for the construction of organized ligands [18–23]. Among such compounds are various modified cyclophanes able to form host–guest complexes, e.g. cyclodextrins [20] and calixarenes [24].

11.1 Water-Soluble Metal Complexes with Modified Poly(ethylene Oxide)s in Catalysis

One of the most popular types of soluble macromolecular metal complexes used in reactions in aqueous media are complexes of terminally modified poly- and oligo(ethylene oxide)s and block copolymers of ethylene oxide and propylene oxide. The main advantages of using these polymers as ligands are the following:

- The high water solubility of poly(ethylene oxide)s, which determines the water solubility of ligands obtained from them. In performing a reaction in a water–organic solvent two-phase system, this enables one, after the reaction, to easily separate the catalyst remaining in the aqueous phase from the reaction products and to reuse the catalyst. In carrying out a reaction in the organic phase, the macromolecular metal complex can readily be removed by adding a solvent (e.g. hexane or ether) that poorly dissolves poly(alkylene ether)s.

- The possibility of easy functionalization of the hydroxyl group of the initial polymers by various complexing groups, which potentially allows us to synthesize a variety of metal complex catalysts for different reactions. Methods of modifying the hydroxyl group of poly(ethylene oxide)s and ethylene oxide–propylene oxide block copolymers are very numerous and are described in a number of reviews [25–30]. Moreover, terminally modified polymers can also be obtained by another technique, which is conventional for producing ethylene oxide-based surfactants, namely, ethylene oxide oligomerization at hydroxyl, amino, etc. groups bound to the low molecular weight ligand.

- Poly(ethylene oxide)s are the only water-soluble polymers which can be terminally functionalized and from which we can obtain complexes bound to the polymer tail. Thereby, several problems encountered in producing conventional polymer-immobilized catalysts can be obviated. The metal complexes synthesized retain the properties of, on the one hand, homogeneous low molecular weight metal complexes, and on the other, poly(ethylene oxide)s or ethylene oxide–propylene oxide block copolymers. Among these properties are, first of all, water solubility and also the ability to concentrate nonpolar substances in polymer globules or micelles formed by polymer ligands.

- The inverse (anti-Arrhenius) temperature dependence of the water solubility of metal complex catalysts based on poly(ethylene oxide). Such a dependence opens up the possibility of using these catalysts as models in developing concepts of "thermoregulated" catalysis and "smart" ligands.

- The possibility of changing the properties of the synthesized macromolecular catalysts by varying the number of ethylene glycol monomer units in the ligand molecule, by introducing nonpolar groups into the oligo(ethylene oxide) block, by varying the ratio between the numbers of monomer units in polar (oligo(ethylene oxide)) and nonpolar (oligo(propylene oxide)) blocks of the ligand, by binding the metal site to the end of a nonpolar or polar block, etc. Thus we can affect the catalyst solubility in various solvents, the cloud point, the substrate concentration near the active site, etc.

Note that, in some studies, poly(ethylene oxide) oligomers, micelle-forming surfactants derived from them, and their complexes were used to perform such reactions in aqueous and alcohol solutions as hydroformylation [31–33], Wacker oxidation [34,35], hydroxylation of aromatic compounds [36–38], carbon dioxide hydrogenation [39], and epoxidation [40]. It was shown that using poly(ethylene oxide)s substantially increases the reaction rate and, in some cases, allows us to separate a metal complex containing oligo(ethylene oxide) [31,40].

Water-soluble macromolecular metal complexes based on terminally functionalized ethylene oxides and ethylene oxide–propylene oxide block copolymers have been used as catalysts for hydroformylation, hydrogenation, Wacker oxidation of unsaturated compounds, hydroxylation of aromatic compounds, oxidation of saturated and alkylaromatic hydrocarbons, metathesis, Heck reaction, and some asymmetric reactions.

Rhodium complexes with oligo- and poly(ethylene oxide)s modified with phosphine groups **1** were proposed as hydroformylation catalysts as far back as

1978 by Wilson et al. [41]. A series of phosphine-containing oligo(ethylene oxide)s **2–6** were synthesized, and their rhodium complexes were investigated in hydroformylation and hydrogenation of unsaturated compounds both as polar substrates in aqueous media and as nonpolar substrates in water–organic solvent two-phase systems [42].

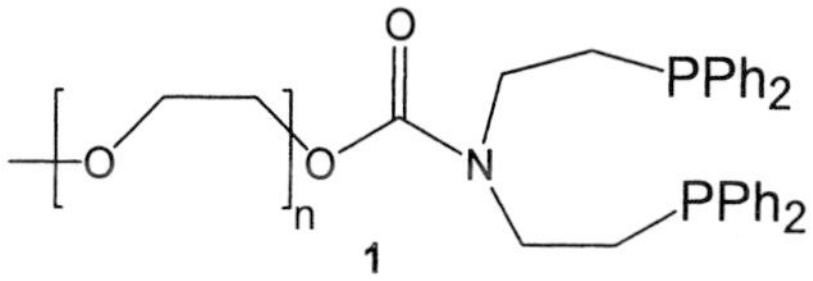

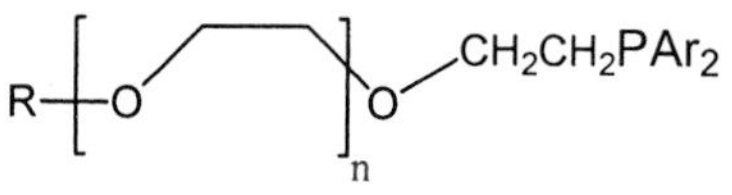

Ar= Ph **2**; Ar= O-2-iso-Pr-5-MePh **3**;
Ar= O-2-*tert*-Bu-4-MePh **4**

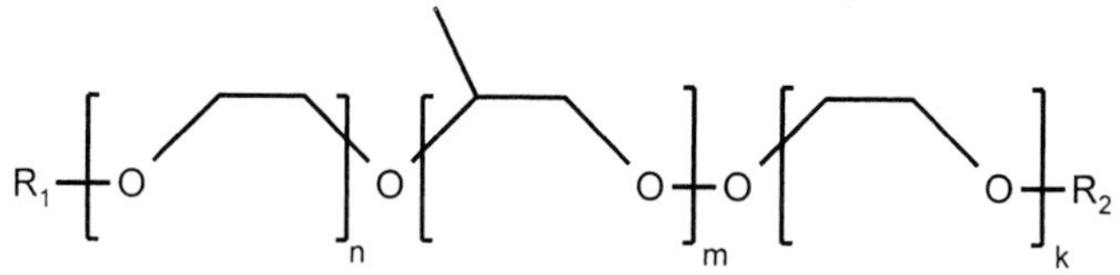

$k = 0$; $R_1 = -CH_2CH_2PPh_2$; $R_2 = -Bu$ **5**; $k = 0$; $R_1 = -Bu$; $R_2 = CH_2CH_2PPh_2$ **6**;
$k = n$; $R_1 = R_2 = -NC(O)[CH_2CH_2PPh_2]_2$ **7**

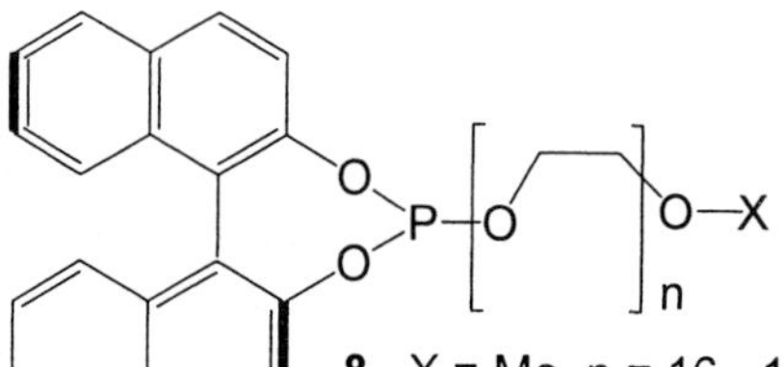

8   X = Me, n = 16 - 17

9   X = PO₂R, n = 4 - 5
[RO₂H₂ = (S)-Binol]

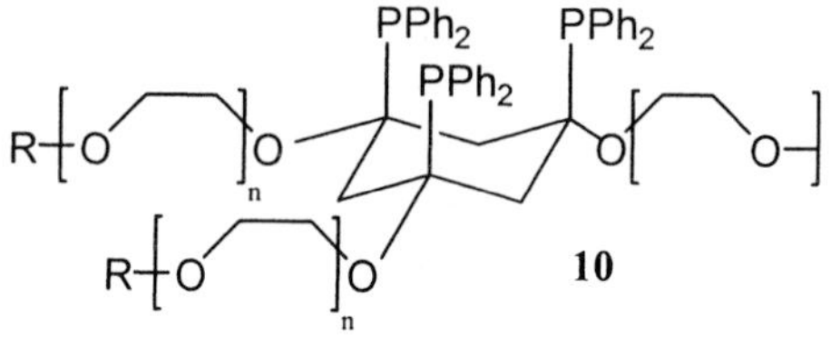

Studying macromolecular complexes based on ethylene oxide–propylene oxide block copolymers **7**, Bergbreiter et al. [42] proposed a concept of "smart" ligands, which would control the catalyst activity as a function of temperature. They showed that a rhodium complex with compound **7** is an active catalyst for allyl alcohol hydrogenation in an aqueous medium at 0 °C. As soon as the temperature rose to 50 °C, the reaction ceased virtually completely (Fig. 11-2). A similar effect was observed in the presence of rhodium complexes with ligand **2** in allyl alcohol hydrogenation (the activity was high at 30 °C and was virtually zero at 70 °C [43]) and a rhodium complex with ligands **8** and **9** in itaconic acid hydrogenation (the reaction took place at 20 °C and ceased at 25 °C [44]). This effect was explained by the fact that all the macromolecular complexes used exhibit a cloud point (lower critical solution temperature) in aqueous solutions, at which there is phase separation in the macromolecular metal complex–water system. As soon as the reaction temperature exceeded the cloud point, the water-soluble substrate and the catalyst were brought into different phases and the reaction ceased.

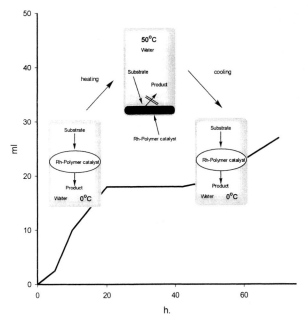

Figure 11-2. Hydrogenation of allyl alcohol using rhodium catalysts that contain smart ligand **7** [42].

In a reaction with nonpolar substrates in two-phase systems, increasing temperature affects the reaction rate differently, depending on the molecular weight of the oligoethylene oxides used and also on whether or not one of the reagents is highly soluble in water. For different types of phosphine- or phosphite-containing ligands involving ethylene oxide oligomers [45–52] with the number of monomer units that are characterized by temperature-dependent water solubility, Jin and coworkers developed the concept of thermoregulated phase-transfer catalysis for hydroformylation and hydrogenation of unsaturated compounds and also for hydrogenation of nitroaromatic compounds.

According to this approach, in a reaction in a water–organic solvent two-phase system at high temperatures, macromolecular rhodium complexes with phosphine-containing ligands bound to ethylene oxide oligomers pass virtually completely into the organic phase where the reaction takes place, and on cooling these complexes return to the aqueous phase and the catalyst can thus be reused (Fig. 11-3).

For this to take place, the ratio between nonpolar (phosphine or phosphite) and polar (ethylene oxide) blocks in phosphine-containing ligands must be sufficient for a cloud point to be exhibited and for the ligand and metal complex obtained to be soluble at high temperature in nonpolar solvents. A series of macromolecular ligands **11-13, 15-20** and metal complexes based on them were synthesized which had different cloud points and different phosphine- and phosphite-containing groups.

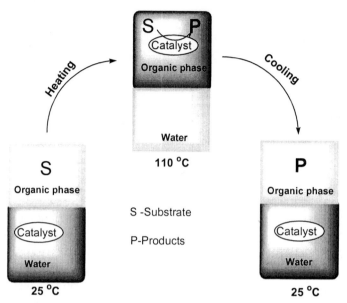

Figure 11-3. General scheme for thermoregulated phase-transfer catalysis.

The catalysts obtained were highly active in the hydroformylation of higher alkenes and styrene at 80–120 °C, whereas their activity below the cloud point proved to be low (Table 11-1). Note also that the activity of the catalysts substantially exceeded the activity exhibited under similar conditions by a catalytic system involving rhodium complexes with sulfonated phosphines and oligo(ethylene oxide). Using phosphine **20**, a *turnover frequency* (TOF) in dodecene hydroformylation of 1280 h^{-1} was obtained All the catalysts could be reused.

Table 11-1. Thermoregulated catalysts based on poly(ethylene oxide)s.

Ligand/metal	Reaction	Substrates	Reference
11-13/Rh(I)	Hydroformylation	Hexene-1, octene-1, dodecene-1, cyclohexene, styrene	43, 45
11-13/Rh(I)	Hydrogenation	Allylic alcohol	43
15/Rh (I)	Hydrogenation	Sodium cinnamate	47
15/Rh (I)	Hydroformylation	Decene-1	47
19/Rh (I)	Hydroformylation	Styrene, 4-isobutylstyrene	49
19/Rh (I)	Hydrogenation	Allylic alcohol	49
20/Rh(I)	Hydroformylation	Decene-1	50
20/Rh(I)	Reduction by CO	Nitroarenes	51

Iridium complexes with triphenylphosphine modified with poly(ethylene oxide)s **14** at one of the phosphine groups were active in two-phase hydrogenation of allylbenzene, although the catalyst activity and selectivity with respect to the hydrogenation product, propylbenzene, was substantially lower than in the case of its low molecular analogue and one of the main products was propenylbenzene. Similar results were obtained for an iridium complex with polyethylene oxide modified with pyridyl groups [53].

We studied the possibility of using rhodium complexes with poly(ethylene oxide)s modified with terminal phosphorus-containing groups with different electron density at the phosphorus atom as hydroformylation catalysts [54–59]. In both two-phase and homogeneous systems, we synthesized a series of ligands with sequentially increasing electron-accepting ability (Experiment 11-2, Section 11.7): $C_4H_9O(CH_2CH_2O)_{111}CH_2CH_2PPh_2$ (*REPPh$_2$*, **2**), $C_4H_9O(CH_2CH_2O)_{111}CH_2CH_2$-$OPPh_2$ (*REOPPh$_2$*, **17**) $(C_4H_9O(CH_2CH_2O)_{111}CH_2CH_2O)_2PPh$ (*REO)$_2$PPh*, **18**); $C_4H_9O(CH_2CH_2O)_{111}CH_2CH_2OP(OPh)_2$ (*REOP(OPh)$_2$*, **21**); then rhodium complexes with these ligands were obtained by ligand exchange. The metal complexes produced allowed us to obtain aldehydes from 1-decene, 1-dodecene, styrene, propenylphenol, and allylphenol in high yields by two-phase hydroformylation.

The complexes are arranged in order of increasing catalytic activity in hydroformylation (substrate conversion after 6-h reaction) as follows: *(REOP(OPPh))$_2$Rh(acac)* < *(REPPh$_2$)$_2$Rh(acac)* < *(REOPPh$_2$)$_2$Rh(acac)* < *((REO)$_2$PPh)$_2$Rh(acac)*. It was shown that the catalysts, which remain in the aqueous phase after the reaction, can be reused (for five catalytic cycles in 6 h, a constant conversion of the initial olefin is attained).

The complexes *(REOPPh$_2$)$_2$Rh(acac)* and *((REO)$_2$PPh)$_2$Rh(acac)* with the macromolecular ligand exhibited an increase in the resistance to hydrolysis, which can be explained by the effect of the polymer chain in the "phosphite" part. Note that Jin also observed [56,57] that phosphite ligands remained stable in the aqueous phase. It was established that, for metal complexes based on both polymers [55,56] and oligomers, the organic solvent used (toluene, hexane, heptane) has no essential effect on the conversion of the initial olefin and on the selectivity with respect to aldehyde of linear structure.

A lower but still rather high activity in two-phase hydroformylation was also exhibited by rhodium complexes with ethylene oxide–propylene oxide block copolymers and butyl ether of poly(propylene oxide)–poly(ethylene oxide) block copolymer modified with diphenylphosphines 3-4 [59].

Noteworthily, the hydroformylation rate as a function of temperature for phosphines and phosphites based on poly(ethylene oxide)s and block copolymers differs substantially from that for catalysts with oligomer ligands. The activity of the system is maximum at 50 °C and is considerably lower above this temperature. Probably, at this temperature, the polymer chain of the ligand has such a conformation that the metal site is most accessible to give the active form of the catalyst. With further increasing temperature, the accessibility of active sites significantly decreases because of the considerable aggregation of polymer coils [58].

Catalysts based on ethylene glycol oligomers functionalized by phosphine and phosphite groups also proved to be active catalysts for hydroformylation in a homogeneous medium. Introduction of anhydrous ether into the reaction medium after completion of the process ensures complete precipitation of the catalyst, which can readily be separated from the reaction products and reused without considerable loss of activity [57–59]. On the basis of ethylene oxides oligomers with phosphine groups, Jin synthesized such hydroformylation catalysts as macromolecular rhodium complexes $P[(p\text{-}CH_2CH_2O)_nH]/RhCl_3$, which exhibited a critical solution temperature in toluene–substrate mixtures, which enabled isolation of the complex after cooling and reuse of it in the reaction [60]. It was demonstrated that, in rhodium complexes with sterically hindered phosphites that are immobilized on butyl ethers of poly(ethylene oxide) 2a and 2b $(R(OCH_2CH_2)_{111}\text{-}OP(O\text{-}2\text{-}iso\text{-}Pr\text{-}5\text{-}MePh)_2$ and $R(OCH_2CH_2)_{111}\text{-}OP(O\text{-}2\text{-}tert\text{-}Bu\text{-}4\text{-}MePh)_2)$, the polymer ligand points out the effect of steric hindrance on the reaction rate and selectivity [61].

1-Hexene hydroformylation was also performed in the presence of water-soluble macromolecular rhodium complexes with the tridentate ligand $cis,cis\text{-}1,3,5\text{-}(PPh_2)_3C_6H_9$ with poly(ethylene oxide) groups 10. It has been shown that the selectivity with respect to aldehydes for this ligand is moderate (75% [62]) because of the rigid arrangement of ligands.

In oxidation of alkylaromatic compounds and cyclohexane, considerable activity is displayed by iron complexes with terminally functionalized acetyl-acetone 22–25, dipyridyl 26–27 and catechol 28–29 ethylene oxide, the monobutyl ether of poly(ethylene oxide) and also the butyl ether of poly(propylene oxide)–poly(ethylene oxide) block copolymers.

L_2–H$_2$CH$_2$C–(–O–)$_n$–O–(–O–)$_m$–CH$_2$CH$_2$–L_1

L_1 = L_2=C$_4$H$_8$ n=111; m=0 **(22)**
          n=22; m=88 **(23)**

L_2 = L_1=C$_4$H$_8$ n=111; m=0 **(24)**
          n=22; m=88 **(25)**

L_1 = OOC– L_2=C$_4$H$_8$ n=67; m=0 **(26)**
               n=54; m=10 **(27)**

L_1 =OOC–   OH OH L_2=C$_4$H$_8$ n=67; m=0 **(28)**
               n=54;m=10 **(29)**

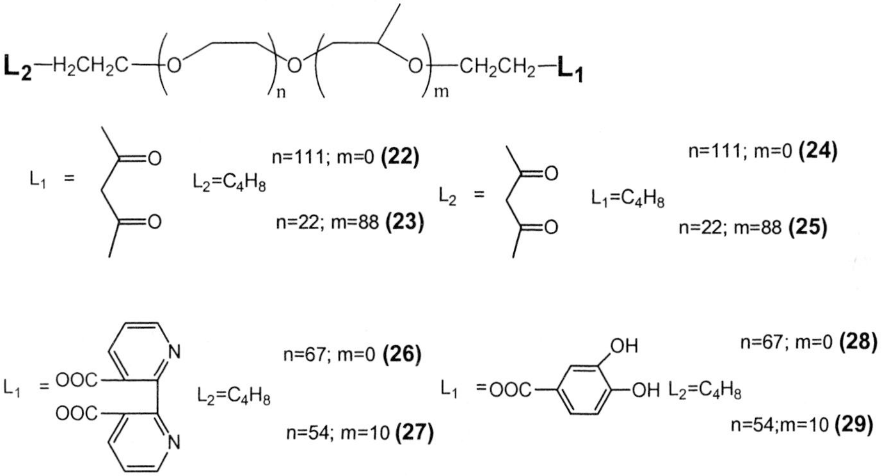

The activity of the catalysts based on terminally modified polyethers proved to be substantially higher than that of similar systems based on a mixture of a polymer and a low molecular weight complex. This can be explained by metal ion binding to the polymer globule in the case of the macromolecular metal complex catalyst [54,58,59]. In cycloxehane oxidation, the activity of all the catalysts is maximum at 30–50 °C. The optimum reaction temperature depends on the active site binding to the hydrophobic (propylene oxide) or hydrophilic (ethylene oxide) parts of the block copolymer: the activity is much higher if the active site is bound to the hydrophilic part of the block copolymer [55]. Similar regularities were observed in the hydroxylation of phenol and benzene by hydrogen peroxide, which is catalyzed by iron complexes with polyethers modified with catechol groups [63].

Palladium complexes with poly(ethylene oxide)s and ethylene oxide–propylene oxide block copolymers modified with iminodipropionitrile groups are catalysts for Wacker oxidation of unsaturated compounds in aqueous and water–alcohol media [64,65]. The yields of the product, methyl ketone, was above 90%, and the catalyst activity considerably exceeded the activity of similar low molecular weight catalytic systems.

The macromolecular ruthenium carbene complex **30** with modified poly(ethylene oxide) demonstrated high activity in ring-closing metathesis of some substrates, which transformed into cyclic compounds, with quantitative yield.

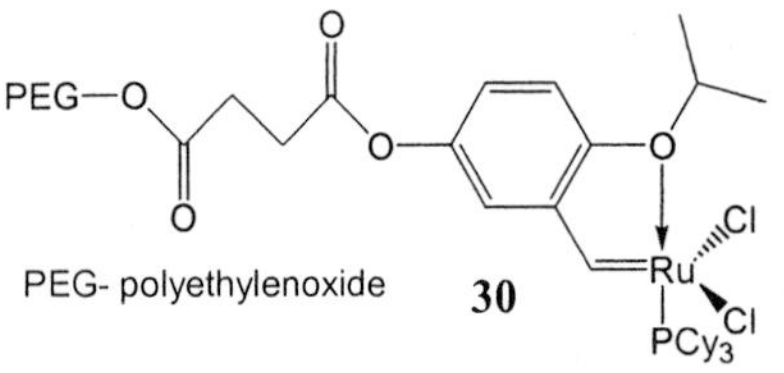

The catalyst was readily separated by precipitation with diethyl ether and used in eight catalytic cycles with almost no loss of activity [66].

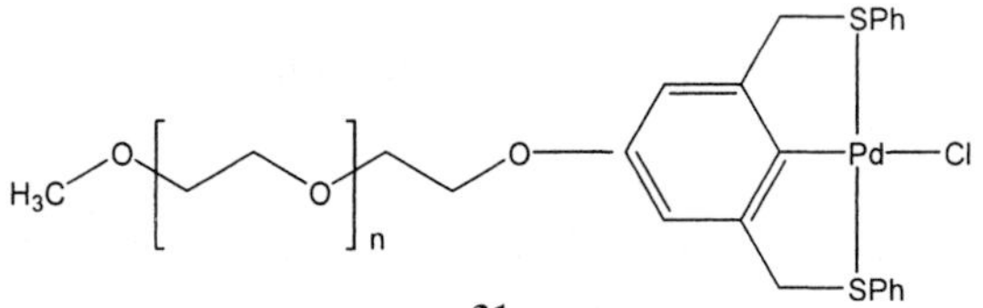

Palladium complexes **31–32** with the tridentate chelate SCS ligands bound to poly(ethylene glycol) of a weight of 5000 was active in Heck reactions in DMFA solution in air. These catalysts were also separated by precipitation with ether and reused without loss of activity [67].

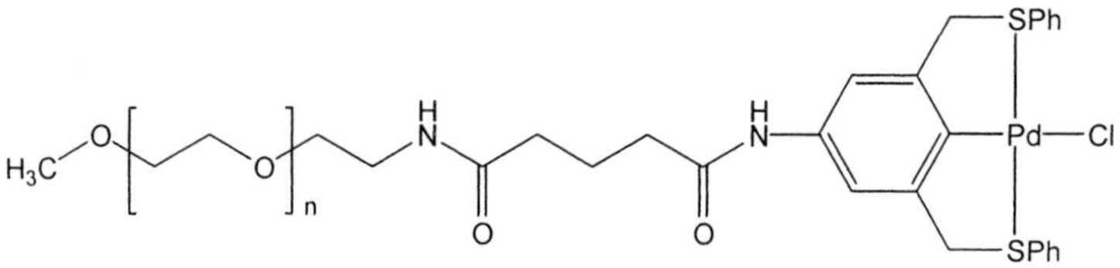

These catalysts were also active in a two-phase system containing 90% aqueous dimethylacetamide and heptane. In reuse of the catalyst in the two-phase system, the catalyst activity essentially increased [68].

Macromolecular metal complexes with modified poly(ethylene oxide)s have also been applied as catalysts for asymmetric reactions: epoxidation, dihydroxylation, hydrogenation, and hydroformylation.

For asymmetric epoxidation of methylstyrene, dehydronaphthalene, and *meta*-chlorobenzoic acid in the presence of *N*-methylmorpholine oxide, Reger and Janda synthesized manganese complexes **33** with the salen ligand bound to the methyl ether of poly(ethylene oxide) [69].

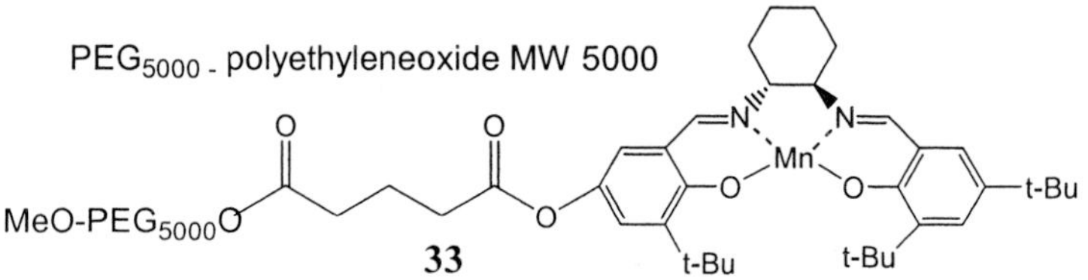

It was shown that the enantioselectivity (76–88%) and product (epoxide) yield (70–79%) obtained for the catalyst are respectively almost equal to those for a low molecular weight analogue. The catalyst can be reused after precipitation with ether without loss of activity and enantioselectivity in only two cycles; later, the enantioselectivity abruptly decreased. Reger and Janda explained it by decomposition of the ligand.

Some researchers analyzed the possibility of using complexes with modified poly(ethylene glycol)s in the Sharpless asymmetric dihydroxylation. The application of alkaloid-modified monomethyl ethers of poly(ethylene oxide)s **33-36** in osmium tetroxide-catalyzed ligand-accelerated asymmetric dihydroxylation ensures considerable enantioselectivity and much higher activity [70].

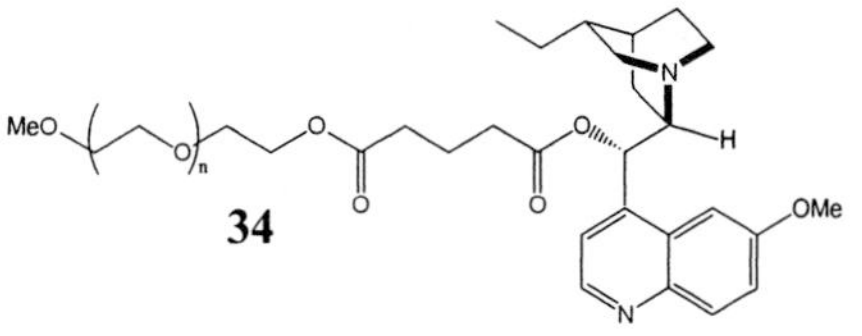

34

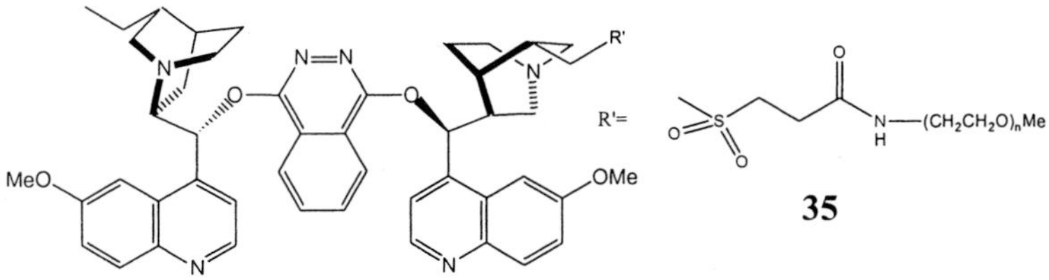

35

For example, in the presence of ligand **34** and *N*-ethylpyrrolidone at conversions of 80 to 90%, the enantioselectivity with respect to *trans*-stilbene was found to be 80%; to *trans*-methylstyrene, 85%; to *trans*-5-decene, 42%; and to styrene, 60%. In using monomethyl ether modified with phthalazine **35** as a ligand and potassium ferrocyanide as an oxidant, the enantioselectivity with respect to the same substrates reached almost 100% [71]. A similar result was also obtained in ethyl *trans*-cinnamate dihydroxylation [72].

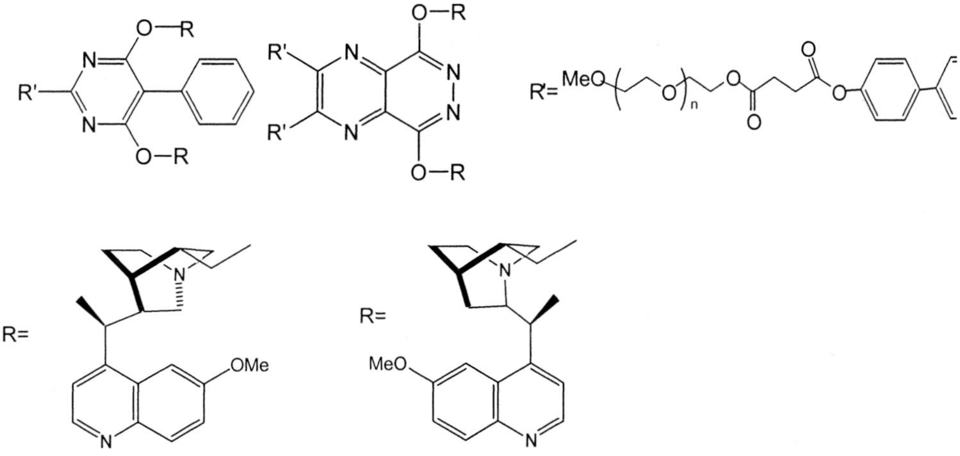

36   **37,38**

High asymmetric induction was demonstrated by ligands **36–38** based on diphenylpyrazinopyridazine (99% for *trans*-stilbene and 95% for toluene) and pyrimidine (87% for 1-decene and 90% for 3,3-dimethyl-1-butene), both immobilized on the methyl ether of poly(ethylene oxide) [73,74]. In all cases, a catalyst could be isolated and reused virtually without decrease in enantioselectivity when a new portion of osmium tetroxide was added.

In asymmetric epoxidation of *trans*-hex-2-en-1-ol by *tert*-butyl hydroperoxide with titanium tetraisopropylate as a catalyst, tartrate esters of methyl ethers of poly(ethylene oxide)s **39** were used as ligands. The epoxide yield under optimum conditions was 90% in enantioselective excess of 93% for dipoly(ethylene oxide) ether and 79% in enantioselective excess of 90% for polyethylene oxide monoether. In an attempt to reuse the ligand, the enantioselectivity was almost halved, with the yield remaining the same [75].

In enantioselective hydrogenation, rhodium and ruthenium complexes with chiral diphosphines bound to poly(ethylene oxide)s **40–42** have been described.

Ruthenium complexes with BINAP-containing polyether ligands **41–42** were active in enantioselective hydrogenation of 2-(6′-methoxy-2′-naphthyl)propionic acid, 2-(*p*-2-methylpropyl)phenylacrylic acid, and *trans*-2,3-dimethylacrylic acid in various solvents: methanol; ethylene glycol; ethyl acetate–water, methanol–water, ethyl acetate–ethylene glycol and other mixtures [76–77]; and complexes with ruthenium **40** were active in enantioselective hydrogenation of β-ketoesters in ethylene glycol [78].

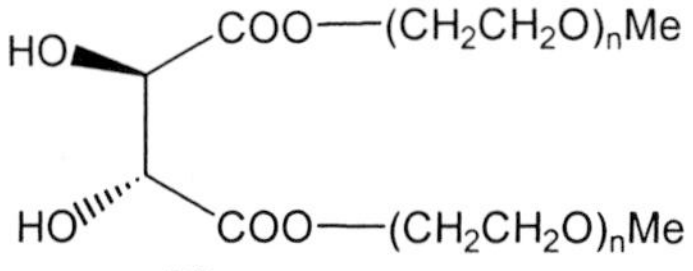

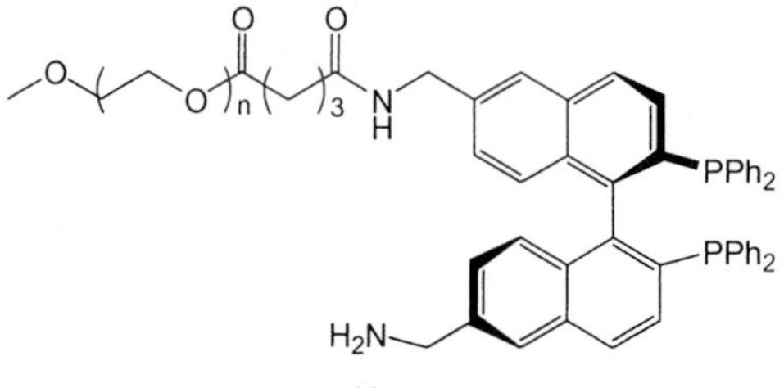

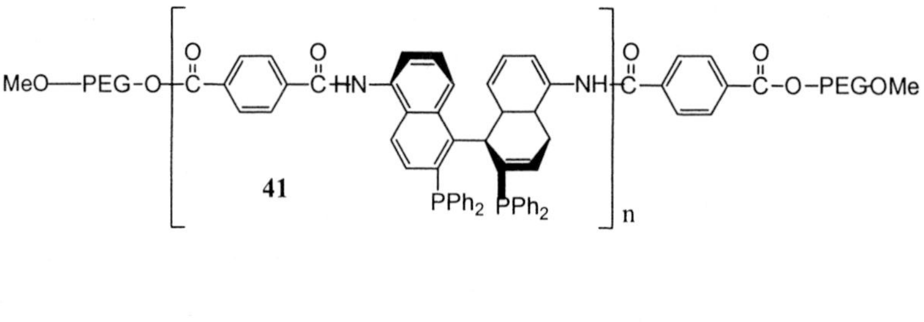

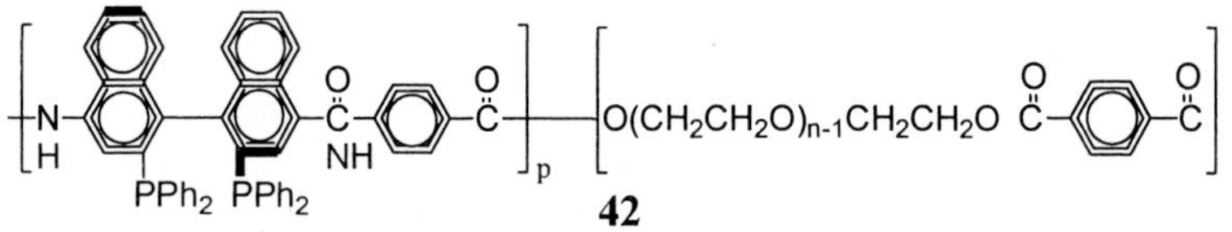

42

The presence of water was shown to essentially decrease the enantioselectivity of the reaction.

A rhodium metal complex with polyethylene glycol modified with Pyrphos ((3*R*, 4*R*)-3,4-bis(diphenylphosphino)-pyrrolidine) **43** was active in hydrogenation of prochiral enamides to amino acids [76]. In all the reactions, the catalysts were used in several cycles without loss of enantioselectivity or activity.

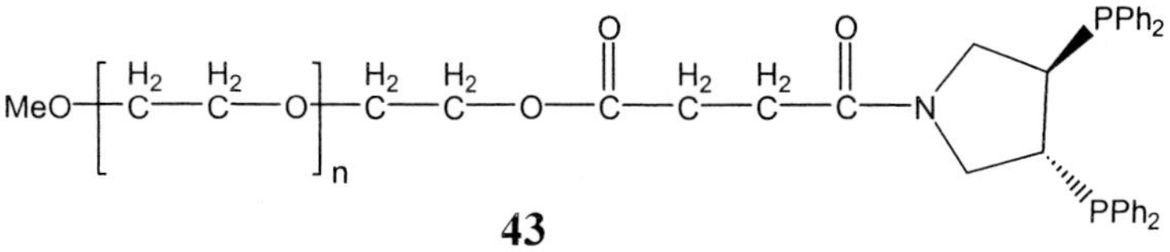

43

In enantioselective hydroformylation, a rhodium complex with chiral polyether phosphite **8-9** has also been used [44]. The system demonstrated high activity and chemo- and regioselectivity in enantioselective hydroformylation of styrene in a two-phase system. The excess of aldehyde in the S-configuration reached 25%.

11.2 Polyamides and Polyacids Based on Soluble Macromolecular Metal Complex Catalysts

One of the types of polymer used to synthesize soluble ligands and complexes is poly(carboxylic acid)s and their derivatives. The presence of acid groups determines the solubility of such polymers and enables us to introduce various ligands into the molecule relatively easily. Several ligands are bound to one polymer molecule. Their arrangement, unlike that in catalysts based on polyethers, generally cannot be determined exactly: after synthesis, ligands are distributed randomly and this affects the activity and properties of the macromolecular metal complex obtained. Moreover, modification of acid or amide groups can enable us to vary the properties of the polymer, such as the ability to transfer nonpolar substrates into the aqueous phase, the dependence of the solubility on temperature, and others.

Poly(acrylic acid) modified with triphenylphosphine groups (PAA-PNH) with a phosphine-to-carboxylic acid ratio 1:5 forms stable water-soluble macromolecular complexes with rhodium, $\{(NBD)RhL_2\}CH_3SO_3$, **44**, which are active in hydrogenation of unsaturated compounds, such as 4-pentenoic acid and but-1-en-4-

ol. At pH < 7, the water solubility of this catalyst is low, and the catalyst can be separated from the solution. At the same time, hydrogenation of higher alkenes, even such as 1-hexene, takes place only on addition of sodium dodecyl sulfonate, which is thought to favor the transfer of water-insoluble alkene into the aqueous phase [79–81]. For hexene, addition of the surfactant increases the product yield from 2.5 to 100%; and for octene, from 10 to 100%. The catalyst activity is more than doubled if the reaction is carried out in a 80:20 methanol–water mixture [79].

A ligand produced by the interaction between bis(diphenylphosphino-ethyl)amine and a copolymer of maleic anhydride with vinyl alkyl ether has been used to synthesize a macromolecular rhodium complex **45**. The catalyst obtained is soluble at pH > 7.5; after acidification of the solution with trifluoroacetic acid, the catalyst precipitates and can be reused. The catalyst in an aqueous medium catalyzes hydrogenation of allyl alcohol, acrylic acid, N-isopropylacrylamide, α-acetamidoacrylic acid, and styrene-p-sodium sulfonate. The activity of the catalyst is in some cases comparable to the activity of a low molecular weight analogue but lower than the activity of ordinary catalysts [82]. The results of studying the activity of rhodium(I) hydrogenation catalysts containing amide groups suggest that the low catalytic activity is because of the coordination of amide groups to the rhodium site. The decrease in the activity in reuse is due to gradual oxidation of phosphine groups [83].

Along with poly(acrylic acid), poly(4-pentenoic acid) has been used to produce phosphine-containing ligands and rhodium complexes **46**. Such complexes proved to be active in two-phase hydroformylation of 1-alkenes without surfactant added. The ability to solubilize the substrate and the activity of the catalyst are much higher than those of poly(acrylamide)-based catalysts. This is likely to be because the length of the chain between the carboxyl group and the polymer chain in this catalyst is greater than that in poly(acrylic acid). In the result, 1-dodecene is hydroformylated at a conversion of 100% and a selectivity with respect to aldehydes of 53% and 1-octene is hydroformylated at the same conversion and a selectivity of 60%. Other products are isomeric alkenes. High selectivity with respect to aldehydes was attained for nonisomerizing alkenes, e.g. 4-vinylcyclohexane, styrene and substituted styrenes; in the last case, the isomeric-to-normal aldehyde ratio exceeded 15:1 [84].

Poly(acrylamide)s were also modified with optically active phosphine-containing ligands - (2S,4S)-4-diphenylphosphino-2-diphenyl-phosphinomethyl-pyrrolidine **47**, its (2R,4R)-isomer and pyrophos ((R,R)-3,4-bis-diphenylphosphino-pyrrolidine) **48** and from them a rhodium catalyst for asymmetric hydrogenation of prochiral amides was synthesized [85–87]. The reaction was performed in aqueous solutions or a water–ethyl acetate two-phase system. The enantioselectivity with respect to the R-isomer using the catalyst with the ligand in the S,S-configuration ranged from 53 to 89%, and the enantioselectivity with respect to the S-isomer using the catalyst with the ligand in the R,R-configuration was from 77 to 83%. For the complex with the ligand in the S,S-configuration, the enantioselectivity was observed to be substantially dependent on the number of phosphine groups: at a sufficiently large number of phosphine groups, a complex forms where the rhodium atom is bound to the phosphine groups of different ligands responsible for nonselective hydrogenation. The selectivity decreases from 90 to 70%. For the complex with the ligand in the R,R-configuration, there is no such change.

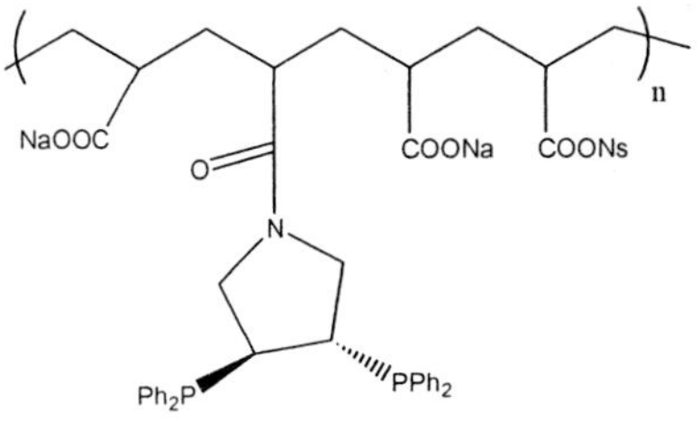

47 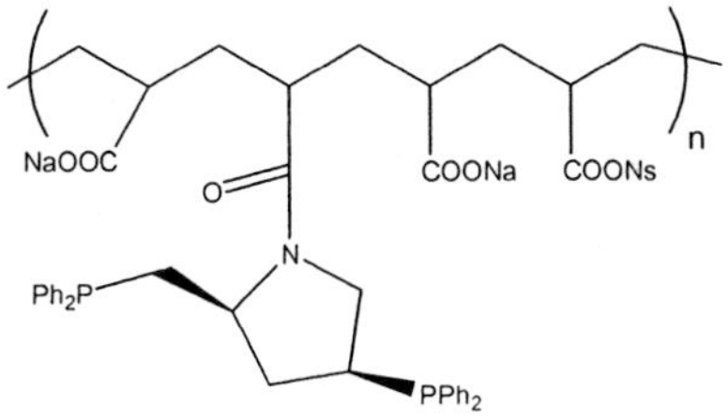   **48**

In cyclohexane oxidation by hydrogen peroxide and hypochlorite in the presence of 4-dimethylaminopyridine, poly(acrylamide)s containing phthalocyanine groups were used. Oxidation for 22 h by hydrogen peroxide gave 1,2-dihydrocyclohexane with 43% yield, and oxidation by hypochlorite gave 1,2-dihydrocyclohexane with 79% yield and 1-hydroxy-2-chlorocyclohexane with 11% yield. The catalyst had high catalase activity [88–90].

The idea of thermomorphic catalysts, which is largely similar to the concept of catalysts with modified poly(ethylene oxide)s, was implemented by the example of modified poly(N-isopropylacrylamides) (PNIPAMs). At low temperatures, PNIPAMs can dissolve in water; and at high temperatures, in two-phase systems, PNIPAMs pass into the organic phase, and also retain partly in water (Fig. 11-4). The upper critical solution temperature depends on substituents at the amide group [83, 91–93]. Several macromolecular metal complexes with PNIPAM ligands have been designed.

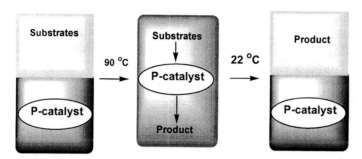

Figure 11-4. General scheme for the use and recovery PNIPAM-based complexes [83].

A rhodium complex with PNIPAM modified with propyldiphosphine **49** groups was more active in hydrogenation of 1-octadecene and 1-dodecene in a system containing heptane and 90% aqueous ethanol (Experiment 11-3, Section 11.7). At 22 °C the catalyst is virtually insoluble in heptane and there is no reaction, but at 70 °C this system forms a homogeneous solution of the polymer and the reaction takes place. The catalyst can be reused without loss of activity [94].

Moreover, the reaction between cinnamyl acetate and cyclohexylamine or dipropylamine, which is catalyzed by a palladium complex with this ligand, shows that the catalyst can be reused even when the product is soluble in the water–ethanol phase. After saturation of this phase with the product in the first catalytic cycle, in the second and subsequent cycles, almost no product is lost. To carry out this reaction in a system composed of 85% aqueous ethanol and heptane, it is necessary for the polymer catalyst to form a single phase at a temperature of 80 °C.

49

A palladium(0) complex with the same ligand was active in the formation of C–C bonds, in allyl substitution, and in cross coupling of aryl iodides with various substrates. Allylic substitution was performed in tetrahydrofuran–water, acetonitrile–water, and ethanol–water systems. The catalyst was separated by precipitation with hexane. In water, the catalyst was removed by heating the mixture to 25 °C, with the lower critical solution temperature being below this temperature [95]. Cross coupling was carried out in the presence of triethylamine in two-phase systems containing heptane and 90% aqueous ethanol or heptane and dimethylacetamide, where the catalyst on heating passed into the organic phase, in which the reaction took place. As in Heck reactions with a low molecular weight analogue [96], the catalyst activity in an aprotic solvent was almost four times higher than that in a protic solvent [97]. In an attempt at reuse, the catalyst proved to be inactive because of oxidation of phosphine groups.

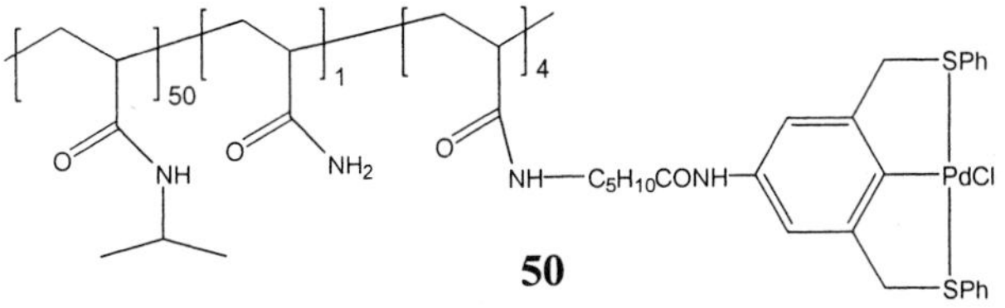

50

The catalyst PNIPAM–SCS–PdCl **50** was also active in C–C coupling. It is air-insensitive and reusable in five cycles, in the first three of which the catalyst activity increases. For this catalyst, reactions were carried out in a system comprising heptane and 90% aqueous dimethylacetamide, where the catalyst exhibited thermophilic behavior, which enabled it to be reused [97].

PNIPAM-modified polymers were used as ligands to synthesize catalysts that are soluble in the nonpolar phase (e.g. alkane) at low temperatures and are miscible with the polar phase (90% aqueous ethanol) owing to the introduction of octadecyl groups into the amide group of the polymer (PNODAM polymers). Rhodium and palladium complexes of polymers modified with diphenylphosphines were active in hydrogenation and Heck reactions, respectively. In Heck reactions, a palladium complex with a PNODAM polymer modified with chelating ligands was used [98].

11.3 Catalysis by Soluble Dendrimers

Recently, the chemistry of dendrimers and macromolecular complexes derived from them has been developed considerably [99–101]. Studies on catalysis by metal complexes with dendrimer-containing ligands is extensive and available in the literature [102–107]; therefore, we consider here only a few examples in this area.

Macromolecular metal complexes of dendrimers have a number of properties, which make them promising for use in catalysis. Firstly, they are relatively readily separable from the reaction products by precipitation with a "poor" solvent or by membrane filtration. Secondly, the introduction of catalytically active sites into dendrimers is controllable; in this they resemble terminally functionalized polymers. The dendrimer structure allows us to purposefully introduce catalytically active groups into the dendrimer core, onto the outer surface, or in intermediate positions. The introduction of catalytically active groups to the outer surface makes them accessible to reaction and enables us to obtain a catalyst that has a high metal content owing to the dendrimer structure. The introduction of catalytically active groups into the inner chains or core of a dendrimer may allow us to affect the catalyst selectivity through the formation of a regular structure near the catalytic site (Fig. 11-5). Thirdly, dendrimers, particularly soluble ones, can bind hydrophobic molecules within themselves much as micelles and can be regarded as "molecular micelles" [108–111].

Macromolecular metal complexes of soluble dendrimers with catalytically active groups on their surface have been used in hydrogenation [112, 113], Kharasch addition [114], Heck reactions and hydroformylation [115–118], metathesis of unsaturated compounds [119–121], oxidation [122], etc. The use of dendrimers in most cases enabled catalyst reuse and also influenced the activity and selectivity of the processes.

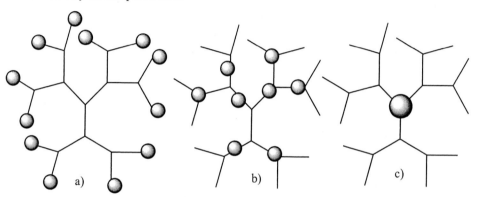

Figure 11-5. Representation of dendrimers with catalytic sites at a) the outer surface; b) the interior; c) the core [106].

In hydrogenation of dienes to monoenes under mild conditions in various solvents (alcohols, tetrahydrofuran, acetone, dimethylformamide), the catalyst DAB-*dendr*-[N(CH$_2$PPh$_2$]$_{16}$ (**51**) and PdCl$_2$(PhCN)$_2$ (DAB dendrimers based on a poly(propylene imine)-diaminobutane core) was much more active than its low molecular weight analogue and proved to be a selective catalyst for hydrogenation of dienes to monoenes [112]

Complexes of different dendrimers with rhodium that contained ferrocenyl phosphine ligands on the surface, **52-53**, were also active in hydrogenation. They catalyzed dimethylacetone hydrogenation at an optical selectivity of 98%, which is comparable to the selectivity of a low molecular weight analogue [113, 123, 124].

Complexes of DAB-*dendr*-[N(CH$_2$PPh$_2$]$_{16}$ **51** with rhodium (I) were active in 1-octene hydroformylation, and their activity was close to that of a low molecular weight analogue [115].

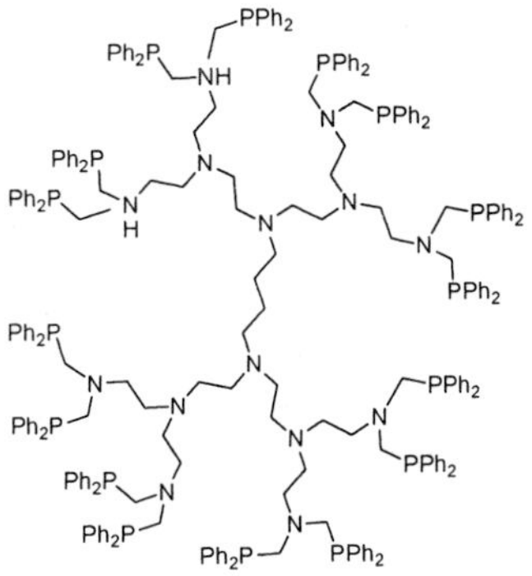

51

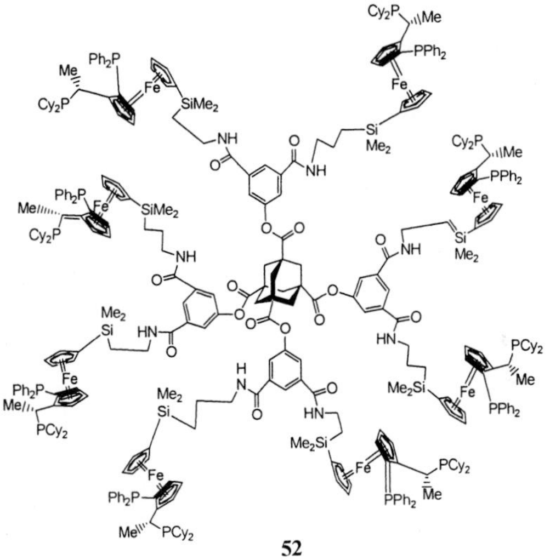

52

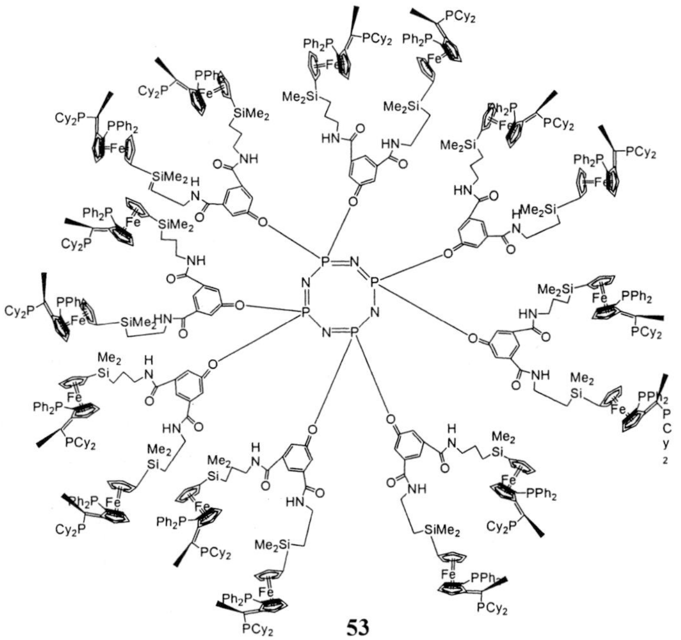

53

Water-soluble metal complexes of rhodium with polyamidoamine-based (PAMAM-based) third-generation (32 terminal groups) dendrimers modified with diphenylphosphine groups have been used in two-phase hydroformylation of 1-octene and styrene. The selectivity in styrene hydroformylation with respect to isoaldehyde was much higher than in the case of trisulfonated triphenylphosphine

(TPPTS), and the isomeric-to-normal aldehyde ratio exceeded 11:1. Introduction of sulfo groups into the dendrimer increased the catalyst activity, whereas introduction of decyl groups decreased this activity because of blocking of the phosphine groups. The loss of rhodium, especially at high temperatures, was high [118].

In some studies, hydroformylation was performed using metal complexes of rhodium with silicon-containing dendrimers modified with diphenylphosphines, for which the dendrimer structure was demonstrated to essentially affect the hydroformylation selectivity. For rhodium complexes with carbosilane-based dendrimers $Si\{(CH_2)_nSi(CH_3)_2(CH_2PPh_2)\}_4$ ($n = 2$, 3; dendrimers of the first to third generation) and $Si\{(CH_2)_nSi(CH_3)(CH_2PPh_2)_2\}_4$, the activity and selectivity in 1-octene hydroformylation were close to those for their monomer analogues [125]. And if the distance along the chain between the terminal silicon and the phosphorus is increased by one methylene group: $(Si\{(CH_2)_nSi(CH_3)-(CH_2CH_2PPh_2)_2\}_8)$, then the selectivity with respect to the linear product abruptly rises: the isomeric-to-normal product ratio changes from 3–4 to 12–14. Such a change is determined by steric hindrance (crowding) that is sufficient for the formation of a six-membered ring, in which the bidentate coordination is preferred [116, 117].

In Kharasch addition to acrylic acid ester that is catalyzed by nickel complexes with silicon-containing dendrimers modified with a chelate aryldiamide ligand, the catalyst activity is lower than that of a low molecular weight analogue. van Koten and coworkers assumed that this is due to the small distance between nickel atoms on the dendrimer surface and to the formation of mixed-valence binuclear complexes [114, 126, 127].

In metathesis, ruthenium catalysts bound to the ends of the polymer chains through the carbene group were used. Although the activity of the complexes remained high in reuse in six cycles, a part of the rhodium was lost in the reaction. The high catalyst activity was assumed to be due to the fact that highly active monophosphine complexes pass into solution [119–121].

One of the interesting examples of catalysis by dendrimers containing active groups within the molecule is a rhodium complex with an organophosphine dendrimer obtained by the reaction between $(CH_3)_3SiNEt_3$ and $P((CH_2)_3OH)_3$. The complexes containing 4 to 46 rhodium atoms were active in 1-decene hydrogenation [128].

The possibility of using the thermotropic properties of macromolecular complexes of dendrimers was demonstrated by the example of oxidation of disulfides in the presence of a macromolecular complex of cobalt phthalocyanine with a polyisopropylacrylamide-based (PIPAAm-based) dendrimer characterized by temperature-dependent solubility. The oxidation rate is almost tripled by increasing the temperature from 34 to 36 °C [129]. These temperatures are close to the lower critical cloud point, and this can explain this phenomenon. At a temperature below 35 °C, the PIPAAm chains are unfolded and prevent the substrate from reaching phthalocyanine sites; with increasing temperature, the chains shrink and the catalyst incorporated in the dendrimer becomes accessible to substrate attack.

Binding of the active site to the dendrimer core essentially changes the properties of the catalyst because of changes in the properties of the microenvironment of the active site. For example, for a palladium complex with a carboxylane dendrimer with the ferrocenyl phosphine group in the dendrimer core, the selectivity of allylic alkylation with respect to the branched product increased with the number of generations of the dendrimer owing to the presence of the nonpolar environment of the active site and to the growth of steric hindrances to nucleophilic attack of the ligand. At the same time, the catalyst activity decreased because of the decrease in the rate of mass transfer from the periphery of the dendrimer to the active site [130].

The idea of steric control of the reaction selectivity has also been used in epoxidation of alkenes with iodobenzene using dendrimers with manganese porphyrin **54** [131, 132].

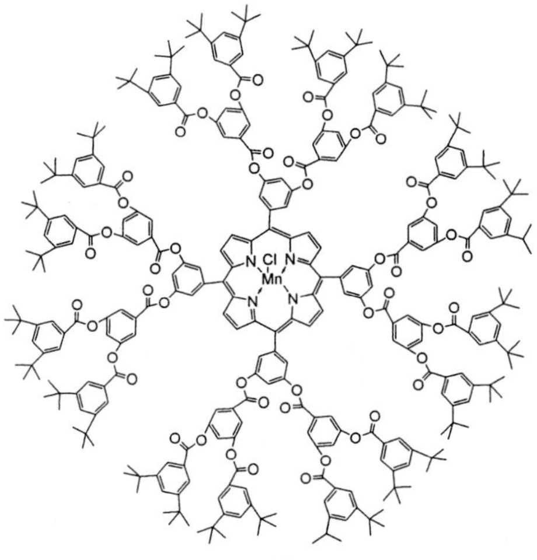

54

It has been shown that this catalyst is selective in epoxidation of linear alkenes: the linear epoxide yield was two to four times higher than in catalysis by ordinary porphyrin. It was also demonstrated that, in catalysis by the dendrimers, cyclic alkenes are oxidized three times more rapidly than similar linear 1-alkenes are. The catalyst activity decreases only by 10% at a turnover number (TON) of 1000, which is much higher than that for the monomolecular analogue. A cobalt complex with dendrimer phthalocyanine was much more stable, while remaining active, in oxidation of 2-mercaptoethanol to the corresponding sulfide [133].

A number of catalysts with the catalytic site in the dendrimer core has also been used in asymmetric hydrogenation. A rhodium complex with dendrimer diphosphine with menthyl groups in its branches as a catalyst in acetamidocinnamic acid hydrogenation exhibited low enantioselectivity. For a catalyst where dendrimers were at the meta positions of phosphine, the reaction rate was substantially higher than that for a low molecular weight analogue; and for a catalyst where dendrimers were at positions 2 and 5 of phosphine, the reaction rate was much lower [134–137].

A ruthenium complex with the dendrimer BINAP ligand obtained from aromatic polyethers **55** catalyzed hydrogenation of 2-[p-(2-methylpropyl)phenyl]acrylic acid to ibuprofen at an optical selectivity of 99% in a methanol–toluene solution; the activity of the dendrimer complex was twice as high as that of a low molecular weight analogue [138].

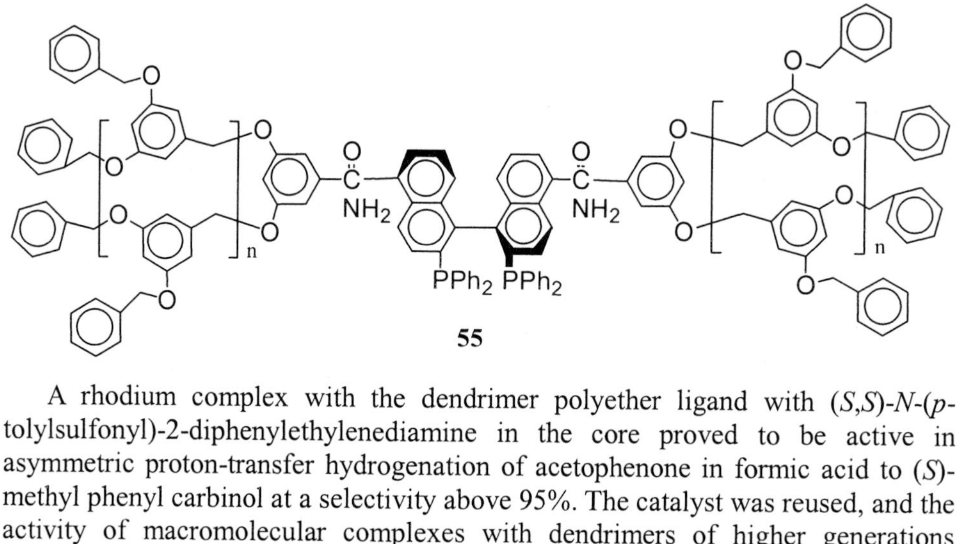

55

A rhodium complex with the dendrimer polyether ligand with (S,S)-N-(p-tolylsulfonyl)-2-diphenylethylenediamine in the core proved to be active in asymmetric proton-transfer hydrogenation of acetophenone in formic acid to (S)-methyl phenyl carbinol at a selectivity above 95%. The catalyst was reused, and the activity of macromolecular complexes with dendrimers of higher generations decreased much more slowly, while the optical selectivity was retained [139]. This also confirms the stabilization of the active site by the dendrimer environment in such macromolecular complexes.

11.4 Macromolecular Metal Complex Catalysts Soluble in Nonpolar Solvents

Among the first soluble macromolecular metal complexes to be used as catalysts were modified linear polystyrenes [140–145]. For example, rhodium complexes with polymer phosphines [146,147] and phosphites [148] were used in hydrogenation and hydroformylation. These catalysts were soluble in nonpolar solvents and could be separated from solutions by adding a "poor" solvent. At the same time, their use provided no perceptible advantages over coated analogues because of changes in the microenvironment of active sites and their possible aggregation. One of the most elegant ways to solve this problem in performing reactions in nonpolar media is to use terminally functionalized polyethylenes characterized by temperature-dependent solubility [149–153]. Metal complexes with such ligands are easily soluble in hydrocarbons on heating to 90–110 °C and precipitate on cooling (Fig. 11-6).

Thus, these catalysts are readily separable from the reaction products and can be reused.

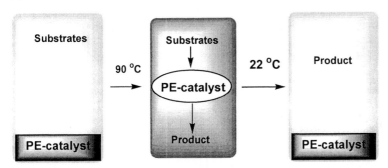

Figure 11-6. Catalyst based on modified polyethylene.

Such complexes have been used in a variety of reactions: hydrogenation [153], hydroformylation [154], carbonylation [154–156], cyclooligomerization [157–159], allylic substitution [160], Kharasch addition [161], reduction of bromides and iodides by sodium borohydride (Table 11-2; see Experiment 11-1, Section 11.7) [162].

In all cases, the catalysts could be reused; in some instances there was a specific effect of the polymer chain in the reaction. For example, in carbonylation of substituted *meta*- and *para*-iodobenzenes, the activity depended on the substituent polarity and the most polar substrates reacted most slowly [154–155].

Another attractive line of application of macromolecular complexes as catalysts is the so-called "fluorous" catalysis, whose idea was proposed in the early 1990s [163–168]. The main idea of such catalysis is that a catalyst is soluble (immobilized) in the perfluorocarbon phase, whereas the product is soluble in an organic solvent. If a suitable solvent for substrates, e.g. toluene or benzene, is used, then the system at sufficiently high temperature is homogeneous, and on cooling, there is phase separation. As a result, the solution of the metal complex in fluorocarbon is readily separable from the reaction products and can be reused.

Naturally, designing such macromolecular complexes require one to use fluorine-containing polymers.

Table 11-2. Macromolecular complexes of terminally functionalized ethylene oligomers in catalysis.

Reaction	Ligand	Metal	Reference
Hydrogenation	PE-PPh$_2$	Rh(I)	153
Hydroformylation	PE-PPh$_2$	Rh(I)	154
Hydroformylation	PE-OPPh$_2$	Rh(I)	156
Carbonylation	PE-PPh$_2$	Pd	155,156
Kharasch addition	PE-PPh$_2$	Ru	161
Cyclooligomerization	PE-PPh$_2$	Ni	157–159
Allylic substitution of allyl ethers with secondary amines	PE-PPh$_2$	Pd	160
Decarboxylation–allylic substitution of allyl ethers of β-keto or cyano carboxylic acids	PE-PPh$_2$	Pd	160
Reduction of alkyl and aryl bromides and iodides by sodium borohydride	PE- SnBu$_2$Cl PE- SnPh$_2$Cl	Sn	162

Perfluorinated oligoethers with the acetylacetonate group at the end of the polymer formed complexes **56** with nickel(I), which catalyzed ethylene oligomerization. The turnover number in the reaction reached 2460 in a perfluorinated ether–toluene two-phase system, and the catalyst, which was in the fluorine-containing phase, was readily separable [169].

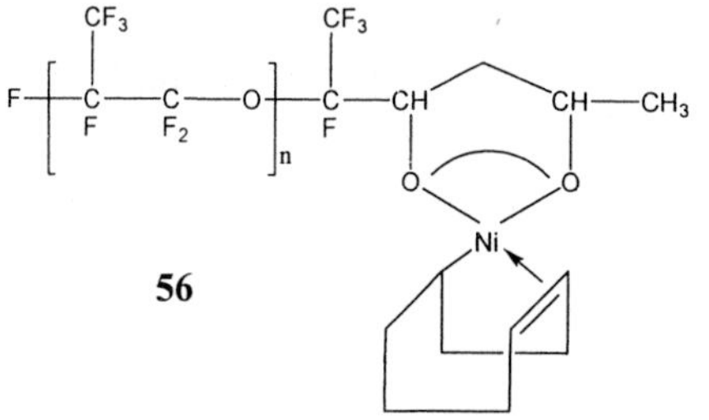

56

Cobalt tetraarylporphyrins with fluorine-containing substituents were active in epoxidation of alkenes using fluorous catalysis in the presence of oxygen and 2-methylpropanal [167,170-171]. Manganese and cobalt complexes of perfluorinated tetraazocyclonone catalyzed allylic oxidation of alkenes with t-BuOOH/O$_2$ [172]. The complex with the salen ligand **57** was active in alkene epoxidation under Mikayama's conditions, and indene was epoxidated at a high stereospecificity [173].

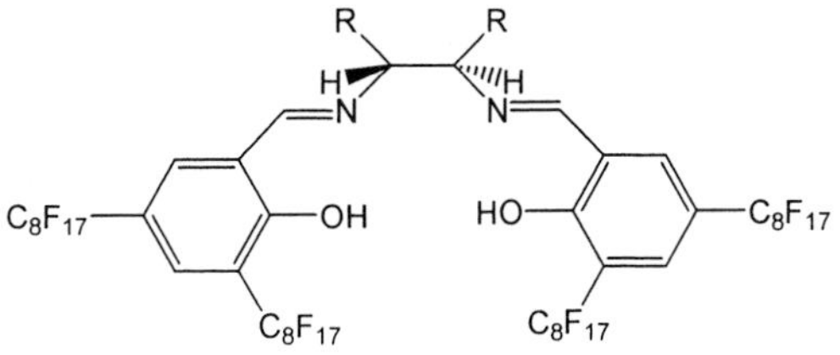

57

Under fluorous biphasic conditions, the activity of rhodium complexes with polymers soluble in organofluorine solvents and containing triphenylphosphine groups has been studied in hydroformylation [174], and the activity of rhodium complexes with polymers containing diphenylphosphine groups was investigated in hydrogenation [175]. In both cases, the catalytic activity was high.

In hydrogenation of various substrates, the activities of rhodium complexes **58–60** with ligands were different. The complex **58** was most active: its activity was comparable to the activity of the Wilkinson complex and exceeded the activity of catalysts based on functionalized polystyrenes, polyethylenes, and amphoteric polymers [175].

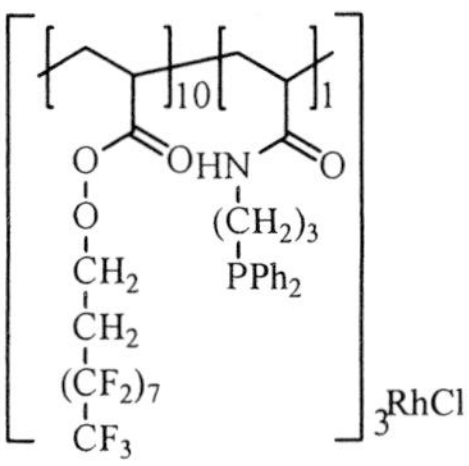

58 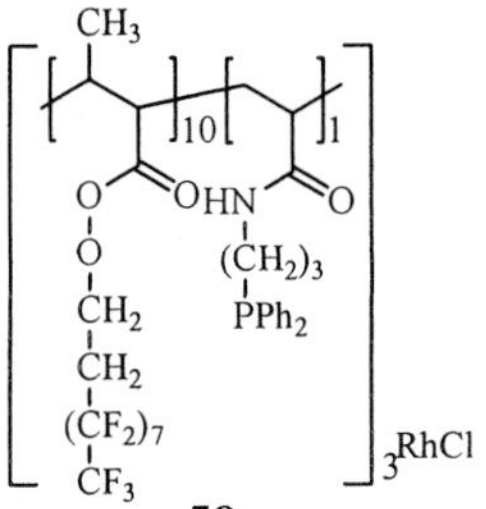   **59** 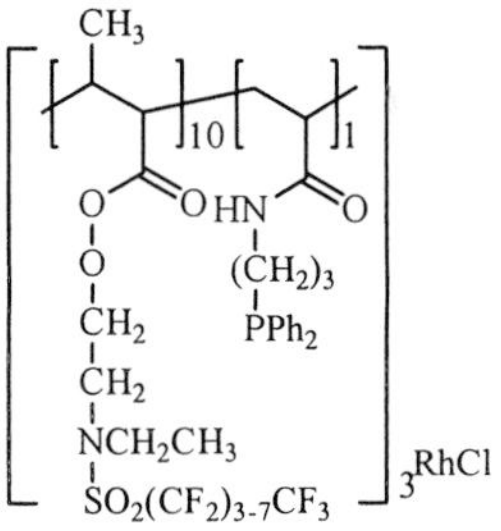   **60**

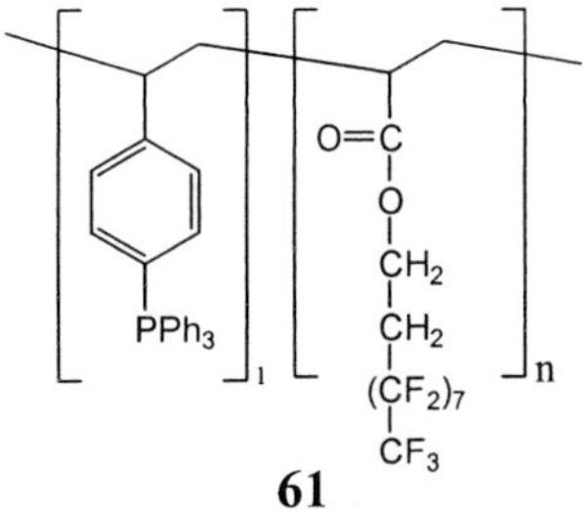

61

A rhodium complex with phosphine **61** in hydroformylation of 1-decene and 1-hexadecene gave linear-to-branched 1-alkene ratios from 4.8:1 to 5.9:1 at a turnover frequency of 136 h^{-1}. For 1-hexene, this quantity under similar conditions was 1456 h^{-1}.and the catalyst remained active up to the substrate-to-rhodium ratio of 200,000:1 [174].

This catalyst was also active in hydroformylation of acrylates in supercritical carbon dioxide; in this solvent, the rate of hydroformylation of linear alkenes was very low [176].

11.5 Metal Complexes with Ligands Based on Other Polymers

Soluble catalysts have also been designed using ligands synthesized on the basis of such polymers as poly(ethyleneimine), copolymer of vinyl alcohol and vinyl acetate, proteins, etc.

Rhodium complexes **62** with polyethyleneimine modified with triphenyl-phosphine groups [79] were studied as two-phase hydrogenation catalysts. The polyacrylic acid hydrogenation rate was high. The reaction was zeroth order with respect to the substrate, and unlike the ordinary Wilkinson complex, the substrate binding here did not determine the reaction rate. The rate of hydrogenation of 4-pentenoic acid and but-1-en-4-ol was lower and was dependent on the substrate concentration. In hydrogenation of nonpolar substrates, such as 1-hexene, the rate was low and cationic or nonionogenic surfactants must be added [80]. It was shown that, in two-phase hydroformylation, the complex decomposes and a part of the rhodium passes into the organic phase [81].

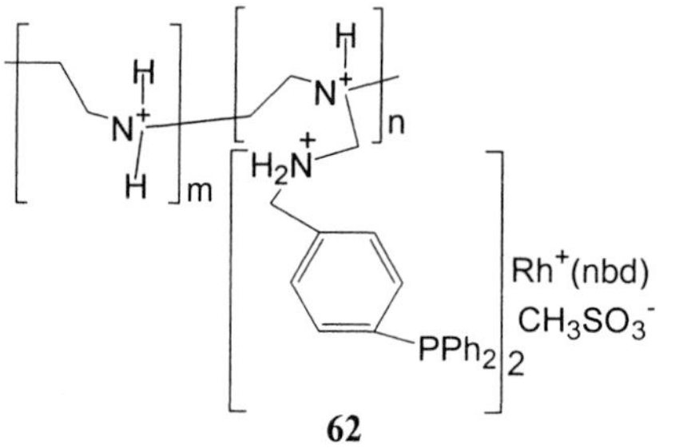

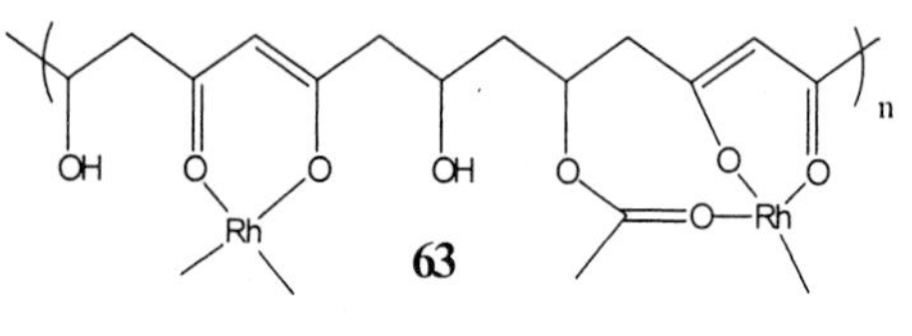

One interesting design of macromolecular soluble metal complexes is binding of a rhodium complex with sulfonated triphenylphosphine to a soluble polyelectrolyte, polydiallyldimethylammonium tetrakis(3,5-bis(trifluorome-thyl)phenyl borate. In this case, the content of phosphine groups enables us to control the catalyst solubility in methanol: at high content of the phosphine counter ion, the polymer is insoluble; and if the ratio of tetraalkylammonium groups to phosphine groups is 4:1 and 10:1, the polymer is soluble in methanol and is an active catalyst of 1-hexene hydroformylation. The catalyst was separated by ultrafiltration [177].

Rhodium complex **63** with oxidized copolymer of vinyl alcohol and vinyl acetate demonstrated high activity in aqueous-phase hydroformylation of linear alkenes, 4-vinylcyclohexane and butyl vinyl ether. In hydroformylation of linear alkenes, the selectivity with respect to aldehyde was from 28 to 77% and substantially decreased with increasing temperature (the second product was 1-octene). In styrene hydroformylation, the product yield was low (13% in 22 h at a synthesis gas pressure of 41.4 bar and 40 °C) but the selectivity with respect to isomeric aldehyde was high [178]. The high isomeric-to-normal aldehyde ratio was also observed in styrene hydroformylation catalyzed by rhodium complexes with human serum albumin, in which there are 30 metal atoms per one protein molecule. Conversion of styrene and other unsaturated compounds – 1-octene, methylstyrene, and methyl vinyl ether – virtually coincided with the theoretically predicted conversion in 22 h at a synthesis gas pressure of 80 bar. For linear alkenes, the isomeric-to-normal product ratio was only 1.2:1 [179].

For asymmetric hydrogenation, soluble polymers containing BINAP ligands **64** obtained by condensation of (S)- or (R)-5,5'-diamino-BINAP with terephthalic anhydride and (2S,4S)-pentandiol have been synthesized.

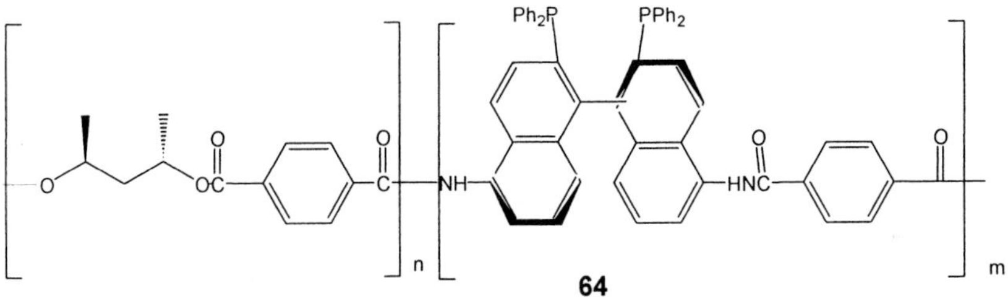

64

Hydrogenation of 2-(6'-methoxy-2'-naphthyl)acrylic acid in water–methanol mixtures, which is catalyzed by a ruthenium complex with this ligand, showed that the activity of the catalyst is much higher than the activity of its monomer analogue and the optical selectivity with respect to the corresponding product exceeds 90%. With increasing methanol content, the catalyst activity decreases. This was explained by conformational changes in the chains and by shrinking of the polymer molecule with increasing methanol content. At high methanol concentrations, these phenomena led to precipitation of the catalyst. As a result, the catalyst could also be reused and retained activity and enantioselectivity after 10 cycles [180].

11.6 Macromolecular Metal Complexes Capable of Forming Host–Guest Complexes

Macromolecules that combine in a single molecule the properties of a metal complex with the ability for molecular recognition and for interfacial transfer of substrate are most interesting in catalysis. [59,63,20,21]. Such catalysts consist of a molecular fragment capable of molecular recognition, a group coordinated to the metal, and a metal site [22,23,181]. In particular, various cyclic isomers, such as cyclodextrins or calixarenes, are used as molecules capable of molecular recognition.

Cyclodextrins are cyclic glucose oligomers consisting of six or more monomer units. The cyclodextrin structure is such that the hydrogens of C–H bonds are directed towards the inside of the cavity of the molecule and hydroxyl groups are directed towards the outside (Fig. 11-7). Therefore, the molecule has a hydrophobic cavity, which owing to its hydrophobicity can bind nonpolar molecules into host–guest complexes and transfer them into the polar phase. This property of cyclodextrins allows us to use them as components of catalytic systems in two-phase catalysis by metal complexes. [20–23, 182–196]; this essentially increased the activity of these catalytic systems.

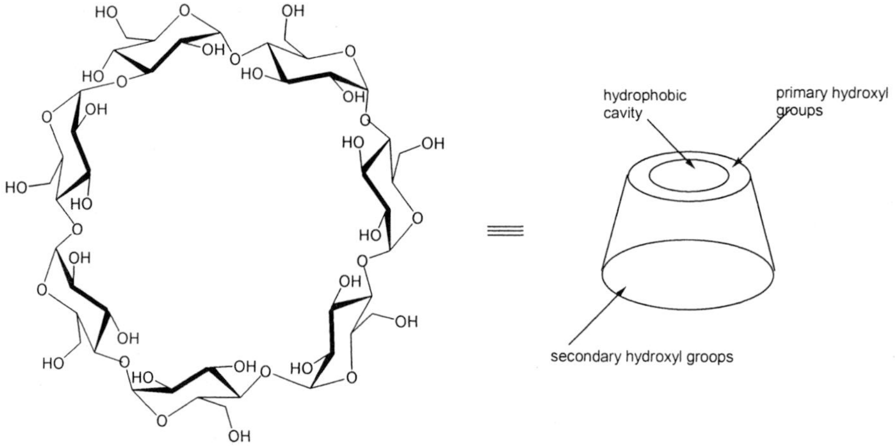

Figure 11-7. Representation of β-cyclodextrin.

Complexes with modified cyclodextrins were initially used as models of hydrolytic enzymes [20,181,197,198]. Then, Breslow and Kato studied such complexes in biomimetic oxidation and epoxidation. Reetz and some other researchers investigated them in hydroformylation. We examined cyclodextrin-based catalysts in hydroxylation and Wacker oxidation. Note that some of these studies demonstrated a substantial increase in the activity of the catalysts in comparison with their analogues prepared as ordinary mixtures, which was due to the cooperative substrate binding.

Breslow synthesized manganese-containing porphyrins **65** and **66**, which selectively catalyzed epoxidation of unsaturated compounds and the benzyl position of C–H bonds in hydrocarbons containing two phenyl groups and steroids owing to the orientation of the substrate in its binding to the catalyst. The activity was maximum for a substrate corresponding to the distance between cyclodextrins in the complex [199–201].

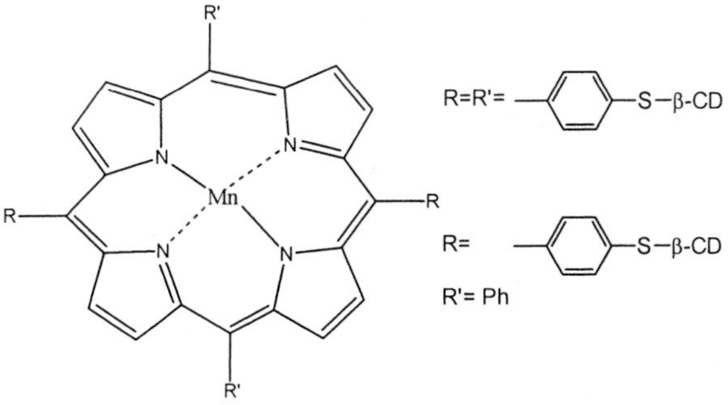

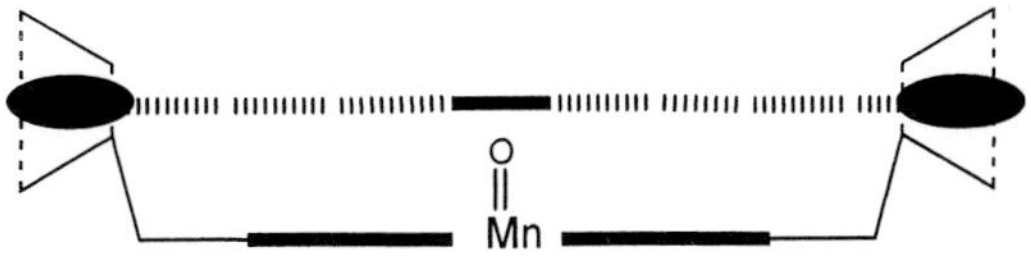

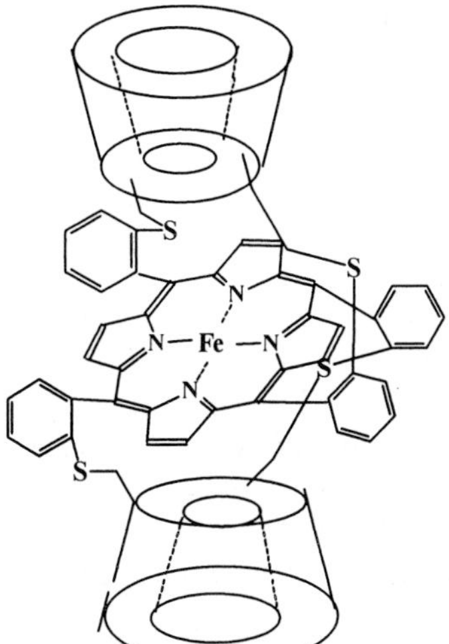

67

For epoxidation of cyclic and branched alkenes by iron-containing porphyrin where cyclodextrins were located above and below the plane of the porphyrin molecule **67**, the substrate selectivity was demonstrated with respect to cyclohexane, for which the reaction rate was several times higher than the rate for styrene and significantly exceeded the rate for a low molecular weight analogue [202].

A variety of rhodium complexes with cyclodextrins modified by various phosphine groups have been synthesized for performing hydroformylation and hydrogenation (Table 11-3).

Reetz showed that these complexes have substrate specificity with respect to aromatic unsaturated compounds in alkene hydrogenation [21,203]. Moreover, rhodium complexes were active in alkene hydroformylation [21,203–205], and so were platinum, rhodium and iridium complexes in hydrogenation of aromatic chloronitro compounds [206].

Table 11-3. Rhodium CD based complexes in hydrogenation and hydroformylation.

Ligand	Reaction	Conditions and yields	References
65,76	Hydrogenation, hydroformylation, hydrogenation of nitroaromatic compounds	25% DMFA in water at 100 atm synthesis gas or 1 atm hydrogen; 25 and 60 °C toluene–water, 20 °C, 20 atm H_2	21,203–206
69–71	Hydrogenation, hydroformylation	25% DMFA in water at 100 atm synthesis gas or 1 atm hydrogen; 25 and 60 °C	203–205
72	Hydroformylation	DMFA at 20 atm synthesis gas, 80 °C	207
77	Hydroformylation	Methanol–water 6:4, 40 atm synthesis gas, 60 °C	208
78	Asymmetric hydrogenation	Methanol, ethanol, tetrahydrofuran, acetone, 1 atm hydrogen	188

Using the complexes in hydroformylation not only increased the reaction rate in comparison with their analogues produced as mixtures but also improved the selectivity with respect to linear aldehyde owing to the orientation of the substrate in cyclodextrin. Unlike catalysis by low molecular weight analogues, the reaction involves even alkenes with an internal double bond (*trans*-3-octene). The catalyst was highly soluble in water, mainly remained in the aqueous phase after the reaction and can be reused.

In hydrogenation of chloronitro compounds, the maximum selectivity was exhibited by platinum complexes with **76** which yielded aminochlorobenzenes at a selectivity above 95% [206].

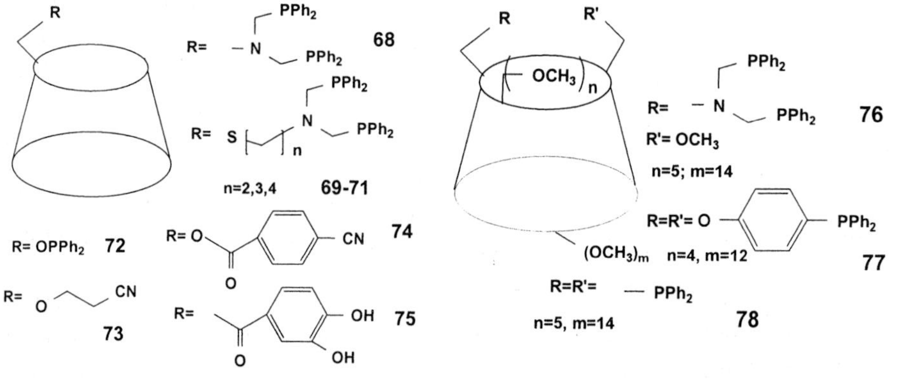

Complexes of cyclodextrin modified with catechol groups **75** proved to be less active in phenol hydroxylation: the main product was catechol and the selectivity with respect to this product exceeded 90%. For copper complexes with unmodified and ethylenediamide-modified cyclodextrins, the selectivity with respect to catechol in this reaction virtually coincided with that predicted theoretically. It was shown that using metal complexes considerably increases the binding constant of phenol, whose orientation in the host–guest complex causes *ortho*-hydroxylation [59,195,209].

A palladium complex with cyclodextrin modified with propionitrile and benzoylnitrile groups **73–74** was active in Wacker oxidation of higher 1-alkenes (Experiment 11-4, Section 11.7), and its activity was much higher than the activity of a catalytic system prepared as a mixture of cyclodextrin and the palladium complex owing to the cooperative substrate binding and to the increase in the stability constant of the catalyst–substrate complex. As in hydroformylation, the catalyst was more active in the reaction with an aromatic substrate, styrene, than with linear alkenes [59,210–211]. The catalyst activity depended on the 1-alkene chain length and was maximum for 1-heptene.

One of the possible advantages of cyclodextrin-based catalysts is related to the difference in the stability constant between optical isomers. For example, in oxidation of a mixture of (*S*)- and (*R*)-pinenes by oxygen under visible light irradiation, iron and copper complexes with porphyrin modified with dimethyl-β-cyclodextrin catalyzed the formation of products primarily from the (*S*)-isomer at an enantioselectivity above 60% [212]. Rhodium complexes of permethylated β-cyclodextrin modified with two triphenylphosphine groups **78** at the neighboring glucose monomer units proved to be active in enantioselective hydrogenation of various substituted α,β-unsaturated acids. The enantioselectivity was from 49 to 92%, depending on the substrate and the solvent [213]. Molybdenum complexes with modified cyclodextrins turned out to be enantioselective catalysts of thioanisole oxidation in water [214].

One more class of compounds capable of forming host–guest complexes are calixarenes, cyclic phenol–formaldehyde oligomers. Water-soluble calixarenes have been used as components of catalytic systems in metal complex-catalyzed Wacker oxidation and cross coupling (Suzuki coupling) [59,215,216]. Macromolecular complexes with modified calixarenes as ligands in two-phase systems have been examined as catalysts of hydroformylation and Wacker oxidation [215,217–218]. Two-phase hydroformylation of 1-octene was carried out in an olefin–water system with no organic solvent added and with rhodium(I) acetylacetonatodicarbonyl and ligand **79** as a catalyst precursor. The reaction was performed at a synthesis gas pressure of 4.0 MPa and 100 °C for 12 h. In this time, 70–95% of the substrate was converted into normal and isomeric aldehydes in the ratio 2:1. It was demonstrated that one and the same portion of the catalyst can be used in several catalytic cycles without loss of activity and selectivity [217].

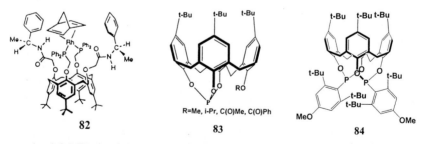

Note that, in alkene hydroformylation, the catalyst was active not only in the reaction with 1-alkenes but also in hydroformylation of such low-activity substrates as 2-octene and 4-octene. In the last case, the reaction yielded a mixture of aldehydes formed as a consequence of isomerization of the initial octene [218].

In Wacker oxidation, palladium complexes with sulfocalixarenes modified with nitrile-containing groups were more active than unmodified sulfocalixarenes. It has been shown that the activity of the catalysts in alkene oxidation from hexene to decene depends on the calixarene cavity size: calix[4]arene is most active in the reaction with 1-hexene and calix[6]arene is most active in 1-octene oxidation [215].

In nonpolar media, complexes with calixarenes modified with phosphine and phosphite groups have been used in hydroformylation [219,220]. Bagatin et al. [221] and Fang et al. [222] described the successful use of the diphenylphosphinated calixarenes **80–81** in hydroformylation of styrene and 1-octene.

Loeber [223] synthesized a complex **82**, which was tested in styrene hydroformylation. Theoretically predicted conversion of the initial compound took place in 48 h at a pressure of 4.0 MPa at a high (95%) selectivity with respect to phenylpropanal. A catalytic system consisting of [Rh(nbd)Cl]$_2$ and phosphine was more active.

At 4.0 MPa and 135 °C, in as little as 6 h, the yield of aldehydes attains its theoretically predicted value, with the normal-to-branched product ratio being close to 1:1. The regioselectivity with respect to 2-phenylpropanal noticeably increases when phosphite of similar structure is applied as a ligand, with the catalytic activity of the system being equally high.

A phosphite-modified calixarene with unsubstituted hydroxyl groups was used as a ligand in 1-hexene hydroformylation catalyzed by rhodium complexes [224]. The reaction was carried out at a synthesis gas pressure of 6.0 MPa and 160 °C. Rh(acac)(CO)$_2$ was a catalyst precursor. In 3 h, the conversion of the initial alkene virtually reached its theoretically predicted value; the yield of aldehydes was 80–85%, and the normal-to-isomeric aldehyde ratio was approximately 1:1. Some similar phosphites **83** were also studied as components of catalytic systems for 1-octene hydroformylation [225]. It was shown that the nature and steric volume of substituent R have no essential effect on the main laws of the process. For example, the conversion was 80–90% at a selectivity with respect to nonanal of about 60% in all cases. The regioselectivity with respect to nonanal was considerably increased to 90–92% by using the chelate biphosphite **84** [220].

11.7 Experimental

Experiment 11-1: Polyethylene Based Catalysts (Section 11.4) [152,153,155,156,160]

Preparation of lithiated ethylene oligomer [153]: A dry 500-mL Fisher–Porter pressure bottle equipped with a magnetic stirring bar was connected to a multiple use (vacuum, nitrogen and ethylene) pressure line through a pressure coupling. The bottle was than evacuated and purged with nitrogen three times. Dry heptane (250 ml), *n*-butyllithium (6.2 mL of a 1.6 N solution in hexane) and *N,N,N′,N′*-tetramethylethylenediamine (2 mL) were added successively by syringe. The bottle was pressurized with ethylene to 30 psig and the oligomerization was carried out at this pressure. The reaction was initially homogeneous and slight warming of the bottle was detectable. After about 30 min., lithiated oligomer precipitated. Depending on the desired size of the final product, the oligomerization was continued for varying lengths of time. Typical times ranged from 36 to 48 h.

Preparation of poly(ethylenediphenylphosphine) PE-PPh$_2$ [152]: To a suspension of oligomer (30 mmol) was added (10.6 mL, 13.3 g, 60 mmol) of chlorodiphenylphosphine at 25 °C. After stirring for 1 h the mixture was heated to 90 °C and vigorously stirred for an additional 12 h. Cooling precipitated the crude product which was placed in Soxhlet apparatus and extracted into the toluene solution. Cooling the hot toluene solution produces a crude phosphine product free of any insoluble salts. This substance was suspended in toluene and allowed to react with trichlorosilane (19.5 g, 145 mmol) for 24 h at 25 °C. The bulk of the excess trichlorosilane was removed under reduced pressure. After the solid oligomer was isolated by filtration, it was suspended in KOH (150 mL of a 1N solution in isopropyl alcohol) and stirred for 24 h. Filtration and washing of the solid product repeatedly with ethanol–water and then ethanol until it was no longer basic led to the product, which was dried in vacuo to yield 18 g of white solid. IR (KBr): 1180, 1120 cm^{-1}; ^{1}H NMR: (toluene d$_8$) d 7.25 (br), 2.1 (m), 1-1.8 (br s), 0.9 (t, CH$_3$) ppm; ^{31}P (toluene) d −16.6 ppm.

Preparation of tetrakis(poly(ethylenediphenylphosphine)) palladium[160]: A 200-mL Schlenk flask was equipped with a magnetic stirring bar, and Pd[P(C$_6$H$_5$)$_3$]. (0.02 mmol) and poly(ethylenediphenylphosphine) (1.25 g, 0.6 mmol diphenylphosphine groups per g) were added. The flask was then purged with argon. Toluene (20 mL) was added, and the suspension was heated to 100 °C. At this point the solution was a light-gold color. Cooling to room temperature precipitated the polyethylene oligomer complex as a light yellow powder. The remaining solution was clear and colorless. The suspension of complexes so formed was then centrifuged, and the supernatant was removed by forced siphon under argon. This process was repeated twice with fresh portions of toluene, and the product was then dried under vacuum. The solid product was somewhat air-sensitive and was therefore stored under argon prior to use.

Preparation of ethylene oligomer ligated rhodium hydrogen [153]: Poly(ethylenediphenylphosphine) (1 g, ~0.4–0.6 mmol of PPh$_2$ per g of polymer) was placed in a 100-mL two-necked round-bottomed flask containing a Teflon-coated magnetic stirring bar. The flask was evacuated and purged several times with argon. Toluene (50 mL) was added, and the contents were briefly heated under argon to produce a clear solution which was then rapidly cooled to room temperature and C$_2$H$_4$ClRh (PPh$_3$)$_2$ (0.1 g, 0.14 mmol) was added. The suspension was stirred under argon at room temperature for 2 h and then heated to 95 °C for 1 h to give an orange-red solution. The solution was cooled to room temperature, and the suspension of the polymeric catalyst was isolated by centrifugation. The product polymeric catalyst was washed with dry toluene until the washings were colorless.

General procedure for carbonylation of aryl iodides [155-156]: Palladium oligomeric catalyst was charged into the reactor and purged three times with argon/vacuum cycle. The reactants and the solvent (o-xylene) were added in a counterflow of argon. The flask was again purged three times with CO/vacuum cycle. The solution was heated to 100 °C and stirred. The course of the reaction with respect to carbon monoxide absorption was followed. After cooling, the catalyst was isolated by centrifugation and supernatant was removed. The catalyst was washed twice with fresh degassed o-xylene, centrifuged and then recycled.

General procedure for hydroformylation of olefins [155]: Rhodium oligomeric catalyst was placed into the reactor and purged three times with argon/vacuum cycles. 1-Dodecene was added and the flask was again purged three times with syngas/argon cycles. The solution was heated to 100 °C and stirred. The reaction was then continued for the indicated time. After cooling the catalyst precipitated and the products remained in the solution. After isolation of the catalyst by centrifugation the products were removed, the catalyst was washed twice with degassed toluene and recycled.

Experiment 11-2: Poly(ethyleneoxide)-Based Catalysts (Section 11.1) [58,226,227]

Synthesis of bromide of monobutyl ether of oligo(ethyleneoxide)[226]: The monobutyl ether of oligoethyleneoxide 5000 (50 g, 0.01 mol) and dry triethylamine (0.93 mL, 0.01 mol) were dissolved in dry toluene (200 mL). Thionyl bromide (2.3 mL, 0.03 mol) was added dropwise with stirring. The mixture stirred under reflux for 4 h. Triethylammonium bromide was removed by filtration, and the filtrate was evaporated. The crude product was dissolved in methylene chloride (40 mL) and added to cold anhydrous ethyl ether (150 mL). The product was collect by filtration.

Synthesis of ligand $Bu(OCH_2CH_2)_n$-$OCH_2CH_2PPh_2$ (17) [227]: The butyl ether of oligo(ethylene oxide) (4 g, 0.8 mmol) in dry benzene (30 mL) was mixed with $PhPCl_2$ (2.5 mL) in toluene (5 mL) and stirred under reflux for 6 h. The solution was added to anhydrous ethyl ether (150 mL) and the product collected by filtration. Further purification was achieved by three precipitations from methylene chloride with ethyl ether. ^{31}P (toluene) δ: –116.

Synthesis of ligand $Bu(OCH_2CH_2)_nCH_2CH_2PPh_2$ (2) [227]: $LiPPh_2$ (1.2 g) was added to the bromide of the monobutyl ether of oligo(ethylene oxide) (1 g) in dry benzene (10 mL) in a glove box under nitrogen and stirred overnight at room temperature. The reaction mixture was brought to reflux and held at this temperature for 2 h and then permitted to cool. Lithium bromide was removed by filtration, and the filtrate added to anhydrous ethyl ether (10 mL). This precipitation process was repeated three times. ^{31}P (toluene) δ: –14 ppm.

Rhodium complexes with 2 or 17 [58]: Acetylacetonate dicarbonyltris rhodium (I) (0.2 g) in toluene (15–50 mL) was mixed with the ligand (1 g) and stirred overnight at reflux. The solution was then added to anhydrous ethyl ether (150 mL) and the product isolated by filtration. Purification was achieved by precipitation from methylene chloride with ethyl ether.

General procedure for hydroformylation using rhodium complexes [58]: Corresponding amounts of complex, toluene (1.5 mL), olefin (1.34 mmol) and water (1.5 mL) were added to a controlled-temperature steel autoclave (volume 30 mL) with a magnetic stirrer. The autoclave was purged of synthesis gas of 0.1–1.5 MPa. The solution was heated to 50 °C and stirred. After the reaction the aqueous layer containing the metal complex was separated and the reaction mixture was analyzed by GC.

Experiment 11-3: PNIPAM-Based Catalysts (Section 11.2) [83,228]

N-Isopropylacrylamide/N-(acryloxy)succinimide co-polymer (PNIPAM/NASI) [83,228]: N-Isopropylacrylamide (5.0 g, 0.044 mol) and N-(acryloxy)-succinimide (0.150 g, 0.88 mmol) were dissolved in *tert*-butyl alcohol (30 mL) at 70 °C under nitrogen. AIBN (30 mg) in *tert*-butyl alcohol (1 mL) was added to the solution. The reaction mixture was stirred at 70 °C for 20 h. It was cooled to room temperature. The solvent was evaporated in vacuo. The polymer (2.2 g) was isolated by succes- sive precipitations from a THF solution into hexane and from a methanol solution into dry diethyl ether: IR (KBr): 3435, 2970, 2924, 1814, 1784, 1734, 1652, 698 cm^{-1}.

Synthesis of PNIPAM-phosphine 49 [83]: PNIPAM/NASI 5 (5 g) was dissolved in THF (150 mL) and (3-aminopropyl)diphenylphosphine (0.75 g) in THF (10 mL) was added under nitrogen. The mixture was stirred at room temperature for 5 h, and an excess of NH$_3$ (saturated aqueous solution) was added to quench any unreacted active succinimide groups. The mixture was again stirred at room temperature for 5 h, and any precipitate was removed by centrifugation. The product polymer was then purified by precipitating the polymer with excess of hexane. Two such reprecipitations yielded a polymer that was then used in the chemistry described below. IR spectroscopy showed that this phosphine-containing polymer **49** did not have any succinimide groups (no signals at 1810, 1780, and 1735 cm^{-1}). The polymer **49** had a poor quality 1H NMR (CDCl$_3$.) spectrum with broad peaks at 1.13,1.20–2.30, 2.60, 3.70, 4.00, and 7.30–7.41 ppm. However, the ^{31}P NMR spectrum of **12** (CDCl$_3$) clearly showed a broad singlet at 15–15.5 ppm and a very minor peak (5%) at 15.34 ppm (which corresponds to the phosphine oxide).

Synthesis of (PNIPAM-PPh$_2$)$_4$Pd(0) [83]: A solution of **49** (250 mg) in a THF:H$_2$O mixture (20 mL of 90:10) was prepared under nitrogen. A solution of Pd(0)(dba)$_2$ (25 mg, 0.044 mmol) in the same solvent (10 mL) was also prepared under notrogen. After the Pd(0) precatalyst had dissolved (~5 min), the precatalyst solution was added to the polymer solution using a double-tipped needle and nitrogen pressure. Within ~5 min the resulting solution changed from dark purple to golden yellow. The solution was then stirred for an additional 15 min. After this time, petroleum ether (~70 mL) was added under nitrogen to the polymer solution, precipitating the complex as a golden-yellow solid. The light-yellow supernatant was then removed, leaving the polymer catalyst ready for dissolution and use in aqueous solution.

Synthesis of (PNIPAM-PPh$_2$)$_3$RhCI [83]: A solution of **12** (75 mg) in 90% EtOH (40 mL) was prepared under nitrogen. To this was added a solution of RhCl((C$_2$H$_4$)$_2$)$_2$ (8 mg, 0.0086 mmol) in dioxane (1 mL). The resulting light-yellow solution was used for hydrogenations after stirring under nitrogen flow for approximately 15 min.

General procedure for hydrogenations using PNIPAM-bound cationic Rh(I) catalysts [83]: A Schlenk reaction flask with an adapter leading to a H$_2$ gas buret was purged with H$_2$ for 10 min. Then allyl alcohol (35 mmol) and the cationic Rh(I) catalyst described above were introduced. Water was added to this flask so that a total volume of 50 mL of reaction solution was reached. The temperature of the reaction was controlled through an ice bath or an oil bath Measurement of the H$_2$ uptake volume at a constant stirring rate was used to follow the rate of the hydrogenation. To confirm that hydrogenation occurred, an aliquot of the reaction mixture was analyzed by GC.

Carbon–carbon coupling reactions by PNIPAM-bound Pd(0) catalyst[83]: 2-Iodophenol (2.2 g, 10 mmol), phenylacetylene (1.03 g, 10 mmol), and triethylamine (2.6 g) were dissolved in THF:water (20 mL, 80:20 by volume) and added to a reaction flask containing a solution of PNIPAM-Pd(0) (30 mL, 0.05 mM in Pd in THF:water, 80:20 by volume) under argon. The mixture was stirred for 5 min and then CuI (50 μmol) was added to the solution. The reaction temperature was raised to ~50 °C, and the mixture was stirred at this temperature for 8 h. After cooling to 25 °C, addition of hexane (100 mL) to the solution led to precipitation of PNIPAM-Pd(0) catalyst. The precipitate was collected by filtration and could be reused in a fresh reaction mixture. The solution was collected, washed with brine (3 × 20 mL), and dried over MgSO$_4$. After the solvent was removed, the pure product was obtained by column chromatography (silica gel, hexane:ethyl acetate, 10:1) in 78% yield.

Experiment 11-4: β-Cyclodextrin-Based Pd Complexes in Wacker Oxidation (Section 11.6) [210]

β-CD-CH$_2$CH$_2$CN ligand 73: Acrylonitrile (0.053 g) was added at 55 °C to a paste compounded of β-CD (1.13 g) and NaOH (0.5 ml of a 10% solution). The mixture was stirred for 30 min and then poured out into methanol (30 mL). A white precipitate formed and was filtered off, dried in vacuum and recrystallized twice from methanol. ^{1}H NMR: 3.19–3.27 – H4 and H2 β-CD, 3.51–3.9 (H3 and H5, H6 β-CD), 4.5–4.7 (H1 β-CD), 3.47, 3.38 ppm (H CH$_2$CH$_2$CN); ^{13}C NMR: 118–119 (CN), 16 and 56 (CH$_2$CH$_2$), 61 and 69.5 (C6 β-CD), 72.6 (C2 and C5 β-CD), 73.9 (C3 β-CD), 81.9 (C4 β-CD), 102.6 ppm (C1 β-CD); FTIR: 2255 (ν CN) cm^{-1}.

Complexes (β-CD-CH$_2$CH$_2$CN)(RCN)PdCl$_2$ (R = –CH$_3$, Ph): Palladium complexes were synthesized from the corresponding ligand and bis(acetonitrile) palladium dichloride or bis(benzonitrile) palladium dichloride at ligand-to-palladium ratios of 1:1 and 2:1. To β-CD-CH$_2$CH$_2$CN (1.2 g) in water (10 mL) Pd(RCN)$_2$Cl$_2$ (1 mmol) was added. The reaction was performed by heating to 80 °C for 6 h under Ar. The solution was cooled and acetone was added. The precipitate formed was filtered off and dried in vacuum. Ratio Pd^{2+}/ligand = 1.

FTIR: 2295 (v CN) cm^{-1}.

Wacker oxidation of an alkene in a two-phase system: A reaction mixture, containing olefin (2 mmol), the corresponding amount of a metal complex (0.05 mmol), the cocatalyst (0.2 mmol) and water (2 mL) was introduced into an autoclave with a volume of 15 mL. The autoclave was purged with oxygen (0.5–1.5 MPa). The solution was vigorously stirred at 65 °C for 4 h. The autoclave was than cooled, pentane (2 mL) was added and the aqueous layer with catalyst was separated from the solution of the product in pentane.

11.8 References

1. A.D. Pomogailo, *Catalysis by Polymer-Immobilized Metal Complexes,* Gordon and Breach Sci. Publ., Amsterdam, 1998.
2. B. Clapham, T.S. Reger, K.D. Janda, *Tetrahedron* **2001**, *57*, 4637.
3. D.C. Sherrington, *J. Poly. Sci. A: Polym. Chem.* **2001**, 39, 2364.
4. S. Bhattacharyya, *Comb. Chem. High. Througput Screen* **2000**, 3, 65.
5. D.C. Sherington, *Catal. Today* **2000**, *57*, 87.
6. A .Choplin, F. Quignard, *Coord. Chem. Rev.* **1998**, *180*, 1679.
7. Ch. Saluzzo, R. ter Halle, F. Touchard, F. Fache, E. Schulz, M. Lemaire, *J. Organomet. Chem.* **2000**, *603*, 30.
8. Ph.L. Osburn, D.E. Bergbreiter, *Prog. Polym. Sci.* **2001**, *26*, 2015.
9. D.E. Bergbreiter, *J. Polym. Sci., Part A: Polym. Chem.* **2001**, 39, 2351.
10. U. Kragl, T. Dwars, *Trends in Biotechnology* **2001**, *19*, 442.
11. P.H. Toy, K.D. Janda, *Acc. Chem. Res.* **2000**, *33*, 546.
12. P. Whentworth, K.D. Janda, *Chem. Commun.* **1999**, 1917.
13. P. Whentworth, *Trends Biotechnol.* **1999**, *17*, 448.
14. D.E. Bergbreiter, *Catal. Today* **1998**, *42*, 389.
15. D.E. Bergbreiter, *Macromol. Symp.* **1996**, *105*, 9.
16. E.A. Karakhanov, E.B. Neymerovets, A.G. Dedov, *Applied. Organometal. Chem.* **1990**, *4*, 1.
17. B. Cornils, H. Hermann (Eds.), *Aqueous Phase Organometallic Catalysis – Concepts and Applications*, Wiley VCH, Weinheim, 1998.
18. K. Takahasho, *Chem. Rev.* **1998**, *98*, 2013.
19. A.E. Rowan, J.A.A.W. Elemans, R.D. Nolte, *Acc. Chem. Res.* **1999**, *32, 995.
20. R. Breslow, S.D. Dong, *Chem. Rev.* **1998**, *98*, 1997.
21. M.T. Reetz, *Topics in Catalysis* **1997**, *4*, 187.
22. E. Rizzarelli, G. Vecchio, *Coord. Chem. Rev.* **1999**, *188*, 343.
23. Robertson, S. Shinkai, *Coord. Chem. Rev.* **2000**, *205*, 157.
24. V. Bohmer *Angew. Chem. Int. Ed.* **1995**, 34, 713.
25. S. Zalipsky, *Bioconjugate Chem.* **1995**, *6*, 150.
26. D.E. Bergbreiter, *Med. Res. Rev.* **1999**, *19*, 439.
27. V.P.J. Torchilin, *Controlled Release* **2001**, *73*, 137.
28. R.B.J. Greenwald, *Controlled Release* **2001**, *74*, 159.
29. R.B. Greenwald, C.D. Conover, Y.H. Choe, *Critical Reviews in Therapeutic*

Carrier System. **2000**, *17*, 101.

30. S. Zalipsky, J.M. Harris, *ACS Symposium Series* **1997**, *680*, 1.
31. Th. Borrmann, H.W. Roesky, U. Ritter, *J .Mol. Cat. A: Chem.* **2000**, *153*, 31.
32. M.V. Sharikova, N.V. Kolesnichenko, N.A. Markova, E.V. Slivinskii, *Russ. Chem. Bull.* **1999**, *48*, 701.
33. M. Haumann, H. Koch, P. Hugo, R. Schomacker, *Appl. Cat. A. General* **2002**, *225*, 239.
34. H. Apler, K. Januszkiewicz, D.J.H. Smith, *Tetrahedron Lett.* **1985**, *26*, 2263.
35. Rico, F. Couderc, E. Perez, J.P. Laval, A. Lattes, *J.Chem.Soc., Chem.Commun.* **1987**, 1205.
36. E.A. Karakhanov, E.A. Ivanova, S.Yu. Narin, A.G. Dedov, *Vestn. Mosk. Univ. Ser. 2 Chem.* **1989**, *30*, 510.
37. E.A. Karakhanov, S. Narin, A.G. Dedov, *Catal. Lett.* **1989**, *3*, 31.
38. E.A. Karakhanov, Ch.R. Pulippurasseril, T.Yu. Filippova, S.V. Egasaryants, A.G. Dedov, *Neftehimiya* **1999**, *32*, 1598.
39. E.A. Karakhanov, A.L. Maksimov, S.V. Egazaeyants, S.V. Kardashev, A.D. Sedih, S.S. Minosyants, *Neftehimiya* **2001**, *41*, 293.
40. K. Dallman, R. Buffon, W. Loh, *J. Mol. Cat. A: Chem.* **2002**, *178*, 43.
41. M.E. Wilson, R.G. Nuzzo, G.M. Whitsides, *J. Am. Chem. Soc.* **1978**, *100*, 2269.
42. D.E. Bergbreiter, L. Zhang, V.M. Mariagnanam, *J. Am. Chem. Soc.* **1993**, *115*, 9295.
43. X. Zheng, J. Jiang, X. Liu, Z. Jin, *Catal. Today* **1998**, *44*, 175.
44. J.A.J. Breuzard, M.L. Tommasino, M.C. Bonnet, M. Lemaire, *J. Organometal. Chem.* **2000**, *616*, 37.
45. Z. Jin, X. Zheng, B. Fell, *J. Mol. Cat. A: Chem.* **1997**, *116*, 55.
46. Y. Wang, J. Jiang, R. Zhang, X. Liu, Z. Jin, *J. Mol. Cat. A: Chem.* **2000**, *157*, 111.
47. J.Jiang, Y.Wang, Ch.Liu, F.Han, Z.Jin *J.Mol.Cat. A. Chem.* **1999**, *147*, 131.
48. Z. Jin, X. Zheng, in: *Aqueous Phase Organometallic Catalysis – Concepts and Applications*, B. Cornils, H. Hermann (Eds.), Wiley VCH, Weinheim, 1998, 233.
49. R. Chen, X. Liu, Z. Jin, *J. Organomet. Chem.* **1998**, *571*, 201.
50. J. Jiang, Y. Wang, Ch. Liu, Q. Xiao, Z. Jin, *J. Mol. Cat. A: Chem.* **2001**, *171*, 85.
51. J. Jiang, J. Mei, Y. Wang, F. Wen, Z. Jin, *Appl. Cat. A: Gen.* **2002**, *224,* 21.
52. Y.H. Wang, X.Z. Liu, C.Y. Jia, Z.L. Jin, *Chinese Chem. Lett.* **2002**, *13*, 468.
53. J.A. Loch, C. Borgmann, R.H. Crabtree. *J. Mol. Cat. A: Chem.* **2001**, *170*, 75.
54. E.A. Karakhanov, Yu.S. Kardasheva, A.L. Maksimov, V.V. Predeina, E.A. Runova, *J. Mol. Cat. A: Chem.* **1996**, *107*, 235.
55. E.A. Karakhanov, A.L. Maksimov, V.V. Predeina, T.Yu. Filippova, A.M. Utukin, *Macromol. Symp.* **1996**, *105*, 67.
56. E.A. Karakhanov, E.A. Runova, Yu.S. Kardasheva, Ye.B. Neymerovets, *Neftehimiya* **1996**, *36*, 321.
57. E.A. Karakhanov, Yu.S. Kardasheva, E.A. Runova, V.A. Semernina, *Neftehimiya* **1998**, *38*, 32.
58. E.A. Karakhanov, Yu.S. Kardasheva, E.A. Runova, V.A. Semernina, *J. Mol. Cat. A: Chem.* **1999**, *142*, 339.
59. E.A. Karakhanov, A.L. Maximov, T.Yu. Filippova, Y.S. Kardasheva, T.S. Buchneva, M.A. Gayevskiy, A.Yu. Zhuchkova, *Polym. Adv. Tech.* **2001**, *12*, 161.

60. R. Chen, J. Jiang, Y. Wang, Z. Jin, *J. Mol. Cat. A: Chem.* **1999**, *149*, 113.
61. E.A. Karakhanov, Yu.S. Kardasheva, E.A. Runova, M.V. Terenina, *Petrochemistry* (Rus) **2000**, *40*, 429.
62. Ph. Stosel, H.A. Mayer, F. Auer, *Eur. J. Inorg.Chem.* **1998**, 37.
63. E.A. Karakhanov, A.L. Maximov, V.V. Predeina, T.Yu. Filippova, S.A. Martynova, I.N. Topchieva, *Catal. Today* **1998**, *44*, 189.
64. E.A. Karakhanov, A.S. Redchin, V.V. Predeina, A.L. Maksimov, *Petrochemistry* **1997**, *37*, 240.
65. E.A. Karakhanov, A.L. Maximov, V.V. Predeina, T.Yu. Filippova, A.Ya. Restakyan, *Macromol. Symp.* **1998**, *131*, 87.
66. Q. Yao, *Angew. Chem. Int. Ed..* **2000**, 39, 3896.
67. D.E. Bergbreiter, Ph.L. Osburn, Y-S. Liu, *J. Am. Chem. Soc.* **1999**, *121*, 9531.
68. D.E. Bergbreiter, Ph.L. Osburn, W.E.M. Sink, *J. Am. Chem. Soc.* **2000**, *122*, 9058.
69. T.S. Reger, K.D. Janda, *J. Am. Chem. Soc.* **2000**, *122*, 6229.
70. H. Han, K.D. Janda. *J. Am. Chem. Soc.* **1996**, *118*, 7632.
71. H. Han, K.D. Janda, *Tetrahedron Lett.* **1997**, *38*, 1527.
72. H. Han, K.D. Janda, *Angew. Chem. Int. Ed. Engl.* **1997**, *36*, 1731.
73. C. Bolm, A. Gerlah, *Angew Chem. Int. Ed. Engl.* **1997**, *36,* 741.
74. C. Bolm, A. Gerlah, *Eur. J. Chem.* **1998**, 21.
75. H. Guo, X. Shi, Z. Qiao, Sh. Hou, M. Wang, *Chem. Com.* **2002**, 119.
76. Q.H. Fan, G.J. Deng, Ch. Ch. Lin, A.S.C. Chan, *Tetrahedron: Assymetry* **2001**, *12*, 1241.
77. Q.H. Fan, G.J. Deng, X.-M. Chen, W.-Ch. Xie, D.-Zh. Jiang, D.-S. Liu, A.S.C. Chan, *J. Mol. Cat. A: Chem.* **2000**, *159*, 37.
78. P. Guerreiro, Ratovelomanana-Vidal, J.-P. Gener Ph. Dellis, *Tetrahedron Lett.* **2001**, *42*, 3423.
79. T. Malmstrom, C. Andersson, *Organometallics* **1995**, *14*, 2593.
80. T. Malmstrom, C. Andersson, *J. Mol. Cat. A: Chem.* **1997**, *116*, 237.
81. T. Malmstrom, C. Andersson, J. Jortkjaer, *J. Mol. Cat. A: Chem.* **1999**, *139*, 139.
82. D.E. Bergbreiter, Y-S. Liu, *Tetrahedron Lett.* **1997**, *38*, 3703.
83. D.E. Bergbreiter, B.L. Case, J.W. Cartaway, *Macromolecules* **1998**, *31*, 6053.
84. N.A. Ajjou, H. Alper, *J. Am. Chem. Soc.* **1998**, *120*, 1466.
85. T. Malmstrom, C. Andersson, *Chem. Commun.* **1996**, 1135.
86. T. Malmstrom, C. Andersson, *J. Mol. Cat. A: Chem.* **1999**, *139*, 259.
87. T. Malmstrom, C. Andersson, *J. Mol. Cat. A: Chem.* **2000**, *157*, 79.
88. M. Kimura, E. Dakeno, E. Adachi, T. Kjoyama, K. Hanabusa, H. Shirai, *Macromol. Chem. Phys.* **1994**, *195*, 3499.
89. M. Kimura, T. Nishigaki, T. Koyama, K. Hanabusa, H. Shirai, *Macromol. Chem.Phys.* **1994**, *195*, 3499.
90. M. Kimura, T. Nishigaki, T. Koyama, K. Hanabusa, H. Shirai, *Reactive & Functional Polymers* **1996**, *29*, 85.
91. D.E. Bergbreiter, J.W. Caraway, *J. Am. Chem. Soc.* **1996**, *118*, 6092.
92. D.E. Bergbreiter, J. Frels, K. Heuz, *Reactive & Functional Polymers* **2001**, *49*, 249.
93. D.E. Bergbreiter, Ph.L. Osburn, C. Li, *Org Lett.* **2002**, *4*, 737.
94. D.E. Bergbreiter, Y-S. Liu, Ph.L. Osburn, *J. Am. Chem. Soc.* **1998**, *120*, 4250.

95. D.E. Bergbreiter, Y-S. Liu, *Tetrahedron Lett.* **1997**, *38*, 7843.
96. F. Zhao, M. Shirai, M. Arai, *J. Mol. Catal. A: Chem.* **2000**, *154*, 39.
97. D.E. Bergbreiter, Ph.L. Osburn, A. Wilson, E. Sink; *J. Am. Chem. Soc.* **2000**, *122*, 9855.
98. D.E. Bergbreiter, Ph.L. Osburn, J.D. Frels; *J. Am. Chem. Soc.* **2001**, *123*, 11105.
99. G.R. Newkome, C.N. Moorefielf, F. Vogtel; *Dendric Molecules: Concepts, Synthesis, Perspectives*, VCH, Weinheim, 1996.
100. Archut, F. Vogtel; *Chem. Soc. Rev.* **1999**, *27*, 233.
101. S. Hecht, J.M.J. Frechet; *Angew. Chem. Int. Ed.* **2001**, *40*, 9174.
102. H.-F. Chow, T.K.-K. Mong, M.F. Nongrum, C.-W. Wan, *Tetrahedron* **1998**, *54*, 8543.
103. L.J. Twyman, A.S.H. King, *J. Chem. Research-S* **2002**, 43.
104. I.P. Beletskaya, A.V. Chuchurjukin, *Uspehi Hmii* **2000**, *69,* 699.
105. D. Astruc, F. Chardac, *Chem. Rev.* **2001**, *101*, 2991.
106. L.J. Twyman, A.S.H. King, I.K. Martin, *Chem. Soc. Rev.* **2002**, *31*, 69.
107. R. Kreiter, A.W. Kleij, R.J.M.K. Gebbink, G.van Koten, *Top. Curr. Chem.* **2001**, *217*, 163.
108. D.K. Smith, F. Diederich, *Chem. Eur. J.* **1998**, *4*, 1353.
109. H.M. Janssen, E.W. Meijer, *Chem. Rev.* **1999**, *99*, 1665.
110. M.E. Piotty, F. Rivers, R. Bond, C.J. Hawker. J.M.J. Frechet, *J. Am. Chem. Soc.* **1999**, *121*, 9471.
111. L.J. Twyman, A.E. Beezer, R. Esfand, M.J. Hardy, J.C. Mitchell, *Tetrahedron Lett.* **1999**, *40*, 205.
112. T. Mizugaki, M. Ooe, K. Ebitani, K. Kaneda, *J. Mol. Cat. A: Chem.* **1999**, *145*, 329.
113. Ch. .Kollner, B. Pugin, A. Togni, *J. Am. Chem. Soc.* **1998**, 120, 10,247.
114. G. van Koten, T.B.H. Jastrzebski, *J. Mol. Cat. A: Chem* **1999**, *146*, 317.
115. M.T. Reetz, G. Lohmer, R. Schickardi, *Angew. Chem. Int. Ed. Engl.* **1997**, *36*, 1526.
116. L. Ropatz, R.E. Morris, D.F. Foster, D.J. Cole-Hamiton, *Chem. Commun.* **2001**, 361.
117. L. Ropatz, R.E. Morris, G.P. Schwartz, D.F. Foster, D. Cole-Hamiton, *Inorg. Chem. Commun.* **2000**, *3*, 714.
118. Gong, Q. Fan, Y. Chen, H. Liu, Ch. Chen, F. Xi, *J. Mol. Cat. A: Chem.* **2000**, *159*, 225.
119. S.B. Garber, J.S. Kingsbury, B.L. Gray, A.H. Hoveda, *J. Am. Chem. Soc.* **2000**, *122*, 8168.
120. T.M. Truka, R.H. Grubbs, *Acc. Chem. Res.* **2001**, *34*, 18.
121. P. Wijkens, J.T.B.H. Jasterzebski, Van der Schaaf, R. Kolly, A. Hafner, G. van Koten, *Organic. Lett.* **2000**, *2*, 1621.
122. H. Zeng, G.R. Newkome, C.L. Hill, *Angew. Chem. Int. Ed.* **2000**, *39*, 1772.
123. R. Schneder, C. Weber, I. Kollner, A. Togni, *Chem. Commun.* **1999**, 2415.
124. Togni, N. Bieler, U. Bruckhardt, C. Kollner, G. Pioda, R. Schneider, A. Schnyfer, *Pure Appl. Chem.* **1999**, *71*, 1531.
125. D. de Groot, P.G. Emmerink, C. Coucke, J.N.H. Reek, P.C.J. Karmer, P.W.N.M. van Leeuwen, *Inorg. Chem. Commun.* **2000**, *3*, 711.

126. L.A.van de Kuil, D.M. Grove, J.W. Zwikker, L.W. Jenneskens, W. Drenth, G. van Koten, *Chem. Mater.* **1995**, *6*, 1675.
127. A.W. Kleij, R.A. Goassage, J.T.B.H. Jasterzbeski, J. Boersma, G. van Koten, *Angew. Chem. Int. Ed.* **2000**, *39*, 176.
128. M. Petrucci-Samija, V. Guillemette, M. Dasgupta, A.K. Kakkar, *J. Am. Chem. Soc.* **1999**, *121*, 1968.
129. M. Kimura, M. Kato, T. Muto, K. Hanabusa, H. Shirai, *Macromolecules* **2000**, *33*, 1117.
130. G.E. Oosterom, R.J. van Haaren, J.N.H. Reek, P.C.J. Kamer, P.W.N.V. van Leeuwen, *Chem. Commun.* **1999**, 1119.
131. P. Bhyrappa, J.K. Young, J.S. Moore, K.S. Suslick, *J. Am. Chem. Soc.* **1996**, *118*, 5708.
132. P. Bhyrappa, J.K. Young, J.S. Moore, K.S. Suslick, *J. Mol. Catal. A: Chem.* **1996**, *113*, 109.
133. M. Kimura, Y. Sugihara, T. Muto, K. Hanabusa, H. Shirai, N. Kobayashi, *Chem. Eur. J.* **1999**, *5*, 3495.
134. H. Brunner, J. Furst, J. Ziegler, *J.Organomet. Chem.* **1993**, *454*, 87.
135. H. Brunner, J. Furst, *Tetrahedron* **1994**, *50*, 4303.
136. H. Brunner, P. Bublak, *Synthesis* **1995**, 36.
137. H. Brunner, M. Janura, S. Stefaniak, *Synthesis* **1998**, 1742.
138. Q.-H. Fan, Y.-M. Chen, X.-M. Chen, D.-Z. Jiang, F. Xi, A.S.C. Chan, *Chem. Commun.* **2000**, 789.
139. Y.-M. Chen, T.-F. Wu, J.-G. Deng, H. Liu, D.-Z. Jiang, M.C.K. Choi, A.S.C. Chan, *Chem. Commun.* **2001**, 1488.
140. E. Bayer, V. Schurig, *Angew. Chem.* **1975**, *6*, 212.
141. E. Bayer, V. Schurig, *Chemtech.* **1976**, *6*, 212.
142. J. Leito, D. Milstein, R. Albright, J. Minklewics, B. Gates, *Chemtech.* **1983**, 46.
143. T. Jongsama, G. Challa, P.W.N.M. Van Leeuwen, *Polymer* **1992**, *33*, 161.
144. D.E. Bergbreiter, *ACS Symp. Ser. No 308, (Polym. Reag. Catal.),* American Chemical Society, Washington DC, 1986, p. 1.
145. G. Challa, *J. Mol. Cat.* **1980**, *21*, 14.
146. K. Ohkudo, K. Fujimoro, K. Yoshinada, *Inorg. Nucl. Chem. Lett.* **1979**, *15*, 231.
147. C. Pan, H. Zong, *Macromol. Symp.* **1994**, *80*, 265.
148. P.W.N.M. van Leeuwen, *Macromol. Symp.* **1984**, *80*, 241.
149. D.E. Bergbreiter, J.R. Blanton, *J.Org. Chem.* **1985**, *50*, 5828.
150. D.E. Bergbreiter, J.R. Blanton, *J. Chem. Soc., Chem. Commun.* **1985**, 337.
151. D.E. Bergreiter, R. Chandran, *J. Am. Chem. Soc.* **1985**, *107*, 4792.
152. D.E. Bergbreiter, J.R. Blanton, R. Chandran, M.D. Hein, K.-J. Huang, D.R. Treadwell, S.A. Walker, *J. Polym. Sci. A: Polym. Chem.* **1989**, *27*, 4205.
153. D.E. Bergbreiter, R. Chandran, *J. Am. Chem. Soc.* **1987**, *109*, 174.
154. E.A. Karakhanov, E.A. Runova, G.V. Berezkin, E.B. Neimerovets, *Macromol. Symp.* **1994**, *80*, 231.
155. E.A. Karakhanov, G.V. Berezkin, E.B. Neimerovets, E.A. Runova, L.E. Rozanceva, *Petrochemistry (Rus.)* **1992**, *32*, 229.
156. E.A. Karakhanov, E.B. Neimerovets, E.A. Runova, Yu.S. Kardasheva, *Vestnik Moskovskogo Universiteta ser. 2 Himiya.* **1996,** *37*, 533.

157. D.E. Bergbreiter, R. Chandran, *J. Chem. Soc., Chem. Commun.* **1985**, 1936.
158. D.E. Bergbreiter, R. Chandran, *J. Org. Chem.* **1986**, *51*, 4754.
159. D.E. Bergbreiter, Z. Chen, R. Chandran, *Macromolecule*s **1985**, *18*, 1055.
160. D.E. Bergbreiter, D.A. Weatherford, *J. Org. Chem.* **1989**, *54*, 2726.
161. J.C. Phelps, D.E. Bergbreiter, G.M. Lee, R. Villani, S.M. Weinreb, *Tetrahedron Lett.* **1989**, *30*, 3915.
162. D.E. Bergbreiter, A.W. Samuel, *J. Org. Chem.* **1989**, *54*, 5138.
163. M. Vogt, *Dissertation Technische Hochschule* Aachen, August 26, 1991.
164. I.T. Horvath, J. Rabai, *Science* **1994**, 266.
165. J.J.J. Juliette, I.T. Horvath, J.A. Gladysz, *Angew. Chem. Int. Ed. Engl.* **1997**, *36*, 1610.
166. B. Cornils, Angew. Chem. Int. Ed. Engl. **1997**, 36, 2057.
167. M. Cavazzini, F. Montanari, G. Pozzi, S. Quici, *J. Fluorine Chem.* **1999**, 94, 183.
168. E.G. Hope, A.M. Stuart, *J. Fluorine Chem.* **1999**, *100, 75.*
169. W. Keim, M. Vogt, P. Wasserscheid, B. Driesen-Holscher, *J. Mol. Cat. A: Chem.* **1999**, *139*, 171.
170. G. Pozzi, F. Montanari, S. Quici, S. Fontana, *Tetrahedron Lett.* **1997**, *38*, 7605.
171. G. Pozzi, F. Montanari, S. Quici, *Chem. Commun.* **1997**, 69.
172. J.-M. Vincent, A. Rabion, V.K. Yachandra, R.H. Fish, *Angew. Chem. Int. Ed. Engl.* **1997**, *36*, 2346.
173. G. Pozzi, F. Cinato, F. Montanari, S. Quici, *Chem. Commun.* **1998**, 877.
174. W. Chen, L. Xu, J. Xiao, *Chem.Commun.* **2000**, 839.
175. D.E. Bergbreiter, J.G. Franchina, B.L. Case, *Org. Lett.* **2000**, *2*, 393.
176. Y. Hu, W. Chen, A.M. Banet, J.A. Iggo, J. Xiao, *Chem. Commun.* **2002**, 789.
177. E. Schwab, S. Mecking, *Organometallics* **2001**, *20*, 5504.
178. H. Alper, J. Chen, *J. Am. Chem. Soc.* **1997**, *119*, 893.
179. M. Marchetti, G. Mangano, C. Botteghi, *Tetrahedron Lett.* **2000**, *41*, 3717.
180. Q.H. Fan, C.Y. Ren, C.H. Yeung, W.H. Hu, A.S.C. Chan, *J. Am. Chem. Soc.* **1999**, *121*, 7407.
181. Tabusi, *Acc. Chem. Res.* **1982**, *15*, 66.
182. H.A. Zahalka, K. Januszkiewcz, H. Apler, *J. Mol. Cat.* **1986**, *35*, 249.
183. Harada, Y. Hu, S. Takahasi, *Chem. Lett.* **1986**, 2083.
184. Y. Shiraishi, H. Tomita, K. Fujiki, H. Hirai, *Reactive & Functional Polymers* **1998**, *36*, 99.
185. Ph. Kalck, L. Miquel, M. Dessoudeix, *Catal. Today* **1998**, *42*, 431.
186. M. Dessoudeix, Ph. Urrutigoity Kalck, *Eur. J. Chem.* **2001**, 1797.
187. J.R. Anderson, E.M. Campi, W.R. Jackson, *Catal. Lett.* **1999**, *9*, 55.
188. E. Monflier, E. Biolet, Y. Barbaux, A. Mortreux, *Angew. Chem. Int. Ed. Engl.* **1994**, *33*, 2100.
189. E. Monflier, S. Tilloy, G. Fremy, Y. Barbaux, A. Mortreux, *Tetrahedron Lett.* **1995**, *36*, 387.
190. E. Monflier, S. Tilloy, G. Fremy, Y. Castanet, A. Mortreux, *Tetrahedron Lett.* **1995**, *36*, 9481.
191. E. Monflier, A. Mortreux, Y. Castanet, S. Tilloy, *Tetrahedron Lett.* **1998**, *39*, 2959.
192. E. Monflier, S. Tilloy, E. Blouet, Y. Barbaux, A. Mortreux, *J. Mol. Catal. A: Chem.* **1996**, *109*, 27.

193. E. Monflier, S. Tilloy, F. Bertoux, Y. Castanet, A. Mortreux, *New J. Chem.* **1997**, *21*, 857.
194. T. Mathivet, C. Meliet, Y. Castanet, A. Mortreux, L. Caron, S. Tilloy, E. Monflier, *J. Mol. Catal. A: Chem.* **2001**, *176*, 105.
195. E. Karakhanov, A. Maximov, P. Poloznicova, K. Suslov, in H. Otsuka (Ed.), *Studies in Surface Science and Catalysis, V.121, Science and Technology in Catalysis*, Elsevier, Amsterdam, **1998**, 127.
196. P. Ravi, R. Ravichandran, S. Divakar, *J. Mol. Catal. A: Chem.* **1999** *148*, 145.
197. B. Zhang, R. Breslow, *J. Am. Chem. Soc.* **1997**, *119*, 1676.
198. D.S. Dong, R. Breslow, *Tetrahedron Lett.* **1998**, *39*, 9343.
199. R. Breslow, X.J. Zhang, R. Xu, M. Maletic, R. Merger, *J. Am. Chem. Soc.* **1996**, *118*, 11678.
200. R. Breslow, X.J. Zhang, Y. Huang, *J. Am. Chem. Soc.* **1997**, *119*, 4535.
201. R. Breslow, J. Yan, S. Belvedere, *Tetrahedron Lett.* **2002** *43*, 363.
202. Y. Kuroda, T. Hiroshige, H. Ogoshi, *J. Chem. Soc., Chem. Commun.* **1990**, 1594.
203. M.T. Reetz ,S.R. Waldwogel, *Angew. Chem. Int. Ed. Engl.* **1997**, *36*, 865.
204. M.T. Reetz, *J. Heterocyclic Compounds* **1998**, 35, 1065.
205. M.T. Reetz, *Catal. Today* **1998**, *42*, 399.
206. M.T. Reetz, C. Frombgen, *Synthesis.* **1999**, 1555.
207. R.M. Dishpande, A. Fukuoka, M. Ichikawa, *Chem. Lett.* **1999**, 13.
208. D. Armspach, D. Matt, *Chem. Commun.* **1999**, 1073.
209. E.A. Karakhanov, Y.S. Kardasheva, A.V. Kirillov, A.L. Maximov, V.V. Predeina, E.A. Runova, *Macromol. Symp.* **2000**, *156*, 137.
210. E.A. Karakhanov, A. Maximov, A. Kirillov, *J. Mol. Catal A: Chem.* **2000**, *157*, 25.
211. E.A. Karakhanov, A.Ya. Guchkova, A.L. Maximov, T.Yu. Filippova, L. Karakpetyan, *Neftehimiya* **2002**, *42*, 233.
212. L. Weber, I. Imiolczyk, G. Hauf, D. Rehorek, H. Henning, *J. Chem. Soc., Chem. Commun.* **1992**, 301.
213. T.Y. Wong, Ch. Yang, K.-Ch. Ying, G. Jia, *Organometallics* **2002**, *21*, 1782.
214. M. Bonchio, T. Carofiglio, F. Difuria, R. Fornasier, *J. Org. Chem.* **1995**, *60*, 5986.
215. T. Buchneva, E. Karakhanov, A. Maksimov, *J. Mol. Cat. A: Chem.* **2002**, *184*, 11.
216. M. Baur, M. Frank, J. Schatz, F. Schildbach, *Tetrahedron* **2001**, *57*, 6985.
217. S. Shimizu, Y. Shirakawa, Y. Sasaki, C. Hirai, *Angew. Chem. Int. Ed.* **2000**, *39*, 1256.
218. Y. Shirakawa, S. Shimizu, Y. Sasaki, *New J. Chem.* **2001**, *25*, 777.
219. R. Paciello, L. Siggel, M. Roper, *Angew. Chem. Int. Ed.* **1999**, *38*, 1922.
220. Z. Csok, G. Szalontai, G. Czira, L. Kollar, *J .Organometal. Chem.* **1998**, *570*, 23.
221. I.A. Bagatin, D. Matt, H. Thonnessen, P.G. Jones, *Inorg. Chem.* **1999**, *38*, 1585.
222. X. Fang, B.L. Scott, J.G. Watkin, C.A.G. Carter, G.J. Kubas, *Inorg. Chim. Acta.* **2001**, *317*, 276.
223. C. Loeber, C. Wieser, D. Matt, A. de Cian, J. Fischer, L. Toupet, *Bull. Soc. Chim. Fr.* **1995**, *132*, 166.
224. C.J. Cobley, D.D. Ellis, A.G. Orpen, P.G. Pringle, *J. Chem. Soc., Dalton Trans.* **2000**, 1109.
225. F.J. Parlevliet, C. Kiener, J. Fraanje, K. Goubitz, A.L. Spek, P.C.J. Kamer, P.W.N.M. van Leeuwen, *J. Chem. Soc., Dalton Trans.* **2000**, 1113.

226. G. Johannson, A. Hartmann, P. Alberston, *Eur. J. Biochem.* **1983**, *254*, 12579.
227. M. Harris, E.C. Struck, M.G. Case, M.S. Paley, M. Yalpani, J.N. van Alstine, D.E.J. Brooks, *J. Polym. Sci., Polym. Chem. Ed.* **1984**, *22*, 341.
228. F.M. Winnik, *Macromolecules* **1990**, *23*, 233.

12 The State of the Art and Perspectives in Catalysis by Heterogenized Polymer-Bound Metal Complexes

Anatolii Pomogailo

In the 1970s functionalized polymer materials received much attention. We may note that advanced research included solid-phase Merrifield's syntheses of peptides [1], polymer-immobilized catalysts [2,3], enzymes [4–6], ligand-exchange chromatography [7] as well as separation and ultrafiltration [8], membrane technology [9], etc. This demonstrates the significant potential and directions of polymer-mediated synthesis. In particular, the use of polymer-supported metal complex catalysts for catalytic organic synthesis (hydrogenation, selective oxidation, hydroformylation, polymerization, etc.) offers many opportunities. Indeed, profitable technologies and processes have been developed from them. In practice two main types of polymer supports are most often used. The first group is highly cross-linked macroporous resins with a high specific surface. The ligand groups are mainly localized on a surface where the contacts with a substrate and a reagent occur. Such systems in their functionalities are similar to catalysts on mineral supports [10]. The second ones include slightly cross-linked microporous (gelatinous) polymers. These swell in solvents. The active sites are localized not only on the surface of the polymer but also in its volume.

Although all of these fields are summarized in the literature, time after time there are different opinions on the subject of the results that may be achieved in catalysis by using MX_n immobilized on a tailor-made polymer. Some researchers believe this approach to be a panacea to eliminate all disadvantages of traditional homogeneous catalysts; others emphasize shortcomings of such systems in comparison with typical heterogeneous catalysts. Here we try to analyze in detail the possibilities and perspectives of polymer-immobilized metal complex catalysts for a wide variety of reactions in organic synthesis.

12.1 The Place of Polymer-Immobilized Metal Complexes in Catalysis

A homogeneous metal complex catalyst can be considered as an individual particle (complex or a combination of complexes) located in the same medium as the reagents (usually solution). The reaction occurs in the transition metal coordination sphere by spatial and electronic rearrangement of complex ligands via the formation of active intermediates [11–13]. All catalytic centers are identical and equally accessible to the reagents (at least theoretically). Their activity and selectivity can be regulated by changes in the metal complex ligand environment.

In heterogeneous catalysis the reaction occurs on the interphase boundary. Adsorption of reagents on the catalyst surface possessing collective electronic properties is followed by their activation. However, not all active centers located on the catalyst surface are equivalent. It is possible that only one of the set is capable of catalyzing the desired reaction, whereas others are inactive and even responsible for some side reactions [14].

The main advantage of homogeneous catalysts is that they provide the opportunity to study the structure of the active center and the mechanism of the catalytic action. By contrast, heterogeneous catalysts are very attractive from a technological viewpoint. They are more efficient and easily separated from reaction products, and sometimes may be regenerated and recycled. This is very important in the case of expensive (for example, rhodium is 300 times more expensive than gold) or toxic metal complexes. This is a very practical problem in the manufacture of foodstuffs, pharmaceutical products, dyes, etc., which is why heterogeneous chemical processes are more widespread in the industry. Much effort has been made to develop new methods for efficient catalyst separation (effective distillation, ultrafiltration, ion exchange, etc.).

The wish to combine the advantages and exclude the shortcomings of homogeneous and heterogeneous catalysts has led to the creation of "hybrid-phase" catalytic systems. These feature a substantial restriction of the mobility of the metal complexes by chemically binding them to a polymeric or mineral support. It seems that this problem can be solved by implanting such complexes into firm elastic granules, fibers, membranes and powders, which can be easily isolated from the reaction mixture and used many times in the same way as a typical heterogeneous catalyst. At the same time other important problems may also be solved.

- Immobilization of metal complexes usually increases their efficiency because their concentration on the polymer support is not limited by solubility (as a rule the set of suitable solvents is restricted by the solubility of MX_n).
- Because of the limited mobility of metal complexes chemically bound with the polymer the state of an infinitely dilute system is imitated.
- Use of metallopolymer catalysts sometimes allows an increase in the operating temperature up to technologically acceptable ranges (homogeneous catalysts usually operate below 100 °C).
- Immobilized complexes are close to homogeneous catalysts in their

reproducibility while the surface structure of heterogeneous catalysts depends strongly on the method of synthesis and the "prehistory" of the catalyst.
- "Hybrid-phase" catalysts are less sensitive to traces of oxygen and moisture as well as to "soft" ligands (poisons) compared with mobile analogues. This is favored by the hydrophobic character of the polymer surface.
- Operating with such catalysts eliminates (or at least reduces) corrosion of apparatus. Also there is no need to cover the wall of the reactor with protective materials.

The advantages of polymer-immobilized catalysts are thus numerous.

However such systems are of interest from more than the practical viewpoint. Experience of synthesis of these catalysts showed that combinations of macroligands and metallocomplexes resulted in unusual properties. Thus, when MX_n binds with a polymer a substitution of some ligands of the metallocomplex by functional groups of a polymer chain takes place. This leads to a change in the catalytic properties of the metallocomplex because the stereochemical surroundings of the metal ion become different. Sometimes by changing these factors (the number of substituting ligands, the kind of functional groups of polymer chain, the structure of the macroligand [15–17], the degree of its swelling, etc.) we may regulate the course of a catalytic process. Moreover, the macroligand may be considered not only as an inert carrier but as favoring a preferred orientation of the substrate with respect to the active center and promoting selectivity. In particular, such a function is the characteristic feature of protein macromolecules in enzyme catalysis.

Much progress has been achieved in homogeneous catalysis by modeling enzyme processes. Therefore a brief comparative summary of relationships between homogeneous and enzyme catalysts as well as immobilized metallocomplexes would be useful. Metal enzymes which catalyze numerous biochemical processes (see Chapter 2) are macromolecular proteins containing metal ions in their active centers. Since in metal enzymes metal ions are incorporated into the structure of biomembranes they may be considered as immobilized complexes embedded in swollen polymer matrices (gels). The high catalytic activity of enzymes is specified by a number of controlling factors, such as structural conformity (some parts of biopolymers possess enhanced affinity to a substrate and form a specific network in the vicinity of the active center whose form and size correspond to those of substrate; this increases substrate concentration, so-called substrate enrichment, and provides favorable substrate orientation), multicenter catalysis (functional groups of the biopolymer coordinate and additionally activate substrate molecules), etc. The same effect can be realized for immobilized catalysts. To describe the kinetics of a number of reactions catalyzed by immobilized metal complexes (for example, hydrogenation, oxidation, etc.) the Michaelis–Menten equation commonly used for biological processes can be applied (Eq. (12-1)) [18].

$$E + S \underset{k_{-1}}{\overset{k_1}{\rightleftharpoons}} ES \xrightarrow{k_2} E' + P \tag{12-1}$$

In Eq. (12-1) E is the catalyst, S is the substrate, P is the products, ES is the Michaelis–Menten complex. E′ is rapidly oxidized to E, the Michaelis constant is equal to $K_M = (k_{-1} + k_2)/k_1$. The initial (w_0) and final (w_M) reaction rates are related by Eq. (12-2).

$$\frac{1}{w_0} = \frac{K_M}{w_M[S]} + \frac{1}{w_M} \tag{12-2}$$

All the statements mentioned above clarify the general direction of development. High-performance catalysts operating as enzyme models should be developed. Consequently, immobilized catalysts combine the main features of homogeneous, heterogeneous and enzymatic catalysts.

The essential role of the polymer consists of the stabilization of catalytically active intermediates which are formed both at the binding of MX_n with a macroligand (for example, monomerization of dimer complexes) and in the course of a catalytic reaction (in particular, establishing isolated low-valence metal ions, prevention of their aggregation, formation of coordinated unsaturated complexes, etc.). The factors determining these effects have been already mentioned and can be summarized as follows:

- the structure and length of chain, and its mobility;
- the ability of a macromolecular helix to constrict and to unwrap depending on the type of solvent and temperature; electrical charges onto chain;
- the kind and part of functional groups bonded with metal (degree of "loading" of chain), etc.

Detailed analysis of these is of paramount interest for establishing the scientific basis for the construction of this type of catalyst because while synthesizing them we can already take into account the main requirements of metallocomplex catalysis. Although a number of original papers, as well as monographs and reviews, have already been published analyzing the different problems on this subject (see [3,17,19–22]) this fruitful approach is still far from realizing its potential. Almost every concrete case is accompanied by a number of complications. It is difficult to create bound complexes of a single type on the polymer support (mobile and immobilized active centers are possible) and overcome the problem of the low stability of the metal–polymer bond under the conditions of the catalyzed reaction. The interaction of MX_n with a macroligand is sometimes complicated by side reactions including their aggregation (see Chapter 3). In addition, some solvents may cause swelling of the macrocomplex.

The aim of the chapter is to indicate the direction of the development of catalysis by polymer-bound metallocomplexes, on the one hand, and to discover the specific role of macroligands in catalytic processes, on the other. Note that in this chapter we will analyze the catalytic properties of complexes bound only to synthetic (or occasionally natural) polymers but not to inorganic carriers.

12.2 The Features of Hydrogenation Reactions Catalyzed by Macromolecular Metal Complexes

Catalytic hydrogenation is a key reaction in organic chemistry. The metal complexes of the Pt group and related metal complexes with polymers are catalysts of the following processes [23,24]:

- hydrogenation of alkenes, aromatic and heteroaromatic substrates;
- partial or exhaustive hydrogenation of oils and unsaturated fatty acids;
- reduction of various functional groups(nitro, carbonyl, nitrile, etc.);
- synthesis of polyols from carbohydrates;
- reduction of alkynes to *cis*- or *trans*-alkenes;

12.2.1 General Kinetic Regularities

The bond energy of H_2 at 437 kJ mol^{-1} is rather high. Consequently, activation of the hydrogen molecule in the intermediate complexes is necessary. Substrate or hydrogen activation may be the controlling reaction factor, depending on the catalyst type, substrate nature and reaction conditions. In the situation where the catalyst surface is filled with hydrogen molecules the value of the activation energy is calculated to be 20–25 kJ mol^{-1}. When the adsorption ability of the substrate increases, there are simultaneously molecules of H_2, substrate and solvent arranged on the catalyst surface (adsorption activation energy for H_2 molecules being 30–40 kJ mol^{-1}). A further rise in the adsorption capacity of the substrate results in the reaction becoming limited by H_2 adsorption (40–60 kJ mol^{-1}). A change in the value of the hydrogen–surface bond energy, even by 4–8 kJ mol^{-1}, may lead to a significant change in hydrogenation selectivity. In some cases the interaction of the substrate with the catalyst may be so great in value (as it is for nitrobenzene) that it is accompanied by the formation of anion radicals, which are capable of interaction with any form of hydrogen. Thus, $PdCl_2$ adsorbed by poly(benzimidazole) (PBI) is reduced to PBI-Pd^0 by hydrogen, producing anchored Pd clusters for which the hydrogenation activity decreases in the sequence: dienes, allyl alcohol > alkene-1 > alkene-2 > 2-methylalkene-2 [25]. The sorption ability of substrates follows the same order. The polymer support probably has an additional effect on surface Pd atoms through moderation of the metal–substrate bond strength. It is important that activation of the catalyst (formation of active centers) is not accompanied by rupture of the chemical bonds between the metal and the macroligand. Special attention is paid to supports prepared from chelate-like polymers, which lead to strong bonding of MX_n. Polymers with pentane-2,4-dione groups, ethylenediimine, 8-amino-quinoline, Schiff bases, etc., are often used for these purposes [26]. These catalysts are stable, and the recycling number in hydrogenation reach 300–1000 or more. $PdCl_2$ complexes immobilized by Indian silk exhibit high activity levels in the reduction of isoprene, hexyne-1, cyclopentadiene and nitrobenzene [27]. The reduced Pd^0 atoms coordinated with protein donor groups ($-NH_2$, $-OH$, $-CONH_2$, $-COOH$) are also active centers of hydrogenation. Metal atoms are retained in the micelles of silk fibers after reduction. Such a catalyst may be recovered a few tens of times without any loss of activity.

Metal complexes immobilized on macroligands with π-donor ability, such as polyacetylene (PA), are active catalysts for acetylene-substrate (S) hydrogenation. The interaction of transition metal ions with PA leads, probably, to unpairing of PA electrons thereby increasing the free radical concentration, and as a consequence there is partial reduction of the transition metal at the stage of its bonding with the support [28]. The transformations of hydrated substrates on such catalysts probably pass through a dissociative chemisorption stage. There are many cases when substrate and products of reaction are irreversibly adsorbed on active centers of the catalyst.

The general scheme of hydrogenation by immobilized catalysts may be represented by two main stages. The first one is the interaction of a substrate molecule with a fixed center containing activated hydrogen. This is the reversible step (Eq. (12-3)).

$$]\text{- MH + S} \underset{k_{-1}}{\overset{k_1}{\rightleftharpoons}}]\text{- M - SH} \tag{12-3}$$

In the second stage the adduct irreversibly reacts with hydrogen yielding the product with subsequent regeneration of the catalytic active center ((Eq. 12-4).

$$]\text{- MSH + H}_2 \xrightarrow{k_2}]\text{- MH + SH}_2 \tag{12-4}$$

The rate of hydrogenation follows the general equation (Eq. (12-5)).

$$w = -\frac{dS}{dt} = \frac{k_1 k_2 [S][H_2]]\text{-MH]}}{k_{-1} + k_1[S] + k_2[H_2]} \tag{12-5}$$

This equation is easily transformed into the Michaelis–Menten form [18] applied for homogeneous enzyme catalysis. The rate of the reaction rises with increasing substrate concentration tending to the constant value w_{max} (Eq. (12-6)), where K_m is the Michaelis factor corresponding to the substrate concentration value at which the reaction rate is one-half of w_{max}.

$$w = - dS/dt = k_2[H_2]~]- MH]$$

(12-6)

From a kinetic analysis of the most frequently observed dependences of hydrogenation rate on the concentration of catalyst, for substrates of different type and H_2 pressure we may reach the following conclusions:

- At low H_2 pressure the rate of hydrogenation increases directly in proportion to the catalyst concentration.
- The dependence of the hydrogenation rate may be either linear in H_2 concentration or not. The latter dependence has been observed for the case when $k_1[S] \gg (k_{-1} + k_2[H_2])$ and becomes linear when $1/w = f(1/P_{H_2})$. This procedure is often employed for calculation of k_1 and k_2 values.
- The hydrogenation rate is ordinarily proportional to the substrate concentration. However some examples are known where the reaction rate is independent of [S], i.e. it is zeroth order with respect to the substrate concentration. Some examples of a non-linear dependence of hydrogenation rate on the substrate concentration are also known. A plot of $1/w$ against $f(1/[S])$ yields a straight line. The values of k_1 and k_{-1} can be calculated from this plot.

While avoiding detailed comparison of immobilized catalysts with others, we shall only note that the parallels sometimes reveal kinetic features that are intrinsic to homogeneous hydrogenation catalysts, and at other times those that resemble heterogeneous catalysts.

At present a lot of evidence demonstrates the similarity of action of homogeneous and heterogeneous hydrogenation catalysts (e.g. poisoning with the same catalytic contaminants, the reduction of olefin hydration rate in the presence of acetylene strongly bound with the metal ion or conjugated diene substrates, compounds of M–H type being detected as intermediates, multi-graded hydrogen addition, etc.). In other words, many processes carrying out in immobilized catalysis are closely similar to those for typical heterogeneous catalysts, for example, as in case of standard 5% Rh/C or the blacks of corresponding metals. Even addition of polystyrene (PS) or copolymers of styrene cross-linked with divinylbenzene (CSDVB) to Rh- or Pt-blacks at wide ratios of active phase/polymer results in an increase in the rate of hydrogenation of nitrobenzene and dimethylethynylcarbinol by 1.5 to 2 times. We shall show later that immobilization of metal complexes leads to the recognition of further similarities between homogeneous and heterogeneous catalysis.

12.2.2 The Influence of the Nature of a Metal Complex on Its Catalytic Properties

The structural parameters of cross-linked polymers are also important in the case of macromolecular metal complexes (MMC) in heterogeneous catalysis. The value of specific surface (S_{sp}, m^2 g^{-1}) and catalyst porosity (v_{poros}, mL g^{-1}) significantly affect the catalytic activity. The S_{sp} and v_{poros} values of copolymers of styrene with divinylbenzene (CSDVB) is a function of the amount of cross-linking agent used (10, 20, 30 and 40 mol%) and amount to 14 and 0.040; 182 and 0.192; 222 and 0.333; 357 and 0.392, respectively [3]. Even simple grinding of CSDVB granules (20% DVB) with fixed titanocene [29] leads to a many-fold rise in catalyst activity (Table 12-1).

Table 12-1. Initial rates of olefin hydrogenation, w_0 (mL of H$_2$/mmol of Ti), by cyclopentadiene (Cp)Ti^{4+} compounds.

Catalyst	Substrate		
	Hexene	Cyclohexene	Cyclooctene
pTiCl$_3$	2.9	2.0	0.5
CSDVB-CpTiCpCl$_2$ (granules)	42	30	7.8
CSDVB-CpTiCpCl$_2$ (powder)	390	194	60.5

The original method of enhancing the value of S_{sp} of such catalysts is to apply metal complexes immobilized on supports of mixed polymer-oxides. These are produced, as a rule, by graft polymerization of different monomers to the surface of oxides (SiO$_2$, MgO, Al$_2$O$_3$, etc.), polymerization of monomers in the presence of oxides, polymer-analogous transformations of grafted polymer (for example, Rh complex contains such groups in the ligand surrounding [30,31]), etc. The specific surface for such catalysts is about 200–700 m^2 g^{-1}. A review of such methods is given in [32]. For illustration let us consider the copolymerization of methacrylate with a mixture of *m*- and *p*-DVB in the presence of SiO$_2$. As a result a cross-linked support SiO$_2$-polymethacrylate (PMAc) is produced. Subsequent fixation of PdCl$_2$·2H$_2$O or H$_2$PtCl$_6$ leads to immobilized Pd^{2+} and Pt^{2+} complex formation, and they reveal high levels of activity in hydrogenation of unsaturated hydrocarbons, nitro compounds and ketones. Such complexes are stable with a rather high turnover number (above 2000) (Figure 12-1) and can easily be isolated from the reaction medium by decantation.

Thus, the supports of mixed type combine the advantages of those of the individual polymer and oxide.

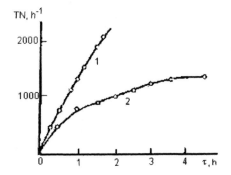

Figure 12-1. Dependence of the turnover number (TN) on time of hydrogenation of nitrobenzene by catalysts SiO_2-PMAc-Pd ([COOH]/[M] = 10; [catalyst]= 0.02 mM; [substrate]=0.05 M).

The mechanism of the influence of ligand properties on the catalytic process is rather different. Both the nature and the concentration of the functional groups (L) are important. The functional groups of the polymer support may not only take part in MX_n binding but also influence the catalytic reaction by promoting a rise in the local concentration of active centers through side reactions or by the protection of active centers from the action of impurities or catalytic poisons. This may lead to catalysts that are essentially different in their activity and selectivity. A similar effect has been observed in styrene hydrogenation by Pd, immobilized on functionalized CSDVB (4–6 mol% DVB) [33]. It has been noticed that the reaction rate increases for supports with functional groups capable of π-complex formation with styrene, thus increasing the lifetime of the absorbed substrate complexes (so called substrate enrichment). In allylacrylate hydrogenation, the catalyst activity correlates with both π-acceptor and polar properties of functional groups of the polymer matrix. The groups capable of associating with substrates are $-NO_2$, and –OMe; others (–OH, $-NH_2$, $-NR_3$, $-CH_2OCH_2-$) may affect the substrate behavior through polar factors. Polymers with a hydrophobic surface cause a decrease in the hydrogenation rate of nonpolar substrates. Even a small amount (7%) of introduced 2,4-dinitrophenyl groups leads to a significant decrease in the 4-vinylcyclohexene hydrogenation rate.

In the optimal variants the ligand groups of polymers possess a dual function: they reveal electron-donor properties if the coordinating transition metal is in a high valence state and electron-acceptor ones for stabilizing the metal in a low or zero valence state (after reduction). Note that the *functional group–substrate* interaction plays a significant role in the case of highly cross-linked polymer supports in which the proportion of vacant groups exceeds those taking part in binding MX_n by scores of times.

The dependence of the hydrogenation rate on the size of the substrate molecule (geometrical or matrix selectivity) is an intrinsic feature of catalysis by immobilized metal complexes. Generally the rate of hydrogenation decreases:

- with elongation of the olefin chain;
- for substrates with inter-chain or inter-ring double bonds (the rate of hexene-1 is 20 times greater than that of hexene-2 or cyclohexene);
- for a substrate change from *cis*-isomers to *trans*-isomers;
- with a rise in the number of substituents in the vicinity of the double bond in both linear and cyclic olefins (donor substituents accelerate alkene reductions while acceptors retard);
- if bulk substituents are introduced into a benzene ring (the ratio of hydrogenation rates of benzene, toluene, *o*-xylene, *tert*-butylbenzene is 1:0.62:0.27:0.01:0.0);
- for nitro-compounds containing substituents.

The latter can be demonstrated by the examples of hydrogenation of different nitrocompounds in the presence of Pd-poly(methyl methacrylate) [34] (Table 12-2).

Table 12-2. The initial rates of hydrogenation of different nitrocompounds in the presence of Pd-poly(methyl methacrylate)(PdPMMA).

Substrate	Initial rate ($mL\ min^{-1}$)	Ratio of the initial rates of substituted substrate and nitrobenzene hydrogenation	Substrate	Initial rate, ($mL\ min^{-1}$)	Ratio of the initial rates of substituted substrate and nitrobenzene hydrogenation
Nitrobenzene	46.1	1.00	*o*-Nitrotoluene	34.3	0.74
p-Nitroaniline	54.5	1.18	*o*-Nitrophenol	48.0	1.04
m-Nitroaniline	50.0	1.08	Nitromethane	9.6	0.21
o-Nitroaniline	28.8	1.63	Nitroethane	4.6	0.10

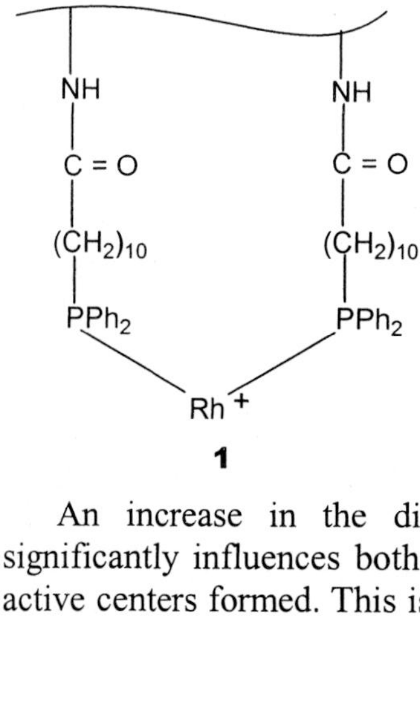

NH   NH
| |
C = O   C = O
| |
(CH₂)₁₀   (CH₂)₁₀
| |
PPh₂   PPh₂
         \   /
           Rh⁺

1

An increase in the distance between the polymer backbone and L significantly influences both the MX$_n$ bonding process and the activity of the active centers formed. This is the result of an increase in active-center mobility facilitating the interaction with substrates. As a rule the activity of such a catalyst is close to that for homogeneous catalysts. This can be attributed to the presence of a rather long "anchor"-like chain between the backbone and the fixed reactive site as exemplified in **1**.

12.2.3 Formation of Coordinatively Unsaturated and Isolated Complexes

As we shall show, the processes of monomerization of di- or multimeric complexes play an essential role in metallocomplex catalysis. This is because there must be coordination vacancies at a metal center for activation of a substrate or H_2. Coordination vacancies may be created either when the catalyst is prepared or during the catalytic reaction (dynamic effect). In this connection, it is important to remember that supported macrocomplexes are dynamic systems, and that unsaturated vacancies of metal centers may either appear or disappear (Fig. 12-2).

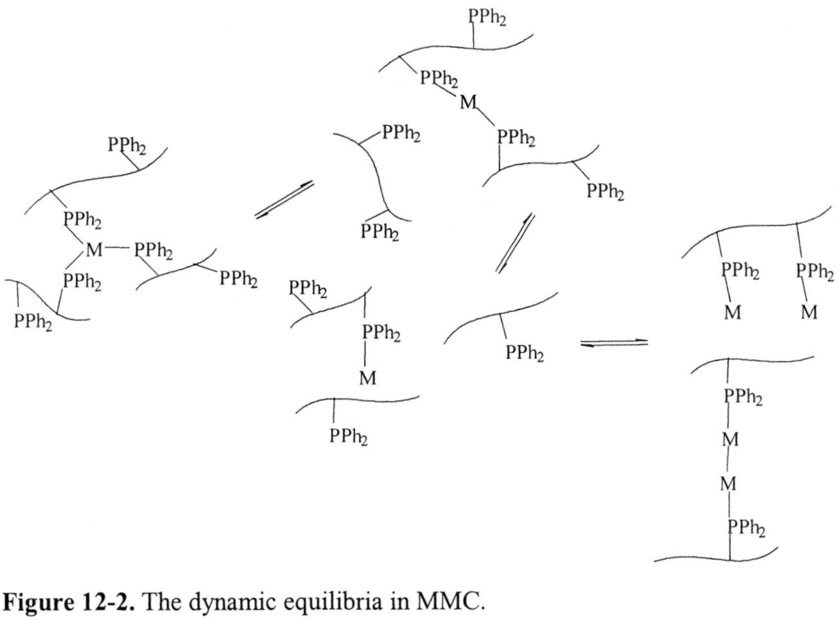

Figure 12-2. The dynamic equilibria in MMC.

As a rule, the formation of macromolecular metallocomplexes include a special design of macroligand, their functional groups, and MX_n with its "own" and "frame" ligands, which are retained in the inner coordination sphere of metal after immobilization. In addition to the halides, carbonyls, nitrosyls, phosphines, acetylacetonates and CH_3COO-, pyridine and Dipy can also be used as frame ligands [35].

At a low degree of cross-linking (2% divinylbenzene) the polymer is rather flexible. The high mobility of polymer segments within a weakly cross-linked polymer matrix may lead to the formation of binuclear complexes (for example, in the case of V_2Cl_8, $[(COD)RhCl_2]_2$ (COD is 1,5-cyclooctadiene). Functional groups in strongly cross-linked polymers (20% DVB) show less mobility, and thus prevent dimer complex formation. Moreover, in such polymers there are also opportunities for bonding coordinatively unsaturated complexes. Mixing of particles of different sizes (macroligand and dimeric complex) may be accompanied by monomerization of associated particles in the vicinity of the chain molecule (see Chapter 3). Either a total monomerization or the formation of dimeric and monomeric metal complexes bound at a polymer may occur. The nature of the functional groups is also very important. Monomeric and polymeric amines give rise to a cleavage of the dimer $[Rh(CO)_2Cl]_2$ while thiols form an addition product $[Rh(CO)_2RS]_2$. The interaction of $[Ru(n\text{-cymene})Cl_2]_2$ with phosphinated polymer matrices has the same effect [36]. It is important that dimeric complexes formed during catalytic reactions have a retarding effect. A classic example of this is the fast deactivation of homogeneous titanocene systems during hydrogenation of olefins, dienes, and acetylenes due to dimerization of complexes, whereas a polymer matrix (polystyrene modified by cyclopentadienyl (Cp) groups) prevents the dimerization process. Heterogenized catalysts therefore not only show a 10–20-fold rise in activity but are also significantly more stable than homogeneous analogues (Table 12-1).

The formation of coordinatively unsaturated complexes is the important factor determining the catalytic activity of immobilized catalysts. Optimization of the metal-to-ligand ratio, M/L, is one of the ways of obtaining these either at the stage of catalyst preparation or during the catalytic reaction. It should also be noted that the constant catalytic action that has been observed for the majority of immobilized catalysts is probably evidence of similar composition of the active forms. Thus, in selective reduction of polynuclear heteroaromatics by rhodium complexes, anchored to phosphynated CSDVB, this P/Rh ratio is only slightly reduced (3.3 to 2.9), even after several thousands of catalytic cycles [2,3], whereas the rate of butadiene, 1,5-COD or acetylene hydrogenation depends in a complex way on both the content of phosphyne groups and the ligand-to-metal (P/Rh) ratio (Table 12-3) [37].

The dispersivity of Rh^0 particles formed on the surface of phosphynated CSDVB (2% DVB) and their fractions in relation to the total metal content also depends on the P/Rh ratio. The general tendency is as follows. The rate of butadiene-1,3 hydrogenation reaches a maximum at P/Rh_s ~1, and does not depend on this ratio at P/Rh_s >2. A decrease in the phosphorus content in the polymer as well as the P/Rh_s ratio leads to an increase of Rh_s particle size at low phosphine-group content and P/Rh_s values approaching the Rh_s particle size formed on the unfunctionalized polymer (nonmodified CSDVB) (Table 12-3).

Table 12-3. Same characteristics of rhodium complexes attached to phosphynated CSSDVB (2% DVB) and their activities in butadiene and acetylene hydrogenation.

P (%)	Rh (%)	Rh$_s$, (%)	Rh particles size (nm)	P/Rh$_s$	$N^a \times 10^2$ butadiene	acetylene
1.6	1.1	1.0	1.5	5.8	1.3	1.0
1.7	0.6	0.5	1.7	10.7	1.0	12
1.7	1.6	1.3	1.9	4.3	0.8	-
1.2	0.5	0.5	1.2	8.4	1.4	-
1.2	1.0	0.9	1.65	4.5	1.3	0.8
1.2	2.0	1.7	1.8	2.3	1.5	-
1.2	5.0	3.3	2.9	1.2	4.2	5.0
1.7	3.0	2.4	1.9	2.3	1.1	-
0.5	0.5	0.4	1.6	4.6	3.9	-
0.5	1.0	0.7	2.9	2.4	3.8	21
0.5	2.0	1.1	4.4	1.6	6.3	-
0.5	5.0	1.7	6.7	0.9	30	-
0(CSDVB)	1.0	0.5	4.8	0	28	42

$^a N$ – number of substrate molecules (at [Rh$_s$] s)$^{-1}$; reaction conditions 348 K; P_{H2} = 67 kPa; P_{C4H6} = 0.11 kPa; P_{C2H2} = 0.11 kPa (368 K).

12.2.4 The Influence of Solvent Properties

The role of the solvent is one of the foremost factors affecting the hydrogenation process. In immobilized systems the solvent causes support swelling and chain flexibility and is responsible for both long-range interactions and the formation of complexes of various compositions. In a solvent in which the polymer does not swell (n-hexadecane) the formation of monomeric complexes is more probable, but dimeric species occur in xylene. In the case of fairly flexible polymers, dimerization of complexes may occur either within one chain or in a local part between the knots of cross-links. Moreover, for polymer species with a modest swelling ability, metal complexes are localized only on the surface of the support, which leads to an increase of S_{sp} (see Section 12.2.2.) of the catalysts formed. Although the diffusion of a substrate into a polymer matrix is not a limiting stage, examples are known when using previously ground particles of catalysts results in an increase of the reaction rate. The swelling ability of nonpolar matrices decreases with a rise in solvent polarity, which consequently reduces the size of the transport channels of catalysts and increases the diffusion hindrances. This last effect is most important for solvated substrate molecules. Similar processes also take place in non-swelling and weakly swelling polymers. A rise in solvent polarity favors the diffusion of polar substrates to the active centers and concurrently acts to prevent the diffusion of non-polar reagents. Comparative analyses of hydrogen solubility in the solvent, the swelling ability of the macrocomplex and the rate of hydrogenation of acrylamide are given in Table 12-4 [38].

Table 12-4. Data on Hydrogen solubility, swelling efficiency, and rate of reduction of acrylamide in various solvents.

| Solvent | H$_2$ solubility (mL of H$_2$/mL of solvent | Pd(II)[a] | | Pd(0)[b] | | Rh(I)[c] | |
		Swelling (%)	TON	Swelling (%)	TON	Swelling (%)	TON
Methanol	0.068	12	974	16	3016	25	1928
Ethanol	0.065	13	772	13	2221	18	1217
2-Propanol	0.045	17	532	20	1801	16	1015
1-Butanol	-	10	478	15	1681	12	846
Acetone	0.0525	7	558	18	2400	24	1420
DMF	0.037	19	532	15	1260	18	676
THF	-	3	551	14	1741	12	973
Water	0.015	14	109	13	540	16	372
Benzene	0.052	2	55	6	240	2	135
Dioxan	-	7	74	13	300	11	338

Supports: [a]Copolymer of 2-hydroxyethyl methacrylate, 2,3-epoxypropyl methacrylate and divinylbenzene; [b]functionalized poly(methacrylic acid); [c]copolymer of 4-vinylpyridine and methylenebisacrylamide. [Acrylamide] = 0.025 mol L^{-1}, 303 K; TON – turn-over number, mol (g atom metal)$^{-1}$.

In general the effects of a solvent on the rate and mechanism of hydrogenation are demonstrated by numerous factors:

- the bond energies of both the substrate and hydrogen with the catalyst surface;
- the adsorption ability of solvent molecules; their participation as specific ligands coordinated with active centers, (such interactions may result in either acceleration or retardation);
- the substrate and product concentration distribution between the catalyst surface and solution;
- the ratio of components; activation and desorption rates (including the formation of hydride complexes of transition metals with participation of solvents, especially alcohols);
- the solubility of hydrogen in the solvent (the Henry constant) and the diffusion rate at the gas-liquid boundary;
- the solvation of the reactive components, especially the support;
- the selective sorption of polar impurities;
- acid–base interactions between solvent, catalyst and functional groups of the macroligand.

The influences of these factors are not equal in strength. In homogeneous catalysis the choice of solvents is restricted by the solubility and stability of the complexes. These restrictions may often be removed in catalysis by immobilized metal complexes. The use of macroporous polymers with 6–75 nm pore size as supports allows us to select most organic solvents. We shall give only one example [39]. The hydrogenation rate of cyclohexene in the presence of polymer-bonded palladium porphyrin [40] depends on the ratio ethanol/water in a complicated way and reaches the maximum value at 80 vol.% ethanol (Fig. 12-3).

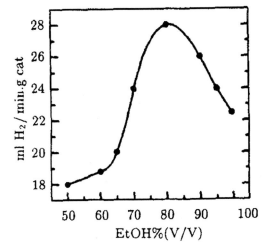

Figure 12-3. Effect of the concentration of EtOH on the hydrogenation catalytic activity of polymer-bound palladium porphyrin.

The examples mentioned above show the multiple influences of the solvent in hydrogenation by immobilized catalysts which are much more versatile than those in homogeneous catalysis. High concentrations of such catalysts can be made in practically any reaction mixture.

12.2.5 The Influence of Reaction Temperature

The influence of temperature on the rate of hydrogenation by heterogenized metal complexes is more complex. This arises because the catalyzed reaction is controlled not only by the temperature but also by the state of the *macroligand–MX_n* system. The curve $w = f(T)$ is often "bell-shaped". There are reasons for this atypical temperature dependence. These include the possible decomposition of polymer matrices, especially at elevated temperatures, which can be initiated or accelerated by metal complexes. It may also be that the curves drop downward because of metal center aggregation; this process becomes especially significant at elevated temperatures.

However, the temperature influence on phase changes in the metal polymer matrices is more significant. Polymer reactions are known to be coupled with the mobility of chain segments, which controls the catalytic properties of immobilized metal complexes. A rise in the chain mobility at a characteristic temperature may lead to changes in the coordinative unsaturation and valence state of the transition metal ion of the active center as well as to the formation of dimeric complexes (see Fig. 12-2).

A particularly interesting phenomenon is the conformation of the polymer matrix that provides the favored configuration for the catalytic center. To illustrate this point let us describe the abnormal Arrhenius dependence of the ethylene hydrogenation rate in the gas phase by rhodium complexes immobilized on phosphynated polystyrene [41]. A sharp increase of hydrogenation rate has been observed when the temperature increases across the glass transition point, T_g, of the polymer (341 K; Fig. 12-4).

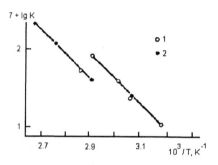

Figure 12-4. Arrhenius plot for the hydrogenation of gaseous ethylene by Rh^+ complexes immobilized on phosphynated CSDVB (2% DVB). The catalyst was evacuated before each experiment at 343 K (1) and 363 K (2); $P_{C2H4} = P_{H2} = 27$ kPa.

The value of the activation energy of the reaction at $T > T_g$ was practically unchanged (33 kJ mol^{-1} for the higher temperatures and 35 kJ mol^{-1} below T_g). Consequently all changes in the rate constant value for this reaction must be connected with changes in the entropy factor. Most probably, the effect is the result of polymer segments unfreezing and promoting the correct configuration of the active centers for catalytic action. Alternatively, such a promotion of lability may also cause the aggregation of metal particles of different sizes composed of zero-valent metal atoms with an activity much higher than that of isolated metal atoms. Another possible reason for the discontinuity in the Arrhenius plot is the existence of a limitation for ethylene coordination. A comparison of olefin adsorptions with their rates of hydrogenation has revealed that both the enthalpy and the entropy of coordinated ethylene molecules correlate with changes in polymer-chain mobilities.

Similarly, the complex dependence $w = f(T)$ in hydrogenation by immobilized complexes may be explained by a variety of effects. The rising branch of the curves is the Arrhenius plot, while the downward drop of the curves results from a decrease of hydrogenation rate, possibly arising from the dimished response of the number of active sites. On the one hand, at $T > T_g$ relaxation of micro Brownian motion may facilitate diffusion and enhance the probability of collisions between the substrate and the active centers but, on the other, movement of neighboring groups of the polymer ligand may sterically hinder coordination and decrease the stability of the coordinated substrate. For these reasons, structural transformations of the active center are also possible. The role of diffusion has been confirmed experimentally: below T_g both reaction rate and amount of adsorbed substrate increase with the rise of olefin molecular size. The phenomenon of an extreme temperature dependence becomes apparent not only in the activity but also in the selectivity of polymer-immobilized catalysts. It is interesting that such behavior is also characteristic of other catalytic reactions (see below). Similar regularities have been observed in C_2H_4 hydrogenation by metallic Pd dispersed on Chromosorb containing a polymer phase, a mixture of poly(ethylene oxide) and PS in 1:1 proportion by mass [42]. The temperature increase above T_g for PS (~368 K) led to a rise in catalytic activity, probably caused by a fusion of PS domains and preferential adsorption of C_2H_4. Target-oriented analysis of the response of polymer-immobilized catalysts to the effect of temperature may lead to the creation of hydrogenation catalysts operating at elevated temperatures, and also to the optimization of their usage.

12.2.6 Surface Density

Not only the total amount of the bound metal complex but also the character of its distribution in the polymer matrices (topography) can have a profound effect on the rate of hydrogenation. As a rule, catalysts in which all active centers are accessible to the substrate show the largest activity. Such MX_n distributions can be obtained either at the stage of catalyst synthesis or sometimes in the course of a catalyzed reaction. Thus, rhodium-containing complexes bound to phosphynated CSDVB with different amounts of cross-linking agent (2–60 mol%) under conditions which exclude polymer swelling gave a catalyst that was 1.5–4 times more active than that produced from polymers that swell easily [3].

The specific catalytic activity depends in a complex way on the bound metal complex content (or the density of the metal complex on the polymer support). As a rule, the specific rate of reaction (w_{sp}, the rate attributed to the total amount of metal in the catalyst) increases, reaches its largest value and then diminishes as the amount of metal complex rises. The principal shapes of such dependences are given in Fig. 12-5.

Such a bell-like dependence of w_{sp} on the surface density of transition metal ions has also been observed in other catalytic reactions (hydroformylation, oxidation, polymerization, etc.), and is probably one of the specific features of catalysis by immobilized metal complexes. While there is no well-founded explanation of the rising branch of the plot, the diminishing trend may be connected with the formation and further growth of low- and/or zero-valent transition metal ion associations, diminishing the catalytic efficiency. Active centers of immobilized catalysts are localized on the boundaries of cluster-like substances with stabilization by their electron systems.

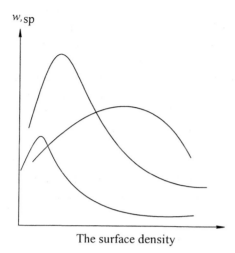

w, sp

The surface density

Figure 12-5. Typical dependences of the specific rate of hydrogenation on the surface density of the metal on a polymer.

As the metal complex surface density rises, the number of such sites suitable for catalysis decreases. It is important that these associations are dynamic; their number may change in the course of bound complexes activation. Thus the reactivity of polymer-bound rhodium complexes in hexene-1 hydrogenation is much greater for catalysts with small amounts of bound rhodium ([Rh] = 0.51 and 1.22%). Obviously in catalytic contacts the highest proportion of metal is localized on the surface of the support and its cross-linked structure prevents particle aggregation both at the stage of activation and in the course of hydrogenation. These processes can be accompanied by a deep interaction of the metal with the polymer matrix. The degree of reduction of $Pd^{2+} \rightarrow Pd^0$ and consequently the catalytic activity for olefin hydrogenation changes significantly even for macroligands of the same type (for example, for polyheteroarylenes) [25]. Two different types of Pd^0 particles have been detected in such reduced forms with electron bond energies $E^b_{Pd\ 3d\ 5/2} = 335.6$ and 338.0 eV. Their sizes were in the range 2–3 nm on average. Such small particles are capable of strong interactions with polymer matrices. Palladium atoms, assigned a positive charge, $Pd^{\delta+}$ due to electron transfer, affect the coordination of the substrate and consequently the activity of such catalysts. The presence of a π-conjugated system in the polymer support, which transfers into an ion-radical state, also enhances the effect. Strong metal ion interactions (in the course of reduction with $NaBH_4$) with the support are followed by electron transfer and metal reduction up to the leuco-base state.

12.2.7 Size Effects in Hydrogenation by Polymer-Immobilized Complexes

As mentioned above, formation of metal complex aggregates composed of zero-valent metal species has often been observed during hydrogenation. This initiated a great number of studies on the dispersion in the polymer of small metal particles or condensation of their ions up to nanoparticles stabilized by a polymer matrix (see reviews in [43,44]). The main advances in this field are as follows. Colloids of Pd^0 (1.8−5.6 nm) bonded with a polymer matrix have been shown to be active and to be selective catalysts for cyclopentadiene reduction to cyclopentane as well as for 1,5-, 1,4-, 1,3- and 2,4-hexadiene hydrogenation. The reduction of conjugated dienes proceeds either at the 1,2 and 1,4 positions depending on the nature of the diene. Palladium particles (diameter 1.8 nm) in PVPr-protected colloids are effective catalysts for hydrogenation of the methyl ester of linolenic acid as well as for partial hydrogenation of soybean oil under ambient conditions. The specific features of such catalysts are a narrow distribution of particle size, colloid stability within a large range of pH values (from 2 to 13), the ease of catalyst separation and long-term stability after many use cycles. Nanoparticle sizes in this range have little influence on the rates of linear olefin hydrogenation but the rate of cyclohexene hydrogenation decreases as the catalyst particle size increases (Table 12-5) [45].

Table 12-5. Sizes of Rh^0 colloid and their activity in olefin[a] hydrogenation.

System	Average Rh particle size (nm)	Activity (mol H_2 per g-at of Rh·s)	
		Hexene - 1	Cyclohexene
$RhCl_3$/PVA[b]/CH_3OH/H_2O	4.0	15.2	3.1
$RhCl_3$/polymethyl ester/CH_3OH/H_2O	4.3	22.5	9.6
$RhCl_3$/PVPr[c]/CH_3OH/H_2O	3.4	15.8	5.5
$RhCl_3$/PVPr[c]/C_2H_5OH	2.2	15.5	10.3
$RhCl_3$/PVPr[c]-iso/C_3H_7OH	2.4	-	9.3
$RhCl_3$/PVPr[c]-CH_3OH/NaOH	0.9	16.9	19.2

[a]Reaction conditions: 303 K, H_2 pressure 0.1 MPa, [olefin] = 25 mM; [Rh] = 0.01 mM.
[b]PVA= poly(vinylalcohol). [c]PVPr=poly(vinylpyrrolidone).

Since the late 1990s, new approaches to the dispersion of noble metal particles in polymer matrixes by means of chemical, photochemical and radiation-chemical reduction, the evaporation of metal atoms (including solvated ones) into different supports, etc. have been developed (see Section 8-1) [44]. Nevertheless, catalysts prepared from individual immobilized metal clusters have more definite, mostly predetermined, structures. For these purposes derivatives of Os_3, Ir_4, Ru_4, Rh_4, Rh_6, etc. clusters are most often used. Earlier studies [46] showed that the rate of ethylene hydrogenation promoted by tetrairidium or tetraruthenium carbonyl clusters bound to phosphynated polymers decreased with an increase of the number of donor phosphyne ligands in the cluster environment (Table 12-6). This effect is also accompanied by a rise of ethylene–catalyst bond strength.

Table 12-6. The influence of the P/Ir ratio on the ethylene hydrogenation[a] rate.

Catalyst	P/Ir	$w \times 10^3$ (mol of C_2H_4/g-at of Ir·s)
]- $PPh_2Ir_4(CO)_{11}$	4.54	21.3
]- $PPh_2Ir_4(CO)_{11}$	14.3	25.3
]- $PPh_2)_2Ir_4(CO)_{10}$	3.4	3.76
]- $PPh_2)_2Ir_4(CO)_{10}$	6.9	3.60

[a]Reaction conditions: ethylene pressure 20 kPa, H_2 pressure 80 kPa, 313 K.

Polymer-immobilized catalysts based on Rh_6 were found to be effective in hydrogenation of cyclohexene [47]. Hydrogenation is fully inhibited by traces of carbon monooxide but the activity is restored after removal of CO. Physico-chemical studies show no M–CO bond rupture during hydrogenation. So coordinative vacancies arise as a result of L–M bond rupture or more probably by breaking of the M–M bond in Ru–Ru, Rh–Rh, Os–Os, etc. In the hydrogenation of unsaturated amino acids in the presence of $Rh_4(CO)_{12}$ or $Rh_6(CO)_{16}$, monomeric particles $HRh(CO)_2·L$ were considered as active species formed through oxidative addition of H_2 and segregation of the cluster [48]. Reduction processes in solutions catalyzed by simple cluster complexes (e.g. of styrene with tetranuclear osmium clusters [49]) also follow the mechanism of cluster fragmentation accompanied by the formation of highly active particles with low nuclearity, possibly mononuclear. The concentration of such species, which are probably active sites for the reaction, is rather low. There are different ways of immobilizing clusters in the case of mixed-type supports as well as for heteroligand and heterometallic clusters. Thus, the heterometallic $RhOs_3$-cluster (derived from $H_2RhOs_3(AcAc)(CO)_{10}$) segregates in the model double reaction of hydrogenation and butene-1 isomerization [50]. This provides separated trinuclear osmium clusters responsible for isomerization, and metallic rhodium particles 1.0 nm in diameter, which are active in the hydrogenation of butene-1.

In recent years, methods for synthesizing such catalysts through copolymerization of cluster-containing monomers with common monomers have been elaborated (see Chapter 4). In these cases the exact ratio *metal–functional group* is determined during the catalyst preparation stage. High activity and selectivity in the action of such catalysts may be expected.

12.2.8 Problems of Selectivity in Hydrogenation Reactions Catalyzed by Macromolecular Metal Complexes

The practical importance of immobilized catalysts is characterized not only by their high catalytic activity [51] and long-term stability but also by their selectivity, i.e. the ability of the reaction to yield one or more desirable products. (The *selectivity factor* is the ratio of the amount of desired product to the sum of all products formed). The main processes that accompany the hydrogenation reaction and decrease its selectivity are isomerization of the substrate or of the intermediate or desired product, alkylation and di- or trimerization, etc. Knowledge of the reaction mechanism and the possibilities of controlling the reaction at the stage of catalyst synthesis and by changing the reaction conditions may allow control of the reaction products.

Four main requirements may be imposed on a heterogeneous catalyst to provide the best selectivity (see, for example, [52]). These are: (i) a balance between bond strengths; (ii) the ability to supply a specific substrate coordination; (iii) the possibility of active ensemble formation; (iv) a template effect. All these parameters depend on both electronic and geometric properties of the metal particles. As a rule, selectivity is the result of the specific adsorption of one of the reagents or the reaction products on the surface catalyst. In essence those factors that are significant for catalyst activity also affect catalyst selectivity. These include the nature and topography of immobilized complexes, the nature of the polymer matrix and functional groups, the L/M ratio and the reaction conditions.   From the numerous examples we shall consider only two. Thus, the nature of the functional groups (in particular the π-acceptor properties) of modified CSDVB has a very profound influence on 4-vinylcyclohexene hydrogenation [33]. The degree of catalyst selectivity (the ratio of 4-vinylcyclohexene yield to other products formed) being equal to 0.4 may reach 0.97–0.98 after 4-ethylcyclohexene separation. Ruthenium complexes bound to phosphinated PS (10% cross-linking) reveal the same selectivity in partial hydrogenation of cyclododecatriene (CDT) to cyclodecene (CDE) [53,54]. Nevertheless, the dependence of selectivity on the support pore diameter (20 nm on average) and assertions about the sieve effect in polymer-immobilized catalysis, seem to be random. The nature of the substrate and solvent, high temperatures (413 K) and pressures (40 atm) are important for the porous characteristics of catalysts.

Immobilized metal complexes selectively hydrogenate different compounds under mild conditions:

- aromatic and heteroaromatic compounds (for example, phenol and cyclohexanol), in particular selective reduction, is characteristic of aromatic compounds containing $\equiv N$, $-C=O$, $-NO_2$, $-C\equiv N$ functional groups;
- carbohydrates to polyols;
- inner alkenes to *cis-* and *trans-*alkenes whereas hydrogenation of allylbenzene and 1,5-cyclohexadiene is often accompanied by isomerization along with migration of a double bond (formation of more stable *cis-* and *trans-*propenylbenzene and 1,4-cyclooctadiene);
- vinylacetylene to 1,3-butadiene (however selectivity decreases with increasing conversion); the selectivity of tetrahydrofurylacetylene alcohol

reduction reaches 98–99%, other acetylene substrates (including those with internal triple bonds) react with 76–84% selectivity;
- partial hydrogenation of 1,5-cyclooctadienes (93–94% selectivity) and cyclododecatriene (about 85%).

Steric hindrance in the vicinity of active centers enhances the selectivity to butene-1 in the hydrogenation of 1,3-butadiene whereas additive coordinative vacancies at the metal atom favor olefin isomerization. The character of the metal center distribution can also influence the catalyst activity and selectivity. This has been demonstrated by comparison of both the activity and selectivity in the hydrogenation of butyne-2 of trinuclear Pd clusters bound to phosphynated PS and incorporated in the volume of the polymer bead or located only in the presurface catalyst layer. Catalytic turnover numbers for the last catalyst were 35 times greater than that for the first. However, the selectivity (with respect to cis-butene-2 formation) was 1.4–3.7 times higher for the first catalyst. This seems to be the result of differences in the structures of the active metal centers, i.e. differences in coordinate unsaturation of palladium ions in these two cases, or in the P/Pd ratio responsible for selectivity. An enhancement in selectivity may be caused in the hydrogenation of one substrate in the presence of another or in a mixture of unsaturated compounds. Thus, a macromolecular complex of $PdCl_2$ with polyethyleneimine is highly selective in the hydrogenation of 1,3- and 1,4-cyclopentadienes: only the 1,3-diene reacts in the mixture of dienes from the initial stages up to 95% conversion and after that the 1,4-diene undergoes reaction. Diene disproportionation is inhibited by acetylene additives.

12.2.9 Perspectives in the Development of Hydrogenation Catalysis by Macromolecular Complexes

Let us consider the directions which, in our opinion, will be further developed in this field. The first of these is membrane catalysis. In heterogeneous catalysis for making semihydrogenated products which form active intermediate products with H_2, special membrane catalysts fabricated from palladium and its alloys [55] are used. Recently, an alternative to these catalysts has been developed through special composite membrane catalysts. These usually consist of an inert polymer support permeable to H_2 with a supported metal layer (which may also be chemically bound). Cation-exchange resins are convenient supports for such contacts. In such films (thickness 0.4 μm) hydrogen ions show an abnormally high mobility (the value of the proton diffusion coefficient in such cation-exchange membranes is greater by an order of magnitude than that for metallic palladium membranes) [56]. It will probably be possible to control the rate of substrate migration through the membrane and in such a way to enhance the selectivity, in other words, to separate zones of substrate and reagent.

We also expect great progress in the study of asymmetric hydrogenation in the presence of immobilized polymer catalysts [57–60]. Considerable success has been achieved with rhodium phosphine polymer catalysts involving a range of macroligands including optically active residues such as 4,5-bis(diphenylphosphyne)methyl-1,3-dioxolane (**2**) (DIOP) or dibenzophosphonic (**3**) groups as well as polymers formed from (2*S*,4*S*)-*N*-acryloil-4-diphenylphosphyno-2-[(diphenylphosphyno)-methyl]pyrrolidone (**4**).

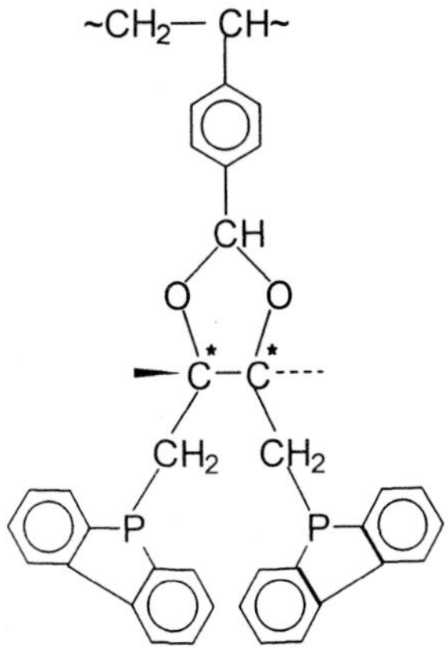

2	**3**	**4**

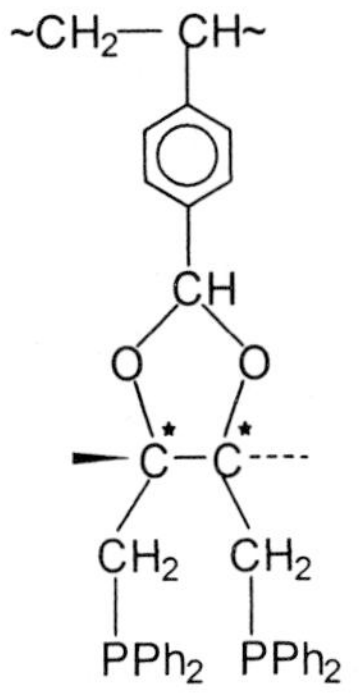

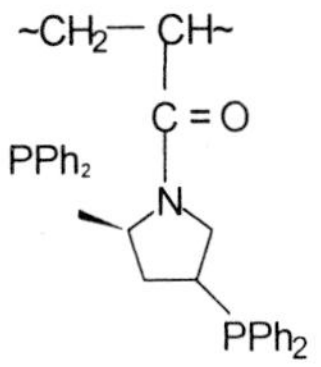

The efficiency of an asymmetric catalytic reaction is characterized either by excess of the predominantly formed enantiomer [61], for example of the *R*-configuration and its optical yield (%), $P = ([R]-[S])/([R]+[S])$, or by the value of the enantioselectivity, $E = [R]/[S]$. Thus, among the amino acids (prochiral enamides) with higher optically active products, the yield (for example, 2-acetamidoacrylic acid) reaches 67%. The same is true of the asymmetric hydrogenation of prochiral ketones.

Complete clarity in the subject of asymmetric hydrogenation by immobilized metal complexes has not, as yet, been achieved. Each of three aspects has to be understood in this connection: (i) the coordination chemistry of the triple complexes, (ii) the mechanism of catalytic hydrogenation and (iii) the reason for enantioselectivity.

A mechanism involving correlation between the stability of the alter-ligand complex on the catalyst surface containing a chiral ligand and the enantioselectivity of the catalyst appears to be the most acceptable explanation. This is the "lock and key" mechanism in which enantioselectivity is specified by the first prochiral substrate bonding with chiral catalysts. Another approach to the reaction mechanism is based not on a consideration of the initial triple complex formation but takes into account the high reactivity of the lesser diastereomeric adduct, catalyst–substrate, corresponding to less favorable substrate bonding. The idea of optical activity transfer from chiral ligand to substrate can also be applied. The chiral information is passed from asymmetric centers in the chelate framework through functional groups to the catalytic metal center and then to the substrate. Recently, a polymer-supported [Ru(cymene)Cl$_2$]$_2$ complex containing chiral-phosphine BINAP groups was found to be active in asymmetric hydrogenation of 2-(6'-methoxy-2'-naphthyl)acrylic acid as the model reaction [62,63]. The scheme of the reaction is shown in Eq. (12-7).

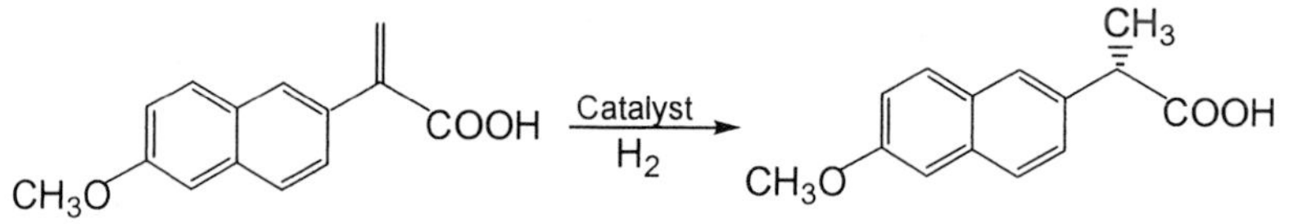

$$\text{(12-7)}$$

The product in this particular reaction (naproxen) is a high-value anti-inflammatory drug. Even a simple microcapsulation of OsO$_4$ into acrylonitrile-butadiene-styrene (ABS) polymer along with addition of a chiral ligand (N-methyl-morpholine N-oxide or 1,4-bis(9-O-dihidroquinidinyl)phthalazine led to a considerable increase in the rate and in the chiral yield during the asymmetric dihydroxylation of olefins (Table 12-7) [64].

Sol–gel physically entrapped chiral rhodium and ruthenium complexes are used as recyclable and effective catalysts in enantioselective hydrogenation of itaconic acid: up to 78% chiral methylsuccinic acid is obtained [65].

Immobilized metal complexes are potential catalysts in the activation of H$_2$, hydrogen/deuterium exchange in water and alcohols, in para–ortho conversion and in hydrogen transfer from one substrate to another, i.e. catalytic hydrogenation of organic compounds using a chemically bonded hydrogen. This also applies to reactions which include hydrosilylation, that is the process whereby alkenes, alkynes and ketones are reduced by hydrogen chemically bound to a silane molecule (X$_3$SiH type). Rhodium complexes bonded with polyamide polymers are the most effective catalysts in such reactions (see, for example, [66,67]).

The mechanism of activated hydrogen transfer in the presence of polymer-immobilized complexes seems to be similar to those involved in enzyme catalysis.

Table 12-7. Asymmetric Dihydroxylation Using ABS-MC OsO$_4$.

Olefin	Product	ABS-MC OsO$_4$ (mol%)	Chiral ligand (mol%)	Yield(%)	ee (%)
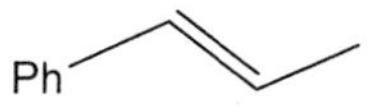 Ph olefin	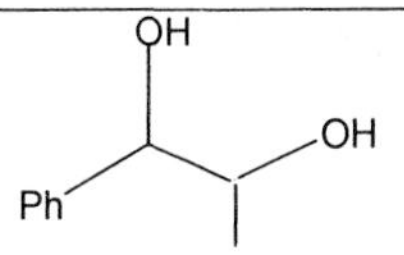 Ph diol (OH, OH)	5	10	75	91
		2.5	5	90	92
		1	2	97	86[a]
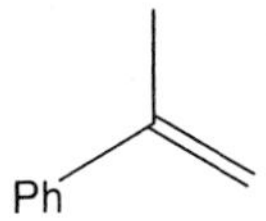 Ph isopropenyl	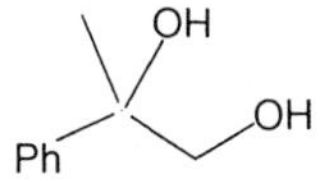 Ph diol	5	5	98	78
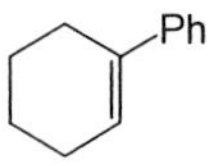 cyclohexenyl-Ph	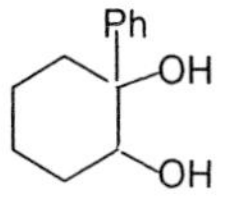 cyclohexane diol, Ph	5	5	64	86
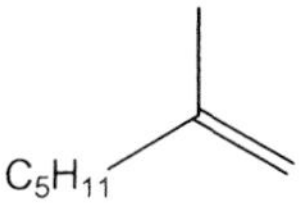 C$_5$H$_{11}$ isopropenyl	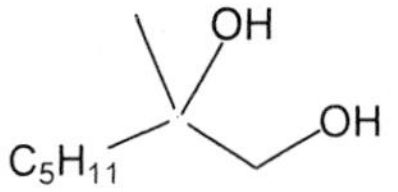 C$_5$H$_{11}$ diol	5	5	90	60
 C$_4$H$_9$-CH=CH-C$_4$H$_9$	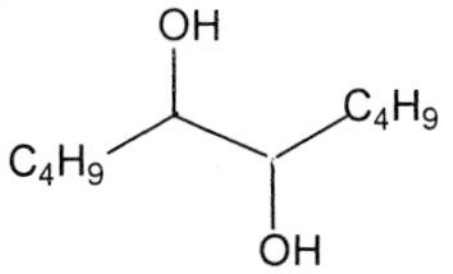 C$_4$H$_9$ diol C$_4$H$_9$	2.5	5	85	63
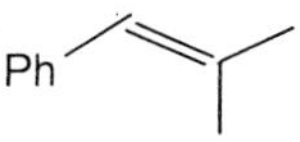 Ph olefin	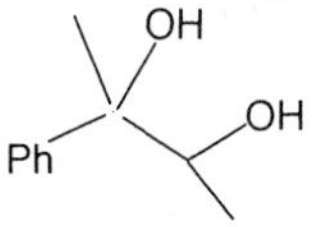 Ph diol	5	5	36	85

12.3 Polymerization Processes Initiated by Macromolecular Complexes

In contrast to hydrogenation and oxidation reactions, n-merization (di-, *iso*-, oligo-, poly- and copolymerization) processes proceed via a single type of substrate activation, namely using reagents containing multiple bonds. Addition polymerization of one or several monomers leads to the formation of one, sometimes huge, molecule. There are at least two entities in the transition metal ion coordination sphere. They are the growing macrochain and the coordinated monomer (for ionic coordinative polymerization).

If the immobilization of transition metal complexes under optimal conditions for hydrogenation and oxidation leads to an increase in catalyst activity and selectivity (as well giving the opportunity to regenerate the catalyst) this procedure causes even larger scale changes in the case of polymerization. This may affect not only catalyst activity but also the length of the macrochains formed, the molecular-mass distribution, the stereoregularity and the composition of the polymer products in the case of copolymerization.

Metal complexes bound to a polymer support most frequently induce ionic polymerization of olefins, dienes and acetylenes, and less commonly radical polymerization of vinyl-type monomers, acting at all reaction stages: initiation, chain propagation and termination. Active sites for the addition of monomer molecules to the growing polymer chain can in many cases be regenerated yielding new polymer chains (catalysis via a polymer chain).

It should be mentioned that simple metal complexes immobilized on polymer supports were initially used for polymerization (1965/1966) in the Solvay catalysts based on titanium complexes bound to macromolecular ligands with C=O, C≡N and C=N groups. Until now the data are mainly available in patent literature, and there are few kinetic studies of polymerization processes involving the action of macromolecular complexes. At the same time the use of metal complexes bound to inorganic supports has been extensively developed in polymerization catalysis. This indicates that there has been inadequate study of the application of metal complexes immobilized on polymer supports to the catalysis of polymerization and copolymerization of different monomers, mainly olefins.

12.3.1 Peculiarities of Olefin Polymerization

Olefin polymerization in the presence of bound complexes is one of the most extensively investigated reactions and is also of commercial importance, especially in the case of ethylene and propylene and their copolymers [68]. This has been preceded by detailed studies over many years of the mechanism of ethylene polymerization with homogeneous and pseudo-homogeneous Ziegler-type catalytic systems $MX_n - AlR_xCl_{3-x}$ (MX_n is a transition metal compound ($TiCl_4$, $TiCl_3$, $Ti(OR)_4$, $(C_5H_5)_2TiCl_2$, VCl_4, $VOCl_3$, $ZrCl_4$, $(C_5H_5)_2ZrCl_2$); AlR_xY_{3-x} is an organometallic compound; R is a hydrocarbon radical, $x = 1, 1.5, 2$, Y is halide or alkoxide). It was established that the active centers of such systems (denoted here as M–R) are organometallic compounds formed *in situ* and stabilized by organoaluminum derivatives in the bimetallic complex. Addition of activated monomer (chain propagation stage) occurs at this bond. The non-stationary character of the action is characteristic of such systems. At the processing temperatures (333–363 K) their activity sharply reduces with time. This is the result of metal–carbon bond formation and cleavage. The reduction of alkylated transition metal ions ($Ti^4 \rightarrow Ti^3$, $V^5 \rightarrow V^4 \rightarrow V^3$, etc.) is one of the reasons for the catalyst deactivation.

The processes of reduction in such systems are often complicated and have

only been clarified since about 1980. These are: bimolecular reactions of secondary alkylation, disproportionation of alkyl groups in the coordination sphere of the transition metal or disproportionation of alkyl radicals of two active centers. The general scheme of these reaction can be presented as follows:
- Formation of active centers (1) (Eq. (12-8)).

$$-\overset{|}{\underset{|}{M}}-Cl + AlR_xCl_{3-x} \longrightarrow -\overset{|}{\underset{|}{M}}-R + AlR_{x-1}Cl_{3-x} \qquad (12\text{-}8)$$

- Chain propagation (2) (Eq. (12-9)).

$$-\overset{|}{\underset{|}{M}}-R + nCH_2 = CH_2 \longrightarrow -\overset{|}{\underset{|}{M}}-(-\,CH_2 - CH_2\,-)_n\text{-}\,R \qquad (12\text{-}9)$$

- Secondary alkylation (3) (Eq. (12-10)).

$$-\overset{|}{\underset{|}{M}}-R + AlR_xCl_{3-x} \longrightarrow \,\diagdown\!\!\underset{\diagdown R}{\overset{\diagup R}{M}}\!\diagup\, + AlR_{x-1}Cl_{3-x} \qquad (12\text{-}10)$$

- Monomolecular disproportionation (4) (Eq. (12-11)).

$$\diagdown\!\!\underset{\diagdown R}{\overset{\diagup R}{M}}\!\diagup \longrightarrow -\overset{|}{\underset{|}{M}} + R_{(+H)} + R_{(-H)} \qquad (12\text{-}11)$$

- Bimolecular disproportionation (5) (Eq. (12-12)).

$$-\overset{|}{\underset{|}{M}}-R + \overset{|}{\underset{|}{M}}-R \longrightarrow 2-\overset{|}{\underset{|}{M}} + R_{(+H)} + R_{(-H)} \qquad (12\text{-}12)$$

The non-stationary character of the polymerization may be the result of changes both in the concentration of active centers (n_p) and in their nature during the polymerization. It is difficult to maintain a high and constant concentration of active centers (above a few percent) in solution because their high reactivity results in side reactions which are competitive with chain propagation (reaction 2).

Therefore immobilization of active centers on the supports is perhaps one possibility of diminishing the prevailing role of side reactions 4 and 5, and thereby of enhancing the efficiency of metal complex catalysts for polymerization of olefins. It was expected that spatial isolation of MX_n (as immobilization of enzymes prevented their deactivation) would lead to a decrease in bimolecular deactivation of active centers and in turn, to a cooperative stabilization preventing monomolecular termination. Instead, as earlier studies have shown (Fig. 12-6) [69] polymer-immobilized complexes are stable over time. Macromolecular metal complexes for polymerization processes can be used as powders, films, fiber materials (for example, poly(vinylacetate) and polyamide fibers, cotton, flax, viscose) with functional groups to which MX_n is bound.

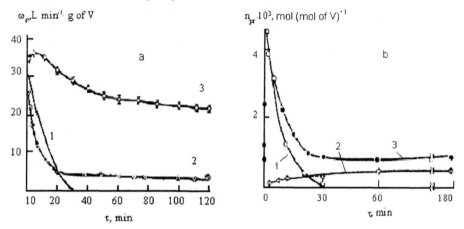

Figure 12-6. The dependence of the reaction rate and active site number on the time of ethylene polymerization by catalytic systems: a: *1*- VCl_4-$(isoBu)_2AlCl$, *2*- VCl_4-PVA$(isoBu)_2AlCl$, *3*- VCl_4-hydrolyzed copolymer of styrene with VA$(isoBu)_2AlCl$; b: *1*- VCl_4-$(isoBu)_2AlCl$, *2*- VCl_4-MMAc$(isoBu)_2AlCl$, *3*- VCl_4-PMMAc-$(isoBu)_2AlCl$; Polymerization conditions: benzene, 313 K, [Al]/[V] = 50, $[C_2H_4]$ = 0.04 mol L^{-1}.

The polymerization is a first-order reaction with respect to monomer and transition metal concentrations. The chain transfer reaction involving monomer competes with this with participation of an organoaluminum compound. The activation energies for the chain propagation step are similar. Along with another kinetic investigations (the constants of the rate reaction of polymerization $(2\div3)\cdot10^8$ $\exp(-4600/RT)$) this indicates that the same active centers are responsible for catalysis in immobilized and homogeneous systems. They differ only in quantitative parameters. If in the standard system the value of n_p reduced by a factor of 45 after 30 min polymerization, for the immobilized catalyst it decreased only by a factor of 5 remaining constant for 3 h (Fig. 12-6 (b)). The polymer matrix stabilizes the active centers of polymerization, the value of the rate constant of bimolecular deactivation reduces from $1.93\cdot10^3$ (for the homogeneous system VCl_4–$Al(C_2H_5)_2Cl$) to $0.75\cdot10^3$ (for $VCl_4\cdot PMMA$ (poly(methylacrylate))–$Al(C_2H_5)_2Cl$). The same is true for the processes of monomolecular deactivation; their rate constants fall almost by a factor of 3 ($2.2\cdot10^{-3}$ to $0.8\cdot10^{-3}$ s^{-1}).

Numerous studies have shown that to increase the catalytic activity of immobilized systems by a factor of 50–250, it is necessary to replace the active sites on the surface of the support or within a thin presurface layer (due to the heterophase character of the reaction). As the topochemistry of bound MX_n is specified by the polymer functional group distribution, special polymer carriers with functional groups grafted to the surface of inert polymer materials, such as PE, polypropylene (PP) and polystyrene (PS), have been developed [70]. The general scheme of formation of such supports as powders is to subject the polymer to some mechanical, chemical or irradiation action in the presence of the desirable function-bearing monomer (allylic alcohol, acrylic acid, vinylpyridine, vinyldiphenylphosphine, etc.). Graft polymerization proceeds on active sites (free radicals, ions) yielding the grafted polymer layer (Eq. (12-13)).

$$
\Big]\ \xrightarrow{\text{initation}}\ \Big]\ \xrightarrow{\ \overset{\displaystyle CH_2=CH}{\underset{Y}{|}}\ }\ \Big] - (-CH_2 - \underset{Y}{\overset{}{CH}} -)_n - \Big] \longrightarrow
$$

$$
\xrightarrow{+MX_n}\ \Big] - (-CH_2 - \underset{Y \cdot MX_n}{\overset{}{CH}} -)_n
$$

$$(12\text{-}13)$$

It has been shown that the grafted polymer layer (at the optimal content of functional groups, 1–2 wt.%), of thickness approx. 10–30 nm, is completely accessible to the reagents. The transition metal compound can penetrate and bind with functional groups at a depth of 10 monolayers of grafted monomer [71]. Data on the catalytic properties and kinetic parameters of such macromolecular complexes are given in Table 12–8 [72]. It is most probable that immobilization of the MX_n molecule is favorable for the substrate entering structures, reducing the insertion barrier. The values of the constants for immobilized systems indicates, in turn, a sharp drop in the rate constant pre-exponential factor. The observed compensating effect indicates that polymer fragments adjacent to the active site are involved in the reaction, resulting in a decrease of entropy for active site reactions.

With a rise in temperature, the ethylene polymerization rate follows the Arrhenius law only over a certain temperature range: up to the glass (or softening) temperature of the polymer support. It has been shown for gas-phase ethylene polymerization catalyzed by $TiCl_4 \cdot P4VP–AlEt_2Cl$ (4VP = poly(4-vinylpyridine), that Arrhenius-dependence of the polymerization rate is valid up to 383 K [73]. The effective activation energy in this range is typical for anionic-coordinate polymerization, 16 ± 2 kJ mol^{-1} (Fig. 12-7). At these temperatures the polymer segments show higher mobility and relaxation relative to reorientation of support macromolecules, which becomes the controlling factor. At higher temperature (>110 °C) the temperature coefficient becomes negative (-130 ± 10 kJ mol^{-1}). If we suppose that the decrease of polymerization rate in this temperature range is the result of irreversible deactivation of active sites caused by enhanced mobility of polymer segments, then $w_p = Kn_{pr}(1 - \Delta n/n_{pr}) = Kn_0(1 - w_d\Delta t)$, where w_d is the specific value of the deactivation rate by any of the possible mechanisms. It therefore follows that $w_d\Delta t = 1-w_p/w_0$, where w_0 is the value of polymerization rate without deactivation, which can be calculated from the temperature dependence (see Fig. 12-7 b). The shape of curve 1 corresponds to those observed for time-relaxation dependences of segment mobility; the apparent value of the activation energy sharply increases in the approach to the temperature of mobility freezing (T_g = 360 K is the temperature which releases the brake on mobility of P4VP leading to a decrease of polymerization). The temperature dependence of the relaxation rate is linearized on the coordinates $\lg w_d$ against $(\Delta T)^{-1}$ (Fig. 127 b, curve 2). Thus, the restriction of transport diffusion of active sites is a significant factor responsible for the activity and stability of immobilized catalysts.

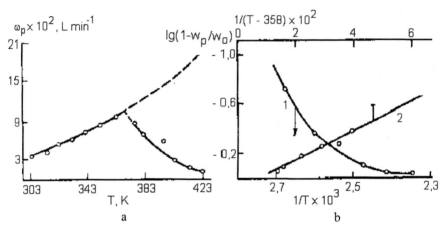

Figure 12-7. Temperature dependence (a) of gas-phase ethylene polymerization rate for the catalytic system $TiCl_4 \cdot P4VP-AlEt_2Cl$ and this (b) in coordinates $1-w_p/w_0-1/T$ (1) and $-1/(T-358)$. The calculated polymerization rate (a) is shown by dashed line, the observed polymerization rate is represented by solid line.

Table 12-8. Characteristics of MX_n immobilized on polymers with grafted functional layers and kinetic parameters of ethylene polymerization (343 K).

Polymer carrier	Content of grafted fragments × 10^4 (mol g^{-1})	MX_n	Content of fixed metal × 10^4 (g-at/g)	n_p, (mmol/mol)	$k_{pr} \times 10^{-4}$ (l/(mol s))	E_a^{eff} (kJ mol^{-1})
PE-gr-poly(acrylonitrile) (PE-gr-PAN)	11.9	VCl$_4$	0.70	4.0	2.8 ± 0.6	38 ± 4
PE-gr-poly(methylmethacrylate) (PE-gr-PMMAc)	12.0	VCL$_4$	1.14	4.5	3.4 ± 0.7	38 ± 4
PE-gr-poly(methylvinylketone) (PE-gr-PMVK)	6.0	VCl$_4$	1.50	4.5	4.0 ± 0.8	31 ± 4
PE-gr-poly(dibutyl]ester of vinylphosphonic acid (PE-gr-PDBEVPAc)	1.1	VCl$_4$	0.50	3.8	2.9 ± 0.6	32 ± 4
PE-gr-poly(vinylpyrrolidone) (PE-gr-PVPr)	8.6	VCl$_4$	2.6	4.2	3.1 ± 0.6	29 ± 4
PE-gr-poly(diallylamine) (PE-gr-PDAA)	2.6	VCl$_4$	2.50	5.0	4.2 ± 0.8	27 ± 4
PE-gr-polya(crylic acid) (PE-gr-PAAc)	11.0	VCl$_4$	1.90	2.1	5.1 ± 1.0	38 ± 4
PE-gr-poly(allyl alcohol) (PE-gr-PAAl)	6.2	VCl$_4$	3.00	12.5	3.4 ± 0.7	34 ± 4
	6.2	Cp$_2$TiCl$_2$	0.10			

12.3.2 Ethylene Polymerization by Titanium–Magnesium Catalysts

At present, immobilized titanium–magnesium catalysts are widely used for ethylene polymerization. These are obtained either by reduction of titanium compounds ($TiCl_4$, $TiCl_3(OR)$, etc.) with Grignard reagents, by supporting Ti^{4+} complexes from solutions on inorganic carriers or by joint grinding of MgO, $MgCl_2$, $Mg(OH)Cl$, $Mg(OR)_2$ with $TiCl_4$, $TiCl_3$, etc. For instance, titanium–magnesium compounds grafted to polymers with a covering of functional groups have been obtained according to a general scheme [74]. Equation (12-14) gives an example with PE-gr-PAAl).

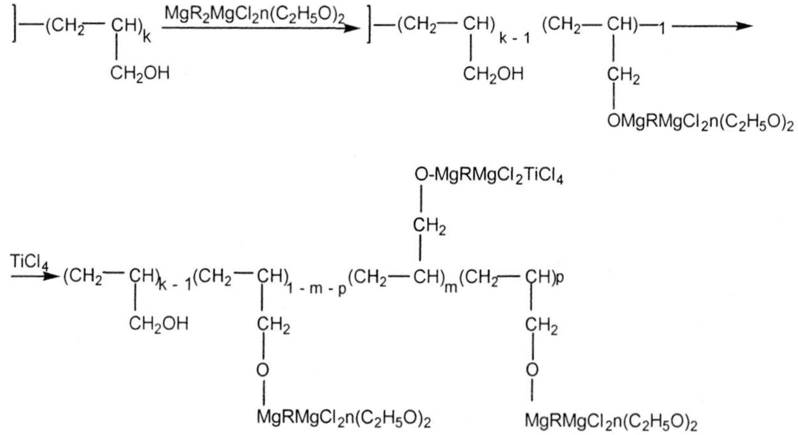

$$(12-14)$$

There are a number of different functional sites on such catalysts, such as unreacted functional groups of the polymer support, resting Mg centers and bound Ti^{4+} and Ti^{3+} species. Species containing about 50% titanium in the form of isolated Ti^{3+} ions showed the highest activity for ethylene polymerization (550 kg of PE/(g of Ti MPa h)$^{-1}$). The strong dependence of catalytic activity on isolated Ti^{3+} ion content leads to the conclusion that these appear in active centers (probably stabilized with $TiCl_3$ or $TiCl_2$ crystallites). The fraction of active sites in immobilized catalysts reaches 10% of the overall titanium content. The same approach has been used [75] to obtain titanium–magnesium catalysts from both CSDVB species and chloromethylated derivatives. Among another examples note immobilized $NdCl_2$ [76] and other rare earth chlorides [77], as well as zirconium-containing polymers [78] and other [79] components of ethylene polymerization catalysts.

12.3.3 Gel-Immobilized Systems

If the main goal of the formation of grafted polymer catalysts was to dispose active sites on the surface of the support so as to increase its specificity, gel-immobilized ones have the opposite goal. In catalysis by gel-immobilized systems we aim to have all active centers dispersed throughout the total polymer volume. This can be achieved by the use of swelling, rather than insoluble polymers in the reaction mixture, so that they are permeable to catalyst components, substrate and solvent molecules, so-called "mosaic" gel polymers [80]. Basically these are rubbers (ethylene–propylene, tercopolymers of ethylene, propylene and unconjugated dienes, etc.) which are functionalized with radical grafting of different monomers. The subsequent rubber linkage gives three-dimensional networks preventing the removal of catalyst particles. Thus, the structure of gel-immobilized catalysts prepared (in the form of granules, fibers, films, etc.) favors the diffusion of reagents into (and out of) the polymer volume, so stabilizing the bound active centers. However, polymerization in the presence of such systems can be accompanied by pore-diffusion resistance even at "superequilibrium" swelling.

12.3.4 Stereospecific Propylene Polymerization

The creation of stereospecific propylene polymerization catalysts is a complex problem. The formation of submicrocrystals of MCl_3 (M = Ti, V) on polymer supports is one of the ways of solving it. Coprecipitation of MCl_3 *in situ*, MCl_4 reduction in the presence of macroligands, including chemisorbed species, and joint grinding of MCl_3 with polymers (functionalized PS, CSDVB, ion-exchange resins, both in the form of NH_4^+ salts and cationites containing carboxylic and sulfoacid groups) are some of the suitable approaches. The ease of MCl_3 precipitation on polymer surfaces, especially on those with excess electrons, is well known. Electron donors favor powdered MCl_3 crystals and significantly affect the morphology of the metal-containing polymer produced. Data concerning the stereospecific polymerization of olefins by immobilized catalysts are available mainly in patents (for instance, it has been reported that the stereoregular fraction (insoluble in boiling heptane) of PP reached 95–98%). It is also important that such systems combine enhanced activity with high stereospecificity. However, the optimism which accompanied the early stages of these investigations is now rather reduced because of the discovery of so-called zirconocene catalysts for ethylene polymerization.

12.3.5 The Heterogenization of Homogeneous Metallocene Catalysts for Olefin Polymerization

Metallocene complexes of Ti, Zr and Hf have revolutionized the world of polyolefins. They represent a new generation of Ziegler–Natta catalysts. It is only necessary to note the record productivity for ethylene polymerization by bridged bis(fluorenyl)complex [$Zr(C_{13}H_4C_2H_4C_{13}H_8)Cl_2$], which can produce 300 tons of PE per g Zr per h [81], the great number of parameters to control the polymerization reactions in terms of stereospecificity, long-chain and short-chain branching and the generation of block copolymers (oscillating catalysts). These catalysts allow production of tailored polyolefins for nearly every purpose and the generation of new materials. The first reports on metallocene catalysts appeared relatively long ago [82]. The composition of such systems can be represented as follows (Eq. (12-15)):

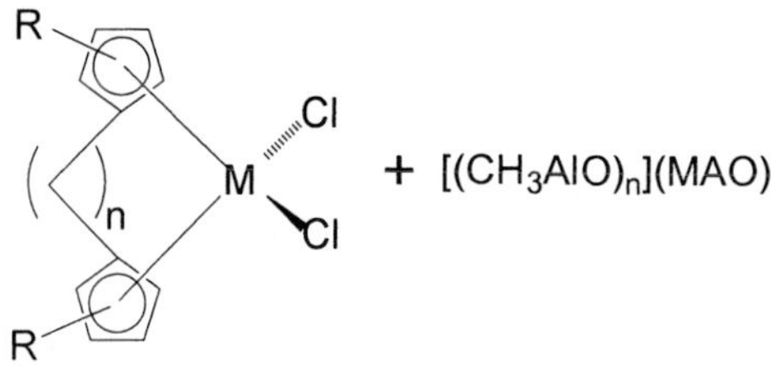

$$\text{+ } [(CH_3AlO)_n](MAO)$$

$$(12\text{-}15)$$

M = Ti, Zr, Hf,
n = bridging unit, consisting of 1–3 atoms in the backbone,
R = hydrocarbyl substituents of fused ring systems (indenyl, fluorenyl and substituted derivatives).

Immobilization of metallocene catalysts is not carried out with a view to enhance their activity because they now possess excellent activities and stereospecificities. However, they are not suitable for technical application, and improvement of their technological characteristics is most likely necessary (raising the operating temperature, changing commonly used aromatic solvents for aliphatic ones, developing the gas-phase processes and preventing the deposition of the polyolefin formed on the reactor walls). Furthermore, manipulating the functional groups in the polymer support opens the possibility of controlling the molecular structure of the polymer product. Inorganic supports such as silica, magnesium dichloride, organosilanes were initially used for these catalysts. Fixation of the catalyst can either be performed by an adsorption process or by forming a chemical bond. Another method is to functionalize the surface of silica by cene ligands or to immobilize the cocatalyst methylalumoxane (see, for instance, [83–89]). Thus, only the first order of the reaction of active center deactivation (Eq. 12–11) is observed in the gas-phase olefin polymerization by immobilized metallocene catalysts [90–92]. The nature of such catalytic systems has been analyzed in a review [93]. Polymers based on cross-linked PS are among the small number of macroligands used to support metallocene complexes [94,95]. An elegant method that allows the avoidance of polar components on the support surface, which could decrease the catalyst activity, was used in the assembly of zirconocene from a polymer precursor [96] (Fig. 12-8).

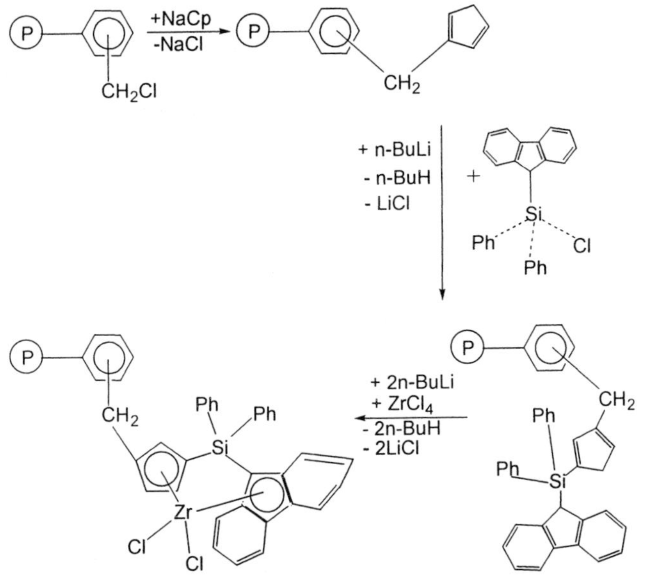

Figure 12-8. Fixation of a catalyst on polystyrene.

However, we should note that such a functionalization of polystyrene by η^5-cyclopentadienyl groups (Cp) to obtain a polymer-immobilized dicyclopentadienyl titanium had essentially been demonstrated earlier [97]. A novel polymer, cross-linked poly(styrene-*co*-4-vinylpyridine), as a support to anchor Cp_2ZrCl_2, and the influence of different factors on ethylene polymerization by such a catalyst, have been studied in detail [98]. The polymer was pretreated with MAO or AlR_3 and was then bound with the zirconium compound. The activity of the catalyst was found to increase with an increase in both the content of 4-VPy units and the degree of cross-linking in the support. The catalyst based on *i*-Bu$_3$Al possessed higher activity than the MAO-based catalyst. It not only retained a relatively higher activity but a narrow molecular weight distribution of the resulting polymer, and a high ability to incorporate an α-olefin (hexene) was also characteristic. The same results were obtained in the case of a divinylbenzene cross-linked copolymer of styrene and acrylamide [99]. Cocatalyst MAO is coordinated with the support via oxygen and amide nitrogen atoms. Adding α-octene did not influence the reaction rate. The comonomer is homogeneously distributed in the polymer chain, and no blocks are observed. Copolymer styrene with acrylamide [100] and saponated copolymer styrene with vinylacetate containing 3 to 19 mol% of hydroxyl groups [101] have been used as supports to obtain polymer-supported half-titanocene catalysts for the syndiospecific polymerization of styrene. The catalyst based on polymer-bound $CpTiCl_3$ and activated MAO revealed higher activity and resulted in polystyrene with a higher proportion of the syndiotactic fraction then the homogeneous analogue. Nevertheless, the same active centers are proposed for both the homogeneous and the immobilized systems. Polymer and inorganic supports differ in their influence on the active centers, which results in the different structures of the polymers formed. One reason for this may be the flexibility of a macroligand in comparison with an inorganic support. The active center of the polymer support unites several MAO molecules, which facilitates stereocontrol during polymerization. The original approach to supporting Cp_2ZrCl_2 is based on its self-immobilization [96,102]. This means the following. Using special methods, fluorenylidene dichloride complexes of Zr were obtained, containing one ω-alkenyl substituent either in the C_2-bridge or in position 3 of the indenylidene ligand. After activation with methylaluminoxane the metallocene complexes not only polymerize ethylene but incorporate themselves into the growing polymer chain as a comonomer. In this way, this homogeneous metallocene catalyst becomes self-immobilizing, and the further formation of polyolefin is catalyzed heterogeneously, the polymer chain serving as an organic support (Fig. 12-9).

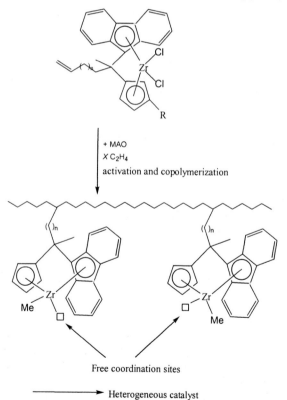

Free coordination sites

Heterogeneous catalyst

Figure 12-9.. A scheme of the self-immobilization of a metallocene catalyst.

In essence, this is one of the variants of the copolymerization of metal-containing monomers with ethylene as analyzed above (see Chapter 4).

Kinetic data on olefin polymerization by polymer-immobilized zirconocene are scarce. It is generally accepted that homogeneous metallocene catalysts contain uniform active sites; however, if they are immobilized on a polymer support, the MWD polymer production becomes broader compared with a homogeneous catalyst [103]. Kinetic analysis of gas-phase ethylene polymerization catalyzed by $(CH_3)_2[Ind]_2ZrCl_2$ bound at a hydroxylated copolymer of styrene with divinylbenzene and previously activated with MAO (0.17 wt.% Zr) has been carried out [104]. The influence of temperature (333 to 353 K), ethylene partial pressure (2 to 6 atm) and MAO level (molar ratio of MAO to zirconium from 2600 to 10,700) were studied. The activity of the catalyst in the gas-phase process changed from 5 to 32 kg PE (g of Zr atm h) $^{-1}$. It is possible that there are two types of active site. They are stable to temperature and deactivated by the same mechanism. A first-order reaction takes place. The propagation rate constants of two active sites show a similar dependence on temperature.

In recent years, novel so-called postcene catalysts for olefin polymerization, i.e. Ni(II) and Pd(II) complexes incorporating bulky α-dimine ligands (see for instance [105,106]), as well as those based on Fe^{2+} and Co^{2+} bearing 2,6-bis(imino)pyridyl ligands [107,108] have been developed. However, data concerning their immobilization on macromolecular ligands are unknown with the exception of self-immobilized catalysts for ethylene polymerization based on salicylaldiminonato phenyl nickel(II) complexes bearing allyl substituents [109].

12.3.6 Polymerization of Vinyl Monomers in the Presence of Macromolecular Complexes

Vinyl polymerization using metallocomplexes commonly proceeds by a radical pathway and rarely involves an ionic mechanism. For instance, metal chelates in combination with promoters (usually halogenated hydrocarbons) are known as initiators of homo- and copolymerization of vinylacetate. Similar polymer-bound systems are also known [3]. The polymerization mechanism is not well understood, but it is believed to be not exclusively radical or cationic (as polymerization proceeds in water). The macrochelate of Cu^{2+} with a polymeric ether of acetoacetic acid effectively catalyzes acrylonitrile polymerization. Meanwhile, this monomer is used as an indicator for the radical mechanism of polymerization. Mixed-ligand manganese complexes bound to carboxylated (co)polymers have been used for emulsion polymerization of a series of vinyl monomers. Macromolecular complexes of $Cu(NO_3)_2$ and $Fe(NO_3)_3$ with diaminocellulose in combination with CCl_4 are active in polymerization of MMA, etc.

Complexes of Cu(II), Fe(III), V(V), etc., immobilized on polymers, are often used as components of redox systems which are effective as initiators of vinyl-type monomer polymerization. For instance, the system vanadylpolycarboxylate–thiocarbamide–water solution of HNO_3 initiates acrylonitrile polymerization. The complex of Cu(II) with poly(vinyl alcohol) acts as an initiator of the (co)polymerization of allyl-type and methacrylic monomers. The process is considerably accelerated in the presence of carbon tetrahalides. Redox polymerization of acrylamide proceeds in the presence of the complexes V(V) bound with chelate copolymer of glycol methacrylate and ethylene glycol dimethacrylate in combination with cyclohexanone at 303 K [110]. The remarkable characteristics of the polymerbound complexes are their ability to be recovered and also to form polymers that contain no metal ion contaminants. The molecular mass of polymers formed reaches values of $6.3–13.3 \cdot 10^5$ ($3.5 \cdot 10^5$ Da in the case of the homogeneous system).

Ionic polymerizations catalyzed by immobilized complexes, in particular Lewis acids on polymer supports, so-called solid superacids, are now extensively developed. A variety of cationic catalyst (protonic and Lewis acids), their combinations (complexes) and complexes with different supports are summarized in a review [111] (Table 12-9).

Table 12-9. The main types of immobilized cationic catalysts.

Types of acids	Types of supports
Bronsted acids	
H_2O, HCl, HF, H_2SO_4, H_3PO_4, $HClO_4$, FSO_3H, heteropolyacids – $H_4SiW_{12}O_{40}$, $H_3PMo_{12}O_{40}$	Oxides, aluminosilicates, zeolites, salts, silica, copolymers of styrene with divinylbenzene, polyphenylketone, perfluorineolefins and their copolymers, net copolymers of alkoxy- and alkoxyorganosilanes
Lewis acids	
R_nAlCl_{3-n}, $AlCl_3$, $AlBr_3$, BF_3, BCl_3, BBr_3, $ZnCl_2$, $CuCl_2$, $FeCl_3$, $TiCl_4$, TiF_4, $SnCl_4$, $SbCl_5$, NbF_5	Oxides, mixed oxides, aluminosilicates, silica, metals, polystyrene, copolymers of styrene with divinylbenzene and 4-vinylpyridine, copolymers of acrylic monomers, coal, graphite
Complex acids	
HX (H_2O, HCl, HF) – $MeX_n(R_nAlCl_{3-n}$, $AlCl_3$, $FeCl_3$, BF_3)	Oxides, aluminosilicates, polystyrene, copolymers of styrene with divinylbenzene, copolymers of acrylic monomers, sulfonated copolymers of styrene with divinylbenzene
Carbene ions	
MX_n – R_nAlC_{3-n}, $AlCl_3$, $TiCl_4$, VCl_4, $SnCl_4$, $AgClO_4$	Poly(vinyl chloride), chlorine (bromine)-butylrubber resins, poly(chloroprene)

All types of acid (Bronsted and Lewis acids, complex acids, including Hustavson acids ($HCl \cdot 2AlCl_3 \cdot 3C_6H_5CH_3$) are anchored to a polymer matrix. Such catalysts are effective in olefin (isobutylene) polymerization and carbonium-ionic polymerization of styrene, α-methylstyrene and β-pinene. As a rule, the activity of a complex depends on its Lewis acid strength and decreases as follows: $SbF_5 > AlCl_3 > ZnCl_2$.

Polymerization of N-vinylcarbazole catalyzed by dimethylglyoxime complexes of different metals immobilized on PVC follows the cationic mechanism. Lewis acids immobilized in a volume of swollen polymer gel catalyze cationic polymerization and oligomerization of vinyl ethers, etc. Cationic complexes of Pd(II) bound to modified PS initiate alternative copolymerization of fluorinated olefins $(C_nF_{2n+1})(CH_2)_mCH=CH_2$ with carbon monoxide [112,113]. The product thus obtained was polyspiroketal rather than polyketone.

12.3.7 Other Perspectives in Polymerization Catalysis by Macromolecular Complexes

Let us briefly summarize the state of the art in the polymerization of other monomer groups in the presence of polymer-bound metal complexes. Considerable progress has been achieved in polymerization of dienes, namely, butadiene and isoprene, catalyzed by macromolecular complexes based on rare earth halides [114]. Co-oligomerization of 1,3-butadiene and CO_2 with immobilized palladium complexes (phosphynated PS as macromolecular ligand) [115] and polymerization of acetylene monomers with immobilized complexes of Mo(V), W(VI), and Pd(II) have been developed though they have not been investigated intensively.

Comparative analysis of kinetic data and the properties of polymers formed by the action of homogeneous and immobilized metal complexes shows the similar nature of the active centers in these systems. The active role of polymer supports consists of regulating the reactions of active-site formation and deactivation and sometimes promoting their regeneration. Under optimal conditions we can clarify the mechanism of each polymerization stage and study intermediates.

The development of bifunctional catalysts for specific catalytic sequences of reactions in which the product of the first reaction can serve as substrate for the second is of great importance. There are many examples of such reactions. They are, for instance, the monomer-isomerizing polymerization of heptene-2, heptene-3 and 4-methyl-2-pentene and the combination of propene disproportionation with oligomerization, etc. Bifunctional catalysts are most widely used for ethylene copolymerization with α-butene *in situ* in the production of so-called low-density linear polyethylene (LDLPE). All general methods for LDLPE production are based on incorporation into a PE backbone of short-chain branches, which can be made by catalytic copolymerization of ethylene with α-olefins $C_3–C_{10}$. A macromolecular ligand offers wide possibilities of joining the different types of active site in the same matrix (see also Section 12.5.2).

The essence of "relay-race" (or concerted) copolymerization thus lies in the coexistence of two different bound active sites in the same polymer matrix, the first being responsible for dimerization and the second for copolymerization of butene with ethylene [116]. The general scheme of the process can be represented as in Eq. (12-16).

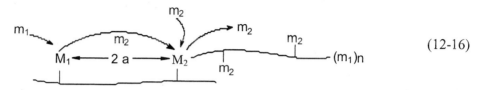

$$(12\text{-}16)$$

The first monomer m_1 is converted into m_2 by some simple reaction, such as di-, oligo-, isomerization, etc. (in this case ethylene is converted into α-butene) at the active center M_1 (for instance, Ni(II)). Copolymerization of m_1 and m_2 proceeds at an active center M_2. The microcell containing both active sites is the size of about half the average distance between two neighboring centers in the bifunctional catalyst. In turn, the distance between M_1 and M_2 can be specified at the stage of catalyst preparation [117]. Sometimes such an approach is called tandem catalysis [118].

Catalysis by immobilized complexes gives opportunities for the production of new composite materials. If the macromolecular complex is used not at catalytic (or initiating) concentrations but as an ingredient, we may obtain polymer–polymer compositions (PPC) from components that are not compatible under common conditions. This approach has already been realized in the case of new type composites, i.e. polyolefin-polyvinyl PPC, chemical hybrids of polyolefins and polydienes, electroconducting PPC, etc. [119].

To solve many of these problems additional efforts are necessary.

12.4 Oxidation Catalysis in the Presence of Macromolecular Complexes

Oxidation by dioxygen and hydroperoxides is an important reaction in organic synthesis. This reaction permits the introduction of oxygen-containing groups into nearly any compound and the preparation of valuable substances. Such reactions with participation of dioxygen are a main part of the complex enzymatic processes occurring in nature (see Chapter 2). One important aspect of partial oxidation is its selectivity. It is important to study the effect of various homogeneous or heterogeneous catalysts, which may be capable of inhibiting, accelerating or increasing the selectivity of the oxidation. The search for and use of new catalysts including immobilized species, as well as the increase of our understanding of the mechanisms and kinetics of oxidation, may help to solve many of these problems.

12.4.1 The Role of Metal Ions in Oxidation

Oxidation processes consist of a large number of consecutive and parallel reactions with participation of initial substances, intermediates and free radicals contributing to chain reactions. The main elementary stages of an oxidation reaction of a hydrocarbon (RH) are the following [120]:

- Chain generation reactions, growth of the chain, generation of chain branching, accompanied by formation of hydroperoxides;
- chain decomposition of hydroperoxide;
- molecular decomposition of hydroperoxide;
- terminations of chains.

The rate of chain generation w_i in the absence of an initiator or catalyst is very low (10^{-9}–10^{-7} mol L^{-1} s^{-1}) and this results in the appearance of a marked induction period for low-temperature non-catalytic oxidation of RH. The ratio of rate constants for chain growth (k_g) and termination (k_t) ($k_g/k_t^{1/2}$) is a characteristic of the ability of hydrocarbons to be oxidized. Radical-chain oxidation can be catalytic or non-catalytic. Ions of the transition metals, M^{n+}, can, in principle, promote or inhibit the direction of any of the above steps of the oxidation process [121–123]. The key role of M^{n+} ions in the liquid-phase oxidation of hydrocarbons is in the acceleration of initiation, as a rule the limiting step for a free-radical chain process. This results in a one-half-order reaction with respect to catalyst concentration and a first-order reaction with respect to hydrocarbon concentration. Reaction acceleration (most often revealed by the disappearance of the induction period) can be achieved in a variety of ways.

- The first is due to catalytic decomposition of hydroperoxides, which are present at very low concentrations and stable under mild reaction conditions [8,12,13]. This process, leading to formation of alkoxy and hydroperoxy radicals, is described by the Haber–Weiss scheme (Eqs. (12-17) and (12-18)) [124]. It has been shown that formation of a complex of ROOH with M^{n+} or M$^{(n+1)+}$ precedes these reactions. Decomposition of coordinated hydroperoxides can proceed by two mechanisms: (i) molecular (heterolytic), leading to the formation of non-radical products, and (ii) radical, the main route to decomposition in developed radical-chain processes.

$$ROOH + M^{n+} \longrightarrow RO^{\cdot} + {}^-OH + M^{(n+1)+} \qquad (12\text{-}17)$$

$$ROOH + M^{(n+1)+} \longrightarrow ROO^{\cdot} + H^+ + M^{n+} \qquad (12\text{-}18)$$

- Another route to acceleration of initiation is through activation of dioxygen by a metal ion and the subsequent interaction between the complex formed and the oxidizable substrate (Eq. 12-19)). It is important that initiation by activation of molecular oxygen is the most promising route from the viewpoint of selectivity: it is possible to introduce O_2 directly into the C–H bond of oxidizable substrates.

$$M^{n+1} + O_2 \longrightarrow {}^{\delta}M^{n+1}...{}^{\delta}O_2 \overset{RH}{\underset{}{\Big\langle}} \begin{matrix} MOOH + \overset{\bullet}{R} \\ \\ M^{(n+1)} + (OH) + \overset{\bullet}{RO} \end{matrix} \tag{12-19}$$

- Peroxides of transition metals are themselves active intermediates in heterolytic and homolytic liquid-phase catalytic oxidation reactions of alkenes, aromatic hydrocarbons and alkanes. Heterolytic oxidations are characterized by a requirement for a free coordination volume near the transition metal atom. Homolytic oxidations proceed via M–O bond cleavage in peroxo complexes.
- Activation of a coordinated substrate involves electron transfer from substrate to metal ion. Most often this transfer proceeds via formation of a cation-radical from the substrate with subsequent loss of a proton. Ions with d^7 (Co^{2+}) and d^5 (Mn^{2+}) configurations are especially active, so it was concluded that ions with an odd number of d-electrons more easily form σ-bonds with coupling of valence electrons between metal and ligand. In many cases the contribution of the σ-bond forms a major part of the total energy of the coordination bond. Participation of uncoupled d-electrons in ligand binding facilitates the formation of the activated substrate–catalyst (especially olefin–catalyst) complex and of intermediate radicals characterized by a high degree of delocalization.

Most salts, including transition metal acetates, used in liquid-phase oxidation of hydrocarbons are incompatible with the medium. Incompatibility diminishes their effectiveness and puts tough restrictions on concentration ratios. To overcome this obstacle, attempts have been made to include MX_n into inverted micelles of ionogenic and non–ionic surface-active substances (SAS).

One may note the deeply rooted idea that metallopolymer systems themselves can be readily oxidized. Thus, hydroperoxide oxidation of 2,3,6-trimethylphenol in the presence of Cu(II) polymer chelates is accompanied by the parallel reaction of oxidation of the polymer matrix [125], which results in transformations of the functional groups. For example, oxidative cleavage of the aromatic ring of trimethylphenol results in the formation of carboxylic acids as well as in oxidative coupling to give 2,2',3,3',5,5'-hexamethyl-4,4'-dioxyphenyl. This was a reason why oxidation catalysis by metallopolymers was developed slowly (see, for instance [2]) although most metallopolymer catalysts are stable in oxygen even at temperatures above 393 K [126]. Meanwhile, the chemical binding of MX_n at a polymer allows us to obtain complexes that have no soluble analogues and to increase the activity and selectivity of the oxidation reaction through some features of their action. First, regeneration of reduced forms of a transition metal is possible, i.e. stoichiometric oxidation becomes catalytic. The nature of the macromolecular ligand and its functional groups influence the dynamics of the oxidation process [127]. The ligands should be rather flexible because part of them is substituted by substrate or oxygen during oxidation. Strong binding of the components may lead to retardation or even inhibition of the reaction. Another feature is the acceleration of hydroperoxide cleavage proceeding both by a heterogeneous–homogeneous mechanism (formation of radicals on the surface of the metallopolymer followed by chain reactions in the bulk liquid phase) and heterogeneously (without the radicals passing into the liquid). In this last case all reactions occur on the surface of catalyst although a mixed mechanism usually takes place: The reaction proceeds simultaneously both on the surface and in the bulk liquid. Metallopolymer catalysts promote oxidations by a heterogeneous mechanism in a "surface cell" that favors activation of only one of the reacting particles and therefore allows us to solve the problem of selectivity. This is also facilitated by fixing oxidants (perbenzoic and other acids), hypochlorite anion, periodates, etc. to polymers [3].

12.4.2 Reactions of Hydrocarbon Oxidation by Oxygen

On the one hand, hydrocarbon oxidation is a model reaction which enables special features of these catalytic processes to be analyzed. In addition, this resembles, to a considerable extent, enzymatic catalysis; it also proceeds at low temperatures with high selectivity and requires small quantities of catalyst [128]. There have been no systematic investigations of catalytic liquid-phase oxidation of paraffins by macromolecular complexes and the scarce data are presented mainly in patents. (Pd complexes bound to ion-exchange resins are highly active in hydrogen oxidation by air (see, for instance [129])).

Among such oxidations, note that liquid-phase oxidations of solid paraffins in the presence of heterogeneous and colloidal forms of manganese are accompanied by a substantial increase (compared with homogeneous catalysis) in acid yield [3]. The effectiveness of n-paraffin oxidations by Co(III) macrocomplexes is high, but the selectivity is low: the ratio between fatty acids, esters, ketones and alcohols is 3:3:3:1. Liquid-phase oxidations of paraffins proceed in the presence of Cu(II) and Mn(II) complexes bound with copolymers of vinyl ether, β-pinene and maleic anhydride (Amberlite IRS-50) [130]. Oxidations of both linear and cyclic olefins have been studied more intensively. Oxidations of linear olefins proceed by a free-radical mechanism; the accumulation of epoxides, ROOH, RCHO, ketones and RCOOH in the course of the reaction testifies to the chain character of these reactions. The main requirement for these processes is selectivity: non-catalytic oxidation of propylene (at 423 K) results in the formation of more than 20 products. Acrylic acid is obtained by oxidation of propylene (in water at 338 K) in the presence of catalyst by two steps: at first to acrolein, then to the acid with a selectivity up to 91%. Oxidation of ethylene by oxygen at 383 K in acetic acid in the presence of metal-polymer catalysts results in formation of methyl methacrylate and a small quantity of ethylene oxide. Methacrolein is obtained at a low temperature (360 K) by oxidation of isobutylene in the presence of Cu(II) macromolecular complexes with a selectivity of 100%. It is important that supported clusters can be used for olefin oxidation. For example, a mixture of clusters $Co_4(CO)_{12}$ and $Rh_4(CO)_{12}$ at a molar ratio between 1:1 and 3:1, immobilized on an amine-containing ion-exchange resin [131], catalyzes the oxidation of 2-olefins to the corresponding alcohols.

Oxidation of cyclohexane, both uncatalyzed and catalyzed by soluble complexes of transition metals, has been used as a model reaction for the investigation of the mechanism of hydrocarbon oxidation. Immobilized complexes are also used in model systems. Cyclohexane oxidation proceeds in two ways, with participation of C=C or C–H bonds (Eq. (12-20)). Cyclohexyl hydroperoxide (CHHP), cyclohexene oxide, cyclo-1-hexene-ol-2 and cyclohexene-on-2 are the major products.

$$(12-20)$$

Oxidation of cyclohexene by polymeric cobalt-containing catalysts [132,133] has many significant features. In the presence of $Co(AcAc)_2$ (AcAc = acetylacetonate) the process is characterized by an induction period, after which the rate rapidly reaches its highest value and decreases thereafter (Fig. 12-10, plot 1).

In the case of heterogenized catalysts there is no induction period and the oxidation rate is constant (Fig. 12-10, plots 2–4). It is important that metal polymer catalysts retain their activity at high degrees of oxidation and can be reused many times after isolation from the reaction mixture. The products formed in the presence of most such catalysts are approximately the same (Fig. 12-11): major products are CHHP, cyclohexenone, cyclohexenol and cyclohexene oxide (CHO).

Oxidation of cyclohexene is a heterogeneous–homogeneous chain process accompanied by escape of radicals into the bulk liquid. The latter was confirmed by introducing acceptors of free radicals (dimer 1,2-bis(4,4′-dimethylaminophenyl)-1,2-diphthaloylethane) (Ph-Ph) into the system. The addition of an inhibitor in the stationary phase of the process results in large induction periods which do not depend on the extent of oxidation (Fig. 12-12).

The rate of initiation at the beginning of the process is 30 times lower than that in the stationary phase. Thus, the processes on the surface must determine the rate of chain generation, and free radicals are not formed in the bulk solution. At the beginning of the process (in the absence of ROOH) Co(II) compounds can interact with O_2 with formation of free radicals (Eq. (12-21))

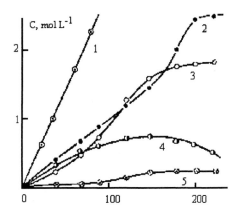

Figure 12-10. Oxygen binding for cyclohexene oxidation in the presence of Co-containing catalysts (333 K, without solvent). (1) Co(AcAc)$_2$, (2) polyethylene-grafted -poly(cobalt acrylate), (3) copolymer of styrene with cobalt acrylate, (4), polyethylene-grafted - poly(acrylamide –Co(II)).

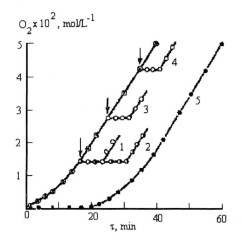

Figure 12-11. Kinetics of product accumulation during cyclohexene oxidation in the presence of PE-gr-poly Co(acrylate)$_2$ at 333 K. (1) Sum of the products, (2) cyclohexenone, (3) CHHP, (4) cyclohexenol, (5) CHO.

$$\text{—Co}^{2+} + O_2 \rightleftharpoons \text{—[Co}^{3+}...O\cdot - O^-]^{2+} \xrightarrow{RH} \text{—Co}^{2+} + R^\cdot + HO_2^\cdot \qquad (12\text{-}21)$$

In the developed process the main contribution to the chain initiation results in the decay of coordinated ROOH (Eq. (12-22)).

$$\text{—Co}^{2+} + n\text{ROOH} \xrightarrow{K} \text{—[Co}^{2+}...n\text{ROOH]} \xrightarrow{ki} RO^\cdot + \text{—[Co(OH)]}^{2+} + (n-1)\text{ROOH} \quad (12\text{-}22)$$

The rate of the initiation is given by Eq. (12-23).

$$w_i = \frac{k_i K [ROOH]_0^n [Co^{2+}]_0}{1 + K[RCOOH]_0^n} \tag{12-23}$$

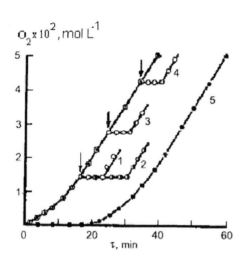

Figure 12-12. Inhibition of cyclohexene oxidation by dimer Ph–Ph. Concentration of cyclohexene 5 mol L^{-1}, catalyst PE-gr-PAA-Co(II), 328 K. Concentration of Ph–Ph 1.7×10^{-4} (1, 3, 4), 3.4×10^{-4} (2), 1.7×10^{-5} mol L^{-1} (5). Arrows indicate the addition of inhibitors.

The main kinetic parameters for the oxidation promoted by Co(II)–polymer complexes are given in Table 12-10.

The small increase of $k_2 k_6^{-1/2}$ cannot lead to the formation of the substantial quantities of products which are characteristic of catalytic oxidations, so it is plausible that the oxidation products are also formed as a result of the first-order termination reaction with participation of immobilized Co^{2+} ions.

The value of K ($>10^2$ L mol^{-1}) and the independence of w_i from [ROOH] indicate that near the surface of the heterogeneous metal-polymer catalyst the local concentration of hydroperoxide is high. It has been shown that in the vicinity of catalytic centers there are always vacant carboxyl groups able to bind ROOH molecules by hydrogen bonds. Probably, the transfer of ROOH molecules to the active center is carried out initially via their migration across the two-dimensional catalyst surface, and only after their decay do the radicals formed escape into the bulk liquid where the chain radical oxidation of hexene takes place.

We can control the process by changing the ratio between vacant functional groups of polymeric support and immobilized metals ("the loading degree"); thus, in such systems there is an additional control lever that is absent from homogeneous catalysis.

Table 12-10. Kinetic parameters of cyclohexene oxidation in the presence of immobilized catalysts (333 K, without solvent).

Catalyst	$[Co(II)]$ $\times 10^3$, $g\,L^{-1}$	$w_i \times 10^6$, $mol\,L^{-1}\,s^{-1}$	$w \times 10^4$, $mol\,L^{-1}\,s^{-1}$	$w[Co^{2+}]$ $\times 10^2\,(s^{-1})$	$k_g k_t^{-1/2}$ $L(mol\,s)^{-1/2}$
PE-gr-poly(Co(II)-acrylate)	0.88	6.5	1.4	160	5.4
Copolymer of styrene and Co(II)-diacrylate	10	2.8	0.97	9.7	5.8
PE-gr-poly(Co(II)-acrylate)	1.0	1.8	0.62	6.2	4.6
Without catalyst (AIBN as an initiator)	-	-	-	-	3.7

Oxidation of alkylaromatic hydrocarbons only affects the alkyl substituent, although benzene itself may be oxidized. Among substrates of this type oxidations of toluene, ethylbenzene and cumene by immobilized complexes of transition metals have been studied intensively [3].

In the direct oxidation of aromatic hydrocarbons in heterogenized systems, triple complexes are formed, which are more stable than in homogeneous systems; these complexes decompose slowly. Sometimes special additives (for instance, dioxane) are used to prevent such decomposition. The kinetics of the accumulation of α-phenylethyl hydroperoxide, methylphenylcarbinol and acetophenone (AP) confirm that the oxidation of ethylbenzene in the presence of immobilized complexes proceeds by the consecutive-parallel scheme. Under optimal conditions selectivity reaches up to 96% with respect to AP. It is known that styrene can be oxidized in two ways: to olefin oxide and RCHO (or ketone), or to the copolymer of the olefin and O_2 (polyperoxide). To improve the selectivity, we must suppress the second alternative, an oxidative polymerization. The rate of formation of the monomeric products (benzaldehyde, styrene oxide) increases and the rate of polymerization decreases in the presence of the metal polymer. The main products of liquid-phase oxidation of cumene in the presence of transition metal complexes with ion-exchange resins are dimethylphenylcarbinol and hydroperoxide cumene. Selectivity depends on the nature of the functional groups. Among phosphorus-containing complexes the selectivity increases in the sequence $Co(II) < Mn(II) < Cu(II) < Ni(II)$. Iron ionite complexes are inactive in these reaction. It should be noted that polymereric Schiff bases containing N and S donor atoms inhibit cumene oxidation because they are acceptors of peroxide radicals [125].

Numerous interesting features were found for the catalytic oxidation of tetralin, a product of the partial hydrogenation of naphthalene, in the presence of immobilized catalysts. Tetralin is oxidized mainly to tetralin hydroperoxide, tetralin-1 and tetralol-1. Firstly, the initial rate of liquid-phase catalytic oxidation of tetralin (O_2, 393 K) is higher by nearly four orders of magnitude than that of non-catalytic oxidation. Secondly, the hydroperoxide accumulation decreases in the presence of immobilized complexes. This testifies to catalytic decomposition of hydroperoxide and free-radical generation by a scheme of secondary catalysis. The traditional scheme of catalytic autoxidation of tetralin, in which the role of metal complexes is only in the radical decomposition of the peroxide formed, does not describe the kinetics of the process. Finally, in this reaction we encounter a new effect in catalysis, the effect of the energy of the non-equilibrium conformations of the polymer that can arise during the course of the reaction through interaction of substrates with the immobilized metal ions (see Chapter 3) [134]. The following example demonstrates the effect of non-equilibrium conformations. Macrocomplexes of $CoCl_2 \cdot PEG$ (poly(ethyleneglycol) with different chain lengths) in tetralin in the absence of O_2 exist as globules that are smaller than free PEG molecules. The activation of O_2 on a metal ion can be accompanied by a temporary breaking of one of the M–O bonds (like the ousting of a ligand from a complex). This results in a non-equilibrium conformation which tends to unfold (Eq. (12-24)).

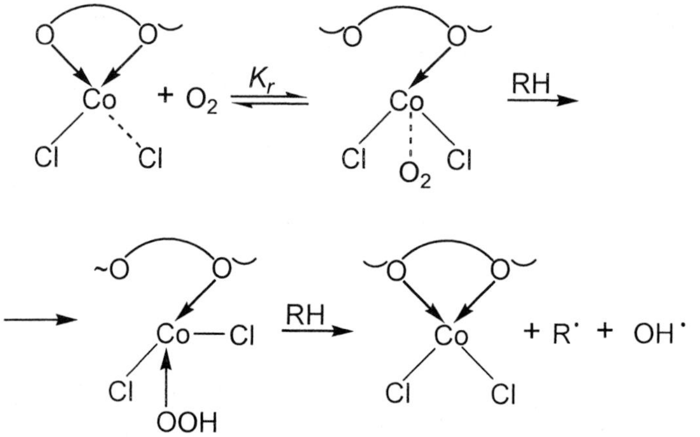

$$(12\text{-}24)$$

This process is reversible and characterized by the value of K_r: $-RT\ln K_r = \Delta G$, where ΔG is the change in the free energy of macrocomplex formation. ΔG can be divided into two parts: $\Delta G = \Delta G_x + \Delta G_c$, where ΔG_x is the contribution from the rupture of old and formation of new chemical bonds, and ΔG_c is the work of chain stretching that was "stored" during production of the macrocomplex. The values of K_r for polymeric complexes and their low molecular mass analogues differ by the factor $\exp(\Delta S_c/R)$, which is a part of the preexponential factor and depends on the molecular mass of the macroligand. In other words, the proportion of complexes bound with O_2 increases with increasing macrochain length; the value of this proportion is contained in the expression for the effective constant of the initiation step. Indeed, with an increase of molecular mass of PEG from 4000 to 40,000 Da, the rate of tetralin oxidation increases by 10 times.

The conformational prearranged effect of both the macromolecular ligand with respect to a certain metal ion and the active sites of metallopolymer catalysts to a substrate was observed in liquid-phase oxidations of alkylaromatic hydrocarbons; recently the same was found in the case of hydroxyarenes [135–137]. Initially, a soluble metallocomplex is formed in the presence of metal ions or substrate. A structural rearrangement, specific to the reacting compounds, is further fixed in the process by cross-linking of the polymer support by *N,N'*-methylenediacrylamide with a subsequent removal of pattern substrates from cross-linked formations. Such an approach leads to a substantial increase of selectivity and activity of the catalyst in comparison with a catalyst obtained without directed generation of its structure.

12.4.3 Peroxide Oxidation of Alkylarenes and Hydroxyarenes

There is no doubt that as well as air oxygen dilute solutions of hydrogen peroxide may also serve as ecologically useful oxidants for large-scale processes. Hydrogen peroxide disproportionation is a model reaction for comparing the activity of homogeneous and heterogeneous complexes. Because there are parallel reactions in the non-catalytic and catalytic decompositions of H_2O_2 in the system studied, the reaction rate may be expressed as shown in Eq. (12-25), where B is the overall decomposition rate in the homogeneous medium on the walls of a glass reactor, or on the polymer surface. Thus, the contributions of each process to the overall reaction rate of H_2O_2 decomposition (1.185 M, 303 K) are equal to 1.0 (in the homogeneous medium), 18 (on the walls), and 81% (on the surface of a catalyst (0.01 g, 0.32 mg-eq. of Cu g^{-1})) [125].

$$W_0 = B + k[M^{n+}][H_2O_2] \qquad (12\text{-}25)$$

Syntheses of 2,3,5-trimethylquinone, a key product in the fabrication of tocopherol, as well as of vitamin K_3 (2-methyl-1,4-naphthoquinone) using this method are of practical importance. Peroxo-acid oxidation models xenobiotic oxidation by means of monooxygenase. Peroxo-acid oxidation of 1,2,4-trimethylbenzene (pseudocumene) to trimethylquinone using peracetic acid also yields acetic acid, which reduces the price of the products obtained (Eq. (12-26)).

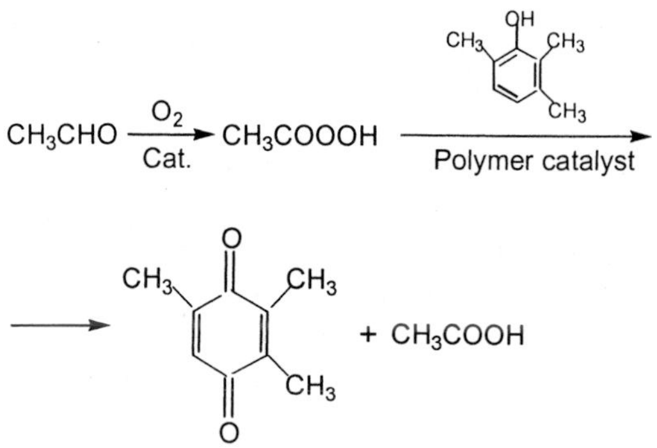

$$\text{(12-26)}$$

Without going into detail we simply note that oxidation of alkylaromatic hydrocarbons with polymer-immobilized catalysts can be considered as a system of successive-parallel reactions including mononuclearic hydroxylation to phenols, oxidation of phenols, oxidative dehydrogenation of hydroquinones, oxidative decomposition of quinones and thermal and catalytic decomposition of peracids.

Oxidation of anthracene is interesting from the viewpoint of obtaining anthraquinone, a valuable product. The oxidation can actually be carried out by a non-catalytic method using any oxidant, but the non-catalytic process requires rigid conditions and is accompanied by formation of a number of side products, especially anthrone and bianthrone (Eq. 12-27).

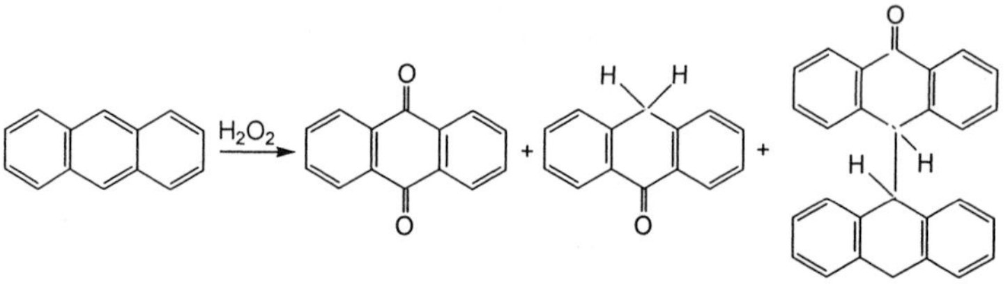

$$\text{(12-27)}$$

Polymeric VO(II) acetylacetonates are effective in the selective oxidation of anthracene by hydrogen peroxide to anthraquinone (up to 85%) in ethylacetate solution at 293–345 K. In spite of the fact that polymer-immobilized catalysts possess a high catalase-like activity in the decomposition of H_2O_2 [138] it has been proposed [125] that such metallopolymers may be valuable in the industrial catalytic oxidation of anthracene.

12.4.4 Catalysis of Olefin Epoxidation by Alkylhydroperoxides

The chemistry of oxiranes is developing in the direction of the industrial production of epoxide compounds (Eq. 12-28).

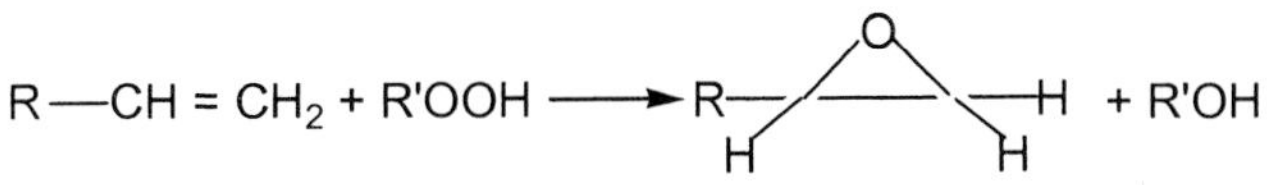

$$R—CH = CH_2 + R'OOH \longrightarrow R—\underset{H}{\overset{O}{\triangle}}—H + R'OH \qquad (12\text{-}28)$$

Olefin epoxidation has many features in common with biological mono-oxygenation. This reaction requires heterolysis of the peroxide bond. The presence of carbonyl groups facilitates heterolysis in peracids. Epoxidation occurs through electrophilic attack, so di- and tri-substituted double bonds are more easily epoxidized. In the presence of metal complex catalysts, epoxidation proceeds under mild conditions because in the transition state the O–O bond is more easily heterolyzed (Eq. 12-29).

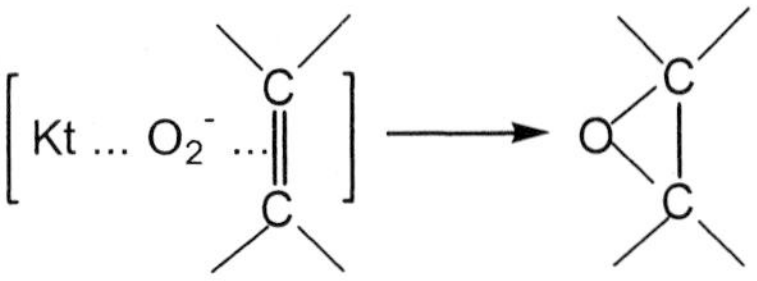

$$\left[Kt \ldots O_2^- \ldots \right] \longrightarrow \qquad (12\text{-}29)$$

The best catalysts are compounds of Mo, V and W in a high-valence state. These compounds have a large positive charge and are able to accept electron pairs on vacant orbitals and to form complexes of varying stability with olefins and hydroperoxides. (Compounds of Fe, Co, Mn, Ni and Cu possess diametrically opposite properties; their single function is homolysis of the peroxide bond by a one-electron mechanism with formation of free radicals.)

One of the indications of the catalytic activity of the complexes is their ability to form mixed-ligand mono- and di-peroxo complexes, active agents of direct epoxidation (Eq. 12-30).

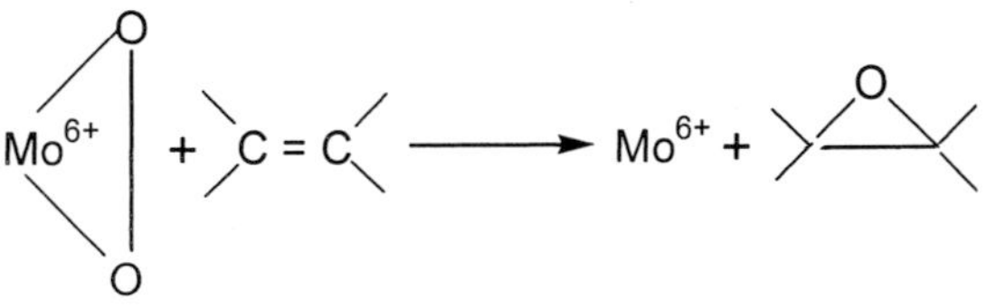

$$Mo^{6+} + C = C \longrightarrow Mo^{6+} + \qquad (12\text{-}30)$$

Formation of diolate complexes is the first step for homogeneous catalytic oxidation of hydroperoxide, monoperoxo complexes being more active than diperoxo complexes by two orders of magnitude.

What could be expected in catalysis of the reaction by fixing the metal complexes with polymers? Firstly, an increase of their activity. This is due to the formation of a higher proportion of monoperoxide complexes and their greater stability. Catalysts obtained by immobilization of molybdenyl groups on modified polyvinylalcohol or furfurolidine resin, are active and stable over 500 h of a continuous process [139]. In this way effective catalysts have been made for the epoxidation of cyclopentane, cyclohexene, styrene, etc. by tert-butyl hydroperoxide

(TBHP) or 1-phenylethyl. The reaction is carried out both with swelling (the effect of hydroperoxide or reaction products) and non-swelling catalysts. Kinetic data show that epoxidation occurs by the interaction of olefins with a HP adsorbed onto a catalyst. The rigid structure of the heterogenized catalyst impedes formation of a bond between Mo(VI) and ROOH. This is proven by the low values of the sorption coefficient K_s (Table 12-11) in comparison with the formation constants of complexes of molybdenyl diolates with hydroperoxides in homogeneous media (3– 5 L mol^{-1}). At the same time, such impediments are insignificant for sorption of reaction products, and values of K_{in} are close to those for a homogeneous catalyst.

Table 12-11. Kinetic parameters for cyclohexene epoxidation (353 K).

Hydroperoxide	K_f, L mol^{-1}	K_{in}, L mol^{-1}	E_a, kJ mol^{-1}
EBHP	0.25 ± 0.10	6.0 ± 0.6	100 ± 2
HPC[a]	0.40 ± 0.10	15.0 ± 2.0	85 ± 4
TBHP	0.10 ± 0.05	18.0 ± 2.0	60 ± 2

[a]Hydroperoxide of cumene.

The selectivity of direct epoxidation (the ratio of the yield of olefin oxide (moles) to equivalents of oxygen consumed) increases with increasing temperature because of the higher value of E_a for this reaction in comparison with non-selective oxidation. The selectivity of the process is determined mainly by the ability of the initial and required products to form complexes. In addition, free-radical decomposition of HP occurs only through low-valence forms of molybdenum, Mo(V), its proportion being small. An interesting comparison of homogeneous and heterogenized catalysts of epoxidation has been reported in [140] for Mo(V) dithiocarbamate complexes immobilized on chloromethylated PS. In the presence of TBHP Mo(V) is oxidized to Mo(VI), which carries out the reaction. An analogous homogeneous system is degraded rapidly in the presence of TBHP due to a side process, ligand oxidation. This leads to a low yield of cyclohexene oxide (20% vs. 80% for the heterogenized system). The olefin oxides formed are reactive compounds and undergo, though to a lesser extent on immobilized catalysts, various chemical transformations with formation of glycols, ethers and other oxygen-containing products that decrease the selectivity of the reaction. In immobilized systems there are diffusional barriers restricting ligand oxidation. In the presence of polymer-immobilized Mn(III) epoxidating agents such as sodium hypochlorite (NaOCl), potassium peroxymonosulfate (KHSO$_5$), NaBO$_3$ and NaIO$_4$ possess a still lower selectivity [141]. Investigation of the epoxidating properties of Mo(VI) immobilized on cellulose phosphate (CH as a substrate, 333 K, TBHP) led to interesting conclusions [142]. A comparison of rate constants and the enthalpies of formation of the intermediate complexes of the catalyst with reagents leads to an important conclusion. The decomposition of a triple complex [TBHP···Mo···CH] on the surface of the support is characterized by a substantially higher activation energy and a higher positive activation entropy than decomposition in solution (E_a and $\Delta S^{\neq}$ are equal to 136 kJ mol^{-1}, 290 J (mol deg)$^{-1}$ and 72 kJ mol^{-1}, 166 J (mol deg)$^{-1}$, respectively. This difference is caused by the heterogeneous character of the system. The triple biligand complex immobilized on a polymeric support is stabilized in an ordered state by an additional interaction with the polymer surface. On the one hand, a higher activation energy is required for decomposition of the ordered complex but, on the other hand, release of the reaction product into solution is accompanied by an abrupt disordering of the system. This phenomenon is common for catalysis by immobilized complexes and was quantitatively analyzed for the first time for this reaction.

Among other metal complexes bound with polymers which effectively catalyze epoxidation reactions we may note VO(AcAc)$_2$ complexes with poly(vinylbenzoylacetone), microporous polystyrene matrixes with chelating residues of iminoacetic acid or bis(2-hydroxyethyl)amine, etc. Oxovanadium(IV) immobilized on PS by chelate groups has a high activity for chalcone (benzylidene-acetophenone) with formation of *trans*-epoxy-1,3-diphenyl-2-propene-1-one [143]. In the presence of TBHP, polymeric complexes are active catalysts for oxidation of *E*-geraniol and linalool, an isomer of geraniol. 100% conversion and high regioselectivity are achieved. Both soluble and immobilized catalysts yield products with the same content: up to 98% of 2,3-epoxygeranyl acetate and 2% of its isomer 6,7-epoxygeranyl acetate [144]. In this case, on the one hand, the heterogenized catalyst can be regenerated many times. On the other hand, macrocomplexes should be used for production of fragrant compounds (derivatives of *E*-geraniol and linalool are fragrant compounds) because in this case complete extraction of the residual catalyst from oxidates is more important than activity.

Ti(O-*iso*-Pr)$_4$ complexes on polystyrene with chiral groups are active in asymmetric epoxidation of geraniol by TBHP to (*S,S*)-2,3-epoxygeraniol [145]. The enantiomeric excess reaches 66%. Recycling of the catalyst leads to a small decrease of asymmetric induction (loss of one-third of activity after five regeneration cycles). There are sparse data on epoxidation of cyclohexene and cyclooctene in the presence of immobilized titanocene, zirconocene and hafnocene [3].

12.4.5 Catalytic Oxidation of Oxygen-Containing Substrates and Prospects for Oxidation Catalysis

Oxidation of the oxygen-containing products of partial oxidation of hydrocarbons is a promising method for obtaining synthetic fats, ionogenic SAS, herbicides, drugs, monomers for polymers, etc. Two problems are most important for production of these compounds. Selectivity is the first. The oxidation rates of alcohols, aldehydes and ketones are, as a rule, higher than those of the initial hydrocarbons. Therefore, it is difficult to stop the oxidation before formation of the thermodynamically stable final products, CO_2 and H_2O, and kinetic control of the transformation degree is very difficult. Extraction of the remaining catalyst from the reaction products is the second problem. Immobilization of metal complexes can solve this problem.

For oxidation of alcohols to the corresponding aldehydes, chromium complexes immobilized on a polymeric support (quaternized ammonium resin, polyvinyl-pyridine) are used most often. It is important that these catalysts are easily regenerated (by washing with NaOH and HCl) without loss of activity in subsequent cycles. In the presence of immobilized Cu(II) complexes, oxidation of sterically hindered phenols by hydroperoxide proceeds by a radical mechanism, in contrast to oxidation by dioxygen. Chelate complexes of Co(II) with tetradentate Schiff bases, derived from salicylaldehyde and diamines (salcomines), are widely used for homogeneous oxidation of sterically hindered phenols by dioxygen. Oxygenated complexes of cobalt, Co(II)¨O_2, are plausible intermediates in the process. The oxidation proceeds via the formation of phenoxyl radicals whereas the formation of a peroxo bridge Co–O–O–Co inhibits the process.

It was supposed that the immobilization of salcomine (N_3O_2 chelate) on a polymer prevents may prevent the formation of the peroxo bridge, because the rigid backbone of the polymer facilitates formation of monomeric oxygenated Co(II) complexes.

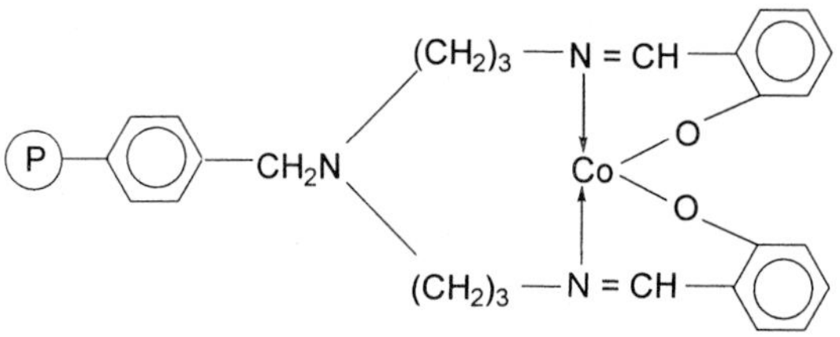

5

2,3,6-Trimethylphenol is effectively oxidized to trimethylquinone on the same mechanism (Eq. (12-31)) [125].

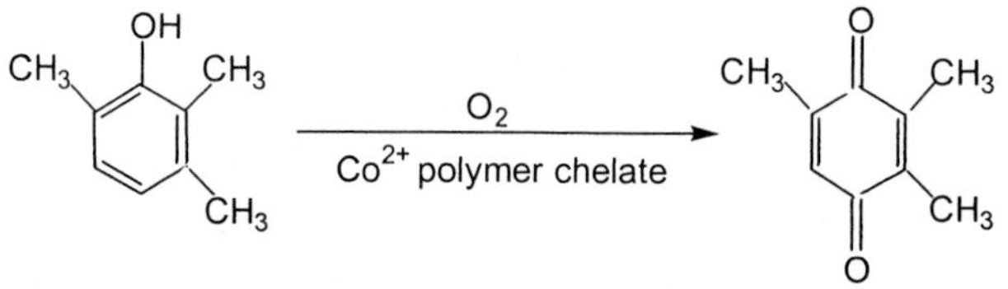

$$(12\text{-}31)$$

Oxidation of sterically hindered phenols by oxygen results in the formation of a number of products. Thus, the oxidative coupling of 2,6-xylenol leads to polyphenylene oxide (C–O coupling) or diphenoquinone (C–C coupling). The first step of their oxidation is the formation of phenoxy radicals. The radicals are then transformed to dimeric and polymeric products linked by C–C and C–O bonds (Eq. (12-32)).

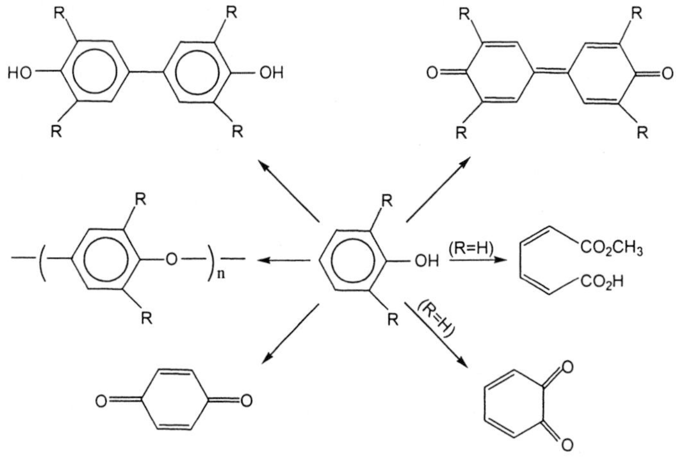

$$(12-32)$$

Recently the kinetics of this reaction has been studied in detail [146] The copolymers containing tertiary nitrogen with anchored Cu(I) or Cu(II) can be used in such reactions [147]. The oxidation of methacrolein dimer and unsaturated 1,3-dioxane derivatives in the presence of polymer-supported Co(II) and Mn(II) complexes [148] also leads to a set of the products. Immobilized Pt complexes are also active in the oxidation of oxygen-containing substrates.

The oxidative dimerization of dialkylphenols proceeds in the presence of immobilized Cu(II) complexes. One-electron reduction of the Cu(II)–phenolate formed results in the production of a phenoxy radical which is oxidized via the intermediate dihydroxybiphenyl to diphenoquinone. At the same time the active centers are regenerated by oxidation of the Cu(I) formed to Cu(II) (the "ping-pong" mechanism). Such cycles – reduction of metal ions at the stage of the reaction with the substrate and its reoxidation by molecular oxygen – are often observed.

Asymmetric induction by polymer-immobilized complexes is an important reaction in oxidation processes (this has already been demonstrated for the hydrogenation transformations described in Section 12.2.9). There are three different methods of synthesis of optically active compounds from optically inactive racemic mixtures: spontaneous, biochemical and chemical. The chemical method is the most common. Immobilized metal complexes are the best models of asymmetric induction by enzymes. They produce large quantities of enantiomeric products from small quantities of chiral compounds. (Ascorbate oxidase is a copper-containing enzyme catalyzing aerobic oxidation of vitamin C. Its polypeptide chain has a relative molecular mass of about 140 000 and contains 8–10 Cu^{2+} ions.) Two types of catalysts can be used for asymmetric induction: (i) catalysts supported on macroligands with optically active groups, (ii) metal complexes that have their own optically active ligands, immobilized on optically inactive polymers. The first type of catalyst is more widespread. For example, complexes of Cu(II) with poly(L-histidine) are active in oxidation of ascorbic or 2,5-dihydroxyphenylacetic acid; those bound with poly(L-lysine) oxidize 3,4-dihydroxyphenylalanine. In addition to polyamine acids, other macromolecules, most often poly(vinylpyridine), with copper, cobalt or iron complexes are used for asymmetric oxidation. Thus, the influence of chain conformation on asymmetric induction can be caused, on the one hand, by formation of a proper chain structure and, on the other hand, by differences in the affinities of substrate enantiomers or the stabilities of intermediates. Oxidation of monosaccharides, especially glucose, aldoses by molecular oxygen as well as transformations of poorly biodegradable molecules such as polychlorinated phenols, catechols and polycondensed aromatics has attracted considerable attention [148–151]. Macromolecular complexes are likely to play a significant role in these processes [152-156].

Catalytic oxidation of CO has been studied intensively, mainly for the purpose of removing CO from exhaust gases. The main information is provided in patents. Polymer chelates of copper and platinum are most often used for this purpose, as well as platinum metals bound with a completely fluorinated polymer "Nafion".

Organic and inorganic sulfur-containing compounds, together with those containing oxygen, are the most extensively studied substrates for oxidation in the presence of immobilized catalysts. Oxidation of sulfur-containing compounds is used not only for production of valuable products, but also for removal of mercaptan from oil cracking products, natural gas and gas condensate, for neutralization of thio salts and waste waters and for resolution of other ecological problems. The most investigated process is the liquid-phase autoxidation of thiols by molecular oxygen. The features of the reaction by metallopolymer catalysts is the same as those considered above. In addition, extended electron-transfer chains are formed in the metallopolymer. These chains facilitate the electron transfer from sulfide ions to oxygen (Eq. (12-33)).

$$ (12\text{-}33) $$

Many aspects of this problem has been already discussed in the book, especially the case of coordinatively bound and polymeric phthalocyanines (Sections 5.2.2 and 6.2.1).

12.5 Hydroformylation in the Presence of Polymer-Immobilized Metal Complexes

Incorporation of oxygen-containing fuctional groups into olefins can be carried out not only by molecular oxygen or peroxides but also with hydroformylation – addition of CO and H_2 to double bonds of hydrocarbons. This reaction utilizes synthesis gas in the presence of catalytic compounds of Co or Rh, and sometimes Ru, Mn and Fe. Industrial oxosynthesis is carried out in the presence of cobalt carbonyls (390–410 K, 20 MPa) for obtaining aldehydes (propion- and butyraldehydes are the most valuable).

12.5.1 Hydroformylation of Olefins

The most important characteristic of selectivity in this reaction is the ratio between the n- and iso-forms of the aldehydes formed (in industrial conditions the ratio is 4:1 to 8:1). The ratio is determined by the rates k_n and k_{iso} of incorporation of monooxide into n- and iso-metal–alkyl intermediates of catalysts (Eq. (12-34)).

$$S = k_n/k_{iso} = \exp[(E_a^n - E_a^{iso})/RT] \qquad (12\text{-}34)$$

In the course of the reaction the double bond is transferred in butene-2, pentene-2, hexene-2, hexene-3 and others.

Although there are numerous examples of hydroformylation of ethylene, propylene, butene by polymer-immobilized complexes (including gas-phase processes and even at atmospheric pressure) metal polymers are widely used for oxosynthesis of olefins C_5–C_7. For instance, hydroformylation of pentene-1 in the presence of Rh complexes on phosphorylated CSDVB is more selective with respect to linear aldehyde ($S = 12$) than hydroformylation with homogeneous systems ($S = 3.3$). However, immobilization of Rh complexes on the ion-exchange resin *Shirotherm* does not result in an increase of selectivity because at high temperatures most of the Rh complexes dissociate from the support and carry on the process in solution, whereas at decreasing temperature Rh complexes bind back to the support [157]. (Thermally reversible catalytic systems are analyzed in detail in Chapter 11.) Rhodium carbonyl supported on polyquinones, modified by phosphine groups and characterized by high thermal and chemical stability, is active at 373–383 K and selective to aldehydes (91%) [158]. From the numerous variants of construction of metal polymeric complexes for hydroformylation [3] we may note $Rh_2Cl_2(CO)_4$ complexes, immobilized on supports of mixed type, which are highly effective and stable catalysts (the number of catalytic cycles occurring at one metal atom reaches 8,000). Polyphosphazene polymer-immobilized complexes containing trinuclear clusters of osmium and iron, as well as the bimetallic clusters $RuOs_3$, $Co_2Rh_2(CO)_{12}$, can be used as high-temperature catalysts (up to 468 K).

Hydroformylation of hexene and high α-olefins C_8–C_{16} has been carried out in the presence of rhodium(I)-hybrid catalysts obtained by sol–gel synthesis [159] as well as with CO/H_2O mixtures in which water is both solvent and the source of protons.

The advantages of localizing the reaction centers of heterophase catalytic systems on the surface or in a thin surface layer of polymer were emphasized in Section 12.3.1. The same conclusion is valid for the hydroformylation of hexene-1 in the presence of $Rh(AcAc)(CO)_2$ supported on the surface of polypropylene-gr-poly(styryldiphenylphosphine) [160]. Under mild conditions (338 K, 1.6 MPa) the ratio between *n*- and *iso*-aldehydes for this system is at least 3.5 times higher than that for the homogeneous analogue.

Polymer-immobilized catalysts can be effectively used in the oxosynthesis of aldehydes (especially from high olefins) followed by their transformations into alcohols and acids including conjugated processes.

12.5.2 Problems of Polyfunctional Catalysis by Immobilized Metal Complexes

Systems consisting of several transition metal compounds (including those bound by a united macroligand) can be used in different variants in catalysis. They may act:

- in parallel homogeneous analogue catalytic cycles,
- independently in successive stages or reactions,
- as combinations (coordinatively) in the same reaction.

Polyfunctional catalysts that initiate successive stages of one multistage reaction are widely used in commercial systems. These include syntheses of CH_3CHO from ethylene (Pd-Cu catalysts), 2-ethylcyclohexene from propylene and CO/H_2O (Co-Rh catalysts) and hydroformylation of allyl alcohol on Rh-containing catalysts followed by hydrogenation of 4-oxybutyric aldehyde formed to 1,4-butandiole (skeletal Ni). The synthesis of alcohols including industrial ones may be carried out by a combination of two catalytic processes (hydroformylation and hydrogenation).

The role of additionally introduced metal (M_1) is many-sided. On the one hand M_1 (usually a non-transition metal) may serve as a "trap" for active centers, binding them more firmly with the support and preventing migration and aggregation. On the other hand these effects can be achieved as a result of the insertion of this metal ion between attached complexes ($-M-M_1-M-$ type) promoting their matrix isolation. In addition, chemical interactions proceeding on the support (formation of weak $M-M_1$ bonds, co-crystallization, formation of intermetallic compounds and alloys, etc.) may promote an increase of catalyst productivity. Both types of active center should react with the substrate molecule, so that a complementary mechanism is realized and diffusional limitations are reduced. This may be shown in the case of the three-stage synthesis of 2-ethylhexanol from propylene and synthesis gas [3]. The process includes the formation of butyric aldehyde (stage 1), its aldol condensation (stage 2) and hydrogenation (stage 3) (Eq. (12-35)).

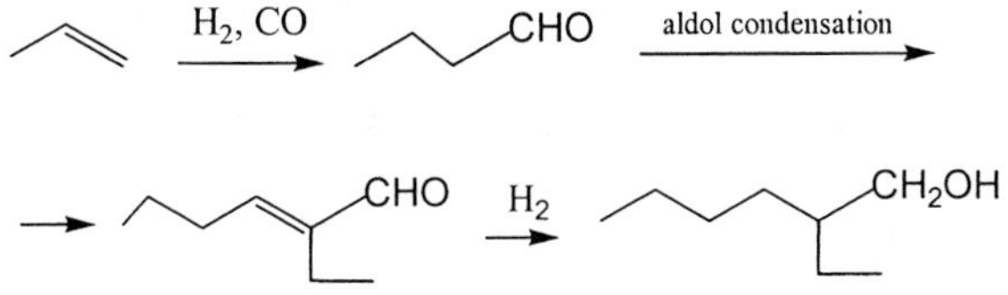

(12-35)

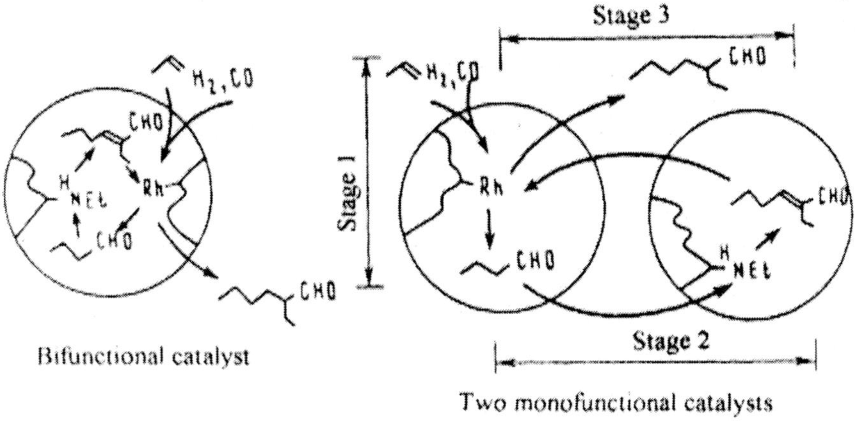

Figure 12-13. Comparison of diffusional inhibition in reactions catalyzed by one bifunctional and two monofunctional catalysts in the case of the three-stage synthesis of 2-ethylhexanol.

This reaction may be achieved by the use of two monofunctional catalysts (hydroformylation and hydrogenation) or one bifunctional catalyst (chloromethylated CSDVB, modified by secondary amino groups with bound $RhCl(PPh_3)_2(CO)$) (Fig. 12-13). The rates of the first, second and third stages of the bifunctional catalyst are 5-, 15- and 30-fold higher respectively than those of the two mono-functional immobilized catalysts. We should also mention the oligomerization of butadiene with the hydroformylation of vinyl cyclohexene by Ni-Rh complexes bound to phosphorylated CSDVB and the preparation of 4-methyl-2-pentanone on polymer-immobilized heterometallic complexes, etc.

The most complicated task in constructing polyfunctional systems is the coordination of the action of the diverse centers and prevention of their autonomous operation. The main approach to resolve this problem is through control of the spatial arrangement of the active centers, which determines the extent of the transfer pathways of reagents between them, regulates diffusional limitations and promotes the existence of the bifunctional catalysis itself.

12.5.3 Asymmetric Hydroformylation Reactions

Asymmetric hydroformylation with participation of immobilized chiral complexes is the most promising reaction for catalytic synthesis of optically active compounds. The reaction proceeds by the scheme shown in Eq. (12-36):

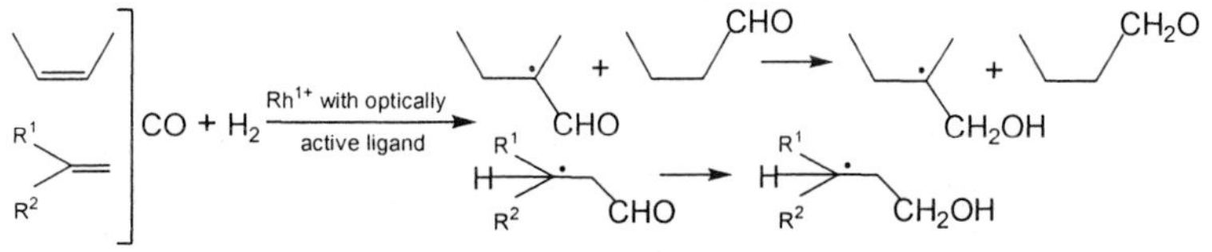

$$ (12\text{-}36) $$

The complexes $[Rh(CO)_2Cl]_2$ and $HRh(CO)(PPh_3)_3$ are immobilized by the same macroligands as those used for hydrogenation (see Section 12.2.9). Immobilization of Rh results in retention of both Rh and the chiral ligand. Examples are not numerous [161,162]. Hydroformylation of cis-2-butane under mild conditions results in the formation of 2-methylbutanol-1 with the same yield (28%) as in analogous homogeneous systems. However, hydroformylation of styrene at 298–353 K is characterized by an enhanced ratio of branched to linear aldehydes (S = 6–20 instead of 2 for homogeneous systems) and by a high yield of (–)(R)-aldehyde. $PtCl_2/SnCl_2$ complexes immobilized on phosphinated macrochelators of 2 and 4 types (see Section 12.2.9) are the best catalysts for asymmetric synthesis. These catalysts promote asymmetric hydroformylation of vinyl acetate, N-vinylphthalimide, norbornadiene and styrene. In the case of styrene under certain conditions an enantiomer is formed, yield 98%, its hydrolysis proceeding also without racemization. The same results were obtained in hydroformylation of vinylacetate and N-vinylphthalimide. Hydroformylation of norbornene proceeds with a very high rate but a low (10–26%) yield of asymmetric 2-formylnorbornane (Eq. (12-37)).

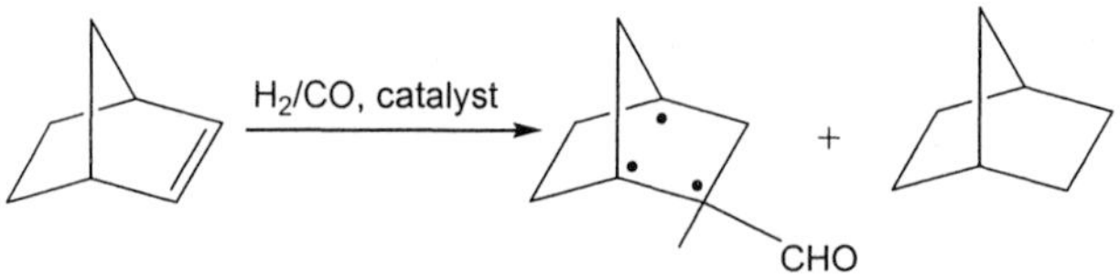

$$ (12\text{-}37) $$

These reactions are characterized by no worse parameters than hydroformylation in homogeneous systems. Recycling of the catalysts leads to only a slight deterioration of the parameters.

12.5.4 Other Prospective Oxo-Processes

Although we do not share the optimistic opinions that predict the achievement of hydrogenation of CO (the Fischer–Tropsch synthesis) (Eq. (12-38)) as the first large-scale project using immobilized metal complexes [2], nevertheless this is a well-developed reaction. The chemistry of the hydrogenation of CO is rather well studied [163, 164] including reaction in the presence of transition metal complexes of the first row on cross-linked polymers. Most likely, metal nanocrystallites or heterogeneous clusters are responsible for their activity whereas homogeneous systems are inactive in this process.

$$nCO + (2n +1)H_2 \rightarrow C_nH_{2n} + nH_2O \tag{12-38}$$

The most interesting reaction of this kind is methanol carbonylation leading to the formation of acetic acid and methyl acetate, products of etherification of acetic acid (in the presence of cobalt or rhodium catalysts promoted by CH_3I). This process is one of the few homogeneous processes realized on an industrial scale. Immobilization of the metal complexes leads to an increase of their activity but the selectivity is not markedly changed; sometimes it is possible to regenerate the catalyst. Consecutive reduction Rh(III) $\rightarrow$ Rh(II) $\rightarrow$ Rh(I) of bound rhodium with poly(ethyleneimine) becomes slower. Water-gas shift reactions are carried out to increase the yield of hydrogen in synthesis gas ($CO + H_2O \rightleftarrows CO_2 + H_2$). Immobilized metal clusters such as $Os_3(CO)_{12}$, $Rh_6(CO)_{16}$ bound with functionalized PS, poly(4-vinylpyridine), etc. are used as heterogenized catalysts. Substantial progress should be expected in the application of macromolecular complexes in catalytic hydrolysis, dehydration and interesterification. These reaction can be considered as activation of a C=O group and the stable molecule H_2O. Thus, hydrolysis of 2,4-dinitrophenylacetate to 2,4-dinitrophenol (using complexes of Cu(II) with poly(vinylalcohol)) may serve as a model of enzymatic catalysis. Catalysis of enantioselective hydrolyses of mixtures of esters of D- and L-amino acids by immobilized complexes is especially interesting as a method of isolating the more valuable L-amino acid from the mixture. Perhaps we can expect the wide use of heterogenized systems in asymmetric reactions (including more complicated stereochemical molecules such as ethers of geranium acid, intermediate products for synthesis of citronellaldehyde, vitamin A and others), at present applied in the field of homogeneous catalysts. Nevertheless, the 1,3-dipolar cycloaddition reactions of $(C_2H_5)_2Zn$ to PhCHO on the enantioselective Lewis acid catalyst $Ti(OPr)_4$, dendritically incorporated in polystyrene beads, have been intensively studied [165,166] This is also true for the activation of alkanes in the presence of polymer-immobilized complexes, an area that has been intensively developed in recent years.

12.6 Conclusion

A comparative analysis of homogeneous and polymer-immobilized systems is mainly based on the discovery of particular features that subsequently indicate the general pattern of evolution of catalysis by macromolecular complexes. At present the actual materials continue to accumulate. However, in contrast to the initial stage, which was more often phenomenological, this field of catalysis now has a quantitative character. Because we only give a brief summary of each part of the chapter here, we mention only some noticeable tendencies. Thus, hybrid organic-inorganic catalysts have been developed in the last few years. Various approaches for the immobilization of molecular homogeneous catalysts have been summarized in a recent review [167]. Laboratory modeling of reactions with polymer-immobilized catalysts to scale-up the catalytic processes is of interest. A promising approach is to transform cyclic processes into continuous ones (for example, using continuously stirred tank reactors and plug flow reactors for better control of exothermic processing, creating a larger margin of safety for scale-up) [168]. The methodology of library synthesis in combinatorial chemistry can be used in catalysis by immobilized metal complexes [169]. The use of supports and macromolecular complexes as nanoparticle species has been outlined (see for instance [170]). Thus, a metallocene/MAO-based solid olefin polymerization catalyst has been developed using chemically functionalized nanoparticles (*p*-hydroxybenzoate-alumoxanes) as a well-defined substrate. Such a catalyst has a high initial activity (~3000 kg of PE (mol $Zr)^{-1} h^{-1}$) [171]. Until now polymerization by organometallic compounds in water has not been realized [172,173]. Probably, hydrophobic supports would be useful for these purposes. Recyclable polymer-supported olefin metathesis catalysts are also of interest. Thus, ruthenium complexes bound with polymer were found to be efficient catalysts for ring-closing metathesis [174].

12.7 Experimental

Experiment 12-1: Typical Procedure for the Catalytic Transfer Hydrogenation of Ketones (Section 12.2) [N.E. Leadbeater in ([22], p.182)

Preparation of a polymer-bound Ru complex and transfer hydrogenation of cyclohexanone: The ruthenium complex $RuCl_2(PPh_3)_3$ can be immobilised by agitating a dichloromethane solution of $RuCl_2(PPh_3)_3$ with resin-bound triphenyl-phosphine (polymer support – polystyrene crosslinked with 2% divinylbenzene, 3 mmol P per g of resin) overnight using a mechanical shaker. Filtration, washing and drying of the polymer gave a black powder of the macromolecular complex. To the complex obtained was added degassed propane-2-ol (10 mL). The mixture was heated to reflux, and then the ketone (cyclohexanone, 10 mmol) was added. The resulting mixture was stirred for 15 min and then a solution of NaOH in propane-2-ol (10 mg in 2 mL) was added dropwise. After 1 h at reflux the mixture was cooled, the supported ruthenium complex filtered off and the product analyzed. The yield of cyclohexanol was 89%.

Experiment 12-2: Hydrogenations of Alkenes (Section 12.2.7) [39]

Preparation of polymer-supported metal complex. An ethanol solution (20 mL) of palladium dichloride (0.42 mmol) was added to the polymer ligand (chloromethylated polystyrene-bound porphyrin). The mixture was refluxed and stirred for 15 h. After cooling, the complex was filtered off, washed thoroughly with water and ethanol, and then dried *in vacuo*.

Hydrogenation of cyclohexene: The substrate cyclohexene (1 mmol) was added to ethanol (10 mL) and the system was heated to an appropriate temperature. The polymer catalyst (5 mg) was added and the system was purged with hydrogen. Since stirring of the mixture was started, the hydrogen consumption was monitored using a gas burette. After the reaction the catalyst was filtered. The products were identified by gas–liquid chromatography. The relative rate for hydrogenation was 23.5 mL of H_2 min^{-1} (g cat)$^{-1}$.

Experiment 12-3: Typical procedure for the Polymerization of α-Olefins (Section 12.3) [175]

Polymerization of alkenes: The polymerization of ethylene, propylene and butene or their copolymerization was carried out in a special facility. It includes a thermostatically controlled stainless-steel reactor and special equipment for stirring and inlet of solvent and components of the catalyst (including organoaluminum compounds). The reactor has equipment for sampling during the reaction for chromatography analysis (gas phase) and for studying the properties of the product (suspension). The copolymerization of ethylene with propene or butene was controlled by chromatographic analysis and on the kinetics of the monomer mixture consumption (followed with a manometer). The polymer obtained was filtered, washed with ethanol and dried *in vacuo*. Molecular-weight and physico-mechanical characteristics, stereoregularity and composition of the copolymers were studied by standard methods.

Experiment 12-4: Typical Procedure for the Oxidation of Hydrocarbons (Section 12.4.3) (N.E. Leadbeater in [22], p.182)

Oxidation of ethylbenzene: To a stirred mixture of hydrocarbons (ethylbenzene) and the supported complex $RuCl_2(PPh_3)_3$ (100 mg) in 1,2-dichloroethane/ethyl acetate (7:1) was added a 30% solution of peracetic acid in ethylacetate (6 mL in 4 mL) dropwise at reflux over a period of 2 h. After 2 h at reflux the mixture was cooled, the supported ruthenium complex filtered off and the product analyzed.

12.8 References

1. R.B. Merrifield, *J. Am. Chem. Soc.* **1963**, *85*, 2149.
2. F.R. Hartly, *Supported Metal Complexes*, Reidel, Dordrecht, 1985.
3. A.D. Pomogailo, *Catalysis by Polymer-Immobilized Metal Complexes*, Gordon & Breach Sci. Publ., Amsterdam, 1998.
4. S.K. Sahmi, J. Reedjick, *Coord. Chem. Rev.* **1984**, *59*, 1.
5. B.J. Abbott, *Adv. Appl. Microbiol.*, **1976**, *20*, 203.
6. R.V. Bahulekar, A.A. Prabhune, H. SivaRaman, S. Ponrathnam, *Polymer* **1993**, *34*, 163.
7. V.A. Davankov, Y.A. Zolotarev, *J. Chromatogr.* **1978**, *155*, 285.
8. D.C. Sherrington, P. Hodge (Eds.), *Syntheses and Separations using Functional Polymers*, Wiley, Chichester, 1988.
9. *Membranes and Membrane Separation Processes*, Proc. Intern. Symp. Torun (Poland), 1989.
10. A. Chopin, F. Quignard, *Coord. Chem. Rev.* **1998**, *178–180*, 1679.
11. R.S. Dickson, *Homogeneous Catalysis with Compounds of Rhodium and Iridium*, Reidel, Dordrecht, 1985.
12. L.S. Hegedus (Ed.), *Catalyst Design Progress and Perspectives*, Wiley, New York, 1987.
13. A. Mortreux, F. Petit (Eds.), *Industrial Application of Homogeneous Catalysis*, Reidel, Dordrecht, 1988.
14. Yu.I. Ermakov, V.A. Likholobov (Eds.), *Homogeneous and Heterogeneous Catalysis*, VNU Science Press, Utrecht, 1987.
15. L.A. Thompson, J.A. Ellman, *Chem. Rev.* **1996**, *96*, 555.
16. D.C. Sherrington, *Chem. Commun.* **1998**, 2275.
17. S.J. Shuttleworth, S.M Allin, P.K. Sharma, *Synthesis* **1997**, 1217.
18. L. Michaelis, M.L. Menten, *Biochem. Z.* **1913**, *49*, 333.
19. W.T. Ford (Ed.), *Polymeric Reagents and Catalysts, ACS Symp.Ser.*, No 308. Washington, DC, 1986.
20. D. Wöhrle, A. Pomogailo, in: *Advanced Functional Molecules and Polymers*, H.S. Nalva (Ed.), Gordon & Breach Sci. Publ., Amsterdam, 2001, Vol. 1, p. 87–161.

21. F. Ciardelli, in: *Macromolecule-Metal Complexes*, F.Ciardelli, E. Tsuchida, D. Wöhrle, (Eds.), Springer, Berlin, 1996, p. 212–233.
22. D.C. Sherrington, A.P. Kybett (Eds.), *Supported Catalysts and their Applications*, Royal Society of Chemistry, Cambridge, 2001.
23. E. Bayer, W. Schumann, *J. Chem. Soc., Chem. Commun.* **1986**, 949.
24. D.J. Bayston, J.L. Fraser, M.R. Ashton, A.D. Baxter, M.E.C. Polyvka, E. Moses, *J. Org. Chem.* **1998**, *63*, 3137.
25. A.A. Belyi, L.G. Chigladze, A.L. Rusanov, M.E. Vol'pin, *Izv. AN SSSR, Ser. Khim.* **1989**, 1961.
26. A.D. Pomogailo, I.E. Uflyand, *Macromolecular Metallochelates*, Khimiya, Moscow, 1991.
27. S.-L- Zhang, Y. Xu, S. Ziao, *J. Catal.* **1986**, 7364.
28. A.V. Korolev, A.R. Brodskii, N.F. Noskova, D.V. Sokol'skii, *Dokl. AN Nauk* **1987**, *296*, 379.
29. E.S. Chandrasekaran, R.H. Grubs, C.H. Brubaker, *J. Organomet. Chem.* **1976**, *120*, 49.
30. C. Bianchini, D.G. Burnaby, J. Evans, P. Frediani, A. Meli, W. Oberhauser, R. Psaro, L. Sordelli, F. Vizza, *J. Am. Chem. Soc.* **1999**, *121*, 5961.
31. C. Merckle, S. Haubrich, J. Blumel, *J. Organomet. Chem.* **2001**, *627*, 44.
32. A.D. Pomogailo, in: *Kompleksnye metalorganicheskii katalizatory polimerizatsii olefinov. Sintez i svoistva polimerizatsionno-napolnennykh polyolefinov*, Vol. 10, Institute of Chemical Physics AN USSR, Chernogolovka, 1986, p. 63.
33. G. Bar-Sela, A. Warchawsky, *React. Polym.* **1983**, *1*, 149.
34. X.-Y. Guo, H.-J. Zang, Y.-J.LiY.-Y. Jiang, *Macromol. Chem., Rapid Commun.* **1985**, *5*, 507.
35. A.D. Pomogailo, E.F. Vainstein, in *Kompleksnye metalorganicheskii katalizatory polimerizatsii olefinov. Metallopolimernye katalizatory,* Vol. 11, Institute of Chemical Physics AN USSR, Chernogolovka, 1991, p. 9.
36. N.E. Leadbeater, K.A. Scott, L.J. Scott, *J. Org. Chem.* **2000**, *65*, 3231, 4770.
37. S. Sgorlon, F. Pinna, G. Strukul, *J. Mol. Catal.* **1987**, *40*, 211.
38. P.C. Selvaraj, V. Mahadevan, *J. Polym. Sci. Part A: Polym. Chem.* **1997**, *35*, 105.
39. R. Wang, Y. He, Z. Lei, Y. Wang, S. Li. Chin, *J. Polym. Sci.* **1998**, *16*, 91.
40. N. Datta-Gupta, T.J. Bordes, *J. Heterocyclic Chem.* **1996**, *3*, 395.
41. T. Uematsu, Y. Nakazawa, F. Akutsu, S. Shimazu, M. Miura, *Makromol. Chem.* **1987**, *188*, 1085.
42. T. Inui, Y. Murakami, T. Susuki, Y. Takegami, *J. Mol. Catal.* **1983**, *22*, 93.
43. A.D. Pomogailo, *Russ. Chem. Bull.* **1997**, *66*, 679.
44. A.D. Pomogailo, A.S. Rozenberg, I.E. Uflyand, *Metal Nanoparticles in Polymers*, Khimiya, Moscow, 2000.
45. H. Hirai, N. Toshima, in *Tailored Metal Catalysts*, Y. Iwasawa, (Ed.), Reidel, Dordrecht, 1986.
46. J. Lieto, J. Rafalko, B.C. Gates, *J. Catal.* **1980**, *62*, 149.
47. S.I. Pomogailo, G.I. Dzhardimalieva, V.A. Ershova, S.M. Aldoshin, A.D. Pomogailo, *Macromol. Symp.* **2002**, *186*, 155.
48. R. Mutin, W. Abboud, J.M. Bassel, D. Sinou, *J. Mol. Catal.* **1985**, *33*, 47.
49. R.A. Sancez-Delgado, A. Andriollo, J. Pupa, G. Martin, *Inorg. Chem.* **1987**, *26*,

1867.

50. J. Lieto, M. Wolf, B.A. Matrana, M. Prochazka, B. Tesche, H. Knozinger, B.C. Gates, *J. Phys. Chem.* **1985**, *89*, 991.
51. Y.-D. Jo, J.-H.Ahn, S. Ihm, *Polym. Intern.* **1997**, *44*, N1.
52. W.M.H. Sachtler, *Chem. Tech.* **1983**, 13, 434.
53. H. Li, B.L. He, *React. Func. Polym.* **1995**, *25*, 61.
54. H. Li, B. He, *Chin. J. Polym. Sci.* **1998**, *16*, 362.
55. V.M. Gryaznov, *Metals and Alloys as Membrane Catalysts*, Nauka, Moscow, 1981.
56. N.N. Gudeleva, B.Yu. Nogerbekov, R.G. Mustafina, *Kinetika i kataliz* **1988**, *29*, 1488.
57. I. Ojima (Ed.), *Catalytic Asymmetric Synthesis*, VCH, Weinheim, 1993.
58. P. Salvadori, D. Pini, A. Petri, *Synlett.* **1999**, 1181.
59. D.J. Gravert, K.D. Janda, *Chem. Rev.* **1997**, *97*, 489.
60. K. Tollner, R. Popovich-Biro, M. Lahav, D. Milstein, *Science* **1997**, *278*, 2100.
61. G.W. Parshall, S.D. Itel, *Homogeneous Catalysis*, Wiley, New York, 1992.
62. I.F.J. Vankelecom, D. Tas, R.F. Parton, V. van de Vyver, P.A. Jacobs, *Angew. Chem., Int. Ed. Engl.* **1996**, *35*, 1346.
63. Q.Fan, C. Ren, C. Yeung, W. Hu, A.S.C. Chan, *J. Am. Chem. Soc.* **1999**, *121*, 7407.
64. S. Kobayashi, M. Endo, S. Nagayama, *J. Am. Chem. Soc.* **1999**, *121*, 11229; *J. Org. Chem.* **1998**, *63*, 6094.
65. F. Gelman, D. Avnir, H. Schumann, J. Blum, *J. Mol. Cat. A:. Chem.* **1999**, *146*, 123.
66. Z.M. Michalska, B. Ostaszewski, J. Zientarska, J.W. Sobczak, *J. Mol. Catal.* **1998**, *129*, 207.
67. Z. M. Michalska, K. Strelec, *React. Funct. Polym.* **2000**, *44*, 189.
68. L. Sun, C.C. Hsu, D.W. Bacon, *J. Polym. Sci. Part A: Polym. Chem.* **1994**, *32*, 2127.
69. A.D. Pomogailo, *Immobilization of Metal Complexes on Macromolecular Supports. Catalytic Properties of Immobilized Systems in Polymerization Processes*, Doctoral Dissertation, Institute of Chemical Physics, Moscow, 1981.
70. D.A. Kritskaya, A.D. Pomogailo, A.N. Ponomarev, F.S. Dyachkovskii, *J. Appl. Polym. Sci.* **1980**, *25*, 349; *J. Polymer Sci., Polym. Symp.* **1980**, *68*, 23.
71. N.M. Bravaya, A.D. Pomogailo, *J. Inorg. Organomet. Polymers.* **2000**, *10*, 1.
72. A.D. Pomogailo, F.S. Dyachkovskii, *Polymer Sci.* **1994**, *36*, 535.
73. A.D. Pomogailo, V.I. Irzhak, V.I. Burikov, F.S. Dyachkovskii, N.S. Enikolopyan, *Dokl. AN USSR* **1982**, *266*, 1160.
74. A.M. Bochkin, A.D. Pomogailo, F.S. Dyachkovskii, *React. Polym.* **1988**, *9*, 99.
75. H. Fuhrmann, F.W. Wilcke, I. Bredereck, *Plaste und Kautschuk* **1990**, *37*, 145.
76. S.Xiao, H. Wang, S. Gai, *Macromol. Chem.* **1991**, *192*, 1059.
77. G.Q.Yu, Y.L. Li, in: *Polymeric Materials Encyclopedia*, J.C. Salamone, (Ed.), CRC Press, Boca Raton, FL, 1996.
78. E. Negishi, T. Takagashi, *Acc. Chem. Res.* **1994**, *27*, 124.
79. P. Sobota, S. Szafert, *J. Chem. Soc., Dalton Trans.* **2001**, 1379.
80. V.I. Smetanyuk, A.V. Ivanyuk, F.I. Prudnikov, *Neftekhimiya* **2000**, *40*, 22.

81. H.G. Alt, S.J. Palackal, *J. Organomet. Chem.* **1994**, *472*, 113.
82. H. Sin, W. Kaminsky, *Adv. Organomet. Chem.* **1980**, *18*, 79.
83. W. Kaminsky, F. Renner, *Makromol. Chem., Rapid Commun.* **1993**, *14*, 239.
84. W. Kaminsky, *Macromol. Chem. Phys.* **1996**, *197*, 3907.
85. M. Jezequel, V. Dufaud, M.J. Ruiz-Garcia, F. Carrillo-Hermosilla, *J. Am. Chem. Soc.* **2001**, *123*, 3520.
86. D. Fregouese, S. Mortana, S. Bresadola, *J. Mol. Catal.* **2001**, *172, 89*.
87. M.L. Ferreira, P.P. Greco, J.H. dos Santos, D.E. Damiani, *J. Mol. Catal.* **2001**, *172*, 97.
88. K. Soga, M. Kaminaka, *Macromol. Chem., Rapid Comm.* **1991**, *12*, 367; **1992**, *13*, 221.
89. N. Suzuki, J. Yu, N. Sinoda, H. Asami, T. Nakamura, T. Huhn, A. Fukuoka, M. Ichikawa, M. Saburi, Y. Wakatsuki, *Appl.Cat. A, General* **2002**, *224*, 63.
90. P. Roos, G.B. Meier, J.J.C. Samson, G. Weickert, K.R. Westerterp, *Macromol. Chem. Rapid Commun.* **1997**, *18*, 319.
91. Z.-G. Xu, S. Chakravarti, *J. Appl. Polym. Sci.* **2001**, *80*, 81.
92. J.C.W. Chien, Z.T. Yu, M.M. Margues, J.C. Flores, M.D. Rausch, *J. Polym. Sci. A: Polym. Chem.* **1998**, *36*, 319.
93. T.F. McKenna, J.B.P. Soares, *Chem. Eng. Sci.* **2001**, *56*, 3981.
94. L. Sun, C.C. Hsu, D.W. Bacon, *J. Polym. Sci. A: Polym. Chem.* **1994**, *32*, 2127.
95. S.B. Roscoe, J.M. Frechet, J.F. Walzer, A.J. Dias, *Science* **1998**, *280*, 270.
96. H.G. Alt, *J.Chem. Soc., Dalton Trans.* **1999**, 1703.
97. R.H. Grubbs, C. Gibbons, L.C. Kroll, W.D. Bonds, Jr, C.H. Brubaker, Jr., *J. Am. Chem. Soc.* **1973**, *95*, 2373.
98. F. Meng, G. Yu, B. Huang, *J. Polym. Sci. A: Polym. Chem.* **1999**, *37*, 37.
99. S.Liu, F. Meng, G. Yu, B.Huang, *J. Appl. Polym. Sci.* **1999**, *71*, 2253.
100. G.Xu, H.Chen, X. Zhang, Z. Jiang, B. Huang, *J. Polym. Sci. A: Polym. Chem.* **1996**, *34*, 2237.
101. J. Xu, J. Ouyang, Z. Fan, D. Chen, L. Feng, *J. Polym. Sci. A: Polym. Chem.* **2000**, *38*, 127.
102. H.G. Alt, M. Jung, *J. Organomet. Chem.* **1999**, *580*, 1.
103. J.D. Kim, B.P. Soares, G.L. Rempel, *J. Polym. Sci. A: Polym. Chem.* **1999**, *37*, 331.
104. J.S. Chung, J.C.Hsu, *Polymer* **2002**, *43*, 1307.
105. L.K. Johnson, C.M. Killian, M. Brookhart, *J. Am. Chem. Soc.* **1995**, *117, 6414*.
106. C.M. Killian, D.J. Tempel, L.K. Johnson, M. Brookhart, *J. Am. Chem. Soc.* **1996**, *118, 11664*.
107. B.L. Small, M. Brookhart, A.M.A. Bennett, *J. Am. Chem. Soc.* **1998**, *120, 4049*.
108. L. Deng, P. Margl, T.Ziegler, *J. Am. Chem. Soc.* **1999**, *121*, 6479.
109. D. Zhang, G.X. Jin, N.H. Hu, *Chem. Commun.* **2002**, 574.
110. S. Skaria, C.R. Rajan, S. Ponrathnam, *Polymer* **1997**, *38*, 1699.
111. Yu. A. Sangalov, *Bashkirskii khim. Zh.* **1995**, *2*,6.
112. K. Nozaki, F. Shibahara, T. Elzner, T. Hiyama, *Can. J. Chem.* **2001**, *79*, 593.
113. K. Nozaki, N. Kosaka, S. Muguruma, T. Hiyama, *Macromolecules* **2000**, *33*, 5340.
114. Y.L. Li, G.Q. Yu, *J. Macromol.Sci.-Chem.* **1990**, *27*, 1335.

115. N. Holzhey, S. Pitter, E. Dinjus, *J. Organomet. Chem.* **1997**, *541*, 243.
116. N.S. Enikolopyan, L.N. Raspopov, A.D. Pomogailo, *Dokl. AN USSR* **1984**, *278*, 1393; *Vysokomol. Soedin.* A. **1989**, *31*, 2624.
117. N.F. Surkov, S.P. Davtyan, A.D. Pomogailo, *Kinetika i Kataliz* **1986**, *27*, 714.
118. Z.J.A. Komon, G.C. Bazan, *Macromol. Rapid Commun.* **2001**, *22*, 467.
119. A.D. Pomogailo, *Rus. Chem. Rev.* **2002**, *71*, 1.
120. E.T. Denisov, V.V. Azatyan, *Inhibition of Chain Reactions*, Gordon & Breach, London, 2000.
121. R.A. Sheldon, J.K. Kochi, *Metal Catalysed Oxidation of Organic Compounds*, Academic Press, New York 1988.
122. T.G. Almazov, L.Ya. Margolis, *High Selective Catalysts for Oxidation*, Khimiya, Moscow, 1988.
123. G.B. Maravin, M.V. Avdeev, E.I. Bagrii, *Neftekhimiya* **2000**, *40*, 3.
124. F. Haber, J. Weiss, *Proc. Roy. Soc.*, *A*, *London*, **1934**, *147*, 332.
125. L.A. Petrov, *Oxidation Catalysis in the Synthesis of Carbonyl Compounds*, Doctoral Dissertation, Institute of Organic Synthesis Ural Branch Russian Academy of Sciences, Ekaterinburg, 2002.
126. Sh. A. Kayumova, U.N. Azizov, S.I. Iskandarov, *Kinetika i Kataliz* **1986**, *27*,1141.
127. L. Canali, D.C. Sherrington, *Chem. Soc. Rev.* **1999**, *28*, 85.
128. S.-H. Chen, *Chin. J. Chem.* **1999**, *17*, 309.
129. V.Z. Radkevich, A.A. Shunkevich, I.E. Kistanova, V.S. Soldatov, Yu. G. Egizarova, *Zh. Prikl. Khimii.* **2000**, *73*, 1861.
130. J. Maslinska-Solich, U. Szaton, *React. Polym.* **1993**, *19*, 191.
131. Finnish. Patent, 71927.
132. A.V. Nikitin, A.D. Pomogailo, V.L. Rubailo, *Izv. Acad. Nauk USSR. Ser. Khim.* **1987**, 36.
133. A.V. Nikitin, A.D. Pomogailo, S.A. Maslov, V.L. Rubailo, *Neftekhimiya* **1987**, *27*, 234.
134. E.F. Vainstein, G.E. Zaikov, in: *Polymer Yearbook*, R.A. Pethric, (Ed.), Harwood Academic Publishers, London, Vol. 10, 1993, p. 231.
135. A.A. Efendiev, *Macromol. Symp.* **2000**, *156*, 155.
136. S.Yu. Men'shikov, A.V. Vurasko, L.A. Petrov, L.S. Molochnicov, E.G. Kovalyova, A.A. Efendiev, *V Intern. Scientific Conference Proceedings "High-tech. in Chemical Engineering 98"*, INTAS, Yaroslavl, 1998, Vol. 2, p. 455.
137. A.I. Kokorin, Izv. *Akad. Nauk. Ser. Khim.* **1997**, 1824.
138. L.Jose, V.N. Rajasekharan Pillai, *Polymer* **1998**, *39*, 229.
139. V.N. Sapunov, *Proceed. Moscow Chemical Technology Institute* **1986**, *N141*, 42.
140. S. Bhaduri, H. Khwaja, *J. Chem. Soc., Dalton Trans.* **1983**, *25*, 415.
141. W. Zhu, W. T. Ford, *J. Polym. Sci. A: Polym. Chem.* **1992**, 30, 1305.
142. A.P. Fillipov, O.A. Polishuk, *Kinetika i Kataliz* **1984**, *25*, 1348.
143. T.H. Kamaluddin, *Oxid. Commun.* **1999**, *22*, 519.
144. E.C. Chapin, E.F. Twohig, L.D. Keys, K.M Corski, *J. Appl. Polym. Sci.* **1982**, *27*, 811.
145. M.J. Farrall, M. Alexis, M. Trecarlen, *Nouv. J. Chim.*, **1983**, *7*, 449.
146. P.C. Selvaraj, V. Mahadevan, *Polymer* **1998**, *39*, 1741.
147. R. Xavier, V. Mahadevan, *J. Polym. Sci. A: Polym. Chem.* **1992**, *30*, 2665.

148. J. Maslinska-Solich, A. Macionga, R. Turczyn, *React. Funct. Polym.* **1995**, *26,* 35.
149. M. Sanchez, N. Chap, J.-B. Cazaux, B. Meunier, *Eur. J. Inorg. Chem.* **2001**, 1775.
150. A. Hadasch, A. Sorokin, A. Rabion, B. Meunier, *New J. Chem.* **1998**, *22*, 45.
151. A. Sorokin, L. Fraisse, A. Rabion, B. Meunier, *J. Mol. Catal. A* **1997**, *117*, 103.
152. F. Minutolo, D. Pini, A. Petri, P. Salvadori, *Tetrahedron Asymm.* **1996**, *7*, 2293.
153. H. Han, K.D. Janda, *Tetrahedron Lett.* **1997**, *38*, 1527.
154. B.B. Lohray, E. Nandanan, V. Bhushan, *Tetrahedron Asymm.* **1996**, *7*, 2805.
155. A. Vidal-Ferran, N. Bampost, A. Moyano, M.A. Pericas, A. Riera, J.K.M. Sanders, *J. Org. Chem.* **1998**, *63*, 6309.
156. D.A. Annis, E.N. Jacobsen, *J. Am. Chem. Soc.* **1999**, *121*, 4147.
157. Y. Kowabata, C.U. Pittman, Jr., R. Kobayashi, *J. Mol. Catal.* **1981**, *12*, 113.
158. M. Ding, J.K. Stille, *Macromolecules* **1983**, *16*, 839.
159. E. Lindner, F. Auer, A. Baumann, P. Wegner, H.A. Mayer, H. Bertagnolli, U. Reinohl, T.S. Ertel, A. Weber, *J. Mol. Catal., A: Chem.* **2000**, *157*, 97.
160. F.R. Hartley, S.G. Murray, A.T. Sayer, *J. Mol. Catal.* **1986**, *38*, 295.
161. J.K. Stille, *J. Macromol. Sci. A* **1984**, *21*, 1689.
162. G. Parrinello, R. Deschenaux, J.K. Stille, *J. Org. Chem.* **1986**, *51*, 4189.
163. G. Henrici-Olive, *Chemistry of CO Catalytic Hydrogenation*, Mir, Moscow, 1987.
164. P.M. Maitlis, *Pure and Appl. Chem.* **1989**, *61*, 1747.
165. H. Sellner, P.B. Rheiner, D. Seebach, *Helv. Chim. Acta* **2002**, *85*, 352.
166. K. Hosoya, S. Tsuji, K. Yoshizako, K. Kimata, T. Akai, N. Tanaka, *React Func. Polym.* **1996**, *29*, 159.
167. M.H. Valkenberg, W.F. Holderich, *Cat. Rev.– Sci. Eng.* **2002**, *44*, 321.
168. N.G. Anderson, *Org. Process Res.Devel.* **2001**, 5, 613.
169. S. Kobayashi, *Current opinion Chem. Biol.* **2000**, 4, 338.
170. S. Amigoni-Gerbier, S. Desert, T. Gulik-Kryswicki, C. Larpent, *Macromolecules* **2002**, *35*, 1644.
171. S.J. Obrey, A.R. Barron, *Macromolecules* **2002**, *35*, 1499.
172. A. Held, S. Mecking, *Chem. Eur. J.* **2000**, *6*, 4623.
173. S. Mecking, A. Held, F.M. Bauers, *Angew. Chem. Int. Ed.* **2002**, *41*, 544.
174. L. Jafarpour, M.-P. Heck, C. Baylon, H.M. Lee, C. Mioscowski, S. Nolan, *Organometallics* **2002**, *21*, 671.
175. A.D. Pomogailo, F.S. Dyachkovskii, *Acta Polym.* **1984**, *35*, 41.

13 Photocatalytic Properties

Masao Kaneko and Dieter Wöhrle

Photocatalysis is a catalysis taking place under photoexcitation. In photo-catalysis, a sensitizer molecule or semiconductor absorbs irradiated light energy, through which an electron transition takes place from the ground state to an excited state of the sensitizer, or from the valence band to the conduction band of the semiconductor. Electron transfer takes place between the excited state and a molecule present nearby, which induces further chemical process(es) leading to a final reaction product. After this photochemical process the sensitizer or semi-conductor is ready to be excited again by light and the same chemical process takes place, i.e. the sensitizer or the semiconductor works as a photocatalyst. In the photosynthesis of green plants such photocatalysis (taking place in chlorophyll molecules, see also Chapter 2) is the most important central process, which provides almost the total energy resource for biological activities. In order to solve the greenhouse-effect problem brought about by fossil fuels, photocatalysis is very attractive for creating energy resources from solar energy and water, which we can call an artificial photosynthesis. It is now established that such a sensitizing process can also lead to the construction of an efficient solar cell. In this chapter only photocatalysis by sensitizer molecules will be described. Please refer to references [1–4] for photocatalysis by semiconductors. Since artificial photosynthetic systems and solar cells are important and an attractive research field in macromolecule-metal complexes, the main part of this chapter is devoted to these two subjects followed by another important photocatalytic process of energy transfer.

13.1 Photocatalysis

Photocatalysis [1–5] is very attractive for creating new energy resources from solar energy to solve global environmental problems such as the greenhouse effect by carbon dioxide, acid rain, and so on. Photocatalysis is catalysis induced by light irradiation. Semiconductors such as TiO_2 are well-known typical materials for achieving photocatalysis. Sensitizers are other candidates to achieve photocatalysis, which should be used in combination with other catalyst(s).

Intensive investigations have been reported aimed at the construction of an artificial photosynthetic system in order to obtain solar energy conversion such as water cleavage by visible light, by utilizing a photocatalyst to produce fuels by sunshine [2–8]. However, a total system for chemical conversion of solar energy (artificial photosynthesis) has not yet been achieved. The first remarkable report of water cleavage by UV light with a TiO_2 photoanode and platinum counter electrode [6] has evoked great attention towards sensitizing large bandgap semiconductors by coating with dye sensitizer. Successful sensitization of a TiO_2 nanoparticle layer by coating with a metal complex sensitizer has led to a sensitized photoregenerative solar cell with 10% conversion efficiency [7], which is now attracting a great deal of attention for practical use as a high cost–high performance device.

In spite of the remarkable development of the photoenergy conversion research field, it is still difficult to establish an artificial photosynthetic system creating energy resources from solar energy by utilizing photocatalysis. In this chapter, we describe some photocatalytic reactions that may lead to artificial photosynthetic systems and solar cells.

13.2 Photocatalysis towards Future Artificial Photo- synthesis

13.2.1 Photosynthesis and the Energy Cycle on the Earth

It is important to understand the energy cycle on the earth in order to design future energy resource systems for our civilization compatible with the global environment. Almost all biological energy sources, as well as the main energy to support our society, are provided directly and indirectly by photosynthesis. Photosynthesis is a kind of photocatalytic reaction of carbon dioxide with water driven by solar visible light energy to produce carbohydrate as a main product (Eq. (13-1), Section 2.7).

$$CO_2 + 2H_2O + 8 \text{ photons} \rightarrow (C_6H_{12}O_6)_{1/6} + O_2 + H_2O \tag{13-1}$$

This reaction is represented by the following three major processes (Eqs. (13-2), (13-3) and (13-4)), where water is an electron donor providing electrons to the whole system (see also Section 2.6). The electrons obtained from water at the Mn protein complex (Eq. (13-2)) are excited by solar photons at the chlorophyll reaction center (two steps) to form high-energy electrons (e^-*) (Eq. (13-3)), and the e^-* reduces CO_2 to carbohydrate (Eq. (13-4)) via reduction of NADP to NADPH.

$$2H_2O \rightarrow 4e^- + 4H^+ + O_2 \tag{13-2}$$

$$4e^- + 8 \text{ photons} \rightarrow 4e^-* \tag{13-3}$$

$$4e^-* + 4H^+ + CO_2 \rightarrow (C_6H_{12}O_6)_{1/6} + H_2O \hspace{2cm} (13\text{-}4)$$

For the oxidation of water ($H_2O \leftrightarrows 0.5\ O_2 + H^+ + 2e^-$) the redox potential is +0.81 V and for the reduction of water ($2H^+ + 2e^- \leftrightarrows H_2$) the redox potential is 0.42 V vs. NHE, respectively. The redox potential of $NADP^+ + H^+ + 2e^- \leftrightarrows$ NADPH with −0.33 V is only ~0.1 V smaller than the reduction potential of water. Therefore through the light reaction in photosynthesis the unidirectional electron transport in the macromolecular protein matrix leads to high-energy storage.

The energetic process of photosynthesis is represented by Fig. 13-1 where the electron from water is driven to higher energy in two steps (at photosystems II and I, abbreviated to PSII and I) at the reaction center of chlorophyll, and finally reduces CO_2 to produce carbohydrate. Metal complexes and metal clusters play a decisive role in photoinduced energy transfer, photoinduced electron transfer and catalysis.

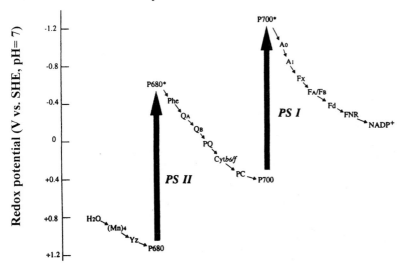

Figure 13-1. Electron flow in photosynthesis: PSI and II are photosystems I and II; P_{680} and P_{700} are chlorophylls; Phe is pheophytine for which Mg is absent in the chlorophyll; Q_A and Q_B are quinones, PQ is plastoquinone, and PC is plastocyanine; Ao and A_1 are electron acceptors; Fx, F_A/F_B, and Fd are iron–sulfur centers. NADP is nicotinamide adenine dinucleotide phosphate; $(Mn)_4$ is a Mn cluster; Y_Z is a tyrosine residue.

The efficiency of this electron flow is almost 100% with negligible back electron transfer (charge recombination), which is indicated by the dynamics of the electron transfer where the forward electron transfer rate is higher than the backward one by almost two to three orders of magnitude.

The photosynthetic product is regarded as a store of high-energy electrons.

We take the products as foods and the high-energy electrons in the foods are accepted by O_2 (produced by the photosynthesis) in respiration, thereby liberating free energy for biological activities to reproduce water and CO_2. Combustion of fossil fuels (oxidation by O_2) also liberates energy and reproduces water and CO_2. Thus the energy cycle on the earth is represented by circulation of electrons coming from water and driven by solar photon energy as shown in Fig. 13-2 [8], which is supported by photosynthesis.

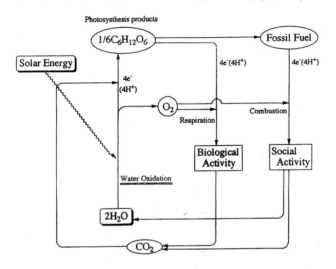

Figure 13-2. Energy cycle on the earth represented by electron circulation driven by solar photons.

The earth and its living organisms are in a non-equilibrium state, but should maintain a steady state (homeostasis). In order to maintain a non-equilibrium steady state, it is important that the system is not closed but open to the outside. Only under open conditions can such non-equilibrium but steady states be maintained. If the earth were closed, all living things and our civilization would come to a catastrophe, but the earth is fortunately open to the universe, especially to solar irradiation. Based on this and on the first and second laws of thermodynamics, it was inevitable that we should meet a problem of environmental and energy resources while we rely on energy cycles conducted in the closed earth. It is important for our future civilization that the earth is open to the universe, i.e. to the sun.

13.2.2 A Model for Artificial Photosynthesis

Taking photosynthesis as a model for photochemical solar energy conversion, an artificial photosynthetic system has been proposed for the creation of fuels from solar energy and water by utilizing photocatalytic reactions (Fig. 13-3) [9]. This scheme considers the minimum requirements for achieving photochemical energy conversion.

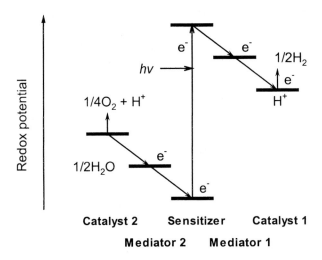

Figure 13-3. Artificial photosynthesis [9] that combines a water oxidation catalyst (C_2), a photoexcitation center (sensitizer) and a proton reduction catalyst (C_1) through mediators (M_2 and M_1).

In comparison to natural photosynthesis only one photoexcitation center (P) was designed for the first stage of the scheme. Dark catalyses of water oxidation (C_2) and proton (or CO_2) reduction (C_1) are designed, which should be coupled electrically with the photoexcitation center via mediators (M_2 and M_1). In order to decompose water by this system, the energy gap between the ground and the photoexcited states of P should be less than 3 eV (wavelength longer than 400 nm), the redox potential of C_2 should at least be more positive than 0.81 V (vs. SHE, at pH 7), and that of C_1 more negative than –0.41 V (vs. SHE, pH 7). In addition to these energy level requirements, it is important to suppress back electron transfer by (1) kinetic design of the electron transfer reactions, and (2) adoption of a heterogeneous phase to achieve uni-directional electron transfer. For this purpose, not only the irradiated reaction but also the dark reactions such as water oxidation and proton reduction are important. As for the photoexcitation center, either dye sensitizers (often metal complex), small bandgap semiconductors or dye-sensitized large bandgap semiconductors are candidates.

To realize this model system, it is essential to establish each unit and then to combine the units by utilizing molecular aggregates or various assemblies. Some progress towards this goal will be described in Section 13.2.3 including dark catalysis and primary photoinduced processes.

13.2.3 Dark Catalysis and Photocatalysis for Artificial Photosynthesis

In this section we describe dark catalysis (water oxidation, proton reduction and CO_2 reduction) and photoexcited state electron transfer carried out in polymer matrixes towards artificial photosynthesis.

13.2.3.1 Dark Catalysis for Water Oxidation

Catalytic water oxidation (dark reaction) is the first important reaction of the electron flow in the photosynthesis represented by Fig. 13-1, in which water is used as a source of electrons provided to the whole system. Although water oxidation is so important for biological activities, its catalyst and reaction mechanism are not yet established [10–13]. In photosynthesis, a Mn-protein complex works as a catalyst for the difficult four-electron oxidation of two molecules of water to form one O_2 molecule (Eq. (13-2)) (see also Section 2.6.1). It is inferred that at least four Mn ions are involved in the active center, but its structure is not yet completely elucidated.

Molecular metal-containing catalysts for water oxidation have been attracting attention a great deal as models for the photosynthetic catalyst. Many structural models have been synthesized. Tetrakis(2,2′-bipyridine)(di-μ-oxo)di-Mn complex (**1**) [11] and tetrakis(2,2′-bipyridine)(μ-oxo)di-Ru complex (**2**, Meyer's complex) [14] are typical examples, but most of them did not show high activity for water oxidation.

$[(bpy)_2Mn(\mu\text{-}O)_2Mn(bpy)_2]^{3+}$, **1**

$[(bpy)_2(H_2O)Ru(\mu\text{-}O)Ru(H_2O)(bpy)_2]^{4+}$, **2**

Water oxidation catalysis has been studied by both chemical (Fig. 13-4, using Ce(IV) as oxidant) and electrochemical (Fig. 13-5, using a polymer-coated electrode) methods, and it has been found that ammine ligand-based Ru complexes, whether trinuclear, dinuclear or mononuclear, show high activity as catalysts for water oxidation.

$4Ce^{IV}$ — Complex^{n+} — $O_2 + 4H^+$

$4Ce^{III}$ — Complex$^{(n+4)+}$ — $2H_2O$

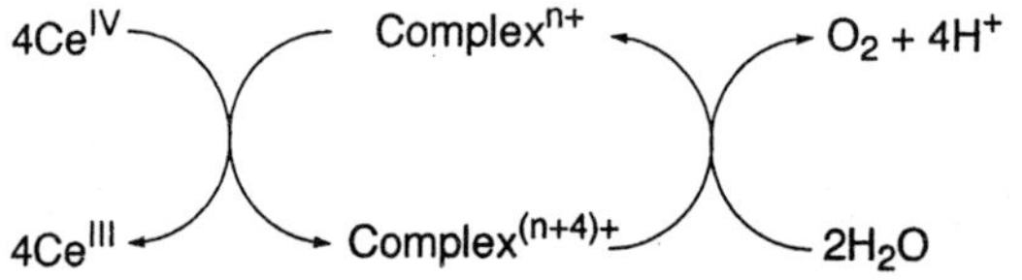

Figure 13-4. Chemical water oxidation catalysis.

Complex^{n+} — $O_2 + 4H^+$

$4e^-$ — Complex$^{(n+4)+}$ — $2H_2O$

| Electrode | Water oxidation catalyst layer | Bulk electrolyte aqueous solution |

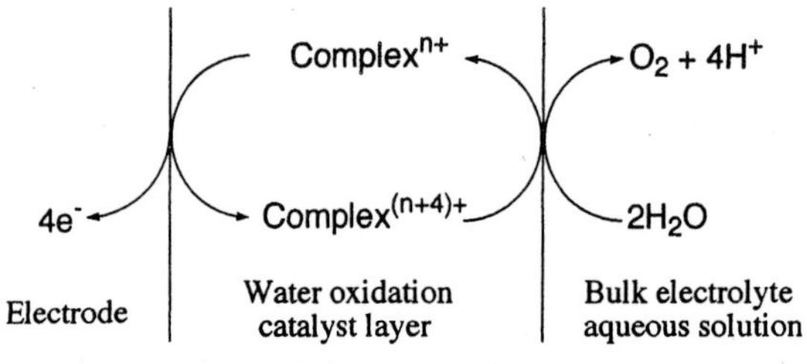

Figure 13-5. Electrocatalytic water oxidation.

Because the catalytic trinuclear Ru-ammine complex called Ru-red (**3**) is very active, it can evolve dioxygen bubbles in the presence of a strong oxidant such as Ce(IV) ion [15–18] (for experiments using **3** in the catalytic and electrocatalytic water oxidations see Section 13.5, Experiments 13-1 and 13-2).

$[(NH_3)_5Ru–O–Ru(NH_3)_4–O–Ru(NH_3)_5]^{6+}$, **3** (Ru-red)

It has been established that other ammine ligand-based Ru complexes including mononuclear (e.g. $[Ru(NH_3)_5Cl]^{2+}$) and dinuclear ones are all active catalysts for water oxidation both in a homogeneous aqueous solution and in a heterogeneous phase such as a polymer membrane and clay [18].

Important results for these catalysts are as follows:

1. Trinuclear (Ru-red) and dinuclear complexes are capable of 4-electron oxidation of water by one molecule. Mononuclear complexes are capable of either 4-electron or 2-electron oxidation of water, two catalyst molecules being required in the latter case.
2. Bimolecular decomposition of the catalyst in its high oxidation state occurs, forming N_2 by oxidation of the ammine ligands in competition with water oxidation. Such bimolecular decomposition is inhibited remarkably by isolating catalyst molecules from each other in a polymer matrix (e.g. Nafion, **4**). The catalytic activity itself remains similar in either a heterogeneous polymer phase or in a homogenous aqueous solution.

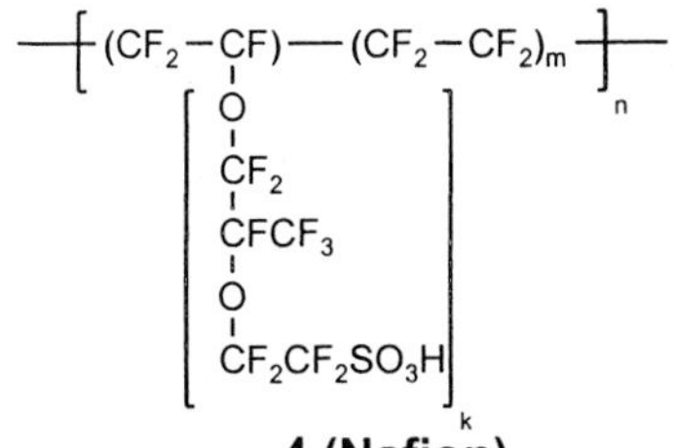

4 (Nafion)

3. By assuming a random distribution of the catalyst in a polymer matrix, the bimolecular decomposition distance was estimated; it was almost the contact distance between molecules [18].
4. When two catalyst molecules are needed for 4-electron water oxidation, the catalytic activity shows an optimum catalyst concentration in a polymer matrix since bimolecular decomposition still takes place, and the cooperative distance between the catalysts can again be determined by assuming a random distribution.
5. For electrocatalytic water oxidation using polymer (Nafion)-coated electrode dispersing catalysts in the polymer membrane, charge transport from the electrode to the catalyst taking place by charge hopping is important. This charge transport by hopping is facilitated by a high concentration of the catalyst in the polymer matrix, which also favours the bimolecular decomposition, so that optimum and delicate concentration

conditions exist for the electrocatalytic system. The charge-hopping distance (r_o) and the bimolecular decomposition distance (r_d) can also be determined by assuming a random distribution.

6. Amino acid residue models such as a tyrosine residue model (*p*-cresol) increase the charge-hopping distance to a remarkable extent, which can solve the electrocatalysis problem mentioned in item 5. above, and significantly enhance the catalytic activity.

Some details of the above items are explained below.

When Ru-red was used as a catalyst in the presence of a large excess of Ce(IV) oxidant (Fig. 13-4), the rate of O_2 evolution was first order with respect to the catalyst concentration, showing that Ru-red is capable of 4-electron oxidation of water. During the catalysis, N_2 was formed as a decomposition product of the Ru-red, and its formation rate was second order with respect to the catalyst concentration, showing that the decomposition is bimolecular. The decomposition distance in a polymer (Nafion) matrix was estimated by assuming a random distribution of the catalyst molecule in the matrix. The probability density $P(r)$ of the distance between the nearest neighbor molecules (r in nm) is represented by Eq. (13-5) according to Poisson statistics where N_A is Avogadro's number (mol^{-1}), c is the concentration of the catalyst molecule in the matrix ($mol\ dm^{-3} = M$), α is the localization factor of the molecule estimated by considering the hydrophilic region fraction in the matrix, and s (nm) is the contact distance of the catalyst molecules.

$$P(r) = 4\pi r^2 N_A \alpha c \times 10^{-24} \exp[-4\pi(r^3 - s^3)N_A \alpha c \times 10^{-24}/3] \tag{13-5}$$

When r_d (nm) is the decomposition distance (center to center) of the catalyst molecules within which they undergo decomposition, the fraction of decomposing catalyst (R_{dec}) is calculated by integrating $P(r)$ from s to r_d. The apparent rate of O_2 formation (k_{app}) should be proportional to $1 - R_{dec}$, so that k_{app} is expressed by Eq. (13-6) when k is an intrinsic rate constant of O_2 formation.

$$k_{app} = k\exp[-4\pi(r_d^3 - s^3)N_A \alpha c \times 10^{-24}/3] \tag{13-6}$$

By applying Eq. (13-6) to the data of k_{app} versus c, r_d was determined almost as a contact distance of the catalyst molecule [18,19]. The results are summarized as follows:

1. Trinuclear and dinuclear Ru complexes are capable of 4-electron oxidation of water but, for mononuclear complexes, only the *cis*-tetraammine Ru complexes are capable of 4-electron oxidation, and other mononuclear ones are 2-electron catalysts by one molecule. For the 2-electron catalysts, the cooperative distance (r_{co}) for bimolecular water oxidation was also estimated by a similar procedure to r_d. the ammine complex shows much higher activity than the bipyridine complex.

2. The bimolecular decomposition distance is nearly the contact distance between the molecules.

3. The activity of the molecular complexes is much higher than that of conventional noble metal oxide catalysts such as PtO_2 or RuO_2, by one to two orders of magnitude based on one molecule and one repeating unit.

In relation to item 6. mentioned above on electrocatalysis, the coexistence of a tyrosine residue model, *p*-cresol (*p*-Cre), greatly enhanced the catalytic activity of Ru-red confined in a Nafion membrane coated on an electrode (Fig. 13-6) [20], which was interpreted as a lengthening of the charge-hopping distance by *p*-cresol by a factor of almost two (from 1.28 nm to 2.25 nm).

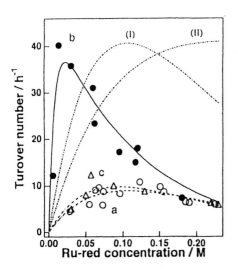

Figure 13-6. Electrocatalytic water oxidation utilizing a polymer-coated electrode as a function of the Ru-red catalyst concentration confined in a Nafion membrane coated on an electrode. Applied potential 1.4 V vs. Ag–AgCl. (a) In the absence of amino acid model compound, (b) in the presence of 5.0×10^{-2} M *p*-cresol (*p*-Cre), and (c) with toluene. The solid line is a simulated curve considering charge mediation, and the dashed lines are the calculated values without mediator. The dash-dotted curve (I) is a simulated curve in the presence of *p*-Cre when r_o and r_d are the same as those in the absence of *p*-Cre, and curve II is a simulated curve in the presence of *p*-Cre when r_o and k_{O2} are the same as those in the absence of *p*-Cre. The mediator lengthens the charge-hopping distance (r_o) between catalyst molecules.

13.2.3.2 Dark Catalysis for Proton Reduction.

Proton reduction is also an important catalysis in water photolysis. Pt and PtO_2 have been the most well-known and active catalysts for proton reduction to produce dihydrogen. However, these colloidal or powdered catalysts are not suited to the construction of a conversion system based on molecules, and, moreover, these strongly colored materials can bring about a filter effect that suppresses photochemical reactions. It is therefore desirable to use a molecular catalyst for constructing an artificial photosynthetic system.

It was reported that cobalt tetraphenylporphyrin complex (CoTPP) coated on an electrode catalyzes electrocatalytic proton reduction [21], but the activity was not very high. It has been found that metal porphyrins (such as metal tetraphenylporphyrin MTPP, **5**) and metal phthalocyanines (**6**, MPc) when incorporated into a polymer membrane coated on an electrode exhibit high activity in electrocatalytic proton reduction to produce H_2 [22,23]. These polymer metal complex catalysts can be much more active than a conventional platinum-based electrode (Fig. 13-7 [23], see Experiment 13-3, Section 13-5).

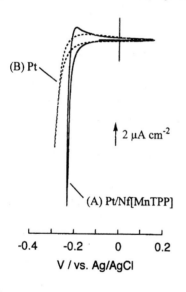

(B) Pt

$2 \mu A \ cm^{-2}$

(A) Pt/Nf[MnTPP]

-0.4   -0.2   0   0.2

V / vs. Ag/AgCl

Figure 13-7. Cyclic voltammogram at a Pt electrode coated with [MnTPP] in Nafion (A) and at a bare Pt electrode (B) in a phosphate buffer aqueous solution (pH 1.0) under Ar. Scan rate, 1 mV s^{-1}.

5

6

The concentration of catalyst in the matrix was especially important for this catalysis. When the catalyst concentration exceeded ~0.05 mol dm^{-3} (M) in a matrix, the activity decreased drastically. Although this result is not yet understood, it has been suggested that since the catalysis is a bimolecular process, a low concentration of the complex catalyst is important in order that two molecules of a H^+-coordinated intermediate can take the best interacting configuration for cooperative H_2 formation.

13.2.3.3 Dark Catalysis for Carbon Dioxide Reduction

Water photolysis is the simplest photochemical solar energy conversion system for which proton reduction catalysis is essential, but carbon dioxide reduction is still an attractive research subject as a synthetic model for CO_2 reduction in photosynthesis. There has been much work on chemical and photochemical CO_2 reduction [24], but it is not the aim of this section to review those projects that mainly involve low molecular weight compounds.

It has been found that when a metal phthalocyanine (MPc) is incorporated into a poly(vinylpyridine) membrane, it is a very active catalyst for electrocatalytic CO_2 reduction to produce carbon monoxide (see Experiment 13-4, Section 13.5) [25]. For CO_2 reduction, selectivity between CO_2 and H^+ reduction is important since in water H^+ reduction is an easier process than CO_2 reduction. It has been shown (Fig. 13-8) that when CoPc is incorporated into a poly(4-vinylpyridine) (PVP) membrane coated on an electrode, the selectivity for CO_2 reduction is much higher than that of a neat CoPc-coated electrode.

By investigating the visible spectral change of CoPc in a PVP membrane coated on a graphite electrode by *in situ* potential step chronospectrometry (PSCS) during CO_2 reduction, it has been shown that, after CO_2 coordination to the two-electron reduced CoPc, one more electron is injected from the electrode to form CO, recovering one-electron reduced CoPc (Path II of Fig. 13-9) [25], different from the previously proposed reaction scheme (Path I).

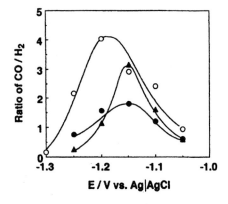

Figure 13-8. Selectivity for CO_2/H^+ reduction forming CO and H_2, respectively, against applied potentials at the graphite electrode coated with CoPc/PVP ($\circ$), CoPc/poly(2-vinylpyridine) ($\blacktriangle$), and neat CoPc ($\bullet$).

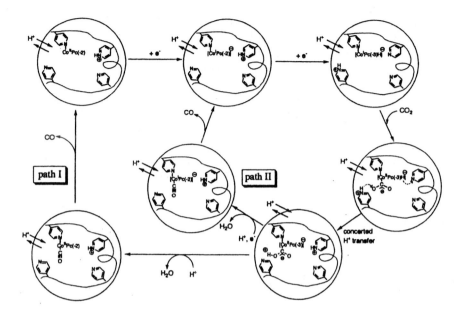

Figure 13-9. Mechanism of CO_2 reduction by CoPc.

The mechanism was different depending on the electron-donating or electron-accepting substituents on the CoPc ring (see reference [26] for the details).

13.2.3.4 Photoexcited-State Electron Transfer in Polymer Matrixes

Electron transfer of sensitizers at photoexcited states present in heterogeneous systems is of essential importance for constructing a photochemical conversion system since energy conversion such as shown in Fig. 13-3 can be realized only in heterogeneous phase(s). In a homogeneous phase, back electron transfer always occurs, so that homogeneous systems cannot be used for photoenergy conversion. Heterogeneous electron transfer is important also in a dye-sensitized solar cell that is now attracting many researchers, as mentioned later, in which TiO_2 is sensitized by a coated dye layer.

Photoinduced electron relay in a polymer solid phase was first found to take place for an ethylenediaminetetraacetic acid (EDTA)-tris(2,2'-bipyridine)-ruthenium(II) $(Ru(bpy)_3^{2+}(7)$-methylviologen $(MV^{2+}, (8))$ system [27] incorporated into a cellulose paper (Fig. 13-10). Electron transfer from EDTA to the oxidized $Ru(bpy)_3^{3+}$ formed after electron transfer from its excited state to MV^{2+} accumulated blue-colored $MV^{\cdot+}$ on the cellulose paper (see Experiment 13-5, Section 13.5).

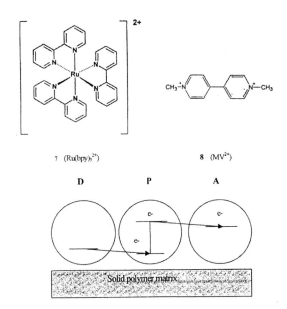

7 (Ru(bpy)₃²⁺)   8 (MV²⁺)

Figure 13-10.
Photoinduced electron
transfer among the
donor (D) photoexcited
sensitizer (P), and
acceptor (A) in a
polymer solid phase.

In a solid phase, since the reaction components cannot move or move only slightly during a photoexcited state, the electron transfer shows a specific aspect entirely different from that in a homogenous solution, for which reaction takes place commonly by a dynamic mechanism. In a solid phase, on the contrary, electron transfer takes place usually by a static mechanism. Depending on the mobility of the components the mechanism can be partly dynamic. The mechanism can be determined by investigating the emission intensity and emission decay based on quenching of the emission by electron transfer. When the mechanism is dynamic, the reaction is expressed by the Stern–Volmer equation (Eq. (13-7)) [28], where I_o is the emission intensity in the absence of quencher (Q, here an electron acceptor or donor), I is that in the presence of Q (concentration [Q] mol dm^{-3}), τ is the corresponding lifetime of the excited state determined by its emission decay, and k_{sv} is the Stern–Volmer constant (Fig. 13-11 (a)).

$$\text{Io}/I \;=\; \tau_0/\tau = \; 1 \;+\; k_{sv}[Q] \tag{13-7}$$

For a static mechanism the τ_0/τ plots do not show a slope (the lifetime is independent of the Q concentration) (see Fig. 13-11 (b)), and very typically the I_o/I plots follow the Perrin mechanism (Eq. (13-8)) [29], where $K_1 = VN_A$ (where V is the quenching sphere volume).

$$I_o/I \;=\; \exp(K_1[Q]) \tag{13-8}$$

In this mechanism the probe (sensitizer) containing Q in the quenching sphere around the probe is quenched entirely, but the probe containing no Q is not quenched at all (see Fig. 13-11 (b)). When the process involves both dynamic and static mechanisms, the I_0/I and τ_0/τ plots are depicted as Fig. 13-11 (c).

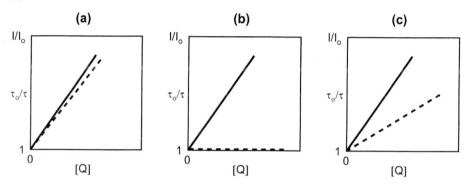

Figure 13-11. Mechanism of photoinduced electron transfer as studied by quenching plots, I_0/I (line) and τ_0/τ (dashed line).

These are only typical cases, and actually the electron-transfer mechanism in a solid state is often more complicated [30]. Possible electron-transfer mechanisms have been investigated in detail based on the following features (more details are described in [30,31]):
- The number of different sites: 1 or 2 sites.
- Dynamic or static mechanisms as well as possible combinations.
- The incorporation equilibrium of Q into the quenching sphere around a probe: is it a one-step equilibrium or a multi-step one?
- Is the rate of static quenching very fast so that emission is negligible, or should the emission be considered since the emission is competitive with the quenching?
- For a multi-step equilibrium between Q and probe, is the static quenching rate is proportional to the number of Q in the quenching sphere, or independent of the number of Q?
    See references [30–33] for details.
    A polysiloxane film containing dispersed $Ru(bpy)_3^{2+}$ and MV^{2+} was prepared, and photoinduced electron transfer from the photoexcited Ru complex to the MV^{2+} was investigated [32]. As shown in Fig. 13-12, the plots based on relative emission intensity (I_0/I) and relative emission lifetime (τ_0/τ) show the typical static quenching behavior shown in Fig. 13-11 (b) (see Experiment 13-6, Section 13.5). By using Eq. (13-8), the electron-transfer distance for static quenching was estimated as 1.4 nm (center to center). In the presence of p-cresol, a tyrosine residue model, in the film, the electron-transfer distance was almost doubled (2.7 nm). Electron hopping via the tyrosine residue model was proposed [32].

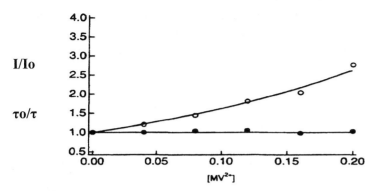

Figure 13-12. Stern–Volmer plots for the electron transfer quenching of $Ru(bpy)_3^{2+}$ by MV^{2+} in a polysiloxane film: I_0/I (○) and τ_0/τ (●).

13.2.3.5 Photochemical H_2 and O_2 Evolution as Half-Reaction Models for Future Artificial Photosynthesis

The photoinduced electron transfer described above has been extended to photochemical H_2 and O_2 evolution to investigate half-reactions for water reduction and oxidation to use for artificial photosynthetic system in the future. These model reactions are represented by Figs. 13-13 and 13-14 in which so-called sacrificial electron donors (such as EDTA and triethanolamine) and acceptors (such as $K_2S_2O_8$ and $[Co(NH_3)_5Cl]^{2+}$) are used, respectively.

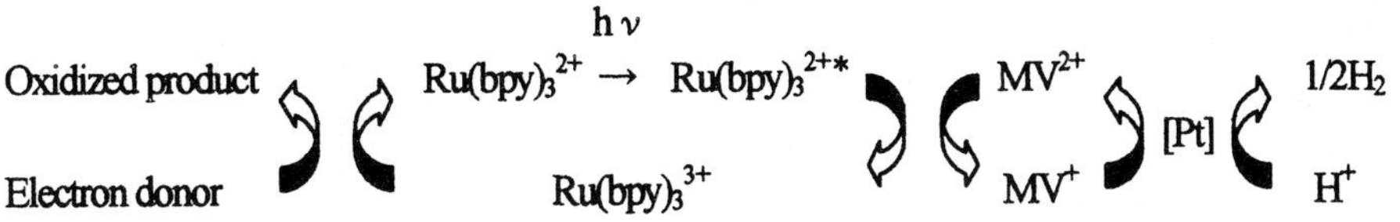

$$h\nu$$

Oxidized product $Ru(bpy)_3^{2+} \rightarrow Ru(bpy)_3^{2+*}$ MV^{2+} $1/2H_2$

Electron donor $Ru(bpy)_3^{3+}$ MV^+ [Pt] H^+

Figure 13-13. A model reaction for photochemical H_2 evolution with a sacrificial electron donor and Pt colloidal catalyst.

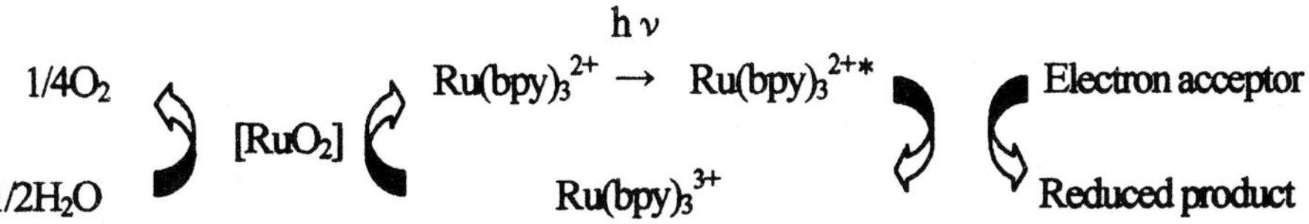

$$h\nu$$

$1/4O_2$ $Ru(bpy)_3^{2+} \rightarrow Ru(bpy)_3^{2+*}$ Electron acceptor

$1/2H_2O$ [RuO_2] $Ru(bpy)_3^{3+}$ Reduced product

Figure 13-14. A model reaction for photochemical O_2 evolution with a sacrificial electron acceptor and colloidal RuO_2 catalyst.

According to the thermodynamics, it should theoretically be possible to combine the reactions in Figs. 13-13 and 13-14 by eliminating both the electron donor and acceptor to achieve simultaneous photochemical H_2 and O_2 evolution from water (i.e. water cleavage). Investigations have been carried out to achieve this with the system, $H_2O/RuO_2/$ $Ru(bpy)_3^{2+}/MV^{2+}/Pt$, by using homogeneous and heterogeneous systems including molecular aggregates and polymers. However, photochemical water cleavage using such a sensitizer/electron relay/catalyst system has not been achieved yet because of charge recombination (back electron transfer). Kinetic requirements should be introduced, as in a photosynthetic system, in order to achieve photolysis of water.

13.3 Photocatalysis for Solar Cells and Water Photolysis by TiO_2

Commercial solar cells are made of inorganic semiconductor junctions, mostly pn-junctions, that combine, for example, n- and p-doped silicon. Although high cost-high performance solar cells made from amorphous semiconductors were developed after the oil crisis in 1973, the cost of these solar cells is still too high for them to be widespread in society, so that development of inexpensive solar cells is an important research subject if we are to utilize renewable energy resources for our civilization. Photochemical solar cells utilizing photocatalytic processes have been attracting attention to develop high cost-high performance cells, but up to 1991 it was difficult to construct cells with high conversion efficiency. A new type of photochemical solar cell called Graetzel's cell, which utilizes sensitized titanium dioxide (TiO_2) film, is now attracting a great deal of attention for the future high cost-high performance solar cells.

After the first report on UV light water photolysis with a n-TiO_2 photoanode [6], sensitization of this large bandgap semiconductor ($E_g = 3$ eV) to utilize visible light has been an important research subject, and dye-sensitization of a TiO_2 photoanode was tried in the 1980s. Adsorption of tris(4,4'-dicarboxy-2,2'-bipyridine)ruthenium(II) ($Ru(dcbpy)_3^{2+}$) onto a TiO_2 photoanode generated a photocurrent with monochromatic 460 nm visible light (intensity 0.22 mW cm^{-2}) with short-circuit photocurrent (I_{sc}) up to 36 µA cm^{-2} and conversion efficiencies of 44% [34]. For the dye-sensitized systems, the photocurrent has been of the order of tens of µA cm^{-2}.

Dye sensitization of a nanometer-sized TiO_2 powder film soaked in an organic medium containing iodine/iodide redox electrolytes successfully generated an open-circuit photovoltage (V_{oc}) of 0.68 V, I_{sc} of 11.2 mA cm^{-2}, Fill Factor (FF) of 0.68 and conversion efficiency of 7.1% under AM 2 irradiation conditions (100 mW cm^{-2}) [7], which evoked great attention to produce high cost-high performance solar cells with an efficiency of about 10%, comparable to the efficiency of amorphous silicon photovoltaic cells. A typical dye used is $Ru(dcbpy)_2(SCN)_2$ (9) where dcbpy is 4,4'-dicarboxyl-4,4'-bipyridine.

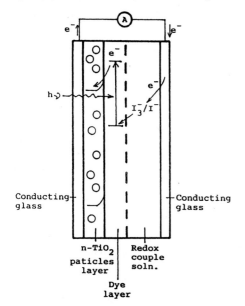

9

The primary reaction of this cell is electron injection from the photoexcited state of the adsorbed dye into the conduction band of TiO_2, which takes place very rapidly (femtoseconds). The concept of this cell is depicted in Fig. 13-15.

Figure 13-15. Photochemical solar cell, a kind of photoregenerative cell, called Graetzel's cell.

This cell functions in photoregenerating the I_3^-/I^- redox couple, so that it is a kind of photoregenerative solar cell. The characteristics of this cell are summarized as follows [3,4,7]:

1. The light-to-electricity conversion efficiency reaches nearly 10% under 100 mW cm^{-2} visible light irradiation, which corresponds to that of an amorphous silicon solar cell. The efficiency at the absorption peak wavelength reaches 80%. The absorption edge of the dye (800 nm) is not sufficient to utilize the solar light efficiently. The development of a dye with absorption at higher wavelength should lead to a higher conversion efficiency.

2. The roughness factor of the nanometer-size (20–30 nm) TiO_2 film is almost 1000, which is an important reason for the high efficiency.
3. The Ru complex dye (**9**, called N3) mentioned above is chemically attached at the –OH group on the TiO_2 surface, which allows rapid electron injection (tens of femtoseconds) from the photoexcited dye to the conduction band of the semiconductor. When the interaction of the dye and the acceptor is very strong and the density of the electron-accepting level is high, Marcus theory is not applicable, and an electron transfer is considered to take place without through vibrational relaxation.
4. The injected electron is believed to be transported by random-walk diffusion among trap levels in the TiO_2. The diffusion coefficient is high, of the order of 10^{-5} cm^2 s^{-1}.
5. Electron transfer from I^- to the oxidized dye takes place in 100 ns, which is faster than the charge recombination between the injected electron and the oxidized dye (μs to ms) by three orders of magnitude.
6. Diffusion of the redox electrolytes in the solution is 10^{-6} to 10^{-5} cm^2 s^{-1}. Specific charge transport via the interaction of the iodide/iodine might also be possible. Other redox electrolytes are not effective.
7. Pt is best for the counter electrode. Some catalytic reaction on the Pt surface to reduce iodine to iodide might be involved.
8. The solvent is limited to organic liquids such as acetonitrile, propionecarbonate, etc. Water drastically decreases the performance of the cell. Water is considered to prohibit the bonding of the dye onto the TiO_2.
9. Attempts have been made to solidify the liquid using polymer gels or p-semiconductors, and a high efficiency of more than 7% has been reported.
10. Continuous operation of the cell for more than 10,000 h has been achieved, and commercialization has started in Australia. Many chemical industries are developing this type of solar cell for possible future commercialization.

In connection with item 9., a precursor mixture composed of polycarbonate (medium), oligomer, redox electrolyte, and radical initiator was injected into a sandwich cell, and heated at 80 to 100 °C to solidify the electrolyte solution, which attains 7.3% conversion efficiency with the solid cell [35].

As mentioned in item 8., this cell can work efficiently only by using an organic medium, and the efficiency is very low in an aqueous medium. Attempts have been made to investigate dye sensitization of TiO_2 film as well as powders in aqueous media by various metal complex dyes in order to utilize it for future artificial photosynthesis, and are described below.

First, fundamental photochemical catalysis by a TiO_2 powder suspension in water should be mentioned as an attempt to cleave water by UV light. Platinized TiO_2 powders (Pt/TiO_2) suspended in water were expected to photolyze water to produce H$_2$ and O$_2$ by UV light, and some reports appeared. However, it has been established that true water photolysis by UV light takes place only under special conditions, e.g. with high concentrations of carbonate anions in water, or with the addition of NaOH in a gas-phase reaction.

It has been reported that platinized TiO$_2$ (TiO$_2$/Pt) suspended in water produced only H$_2$ under irradiation [36] without giving any detectable oxidized product. After complete deactivation of the catalyst in a long-term photoreaction, the recovered TiO$_2$/Pt did not photoproduce H$_2$ even with addition of alcohol (electron donor), showing that the TiO$_2$ underwent some structural change instead of the H$_2$ evolution being caused by the presence of some reductive impurities. As for the change of the catalyst, oxidation of the Ti^{4+} to Ti^{5+} was suggested from XPS data giving an average structure of (Ti$_2$O$_5$)(TiO$_2$)$_2$.

Sensitization of TiO$_2$/Pt powders in water by dissolved sensitizer was carried out under visible light irradiation in the presence of sacrificial electron donor EDTA to produce H$_2$. It was found that tris(bipyrimidine)Ru(II) is an efficient sensitizer for visible light H$_2$ formation. A high concentration of dissolved Ru(bpy)$_3$$^{2+}$ was required for the sensitization, suggesting adsorption of the sensitizer onto the TiO$_2$. Tetrakis(4-carboxyphenyl)porphine and tetrakis(4-sulfonic phenyl)porphine sensitizers dissolved in water were also effective for H$_2$ evolution. Zinc and aluminum complexes of sulfonated and carboxylated phthalocyanines were also investigated [37]. The presence of water oxidation catalysts such as Ru ammine complexes or RuO$_2$ attached to the Pt/TiO$_2$ in place of EDTA was not effective for the H$_2$ formation showing that catalytic electron donation from water is still the most difficult and unsolved process for complete water photolysis by visible light [38].

Sensitization of TiO$_2$ nanosized particle films soaked in water was tried by dissolving a sensitizer and a sacrificial electron donor (EDTA) in the water phase. The photocurrent was strongly dependent on the concentration of Ru(bpy)$_3$$^{2+}$, reaching a saturation at concentrations greater than 2 mM. Analysis of the photocurrent vs. concentration curve, suggested a Langmuir-type adsorption of the dye.

Sensitization of TiO$_2$ powders and films for water photolysis is still an attractive and yet unsolved problem towards constructing an artificial photo-synthetic system for creating an energy resource from solar energy and water.

13.4 Photoinduced Energy Transfer

As pointed out in Section 2.7, in photosynthesis energy transfer occurs between the light-harvesting LH antenna systems and the photosynthetic reaction centers. This process is described as a singlet–singlet energy transfer after the Förster mechanism [39,40]. After excitation the donor in the excited singlet state $^1D^*$ transfers energy to an acceptor with the formation of the acceptor in the excited singlet state $^1A^*$ (Eq. (13-9)). It is necessary that the emission spectrum of D overlaps the absorption spectrum of A.

$$^1D^* + {}^1A \rightarrow {}^1D + {}^1A^* \tag{13-9}$$

Various artificial systems have been synthesized to mimic light harvesting and energy transport. As one example, consider the oligomeric molecule **10**, consisting of a carotenoid C, a Zn-porphyrin P_{Zn}, a metal-free porphyrin P, a naphthoquinone Q_A and a quinone Q_B [41]. With **10** dissolved in an organic solvent, excitation of P_{Zn} at 650 nm leads to singlet–singlet energy transfer to P ($k_1 = 2.3 \cdot 10^{10}$ s^{-1}) (Eq. (13-10)). The subsequent electron transfer to Q_A results in electron–hole separation ($k_Q = 7.1 \cdot 10^8$ s^{-1}) (Eq. (13-11)). Also subsequent further separation of charges to the donor C and the acceptor Q_B are very rapid (Eq. (13-12)). Altogether, molecule **10** combines photoinduced energy transfer and electron transfer. The quantum yield of the whole process is >0.7. A further example of a large multiporphyrin array was discussed in Section 6.3 (dentritic porphyrin **51**).

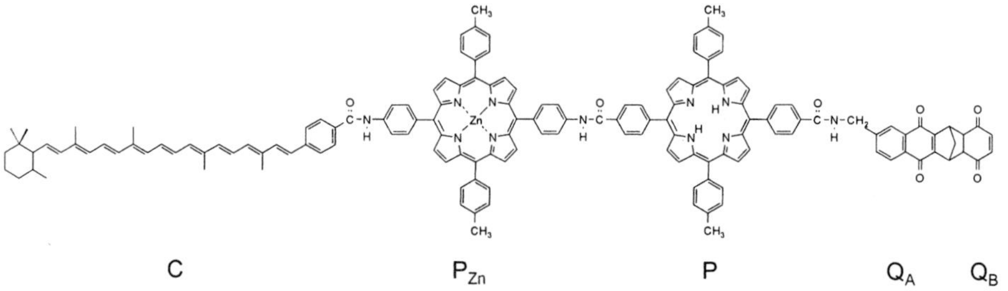

$$C-P_{Zn}-P-Q_A-Q_B \xrightarrow{h\nu} C-{}^1P_{Zn}-P-Q_A-Q_B \tag{13-10}$$

$$\xrightarrow{k_1} C-P_{Zn}-{}^1P-Q_A-Q_B$$

$$C-P_{Zn}-{}^1P-Q_A-Q_B \xrightarrow{k_Q} C-P_{Zn}-P^{\bullet+}-Q_A^{\bullet-}-Q_B \tag{13-11}$$

$$C-P_{Zn}-P^{\bullet+}-Q_A^{\bullet-}-Q_B \rightarrow C^{\bullet+}-P_{Zn}-P-Q_A-Q_B^{\bullet-} \tag{13-12}$$

Another possibility is photoinduced energy transfer to a separate species, which can further react with different compounds. A known example is photoinduced energy transfer from an excited photosensitizer to triplet oxygen, with formation of singlet oxygen, which is spin-allowed and takes place via the excited triplet state of the photosensitizer (PS) (Eqs. (13-13)– (13-15)) [42–46].

$$PS \xrightarrow{\ h\nu\ } {}^1PS^* \xrightarrow{\ ISC\ } {}^3PS^* \qquad\qquad (13\text{-}13)$$

$${}^3PS^* + {}^3O_2 \rightarrow PS + {}^1O_2 \qquad\qquad (13\text{-}14)$$

$${}^1O_2 + substrate \rightarrow oxidized\ substrate \qquad\qquad (13\text{-}15)$$

Suitable photosensitizers absorbing in the visible region of light are Rose Bengal, methylene blue, porphyrins and phthalocyanines leading to singlet oxygen by irradiation with visible light or solar radiation with quantum yields between 0.2 and 0.8 [42,47]. Sulfonated metal phthalocyanines **11** are particularly active and stable photosensitizers [48]. The following reactions, which are of technical interest in the petroleum industry and for wastewater cleaning, are examples. In the dark using the Co(II) complex of the phthalocyanine **11** as catalyst the toxic sulfide is oxidized to sulfur or thiosulfate, thiols to disulfide and phenol to quinone. However, by irradiation with visible or solar light, using Zn(II), Al(III)OH or Si(IV)(OH)$_2$ complexes, these substrate are more efficiently photooxidized: sulfide to sulfate (Eq. (13-16)), thiols to sulfonic acids (Eq. (13-17)) and phenols partially to carbonate (Eq. (13-18)) [42,48].

$$HS^- + 2O_2 + HO^- \rightarrow SO_4^{2-} + H_2O \qquad\qquad (13\text{-}16)$$

$$2RS^- + 3O_2 \xrightarrow{\ h\nu\ } 2RSO_3^- \qquad\qquad (13\text{-}17)$$

$$C_6H_5O^- + 3.5O_2 + 4OH^- \rightarrow CO_3^{2-} + HCOOH$$
$$+\ {}^-OOC\text{-}CH=CH\text{-}COO^- + 3H_2O \qquad\qquad (13\text{-}18)$$

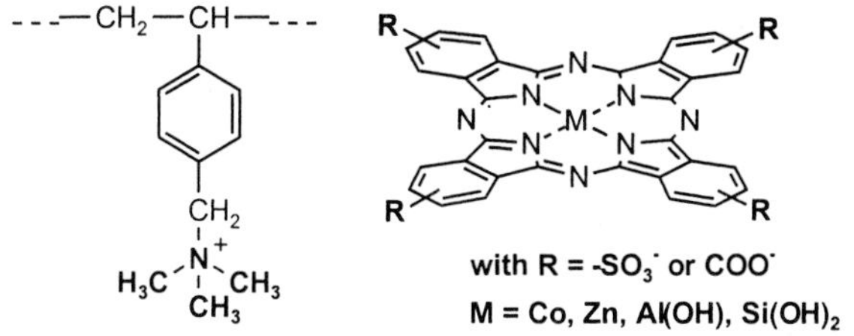

with R = -SO$_3^-$ or COO$^-$
M = Co, Zn, Al(OH), Si(OH)$_2$

11

These photooxidations were carried out with low molecular weight photosensitizers dissolved in aqueous alkaline solution. Heterogeneous photosensitizers are more useful for practical applications. Therefore the phthalocyanines were ionically bound to commercially available ion-exchange resins like Amberlite IRA 400 (see Section 5.4.13) [48].

Figure 13-16 shows an efficient photooxidation of phenol with the heterogeneous **11** (M = Si(OH)$_2$ demonstrated by the oxygen consumption over time. The polymer-bound photosensitizers shows no loss in activity when reused (Fig. 13-17). The photooxidation of phenol using **11** (M= Al(OH)) is described in Section 13.5, Experiment 13-7.

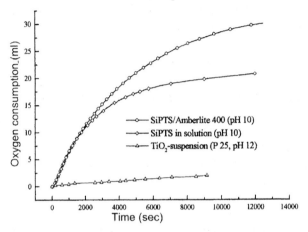

Figure 13-16. Photooxidation of 0.36 mmol phenol in 50 mL aqueous solution (irradiation with 180 mW cm^{-1} of a slide projector). −○− 0.25 μmol **11** and −◇− 10 μmol **11** on Amberlite. −△− 160 mg TiO$_2$ for comparison.

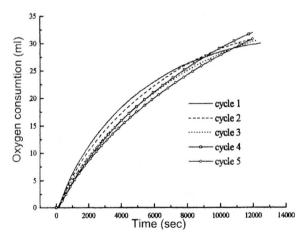

Figure 13-17. Photooxidation of phenol with **11** on Amberlite under conditions described in Fig. 13-16. Employment of the same polymer photosensitizer 5 times (cycles).

13.5 Experimental

Experiment 13-1: Catalytic Water Oxidation by a Trinuclear Ru-ammine Complex (Ru-red) Incorporated into a Nafion Membrane (Section 13.2.3.1) [19]

Preparation of Ru-red in Nafion: A commercially available Nafion membrane (4, thickness 180 μm, purchased from Aldrich Chemical Co.) was swollen sufficiently by soaking in distilled water for ~24 h. The swollen membrane (2 × 3 cm) was immersed in an aqueous solution containing of Ru-red (3, purchased from Wako Pure Chemical Industries Ltd.) (3 mL of $5 \cdot 10^{-3}$ mol dm^{-3} solution, $1.5 \cdot 10^{-5}$ mol) for 5 h, so that the complex was incorporated into the membrane by cation exchange. The amount of the complex incorporated into the membrane was estimated from the visible absorption spectral change of the aqueous solution before and after the incorporation of the complex. The molar concentration of the complex in the membrane (M) was calculated from the amount of the adsorbed complex and the membrane volume.

Catalytic water oxidation: This membrane with adsorbed Ru-red was placed in distilled water (5 mL) in a cell equipped with a rubber septum, Ar gas was bubbled into the water for 1 h, and then cerium(IV) diammonium nitrate powder ($6 \cdot 10^{-3}$ mol) was added quickly. The Ce salt dissolves in water rapidly, and dioxygen bubbles were seen on the surface of the Nafion membrane. The evolved O_2 was analyzed on a gas chromatograph equipped with a 5-Å molecular sieve column using Ar carrier gas (flow rate 40 mL min^{-1}) at 50 °C.

Experiment 13-2: Electrocatalytic Water Oxidation by a Trinuclear Ru-ammine Complex (Ru-red) Incorporated into a Nafion Membrane (Section 13.2.3.1) [20]

Preparation of Ru-red in Nafion: A commercially available alcoholic solution of Nafion 117 (Aldrich Chemical Co.; 5 wt%) was diluted to half by methanol, and this solution (30 μL) was cast onto a Pt plate electrode (1 cm^2). The Nafion-coated electrode was immersed in aqueous Ru-red (Wako Pure Chemical Ind. Ltd.) solution (2 mL of $1 \cdot 10^{-4}$ mol dm^{-3}) to adsorb the complex into the membrane. The amount of the complex incorporated into the membrane was estimated from the visible absorption spectral change of the aqueous solution before and after the incorporation of the complex. The molar concentration of the complex in the membrane (M) was calculated from the amount of the adsorbed complex and the membrane volume.

Electrocatalytic water oxidation: A conventional single-compartment cell was equipped with the modified working electrode, a Ag/AgCl reference electrode and a platinum wire counter electrode. A supporting electrolyte solution (pH 5.4) of 0.1 mol dm^{-3} KNO$_3$ was deaerated by bubbling argon gas for 1 h. Potentiostatic electrolysis at an applied potential of 1.4 V vs.Ag/AgCl was carried out at room temperature, and the evolved O$_2$ was analyzed on a gas chromatograph equipped with a 5-Å molecular sieve column using Ar carrier gas (flow rate 40 mL min^{-1}) at 50 °C. The amount of O$_2$ evolved through catalysis by the complex was calculated by subtracting the amount of O$_2$ detected for a blank experiment without the complex. Although O$_2$ can also evolve on the electrode surface without the catalyst, at least two-thirds of the evolved O$_2$ was obtained via the catalysis of the complex. The Ru-red concentration of 0.1 mol dm^{-3} in the Nafion gave an optimum activity with the turnover number (TN) of the catalyst 8 h^{-1}. When p-cresol, a tyrosine-residue model compound, was added as a charge mediator in the Nafion solution at a concentration in the coated Nafion of 5·10^{-2} mol dm^{-3}, the O$_2$ evolution was enhanced by four times, giving a TN of 36 h^{-1} at a Ru-red concentration of ~0.02 mol dm^{-3} (refer to Fig. 13-6).

Experiment 13-3: Electrocatalytic Proton Reduction to H$_2$ Evolution by Cobalt Tetraphenylporphyrin 5 (CoTPP) Incorporated into a Nafion Membrane Coated on an Electrode (Section 13.2.3.2) [23(b)].

Preparation of CoTPP in Nafion: A DMF solution of CoTPP (Aldrich Chemical Co.; 0.2 mmol dm^{-3}) containing Nafion (0.5 wt%) was prepared, and this solution (50 µL) was cast onto a Pt electrode followed by solvent evaporation under vacuum. The membrane thickness was ~1 µm.

Electrocatalytic proton reduction: An electrochemical cell was equipped with the working Nafion[Co-TPP]/Pt electrode, a Ag/AgCl reference electrode and a spiral Pt counter electrode. After bubbling Ar gas into the aqueous phosphate buffer solution (pH 1), potentiostatic electrolysis to reduce protons to form H$_2$ was carried out, and the H$_2$ evolved was analyzed on a gas chromatograph with a 5-Å molecular sieve column using Ar carrier gas (flow rate 40 mL min^{-1}) at 50 °C. The Nafion[Co-TPP]/Pt electrode (coated amount of CoTPP is 1·10^{-8} mol cm^{-2}) gave 90.8 µL H$_2$ h^{-1} at an applied potential of –0.3 V vs. Ag/AgCl, while a bare Pt electrode gave only 14.4 µL H$_2$ h^{-1} under the same conditions. The TN of the complex was 4·10^3 h^{-1} when the CoTPP concentration in the Nafion was 0.05 mol dm^{-3}.

Experiment 13-4: Electrocatalytic Carbon Dioxide Reduction by Cobalt Phthalocyanine 6 Incorporated into a Poly(4-vinyl-pyridine) (PVP) Membrane Coated on an Electrode (Section 13.2.3.3) [25]

Preparation of CoPc in PVP: A mixed solution of DMF CoPc (Kanto Chemical Co.Inc.; 5 to 50 μmol dm^{-3}) and PVP (1 wt.%, prepared by radical polymerisation of 4-vinylpyridine) was prepared, and this solution (2 μL) was cast on a basal plane pyrolytic graphite electrode (BPG) (effective area 0.17 cm^2) followed by solvent evaporation under ambient conditions.

Electrocatalytic carbon dioxide reduction: An electrochemical cell was equipped with this modified working electrode, an Ag/AgCl reference electrode and a spiral counter electrode. An aqueous phosphate buffer solution (pH from 2.3 to 6.8) was placed in the cell and CO_2 gas was bubbled through for 1 h. Then potentiostatic electrolysis was carried out at an applied potential of -1.05 to -1.30 V vs. Ag/AgCl at room temperature, and the gas evolved was analyzed on a gas chromatograph with a 5-Å molecular sieve column using Ar carrier gas (flow rate 40 mL min^{-1}). CO and H_2 were detected as reduction products of CO_2 and protons, respectively. 20 μL CO and 5 μL H_2 were obtained with a CoPc coverage of 10^{-10} mol cm^{-2} at pH 4.4 at an applied potential of -1.20 V vs. Ag/AgCl for 2 h, and the selectivity ratio for CO_2 and H$^+$ reduction was 4.

Experiment 13-5: Photoexcited State Electron Transfer: Photochemical Formation of MV$^{\bullet+}$ on a Cellulose Paper with Adsorbed Ru(bpy)$_3^{2+}$, MV^{2+} and EDTA (Sacrificial Reducing Agent) (Section 13.2.3.4) [27]

Adsorption on cellulose paper: A mixture of MV^{2+} 8 (Aldrich Chemical Co.; $5 \cdot 10^{-2}$ mol dm^{-3}), Ru(bpy)$_3^{2+}$ 7 (Aldrich Chemical Co.; 10^{-3} mol dm^{-3}), and disodium EDTA ($5 \cdot 10^{-2}$ mol dm^{-3}) in water was adsorbed on a cellulose paper by dipping the paper in the solution for 10 min, and the paper was dried *in vacuo* at room temperature. The masses of adsorbed MV^{2+}, Ru(bpy)$_3^{2+}$, and EDTA were 64.4, 1.88 and 92.4 mg per g cellulose, respectively, which corresponds to a mole ratio of 100/1/100.

Photochemical formation of MV^{2+}: This cellulose paper was irradiated with 470-nm monochromatic light *in vacuo* at 30 °C, the paper turned blue by the photochemically formed MV$^+$, and the initial rate of MV$^+$ formation was determined from the increase of the absorbance maximum at 620 nm as 0.010 absorbance unit min^{-1}. The visible absorption spectrum was measured by placing a blank cellulose paper at the reference side of the spectrophotometer. In contrast to a solution reaction, MV$^{\bullet+}$ can be accumulated even in the presence of O_2 (under air) since oxidation of MV$^{\bullet+}$ by O_2 is prohibited in the solid phase.

Experiment 13-6: Photoinduced Electron Transfer from $Ru(bpy)_3^{2+}$ to MV^{2+} Incorporated in a Polysiloxane Film (Section 13.2.3.4) [32]

Incorporation in polysiloxane film: Fixed amounts of polysiloxane containing carboxyl groups at side chains (prepared by the author, refer to reference [32]), $Ru(bpy)_3^{2+}$ **7** and MV^{2+} **8** were dissolved in ethanol to obtain a concentration of the total solutes of 0.02 g mL^{-1}. The solution was cast on a glass plate and allowed to dry under vacuum at 35 °C to prepare a polymer film. The film thickness was estimated to be ~1 μm. The Ru complex concentration in the film was fixed at 0.05 mol dm^{-3} (estimated by using a film density of 1.3 g mL) to minimize the concentration quenching of the photoluminescence from $Ru(bpy)_3^{*2+}$. At this concentration only 10% of the photoexcited Ru complex is subject to concentration quenching. The concentration of MV^{2+} in the film was from 0.04 to 0.20 mol dm^{-3}.

Photoinduced electron transfer: All the measurements of emission intensity were carried out on a fluorospectrophotometer at 25 °C under Ar. The front surface of the film was irradiated with monochromatic light of 450 nm, and the emission (maximum around 600 nm) was monitored from the back side of the glass at a right angle from the excitation light to minimize the excitation light-scattering effect. The emission decay was measured by a time-correlated single photon counting apparatus equipped with a 10 atm hydrogen lamp; the emission was monitored through a cutoff filter (O-58) to minimize the scattering. The Stern–Volmer plots (Eq. (13-7)) based on emission intensity (I_0/I) and emission lifetime (τ_0/τ) were plotted against MV^{2+} concentration (see Fig. 13-12) and the results showed a typical example of a static mechanism as shown in Fig. 13-11 (b).

Experiment 13-7: Photooxidation of Phenol (Section 13.4) [48]

Photooxidation experiment: A thermostated glass apparatus consisting of a 50-mL gas buret and 100-mL reaction vessel which are filled with oxygen were used. An aqueous solution (50 mL) containing the following compounds was employed: the low molecular weight phthalocyanine **11** (0.25 μmol) as photosensitizer; for measurements at pH 13 NaOH (0.12 mol dm^{-3}), at pH 10 borate buffer (25 mL), at pH 7 phosphate buffer (25 mL). In all experiments the whole apparatus was thermostated at 25 °C. After filling the reaction flask with the solution (50 mL), the apparatus was flushed with pure oxygen for 10 min. Then an ethanolic phenol solution (0.25 mL of a solution of phenol (72 mmol) in ethanol (50 mL), 0.36 mmol) was added (chlorophenols were also employed). The reaction vessel was closed immediately. The mole ratio substrate to PS amounts to 1430:1. In the case of ionically polymer-bound phthalocyanine **11** an aqueous buffer solution (50 mL) with polymer (2.5 g containing 10 µmol PS (mole ratio substrate:PS = 36:1)) was first added to the reaction vessel. The reaction flask was irradiated with intensive magnetic stirring and the oxygen consumption over time was measured volumetrically.

13.6 References

1. A. Fujishima (Ed.), *TiO₂ Photocatalysis – Fundamentals and Application*, BKC Publishers , Tokyo, 1999.
2. K. Kalyanasundaram, M. Graetzel (Eds.), *Photosensitization and Photocatalysis Using Inorganic and Organometallic Compounds*, Kluwer Academic Publishers, Dordrecht, 1993.
3. M. Kaneko, I. Okura (Eds.), *Photocatalysis – Science and Technology*, Springer-Verlag, Berlin, 2002.
4. *Photochemical Conversion and Storage of Solar Energy, Proceedings of the 12ᵗʰ International Conference*, Oldenbourg Wissenschaftsverlag, München, 2000.
5. G.J. Meyer (Ed.), *Molecular Level Artificial Photosynthetic Materials*, Progress in Inorganic Chemistry, Vol. 44, Interscience Publication, New York, 1997.
6. A. Fujishima, A. Honda, *Nature* **1972**, *238*, 37.
7. B. O'Regan, M. Graetzel, *Nature* **1991**, *353*, 737.
8. M. Kaneko, D. Woehrle, *Adv. Polym . Sci.* **1987**, *84*, 141.
9. M. Kaneko, *11ᵗʰ Symposium on Unsolved Problems of Polymer Chemistry*, The Society of Polymer Science, Japan, 1976, p. 21.
10. D.O. Hall, K.K. Rao, *Photosynthesis*, 6th Edn., Cambridge University Press, Cambridge, 1999.
11. W. Ruetinger, G.C. Dismukes, *Chem. Rev.* **1997**, *97*, 1.
12. R. Manchanda, G.W. Brudvig, *Coord. Chem. Rev.* **1995**, *144*, 1.
13. V.L. Pecoraro, M.J. Baldwin, A. Gelasco, *Chem. Rev.* **1994**, *94*, 807.
14. J.A. Gilbert, D.S. Eggleston, W.R. Murphy, Jr., D.A. Geselowitz, S.W. Gersten, D.J. Hodgson, T.J. Meyer, *J. Am. Chem. Soc.* **1985**, *107*, 3855.
15. R. Ramaraj, A. Kira, M. Kaneko, *Angew. Chem. Int. Ed.* **1986**, *25*, 1009.
16. R. Ramaraj, A. Kira, M. Kanneko, *J. Chem. Soc., Faraday Trans. 1* **1987**, *83*, 1539.
17. R. Ramaraj, M. Kankeo, *Adv. Polym. Sci.* **1995**, *123*, 215.
18. M. Yagi, M. Kaneko, *Chem. Rev.* **2001**, *101*, 21.
19. M. Yagi, S. Tokita, K. Nagoshi, I. Ogino, M. Kaneko, *J. Chem. Soc., Faraday Trans.* **1996**, *92*, 2457.
20. M. Yagi, K. Kinoshita, M. Kaneko, *J. Phys. Chem. B* **1997**, *101*, 3957.
21. R.M.Kellett, T.G. Spiro, *Inorg. Chem.* **1958**, *24*, 2373, 2378.
22. T. Abe, H. Imaya, S. Tokita, D. Woehrle, M. Kaneko, *J. Porphyrins Phthalocyanines* **1997**, *1*, 215.
23. (a) F. Taguchi, T. Abe, M. Kaneko, *J. Mol. Cat. A: Chem.* **1999**, *140*, 41. (b) T. Abe, F. Taguchi, H. Imaya, F. Zhao, J. Zhang, M. Kaneko, *Polym. Adv. Tech.* **1998**, *9*, 559.
24. H. Tanaka, B.C. Tzeng, H. Nagao, S.M. Peng, K. Tanaka, *Inorg. Chem.* **1993**, *32*, 1508.

25. T. Abe, T. Yoshida, S. Tokita, F. Taguchi, H. Imaya, M. Kaneko, *J. Electroanal. Chem.* **1996**, *412*, 125.
26. T. Abe, H. Imaya, T. Yoshida, S. Tokita, D. Schlettwein, D. Wöhrle, M. Kaneko, *J. Porphyrins Phthalocyanines* **1997**, *1*, 315.
27. M. Kaneko, J. Motoyoshi, A. Yamada, *Nature* **1980**, *285*, 468.
28. O. Stern, M. Volmer, *Phys. Z.* **1919**, *20*, 183.
29. J. Perrin, *Comp. Rend. Acad. Sci. Paris* **1927**, *184*, 1097; **1924**, *178*, 1978 .
30. M. Kaneko, *Progress in Polymer Science* **2001**, 26, 1101.
31. K. Nagai, N. Yakamiya, M. Kaneko, *Macromol. Chem. Phys.* **1996**, *197*, 2883.
32. K. Nagai, J. Tsukamoto, N. Takamiya, M. Kaneko, *J. Phys. Chem.* **1995**, *99*, 6648.
33. T. Abe, T. Ohshima, K. Nagai, S. Ishikawa, M. Kaneko, *Reactive Functional Polymers* **1998**, *37*, 133.
34. J. Desilvestro, M. Grätzel, L. Kavan, J. Moser, *J. Am. Chem. Soc.* **1985**, *107*, 2988.
35. S. Hayase, *Recent Advances in Research and Development for Dye-Sensitized Solar Cells*, H. Arakawa (Ed.), CMC, Tokyo 2001, p. 270.
36. T. Abe, E. Suzuki, K. Nagoshi, K. Miyashita, M. Kaneko, *J. Phys. Chem. B* **1999**, *103*, 1119.
37. MD.K. Nazeeruddin, R. Humpfry-Baker, M. Grätzel, D. Wöhrle, G. Schnurpfeil, G. Schneider, A. Hirth, N. Trombach, *J. Porphyrins Phthalocyanines* **1999**, *3*, 230.
38. M.Kaneko et al., Unpublished data.
39. D.R. Ort, C.F. Yocum (Eds.), *Oxygenic Photosynthesis: The Light Driven Reactions*, Kluwer Academic Publisher, Dordrecht, 1996.
40. H.T. Witt, *Ber. Bunsenges. Phys. Chem.* **1996**, *100*, 1923.
41. D. Gust, T.A. Morre, A.L. Morre, *Acc. Chem. Res.* **1993**, *26*, 198; H. Kureck, M. Huber, *Angew. Chem.* **1995**, *107*, 929.
42. D. Wöhrle, M. Tausch, W.-D. Stohrer, *Photochemie — Konzepte, Methoden, Experimente*, Wiley-VCH, Weinheim, 1998.
43. F. Wilkinson, W.P. Helman, A.B.P. Ross, *J. Phys. Chem.. Ref. Data* **1993**, *22*, 113.
44. E.A. Lissi, M.V. Encinas, E. Lemp, M.A. Rubio, *Chem. Rev.* **1993**, *93*, 699.
45. M. Prein, W. Adam, *Angew. Chem.* **1996**, *168*, 519.
46. P. Esser, B. Pohlmann, H.-D. Scharf, *Angew. Chem.* **1994**, *106*, 2093.
47. W. Spiller, H. Kliesch, D. Wöhrle, S. Hackbarth, B. Roeder, *J. Porphyrins Phthalocyanines* **1998**, *2*, 145.
48. R. Gerdes, O. Bartels, G. Schneider, D. Wöhrle, G. Schulz–Ekloff, *Intern. J. Photoenergy* **1999**, *1*, 41; *Polym. Adv. Technol.* **2001**, *12*, 152.

14 Electron- and Photon-Induced Processes

Masao Kaneko

One of the important functions of metal complexes is charge transfer. This is realized mainly by redox reactions of the central metal ion in the ground as well as in excited states. When metal complexes are incorporated in a polymer matrix, the polymer matrix can take on the function of transporting charges in the dark or under light irradiation, which can lead to electronic and photonic devices. Please refer also Chapter 13. Another possibility where metal complexes are involved are electroluminescent devices in which a material is excited by applying a voltage, and luminescence is observed.

Parts of this chapter deal with charge transfer in solid polymer matrixes with incorporated metal complex redox centers in their ground and photoexcited states. It is important to consider electron transfer in both the ground and excited states if we are to obtain a complete understanding of electron transfer events. However, charge transfer in the ground and photoexcited states have been studied separately by scientists in the fields of electrochemistry and photo-chemistry, respectively, and have not previously been discussed together. An overview and an outline of the fundamental problems on each topic will first be given, followed by a review of recent results obtained by the present author. Charge transfer takes place by different mechanisms, i.e. physical diffusion of redox molecules, charge hopping between redox molecules, and the combined process of diffusion and charge hopping. For charge hopping, the distance that the charge can hop between molecules is especially important.

Charge transfer in the ground state of metal complexes has been investigated mainly by electrochemical methods, which provide information about integrated charge propagation in a polymer matrix. In the ground state, charges are transported by molecular diffusion, charge hopping between molecules or a combined mechanism involving both. Refer also to Chapter 13 for information on electrocatalysis by polymer-redox-molecule-coated electrodes.

Charge transfer in a photoexcited state has been investigated by photochemical methods such as laser flash photolysis and luminescence quenching, providing information on one single charge-transfer step at an excited state. The mechanism may be both dynamic and static; the former taking place by diffusion and collision of the molecules and the latter taking place without diffusion (see also Chapter 13). Various models have been presented to analyze the mechanism and the charge-transfer distance by luminescence quenching. In solid polymer matrixes charge transfer in an excited state takes place mainly by a static mechanism, and the charge transfer distance is similar to that in the ground state.

Combined systems of charge transfer in ground and excited states will also be briefly reviewed, and approaches to photoelectric devices introduced.

Interesting electrochemical and photochemical electron transfer in a quasi-solid-state polymeric material containing a large excess of water, where the electron transfer takes place in the same way as in liquid water, is also described.

Elsewhere in this chapter, electroluminescent devices including macro-molecular metal complexes as active materials are described. These materials are very attractive as thin and light display devices.

14.1 Charge Transport by Metal Complexes in the Ground State Confined in Polymer Matrixes

14.1.1 Fundamental Aspects

Charge transfer by redox molecules in solid matrixes such as proteins and synthetic polymers is an important process for biological activities and various devices [1–9]. Biological activities are supported by electron transfer among central redox molecules confined in enzymes. These center molecules are often metal complexes such as cytochrome *c*, blue-copper proteins, carboxy-peptidases, etc. (see Chapter 2). In these enzymes electron transfer takes place by electron hopping between active central metal ions. A specific aspect of such enzymatic electron transfer is a long-range electron transfer mediated by proteins. Much research has been published on electron transfer in biological materials [10].

This section focuses on charge transfer in synthetic polymer solid phases (for metal enzymes refer to Chapter 2). In many electronic devices such as electrocatalytic systems, sensors, and electrochromic display or photoconversion systems, charge transfer between redox molecules incorporated in a solid polymer matrix plays an essential role in their functions. From these points of view, investigation of charge transfer between molecules confined in heterogeneous polymer phases is important.

In order to investigate charge transfer in the ground state, an electrochemical technique is usually used, in this case employing polymer-modified electrodes containing redox centers in a polymer matrix (Fig. 14-1). Charges are transported via redox reactions of the metal complexes in the polymer matrix, and this process is called charge propagation. The electrochemical method provides information on integrated charge transfer in a polymer matrix in addition to elucidating one single step of a charge transfer between molecules.

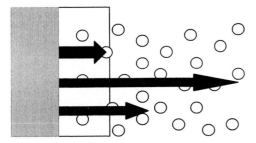

Electrode/Polymer film with metal complexes/Electrolyte solution

Figure 14-1. Charge transfer between metal complexes confined in a polymer matrix coated on an electrode.

A polymer-coated electrode is usually soaked in a solution containing electrolytes, and electrochemical charge transfer in the polymer film is investigated by a three-electrode system with working, counter and reference electrodes under cyclic or step potential scanning.

Redox molecules are incorporated in a polymer matrix by covalent bonding, electrostatic binding or/and hydrophobic interaction. The mobility of the redox-center molecule determines the mechanism of charge transfer; either physical diffusion (or physical replacement) of the molecules, charge hopping between molecules or a combined mechanism (Fig. 14-2). Migration of counter ions also affects the charge transfer. From these points of view the nature of both the metal complex and the polymer matrix is very important for the whole charge-transfer event.

e^- or h^+

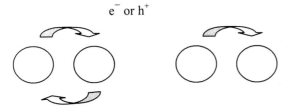

Diffusion (Physical replacement) Charge hopping

Figure 14-2. Mechanisms of charge transfer between metal complexes in a polymer matrix.

Charge propagation is understood as a diffusion of charges in a matrix, so that it is represented by an apparent diffusion coefficient (D_{app} cm^2 s^{-1}). Such charge diffusion was expressed by a Dahms–Ruff equation which combines terms of physical diffusion of the redox-center molecules (D_{phys}) and charge hopping between them (D_e) (Eqs. (14-1), (14-2)) [11,12], where k_{ex} is the rate constant of charge exchange between two molecules, δ is the charge hopping distance and c is the concentration of redox molecule in the polymer matrix.

$$D_{app} = D_{phys} + D_e \tag{14-1}$$

$$D_e = k_{ex}\delta^2 c \tag{14-2}$$

From this it may be seen that, if the D_{app} value is independent of the concentration, the charge transfer takes place only by physical diffusion of the redox center, and if it is dependent on the concentration, the process involves both diffusion and charge hopping. Many papers have reported a dependence of D_{app} upon c which leads to determination of the charge transfer mechanism.

Andrieux and Saveant proposed Eq. (14-3) for a system where only charge hopping takes place[13].

$$D_{app} = (1/6)k_{ex}\delta^2 c \tag{14-3}$$

Monte Carlo simulation was carried out by Blauch and Saveant based on a percolation process, and D_{app} was obtained as shown in Eq. (14-4) considering charge hopping and bounded motion of the redox center [14]. Bounded motion λ is a kind of local oscillation of redox molecules. In this model, charge transfer by molecular diffusion is not taken into account.

$$D_{app} = (1/6)k_{ex}(\delta^2 + 3\lambda^2)c \tag{14-4}$$

In addition to these mechanistic studies, many papers have been published by the groups of, for example, N. Oyama [15,16], F.C. Anson [17,18], R.W. Murray [19,20], A.J. Bard [21,22], M.S. Wrighton [23,24], J.G. Vos [25,26], T.J. Meyer [27,28], D. Abruna [29,30], H. Larsson and M. Sharp [31,32], A. Heller [33,34], K. Aoki [35] and other researchers. A general analysis has been proposed for covalently attached redox polymers by taking into account the influence on the charge-transfer dynamics of both bounded motion and free physical diffusion rates [36a].

Interactions and electric communication between redox centers are of importance for charge transfer. A Ru complex of poly[2-(2-pyridyl)-bisbenz-imidazole] where the redox centers are coordinated directly to a long-range π network has been investigated [36b]. Electron-transfer studies yielded electron diffusion coefficients of over 10^{-8} cm^2 s^{-1} for the Ru(III/II) state, at least one order of magnitude higher than for a non-conjugated Ru(bpy)$_3^{2+}$ type polymers.

In the following sections one approach is described to the mechanism of charge transfer (Section 14.1.2) and to the distance of charge hopping with bounded motion (Section 14.1.3). A percolation theory to treat charge transfer by redox centers without diffusion and bounded motion will be mentioned in Section 14.1.4.

14.1.2 The Mechanism of Charge Transfer

The mechanism of charge transfer by redox centers can be investigated by measuring the dependence of the initial charge transfer rate (V_{ct}) on the concentration of redox centers (c, represented by mol dm^{-3}) in a polymer membrane when the membrane thickness is not changed. If charge propagation takes place by a diffusion mechanism (physical replacement), V_{ct} should be proportional to c. If it takes place by a charge hopping mechanism with sufficient bounded motion, V_{ct} should be proportional to c^2. However, it should be noted here that charge hopping should be treated by a percolation theory as mentioned later if bounded motion does not exist. These processes are represented by Eqs. (14-5)–(14-7) [37].

Diffusion of redox molecules (physical replacement): $V_{ct} = k_1 c$ (14-5)

Charge hopping between redox molecules with sufficient bounded motion:
$$V_{ct} = k_2 c^2 \qquad (14\text{-}6)$$

Combination of diffusion and hopping: $V_{ct} = k_1 c + k_2 c^2$ (14-7)

The mechanism can be determined by these relations or by plotting the value V_{ct}/c against c (Eqs. (14-8)–(14-10)) as shown in Fig. 14-3.

$$V_{ct}/c = k_1 \qquad (14\text{-}8)$$

$$V_{ct}/c = k_2 c \qquad (14\text{-}9)$$

$$V_{ct}/c = k_1 + k_2 c \qquad (14\text{-}10)$$

Instead of the V_{ct} values, D_{app} can also be used for determining the mechanism using a similar treatment.

Besides the charge transfer rate, the fraction of the redox molecule which accepts charge(s) (R_{ct}) is another important factor that represents the characteristics of the charge transfer. In a diffusion mechanism R_{ct} is usually independent of c and reaches a high value of more than 0.8 to 0.9 after saturation, although sometimes it decreases when c is too high, owing to hindrance of molecular diffusion. The R_{ct} value after saturation of the charge propagation is shown in Fig. 14-4. In a hopping mechanism R_{ct} is strongly dependent on c and with sufficient bounded motion (see below), R_{ct} can increase with an increase in c. However, with or without insufficient bounded motion, R_{ct} can increase only when c is high enough (percolation threshold) since charge hopping is possible only when a nearest neighboring molecule is located within a charge-hopping distance.

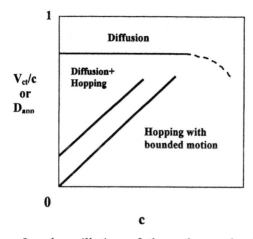

Figure 14-3. Dependence of V_{ct}/c on the metal complex concentration (c) in a polymer matrix for various mechanisms.

Local oscillation of the redox molecule called "bounded motion" is an important factor for charge hopping. Assuming a random distribution of redox molecules confined in a matrix, there will exist fractions of isolated molecules and clusters (a group of molecules present within a charge-hopping distance of each other, but isolated from other molecules or clusters) to which charge propagation does not occur. When the molecules do not oscillate at all (no bounded motion), charge propagation can take place only when the concentration exceeds a certain value called the threshold (or critical) concentration (c_{th}, see Section 14.2.4) because of the isolated molecules and clusters. This process can be treated by percolation theory, and the R_{ct} value (after saturation) against c shows a sigmoidal curve as shown in Fig. 14-4. However, when the molecules oscillate (bounded motion), rearrangement of the isolated molecules and clusters always takes place resulting in easier charge propagation in the matrix than in the system without bounded motion.

The rate of charge transfer is usually measured by the amount of charge passed (coulomb number). However, it happens sometimes that such a coulomb number does not show the true change in the redox molecules because of some charging current of the polymer film. Observation of the true change of the redox molecules by spectroscopy can give a more exact charge-propagation rate than the measurement of the coulomb number. *In-situ* spectrocyclic voltammetry (SCV) or potential-step chronoamperospectrometry [37,38] which combines conventional voltammetry with visible absorption spectroscopy have been adopted to study the mechanism. Typical charge-transfer systems described in this section are summarized in Table 14-1. The details of this Table are mentioned below.

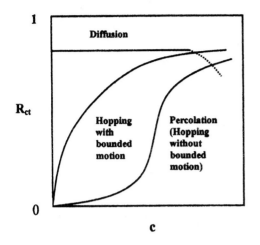

Figure 14-4. Dependence of the fraction of reacted redox center (R_{ct}) on its concentration in a matrix for various charge transfer mechanisms; diffusion, hopping with bounded motion, and hopping without bounded motion (percolation).

Table 14-1. Summary of charge transfer by redox molecules in a polymer matrix.

Mechanism	Bound motion Examples (redox center/polymer)[a]	$D_{app}(10^{-10}$ $cm^2 s^{-1})$
(1) Physical diffusion Easy	$MV^{2+} \rightarrow MV^+$ /Nf	11
	$Ru(bpy)_3^{3+} \rightarrow Ru(bpy)_3^{2+}$/Nf	0.85
	$[Co^{III}TPP(-2)]^+ \rightarrow Co^{II}TPP(-2)$ /Nf	4.4
	$Co^{III}TPP(-2) \rightarrow [Co^{II}TPP(-2)]^+$ /Nf	0.73
(2) Charge hopping Easy	$Ru(bpy)_3^{2+} \rightarrow Ru(bpy)_3^{3+}$ /Nf	0.24
	$[Co^{III}TPP(-2)]^+ \rightarrow [Co^{III}TPP(-1)]^{2+}$ /Nf	0.03
	$Zn^{II}TPP(-2) \rightarrow [Zn^{II}TPP(-1)]^+$ /P-2VP	0.13
Difficult (but possible)	$[(bpy)_2RuORu(bpy)_2]^{3+} \rightarrow$ $[(bpy)_2RuORu(bpy)_2]^{4+}$ /Nf	
Very difficult (Percolation)	$Poly-MV^{2+} \rightarrow Poly-MV^+$ $Poly- Ru(bpy)_3^{2+} \rightarrow Poly- Ru(bpy)_3^{3+}$	
(3) Diffusion + Hopping Easy	Ru-red $\rightarrow$ Ru-brown (oxidation) /Nf	

[a] Nf = Nafion, P2VP = Poly(2-vinylpyridine).

In Table 14-1 a Nafion **1** (sulfonated perfluoroalkyl polyanion polymer) membrane is often used since it is a stable cation-exchange polymer able to adsorb cationic compounds easily. A Nafion membrane is composed of hydrophilic columns incorporating anionic sulfonate groups, hydrophobic columns composed of main chains, and interlayer regions between the hydrophilic and hydrophobic columns. When a cationic substrate is adsorbed from an aqueous solution, the adsorbed material is located only in the hydrophilic columns or interlayer regions, so that a real local concentration of the adsorbed material should be calculated by adopting a localization factor (α) which shows the degree of localization. The α value was estimated as 5.1 for a Nafion film cast from its alcoholic solution, and 40 for a commercially available film [37]. In many cases the thickness of a membrane was taken as 1 μm, the concentration of redox-center molecules (c) 0.01–1 M, and the coverage 10^{-9}–10^{-7} mol cm^{-2}.

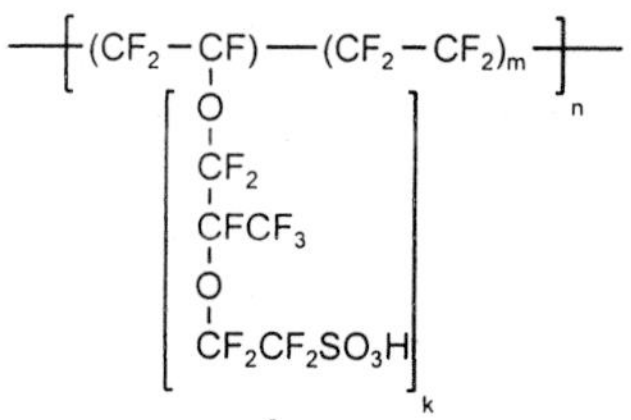

1

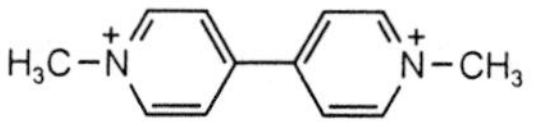

2 (MV^{2+})

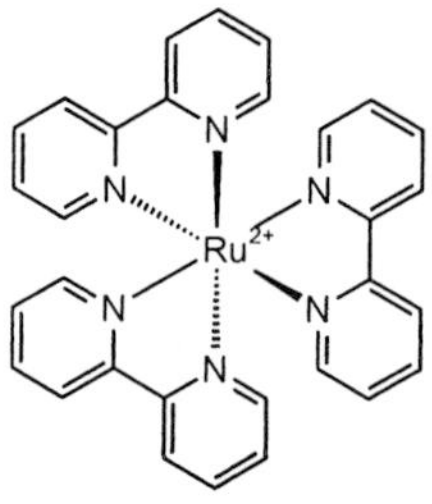

3 Ru(bpy)$_3^{2+}$)

    A typical example of the diffusion mechanism is charge transfer by methylviologen **2** (MV^{2+}) confined in a Nafion membrane [39] by adsorption of MV^{2+} from its aqueous solution into a precast membrane.

    The rate of visible absorption spectral change of MV^{2+} (V_{ct}) confined in a Nafion membrane coated on a transparent indium tin oxide (ITO) electrode after potential step from 0 to –0.75 V vs. Ag/AgCl was proportional to the MV^{2+} concentration in the Nafion, where the increase of the absorption at 550 nm (dimeric species) indicates formation of the cation radical (MV$^+$). By this charge transfer the membrane changes from colorless to blue-colored, and this color change is reversible. The linear dependence of V_{ct} on the MV^{2+} concentration shows clearly that the charge transfer takes place by a diffusion mechanism [39]. The first-order rate constant for the charge propagation by diffusion was obtained as $k_1 = 1.4 \cdot 10^{-1}$ s^{-1}.

    The cationic MV^{2+} adsorbed from its aqueous solution is incorporated electrostatically into the hydrophilic column, and the reduced species diffuse there. This is an important and typical feature of charge transfer taking place by diffusion in contrast to a charge-hopping mechanism for which R_{ct} can increase only with the increase of c as mentioned later. For the diffusion mechanism R_{ct} sometimes decreases in highly concentrated regions due to steric hindrance of diffusion. The dependence of R_{ct} on concentration is also a clear indicator of the mechanism in addition to V_{ct}.

    When MV^{2+} is anchored onto a polymer by a covalent bond, the charge transfer is converted into a hopping mechanism as shown by the strong dependence of R_{ct} on c [39] because physical diffusion is prohibited.

A typical example of the charge hopping mechanism is shown by the tris(2,2'-bipyridine) ruthenium(II) complex **3** (Ru(bpy)$_3{}^{2+}$) incorporated into a Nafion membrane on its oxidation from a 2+ to a 3+ complex. The visible absorption spectral change of the Ru(bpy)$_3{}^{2+}$/Nafion after potential steps from 0 to 1.3 V vs. SCE (Fig. 14-5) shows a decrease of the 2+ complex (453 nm) with a concomitant increase of the absorption by the 3+ complex (around 420 nm), whose rate was second order with respect to the concentration [40]. The second-order rate constant for the charge hopping was $k_2 = 1.1 \cdot 10^{-1}$ mol^{-1} s^{-1} (see Experiment 14-1, Section 14.5).

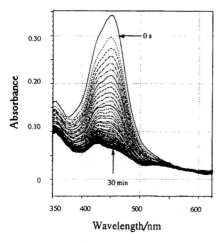

Figure 14-5. Visible absorption spectral change of Ru(bpy)$_9{}^{2+}$ (0.3 M) in Nafion coated on an ITO electrode soaked in a 0.1 M NaClO$_4$ aqueous solution at pH 1.2 after potential step from 0 to 1.3 V (vs. SCE) as measured by a potential-step chronospectrometry.

In contrast to the hopping mechanism in the oxidation to Ru(bpy)$_3{}^{3+}$ in Nafion, its reverse reduction from the 3+ to the 2+ complex takes place by a diffusion mechanism [41]. This is interpreted as an interaction of the charged species (product) with the polymer matrix. On oxidation of the 2+ complex to the 3+ one, the electrostatic interaction between the cationic complex and the anionic polymer becomes stronger, which suppresses diffusion of the molecule. However, on reduction of the 3+ complex to the 2+ one, the interaction becomes weaker, which makes diffusion of the product molecule possible. The first-order rate constant for the diffusion was $k_1 = 7.5 \cdot 10^{-2}$ s^{-1} ,which is about half that for the reduction of MV^{2+}/Nafion.

The question now is how large molecules such as metal porphyrins and metal phthalocyanines behave in a polymer matrix. Kinetic investigation of the electrochemical process of these molecules confined in a polymer matrix gives us important information about this. Metal tetraphenylporphyrin **4** (MTPP) or metal phthalocyanine **5** (MPc) has been incorporated into a polymer membrane by a mixture casting method (cast from their mixed solution onto an electrode), and an electrochemical study was carried out.

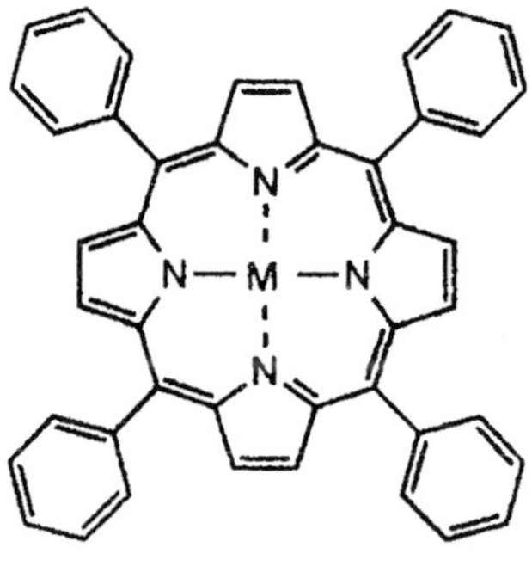

4 (MTPP) 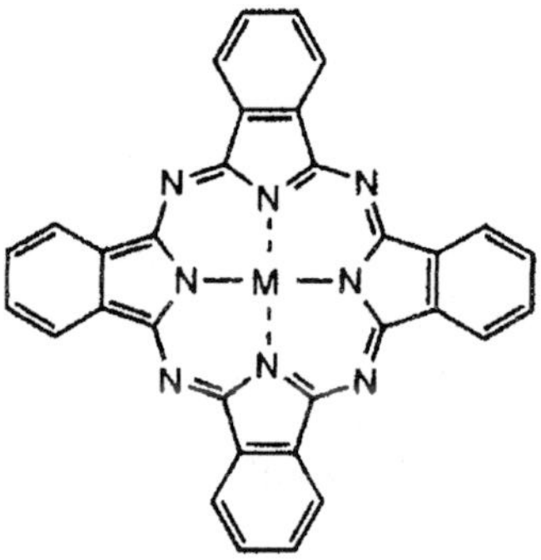   5 (MPc)

CoTPP confined in a Nafion membrane showed only monomeric species. The visible absorption spectral change of the $[Co(III)TPP (-2)]^+$ incorporated in a Nafion membrane has been investigated when it is reduced to $Co(II)TPP(-2)$ [42]. It was interesting to find out by analysis of the rate that, when $Co(II)TPP(-2)$ (non-charged) is oxidized to $[Co(III)TPP(-2)]^+$ in Nafion, or when $[Co(III)TPP(-2)]^+$ is reduced to $Co(II)TPP(-2)$, the charge-transfer mechanism is physical diffusion [42]. The R_{ct} value after saturation was not dependent on c and reaches 1 when c is low, but it started to decrease as c increased. This decrease is ascribable to the steric hindrance of diffusion due to the large size and hydrophobicity of the CoTPP. On the contrary, when $[Co(III)TPP(-2)]^+$ is oxidized to $[Co(III)TPP(-1)]^{2+}$, holes are transported by a hopping mechanism. These results are again interpreted by the degree of interaction between the product (species after accepting charges) and the matrix. When $Zn(II)TPP(-2)$ was dispersed in a poly(2-vinylpyridine) membrane, its oxidation to $[Zn(II)TPP(-1)]^+$ took place by a hopping mechanism [43a], which was explained by coordination of the complex to the pyridine residue at its fifth coordination site.

The visible absorption spectral change of tetrasulfonated CoPc incorporated into poly(4-vinylpyridine-co-styrene) (P(VP-St)) is shown in Fig. 14-6 after a potential step from 0 to –0.50 V (vs. Ag–AgCl) [43(b)]. The rate of initial charge propagation in this reduction process (V_{ct}) was proportional to the concentration (inset of Fig. 14-7) showing that the mechanism is diffusion of the complex.

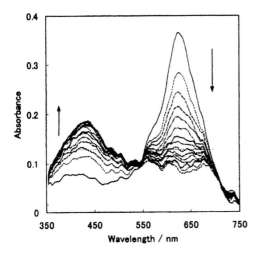

Figure 14-6. Visible absorption spectral change of $(CoPc(SO_3H)_4)/P(VP-St)$ coated electrode after a potential step from 0 to -0.50 V vs. Ag-AgCl.

ZnPc incorporated in a Nafion membrane gave similar but slightly different results. The complex was in equilibrium between monomeric and dimeric species in Nafion. From a kinetic analysis it was shown that only the monomeric ZnPc contributes charge transfer (by a diffusion mechanism), and the contribution of dimeric species to the charge transfer was negligible [44]. The k_1 value for the monomeric complex was $3.3 \cdot 10^{-3}$ s^{-1}, a fairly low value due to the large and hydrophobic molecule.

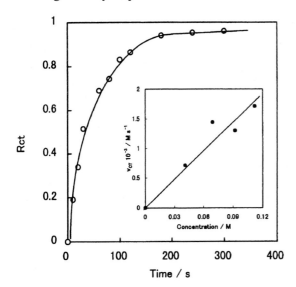

Figure 14-7. Correlation of R_{ct} and V_{ct} (inset) against time in the reduction of $(CoPc(SO_3H)_4)/P(VP-St)$ exhibited in Fig. 14-6.

The above MTPP or MPc/Nafion system gives us important information about the behavior of such large and hydrophobic molecules in Nafion. If the complex is located in hydrophobic columns composed of the Nafion main chain, it could not diffuse because of its large size and hydrophobicity. The high R_{ct} value (nearly 1) even at low concentrations is a clear indication that the complex is located in hydrophilic columns or interlayer regions. After casting an alcoholic/DMF mixture solution of the complex and the Nafion onto an electrode, the solvent is evaporated slowly. During this evaporation, the large molecules would be expelled from the hydrophobic columns into either the hydrophilic columns or the interlayer regions. Since such large and hydrophobic molecules would be unstable and trapped under a delicate balance in these locations, an electrochemical change of its charged state would force it to move from its trapped location looking for another stable site, resulting in diffusion.

Because it is affected by the interaction of a redox molecule and a matrix, charge transfer can involve both physical diffusion and charge hopping. A typical example is an oxo-bridged trinuclear ammine ruthenium complex called Ru-red **6**.

Ru-red

6

This complex has been incorporated by adsorption from its aqueous solution into a precast Nafion membrane, and oxidized electrochemically from the 6+ valent state to the 7+ state (called Ru-brown) [45]. The rate of charge (hole) transfer divided by c (= k_{app}) provided evidence that the charge propagation process is composed of molecular diffusion and charge hopping (see Eq. (14-10)). Although this molecule is large and its charge is high (7+), the hydrophilicity of the ammine ligand would allow its diffusion in the hydrophilic column of the Nafion. The first-order rate constant for the physical diffusion was calculated as $k_1 = 1.0 \cdot 10^{-2}$ s^{-1}, and the second-order one for charge hopping was $k_2 = 6.7 \cdot 10^{-2}$ mol^{-1} s^{-1}. These values are not high because of the large molecular size.

In contrast to the partial diffusion of Ru-red in Nafion, di-μ-oxo-tetrakis(2,2'-bipyridine)diruthenium (7, Meyer's complex) incorporated in a Nafion membrane transported positive charges by a hopping mechanism in its oxidation from Ru(II)-Ru(III) to Ru(III)-Ru(III). The concentration dependence of the k_{app} value (= V_{ct}/c) did not show a simple hopping behavior, but showed typical percolation characteristics for which charge transfer starts to take place beyond a critical concentration (see below). This percolation system is described in Section 14.1.4.

$[(bpy)_2\text{-}Ru\text{-}O\text{-}Ru(bpy)_2]^{3+}$
7 (Meyer's complex)

The rate of charge transfer has been represented by an apparent diffusion coefficient of the charges (D_{app}), which is obtained from the slope of a conventional Cottrell plot (Eq. (14-11)) by chronoamperometry or chronospectrometry [47], where i is the current density, n the number of charges involved in the reaction, F Faraday's constant, c the redox-center concentration and t the reaction time.

$$i = nFcD_{app}^{1/2}/(\pi t)^{1/2} \tag{14-11}$$

This parameter can also be an indication of the mechanism (see also Eqs. (14-1) and (14-2)). Typical D_{app} values for each mechanism are shown in Table 14-1. For a diffusion mechanism, D_{app} is usually somewhat larger (of the order of 10^{-8}–10^{-10} cm^2 s^{-1}) than that for a hopping one ($< \sim 10^{-11}$ cm^2 s^{-1}). In addition, D_{app} is independent of c for a diffusion mechanism, but strongly dependent on c for a hopping mechanism. If D_{app} decreases with an increase of c, it follows a diffusion mechanism. Some behaviors of D_{app} values are shown below.

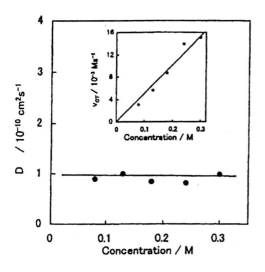

Figure 14-8. Apparent diffusion coefficient of Ru(bpy)$_3^{3+}$ on reducing to 2+ complex by a potential step from 1.5 to 0 V vs. SCE. The inset shows the initial reduction rate vs. c.

In the reduction of a Ru(bpy)$_3^{3+}$ to a 2+ complex in Nafion, D_{app} values are independent of c because of its diffusion mechanism (Fig. 14-8) [41]. However, in the oxidation of Ru(bpy)$_3^{2+}$ to a 3+ complex D_{app} is strongly dependent on c, typical for a hopping mechanism [41].

In the reduction of [Co(III)TPP(-2)]+ to Co(II)TPP(-2), D_{app} decreased with an increase of pH in Nafion, but in poly(4-vinylpyridine-co-styrene) it was almost independent of pH [42].

The following conclusions have been obtained for charge transfer:

1. **Charge-transfer mechanism**: The mechanism is determined by the interaction of the matrix with the redox species that accepted charges. When this interaction is weak and the product can diffuse in the matrix, charge is transported by a diffusion mechanism, but when the interaction is strong and the product cannot diffuse, charge is transported by a hopping mechanism. The mechanism can therefore be different for transport of either positive or negative charges even for the same redox center/matrix system. A typical example of this is the $Ru(bpy)_3^{2+}$/Nafion system.

2. R_{ct} (the fraction of redox molecules accepting charges): For a diffusion mechanism, the R_{ct} value after saturation of charge transfer is independent of the redox-center concentration (c), and usually high, reaching 0.8 to 1. For a hopping mechanism, R_{ct} is strongly dependent on c and can increase only at high concentrations.

3. D_{app}: For a diffusion mechanism, the apparent diffusion coefficient of the charge (D_{app}) is independent of c, and usually shows a high value of the order of 10^{-8}–10^{-10} cm^2 s^{-1}. For a hopping mechanism D_{app} is dependent on c, usually shows a low value of the order of less than 10^{-11} cm^2 s^{-1}, and can increase only at high concentrations.

4. **Diffusion of large molecules**: Large and hydrophobic molecules such as metal tetraphenylporphyrins and metal phthalocyanines incorporated by a mixture-casting method are located mainly in the hydrophilic columns or interlayer regions of Nafion, and can diffuse there exhibiting fairly high D_{app} values.

5. **Redox molecules covalently attached to polymers**: Charge transfer takes place only by a hopping mechanism even when the corresponding monomeric molecule transports charges by a diffusion mechanism.

No example has yet been found in which charge is transported first by diffusion of redox molecules and then by charge hopping between them.

14.1.3 Distance of Charge Hopping with Bounded Motion

The charge-hopping distance is an important parameter for a charge-hopping mechanism. This is true not only for the present polymer systems but also for biological ones for which the electron-hopping distance is a crucial parameter for investigating the process. As mentioned before, the charge-hopping distance can be estimated from either Eq. (14-1) or (14-2) (in the presence of a diffusion mechanism) or from Eq.(14-3) or (14-4) (in the absence of diffusion).

Charge transfer by a hopping mechanism is strongly influenced by bounded motion of the redox molecule, which is an oscillation of the molecule at its confined position in the matrix. If such bounded motion does not take place at all, there will always be isolated molecules and clusters (a group of molecules present within charge-hopping distance of each other) to which charges cannot be transported. In this case charge transfer follows a percolation process as mentioned in Section 14.1.4 below.

On the contrary, if bounded motion of the molecules occurs, the problem of isolated molecules and clusters becomes negligible (depending on the degree of the bounded motion) because of a continual reorganization of the isolated molecules and clusters, so that the system can be treated approximately by a probability function.

When assuming a random distribution, the probability density $P(r)$ $(nm^{-1})$of the distance (r) between the nearest-neighbor molecules (center-to-center) is represented by Eq. (14-12),

$$P(r) = 4\pi r^2 N_A c 10^{-24} \exp[-4\pi(r^3 - s^3)N_A c 10^{-24}/3] \tag{14-12}$$

where N_A is Avogadro's number (mol^{-1}), and s (nm) the contact distance (center-to-center) between molecules. When the redox center is incorporated only in the hydrophilic columns of Nafion, for instance by adsorption from an aqueous solution, αc is used instead of an average concentration c by considering the fraction of the hydrophilic columns (see above).

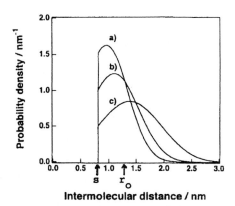

Figure 14-9. Probability density $P(r)$ versus the distance between the nearest-neighbor molecules in a matrix as obtained from Eq. (14-12).

$P(r)$ vs. r can be depicted as a function of c and is shown in Fig. 14-9. When assuming a charge-hopping distance (r_o) between molecules (center-to-center) within which charges can hop, the fraction of molecules that can accept charge (R_{ct}) is obtained by integration of the $P(r)$ function from s to r_o. The fraction of molecules that cannot accept charge is then represented by $1 - R_{ct}$, which is obtained as Eq. (14-13) [37,38].

$$1 - R_{ct} = \int_{r_o}^{\infty} P(r)dr = \exp{-4\pi(r_o^3 - s^3)N_A c 10^{-24}/3} \tag{14-13}$$

This equation can be rewritten as Eq. (14-14).

$$-\ln(1 - R_{ct}) = 4(r_o^3 - s^3)N_A c 10^{-24}/3 \qquad (14\text{-}14)$$

In the oxidation of Ru(bpy)$_3^{2+}$ to Ru(bpy)$_3^{3+}$ in Nafion the change of $1 - R_{ct}$ with c was obtained as a linear relationship [41], indicating that this treatment is reasonable. From the slopes of these lines r_o was calculated as a function of reaction time. From a saturated value of r_o after a long time, the final charge-hopping distance was obtained as 1.74 nm. In order to see the change of r_o with time more clearly, log r_o was plotted against time. The plots can be divided into three major lines; the first linear part is considered to represent charge propagation in a film by semi-infinite diffusion, the second part should be dominated by a finite diffusion, and the third part would be contributed mainly by bounded motion of the molecules. The r_o values in the first and second part of the linear relationship do not have any meaning since charges propagate from the electrode surface up to the film/liquid interface in these regions. Extrapolation of the third line to the y-axis can give an estimated charge-hopping distance (1.47 nm in this case) without bounded motion. The bounded motion distance is then calculated to be 0.27 nm [41].

Charge hopping between redox centers in biological systems is mediated by protein main chains or aromatic amino acid residues. It is an interesting question whether such mediation can also take place in synthetic systems. In order to investigate this, charge hopping was studied in the presence of amino acid residue models. In the presence of a model compound, either 3-methylindole (IND, tryptophan residue model), 2-methylimidazole (IMD, histidine residue model) or p-cresol (CRE, tyrosine residue model) for a system Ru(bpy)$_3^{2+}$/Nafion, the charge-hopping distance was remarkably increased (in the absence of mediator 1.3 nm, and in the presence of mediator 1.8 to 2.1 nm) [40], while the bounded motion distance was not changed (0.2 to 0.3 nm). The amino acid itself was not effective for mediation. It was inferred that the residue models mediate charge hopping by a kind of superexchange mechanism. Such mediation effects of amino acid residue models were also observed in electrocatalysis as described in Chapter 13 and in photoexcited charge transfer as described in Section 14.2.

14.1.4 The Percolation Process for Charge Hopping without Bounded Motion

When charge hopping takes place with sufficient bounded motion or diffusion of central redox molecules, the initial rate of charge transfer shows second-order dependence on the concentration as described in the last section. However, if charge hopping takes place without any bounded motion or with only a small degree of it, the process cannot be analyzed by a simple bimolecular process like the above treatment since there are isolated molecules and clusters that do not contribute to the charge transfer depending on the redox-center concentrations. Such process has been analyzed by a so-called percolation theory [48–50].

The probability of charge propagation (percolation) (P) against redox-center concentration is expressed by a sigmoidal curve according to the percolation theory. In order to design devices it is important to know the distribution of propagated charges especially in a percolation process when the propagated charge density is low. For this purpose R_{ct} was calculated for a static percolation process by a Monte Carlo simulation instead of a conventional P value (Fig. 14-10) [51]. When the concentration (represented by occupied probability p of the redox center) is low, charge propagation does not take place because the distance between molecules is too large for the charge to hop. When the concentration increases, the charge starts to propagate, and R_{ct} increases steeply at a threshold concentration c_{th}.

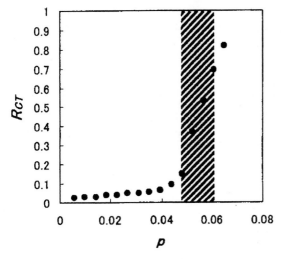

Figure 14-10. R_{ct} versus redox-center concentration (represented by occupied site, p) calculated by percolation theory. The p value of the middle point of the shaded area is regarded as c_{th}.

According to the simulation, the charge propagation was visualized as shown in Fig. 14-11 [51]. These figures show the distribution of propagated charges after saturation at low and high redox-center concentrations. When c is low, the propagated charges are localized only near the electrode surface, and when the redox-center concentration increases the charges start to grow towards the matrix/liquid interface.

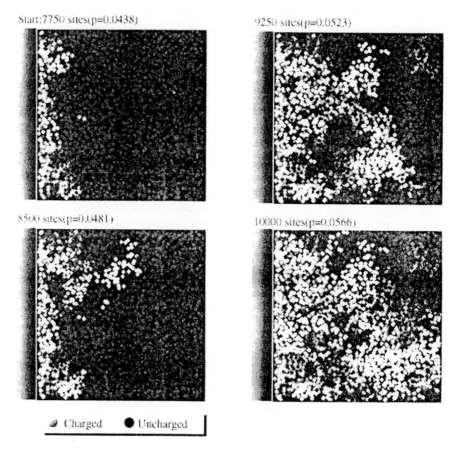

Figure 14-11. Images of the distribution of propagated charges in a polymer membrane with increasing redox-center concentrations as visualized by a Monte Carlo simulation. The left side is the electrode and the right side the solid/liquid interface [51].

However, unlike the behavior expected from the percolation theory, charge propagation can start slowly from low concentration regions in actual systems [48,49] because of a dynamic percolation behavior [50] or bounded motion in which limited site motion exists.

Charge transfer has been investigated for ionically conductive room temperature melts composed of derivatized metal-tris(2,2'-bpy)complexes with oligo(ethylene glycol) **8** tails and dissolved Li electrolytes [48].

$[M(bpy(CO_2MPEG350)_2)_3](ClO_4)_2$

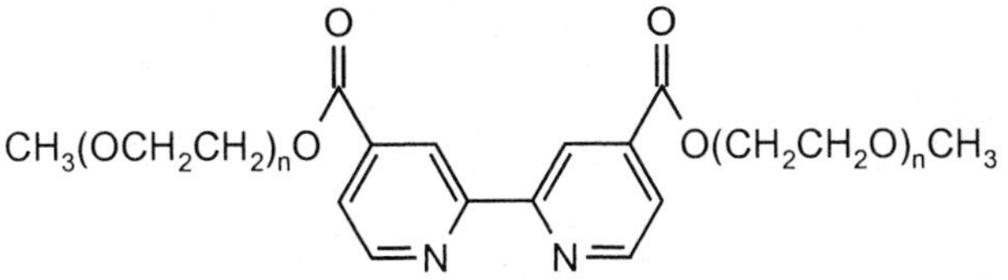

$CH_3(OCH_2CH_2)_nO$ ... $O(CH_2CH_2O)_nCH_3$

n = 7 on average

M = Fe(II), Co(II), Ni(II)

8

The D_{app} value for the $Fe^{2+/3+}$ couple against mole fraction of Fe in 8 exhibited a distinct deviation from percolation-theory behavior at low Fe concentrations, which was ascribed to physical motion of the melt lattice over a small number of molecular dimensions.

It should be noted here that dynamic percolation does not simply mean that the charge-hopping distance is lengthened by bounded motion or by diffusion of redox-center molecules. If such dynamic motion only lengthens the hopping distance, the main feature of Fig. 14-10 does not change except to decrease the c_{th} value. Rather, such dynamic motion brings about rearrangement of isolated molecules and clusters of the redox molecules with time, leading to deviation of the R_{ct} against c curve from theory. It is inferred that this effect causes deviation especially at low concentration regions, as reported for many actual systems. A theoretical analysis of such dynamic percolation behavior has not yet been reported.

14.1.5 Electrocatalysis – a Combination of Charge Transfer and Catalysis

Electrocatalysis is an important application of electron-induced processes. In electrocatalysis the catalyst has at first to accept charge(s) from the electrode, and thereafter catalysis can take place. Enzyme-immobilized electrodes are typical examples used for various biosensors as well as for investigation of fundamental biocatalysis. The enzymatic active center is often located inside a protein molecule, so that mediation of charges from the electrode surface to the active center is important. Viologens, metallocenes and other metal complexes have been used as such mediators.

Poly(N-isopropylacrylamide-co-vinylferrocene) gel **9** acted as an electron mediator between an oxidase and an electrode [52(a)]. Swelling and shrinking of this polymer gel is temperature dependent; it swells below 20 °C and shrinks above 28 °C because of a temperature-dependent hydrophobic interaction of the isopropylacrylamide groups.

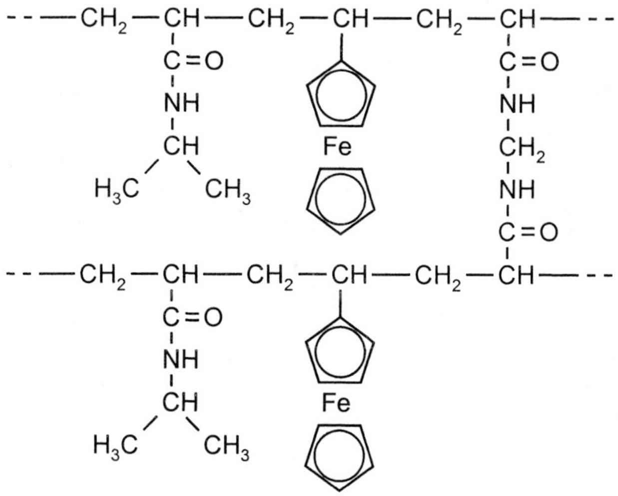

$$--\ --CH_2-CH-CH_2-CH-CH_2-CH--\ --$$

(structure of polymer 9, showing repeating units with C=O, NH, CH(CH$_3$)$_2$ isopropylamide side groups and ferrocene (Fe) units)

9

This polymer was coated on a glass electrode, and glucose oxidase was loaded from its aqueous solution at the temperature where the gel swells (below 10 °C). At a temperature above 35 °C the gel shrank and the enzyme was unloaded. When the enzyme is loaded, the modified electrode gives an anodic current in the presence of glucose in the aqueous phase, owing to electrocatalytic glucose oxidation catalyzed by the enzyme.

A fructose biosensor based on D-fructose dehydrogenase (FDH) was developed by using tris(1,10-phenanthroline)cobalt (II) complex (Co(phen)$_3$) as a redox mediator [52(b)]. A self-assembled monolayer (SAM) of cystamine was formed on a gold electrode and then FDH was covalently bound on the SAM using glutaraldehyde which cross-links the amino acid groups of cystamine with the FDH. With the use of the Co(phen)$_3$ mediator, the catalytic current was increased by the addition of D-fructose.

A poly(allylamine) hydrogel containing Os(bpy)$_2$Cl groups and cross-linked with glucose oxidase was used as an amperometric enzyme electrode; volume, mass and viscoelastic changes of the hydrogel were investigated by electroacaustic impedance at 10 MHz [53]. Swelling of the film during Os(II) oxidation resulted in mass uptake and a decrease of the shear modulus. These quantities depend only on the degree of polymer oxidation and not on film history or kinetics.

Nanometer-sized monolayer-protected gold clusters functionalized with anthraquinone by using 1-(1,3-dithiapropyl)anthracene-9,10-dione were used to mediate electroreduction of 1,1-dinitrocyclohexane [54a]. The enhancement in catalytic efficiency for cluster-bound anthraquinone compared to its monomer was ascribed to the smaller diffusion coefficient of the cluster-bound anthraquinone, resulting in spatial compression of the reaction zone next to the electrode. Electrocatalytic oxidation of ascorbic acid at an [Os(bpy)$_2$(poly-4-vinylpyridine)Cl]Cl-modified electrode has been achieved [54b], and this was studied as an electrochemical sensor for biological substances.

Water oxidation in the photosynthesis of green plants is the first chemical reaction of most biological activity since photosynthesis is the energy source for almost all living things. Artificial water oxidation is important not only for simulating the photosynthetic oxygen-evolving center (OEC) but also for future artificial photosynthesis which is very attractive for creating energy resources from water and solar energy [55–61]. Molecular catalysts for water oxidation have been studied both by an electrocatalytic method using a polymer-coated electrode incorporating the water oxidation catalyst and by chemical methods using Ce(IV) as oxidant [55]. Please refer to Section 13.2.3.1 for details.

Dihydrogen (H$_2$) formation by proton reduction is another fundamental electrocatalytic reaction important for water electrolysis as well as for future solar-energy conversion systems [3,9,62]. Electrocatalysis by neat CoTPP deposited on a graphite electrode for proton reduction has been reported [63]. It was found that metal-TPP complexes show high electrocatalytic activity when incorporated into a polymer membrane such as Nafion coated on an electrode [64–66]. An increase of the catalyst concentration in the polymer matrix reduced the catalytic activity. This is ascribed to (1) unfavorable charge transfer with high concentration since the electron transfer takes place by a diffusion mechanism (see above), and (2) an unfavorable situation for intermediate formation such as a catalyst-H-H-catalyst structure at high concentrations. CoPc and its octacyano and tetra(sulfonic acid) derivatives incorporated into a poly(4-vinylpyridine-co-styrene) also showed high activity (see Experiment 14-2, Section 14.5) [66]. Refer also to Section 13.2.3.2.

14.2 Charge Transfer in a Photoexcited State – Photochemistry

14.2.1 Fundamental behavior

Electron transfer between molecules is the most essential process in biological systems [1,2]. Electron transfer in biological substances such as enzymes is very rapid, so that it is in many cases investigated by utilizing electron transfer in a photoexcited state with a laser flash technique. By this method one single step of an electron transfer can be measured. For this purpose a specific position of an enzyme is modified with a redox-center molecule such as a metal complex, and electron transfer is studied with laser flash photolysis.

The central metal ion of the iron-heme in cytochrome *c* or myoglobin was substituted by Zn. Ru ion was coordinated to a surface imidazole of a histidine residue, and then the electron transfer from the photoexcited Zn porphyrin to the Ru moiety was investigated [67,68]. The relation between the rate constant for the electron transfer k_{et} and the spatial distance between the redox centers (R) was represented by the known Eq. (14-15) showing that the electron transfer takes place via the spatially shortest pathway.

$$-\ln k_{et} = R \qquad\qquad\qquad (14\text{--}15)$$

Marcus theory [69] predicts that the rate of non-adiabatic electron transfer with weak exergonicity shows a bell-shaped behavior with respect to the free energy difference ($-\Delta G$) between electron donor and acceptor giving a maximum rate when ΔG equals the reorganization energy of the medium. Extensive studies have been published especially on intramolecular electron transfer in this regard, and we shall not review this work here. Instead, this section will review intermolecular electron transfer between a photoexcited redox molecule and the ground state one incorporated in a polymer.

Electron (or charge) transfer reaction of a photoexcited molecule can be investigated by the following methods:

1. Measurement of a steady state UV-vis absorption spectrum when the back electron transfer is slow, so that the equilibrium under irradiation shifts to the product side. Such a reaction is called a photoredox system. There are not many examples of this type. When the back electron transfer is rapid, the presence of some sacrificial donor or acceptor is effective in shifting the reaction to the product side and therefore to monitor the reaction.
2. Monitoring a transient absorption spectrum with a laser flash photolysis method especially when the rates of electron transfer and the back reaction (recombination) are rapid.
3. Measurement of emission intensity or emission decay (lifetime) when the photoexcited state emits light (luminescence, either fluorescence or phosphorescence).

14.2.2 Photoexcited State Electron Transport in Solid Polymer Matrixes

Electron transfer from the photoexcited $Ru(bpy)_3^{2+}$ (3) to MV^{2+} (2) in a polymer solid phase was reported for the first time by utilizing a cellulose (paper) matrix adsorbing the Ru complex, MV^{2+}, and a sacrificial electron donor, ethylenedia-minetetraacetic acid (EDTA) [70]. Irradiation of the cellulose by visible light (λ_{max} 450 nm) gave a blue-colored MV^+ formed by electron relay from $Ru(bpy)_3^{2+*}$ to MV^{2+} and then from EDTA to the produced $Ru(bpy)_3^{3+}$ (Eq. (14-16)). This is therefore a combined electron-transfer system, the first step being excited state electron transfer, and the second step ground state electron transfer.

$$\text{EDTA} \xrightarrow[\text{2}^{nd}\text{ step, e}^-]{} \text{Ru(bpy)}_3^{2+} \xrightarrow{\text{1}^{st}\text{ step, h}\nu,\text{ e}^-} \text{MV}^{2+} \qquad (14\text{-}16)$$

The electron transfer from $\text{Ru(bpy)}_3^{2+}*$ to MV^{2+} can also be monitored through the quenching of the red phosphorescence (E_{max} 603 nm) from the $\text{Ru(bpy)}_3^{2+}*$ by MV^{2+}.

Polystyrene-pendant Ru(bpy)_3^{2+} was prepared, and an electron relay took place from EDTA present in the polymeric Ru-complex solid film to MV^{2+} present in a water phase in contact with the polymer film via an excited state of the Ru complex forming the MV^+ cation radical [71]. Chelating resin beads containing pendant $-\text{CH}_2\text{N(COOH)}_2$ groups and adsorbing Ru(bpy)_3^{2+} and MV^{2+} produced MV^+ on irradiation with visible light, with the imino dicarboxylic acid groups working as electron donors [72]. The interior of the opaque beads also turned blue-colored with the MV^+, showing that the electrons migrate from the outer surface to the inside of the beads via MV^{2+} most probably by hopping.

The behavior and electron transfer of photoexcited Ru(bpy)_3^{2+} incorporated in cellulose [73], cellophane [74], poly(methacrylic acid) [75] and layer silicates [76] matrixes have been investigated by both steady state and time-resolved methods. An important feature of this kind of solid matrix system is that the redox molecules do not move at all or move only slightly during the short lifetime of the photoexcited state, in contrast to free diffusion in a liquid. Long-range electron transfer from a series of aromatic amines to a series of photoexcited $\text{Ru(derivatized bpy)}_3^{2+}$ complexes in a rigid polycarbonate medium has been investigated [77], and a tunneling distance up to 1.2 nm (edge to edge) was reported.

Back electron transfer (charge recombination) is normally as fast as the photoinduced electron transfer, so that the charge separated state is only short-lived. However, long-lived charge-separated states have been reported following photoinduced electron transfer in a macroreticular resin, Amberlite XAD-8, adsorbing poly(vinylcarbazole) (PVCz) and 1,2,4,5-tetracyanobenzene (TCNB) as electron acceptor [78]. The charge-separated state formed by electron transfer from excited PVCz to TCNB was found more than 8 h, which was ascribed to both a hole-migration process along the polymer chain and a hole-trapping process resulting in a large interionic distance. Photoinduced electron transfer along linked porphyrin arrays doped in a poly(methyl methacrylate) film has been investigated [79]. The change in dielectric dipole moment suggested that the electron transfer takes place along the linked porphyrin arrays.

A Ru complex, Ru(dcbpy)$_3^{2+}$ where dcbpy is 4,4'-dicarboxyl-2,2'-bipyridine, was covalently anchored onto an insoluble Sephadex, and heterogeneous electron-transfer quenching of the excited state in the solid phase was studied in water containing MV^{2+} or N,N-dimethylaniline [80]. The electron transfer was very efficient in polar solvents. Electron transfer of excited state Methylene Blue incorporated in a Nafion membrane by Fe^{2+} has been reported [81]. The electron-transfer process involved excitation of the dye dimeric species or excimers.

14.2.3 The Mechanism of Photoexcited State Electron Transport in Polymer Solid Matrixes

For the mechanism of photoexcited state electron transfer in polymer matrixes, refer also to Section 13.2.3.4.

Measurement of the emission intensity and decay of a photoexcited probe gives important information on the mechanism of electron transfer, the mobility of the probe, and the nature of the microenvironment around the probe. Since electron transfer from or to an excited state of a probe quenches the emission, quenching   reflects the electron transfer. The mechanism of electron-transfer quenching can be roughly divided into three extreme cases (Fig. 13-11). One is a dynamic quenching for which donor and acceptor molecules diffuse freely, collide and react. The other is a static quenching for which the molecules do not move, and when a quencher (Q) molecule exists in a quenching sphere around the probe, it quenches the emission (electron transfer takes place). A system involving both the dynamic and static mechanisms is also possible.

Such quenching is analyzed by plotting relative emission intensity (I_0/I, where I_0 is the emission intensity without Q, and I is that with Q) against Q concentration according to the Stern–Volmer equation (Eq. (13-7)) [82,83] (called Stern–Volmer plots [82]).

The Stern–Volmer plots based on the lifetime of the excited state (τ, the reciprocal of the first-order decay constant of the excited state) is also an important means of determining the electron-transfer mechanism. For a complete dynamic quenching case, both the I_0/I and τ_0/τ plots are linear and fall on the same line (Fig. 13-11(a)). For a complete static quenching case, the I_0/I plots are upwardly curved according to the Perrin equation [84] represented by Eq. (13-8) under higher concentration conditions. For this static quenching the plots τ_0/τ do not have a slope (Fig. 13-11(b)), i.e. the excited state involving quencher(s) in its quenching sphere does not emit at all, but the excited state involving no quencher in its quenching sphere emits the same as the system without quencher.

For polymer solid phases there are two major problems as follows:

1. The microenvironment of the excited state is not homogeneous, but differs

depending on the location of the probe, which affects emission decay and quenching of the excited state. This is reflected in a multiexponential decay of the excited state.

2. The (electron-transfer) quenching mechanism can be both dynamic and static, which makes the I_0/I plots complicated.

The equations to analyze electron transfer quenching were derived based on the following principles [85]:

a. The number of different sites; 1-site or 2-sites?
b. Dynamic or static quenching or a combination?
c. The incorporation equilibrium of Q into the quenching sphere around a probe; a one-step equilibrium or a multi-step one?
d. The rate of static quenching; is it very fast so that emission is negligible, or should the emission be considered since it is competitive with the quenching?
e. For a multi-step equilibrium between Q and probe; is the static quenching rate proportional to the number of Q in the quenching sphere, or independent of the number of Q?

Based on these factors 11 models and corresponding equations have been proposed for a 1-site model, and 2 models for a 2-site model [85].

For the completely static quenching case, Eq. (14-17) was derived considering the size of the probe by assuming that the probes are distributed randomly in a matrix[86],

$$I_0/I = \exp[-4\pi(R_0^3 - s^3)N_A 10^{-24}[Q]/3] \tag{14-17}$$

where R_0 is the electron-transfer distance (nm, center-to-center of donor and acceptor), s is the contact distance between donor and acceptor (nm) and [Q] is the Q concentration (mol dm^{-3}) in the matrix.

Electron-transfer quenching of the photoexcited Ru complex by MV^{2+} has been investigated in a polysiloxane film containing dispersed Ru(bpy)$_3^{2+}$ and MV^{2+}, [86] (see Section 13.2.3.4). The Stern–Volmer plots for the relative emission quantum yield (equal to I_0/I) and the relative lifetime of the excited state (τ_0/τ) against MV^{2+} concentration showed that the electron-transfer quenching takes place by a static mechanism. By applying Eq. (14-17) ($s = 0.82$ nm), the electron-transfer distance R_0 was obtained as 1.4 nm. This is a similar value to the ground-state charge-transfer distance obtained by an electrochemical method, and also a reasonable value considering the electron-transfer distance in biological systems is often reported to be around 1.3 nm.

In biological systems proteins mediate electron transfer. This mediation takes place either through bonds in the protein, or through space, when aromatic amino acid residues are considered to mediate the electron transfer in the space.

The aromatic amino acid residue model, 3-methylindole as a model of tryptophan residue, was codispersed in the above $Ru(bpy)_3^{2+}/MV^{2+}$ system. In the presence of the mediator model the electron transfer was greatly enhanced [86]. An equation taking into account mediated electron transfer was derived (see [86]), and applied to the results giving mediated electron transfer of 2.7 nm, which is about twice that without the mediator. Tryptophan itself or a model of tyrosine (*p*-cresol) was not effective for the mediation.

Electron transfer quenching of excited $Ru(bpy)_3^{2+}$ by MV^{2+} incorporated in a poly(ethylene oxide) film takes place by both static and dynamic mechanisms, unlike the above polysiloxane film, and the electron-transfer distance was 1.7 nm [87].

These and other results showed that electron transfer in a polymer solid usually takes place by a static mechanism, but a dynamic mechanism can also be involved when diffusion or bounded motion of the redox molecules in the polymer matrix is possible. The electron-transfer distance in a photoexcited state studied by a photochemical method is in principle comparable to that in the ground state studied by electrochemical methods (see Section 14.1).

14.2.4 Photoelectrochemical Devices Utilizing Charge Transfer in Photoexcited and Ground States

Photoelectric and electronic devices can be made that utilize charge transfer between redox molecules in their photoexcited and ground states taking place in polymer matrixes. These devices are attracting attention for future photoenergy conversion devices aiming at creating new energy resources from solar energy [2,3,4,7,9,62,88,89,90]. For these devices charge transfer in both the excited and ground states is important.

Excited-state reactions can be classified into two major classes as follows depending on the rate of back charge transfer (charge recombination) [89,90]:

1. Photoexcited charge-transfer reactions for which the back charge transfer is rather slow, shifting the equilibrium of the system to the product side under illumination. This type of reaction is called a "photoredox reaction" [89,90], and a typical example is the reaction of the photoexcited state of thionine (dye) with Fe^{2+} as an electron acceptor shifting the equilibrium to the product side (the color of the aqueous solution is violet in the dark and colorless under illumination, and this color change is reversible). For this kind of reaction couple, when two electrodes are dipped in the solution of the reaction couple, a photopotential and photocurrent can be induced by introducing an asymmetric factor in the cell. Such an asymmetric factor is, for instance, two different electrodes or different illumination of the electrodes (e.g. illumination at one electrode while the other electrode is kept in the dark).

2. In the most of the excited-state charge-transfer reactions, back charge transfer is very rapid, so that the equilibrium does not shift to the product side under illumination. In this case a photoresponse cannot be obtained from a simple solution of the reaction couple. The $Ru(bpy)_3^{2+}/MV^{2+}$ couple is a typical example.

However, it has been found that if one or both components are coated on an electrode, a photoresponse can be obtained at the modified electrode since electric communication between the photoreaction product and the electrode becomes possible, competing with the back charge transfer [9,89,90,91]. Thus, polymer-pendant $Ru(bpy)_3^{2+}$ (water insoluble) was coated on a graphite electrode, and a photoresponse was obtained in the presence of MV^{2+} in an aqueous phase in contact with the coated electrode. Bilayer membranes composed of polymer-pendant $Ru(bpy)_3^{2+}$ (10) and polymer-pendant MV^{2+} (11) sandwiched between Ag and conductive glass (NESA) electrodes (configuration Ag/polymer $Ru(bpy)_3^{2+}$ (10)/polymer MV^{2+} (11)/NESA) generated a photoresponse (see Experiment 14-2, Section 14.5 for more details [192]).

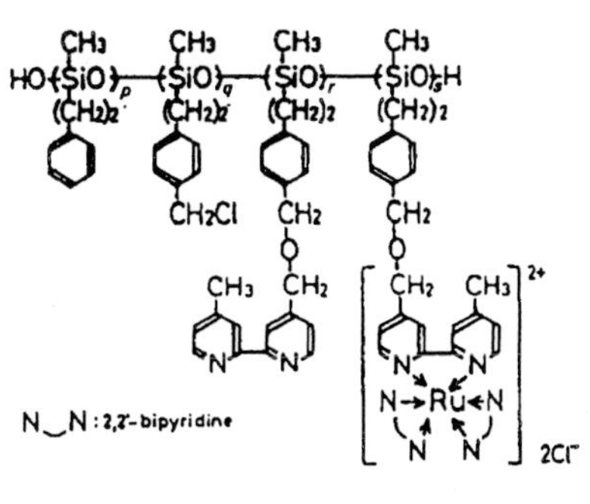

10

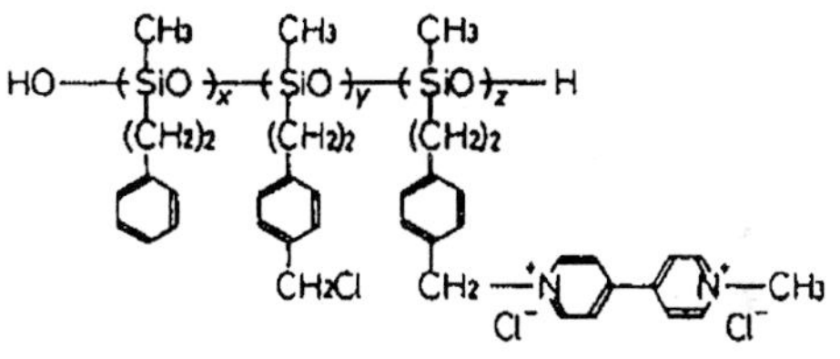

11

In order to obtain a photoresponse by combination of charge transfer in both photoexcited and ground states, it is important to arrange the components in a donor (D)–sensitizer (P)–acceptor (A) sequence. Langmuir–Blodgett (LB) films are promising matrix candidates for arranging the components in a sequential way. A LB film containing donor (ferrocene)–sensitizer (pyrene)–acceptor (viologen) sequence coated on an ITO electrode induced photocurrents [93]. A photoelectrochemical response was obtained by a polymer LB film containing polymer-pendant $Ru(bpy)_3^{2+}$ (12) in the presence of a donor (thiosalythilic acid) in an aqueous solution in contact with the LB film [94(a)]. Since such a LB film can be used only with its monolayer for this photoinduced charge relay, the light absorbance is so small that the photocurrent is only of the order of 20 nA cm^{-2}, the conversion of photon energy into electricity being 0.8–1.1%. It is therefore important to introduce the concept of a roughness factor (see Section 13.3) to enhance the light absorption.

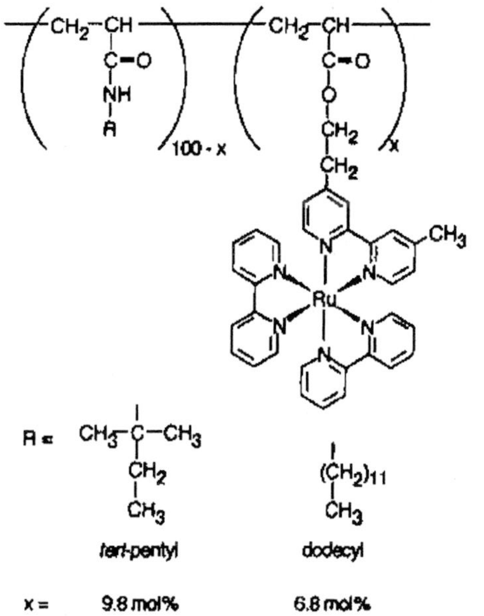

12

A self-assembled monolayer (SAM) composed of a mercaptane compound and gold electrode is attracting attention to organized molecules for investigating the photoelectrochemical response of a D–P–A model coated on an electrode. SAM monolayers of a ferrocene, porphyrin and fullerene triad on a gold electrode (Fig. 14-12) has been investigated [94 (b,c)]. In the presence of an electron acceptor (MV^{2+}) in the liquid phase and in the presence of dioxygen, the quantum yield for the photocurrent reached 20–25%. As shown in the potential diagram of Fig. 14-11, the photoinduced electron flow takes place from the electrode to the O_2.

A thin film composed of fragments of a purple membrane containing bacteriorhodopsin (bR) formed on a conductive glass (NESA) by a LB method generated a transient and rectified photocurrent in the cell configuration, NESA/bR LB membrane/electrolyte aqueous gel/Au [95,96]. Transmembrane charge transfer in a vesicle-based photocatalytic system [97], photo-gated electron transfer in two-component self-assembled monolayers [98], and construction of artificial photosynthetic reaction centers on a protein surface [99], are also interesting and promising topics although they are do not involve synthetic polymers. Details may be obtained from the references.

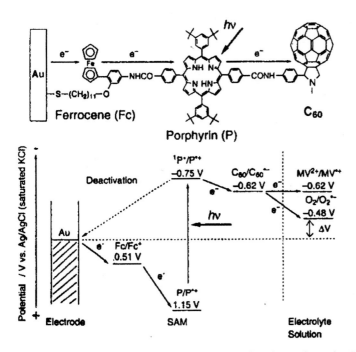

Figure 14-12. Photoinduced electron flow from the electrode to the O_2 in the liquid phase in the SAM triad compound composed of ferrocene, porphyrin and C_{60} in the presence of MV^{2+} and O_2 in the outer aqueous phase.

A suspension of platinized TiO_2 photocatalyst particles in water sensitized by surface-coated polymer-pendant $Ru(bpy)_3^{2+}$ generated H_2 with visible light irradiation in the presence of a sacrificial electron donor (EDTA) [100]. Sensitization of a TiO_2 polycrystalline electrode (see also Section 13.3) was achieved by adsorbed $Ru(bpy)_3^{2+}$, and a highly efficient photocurrent was obtained with more than 40% quantum yield in the presence of hydroquinone in water [101]. A TiO_2 electrode coated with a Nafion film incorporating $Ru(bpy)_3^{2+}$ also gave a sensitized photocurrent [102]. This sensitized semiconductor system was applied to a film composed of nanometer-sized TiO_2 particles adsorbing a $Ru(4,4'$-dicarboxy-$2,2'$-bipyridine)$_2(NCS)_2$ complex soaked in an organic medium dissolving I_3^-/I^- redox couple, which yielded photocurrents with J_{sc} 11.2 mA cm^{-2}, V_{oc} 0.68 V, fill factor FF 0.68, and conversion efficiency 7.1% [103,104] (see also Section 13.3). Later about 10% conversion efficiency was claimed which can compete with an amorphous silicon solar cell. These solar-cell reactions consist of rapid electron injection from the excited dye to TiO_2 and subsequent reduction of the oxidized dye by I^-. It is attracting a great deal as one of the promising approaches to fabricating low-cost solar cells as a way of solving the energy resource problem. The key point for this type of photoelectrochemical cell mimicking a photosynthetic system is a combination of charge transfer in both excited and ground states by utilizing redox molecules confined in a matrix.

The combination of charge transfer in both ground and photoexcited states is one of the most important concepts for designing future artificial photosynthetic systems [1–3,9,62,89,90,105,106] to create renewable energy resources from solar energy and water.

14.3 New Aspects and Future Scope of Charge Transfer in Polymeric Quasi-Solids

The solid phase of polymeric materials is of great use for fabricating devices, as shown in the previous sections. However, one problem remains in utilizing polymeric solid materials, that is, the slow propagation of charges in solid polymer matrixes whether through diffusion of the metal complex or charge hopping between them, D_{app} (the apparent propagation coefficient of the charges) being the order of 10^{-8} to 10^{-13} cm^2 s^{-1}. To overcome this problem, the present author has found that in tight and elastic solids composed of a small amount of polysaccharide and a large amount of water, diffusion of molecules, i.e. diffusion of charges, can take place as rapidly as in pure liquid water [107]. We call this a "molecular reactor", since chemical reactions, including electrochemical and photochemical ones, can take place in this solid just as in liquid water but without any flask or cell.

Chemical reactions can take place in a typically solid, but in such case the reactants do not diffuse at all or move only slightly, so that chemical reactions in a solid are usually entirely different from those in a pure liquid. If chemical reactions can occur in a stable and tight solid in the same way as in a pure liquid, it leads to new and wide applications such as chemical reactors, microreactors, electrochemical/photochemical sensors, microanalysis, combinatorial chemistry, laboratories on chips, etc. We have now succeeded in preparing tight and stable solids from polysaccharides containing a large excess water by carefully choosing the conditions, and found that electrochemical and photochemical reactions in these solids can take place the same as in pure water. κ-Carrageenan[1] (13) and agarose[2] (14) are examples of the polysaccharides we used.

The electrochemical behavior of redox molecules in thin polymer films and gels coated on an electrode has been investigated as mentioned previously, but they have been studied in the presence of an outer electrolyte solution except for the work reported in [108], which uses a micro-electrode in the absence of an outer electrolyte solution.

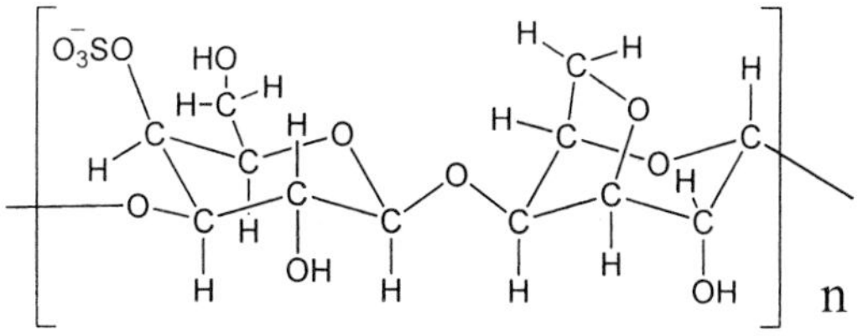

13 (κ-carrageenan)

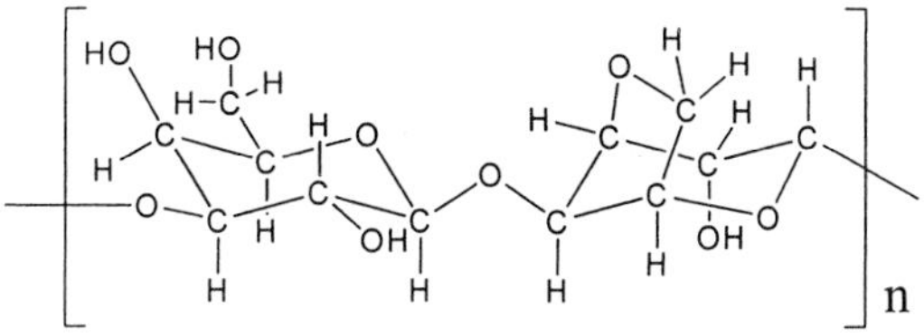

14 (agarose)

The diffusion coefficient (D_{app}) of the redox substrate in these films or solids coated on an electrode has usually been very small. An entirely solid-state voltammogram using a conventional three-electrode system has not been reported yet (see Experiment 14-4, Section 14.5). Electrochemical reactions have been successfully achieved in a tight solid made of polysaccharide and water [107]. In this solid, conventional electrochemical reactions and measurements could be performed using a normal three-electrode system as in pure water without any outer water phase and vessel, and the electrochemical reactivity was almost the same as in pure water.

A few percent κ-carrageenan or agarose was solubilized in water under heat. After cooling the clear solution to room temperature under ambient conditions, a tight, elastic and clear solid was obtained. The solid is tight and elastic when the content is over about 1 wt.% for agarose and over 2 wt.% for carageenan.

In preparing a 1 wt.% agarose/water (4 mL) solution, (Ru(bpy)$_3^{2+}$) **3** and KNO$_3$ were added in the water so as to have their concentrations 1 mmol dm^{-3} and 0.1 mol dm^{-3}, respectively, and the mixture was heated and cooled as above. While the mixture was in a viscous state, before cooling was complete, a transparent indium tin oxide (ITO) working electrode, a platinum counter electrode and an Ag-AgCl reference electrode were soaked in the solution, and the mixture was solidified by cooling to room temperature. A cyclic voltammogram (CV) was measured for this solid, and is shown in Fig. 14-13 [107].

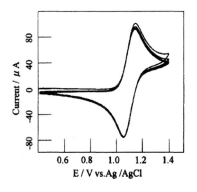

Figure 14-13. Solid-state cyclic voltammogram of Ru(bpy)$_3{}^{2+}$ (1 mmol dm^{-3}) in a 2 wt.% agarose/water solid containing 0.1 mol dm^{-3} KNO$_3$ electrolyte without any outer water phase and vessel. Scan rate 20 mV s^{-1}.

The voltammogram in the solid shows almost the same features as in liquid water including redox potential, peak separation and peak currents. The solid was so stable and self-supporting that the CV could be measured without any outer cell or vessel after being taken out of the vessel used for the preparation of the solid. The D_{app} value in the solid was the same order of magnitude as in liquid water (4.28·10^{-6} cm^2 s^{-1}). Similar results were obtained for the carrageenan solid. In a Nafion film containing adsorbed Ru(bpy)$_3{}^{2+}$ in the water phase in the hydrophilic column (water is present in a ~20 vol.% hydrophilic column of the Nafion **1**) coated on an ITO electrode and soaked in a 0.1 mol dm^{-3} KNO$_3$ aqueous solution, D_{app} was 10^{-10} cm^2 s^{-1}, four orders of magnitude lower than that in the above agarose or carrageenan solid.

To further investigate chemical reactions in the solid, electron transfer from a photoexcited state Ru(bpy)$_3{}^{2+}$ to methylviologen (MV^{2+}) was studied in the polysaccharide solid containing a large excess water. Electron transfer from a photoexcited Ru(bpy)$_3{}^{2+}$ to MV^{2+} in a dry solid matrix was first reported by the present author by adsorbing the compounds on a cellulose sheet [70], and the mechanism of such electron transfer in a dry polymer solid phase has later been established to be a static one in which the reaction components do not move at all for the electron transfer to occur (see above). In a 2 wt.% carrageenan/water solid containing both Ru(bpy)$_3{}^{2+}$ (50 µmol dm^{-3}) and MV^{2+} (from 50 to 400 µmol dm^{-3}), electron transfer from the photoexcited Ru(bpy)$_3{}^{2+}$ to MV^{2+} was investigated under irradiation with the wavelength of 453 nm (λ_{max}) at 25 °C. The visible absorption and emission spectra of the Ru(bpy)$_3{}^{2+}$ (λ_{max} 453 nm and emission max E_{max} 627 nm) were the same as in pure water. The Stern–Volmer plots of the relative emission intensity (I_o/I, where I_o is the emission intensity at 627 nm without MV^{2+} and I with MV^{2+}) from the photoexcited Ru complex as well as relative emission lifetime (τ_o/τ) in the carrageenan solid are shown in Fig. 14-14.

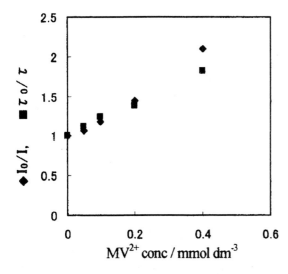

Figure 14-14. Stern−Volmer plots of the relative emission intensity (♦) and lifetime (■) against MV^{2+} concentration for the electron transfer quenching of the photoexcited $Ru(bpy)_3^{2+}$ by MV^{2+} in a 2 wt.% carrageenan/water solid ($1 \times 1 \times 3$ cm size).

Both the plots of I_0/I and τ_0/τ fell on almost the same line showing that the electron transfer takes place by a dynamic mechanism (see Section 13.2.3.4) in which the molecules at first diffuse and react after collision. The second-order rate constant of the electron transfer obtained by dividing the slope of the plots of Fig. 14-14 by τ_0 (447 ns), $4.54 \cdot 10^9$ $dm^3 mol^{-1} s^{-1}$, is almost a diffusion-controlled value. On the contrary, a static mechanism took place in both dry and wet Nafion films containing adsorbed $Ru(bpy)_3^{2+}$ and MV^{2+} for which the Stern−Volmer plots based on τ_0/τ gave no slope.

To conclude, tight and stable solids could be prepared from polysaccharides and water, and chemical reactions including electrochemical and photochemical ones could be carried out in these solids in the same way as in pure water, which offers a new concept of solution chemistry taking place in solids. In addition to its usage for conventional chemical reactions, any application to micro-scale reactors would be promising.

Electron processes of metal complexes in polymer matrixes in ground and photoexcited states have been described. Charge transfer between molecules in the ground state has been studied mainly by electrochemical methods, and that in a photoexcited state by photochemical methods. In the former process integrated charge propagation in a polymer matrix is observed, but in the latter only one charge transfer step at an excited state is usually observed. In this chapter both of them have been described in Section 14.1 to correlate them with each other. They show similarities, for instance in the mechanism (diffusion or dynamic process, charge hopping or static process) as well as in the charge-transfer distance that usually is near 1.4 nm without mediator, and nearly twice as great with a mediator molecule. Electrocatalysis is also one of the topics described relevant to the electrochemical process, as mentioned in Section 13.2.

Combined systems of charge transfer in ground and excited states have been mentioned in Section 14.2. This provides a fundamental idea for designing future electronic and photonic devices to solve global problems such as environment and energy resources. New quasi-solid materials composed of polysaccharide containing a large excess water have been described in which electrochemical and photochemical reactions can be carried out in the same way as in pure water without any outer cell or flask. These new materials could open the way to a new chemistry and devices in the future.

14.4 Organic Electroluminescent Devices

14.4.1 General Overview

Electroluminescent (EL) devices are very attractve as thin and light display devices that can replace the cathode ray tubes (CRT) currently used as the normal display device for televisions (TV) and computers. Electroluminescence is luminescence from an EL material excited by applying a voltage. Mono-chromic display panels for EL devices have already been commercialized. Since a full-color display is also possible, heavy and large CRT monitors are expected to give way to EL devices capable of hanging on a wall. As a light and thin display device, full-color liquid crystal devices (LCD) are already popular for personal computers and small TVs, but the LCD has problems; it does not emit light by itself so that it needs back lighting, it is difficult to see from an angle and the on–off rate is not rapid enough to monitor a quick motion. EL devices can overcome all these problems of LCDs.

Both inorganic and organic EL materials exist. Inorganic EL materials have a long history, and were the first to be commercialized (for example ZnS/Mn giving an orange luminescence). Organic EL materials are also used in practice, and will be especially important for the future fabrication of flexible or large-size display devices. A trial product has already appeared for a full-color organic TV display device. In organic EL devices metal complexes are important as light-emitting and electron-transfer materials. In this section we describe organic EL devices (see Experiment 14-5, Section 14.5 for an example).

14.4.2 Structure and Mechanism of EL Devices

The device proposed by Tang (Fig. 14-15) has opened a new era of organic EL device [109,110].

In this device tris(8-quinolinolato)alminum(III) (15, Alq) was used as an EL compound and an amine compound (**16**, TPAC) as a hole-transport material. The hole-transport layer (**16**) was coated on an indium tin oxide conductive glass (ITO) by vapor deposition, followed by coating of Alq (also by vapor deposition), and finally an MgAg electrode was attached as an electron-injection electrode.

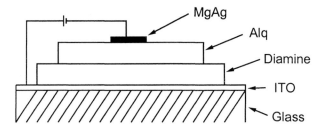

Figure 14-15. Organic EL device reported in [109].

Alq works also as an electron-transfer material. This device gave bright luminescence of over $1000 \ cd \ m^2$ under $10 \ V$ applied voltage with the maximum luminance over $10^4 \ cd \ m^2$. The luminance of a CRT is of the order of $100 \ cd \ m^2$, and that of a fluorescent lamp some thousands of $cd \ m^2$, so that the luminescence of this device is strong enough not only for display devices but also for a light source. Various compounds share the functions (hole transport, electron transfer, and electroluminescence), which is the origin of the high performance.

15 (Alq)

16 (TPAC)

The idea of using a multilayer is important for EL devices. Various multilayer devices have been investigated such as a triple layer composed of HITL (hole-injection and transport layer), EITL (electron-injection and transfer layer), and EML (emitter layer) as shown in Fig. 14-16 [111–113].

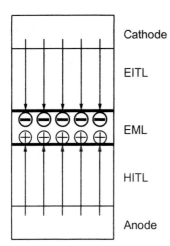

Figure 14-16. Multilayer EL device composed of HITL (hole-injection and transport layer), EITL (electron-injection and transfer layer) and EML (emitter layer).

The thickness of the layer is about 50 nm for HITL, 30 nm for EITL, and 20 nm for EML, in total being around 100 nm. Because of the thin structure, the make-up of the films is very important. Crystalline materials tend to form pinholes that make the device unstable, so that amorphous materials are preferable. For this purpose the introduction of bulky substituents such as *tert*-butyl groups, a low symmetry molecular structure, and a non-planar structure are effective. In order to suppress crystallization due to heat evolution during operation, a high glass temperature (T_g) is desired, for which large molecular weight compounds are effective. Low molecular weight materials are coated by vapor evaporation, and high molecular weight ones by spin coating from a solution [112].

Application of direct current voltage induces injection of electrons and holes at cathode and anode producing radical anions and radical cations, respectively, and their recombination excites the EL compound. The injected charges are transported by hopping between the carrier compounds at a high voltage of 10^6 V cm^{-1}. The carrier mobility is not high, around 10^{-7}–10^{-3} cm^2 V^{-1} s^{-1}, but the time needed to reach the recombination layer is very short, about 10^{-8} s, because of the thin layer of around 100 nm, so that the response of the device is very quick [113].

Multilayer devices are excellent systems for EL, but they also involve some drawbacks [113]. It takes more time and cost to fabricate a multilayer structure than to make a monolayer one. Because there are more interfaces than for a monolayer, separation of the layers and mixing of the components can more easily take place. For these reasons monolayer structures have also been investigated using a bipolar layer where both the carriers can be injected. Although this monolayer structure is superior in its simple fabrication and stability of the layer, the region for exciton formation is not limited to the central region of the layer. Excitons formed near the electrode can be quenched there, so a larger proportion of carriers do not contribute to the formation of excited states of the luminophor, resulting in a possible lower efficiency than a multilayer device.

14.4.3 Materials for EL Devices

Various materials including luminescent compounds, and hole- and electron-transfer materials can be used for EL devices (for the preparation of a device see Experiment 14-5, Section 14.5).

Luminescent Materials

Luminescent compounds are divided into three types: low molecular weight compounds, polymeric compounds and phosphorescent compounds. Luminescent compounds should have also the ability to transport electrons or holes, or bipolar characteristics, since carrier injection is important.

Some typical examples of low molecular weight luminescent compounds are shown in Fig. 14-17 [114]. Metal quinoline complexes are the most typical compounds. The highest luminescence (green colored) from Alq is over 10,000 cd m^{-2}. The benzoquinolinolato complex of Be (BeBq$_3$) gives a much higher luminance (green) of 19,000 cd m^{-2}. The luminescence maximum can be red-shifted when making a device by the formation of a charge-transfer complex or exciplex, which may cause luminescent efficiency to decrease. To avoid this, it is important to prohibit sterically the overlap of the molecular orbitals. For this purpose introduction of bulky groups or non-planar structures is effective; the styryl biphenyl derivative (DPVBi) in Fig. 14-17 is a typical example. The distorted structure of the terminal phenyl groups does not induce complex formation with a hole-transport material, tetraphenylenediamine derivative (TPD), resulting in the same blue luminescence from the DPVBi itself when incorporated into a device structure [113].

Concentration quenching of luminescence is a problem for organic luminescent dyes. It is often effective to dope a small number of luminescent molecules into a host layer to increase the luminescence efficiency. Thus, doping of a small amount of coumalin 6 (0.5–1%) into an Alq layer increased the efficiency from 1% to 2.5% with a bluish-green luminescence from the coumalin 6 [115] different from the color from the Alq (green). Doping can therefore not only increase the luminescence efficiency, but also change the color of the luminescence. The DCM dopant (see Fig. 14-17) gives orange to red luminescence as a guest.

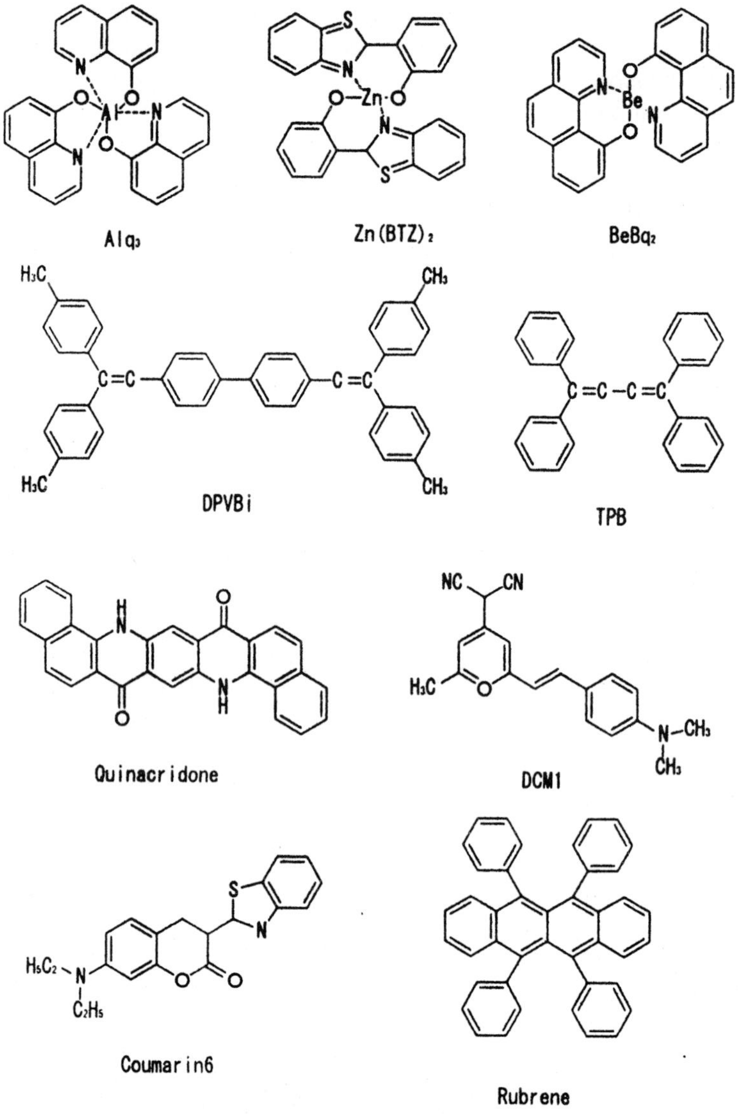

Figure 14-17. Low molecular weight luminescent compounds.

The mechanism may be either energy transfer from the excited host molecule to the guest, or direct trapping and recombination of the charge carriers on the guest molecule; it is not easy to differentiate these two mechanisms. In any case the lowest singlet excited state of the guest should be lower than that of the host molecule, so that the luminescence from a guest is located at higher wavelength than the host. Quinacridone and Rubrene are used as guest molecules with high durability. Luminescence of the three primary colors (red, green and blue) is possible by using Al-quinoline compound with different dopant materials.

Typical polymeric luminescent materials are shown in Fig. 14-18 [116]. The largest merit of polymeric luminescent materials is that a film can be formed by a wet method such as spin coating. The possibility of fabricating a large-area device is another advantage of polymeric materials and flexible devices are also possible. There are some disadvantages in preparation, e.g. control of the molecular weight, molecular distribution, microstructure of the film, and so on [116]. Conjugated polymers such poly(phenylenevinylene) (PPV) are typical luminescent materials. Much research has been carried out since the first report of an EL device using PPV in 1990 [117]. The mechanism is considered the same as for the low molecular weight compound, which involves formation of excitons by the reaction of radical cations and radical anions. For this reason, electron injection is not sufficient in a monolayer device with a hole-transporting PPV, as the luminance is not very high. This problem has been improved by fabricating a bilayer with an electron-transferring layer. A bilayer device composed of electron-transferring CN-PPV with electron-accepting CN groups and hole-transporting PPV gave a high luminescence efficiency of 4% [118]. A soluble polyphenylene derivative (LPPP), polyfluorene with long alkyl chains (PAF), and its copolymer (PAF-copolymer) showed efficient luminescence, and the long-term stability is almost the same as for low molecular weight compounds [116].

Phosphorescent EL materials are shown in Fig. 14-19 [116]. Singlet and triplet excitons are formed in the ratio of 1:3 by the recombination of electrons and holes. For this reason, fluorescent materials can give a maximum luminescence yield of only 25% internal efficiency. Since the yield of transmission of the luminescence through the layer is around 20%, the total external efficiency is around 5%. Phosphorescent compounds can utilize the triplet excited state to emit luminescence, so that the use of both fluorescent and phosphorescent compounds could give theoretically 100% internal quantum efficiency. Actually a device with 15% external luminescence efficiency has been reported. One of the topics in EL-device research is how to realize luminescence of 100% internal quantum efficiency. The efficiency of fluorescence is often near 100%, but that of phosphorescence is usually not very high. Eu or Tb ions give efficient phosphorescence, and so some complexes of Eu or Tb have been studied for EL devices, but there has been no report of efficient luminescence.

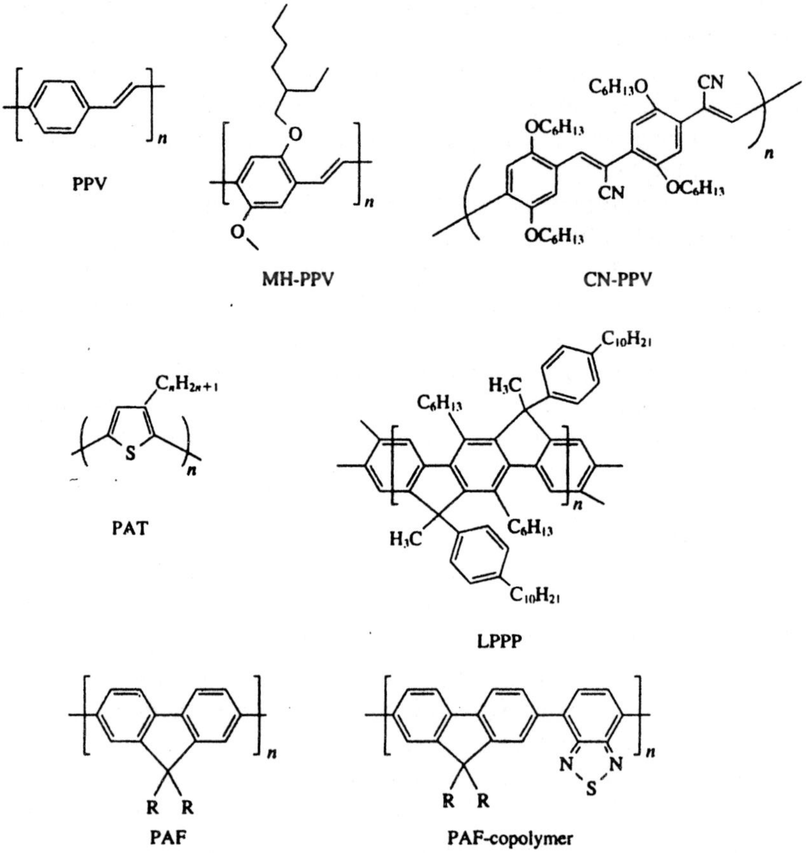

Figure 14-18. Polymeric luminescent materials.

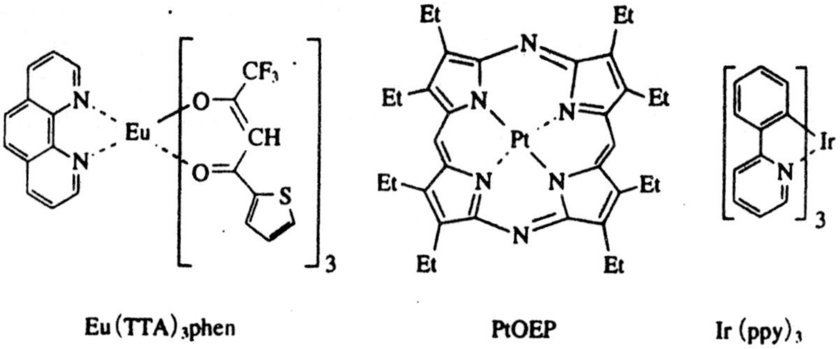

Figure 14-19. Phosphorescent materials.

The Pt complex of octaethylporphyrin (PtOEP) is known to give a relatively high phosphorescence efficiency, and an EL device with it gave fairly high efficiency, but the phosphorescence is saturated at a high luminance. Some iridium complexes, e.g. Ir(ppy)$_3$, give efficient phosphorescence, and saturation behavior is not observed, resulting in 15% quantum efficiency (see Experiment 14-5, Section 14.5) [119].

Emission of white light is possible by combining the three primary colors (blue, green and red) utilizing a multi-layer device [113,120]. In a bilayer device of TPD/Alq (for TPD see hole-transport materials in Fig. 14-20 below), an electron-transferring and hole-blocking 1,2,4-triazole derivative layer was put between the two layers to control the carrier recombination, which induced a bluish-green emission due to the emission both in the TPD and Alq layers, depending on the layer thickness. Doping of Nile red into the Alq layer gave white emission by the three colors from TPD (bluish-green), Alq (green) and Nile red (red) [120].

Hole-Transport Materials

Aromatic amine compounds are usually used as hole-transport materials (Fig. 14-20 [116]). Compounds with smaller ionization potential (I_p) can easily accept holes and hole mobility is higher due to the lower activation energy for hopping transport. In order to decrease I_p, introduction of electron-donating groups such as $-NR_2$ is effective. The diamine compound TPD is widely used as a hole-transport material. The I_p of this compound is small, 5.4 eV, and hole drift mobility is 10^{-3} cm^2 V^{-1} s^{-1} [113]. However, the glass transition temperature (T_g) is low (63 °C), and the evaporated film aggregates as the temperature rises, so that suppression of molecular mobility is required by increasing T_g. Introduction of a naphthyl group (α-NPD) increases T_g to 96 °C, so that durability against heat becomes higher than for TPD. The large molecule *m*-MTDATA (called starburst amine) has a high T_g of 75 °C showing higher heat resistance, smaller I_p (5.1 eV), and so higher hole mobility than TPD. Higher molecular weight favors heat resistance. A polymer containing pendant TPD groups has the same I_p as TPD (5.4 eV), high T_g (180 °C), and therefore showed high luminance of 20,000 cd m^{-2} [113].

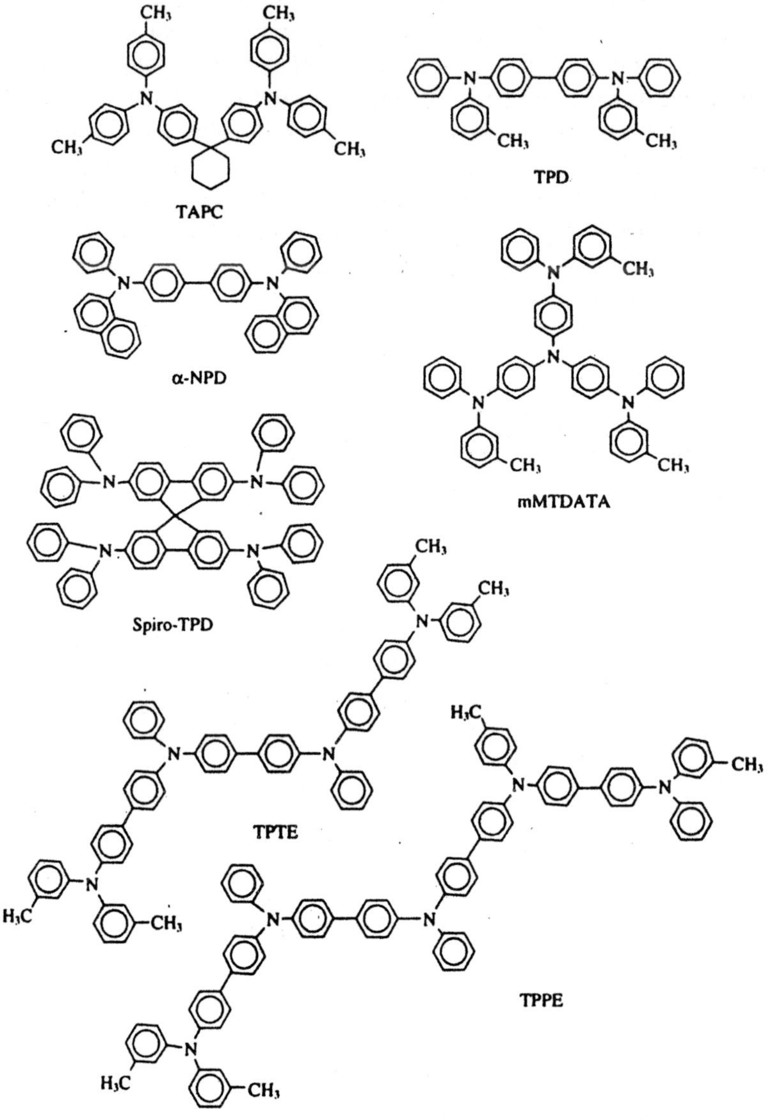

Figure 14-20. Hole-transport materials [8].

Electron-Transfer Materials

Many metal complex luminescent compounds work also as electron-transfer materials. The electron affinity (E_a) is a factor for this function. Alq and Beq luminescent compounds are also good electron-transferring materials. In the Al quinolinol complex E_a is 3.0 eV and electron mobility 10^{-6} cm^2 V^{-1} s^{-1} where the central metal ion is inferred to work as an electron-accepting group. Metal complexes are promising because of their high heat resistance and durability. Introduction of electron-accepting hetero-ring groups such as the oxadiazole ring (PBD, OXD-7) or triazole ring (TAZ) is effective for electron-transfer materials (Fig. 14-21) [116]. The cyano group is also effective as in CN-PPV (Fig. 14-18). For the hole- or electron-accepting materials, not only are I_p and E_a values important, but also a larger energy gap than for luminescent materials, so that the excitons formed in the luminescent layer are not quenched by the hole or electron transporting layers.

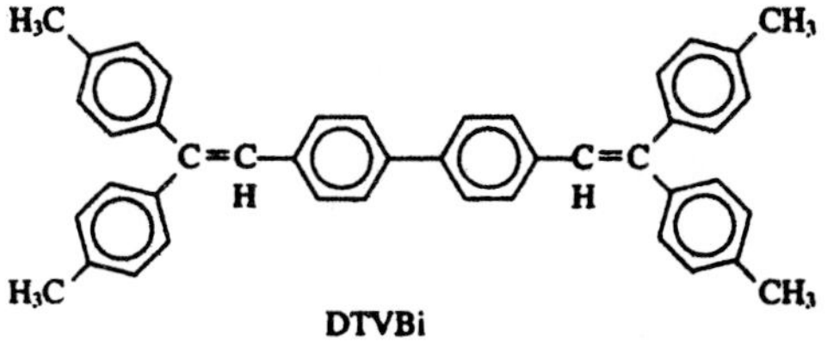

DTVBi

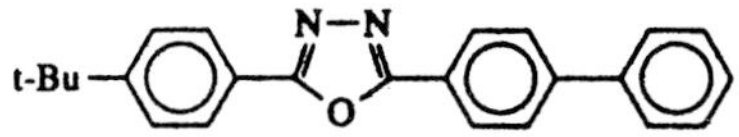

PBD

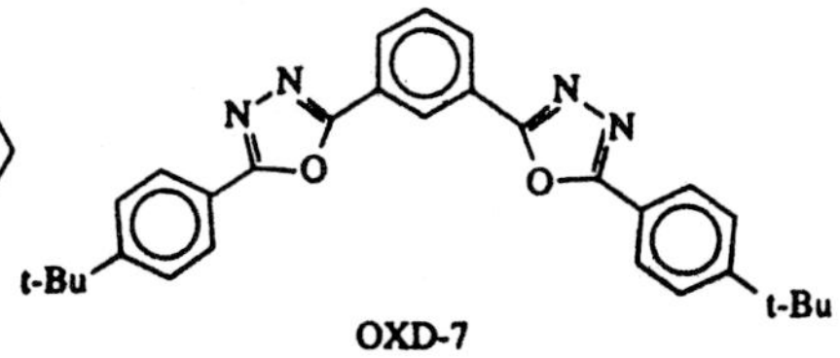

OXD-7

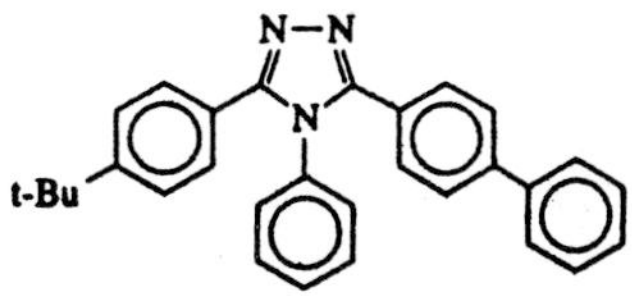

TAZ

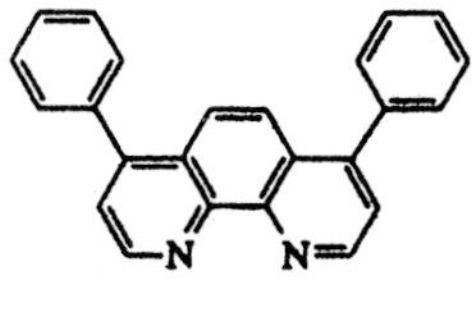

Bphen

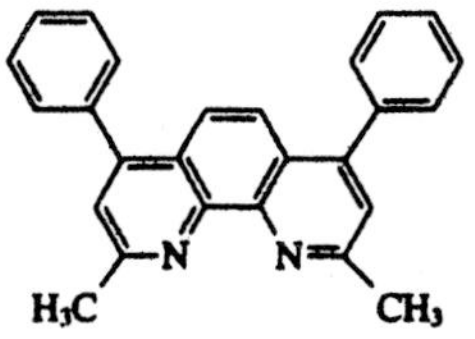

BCP

Figure 14-21. Electron-transfer materials.

Distyrylbenzene derivatives (e.g. DTVBi) with a nonplanar structure show blue luminescent and electron-transferring properties with high durability. Bathophenathrene (Bphen) and its derivatives (e.g. BCP) are also used for electron transfer. The compounds in Fig. 14-21 show good electron-transfer properties, but E_a is not very large, so that they do not decrease the barrier for electron injection at the electrode, that is, they do not have an electron injection ability [116]. They work as ETL materials by blocking holes. When they are used as ETLs, other methods such as the introduction of a buffer layer are required.

14.4.4 Commercialization

A green monocolor display was commercialized in 1997, and a multi-color display with blue, green and orange colors in 1999: both are used for automobile front-panel displays. A full-color demonstration display was made in 1998. A full-color TV monitor which can substitute for the present CRT and be used as a wall-hanging type is expected to be commercialized in a few years. The response time of EL devices, at 10 to 20 ms, can satisfy one of the important characteristics for TV. A prototype full-color TV monitor (13 inch) was demonstrated by the Sony Co. Ltd. in 2001, and attracted much attention at an Electronics Show held in Japan. The Japanese Government started projects on EL devices in 2002; one concerns large-area EL devices and the other flexible ones.

Table 14-2. Characteristics of recent EL materials [13]

Color	Research Company	Efficiency/ cd A^{-1}		Lifetime/10^4 h (Initial luminance/cd m^{-2})
Blue	Pioneer	3.9	1	(100)
	Idemitsu	4.7	1	(200)
Green	Pioneer	16	1	(300)
	NEC	20	1	(300)
	Pioneer (Phosphorescence)	25	0.33	(818)
Red	Pioneer	2.6	1	(250)
	NEC	7	1	(200)
	Pioneer (Phosphorescence)	3.2	3	(135)
White	TDK	-	1	(500)
	Matsushita (Yamagata Univ.)	10	-	
	Idemitsu	11	1	(1000)

Recent data of organic EL materials are shown in Table 14-2, but the details of the materials are not available. The lifetime of 10^4 h (corresponding to about a year) of continuous operation is the first requisite for commercial durability, and this criterion has in principle been achieved. Continuous operation for $3 \cdot 10^4$ h is also possible for some devices. EL devices are also expected to substitute for conventional light sources such as tungsten and fluorescent lamps in the future, because of their high luminance efficiency.

14.5 Experimental

Experiment 14-1: In-situ Potential Step Chronoamperospectrometry (PSCAS) of Ru(bpy)$_3^{2+}$ 3 Confined in a Nafion Film Coated on an ITO Electrode (Section 14.1.2) [40]

Preparation of the film: An alcoholic solution of Ru(bpy)$_3^{2+}$ **3** (containing 3.16 to 1.05 mmol dm^{-3}) and Nafion (2.5 wt.%) was prepared, and this solution (10 µl) was cast on an ITO electrode (effective area 1 cm^2). It was dried under ambient conditions to form a Nafion film containing a known concentration of the Ru complex. The thickness of the film was estimated to be about 1 µm for each modified electrode. The concentration range of the Ru complex in the film was from 0.1 to 0.3 M. The modified electrodes were stored in ion-exchange water before measurement.

Chronoamperospectroscopic measurements: The visible absorption spectral change of the modified electrode was measured in a combination of a multichannel photodiode array spectrophotometer (Otsuka Electronics Co., model IMUC-7000) with a voltammeter (Rikadenki Kogyo Co., model CV-27) and an XY-recorder. A conventional single-compartment quartz cell was equipped with the modified ITO, a calomel reference electrode and a platinum counter electrode. A 0.1 M sodium perchlorate aqueous solution was adjusted to pH 1.2 with perchloric acid. The solution was deaerated by bubbling argon gas for 30 min before measurement. The visible absorption spectral change after a potential step from 0 to 1.3 V vs. SVE is shown in Fig. 14-5. From the initial rate of the spectral change, the initial rate of charge propagation (V_{ct}) was obtained as a function of the concentration, and from these plots (V_{ct} is proportional to c^2), the mechanism was determined as a hopping one.

Experiment 14-2: Charge Transfer by Co(II)Pc(-2)(SO₃H)₄ Confined in a Film of Poly(4-vinylpyridine-co-styrene) (P(VP-St)) Coated on an ITO Electrode (Section 14.1.5) [66]

Film preparation and charge-transfer measurements: P(VP-St), a random copolymer with a molar ratio 9:1 of VP:St, was dissolved in *N,N*-dimethylacetamide to prepare a solution (0.5 wt.%, density 0.8 g cm^{-3}). This solution (30 μL) was cast on an ITO electrode (effective area 1 cm^2), and then it was dried under ambient conditions to form a film (estimated thickness 1 μm). This polymer-coated ITO was soaked in an aqueous solution of Co(II)Pc(-2)(SO₃H)₄ at pH 1 to adsorb the complex into the film by anion exchange. PSCAS measurement was carried out as in the previous experiment, and the spectral change in an aqueous solution of pH 1.2 NaClO₄ (0.07 mol dm^{-3} complex in the film) after the potential step from 0 to -0.50 V vs. Ag-AgCl was obtained as shown in Fig. 14-6. The time dependence of the charge propagation (R_{ct}) and the initial rate of charge propagation (V_{ct}) were obtained and shown in Fig. 14-7. The linear dependence of V_{ct} on the complex concentration (inset of Fig. 14-7) shows clearly that the mechanism of the charge propagation is diffusion of the molecules.

Experiment 14-3: A Photochemical Device Composed of a Bilayer of Polymer-pendant Ru(bpy)₃²⁺ 10 and Polymer-Pendant MV²⁺ 11 (Section 14.2.4) [92]

Preparation of the device and photovoltaic measurements: The polymer-pendant MV^{2+} (**11**, MV^{2+} fraction is 12.8% of the repeating unit) was coated on a SnO₂ conductive glass (effective area 1 cm^2) from its methanol solution, and dried at 353 K *in vacuo* so as to form a film of 5.0-μm thickness. The second layer was formed by coating the polymer-pendant Ru(bpy)₃$^{2+}$ (**10**, Ru complex fraction is 10.5% of the repeating unit) from its THF solution, and dried under the same conditions (film thickness 5.0 μm). Ag was vacuum evaporated on top of the second layer to fabricate the cell, SnO₂/**11**/**10**/Ag. The device was irradiated with visible light (intensity 34 mW cm^{-2}) from a 500 W xenon lamp through UV and IR cutoff filters (Toshiba UV-37 and HA-50), and *I–V* characteristics were obtained by measuring current against voltage. The open circuit voltage V_{oc} was 0.25 V, the short-circuit photocurrent density J_{sc} 0.485 μA cm^{-2}, fill factor FF 0.25 and the conversion efficiency η about 10^{-4}%.

Experiment 14-4: Cyclic Voltammogram of Ru(bpy)₃²⁺ in a Quasi-Solid of Agarose 14 (Section 14.3) [107,108]

Preparation of the quasi-solid and electrochemical measurements: Agarose (2 wt.%) was dissolved in a solution containing Ru(bpy)$_3{}^{2+}$ (1 mmol dm^{-3}) and KNO$_3$ (0.1 mol dm^{-3}) with warming. The mixture was allowed to cool, but before cooling was complete, while the material was still soft, three electrodes, a working electrode (Pt, ITO or graphite), a counter electrode (Pt), and a reference electrode (Ag-AgCl) were dipped in the mixture. After cooling was complete, a cyclic voltammogram (shown in Fig. 14-13) was measured in the agarose quasi-solid. The features of the CV in the solid were almost the same as in liquid water. The voltammogram can be measured in the absence of any outer cell because the solid is stable enough to hold the electrodes.

Experiment 14-5: EL Characteristics of Poly(N-vinylcarbazole) Doped with Phosphorescent Compounds (Section 14.4) [122]

Three-layered EL device, ITO/poly(3,4-ethylenedioxythiophene) (PEDOT)/ dye-doped PVK/Bphen:Cs/Al: On an ITO electrode, PEDOT and then PVK were spin-coated from their toluene solution, and the solvent was removed under reduced pressure. Bphen and Cs were coevaporated under high vacuum below 10^{-5} Torr on top of the PVK layer, and then the Al cathode was evaporated on the top of the Bphen:Cs in the same way. 2,3,7,8,12,13,17,18-octaethyl-21H,23H-porphyrinplatinum(II)(PtOEP) was doped in the PVK layer by using a mixed solution of the Pt complex and PVK on spin-coating. Optimum conditions were obtained at a dopant concentration of 1.0 mol% and a PVK thickness of 100 nm. The EL spectrum is shown in Fig. 14-22.

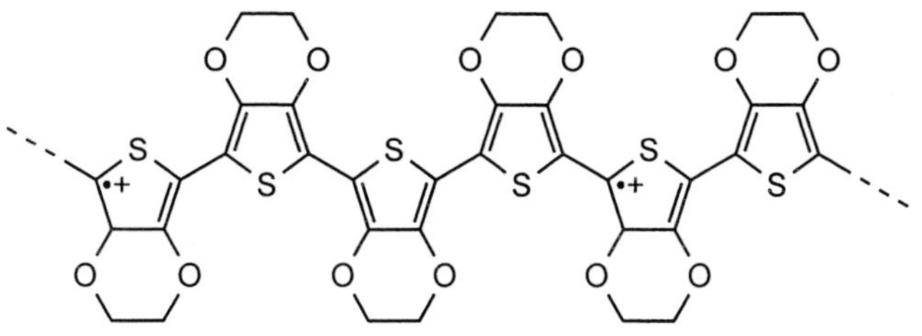

PEDOT

Device with Ir(ppy)$_3$ in PVK: Tris(2-phenylpyridine)iridium, Ir(ppy)$_3$, was doped in the PVK layer by using a mixture of the Ir complex and PVK on spin-coating. The thickness of the anode buffer layer (PEDOT) was taken as constant (60 nm), and the PVK layer thickness and the dopant (Ir complex) concentration were changed to optimize the conditions. A PVK layer thickness of 80 nm and a dopant (Ir(ppy)$_3$) concentration of 1.0 mol% gave the optimum conditions, for which the maximum luminance of 31,000 cd m^2 at 11 V, a luminance efficiency of 17.2 lm W^{-1}, a maximum current efficiency of 33.0 cd A^{-1}, and a maximum external efficiency of 9.58% (at 7 V) were obtained (Fig. 14-23).

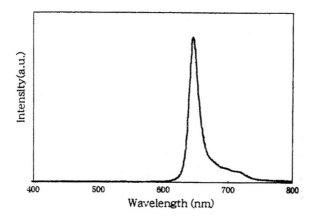

Figure 14-22. EL spectrum of PtOEP.

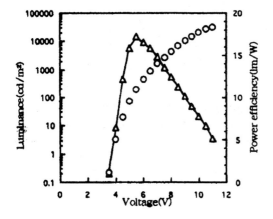

Figure 14-23. Luminance-voltage (○) and power efficiency-voltage (△) characteristics of the device, with Ir(ppy)₃ in PVK.

14.6 References

1. S.S. Isied (Ed.), *Electron Transfer Reactions*, Adv. in Chem. Ser., American Chemical Society, Washington DC, 1997.
2. G.J. Meyer (Ed.), *Molecular Level Artificial Photosynthetic Materials*, Progress in Inorganic Chemistry, Vol. 44, Interscience Publication, New York, 1997.
3. M. Graetzel, *Heterogeneous Photochemical Electron Transport*, CRC Press, Boca Raton, FL, 1989.
4. K. Sienicki (Ed.), Molecular Electronics and molecular Electronic Devices, CRC Press, Boca Raton, FL, 1993
5. R. Arshady (Ed.), *Desk Reference of Functional Polymers*, American Chemical Society, Washington DC, 1997.
6. K. Kalyanasundaram, M. Graetzel, *Coordination Chem. Rev.* **1998**, *77*, 347 .
7. F. Ciardelli, E. Tsuchida, D. Wöhrle (Eds.), *Macromolecule-Metal Complexes*,

Springer Verlag, Berlin, 1996.
8. J.L. Anderson, L.A. Coury, Jr., J. Leddy, *Anal. Chem.* **1998**, *70*, 519R.
9. M. Kaneko, *Organic Conductive Molecules and Polymers*, H.S. Nalwa (Ed.), Wiley, 1997, Chichester, Vol. 4, p. 661.
10. C.C. Page, C.C. Moser, X. Chen, L. Dutton, *Nature* **1999**, *402*, 47. J.M. Gruschus, A. Kuki, *J. Phys. Chem.* **1999**, *103*, 11407. (c) G. Hartwich, D.J. Caruana, T. de L.-Woodyear, Y. Wu, C.N.Campbell, A. Heller, *J. Am. Chem. Soc.* **1999**, *121*, 10803. (d) D. Ly, L. Sanii, B. Schuster, *J. Am. Chem. Soc.* **1999**, *121*, 9400. (e) G. Jones, V. Vullev, E.H. Braswell, D. Zhu, *J. Am. Chem. Soc.* **2000**, *122*, 388. (f) H.-A. Wagenknecht, E.D.A. Stemp, J.K. Barton, *J. Am. Chem. Soc.* **2000**, *122*, 1. (g) D. Burdinski, K. Wieghardt, S. Steenken, *J. Am. Chem. Soc.* **1999**, *121*, 10781. (h) G.J. II, L.N. Lu, H. Fu, C.W. Farahat, C. Oh, S.R. Greenfield, D.J. Gosztola, M.R. Wasielewski, *J. Phys. Chem. B* **1999**, *103*, 572.
11. H.J. Dahms, *J. Phys. Chem.* **1968**, *72*, 362.
12. Ruff, J. Friedlich, *J. Phys. Chem.* **1971**, *75*, 3297.
13. C.P. Andrrieux, J.-M. Savant , *J. Electroanal. Chem.* **1980**, *11*, 377.
14. D.N. Blauch, J.-M. Savant, *J. Am. Chem. Soc.* **1992**, *114*, 3323.
15. N. Oyama, F.C. Anson, *J. Am. Chem. Soc.* **1979**, *101*, 739 , 3450.
16. N. Oyama, T. Takuma, T. Sato, T. Sotomura, *Nature* **1995**, *373*, 598.
17. D. Rong, F.C. Anson, *J. Electroanal. Chem.* **1996**, *404*, 171.
18. F.C. Anson, D.N. Blauch, J.-M. Savant, C.-F. Shu, *J. Am. Chem. Soc.* **1991**, *113* 1922.
19. M. Watanabe, M.L. Longmire, R.W. Murray, *J. Phys. Chem.* **1990**, *94*, 2614.
20. J.J. Pietron, R.W. Murray, *J. Phys. Chem.* **1999**, *103*, 4440: (b) M.E. Williams, R.W. Murray, *J. Phys. Chem.* **1999**, *103*, 10221.
21. M.V. Mirkin, F.-R.F. Fan, A.J. Bard, *Science* **1992**, *257*, 364.
22. C.R. Martin, I. Rubinstein, A.J. Bard, *J. Am. Chem. Soc.* **1982**, *104*, 4817.
23. G.T.R. Palmore, D.K. Smith, M.S. Wrighton, *J. Phys. Chem.* **1997**, *101*, 2437.
24. D.K. Smith, L.M. Tender, G.A. Lane, S. Licht, M.S. Wrighton, *J. Am. Chem. Soc.* **1989**, *111*, 1099.
25. A.P. Clarke, J.G. Vos, H.L. Bandey, A.R. Hillman, *J. Phys. Chem.* **1995**, *99*, 15973.
26. R.J. Forster , J.G. Vos, *J. Electrochem. Soc.* **1992**, *139*, 1503.
27. S. Gould, K.H. Gray, R.W. Linton, T.J. Meyer, *J. Phys. Chem.* **1995**, *99*, 16,052.
28. S. Gould, R.M. Leasure, T.J. Meyer, *Chemistry in Britain* **1995**, 891.
29. J.D. Tirado, D. Abruna, *J. Phys. Chem.* **1996**, *100*, 4556.
30. J.A.R. Sende, C.R. Arana, L. Hernandez, K.T. Potts, M. Keshevarz, D. Abruna, *Inorg. Chem.* **1995**, *34*, 3339.
31. H. Larsson, M. Sharp, *J. Electroanal. Chem.* **1995**, *381*, 133.
32. M. Sharp, H. Larsson, *J. Electroanal. Chem.* **1995**, *386*, 189.
33. W. Schuhmann, T.J. Ohara, H.-L. Schmidt, A. Heller, *J. Am. Chem. Soc.* **1991**, *113*, 1394.
34. Heller, *J. Phys. Chem.* **1992**, *96*, 3579.
35. K. Aoki, J. Electroanal. Chem. **1995**, 395, 51; J. Electroanal. Chem. **1990**, 292, 53.
36. J. Umamaheswari, M.V. Sangaranarayanan, *J. Phys. Chem., B* **1999**, *103*, 5687.

(b) C.G. Cameron, G. Pickup, *J. Am. Chem. Soc.* **1999**, *121*, 11773.

37. J. Zhang, F. Zhao, M. Kaneko, *J. Porphyrins Phthalocyanines* **1999**, *3*, 1.
38. M. Yagi, K. Nagai, T. Onikubo, M. Kaneko, *J. Electroanal. Chem.* **1995**, *383*, 61.
39. J. Zhang, T. Abe, M. Kaneko, *J. Electroanal. Chem.* **1997**, *438*, 133.
40. J. Zhang, M. Yagi, M. Kaneko, *J. Electroanal. Chem.* **1998**, *445*, 109.
41. J. Zhang, F. Zhao, T. Abe, M. Kaneko, *Electrochim. Acta.* **1999**, *45*, 399.
42. F. Zhao, J. Zhang, T. Abe, M. Kaneko, *J. Porphyrins Phthalocyanines* **1999**, *3*, 238.
43. J. Zhang, T. Abe, M. Kaneko, *J. Porphyrins Phthalocyanines* **1998**, *2*, 93: (b) F. Zhao, J. Zhang, T. Abe, D. Wöhrle, M. Kaneko, *J. Mol. Cat., A: Chem.* **1999**, *145*, 245.
44. M. Yagi, H. Fukiya, T. Kaneko, T. Aoki, E. Oizawa, M. Kaneko, *J. Electroanal. Chem.* **2000**, *481*, 69.
45. M. Yagi, K. Kinoshita, K. Nagoshi, M. Kaneko, *Electrochim. Acta* **1998**, *43*, 3277.
46. K. Kinoshita, M. Yagi, M. Kaneko, *Electrochim. Acta* **1999**, *44*, 1771.
47. R.W. Murray, *Molecular Design of Electrode Surfaces*, Wiley, New York , 1992.
48. J.W. Long, C.S. Velazquez, R.W. Murray, *J. Phys. Chem.* **1996**, *100*, 5492.
49. J.S. Facci, R.H. Schmehl, R.W. Murray, *J. Am. Chem. Soc.* **1982**, *104*, 4959.
50. D.N. Blauch, J.-M. Savant, *J. Am. Chem. Soc.* **1992**, *114*, 3323.
51. H. Shiroisi, K. Ishikawa, K. Hirano, M. Kaneko, *Polym. Adv. Tech.* **2001**, *12*, 237.
52. T. Tatsuma, K. Saito, N. Oyama, *J. Chem. Soc., Chem. Commun.* **1994**, *1994*, 1853; (b) S. Watanabe, I. Kubo, *Electrochemistry* **2002**, *70*, 258.
53. E.J. Calvo, R. Etchenique, *J. Phys. Chem.* **1999**, *103*, 8944.
54. J.J. Pierton, R.W. Murray, *J. Phys. Chem.* **1999**, *103*, 4440. (b) A.P. Doherty, M.A. Stanley, J.G. Vos, *Analyst* **1995**, *120*, 2371.
55. R. Ramaraj, M. Kaneko, *Adv. Polym. Sci.* **1995**, *123*, 215.
56. R. Ramaraj, A. Kira, M. Kaneko, *Angew. Chem., Int. Ed. Engl.* **1986**, *25*, 825.
57. R. Ramaraj, A. Kira, M. Kaneko, *J. Chem. Soc., Faraday Trans. 1* **1987**, *83*, 1539.
58. M. Yagi, S. Tokita, K. Nagoshi, I. Ogino, M. Kaneko, *J. Chem. Soc., Faraday Trans.* **1996**, *92*, 2457.
59. M. Yagi, K. Kinoshita, M. Kaneko, *J. Phys. Chem.* **1996**, *100*, 11098.
60. M. Yagi, K. Kinoshita, M. Kaneko, *J. Phys. Chem., B* **1997**, *101*, 5143.
61. K. Kinoshita, M. Yagi, M. Kaneko, *Macromolecules* **1998**, *31*, 6042.
62. M. Kaneko, D. Wöhrle, *Macromolecule-Metal Complexes*, F. Ciardelli, E. Tsuchida, D.Wöhrle (Eds.), Springer-Verlag, Berlin, 1995, pp. 267–307.
63. R.M. Kellett, T.G. Spiro, *Inorg. Chem.* **1985**, *24*, 2373.
64. T. Abe, F. Taguchi, H. Imaya, F. Zhao, J. Zhang, M. Kaneko, *Polym. Adv. Technol.* **1998**, *9*, 559.
65. F. Taguchi, T. Abe, M. Kaneko, *J. Mol. Cat., A: Chem.* **1999**, *140*, 41.
66. F. Zhao, J. Zhang, T. Abe, D.Wöhrle, M. Kaneko, *J. Mol. Cat., A: Chem.* **1999**, *145*, 245.
67. M.J. Therien, M. Selman, H.B. Gray, *J. Am. Chem. Soc.* **1990**, *112*, 2420.
68. D.N. Beratan, J.N. Onuchic, J.N. Betts, B.E. Bowler, H.B. Gray, *J. Am. Chem. Soc.* **1990**, *112*, 7915.
69. R.A. Marcus, *J. Chem. Phys.* **1956**, *24*, 966; ibid, **1965**, *43*, 679.
70. M. Kaneko, J. Motoyoshi, A. Yamada, *Nature* **1980**, *285*, 468.

71. M. Kaneko, M. Ochiai, K. Kinosita, Jr., A. Yamada, *J. Polym. Sci., Chem. Ed.* **1982**, *20*, 1011.

72. Y. Kurimura, M. Nagashima, K. Takato, E. Tsuchida, M. Kaneko, A. Yamada, *J. Phys. Chem.* **1982**, *86*, 2432.

73. B.H. Milosavljevic, J.K. Thomas, *J. Phys. Chem.* **1985**, *89*, 1830.

74. B.H. Milosavljevic, J.K. Thomas, *J. Am. Chem .Soc.* **1986**, *108*, 2513.

75. D.Y. Chu, J.K. Thomas, *J. Phys. Chem.* **1985**, *89*, 4065.

76. V.G. Kuykendall, J.K. Thomas, *J. Phys. Chem.* **1990**, *94*, 4224.

77. T. Guarr, M.E. McGuire, G. McLendon, *J. Am. Chem. Soc.* **1985**, *107*, 5104.

78. S. Kotani, H. Miyasaka, A. Itaya, *J. Phys. Chem.* **1996**, *100*, 19898.

79. N. Ohta, Y. Iwaki, T. Itoh, I. Yamazaki, A. Osuka, *J. Phys. Chem.* **1999**, *103* 11242.

80. J.L. Bourdelande, J. Font, G. Marques, M. Valiente, *J. Photochem. Photobiol.* **1996**, *95*, 235.

81. J. Kiwi, N. Denisov, V. Nadtochenko, *J. Phys. Chem.* **1999**, *103*, 9141.

82. O. Stern, M. Volmer, *Phys. Z.* **1919**, *20*, 183.

83. N.J. Turro, *Modern Molecular Photochemistry*, Benjamin/Cummings Publishing Co., Menlo Park, CA, 1978.

84. J. Perrin, *Comp. Rend. Acad. Sci. Paris* **1927**, *184*, 1097; ibid, **1924**, *178*, 1978.

85. K. Nagai, N. Takamiya, M. Kaneko, *Macromol. Chem. Phys.* **1996**, *197*, 2983.

86. K. Nagai, J. Tsukamoto, N. Takamiya, M. Kaneko, *J. Phys. Chem.* **1995**, *99*, 6648.

87. T. Abe, T. Ohshima, K. Nagai, S. Ishikawa, M. Kaneko, *Reactive Functional Polymers* **1998**, *37*, 133.

88. R.-J. Lin, M. Kaneko, in: *Molecular Electronic and Molecular Electronic Devices*, Vol. 1, K. Sienicki (Ed.), CRC Press, Boca Raton, FL, 1992, pp. 201–241.

89. M. Kaneko, A. Yamada, *Adv. Polym. Sci.* **1984**, *55*, 1.

90. M. Kaneko, D. Woehrle, *Adv. Polym. Sci.* **1987**, *84*, 141.

91. M. Kaneko, M. Ochiai, A. Yamada, *Makromol. Chem. Rapid Commun.* **1982**, *3*, 299.

92. K. Yamada, N. Kobayashi, K. Ikeda, R. Hirohashi, M. Kaneko, *Japan. J. Appl. Phys.* **1994**, *33*, L544.

93. M. Fujihira, *Organic Thin Films and Surfaces*, Vol. 1, Academic Press, Boston, MA, 1994.

94. T. Taniguchi, Y. Fukasawa, T. Miyashita, *J. Phys. Chem.* **1999**, *103*, 1920; (b) H. Imahori, H. Yamada, Y. Nishimura, I. Yamazaki, Y. Sakata, *J. Phys. Chem., B* **2000**, *104*, 2099; (c) H. Yamada, T. Hasobe, H. Imahori, S. Fukuzumi, 2nd *International Conference on Porphyrins and Phthalocyanines*, Kyoto (Japan), P-187, June 2002.

95. T. Miyasaka, K. Koyama, I. Itoh, *Science* **1992**, *255*, 342.

96. Y. Saga, T. Watanabe, K. Koyama, T. Miyasaka, *J. Phys. Chem. B* **1999**, *103*, 234.

97. R.F. Khairutdinov, J.K. Hurst, *Nature* **1999**, *402*, 510.

98. D.G. Walter, D.J. Campbell, C.A. Mirkin, *J. Phys. Chem. B* **1999**, *103*, 402.

99. Y.-Z. Hu, S. Tsukiji, S. Shinkai, S. Oishi, I. Hamachi, *J. Am. Chem. Soc.* **2000**, *122*, 241.

100. T. Nakahira, Y. Inoue, K. Iwasaki, H. Tanigawa, Y. Kouda, S. Iwabuchi, K.

Kojima, *Makromol. Chem., Rapid Commun.* **1988**, *9*, 13.

101. J. Desilvestro, M. Graetzel, L. Kavan, J. Moser, *J. Am. Chem. Soc.* **1985**, *107*, 2988.

102. L. Kavan, M. Graetzel, *Electrochim. Acta* **1989**, *9*, 1327.

103. B. O'Regan, M. Graetzel, *Nature* **1991**, *353*, 737.

104. S. Burnside, J.-E. Moser, K. Brooks, M. Graetzel, D. Cahen, *J. Phys. Chem.* **1999**, *103*, 9328.

105. M. Kaneko, *11th Symposium on Unsolved Problems of Polymer Chemistry*, Society of Polymer Science, Japan, 1986, pp. 21–26.

106. M. Yagi, M. Kaneko, *Chem. Rev.* **2001**, *101*, 21.

107. M. Kaneko, N. Mochizuki, K. Suzuki, H. Shiroishi, K. Ishikawa, *Chem. Lett.* **2002**, *2002*, 530.

108. M. Watanabe, H. Nagasaka, N. Ogata, *J. Phys. Chem.* **1995**, *99*, 12,294.

109. C.W. Tang, S.A. van Slyke, *Appl. Phys. Lett.* **1987**, *51*, 913.

110. S. Tanaka, *Monthly Display* **1998**, *4*, 1.

111. C. Adachi in *Organic Colorants for Electronics* (S.Tokita, Ed.), CMC Co. Ltd., Tokyo, 1998 p. 61.

112. J. Kido, Kobunshi (Monthly Journal from The Society of Polymer Science, Japan) **1994**, 43, 301.

113. J. Kido, *Denki Kagaku* **1995**, *63*, 901.

114. K. Yoshino, A. Fujii, *Electronic Engineering* **2002**, *44*, 12.

115. C.W. Tang, S.A. van Slyke, C.H. Chen, *J.Appl.Phys.* **1989**, *65*, 3610.

116. T. Tsutsui, in: *Organic EL Display, Optical Materials Handbook [New Edition]*, T. Fukumi, (Ed.), Realize Inc., Tokyo, 2000, p. 583.

117. J.H. Burroughes, D.D.C. Bradley, A.R. Brown, R.N. Marks, K. Mackay, R.H. Friend, P.L. Burns, A.B. Holmes, *Nature* **1990**, *347*, 539.

118. N.C. Greenham, S.C. Moratti, D.D.C. Bradley, R.H. Friend, A.B. Holmes, *Nature* **1993**, *365*, 628.

119. T. Tsutsui, M.-J. Yang, M. Yashiro, K. Nakamura, T. Watanabe, T. Tsuji, Y. Fukuda, T. Wakimoto, S. Miyaguchi, *Jpn. J. Appl. Phys.* **1999**, *38*, L1502.

120. J. Kido, M. Kimura, K. Nagai, *Science* **1995**, *267*, 1332.

121. N. Miura, Electronic Engineering **2002**, 44, 17.

122. J. Kido, J. Tanaka, *Polymer Preprints, Japan* **2001**, *50*, 3397.

15 Outlook

Dieter Wöhrle and Anatolii Pomogailo

The prototypes of synthetic combinations of metal complexes and metals with macromolecules (also called macromolecular metal complexes – MMCs) are found in biological systems. The range of metals in living organisms is very large, ranging from the alkali metals to the transition metals (Section 2.1). They play an essential role in both growth and metabolism. Some metals are necessary in gram quantities, others are trace elements and beneficial nutrients at low levels but metabolic poisons at high levels. As pointed out in Section 2.2, metals fulfill several functions essential for life (Fig. 15-1; for some examples see Sections 2.3–2.8):

- Metals as cofactors in proteins and metalloproteins: oxygen binding and transport, electron transfer, energy transfer followed by electron transfer, and several regulating functions.
- Metalloenzymes: hydrolytic enzymes, redox enzymes, redox reactions with several electron pairs, rearrangements.
- Interference of detrimental metal ions with normal cell metabolism: metabolic poisons.
- Communicative functions of metals: magnetic compass, initiator of cell functions, regulation of gene expressions.
- Interaction with nucleic acids: stabilization of DNA and RNA.

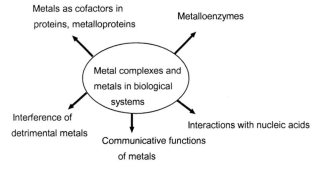

Figure 15-1. Examples of metal complexes and metals in biological systems.

An understanding of the structure and reactivity of biological reaction centers is important for the construction and optimization of artificial systems. Research has shown that the combination of a specific metal in a specific ligand surrounding and in combination with a specific macromolecule is the fundamental prerequisite for activity and selectivity.

Systematic studies on artificial combinations of metal ions, metal complexes and metals with macromolecules began only around sixty years ago. Now this field has reached the first advanced state in detailed synthesis, structure investigation and first optimization of structure–property relationships.

The synthesis of macromolecular metal complexes has come close to the level seen in nature. Many examples and routes are available to obtain different and tailor-made structural features. We have therefore attempted to discuss all the presently available approaches to the preparation of MMCs from different ligands and metal derivatives, mentioning the related chemical or physical interactions involved.

As there are many examples, we have given a classification of four types of combinations of metal complexes and metals with macromolecules in Section 1.2 (Fig. 15-2) and dealt with them in detail in Chapters 4 through 8.

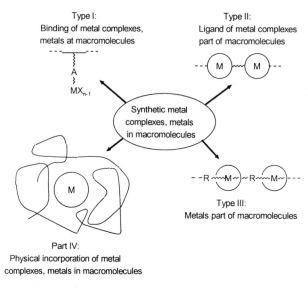

Figure 15-2. Classification of synthetic metal complexes and metals in macromolecules.

Polymerization of metal-containing monomers (Chapter 4) and binding of metal ions and metal complexes to macromolecular carriers (Chapter 5) are ways of having macromolecular metal complexes connected by different bonds to a macromolecular chain or network.

Highly interesting results are obtained from the study of polymerization and copolymerization of metal-containing monomers based on metals in different valence states (Chapter 4), but the correlation of the structure of the monomers and their reactivity is far from being understood. Also side reactions during polymerization processes are not well understood. The copolymerization of polynuclear and cluster-containing monomers is of increasing interest for tailor-made catalytic centers and must be investigated in more detail. Non-traditional methods of polymerization in solution and solid-phase polymerization of metal-containing monomers should be investigated more intensively to overcome problems during their polymerization.

Studies of the binding of metal ions at organic or inorganic macromolecules containing various ligand groups are well advanced, and are interesting for selective metal ion binding and the preparation of catalysts (Section 5.1). The kinetics and thermodynamics of complexation in dilute polymer solutions or at insoluble polymers are described in Chapter 3. Quantitative data (reaction constants, functions of formation, thermodynamic parameters) of the interaction of the metal compounds with macroligands are important for the evaluation of the complexation ability of polymeric ligands. So far, only a few systems have been investigated to give a first look at the influence of the macromolecular chain on the character of the coordination centers formed.

Sections 5.2 and 5.3 describe with a few examples the covalent, coordinative and ionic binding of metal complexes. This is a well-developed field. Well-characterized new polymeric systems can be obtained based on defined polymers and metal complexes.

Polymers containing metal complexes as parts of linear or crosslinked macromolecules (Chapter 6) are often difficulty to characterize because of their limited solubility and heterogeneous nature. The solubility can be improved by the introduction of bulky substituents, and the systems have been characterized in more detail in recent papers (Section 6.1). Progress in the synthesis of well-defined linear polymers has been made by polyreactions of bifunctional porphyrins (Section 6.2.2). In contrast, the electropolymerization of suitably substituted metal complexes results in insoluble polymer films. The combination of conducting pathways and active centers (metal complexes) is a developing field for electrocatalysis and catalysis. Finally, Chapter 6 describes the first results in the developing field of new dendrimers and networks hydrogen bonded via the ligand.

Chapter 7 shows how metals can be part of a polymer in different ways. In work of great originality polymers have been constructed with metals directly in the backbone. As well as homochain and heterochain polymers containing metals (Sections 7.1 and 7.2) the supramolecular organization of coordination polymers is a wide field (Section 7.3). In general, very well-defined structures are obtained. Recent progress has been made in obtaining stable networks with defined cavities and channels for reversible inclusion of guest molecules. Cofacial stacked polymeric metal complexes with high electrical conductivities is a well-developed field (Section 7.4). Finally, more recent reports describe metallodendrimers (Section 7.6).

Chapter 8 concentrates on the physical incorporation of metal complexes or clusters in macromolecules. Techniques for the inclusion of metal complexes in polymers in order to get thin-film electrodes are well-investigated (Section 8.1, and Chapters 9, 10, 13 and 14). There are also several methods for forming and stabilizing metal particles in macromolecules, especially organic polymers (Section 8.3). Up to now it has been difficult to establish a correlation between the molecular structure and the properties of the generated material.

If, on the one hand, preparative approaches have recently reached the level found in nature, the properties and applications of materials based on combinations of metal complexes and metals with macromolecules seem, on the other hand, to be still under development. The future offers the prospect of providing new materials with unique and unexpected features. Chapters 9 through 14 give insight in the most developed fields of binding of small molecules, sensors, catalysis and electron transfer, including processes induced by photoexcited states.

The binding of small molecules, as described in Chapter 9, offers extensive potential applications such as separation membranes, absorbents and sensors. Metal complexes in macromolecules exhibit characteristic behavior in binding reactions with small gaseous molecules as well as unique chemical reactivities and physicochemical properties. In Section 9.2 the olefin (ethene, propene) binding and reversibility at Ag(I)-polymer complexes are discussed. Also the oxygen binding and reversibility at metalloporphyrins embedded in polymers are treated in detail. Selective membrane separation of gaseous small molecules through dense (non-porous) polymeric membrane containing metal complexes is the next step towards practical applications (Section 9.3). The present state shows a high alkene/alkane and oxygen/nitrogen selectivity. Further developments towards the separation of other small gaseous molecules and other applications are pointed out in Sections 9.3.4 and 9.4.

The uses of macromolecular metal complexes as optical and physical sensors are treated in Chapter 10. Optical gas sensors based on luminescence or absorption changes of a metal complex in a polymer matrix have recently been paid much attention, and are now interesting for practical applications (Section 10.1). Very reproducible oxygen sensing is available by luminescence quenching of Pt/Pd porphyrins or triplet–triplet absorption by Zn porphyrins in specific organic or inorganic macromolecules. These results should be transferred to optical sensing of other molecules using other metal complexes in macromolecules. An electrochemical sensor is a chemical sensor with an electrochemical transducer (Section 10.2). A macrocyclic Ni(II) complex self-assembled on a gold electrode has been thoroughly investigated as an electrochemical sensor (anion recognition, H_2O_2, NADH, pH). Based on these results, a superoxide biosensor (superoxide dismutase) has been constructed (Section 10.2.3). This field is now open for the investigation of other enzyme-based sensors.

The nature of the macroligands plays an important role in the construction of macromolecular catalysts: along with the reaction conditions it influences the main properties of the catalysts obtained. Based on their solubility in the reaction medium (both polar and non-polar), macrocomplexes may be classified as soluble (Chapter 11) or heterogenized (Chapter 12). Soluble macromolecule metal complexes based on modified poly(ethylene oxides), polyamides and polyacids, and complexes with polymer-modified cyclodextrins, are the most typical catalysts of the first type. In such systems the polymer forms a specific ligand environment at the active center of the catalyst. Modification of macroligands by polar groups increases their solubility in water or water–organic media and enhances the adsorption of reagents, etc. Note especially the overall and repeated reversibility of the binding of MX_n by these polymers. Such an approach is used for preparing "temperature-controlled" immobilized catalysts. The principle of their operation is as follows: the macromolecular complex is solid under normal conditions; during the catalytic process at a higher temperature it dissolves and operates as a homogeneous catalyst. At a lower temperature the catalyst becomes insoluble and can be easily isolated and used repeatedly. The ability to combine the advantages of homogeneous and heterogeneous catalysts and to overcome their disadvantages led to the creation of "hybrid phase" catalytic systems (Chapter 12). The chemical binding of metal complexes with polymer or inorganic supports as firm and elastic granules, powders, fibers or membranes, which, like common heterogeneous catalysts, can be easily isolated and used in repeated cycles, is a new approach in metallocomplex catalysis. By varying the nature of the functional groups of the polymer chain, the MMC structure, the swelling degree of the polymer support, the surface density of the metallocomplex (topography), and the temperature, we can control the catalytic process. The macroligand can serve not only as an inert support but also to facilitate concentration and a favorable orientation of the substrate at the active center, thereby promoting its selectivity. Such a function is characteristic of protein macromolecules in enzyme catalysis. Thus, catalysis by MMC combines the main features of homogeneous, heterogeneous and enzyme processes (Fig. 15-3).

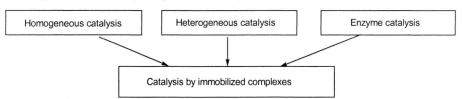

Figure 15-3. Overview of immobilized metal complexes as catalysts.

There are many examples in catalysis by MMCs when the isolation and identification of the active species as well as the mechanism of the catalytic process have been successfully studied. We may expect that investigations in this direction will be further developed. Really, there is increasing interest in this area of science. Recently, in a special issue of *Chemicl Reviews* (2002, vol. 102) 21 articles devoted to the broad interdisciplinary subject "Recoverable Catalysts and Reagents" had been published. The reviews in this issue are concerned with the preparation of macroligands and catalytic systems on the base of macromolecular metal complexes. One can mention especially the review articles by D.E. Bergbreiter as well as T.J. Dickerson and coworkers, C.A. McNamara et al., N.E. Leadbeater and M. Marco. The detailed reviews (C.E. Songa and S. Lee; D. Rechavi and M. Lemaire; Q.-H. Fan et al.) focus on supported chiral catalysts and factors influencing the enantioselectivity. The papers by A.P. Wight and M.E. Davis, D.E. De Vos et al., R. Duchateau offer analysis of the state–of–the art in organic–inorganic catalysts. The catalysis by heterogeneous nanosized particles is also described (A.K. Kakkar, A. Roucoux et al.) as well as dendrimers as supports and reagents (R. Van Heerbeek et al.). All this emphasize that the idea of macromolecular metal complexes is viable and their potentialities are not yet fully realized.

The photocatalytic properties and electron/photon-induced processes related to natural systems treated in Chapters 13 and 14 have been researched in depth. Different single fundamental multi-electron catalytic processes and photoexcited state electron-transfer reactions, both in polymer matrixes, are described in relation to photosynthesis (Section 13.2). It is now necessary to combine these reactions step by step to produce artificial photosynthetic systems. Some photoinduced energy-transfer processes (photooxidations) have now reached the level of practical application for wastewater cleaning (Section 13.4) and should be extended to other reactions induced by irradiation with visible light.

The first part of Chapter 14 concentrates on charge transfer in the ground and photoexcited state by metal complexes in polymer matrixes (Sections 14.1 and 14.2) and this is then extended to quasi-solids (Section 14.3). Detailed investigations show an in-depth knowledge of charge hopping and quenching processes in polymers. Organic electroluminescent devices are in practical use, and will be especially important for fabrication of flexible or large-size display devices in future (Section 14.4). Metal complexes are important as electro-luminescent materials and their use will be further developed.

Figure 15-4 illustrates the properties mentioned and gives additional examples.

Many of the properties described arise from high-order functions of metal complexes and metals in macromolecules, caused by the conjugation of dynamic interactions and electronic processes. The discussed properties and additional advanced functions deriving from molecular-level features are collected in Figure 15-4. Most of the properties of the new materials are under development and promise to have a great impact in related scientific and technological fields.

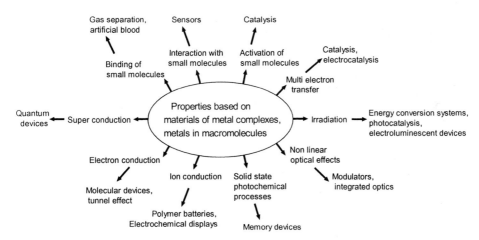

Figure 15-4. Properties based on materials containing metal complexes and metals in macromolecules.

The extraordinary materials "metal complexes and metals in macromolecules" have reached the scientific standard of an interdisciplinary and natural science. Based on the high level of synthetic procedures and detailed structural characterizations, a more intensive investigation of structure–property relationships is now possible. This will open new areas and contribute efficiently to new processes of fundamental technical importance for synthetic materials. The key for all these developments is the metal complex/metal as the active part in a well-defined macromolecular environment.

***

Let us briefly highlight the problems that must be solved in the field of metal complexes and metals in macromolecules that we are considering. First, until now the specific character (e.g. length, stability) of the metal–polymer ligand bond, and its difference from that of low molecular weight analogues, is not entirely clear. The same is true of the role of the molecular motion of the macromolecular chain on all three levels (local, molecular and permolecular) and to the kinetics and thermodynamics of complexing processes in such systems; there are no quantitative data for comparative analysis. The theory of macromolecular reactions has developed rather slowly. Furthermore, MMC preparation requires a high skill in synthesis and is often a time-consuming and unprofitable process. Therefore, the industrial uses of MMCs as heterogeneous catalysts are still at the developmental stage. Nevertheless, we may expect that catalysts of this type will allow us to transform coal into oil and gas fuel, starch into sugar, cellulose and carbohydrate into foodstuffs. Activation of molecular nitrogen and the development of new heat sources may change human life radically. MMCs are a link between living and inorganic nature. Indeed, *in macro veritas*.

Index